D0583422

Handbook of Neurochemistry and Molecular Neurobiology

Practical Neurochemistry Methods

Abel Lajtha (Ed.)

Handbook of Neurochemistry and Molecular Neurobiology

Practical Neurochemistry Methods

Volume Editors: Glen Baker, Susan Dunn, Andrew Holt

With 104 Figures and 28 Tables

Editor
Abel Lajtha
Director
Center for Neurochemistry
Nathan S. Kline Institute for Psychiatric Research
140 Old Orangeburg Road
Orangeburg
New York, 10962
USA

Volume Editors
Glenn Baker, Susan Dunn, Andrew Holt
Department of Pharmacology
University of Alberta
Edmonton, Alberta, T6G 2H7
Canada

Library of Congress Control Number: 2006922553

ISBN-13: 978-0-387-30359-8

Additionally, the whole set will be available upon completion under ISBN-13: 978-0-387-35443-9
The electronic version of the whole set will be available under ISBN-13: 978-0-387-30426-7
The print and electronic bundle of the whole set will be available under ISBN-13: 978-0-387-35478-1

springer.com

Printed on acid-free paper

SPIN: 11417699 2109 - 5 4 3 2 1 0

Preface

When preparing this volume we were well aware that it is not possible to cover all relevant neurochemical techniques in a single book of this length, but we have done our best to gather a group of techniques that will have wide general interest and are applicable to many analytical problems currently faced by neurochemists and other neuroscientists. There are also some techniques, e.g., neuroimaging methods such as magnetic resonance imaging/spectroscopy (MRI/MRS) and positron emission tomography (PET), which are used extensively in the neurosciences but could take up an entire volume in their own right. These are better covered in volumes of this handbook focusing on applications of such techniques. However, we feel that the current volume covers some very interesting techniques and that the chapters, written as they are by active investigators, provide useful details about theory, instrumentation, potential problems, applications, and relevant references that will be of considerable interest to both novice and seasoned investigators.

The first chapter deals with gas chromatography. Although this technique has been replaced by high performance liquid chromatography (HPLC) in many applications, it still represents a useful, relatively inexpensive, and widely accessible technique, which can be applied to numerous neurochemicals, drugs, and their metabolites. The topic of the second chapter is HPLC, which is now among the most frequently used techniques in neurochemistry. In these first two chapters, there is information on basic principles, columns, and detectors, and examples of applications to several neurochemicals and psychotropic drugs are then discussed. Cytochrome P450 (CYP) enzymes are involved in metabolism of a wide variety of endogenous neurochemicals and important drugs, and a comprehensive overview of these enzymes is provided in Chapter 3. Although there are many review papers in the literature on CYP enzymes, most of these do not focus on these enzymes in brain; this chapter should be a particularly valuable resource for neuroscientists interested in CYP enzymes in the central nervous system. Many neurochemists carry out studies on enzymes, but often their knowledge of enzymology is less than desired. Chapter 4 deals with practical aspects of enzymology and enzyme kinetics. Chapter 5 provides an interesting overview of the theory and applications of fluorescence techniques used to provide information about receptor–ligand interactions. Mass spectrometry is a widely used technique, primarily because of its selectivity and large number of neurochemical applications. Chapter 6 in this volume deals with basic principles and applications of mass spectrometry, while the following chapter discusses applications of mass spectrometric techniques to neuroactive steroids, compounds that interact with several important receptors and seem to be important in the etiology and pharmacotherapy of a number of psychiatric and neurologic disorders. Immunochemistry has done much over many years to enhance our knowledge in the neurosciences, and Chapter 8 focuses on immunohistochemistry, Western blotting, and ELISA assays, with many practical suggestions provided.

In vivo microdialysis, discussed in Chapter 9, permits neurochemical sampling of extracellular fluid from brain regions in living, freely moving organisms and has proved to be a popular technique in the neurosciences. The chapter on this topic discusses many practical aspects of this technique and gives examples of several diverse applications, including clinical ones. Radiolabeled ligand-binding techniques have been used for many years to investigate receptors and the drugs that interact with them. Chapter 10 provides information on theory, practical considerations, and limitations and gives some examples of protocols. Receptor autoradiography, the topic of Chapter 11, permits anatomical localization and quantification of receptor density in discrete regions of the brain and other tissues. Both in vitro and in vivo receptor autoradiography are discussed, and the applications section focuses on receptors for insulin-like growth factors. In silico molecular modeling, the topic of Chapter 12, is useful for studying proteins whose structures are not fully characterized. With this technique, homology modeling, i.e., using a related protein whose structure is known as a template for model building, is employed. The chapter introduces basic strategies used with homology models and in investigating the docking of ligands in the binding sites of protein structures generated in silico; applications discussed in the chapter focus on G protein–coupled receptors. Chapter 13 deals with flow cytometry, a technique, which uses fluorescence and light scatter to examine biochemical and biophysical properties of cells in fluid suspension. The technique is used routinely in many disciplines, including neuroscience, and this chapter has sections on principles and instrumentation, practical considerations, data acquisition and analysis, applications, and protocols. Oocytes from *Xenopus laevis*, the topic of Chapter 14, represent a powerful system for studying the transient heterologous expression of proteins and have been used widely for studying receptors associated with ligand-gated ion channels. This chapter provides a comprehensive discussion of practical aspects of using the oocyte expression system.

Polymerase chain reaction (PCR) amplification is a widely used technique for obtaining detectable amounts of DNA or RNA that can be manipulated. Chapter 15 discusses different approaches to PCR and gives examples of protocols and practical aspects to consider. The next chapter deals with in situ hybridization histochemistry, a technique used widely in studies on genes to examine the spatial and temporal distribution of their mRNA. Basic methodology and examples of standard applications and some more novel applications are discussed. Differential display, the topic of Chapter 17, is a technique that employs PCR and DNA sequencing to amplify, visualize, and identify differences in mRNA levels among cell types or tissues. This chapter deals with advantages and limitations of the technique, experimental design, protocols, and verification of differential display products. Chapter 18 deals with gene arrays, which permit simultaneous analysis of gene expression of multiple genes in a single sample. Although the chapter focuses on applications to stroke studies, a great deal of practical information about general experimental design and analysis and interpretation of data is provided. The yeast two-hybrid system, a genetic-based high throughput technique for the study of protein–protein interactions, is covered in Chapter 19. Theory, practical factors to be considered in its use and advantages and disadvantages of the technique are discussed. Two protein engineering techniques, namely chimeragenesis (formation of chimera between proteins encoded by homologous cDNAs) and site-directed mutagenesis, are discussed in detail in Chapter 20. These techniques have been used extensively, often in concert, to identify specific domains or amino acids that confer unique properties to the protein under examination, and this chapter provides a great deal of useful information about their applications. The use of cysteine-scanning mutagenesis for mapping binding sites of ligand-gated ion channels is the topic of Chapter 21. In this technique, cysteines are introduced, one at a time, into a protein region, and thiol-specific reagents are subsequently applied to the mutant proteins to determine if these cysteines are accessible to modification. The volume concludes with a chapter on the theory and applications of protein X-ray crystallography. As indicated by the author, this chapter is intended to provide the readers with a basic understanding of what is required to conduct crystallographic experiments and to permit them to determine if collaboration with a crystallographer might enhance their research.

We hope that this volume will be of use to not only neurochemists but also to many neuroscientists interested in the functioning of the nervous system and the effects of drugs on that functioning. We realize that some relevant techniques may well have been omitted in the volume, and would appreciate feedback from readers about techniques that should be included on the website and in future hard copies of this volume in the Handbook of Neurochemistry and Molecular Neurobiology.

Glen Baker, Susan Dunn, and Andy Holt

Table of Contents

Contributors

A. A. Alomary
Department of Neuropharmacology, Scripps Research Institute, La Jolla, CA, USA

J. Auta
The Psychiatric Institute, Department of Psychiatry, University of Illinois, Chicago, IL, USA

G. B. Baker
Department of psychiatry, University of Alberta, Edmonton, Canada

A. N. Bateson
Department of Biochemistry, University of Leeds, Leeds, UK

Y. Chen
The Psychiatric Institute, Department of Psychiatry, University of Illinois, Chicago, IL, USA

W. J. Costain
Cerebrovascular Research Group, Institute for Biological Sciences, National Research Council of Canada, 1200 Montreal Road, Ottawa, ON, Canada

R. T. Coutts
Novokin Biotech Inc. and Department of Psychiatry, University of Alberta, Edmonton, Canada

C. Czajkowski
Cynthia Czajkowski, Department of Physiology, University of Wisconsin, Madison, WI, USA

M. Davies
Department of Pharmacology, University of Alberta, Edmonton, Canada

E. M. Denovan-Wright
Department of Pharmacology, Dalhousie University, Halifax, NS, Canada

J. M. C. Derry
Department of Pharmacology, University of Alberta, Edmonton, Canada

S. M. J. Dunn
Department of Pharmacology, University of Alberta, Edmonton, Canada

E. T. Everhart
Department of Neuropharmacology, Scripps Research Institute, La Jolla, CA, USA

R. L. Fitzgerald
Department of Neuropharmacology, Scripps Research Institute, La Jolla, CA, USA

K. J. Frantz
Department of Psychology, and Department of Biology, Georgia State University, Atlanta, GA, USA

W. P. Gati
Department of Pharmacology, University of Alberta, Edmonton, Canada

R. W. Gilbert
Department of Pharmacology, Sir Charles Tupper Medical Building, Dalhousie University, Halifax, NS, Canada

K. L. Gilby
Department of Pharmacology, Dalhousie University, Halifax, NS, Canada

D. R. Grayson
The Psychiatric Institute, Department of Psychiatry, University of Illinois, Chicago, IL, USA

R. L. Haining
Department of Basic Pharmaceutical Sciences, West Virginia University, Morgantown, WV, USA

D. R. Hampson
University of Toronto, Toronto, Canada

R. S. Hansen
Department of Pharmacology, University of Alberta, Edmonton, Canada

C. Hawkes
Department of Psychiatry, University of Alberta, Edmonton, Canada

A. Holt
Department of Pharmacology, University of Alberta, Edmonton, Canada

L. L. Jantzie
Department of Psychiatry, University of Alberta, Edmonton, Canada

A. Kapur
Department of Pharmacology, University of Alberta, Edmonton, Canada

S. Kar
Department of Psychiatry, University of Alberta, Edmonton, Canada

L. P. Kotra
Faculty of Pharmacy, University of Toronto, Toronto, Canada

D. L. Krebs-Kraft
Department of Psychology, and Department of Biology, Georgia State University, Atlanta, GA, USA

I. L. Martin
Division of Life Sciences, University of Aston, Birmingham, UK

S. H. Mellon
Department of Neuropharmacology, Scripps Research Institute, La Jolla, CA, USA

J. G. Newell
Department of Physiology, University of Toronto, Toronto, Canada

J. Odontiadis
Department of Psychiatry, University of Alberta, Edmonton, Canada

M. B. Parent
Department of Psychology, and Department of Biology, Georgia State University, Atlanta, GA, USA

L. H. Parsons
Department of Neuropharmacology, Scripps Research Institute, La Jolla, CA, USA

K. Pozo
School of Pharmacy, University of London, London, UK

R. H. Purdy
Department of Neuropharmacology, Scripps Research Institute, La Jolla, CA, USA

G. Rauw
Department of Psychiatry, University of Alberta, Edmonton, Canada

K. A. Rittenbach
Department of psychiatry, University of Alberta, Edmonton, Canada

H. A. Robertson
Department of Pharmacology, Sir Charles Tupper Medical Building, Dalhousie University, Halifax, NS, Canada

W. B. Ruzicka
The Psychiatric Institute, Department of Psychiatry, University of Illinois, Chicago, IL, USA

D. A. R. Sanders
Biomolecular Structure and Research Program, Department of Psychiatry, University of Saskatchewan, Saskatoon, SK, Canada

B. D. Sloley
Novokin Biotech Inc. and Department of Psychiatry, University of Alberta, Edmonton, Canada

M. J. Smith
School of Pharmacy, University of London, London, UK

F. A. Stephenson
School of Pharmacy, University of London, London, UK

V. A-M. I. Tanay
Department of Psychiatry, University of Alberta, Edmonton, Canada

K. G. Todd
Department of Psychiatry, University of Alberta, Edmonton, Canada

M. Wang
University of Toronto, Toronto, Canada

1 Gas Chromatography

K. A. Rittenbach · G. B. Baker

Abstract: Gas chromatography (GC) is commonly used for the analysis of a myriad of compounds in neurochemistry. In this chapter various aspects of GC, including inlets, columns and detectors are discussed. Appropriate sample preparation, including extraction and derivatization techniques are also covered. In the latter portion of the chapter, examples of the analysis of specific types of endogenous and exogenous compounds by GC are dealt with.

List of Abbreviations: 5-HIAA, 5-hydroxyindole-3-acetic acid; 5-HT, 5-hydroxytryptamine; 6-MAM, 6-monoacetylmorphine; BDZ, benzodiazepine; DA, dopamine; ECD, electron-capture detector; FID, Flame ionization detector; GABA, γ-aminobutyric acid; GC, gas chromatography; GC x GC, two dimensional gas chromatography; GC-MS, combined gas chromatography-mass spectrometry; GHB, γ-hydroxybutyrate; *m*-, *p*-HPPA, *m*-, *p*-hydroxyphenylacetic acid; HPLC, high performance liquid chromatography; HVA, homovanillic acid; LLE, liquid-liquid extraction; MHPG, 3-methoxy-4-hydroxyphenylethylene glycol; NA, noradrenaline; NPD, nitrogen-phosphorus detector; PEA, 2-phenylethylamine; PFBC, pentafluorobenzoyl chloride; PFBS, pentafluorobenzenesulfonyl chloride; PFPA, pentafluoropropionic anhydride; SCOT, support coated open tubular; SPE, solid phase extraction; TCA, tricyclic antidepressant; TCD, thermal conductivity detector; TMS, trimethylsilyl; WCOT, wall coated open tubular

1 Introduction

Gas chromatography (GC) is a versatile and powerful technique which has many applications in neurochemical investigations. The strengths of this technique include: high efficiency of separation; the ability to automate most analyses; negligible waste production; and relatively low cost. Limitations include the following: compounds must be stable and volatile; some samples must be derivatized; and aqueous samples can harm the system. This chapter will cover the basic principles of GC, commonly used methods for derivatization and detection of compounds of interest, and specific applications of GC to endogenous components of the nervous system and exogenous compounds, mainly psychotropic drugs. Combined gas chromatography–mass spectrometery (GC–MS) will not be covered in detail as MS is the topic of a separate chapter in this volume.

2 Basic Principles

This chapter focuses on gas–liquid chromatography, in which compounds in a sample are separated based on vapor pressures and differences in affinity for the stationary phase (a high boiling point liquid) versus the gaseous mobile phase. The time between sample injection and detection of the individual compound eluting from the column is called the retention time. Compounds that have limited solubility in the stationary phase will exit the column quickly as a large proportion will remain in the mobile phase. Compounds with polarity similar to that of the stationary phase will have longer retention times and potentially broader peaks, due to increased interaction with the stationary phase.

2.1 Gas Chromatography Systems

2.1.1 Inlets

Gas chromatograph systems are composed of an inlet, carrier gas, a column within an oven, and a detector (➋ *Figure 1-1*). The inlet should assure that a representative sample reproducibly, and frequently automatically, reaches the column. This chapter will cover injection techniques appropriate for capillary columns. These include direct, split/splitless, programmed temperature vaporization, and cool on-column injection (Dybowski and Kaiser, 2002).

In direct injection, all injected material is carried onto the column by the mobile phase. This eliminates the possibility of sample discrimination in the inlet, but can overload capillary columns and so is more commonly used with packed columns and wide-bore capillary columns.

Figure 1-1
Schematic of a gas chromatograph system

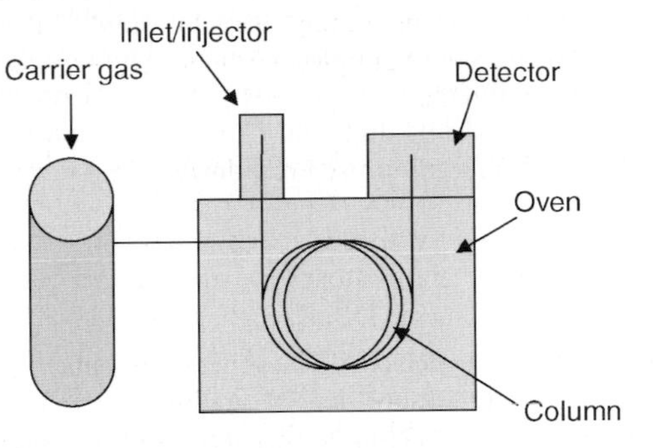

One method that prevents overloading of narrow-bore capillary columns is split injection. The inlets are usually bimodal split/splitless inlets, and either mode can be selected for a given analytical method. In the split mode, the sample is rapidly vaporized in the inlet and a portion is introduced into the column in a narrow band with carrier gas, while the rest of the sample is vented (Dybowski and Kaiser, 2002). The amount introduced can vary for each method and is chosen as a ratio, e.g., 1:100. Easily vaporized compounds may preferentially vent, leading to the introduction of a nonrepresentative sample to the column (Watson, 1999). When the splitless mode is chosen, the entire sample is introduced into the column and the vent is opened after a predetermined period of time, to flush the excess solvent from the injector (Dybowski and Kaiser, 2002).

Introduction of the sample into a cool inlet characterizes programmed temperature vaporization. A temperature program is then utilized to vaporize the sample and introduce it into the column (Dybowski and Kaiser, 2002).

The most reproducible, but most difficult injection technique to automate is cool on-column injection. The sample is injected directly onto a section of column that has been stripped of the stationary phase using a small-bore needle (Dybowski and Kaiser, 2002).

2.1.2 Columns

In the last 25 years a major shift has occurred in the type of columns used in GC, from the original packed columns to capillary columns. Capillary columns provide greater separation efficiency, but sample size must be limited, which can make detection of low-concentration analytes more difficult (Watson, 1999). Capillary columns were originally made from glass or even metal, but are now generally made from fused silica, for inertness, and coated with polyamide, to provide strength and flexibility. The stationary phase is coated along the walls of these columns in thicknesses from 0.1 to 10.0 μm; both the thickness and the polarity of the stationary phase affect separation, and many variations are commercially available. There are two types of capillary columns: wall coated open tubular (WCOT) and support coated open tubular (SCOT). A thin liquid phase layer of stationary phase is coated directly onto the walls of WCOT columns (Coutts and Baker, 1982). In SCOT columns the thin layer of stationary phase is coated onto a thin layer of support material, which can improve stability and allow for higher temperatures (Coutts and Baker, 1982).

A wide variety of stationary phases are available commercially; the principal difference between them is the polarity of the phase, which changes separation characteristics. Polarity differences can change elution order dramatically, but are not always sufficient to separate the compounds of interest in an efficient manner.

To efficiently separate mixtures with varying boiling points, the column temperature is frequently manipulated, which can also improve the resolution of higher boiling points/later eluting peaks

(Burtiset al., 1987). Current gas chromatographs have the capability of very complex temperature programs that can minimize the time per run and maximize the resolution of the compounds of interest. The minimum and maximum temperatures are usually the only nonvariable parameters. The solvent usually determines the minimum temperature because it must be volatized for sample introduction. The maximum temperature is determined by the stationary phase, because it may break down at high temperatures, which can result in an elevated baseline of the chromatograph. Thus, the manufacturer recommends a maximum temperature to maximize column life and minimize breakdown.

2.1.3 Detectors

A major component of a GC system is the detector. The general principles of detectors commonly used in neurochemical studies will be discussed here, and specific applications will be covered later in the chapter.

The thermal conductivity detector (TCD) is a universal detector that is nondestructive, which is a major advantage for preparative work (Dybowski and Kaiser, 2002). However, it is not sensitive enough for many of the analyses discussed later. This detector operates on the principle that a hot body loses heat at a rate dependent on the composition of the material surrounding it (Burtis et al., 1987). In a TCD, two filaments are heated, one in carrier gas, and the other in the column effluent. The voltages required to maintain the filament at a constant temperature are measured and compared. When compounds elute from the column the voltage of the sample filament is different from that of the filament in carrier gas and is recorded as a peak (Burtis et al., 1987).

The flame ionization detector (FID) burns the effluent from the column in a mixture of hydrogen and air and measures the ionization products as an increase in current between the jet and the collector (Dybowski and Kaiser, 2002). Carbon atoms attached to oxygen, nitrogen, or chlorine are not sensed, but carbon–hydrogen containing compounds are (Watson, 1999). This detector is linear over large ranges but is neither as sensitive nor selective as some other detectors. It is, however, used in many laboratories and for a large variety of analyses partially because it is less selective and hence more flexible; it is also very reliable (Burtis et al., 1987).

The electron-capture detector (ECD) is used extensively for the analysis of drugs in biological samples. A radioactive source, usually nickel 63, interacts with an appropriate carrier gas to create a constant stream of electrons that is measured by the collector; this is the "standing current" (Dybowski and Kaiser, 2002). When an electrophoric substance exits the column into this stream of electrons, some of the electrons are absorbed and the standing current decreases, which is recorded as a peak. This detector is very sensitive to halogenated compounds and to compounds containing ketone or nitro groups (Burtis et al., 1987). With technological advances, linearity has been increased since the first development of the ECD, making it useful for many analyses (Coutts and Baker, 1982).

The nitrogen–phosphorus detector (NPD) is, as the name implies, preferentially sensitive to nitrogen and phosphorus atoms. The detector contains a ceramic bead formed with an alkali salt, such as rubidium or cesium. The bead is heated in the presence of hydrogen, but the flow of gas is limited so that no flame is present. As the column effluent passes around the bead, nitrogen- and phosphorus-containing compounds produce electronegative moieties, which are treated as an increase in current and recorded as peaks (Carlsson et al., 2001).

2.2 Two-Dimensional Gas Chromatography

Comprehensive two-dimensional GC (GC × GC) has been applied to the analysis of many complicated systems including petroleum and air samples (Ryan and Marriott, 2003). This technique is more powerful than previously developed heart-cutting methods, in which a small section of eluent from one column is injected onto a second for further separation (Ryan and Marriott, 2003). In GC × GC, two columns with fundamentally different separation mechanisms are connected by a modulator (➋ *Figure 1-2*). The modulator continuously refocuses portions of the eluent from the first column onto the second column. The

Figure 1-2
Schematic of two dimensional GC

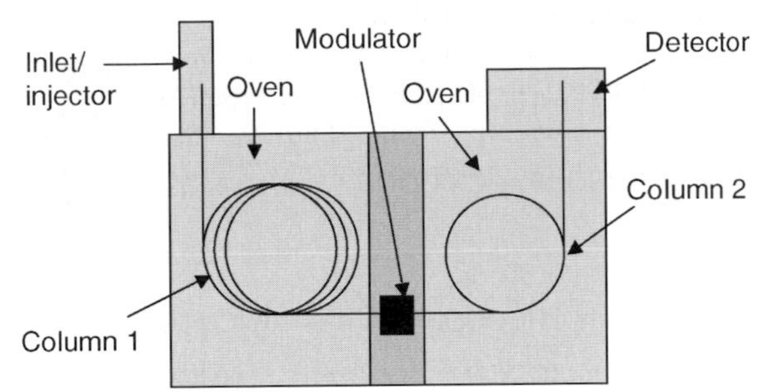

second column gives very quick separation; ideally all components should elute before the next portion from column one enters. The size of the portions refocused by the modulator should be such that each peak from the first column is split into four segments on the second column (Dalluge et al., 2003). The columns are sometimes housed in the same oven, sometimes separate ones (Dalluge et al., 2003). For true orthogonality, which allows for ordered chromatograms, the first column should be truly nonpolar and of a regular length, whereas the second column should be polar or shape-selective and much shorter and narrower (Dalluge et al., 2003). The resulting chromatogram requires more interpretation as peaks are split and must be properly identified. This complication is compensated for by the increase in separation efficiency (Dalluge et al., 2003). An excellent review by Dalluge et al. in 2003 gives examples of various applications including steroids in microalgae. Another review from 2003 covers the application of GC × GC to drug analysis, specifically 27 different primary, secondary, and tertiary amines from urine (Kueh et al., 2003). One disadvantage of GC × GC is that software packages for interpretation of results have not been commercialized and can be quite complicated (Ryan and Marriott, 2003).

2.3 Derivatization

Some compounds are not easily volatized, making them unsatisfactory for GC; these may be good candidates for HPLC, which does not require volatization. However, another option may be derivatization. This often involves the replacement of an active hydrogen atom by a moiety that improves the chromatography of the original compound. Improvement of the chromatograph results from the increased volatility of the compound, reduced polarity (which can result in sharper peaks), and stabilization of thermally labile structures (Burtis et al., 1987). Derivatization is also used in some cases to improve sensitivity to specific GC detectors; for example, adding halogenated moieties can make compounds detectable at much lower concentrations with ECDs. These improvements generally outweigh the disadvantage of extra steps in the assay.

The main derivatization techniques used are alkylation, acylation, silylation, or condensation (Drozd, 1975; Ahuja, 1976; Perry and Feit, 1978; Coutts and Baker, 1982). One of the few fundamental additions to derivatization reagents since the 1970s is the use of chloroformates (Wells, 1999). Chloroformates are useful in aqueous media, and a reaction with them proceeds rapidly at room temperature (Hušek, 1998). Some examples of derivatization techniques are shown in *Figure 1-3*.

Derivatization can also be used to increase the yield of the extraction step of an assay. For example, when phenolic amines are first acetylated and then extracted into nonpolar solvents, the yield increases significantly (Baker et al., 1994).

A review of sample preparation techniques written recently by Smith (2003) purports that derivatization is not very useful and that with advances in separation techniques or by using a different analytical technique, such as HPLC, derivatization can be avoided. While a "laboratory will examine almost any alternative to avoid derivatization" according to Smith (2003), switching equipment can be prohibitively expensive, and so derivatization certainly still has an important role in most laboratories.

Figure 1-3
Examples of the principal types of derivatizations used for GC analysis: (a) silylation; (b) alkylation; (c) acylation; (d) condensation; and (e) chloroformation

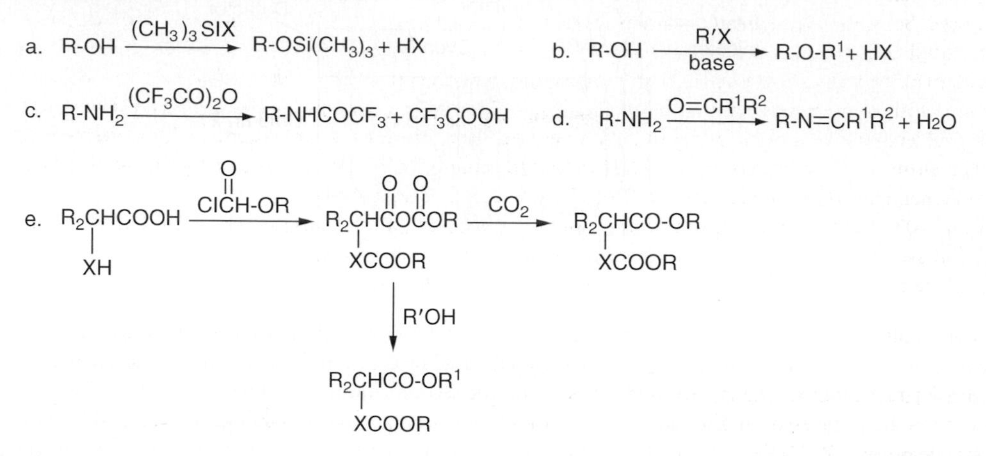

3 Analysis of Specific Types of Compounds by Gas Chromatography

Since the last edition of this handbook, HPLC has become more common and many compounds are now routinely analyzed with HPLC. However, GC is still very practical for the analysis of many compounds of interest to neurochemists. Methods published since 1982 are the primary focus of this chapter, and representative examples for many compounds, both endogenous and exogenous, are included here. For a review of information prior to 1982, readers are referred to Coutts and Baker (1982).

3.1 Chirality

Since the last edition of the Handbook of Neurochemistry, considerable advances have been made in the separation of enantiomers of compounds. Enantiomers are nonsuperimposable mirror images that may differ markedly in pharmacodynamics and pharmacokinetics (Jamali et al., 1989; Tucker, 2000; Baker et al., 2002). Their separation can be accomplished either by the use of chiral derivatizing agents and the usual capillary columns (Srinivas et al., 1995; Van Bocxlaer et al., 1997) or by using capillary columns that react preferentially with one enantiomer (Caldwell, 2001). One concern in using chiral derivatizing agents is that the reaction may be stereoselective, and thus the results would not reflect the original enantiomeric ratio (Smith, 2003). An excellent review of GC enantiomeric separation that explains the different forces that influence the separation of enantiomers and the main types of stationary phases used has been published recently (Schurig, 2001).

Williams and Wainer (2002) use examples of two chiral separations to demonstrate their utility in research. In one example, the difference between enantiomers in the competitive displacement of cyclosporine from immobilized P-glycoprotein was studied. In the other, the pharmacokinetic profiles of (+) and (−)-ketamine and (+) and (−)-norketamine were determined (Williams and Wainer, 2002).

Examples of specific methods important to neurochemists include separation and quantification of R- and S-fluoxetine and R- and S-norfluoxetine in brain tissue and body fluids using derivatization with (−)-(S)-N-(trifluoracetyl)prolyl chloride, a chiral derivatizing agent (Torok-Both et al., 1992; Aspeslet et al., 1994). A similar method has been used to separate the enantiomers of 3,4-methylenedioxyamphetamine (MDA) and 3,4-methylenedioxymethamphetamine (MDMA) (Hegadoren et al., 1993). Fluoxetine and norfluoxetine enantiomers have also been separated on a chiral column in series with a nonchiral column with NPD detections (Ulrich, 2003). Reviews of the analysis of enantiomers of several drugs of abuse are available (Jirovský et al., 1998; Tao and Zeng, 2002; Liu and Liu, 2002).

3.2 Sample Preparation

A pivotal step in the analytical process is sample preparation. Frequently liquid–liquid extractions (LLEs) are used. Solvents, pH, and multiple back extractions are all manipulated to increase selectivity and decrease unwanted contaminants before injection on the GC system. Solid phase extraction (SPE) is more convenient than it used to be because of an increase in commercially available SPE columns. SPE columns are packed with an inert material that binds the drug of interest, allowing impurities to pass through. As with LLE, solvent choices and pH affect retention and recovery. There are three commercially available types of SPE columns, diatomaceous earth (which uses the same principles as LLE), polystyrene–divinylbenzene copolymer, and mixed mode bonded silica (Franke and de Zeeuw, 1998).

In 2003, Smith reviewed newer sample preparation techniques, including pressurized liquid extraction, solid phase microextractions, membrane extraction, and headspace analysis. Most of these techniques aim to reduce the amount of sample and solvent required for efficient extraction.

3.3 Endogenous Compounds

2.3.1 Biogenic Amines

Biogenic amines are of great interest to researchers because of their potential roles in several psychiatric and neurological disorders. They include dopamine (DA), noradrenaline (NA), 5-hydroxytryptamine (5-HT, serotonin), histamine, and "trace amines" such as 2-phenylethylamine (PEA), tyramine, octopamine, phenylethanolamine, and tryptamine (Coutts and Baker, 1982). Although GC assays for DA, NA, and 5-HT are available, HPLC analysis with electrochemical detection has for many years now been the method of choice for analysis of these neurotransmitter amines.

As described in the chapter on GC in the previous edition of the Handbook of Neurochemistry (Coutts and Baker, 1982), analysis of biogenic amines in brain using GC usually involves analysis by GC–ECD following extraction and derivatization (usually with halogenated reagents). Amphoteric amines such as tyramine, which contain both acidic (phenol) and basic (amine) moieties present a special problem because they are difficult to extract from aqueous solutions at any pH value. In our laboratories, we overcame this problem by derivatizing both phenol and amine groups with acetic anhydride under slightly basic aqueous conditions, readily extracting the resultant neutral compounds (the phenols had been converted to esters and the amines to amides) into organic solvents, hydrolyzing the esters, and reacting the resultant free phenol and the amide with a reagent such as trifluoro- or pentafluoropropionic anhydride (PFPA) (➋ *Figure 1-4*) (Baker et al., 1981).

Another way to readily extract phenol- and amine-containing compounds from aqueous solution is to shake the aqueous mixture (under basic conditions) with reagents such as pentafluorobenzoyl chloride (PFBC) or pentafluorobenzenesulfonyl chloride (PFBS) in organic solvents. These reagents, unlike the perfluoronated anhydrides such as PFPA, are stable under aqueous conditions, and allow the derivatization and extraction into organic solvents in one step (Baker et al., 1994). Such procedures have been applied to PEA (Baker et al., 1985; Bakeret al., 1986a; Nazarali et al., 1987), but are also applicable to other amine- and phenol-containing compounds, including drugs and their metabolites.

Metabolites of biogenic amines have also been analyzed by GC–ECD. For the simultaneous analysis of 5-hydroxyindole-3-acetic acid (5-HIAA), homovanillic acid (HVA) and *m*- and *p*-hydroxyphenylacetic acid (*m*-, *p*-HPAA), (metabolites of 5-HT, DA, and *m*- and *p*-tyramine acid respectively) in urine, a simple acidic extraction followed by derivatization with PFPA (derivatizes phenols) and hexafluoroisopropanol (derivatizes carboxylic acid groups) has been used (Davis et al., 1982; Baker et al ., 1987).

The extraction of 3-methoxy-4-hydroxyphenylethylene glycol (MHPG), a metabolite of NA, is difficult from polar matrixes such as urine and blood. However, Baker et al. (1986b) found good results when MHPG was acetylated in the aqueous phase and then extracted and further derivatized with a perfluoroacylating reagent. Sensitivities of 10 ng/ml were achieved using ECD detection.

Figure 1-4
Comparison of the derivative of *p*-tyramine formed when the sample is acetylated, hydrolyzed under moderately basic conditions, and reacted with TFAA (a) and when the hydrolysis is omitted (b) AA = acetic anhydride; R = $COCH_3$; TFAA = trifluoroacetic anhydride

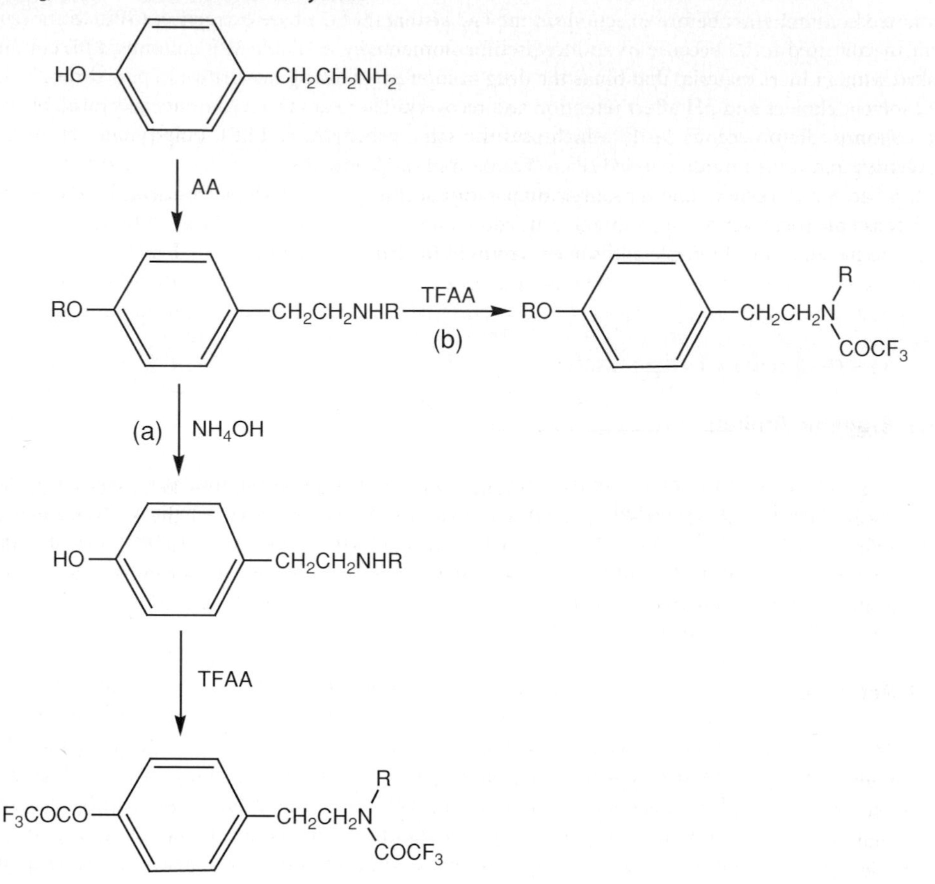

Wong et al. (1988) published a method for analysis of phenylacetic acid (the major metabolite of PEA) in urine samples that involves derivatization under aqueous conditions using pentafluorophenol in the presence of the condensing agent dicyclohexylcarbodiimide.

3.3.2 Amino Acids

The volatility of amino acids is not sufficient for GC analysis, which is why HPLC is frequently used. Furthermore, the zwitterion characteristics of amino acids cause them to be difficult to extract using common techniques. Frequently they are extracted from aqueous medium using ion-exchange techniques followed by derivatization of the amino group by acylation and the carboxylic group by esterification (Coutts and Baker, 1982). Another method is a dual derivatization using isobutyl chloroformate as the first step and pentafluorophenol as the second (Wong et al., 1990a). This method has been used for analysis of γ-aminobutyric acid (GABA) levels in rat brain. Advantages of this method include the use of only 10 μL of homogenate and a single tube for the extraction and derivatization (Wong et al., 1990a). This method can also be used for the analysis of alanine (Yamada et al., 1993). Phenylalanine, a precursor of *p*-tyrosine, can be analyzed by a similar procedure using benzoyl chloride and pentafluorobenzyl alcohol as the derivatizing agents (Wong et al., 1990b). Both of these methods use GC–ECD and attain good sensitivity. The latter method can also be used for analysis of other amino acids such as leucine, isoleucine, and valine in tissue and body fluids.

Kataoka (1997) describes a method for the analysis of 21 protein amino acids and 33 nonprotein amino acids with NPD detection. One disadvantage of this method is the use of diazomethane, which is explosive and toxic (Kataoka, 1997). A method for homocysteine with GC–FID analysis uses a one-step derivatization with ethyl chloroformate and an extraction procedure (Hušek et al., 2003).

Amino acid enantiomers can be separated on a chiral stationary phase after derivatization with chloroformates (Abe et al., 1996). The derivatization procedure is quite simple and rapid, but the derivatizing reagent must be synthesized, which complicates the assay. Another method for the analysis of amino acid enantiomers uses N,O-pentafluoropropionyl isopropyl derivatives and a chiral column with NPD detection (Hashimoto et al., 1992).

In recent years, the abuse of γ-hydroxybutyrate (GHB) has risen dramatically (LeBeau et al., 2000). This metabolite of GABA can be difficult to analyze because of its short half-life and endogenous presence. However, a simple and fast head space analysis using GC–FID with MS confirmation has been published (LeBeau et al., 2000).

3.3.3 Acetylcholine and Choline

Sensitivities of 1 ng can be achieved by demethylation of the quarternary amines to tertiary amine analogs and analysis with FID detection (Tsai, 2000). However, the lack of volatility of these compounds makes them difficult to analyze satisfactorily with GC. Most work since the last edition of the *Handbook of Neurochemistry* on these compounds has utilized HPLC because these substances can be analyzed without demethylating or derivatizing (Tsai, 2000).

3.3.4 Polyamines

While GC has not been used extensively to analyze polyamines, a review in 2002 comprehensively covers useful methods (Teti et al., 2002) and a review in 2001 includes specific methods for various polyamines that are used to monitor cancer remission (Khuhawar and Qureshi, 2001). Using activated Permutit for cleanup and heptafluorobutyric anhydride for derivatization, sensitivities of 0.02–0.1 pmol were achieved for commonly analyzed polyamines using GC–ECD and human plasma samples (Teti et al., 2002). NPD also has utility because low detection limits are easily attained (Teti et al., 2002). PFB derivatives have shown some utility with GC–ECD (Coutts and Baker, 1982; Clements et al., 2004), and heptafluorobutryic anhydride with GC–ECD has also been used with success for analysis in human blood samples (Fujihara et al., 1983).

3.3.5 Steroids

Extensive reviews have been published on the analysis of steroids (e.g. Shimada et al., 2001; Andrew, 2001). GC is still used to measure levels of some volatile steroids, but LC–MS (Shimada et al., 2001) and GC–MS (Kim et al., 2000; Vallée et al., 2000; Purdy et al., 2004 and chapter in this volume) are now used for most analyses because of versatility and selectivity.

Endogenous and exogenous androgens can be derivatized with trimethylsilyl (TMS) for hydroxy functions and by O-methylation for ketones, and analyzed with GC–FID or GC–MS (Shimada et al., 2001). MS is more prevalent due to unequivocal identification and greatly increased sensitivity but FID is still used in laboratories for some steroids. Sterols have typically been analyzed by GC–FID and GC–MS with derivatization to optimize peak shape (Shimada et al., 2001), and bile acids can be derivatized with *n*-butyl ester-TMS ether and analyzed by GC–FID from plasma samples (Batta et al., 1998). Juricskay and Telegdy (2000) reported an assay capable of analyzing 28 steroids in urine samples using GC–FID.

3.4 Exogenous Compounds

Drug screens, used generally in toxicology laboratories (both clinical and forensic), are capable of separating a multitude of drugs in a variety of matrices using primarily GC–NPD or GC–FID. A review in 2002 updated the status of drug testing using hair and saliva samples employing various analytical techniques including GC, although primarily for drugs of abuse and doping agents (Kintz and Samyn, 2002). Hair analysis generally uses GC–MS for detection, but nitrazepam was analyzed using GC–NPD (Sachs and Kintz, 1998). Examples of methods using blood samples (both whole blood and plasma) are extensive (Ojanperä et al., 1991; Drummer et al., 1994; Chen et al., 1994; Moeller et al., 1998). Liver can also be an important matrix (Huang et al., 1996) and saliva is being used more commonly (Schramm et al., 1992).

Some methods are capable of screening for over 60 commonly prescribed drugs, including: antidepressants such as tricyclics; anxiolytics such as benzodiazepines; neuroleptics such as haloperidol; drugs of abuse, such as cocaine; and licit drugs that can be abused, such as codeine and morphine (Drummer et al., 1994; Huang et al., 1996; Drummer, 1999). These procedures are usually simple, rapid, and can estimate the concentration of the compound. Confirmation of concentration is generally done on a second instrument (usually MS). However, in general the methods used can also quantitate the compound of interest on the non-MS detector if a standard curve is included with the samples.

3.4.1 Antidepressants

GC is used extensively for analysis of antidepressants (Orsulak et al., 1989), but HPLC assays and enzyme immunoassays have become more popular in recent years. However, GC has advantages such as economy and ready availability. ECD and NPD generally are the detectors of choice (Coutts and Baker, 1982). NPD is relatively efficient for the analysis of tricyclic antidepressants (TCAs) as derivatization is not necessary, although the secondary, demethylated amines are sometimes derivatized to improve resolution and peak shape (Coutts and Baker, 1982). Acetylation, under aqueous or anhydrous conditions, followed by GC–NPD, has been used extensively for analysis of TCAs and the tetracyclic antidepressant maprotiline in plasma samples (Drebit et al., 1988). ➋ *Table 1-1* summarizes GC assays for some commonly prescribed antidepressants and their metabolites.

In 1986, Baker et al. published an article describing the analysis of pentafluorobenzoyl and pentafluorobenzenesulfonyl derivatives of several antidepressants and their metabolites (Baker et al., 1986c). The original article documented the derivatization of desipramine, 2-hydroxydesipramine, maprotiline,

Table 1-1
Representative GC assays for several antidepressants

Drug/Metabolite	Detector	Reference
Amitriptyline	NPD	Coutts et al., 1997
m-Chlorophenylpiperazine	NPD	Rotzinger et al., 1998
Desipramine	NPD	Goodnough et al., 1995
Fluoxetine and *p*-trifluoromethylphenol	ECD	Liu et al., 2002
Fluoxetine and Norfluoxetine	NPD	Goodnough et al., 1995
Fluoxetine and Norfluoxetine	ECD	Dixit et al., 1991; Lantz et al., 1993
Fluvoxamine	ECD	Rotzinger et al., 1997
Imipramine and metabolites	NPD	Coutts et al., 1993; Su et al., 1993
Maprotiline and desmethylmaprotiline	NPD	Drebit et al., 1988
Paroxetine	ECD	Lai et al., 2000
p-Trifluoromethylphenol	ECD	Urichuk et al., 1997
Sertraline and desmethylsertraline	ECD	Tremaine and Joerg, 1989
Trimipramine	NPD	Bolaji et al., 1993

nortriptyline, desmethylclomipramine, desmethyltrimpramine, 2-hydroxyimipramine, tranylcypromine, 4-hydroxytranylcypromine and nomifensine. However, the methods have been shown to be transferable and have been used since then to analyze other antidepressants and their metabolites (Urichuk et al., 1997; Rotzinger et al., 1997; Rotzinger et al., 1998; Liu et al., 2002).

Solid phase extraction (SPE) has been used to efficiently extract several types of antidepressants, which can then be conveniently analyzed on GC–NPD. One assay extracted and analyzed viloxazine, venlafaxine, imipramine, desipramine, sertraline, and amoxapine from whole blood in one procedure (Martinez et al., 2002). The same laboratory analyzed fluoxetine, amitriptyline, nortriptyline, trimipramine, maprotiline, clomipramine, and trazodone in whole blood in one assay (Martinez et al., 2003). SPE has also been used for the simultaneous analysis of TCAs and their metabolites by de la Torre et al. (1998).

3.4.2 Anxiolytics

3.4.2.1 Benzodiazepines Benzodiazepines (BDZs) are commonly used as anxiolytics, antipanic drugs, hypnotics, muscle relaxants, and anticonvulsants; they are also commonly abused in many segments of the population (Drummer, 1998). A major review of methods for quantitating BDZs in biological samples was published in 1998 and covered much of the progress in analysis (Drummer, 1998). While there are several alternative methods for analyzing BDZs, many can be analyzed on GC–ECD after a simple extraction because they have several electrophoric substituents in their structure (Cooper, 1988). An SPE followed by GC–MS detection separated and identified 19 BDZs and 2 thienodiazepines from whole blood (Inoue et al., 2000). This SPE could perhaps be used to clean up samples for analysis with other detectors also.

3.4.2.2 Nonbenzodiazepines GC–NPD was used by Kivistö et al. (1999) to quantitate the nonbenzodiazepine anxiolytic buspirone and its major metabolite, 1-(2-pyrimidinyl)-piperazine, using separate extraction methods and separate assays. The limit of quantification in plasma for both compounds was 0.2 ng/mL, which makes this assay useful for pharmacokinetic studies of this compound (Kivistö et al., 1999). A rapid, simple method for analysis of buspirone in rat brain requiring a single extraction step followed by GC–NPD has also been described (Lai. et al., 1997).

3.4.3 Antipsychotics

Many antipsychotics show great interindividual variation in plasma levels and so analysis of therapeutic levels can be important clinically as well as in the research laboratory. In addition, nonresponse to the drugs may actually be due to excessive levels of neuroleptics, a paradoxical situation that requires analysis to identify (Rockland, 1986). Several methods using FID were cited in the previous edition of the Handbook of Neurochemistry, but ECD and NPD have both shown utility for the typically low therapeutic levels (Cooper, 1988). GC–FID has been used to analyze levels of clozapine in blood, gastric, and urine samples in fatal cases of overdose with this drug (Ferslew et al., 1998), and olanzapine has been measured in blood and urine samples by GC–NPD in overdoses (Stephens et al., 1998). 4-(4-Chlorophenyl)-4-hydroxypiperidine, a metabolite of haloperidol, was analyzed in urine, plasma, brain, and liver from haloperidol-treated rats by GC–ECD, after derivatization with PFBC under aqueous conditions (Fang et al., 1996).

3.4.4 Stimulants

Many stimulants, such as amphetamine, methamphetamine, and caffeine contain nitrogen atoms, which makes NPD analysis fairly straightforward (eg. Koide et al., 1998; Bach et al., 1999). Enantiomeric separation can be of particular importance for these drugs. A review by Liu and Liu (2002) provides extensive examples of methods for the determination of amphetamine and methamphetamine enantiomers and includes examples using FID, NPD, ECD, or MS. A review of GC analysis of "performance enhancing drugs" by Müller et al. (1999) also covers stimulants.

Urine samples from apparent cases of fatal overdoses have been analyzed using GC–FID in a method that separated the enantiomers of amphetamine using a homochiral derivatizing reagent and an achiral column (Van Bocxlaer et al., 1997). GC–ECD has been used for analysis of amphetamine in brain, liver, heart, and plasma from rats and urine from humans (Paetsch et al., 1991; Sherry et al., 2000; Asghar et al., 2002). Analysis of stimulants is frequently a legal matter and so MS is often used because of its high degree of selectivity; however, caffeine can be easily quantified from human plasma using GC–NPD (Carregaro et al., 2001).

3.4.5 Alkaloids

The analysis of codeine, morphine, 6-monoacetylmorphine (6-MAM, a metabolite of heroin), and cocaine is important for many toxicology labs to determine illicit drug use. When analyzing opiates in urine samples, frequently the matrix chosen for drug screening, the conjugated metabolites must be hydrolyzed; however, this process can break down 6-MAM (Christophersen et al., 1987). These compounds can be derivatized to increase sensitivity, and both ECD and NPD are used for these assays. Derivatizations used include reaction with N-methyl-N-trimethylsilyltrifluoroacetamide followed by GC–FID (Lin et al., 1994) or with N,O-bis(trimethylsilyl)trifluoroacetamide (Christophersen et al., 1987; Lee and Lee, 1991), PFPA (Christophersen et al., 1987), or heptafluorobutyric anhydride (HFBA) followed by GC–ECD. All these methods show good sensitivity and selectivity.

Cocaine and its ethyl homolog, cocoethylene, have been assayed simultaneously using GC–NPD (Hime et al., 1991). These compounds can be derivatized by reducing the ester groups to alcohols, which can then be derivatized with a perfluoro acylating reagent to provide sensitivity for GC–ECD (Blake et al., 1974).

4 Conclusions

The use of GC with capillary columns and sensitive and selective detectors permits the analysis of a wide variety of compounds, both endogenous and exogenous, that are of interest to neurochemists. The detectors covered in this chapter do not provide the same degree of selectivity and sensitivity that MS does; however, when used with reference samples for verification they are more than satisfactory for routine analysis.

Acknowledgements

The authors are grateful for funding from the Canadian Institutes of Health Research (CIHR), the Canada Research Chairs program, the Davey Endowment, the Berger Endowment, and the Faculty of Graduate Studies and Research, University of Alberta.

References

Abe I, Fujimoto N, Nishiyama T, Terada K, Nakahara T. 1996. Rapid analysis of amino acid enantiomers by chiral-phase capillary gas chromatography. J Chromatogr A 722: 221.

Ahuja S. 1976. Derivatization in gas chromatography. J Pharm Sci 65: 163.

Andrew R. 2001. Clinical measurement of steroid metabolism. Best Prac Res Clin Endocrinol Metab 15: 1.

Asghar SJ, Baker GB, Rauw, GR, Silverstone PH. 2002. A rapid method of determining amphetamine in plasma samples using pentafluorobenzenesulfonyl chloride and electron-capture gas chromatography. J Pharmacol Toxicol Methods 46: 111.

Aspeslet LJ, Baker GB, Coutts RT, Torok-Both GA. 1994. The effects of desipramine and iprindole on levels of enantiomers of fluoxetine in rat brain and urine. Chirality 6: 86.

Bach MV, Coutts RT, Baker GB. 1999. Involvement of CYP2D6 in the in vitro metabolism of amphetamine, two N-alkylamphetamines and their 4-methoxylated derivatives. Xenobiotica 29: 719.

Baker GB, Coutts RT, Bornstein RA, Dewhurst WG, Douglass AB, et al. 1986b. An electron-capture gas chromatographic method for analysis of urinary 3-methoxy-4-hydroxyphenylethylene glycol (MHPG). Res Commun Chem Pathol Pharmacol 54: 141.

Baker GB, Coutts RT, Holt A. 1994. Derivatization with acetic anhydride: applications to the analysis of biogenic amines and psychiatric drugs by gas chromatography and mass spectrometry. J Pharmacol Toxicol Methods 31: 141.

Baker GB, Coutts RT, Martin IL. 1981. Analysis of amines in the central nervous system by gas chromatography with electron-capture detection. Prog Neurobiol 17: 1.

Baker GB, Koilpillai M, Nazarali AJ, Rao TS, Coutts RT. 1986c. Gas chromatography of antidepressants and their metabo lites as pentafluorobenzoyl and pentafluoro benzenesulfonyl derivatives. Proc West Pharmacol Soc 29: 291.

Baker GB, Nazarali, AJ, Coutts RT. 1985. A rapid and sensitive procedure for the simultaneous analysis of β-phenylethylamine and tranylcypromine in rat brain using trichloroacetylation and gas chromatography. Res Commun Chem Pathol Pharmacol 5: 317.

Baker GB, Prior TI, Coutts RT. 2002. Chirality and drugs used to treat psychiatric disorders. J Psychiatry Neurosci 27: 401.

Baker GB, Rao TS, Coutts RT. 1986a. Electron-capture gas chromatographic analysis of β-phenylethylamine in tissues and body fluids using pentafluorobenzenesulfonyl chloride for derivatization. J Chromatogr Biomed Appl 381: 211.

Baker GB, Yeragani VK, Dewhurst WG, Coutts RT, Mac Donald RN, et al. 1987. Simultaneous analysis of urinary *m*- and *p*-hydroxyphenylacetic acid, homovanillic acid and 5-hydroxyindole-3-acetic acid using electron-capture gas chromatography. Biochem Arch 3: 257.

Batta AK, Salen G, Rapole KR, Batta M, Earnest D, et al. 1998. Capillary gas chromatographic analysis of serum bile acids as the *n*-butyl ester–trimethylsilyl ether derivatives. J Chromatogr B 706: 337.

Blake JW, Ray RS, Noonan JS, Murdick PW. 1974. Rapid, sensitive gas–liquid Chromatographic screening procedure for cocaine. Anal Chem 46: 288.

Bolaji OO, Coutts RT, Baker GB. 1993. Metabolism of trimipramine in vitro by human CYP2D6 isozyme. Res Commun Chem Pathol Pharmacol 82: 111.

Burtis CA, Bowers LD, Chattoraj SC, Ullman MD. 1987. Chromatography. Fundamentals of Clinical Chemistry 3rd edition, NW Tietz, editors. Philadelphia: W.B. Saunders Company; pp. 105–123.

Caldwell J. 2001. Do single enantiomers have something special to offer? Hum Psychopharmacol 16: S67.

Carlsson H, Robertsson, G, Colmsjö. 2001. Response mechanisms of thermionic detectors with enhanced nitrogen selectivity. Anal Chem 73: 5698.

Carregaro AB, Woods WE, Tobin T, Queiroz-Neto A. 2001. Comparison of the quantification of caffeine in human plasma by gas chromatography and ELISA. Braz J Med Biol Res 34: 821.

Chen X-H, Franke J-P, Wijsbeek J, de Zeeuw RA. 1994. Determination of basic drugs extracted from biological matrices by means of solid-phase extraction and wide-bore capillary gas chromatography with nitrogen–phosphorus detection. J Anal Toxicol 18: 150.

Christophersen AS, Biseth A, Skuterud B, Gadeholt G. 1987. Identification of opiates in urine by capillary column gas chromatography of two different derivatives. J Chromatogr B Biomed Appl 422: 117.

Clements RLH, Holt A, Gordon ES, Todd KG, Baker GB. 2004. Determination of rat hepatic polyamines by electron-capture gas chromatography. J Pharmacol Toxicol Methods 50: 35.

Cooper TB. 1988. Gas–liquid chromatography of antidepressant, antipsychotic and benzodiazepine drugs in plasma and tissues. Neuromethods, Vol. 10: Analysis of Psychiatric Drugs. Boulton AA, Baker GB, Coutts RT, editors. New Jersey: Humana Press; pp. 65–98.

Coutts RT, Bach MV, Baker GB. 1997. Metabolism of amitriptyline with CYP2D6 expressed in a human cell line. Xenobiotica 27: 33.

Coutts RT, Baker GB. 1982. Gas chromatography. Handbook of Neurochemistry, second edition. Volume 2: Experimental Neurochemistry. A. Lajtha editor. New York: Plenum Press; pp. 429–448.

Coutts RT, Su P, Baker GB, Daneshtalab M. 1993. Metabolism of imipramine in vitro by isozyme CYP2D6 expressed in a human cell line, and observations on metabolite stability. J Chromatogr B Biomed Appl 615: 265.

Dallüge J, Beens J, Brinkman UAT. 2003. Comprehensive two-dimensional gas chromatography: a powerful and versatile tool. J Chromatogr A 1000: 69.

Davis BA, Yu PH, Carlson K, O'Sullivan K, Boulton AA. 1982. Plasma levels of phenylacetic acid, *m*- and *p*-hydroxyphenylacetic acid, and platelet monoamine oxidase activity in schizophrenic and other patients. Psychiatry Res 6: 97.

de la Torre R, Ortuño J, Pascual JA, González, S, Ballesta J.1998. Quantitative determination of tricyclic antidepressants and their metabolites in plasma by solid-phase extraction (Bond-Elut TCA) and separation by capillary gas chromatography with nitrogen–phosphorus detection. Ther Drug Monit 20: 340.

Dixit B, Nguyen, H, Dixit VM. 1991. Solid-phase extraction of fluoxetine and norfluoxetine from serum with gas-chromatography–electron-capture detection. J Chromatogr B 563: 379.

Drebit R, Baker GB, Dewhurst WG. 1988. Determination of maprotiline and desmethylmaprotiline in plasma and urine by gas chromatography with nitrogen–phosphorus detection. J Chromatogr B Biomed Appl 432: 334.

Drozd J. 1975. Chemical derivatization. J Chromatogr 113: 303.

Drummer OH, Horomidis S, Kourtis S, Syrjanen ML, Tippett P. 1994. Capillary gas chromatographic drug screen for use in forensic toxicology. J Anal Toxicol 18: 134.

Drummer OH. 1998. Methods for the measurement of benzodiazepines in biological samples. J Chromatogr B 713: 201.

Drummer OH. 1999. Chromatographic screening techniques in systematic toxicological analysis. J Chromatogr B 733: 27.

Dybowski C, Kaiser MA. 2002. Chromatography. Kirk-Othmer Encyclopedia of Chemical Technology, John Wiley and Sons, Inc; http://www.mrw.interscience.wiley.com/kirk/articles/chrokais.a01/frame.html.

Fang J, Baker GB, Coutts RT. 1996. Determination of 4-(4-chlorophenyl)-4-hydroxypiperidine, a metabolite of haloperidol, by gas chromatography with electron-capture detection. J Chromatogr B 682: 283.

Ferslew KE, Hagardorn AN, Harlan GC, McCormick WF. 1998. A fatal drug interaction between clozapine and fluoxetine. J Forensic Sci 43: 1082.

Franke JP, de Zeeuw RA. 1998. Solid-phase extraction procedures in systematic toxicological analysis. J Chromatogr B 713: 51.

Fujihara S, Nakashima, T, Kurogochi Y. 1983. Determination of polyamines in human blood by electron-capture gas–liquid chromatography. J Chromatogr B Biomed Appl 277: 53.

Goodnough DB, Baker GB, Coutts RT. 1995. Simultaneous quantification of fluoxetine, norfluoxetine and desipramine using gas chromatography with nitrogen–phosphorus detection. J Pharmacol Toxicol Methods 34: 143.

Hashimoto A, Nishikawa T, Hayashi T, Fujii N, Harada K, et al. 1992. The presence of free D-serine in rat brain. Federat Europ Biochem Soc 296: 33.

Hegadoren KM, Baker GB, Coutts RT. 1993. The simultaneous separation and quantitation of the enantiomers of MDMA and MDA using gas chromatography with nitrogen–phosphorus detection. Res Commun Subst Abuse 14: 67.

Hime GW, Hearn WL, Rose S, Cofino J. 1991. Analysis of cocaine and cocaethylene in blood and tissues by GC–NPD and GC-ion trap mass spectrometry. J Anal Toxicol 15: 241.

Huang Z-P, Chen X-H, Wijsbeek J, Franke J-P, de Zeeuw RA. 1996. An enzymic digestion and solid-phase extraction procedure for the screening for acidic, neutral, and basic drugs in liver using gas chromatography for analysis. J Anal Toxicol 20: 248.

Hušek P. 1998. Chloroformates in gas chromatography as general purpose derivatizing agents. J Chromatogr B 717: 57.

Hušek P, Matucha P, Vránková A, Šimek P. 2003. Simple plasma work-up for a fast chromatographic analysis of homocysteine, cysteine, methionine and aromatic amino acids. J Chromatogr B 789: 311.

Inoue H, Maeno Y, Iwasa M, Matoba, R, Nagas M. 2000. Screening and determination of benzodiazepines in whole blood using solid-phase extraction and gas chromatography/mass spectrometry. Forensic Sci Int 113: 367.

Jamali R, Mehvar, R, Pasutto RM. 1989. Enantioselective aspects of drug action and disposition: therapeutic pitfalls. J Pharm Sci 78: 695.

Jirovský D, Lemr K, Ševčík J, Smysl B, Stránský Z. 1998. Methamphetamine – properties and analytical methods of enantiomer determination. Forensic Sci Int 96: 61.

Juricskay S, Telegdy E. 2000. Urinary steroids in women with androgenic alopecia. Clin Biochem 33: 97.

Kataoka H. 1997. Selective and sensitive determination of protein and non-protein amino acids by capillary gas chromatography with nitrogen–phosphorus selective detection. Biomed Chromatogr 11: 154.

Khuhawar MY, Qureshi GA. 2001. Polyamines as cancer markers: applicable separation methods. J Chromatogr B 764: 385.

Kim Y-S, Zhang H, Kim H-Y. 2000. Profiling neurosteroids in cerebrospinal fluids and plasma by gas chromatography/electron capture negative chemical ionization mass spectrometry. Anal Biochem 277: 187.

Kintz P, Samyn N. 2002. Use of alternative specimens: drugs of abuse in saliva and doping agents in hair. Ther Drug Monit 24: 239.

Kivistö KT, Laitila J, Mårtensson K, Neuvonen PJ. 1999. Determination of buspirone and 1-(2-pyrimidinyl)-piperazine (1-PP) in human plasma by capillary gas chromatography. Ther Drug Monit 21: 317.

Koide I, Noguchi O, Okada K, Yokoyama A, Oda H, et al. 1998. Determination of amphetamine and methamphetamine in human hair by headspace solid-phase microextraction and gas chromatography with nitrogen–phosphorus detection. J Chromatogr B 707: 99.

Kueh AJ, Marriott PJ, Wynne PM, Vine JH. 2003. Application of comprehensive two-dimensional gas chromatography to drugs analysis in doping control. J Chromatogr A 1000: 109.

Lai C-T, Gordon ES, Kennedy SH, Bateson AN, Coutts RT, et al. 2000. Determination of paroxetine levels in human plasma using gas chromatography with electron-capture detection. J Chromatogr B 749: 275.

Lai C-T, Tanay VA-MI, Rauw GA, Bateson AN, Martin IL, et al. 1997. Rapid, sensitive procedure to determine buspirone levels in rat brains using gas chromatography with nitrogen–phosphorus detection. J Chromatogr B 704: 175.

Lantz RJ, Farid KZ, Koons J, Tenbarge JB, Bopp RJ. 1993. Determination of fluoxetine and norfluoxetine in human plasma by capillary gas chromatography with electron-capture detection. J Chromatogr A 614: 175.

Le Beau MA, Montgomery MA, Miller ML, Burmeister SG. 2000. Analysis of biofluids for gamma-hydroxybutyrate (GHB) and gamma-butyrolactone (GBL) by headspace GC–FID and GC–MS. J Anal Toxicol 24: 421.

Lee H-M, Lee C-W. 1991. Determination of morphine and codeine in blood and bile by gas chromatography with a derivatization procedure. J Anal Toxicol 15: 182.

Lin Z, Lafolie P, Beck O, 1994. Evaluation of analytical procedures for urinary codeine and morphine measurements. J Anal Toxicol 18: 129.

Liu J-T, Liu RH, 2002. Enantiomeric composition of abused amine drugs: chromatographic methods of analysis and data interpretation. J Biochem Biophys Methods 54: 115.

Liu Z-Q, Tan Z-R, Wang D, Huang S-L, Wang L-S, et al. 2002. Simultaneous determination of fluoxetine and its metabolite *p*-trifluoromethylphenol in human liver microsomes using a gas chromatographic-electron-capture detection procedure. J Chromatogr B 769: 305.

Martínez MA, Sánchez de la Torre C, Almarza E. 2002. Simultaneous determination of viloxazine, venlafaxine, imipramine, desipramine, sertraline, and amoxapine in whole blood: comparison of two extraction/cleanup procedures for capillary gas chromatography with nitrogen–phosphorus detection. J Anal Toxicol 26: 296.

Martínez MA, Sánchez de la Torre C, Almarza E. 2003. A comparative solid-phase extraction study for the simultaneous determination of fluoxetine, amitriptyline, nortriptyline, trimipramine, maprotiline, clomipramine, and trazodone in whole blood by capillary gas–liquid chromatography with nitrogen–phosphorus detection. J Anal Toxicol 27: 353.

Moeller MR, Steinmeyer S, Kraemer T. 1998. Determination of drugs of abuse in blood. J Chromatogr B 713: 91.

Müller RK, Grosse J, Thieme D, Lang R, Teske J, et al. 1999. Introduction to the application of capillary gas chromatography of performance-enhancing drugs in doping control. J Chromatogr A 843: 275.

Nazarali AJ, Baker GB, Coutts RT, Yeung JM, Rao TS. 1987. Rapid analysis of β-phenylethylamine in tissues and body fluids utilizing pentafluorobenzoylation followed by electron-capture detection. Prog Neuro-Psychopharmacol Biol Psychiatry 11: 251.

Ojanperä I, Rasanen, I, Vuori E. 1991. Automated quantitative screening for acidic and neutral drugs in whole blood by dual-column capillary gas chromatography. J Anal Toxicol 15: 204.

Orsulak PJ, Haven MC, Burton ME, Akers LC. 1989. Issues in methodology and applications for therapeutic monitoring of antidepressant drugs. Clin Chem 35: 1318.

Paetsch PR, Baker GB, Caffaro LE, Greenshaw AJ, Rauw GA, et al. 1991. An electron-capture gas chromatographic procedure for simultaneous determination of amphetamine and N-methylamphetamine. J Chromatogr Biomed Appl 573: 313.

Perry JA, Feit CA. 1978. Derivatization techniques in gas–liquid chromatography. GLC and HPLC Determination of Therapeutic Agents, Part 1, Tsuji K. Marozowich W, editors. New York: Marcel Dekker; pp. 137–208.

Purdy RH, Fitzgerald RL, Everhart ET, Mellon SH, Alomary A, Parsons LH. 2004. The analysis of neuroactive steroids by mass spectrometry. Handbook of Neurochemistry and Molecular Neurobiology, 3rd edition, Volume 18: Practical Neurochemistry (Methods). Baker GB, Dunn SMJ, Holt A, editors. New York: Kluwer Academic Publishers.

Rockland LH. 1986. Neuroleptic blood levels and clinical response. Can J Psychiatry 31: 299.

Rotzinger S, Fang J, Baker GB. 1998. Trazodone is metabolized to *m*-chlorophenylpiperazine by CYP3A4 from human sources. Drug Metab Dispos 26: 572.

Rotzinger S, Todd KG, Bourin M, Coutts RT, Baker GB. 1997. A rapid electron-capture gas chromatographic method for the quantification of fluvoxamine in brain tissue. J Pharmacol Toxicol Methods 37: 129.

Ryan D, Marriott P. 2003. Comprehensive two-dimensional gas chromatography. Anal Bioanal Chem 376: 295.

Sachs H, Kintz P. 1998. Testing for drugs in hair, critical review of chromatographic procedures since 1992. J Chromatogr B 713: 147.

Schramm W, Smith RH, Craig PA, Kidwell DA. 1992. Drugs of abuse in saliva: a review. J Anal Toxicol 16: 1.

Schurig V. 2001. Separation of enantiomers by gas chromatography. J Chromatogr A 906: 275.

Sherry RL, Rauw G, McKenna KF, Paetsch PR, Coutts RT, et al. 2000. Failure to detect amphetamine or 1-amino-3-phenylpropane in humans or rats receiving the MAO inhibitor tranylcypromine. J Affect Disord 61: 23.

Shimada K, Mitamura K, Higashi T. 2001. Gas chromatography and high-performance liquid chromatography of natural steroids. J Chromatogr A 935: 141.

Smith RM. 2003. Before injection—modern methods of sample preparation for separation techniques. J Chromatogr A 1000: 3.

Srinivas NR, Shyu WC, Barbhaiya RH. 1995. Gas chromatographic determination of enantiomers as diastereomers following pre-column derivatization and applications to pharmacokinetic studies: a review. Biomed Chromatogr 9: 1.

Stephens BG, Coleman DE, Baselt RC. 1998. Olanzapine-related fatality. J Forensic Sci 43: 1252.

Su P, Coutts RT, Baker GB, Daneshtalab M. 1993. Analysis of imipramine and three metabolites produced by isozyme CYP2D6 expressed in a human cell line. Xenobiotica 23: 1289.

Tao QF, Zeng S. 2002. Analysis of enantiomers of chiral phenethylamine drugs by capillary gas chromatography/mass spectrometry/flame-ionization detection and pre-column chiral derivatization. J Biochem Biophys Methods 54: 103.

Teti D, Visalli M, McNair H. 2002. Analysis of polyamines as markers of (patho)physiological conditions. J Chromatogr B 781: 107.

Torok-Both GA, Baker GB, Coutts RT, McKenna KF, Aspeslet LJ. 1992. Simultaneous determination of fluoxetine and norfluoxetine enantiomers in biological samples by gas chromatography with electron-capture detection. J Chromatogr B Biomed Appl 579: 99.

Tremaine LM, Joerg EA. 1989. Automated gas chromatography–electron-capture assay for the selective serotonin uptake blocker sertraline. J Chromatogr B 496: 423.

Tsai T-H. 2000. Separation methods used in the determination of choline and acetylcholine. J Chromatogr B 747: 111.

Tucker GT. 2000. Chiral switches. Lancet 355: 1085.

Ulrich S. 2003. Direct stereoselective assay of fluoxetine and norfluoxetine enantiomers in human plasma or serum by two-dimensional gas–liquid chromatography with nitrogen–phosphorus selective detection. J Chromatogr B 783: 481.

Urichuk LJ, Aspeslet LJ, Holt A, Silverstone PH, Coutts RT, et al. 1997. Determination of *p*-trifluoromethylphenol, a metabolite of fluoxetine, in tissues and body fluids using an electron-capture gas chromatographic procedure. J Chromatogr B 698: 103.

Vallée M, Rivera JD, Koob GF, Purdy RH, Fitzgerald RL. 2000. Quantification of neurosteroids in rat plasma and brain following swim stress and allopregnanolone administration using negative chemical ionization gas chromatography/mass spectrometry. Anal Biochem 287: 153.

Van Bocxlaer JF, Lambert WE, Theinpont L, De Leenheer AP 1997. Quantitative determination of amphetamine and α-phenylethylamine enantiomers in judicial samples using capillary gas chromatography. J Anal Toxicol 21: 5.

Watson DG. 1999. Pharmaceutical Analysis A Textbook for Pharmacy Students and Pharmaceutical Chemists. Churchill Livingstone, Edinburgh: pp. 207-235.

Wells RJ. 1999. Recent advances in non-silylation derivatization techniques for gas chromatography. J Chromatogr A 843 : 1.

Williams ML, Wainer IW. 2002. Role of chiral chromatography in therapeutic drug monitoring and in clinical and forensic toxicology. Ther Drug Monit 24: 290.

Wong JTF, Baker GB, Coutts RT. 1988. Rapid and simple procedure for the determination of urinary phenylacetic acid using derivatization in aqueous medium followed by electron-capture gas chromatography. J Chromatogr B Biomed Appl 428: 140.

Wong JTF, Baker GB, Coutts RT. 1990a. A rapid, sensitive assay for γ-aminobutyric acid in brain using electron-capture gas chromatography. Res Commun Chem Pathol Pharmacol 70: 115.

Wong JTF, Paetsch PR, Baker GB, Greenshaw AJ, Coutts RT. 1990b. A rapid procedure for the analysis of phenylalanine in brain tissue utilizing electron-capture gas chromatography. J Neurosci Methods 32: 105.

Yamada N, Takahashi S, Todd KG, Baker GB, Paetsch PR. 1993. Effects of two substituted hydrazine monoamine oxidase (MAO) inhibitors on neurotransmitter amines, γ-aminobutyric acid, and alanine in rat brain. J Pharm Sci 82: 934.

2 High-Performance Liquid Chromatographic Analysis of Psychotropic and Endogenous Compounds

J. Odontiadis · G. Rauw

Abstract: High-performance liquid chromatography (HPLC) is a versatile analytical technique used extensively in the neurosciences to investigate levels and possible functions of a wide variety of neurochemicals and to study drug levels and metabolism and drug-drug interactions. In this chapter, we describe basic principles and instrumentation (including columns, mobile phases and detectors). There are also general discussions about sample preparation and chirality, followed by specific examples of applications of HPLC to analysis of endogenous compounds, drugs and/or their metabolites.

List of Abbreviations: HPLC, High-performance liquid chromatography; ODC, octadecyl silane; Ph, phenyl; CN, cyano; UV/VIS, ultraviolet/visible; PDAD, photodiode array detector; EC, electrochemical; MS, mass spectrometer; SPE, solid phase extraction; CSP, chiral stationary phase; CMPA, chiral mobile phase additive method; OVM, ovomucoid; OVG, ovoglycoprotein; AVD, avidin; FLA, flavoprotein; CBH I, Cellobiohydrolase I; NE, norepinephrine; E, epinephrine; DA, dopamine; 5-HT, 5-hydroxytryptamine; CSF, cerebrospinal fluid; 5-HIAA, 5-hydroxyindole-3-acetic acid; HVA, homovanillic acid; DOPAC, 3,4-dihydroxyphenylacetic acid; FMOC-CL, 9-fluorenylmethyloxycarbonyl chloride; NM, normetanephrine; MN, metanephrine; DANSYL, 1-dimethyl aminonaphthalenesulfonyl chloride; PITC, phenylisothiocyanate; AQC, 6-aminoquinoquinolyl-*N*-hydrocysuccinimidyl carbamate; OPA, *ortho*-phthalaldehyde; NDA, naphthalene-2,3-dicarboxaldehyde; GLU, glutamate; GLY, glycine; TAU, taurine; GABA, γ-aminobutyric acid; CBI, 1-cyanobenz[*f*]isoindole; *D*-Ser, *D*-serine; *D*-Asp, *D*-aspartate; NMDA, N-methyl- D-aspartate; NAC, *N*-acetyl-L-cysteine; Boc-*L*-Cys, *N-tert*-butyloxy-carbonyl- *L*-cysteine; IBLC, *N*-isobutyryl -*L*-cysteine; IBDC, *N*-isobutyryl- *D*-cysteine; BTCC, *N*-(*tert*-butylthiocarbamoyl)- *L*-cysteine; ACh, acetylcholine; Ch, choline; IMER, immobilized enzyme reactor; Pt, platinum; GC, gas chromatography; HRP-GCE, horseradish peroxidase-osmium redox polymer-modified glassy carbon electrode; PFCE, disposable, film carbon electrode; Os-gel-HRP, osmium-polyvinylpyrridine-wired horderadish peroxidase gel polymer; PUT, putrescine; SP, permine; SPD, spermidine; CAD, cadaverine; PSE, 4-(1-pyrene)butyric acid *N*-hydroxysuccinimide ester; LLE, liquid–liquid extraction; RP, reverse phase; NP, normal phase; DNSH, dansylhydrazine; TCAs, tricyclic antidepressants; AMI, amitripyline; IMI, imipramine; NT, nortryptyline; CMI, clomipramine; DMI, desipramine; LLOQ, lower limit of quantification; LLOD, lower limit of detection; NBD-COCL, 4-(N-chloroformylmethyl-N-methyl)amino-7-nitro-2,1,3,-benzoxadiazole; BZD, benzodiazepine; 1-PP, 1-(2 pyrimidinyl) piperazine; CBA, carboxymethyl; SDS, sodium dodecyl sulphate; PTZ, phenothiazine; MDMA, 3-4-methylenedioxymethamphetamine; MDEA, N-ethyl-3,4-methylenedioxyamphetamine; MDA, methylenedioxyamphetamine; MBDB, N-methyl-1-(1,3-benzdioxol-5-yl)-2-butamine

1 Introduction

High-performance liquid chromatography (HPLC) has become a powerful tool of analysis since its appearance in analytical laboratories several decades ago. Improvements in technology have led to superior packing materials with particle diameters of 3 μm to 10 μm, which in turn have led to greatly improved pump engineering and detector flow cell dynamics to provide low flow rates, relatively short retention times, and high throughput of samples. HPLC offers advantages over gas chromatography such as, the ability to analyze nonvolatile or heat labile samples without derivatization; manipulation of a wide range of separation mechanisms; and analysis at near ambient temperature. This chapter presents a brief review of the principles of HPLC as well as its applications.

2 Basic Principles

In HPLC, the separation of a mixture of compounds is performed on an analytical column that is packed with small particles of stationary phase (typically silica 3, 5, or 10 μm in diameter) by elution with a liquid (mobile phase) under pressure.

There are four different mechanisms of separation utilized in HPLC: adsorption, partition, ion-exchange, and size exclusion chromatography.

Adsorption arises from the interaction between solutes and the surface of a solid stationary phase. Ion-exchange chromatography involves a solid stationary phase with either anionic or cationic functional groups on the surface to which solute molecules of opposite charge are attracted. Size exclusion chromatography involves a solid stationary phase with a specific pore size. Solutes are separated according to their molecular size. The large molecules are unable to enter the pores, thus retaining the least and eluting first. Partition chromatography involves a liquid stationary phase, which is immiscible with the eluent (mobile phase) and is coated on an inert support. This is the most commonly used mode of separation for drug analysis.

There are two types of partition chromatography that are distinguishable based upon the relative polarities of the mobile and stationary phases. In normal phase chromatography, a highly polar stationary phase is used with a relatively nonpolar mobile phase. As a general principle in normal phase chromatography, the least polar components are eluted first and, increasing the polarity of the mobile phase decreases the elution time.

In reverse-phase chromatography, the stationary phase is nonpolar (often a hydrocarbon) and the mobile phase is relatively polar (e.g., water, methanol, and acetonitrile). The most polar components elute first, and increasing the mobile phase polarity (i.e., decreasing the organic solvent concentration) increases elution time.

The concept of the four different modes of separation is an oversimplification. In reality there are no distinct boundaries, and several different mechanisms often operate simultaneously.

There are typically six components in an HPLC system: (1) solvent reservoirs; (2) a pumping or solvent management system; (3) an injector, which can be either manual or automated; (4) a column; (5) a detector; (6) a data recorder, which can be an integrator or a computer system.

The pumping or sample management system can be as simple as an isocratic system, consisting of a single piston pump head connected to a manual injector. It can also be as encompassing as a quartenary gradient system with two independently driven pump plungers for optimal flow control connected to an autosampler capable of accurately handling sample injections as low as 5 μl out of a total volume of 10 μl. Choice of HPLC components will depend on the application of the system. Any application that is flow sensitive, such as an electrochemical detector, or the use of a microbore column, will require a pumping system that generates a noise-free stable and flat baseline. Generally, a pump head using two or more reciprocating pump heads will have less baseline noise than a single pump head. A pumping system containing independently driven pump plungers controlled by individual transducers has the ability to produce less pump noise than the reciprocating pump head system. Isocratic systems meet most laboratory needs. Gradient systems are beneficial when separating multiple analytes in one sample, during method development, or for washing the column between injections.

3 HPLC Analytical Columns

The most widely used support substance for the manufacture of packing materials in analytical HPLC columns is silica. Silica can be treated with organochlorosilanes or similar reagents to produce siloxane linkages of any derived polarity similar to what is done for GC columns (stationary phases). The most popular materials are octadecyl silane (ODS), which contains a carbon loading of C18 groups and octyl, which contains C8 groups; materials such C2, C6, and C22 are also available.

Variations in elution order on different packing materials (e.g., ODS-silica) are often attributed to differences in the surface coverage and the presence of exposed residual silanol (SiOH) groups. ODS-silicas are more lipophilic and retain organic compounds more strongly than octyl groups (i.e., C18 > C8 > C2). There is also a vast range of materials that have intermediate surface polarities arising from the loading of the silica with other organic compounds such as phenyl (Ph), cyano (CN), amino (NH_2), and hydroxyl (OH). The pH range for the use of these bonded silica columns is between 2 and 8. Eluents with a pH above 8 will dissolve the silica support material and a pH below 2 will damage or remove the siloxane linkage (stationary phase) on the surface of the silica support.

Newer types of packing materials have emerged that have proven to be stable at a wider pH range (pH 1–12). In one type, silicas and polymers are combined and distributed throughout the particle backbone.

In another type, a high-purity silica is densely bound and end capped. New chiral stationary phases have also been developed over the past decade and are discussed further in this chapter.

Other matters to consider in column choice are column length, column diameter, and particle size. Column efficiency (theoretical plate count) is determined by a ratio of column length to particle size. A shorter column with the same particle size may give a shorter run time but at a loss of resolution. A shorter column with a smaller particle size with a lower flow rate may give a similar resolution in a shorter time. Retention time reproducibility improves in systems where column temperature can be controlled, especially in cases where ambient room temperature varies.

Analysis of biological samples requires an inline precolumn or guard column installed before the analytical column, to protect against the gradual accumulation of particulates or contaminants originating from the sample. They should be packed with a similar material as the analytical column and should be disposed of on the first indication of contamination (i.e., high back pressure or loss of resolution).

4 HPLC Mobile Phase Systems

There are a large number of eluents/packing material combinations that are used for drug analysis. The most commonly used HPLC systems utilize the reverse-phase mode to achieve the required separations and retentions by controlling the partition between the organic stationary phase (bonded silica) and the polar mobile phase. The mobile phase can be modified by changes in polarity, pH, and the use of ion-pairing reagents. Mobile phases typically employed are usually mixtures of methanol, acetonitrile with water, or aqueous buffer solutions. Retention is mainly controlled by the hydrophobic interactions between the drugs and the hydrocarbon chains bonded onto the silica support surface. The retention increases as the solute decreases in polarity (i.e., polar species are eluted first) and vice versa. This can also be affected by adjusting the polarity of the mobile phase by varying the concentration of the organic solvent to water or aqueous buffer.

The pH of the mobile phase and the pKa (dissociation constant) of the analyte are also important because it is the nonionized species that shows greatest retention. Hence, organic acids show an increase in retention as the pH is reduced whereas organic bases show a decrease. It is important to use a buffer of sufficient buffering capacity to cope with any injected sample size volume; otherwise, tailing peaks can arise from changes in ionic form during chromatographic separation. Phosphate buffers are widely used as they have a good pH range of buffering capacity and low ultraviolet absorbance with spectrophotometric detection.

Analytes containing basic nitrogen atoms sometimes show poor efficiencies and give tailing peaks due to interactions with residual silanol groups on the surface of silica support materials. This can often be improved by the addition of an amine compound such as triethylamine to the mobile phase that competes with the solutes for the silanol adsorption sites on the silica surface. Other hydrocarbon bonded packing materials of lesser carbon loading (e.g., C8 or C2 in reverse-phase mode) are generally associated with a decrease in retention times. Analytes bearing positive or negative charges are poorly retained in reverse-phase systems. If the pH of the mobile phase cannot be changed to convert the analyte to its nonionized form, a hydrophobic ion of opposite charge can be added to form a neutral "ion-pair" and increase retention time. For example, an acidic mobile phase or eluent is chosen for a compound that possesses a high pKa value and a hydrophobic anion is added. This technique is referred to as reverse-phase ion-pair chromatography. The sodium salts of alkylsulphonic acids (RSO_3^- Na^+ where R = pentyl, heptyl, or octyl) are widely used as ion-pairing reagents for basic compounds, whereas ammonium compounds such as tetrabutylammonium salts are used for acidic compounds.

The quality of solvents and inorganic salts used for mobile phase preparations is an important consideration. Soluble impurities can generate noisy detector baseline disturbances and spurious chromatographic peaks or can build up on the surface of the packing material, leading to increasingly high analytical column head pressures and changing retention times. Air dissolved in the mobile phase can also lead to various problems. The formation of air bubbles in high-pressure pumps usually reduces or completely stops the eluent flow whereas air bubble formation in detector systems produces erratic and unstable baselines. The remedy is to remove dissolved gases by degassing the eluent. One method of degassing consists of "sparging" in which the dissolved gases are swept or displaced out of the solution by fine bubbles of an inert gas that is not soluble in the mobile phase (e.g., H_2 or He gases). Some HPLC systems have built-in degassers that ensure the removal of any dissolved gases from the eluent before their introduction to the remainder of the HPLC apparatus. Another method involves filtration of the mobile phase under vacuum, which not only removes dissolved gases but also removes any particulate material that may cause additional problems in the HPLC system.

Measurements for pH determinations in a mixture of aqueous and organic solvents should be described as "apparent" pH. The true pH value can only be measured in aqueous solutions. In general, the apparent pH of a buffer solution increases as the proportion of organic solvent in the aqueous mixture increases. When preparing an eluent, it is usually best to dissolve the required buffer salts in distilled water at the appropriate concentration, adjust the pH, then mix this solution with the required organic solvents.

5 HPLC Detectors

5.1 Ultraviolet/Visible Spectrophotometric Detectors (UV/VIS)

This is generally regarded as being the most popular detector for drug analysis. In its simplest form, it consists of a single wavelength source (mercury lamp) set at 254 nanometers (nm). The detector monitors the effluent from the column and will produce a response for all compounds that have an absorption at this wavelength. The limit of detection will depend on the extinction coefficient (ε) of the compound at 254 nm and also upon the separation efficiency of the analytical column. The limits of detection for a particular compound can be extended by the use of a variable wavelength detector, which permits setting the wavelength at the absorption maxima of the compound of interest. However, it is often just as useful to set the wavelength *away* from the wavelength at which endogenous components may interfere. The two factors have to be balanced to make the best use of variable wavelength detectors. The majority of compounds show some absorption at very low wavelengths (220 nm or less) but as selectivity is low, such detection wavelengths should only be used to enhance the sensitivity in the analysis of samples known to contain a particular compound.

The photodiode array detector (PDAD) measures absorption of light waves by a sample. This is considered the most powerful of the ultraviolet spectrophotometric detectors. The optical system focuses light from a deuterium source through the sample flow cell onto several photodiodes. These act as capacitators by holding a fixed amount of charge. When light strikes the photodiodes, they discharge a certain amount of current.

The magnitude of the discharge is proportional to the intensity of light striking the photodiode and related to the intensity of light transmitted through the sample flow cell.

Measurements are taken very rapidly, allowing spectra to be taken at multiple wavelengths during the same analysis (Skoog et al., 1998).

Diode resolution is an important consideration in a PDAD detector. It is calculated by dividing the wavelength range by the number of diodes in the array. This, in combination with the optical resolution, will affect the spectral resolution of the instrument (i.e., the wavelength in nm, between data points in an acquired spectrum). (Operation and Maintenance of the ZQ with MassLynx, Waters Corporation.)

5.2 Fluorescence Detectors

In this detector, the solute is excited with UV radiation and emits radiation at a longer wavelength. The limits of this detection by a spectrofluorometer depend upon the physico-chemical properties of the compound (degree of aromacity), the solvent, and the pH. The eluents (mobile phase) should neither fluoresce nor should they absorb at the excitation and emission wavelengths used. The pH is important because some compounds show fluorescence only in particular ionic forms. There are only a few compounds that have strong native fluorescence, and for these analytes fluorescent detection can achieve better sensitivities in comparison to other spectrophotometric techniques. Reactions with a derivatizing reagent such as *o*-pthaldialdehyde are often employed to enhance the fluorescence species, for example, in the analysis of amino acids.

5.3 Electrochemical (EC) Detectors

These detectors are based upon the ability of the compound to undergo oxidation or reduction at the surface of a carbon or platinum electrode, which has an applied constant potential. The resulting current is measured and is proportional to the concentration of the electrode species present. There are two types of EC detectors commonly available, amperometric and coulometric.

In amperometric detectors, the eluent flows by the surface of the glassy carbon electrode in which only 5–15% of the electroactive species is present and this undergoes electrolytic conversion (oxidation or reduction) as the surface area of the electrode is relatively small.

In coulometric detectors, the eluent flows *through* a porous graphite electrode such that, in theory, 100% of any electroactive species will undergo electrolytic conversion. As a result, this significantly increases the detection sensitivity, as the surface area is relatively large.

In most applications, the electrochemical compounds are usually oxidized, yielding one or more electrons per molecule reacted. The oxidized form is usually unstable and reacts further to form a stable compound that flows past the carbon electrode surface. Unfortunately, this is not always the case, with the stable oxidized form occasionally building up at the surfaces of the carbon electrode. This creates sensitivity problems and decreases the efficiency of the detector. However, the problem is usually overcome by regularly cleaning the carbon electrode surfaces, removing any oxidizable products. Eluents for EC detection must be electrochemically conductive, which is achieved by the addition of inert electrolytes (to maintain a baseline current) such as phosphate or acetate. All solvents and buffers used in preparation of an eluent must be relatively pure and selected so as to not undergo electrochemical changes at the applied electrode potentials.

5.4 Mass Spectrometer (MS)

The mass spectrometric detector separates gas phase ions according to their m/z (mass to charge ratio) value. Introduction of the sample from an HPLC system to an MS detector is usually done under atmospheric conditions and requires special considerations with regard to flow rate, pH, and mobile phase constituents. This is discussed further in the chapter of this handbook by Sloley et al.

5.5 Other Detectors

The refractive index detector operates by comparing the refractive index of the mobile phase prior to the column with the refractive index of the column eluate. This detector responds to nearly all solutes but it is highly temperature-sensitive (Skoog et al., 1998). This type of detector can be used for sugars and fatty acids.

Evaporative light scattering detectors for nonchromophoric compounds are gaining in popularity and are used in analysis of various classes of lipids (LaCourse, 2002). The effluent from the column is nebulized with the aid of nitrogen or air, vaporizing the mobile phase and analyte molecules that pass through a laser beam. A silicon photodiode detects scattered radiation at right angles to the flow. This type of detector is more sensitive than a refractive index detector (Skoog et al., 1998).

Infrared detectors are similar in construction to those used in UV detection. The main difference is that the sample cell windows are constructed of sodium chloride, potassium bromide, or calcium fluoride. A limitation of this type of detector is caused by the low transparency of many useful solvents (Skoog et al., 1998). Recent changes to interface systems that use spraying to induce rapid evaporation of the solvent provide good sensitivity and enhanced spectral quality (LaCourse, 2000).

6 Analysis

6.1 Sample Collection and Preparation

The first step in analysis is appropriate collection of the sample that is to be assayed. In the case of drug analysis there has, historically, been documented interference of analysis of tricyclic antidepressants and neuroleptics by vacutainer stoppers containing the plasticizer tris(2-butoxyethyl)phosphate (LeGatt, 1988). Current literature suggests that serum tubes with gel separators should be avoided when collecting samples for many drugs, especially tricyclic antidepressants or steroids such as progesterone (LeGatt, 1988; Ernst and Ernst, 2003). Mei and colleagues (2003) suggest that lithium heparin should be avoided as an anticoagulant for plasma samples to be assayed by HPLC/MS/MS because it can contribute to matrix effect problems.

Whole blood should be quickly centrifuged and serum or plasma stored at a minimum of −20°C, and preferably at −60 to −80°C. Freezing and thawing of samples should be avoided. Collecting tissue samples from animals, particularly from their brain or brain regions, for neurochemical and drug analysis requires a quick freezing method to prevent postmortem changes. Rapid enzyme inactivation is discussed in a previous edition of this handbook (Lenox et al., 1982).

Our laboratory routinely immerses rodent brain tissue in ice cold 2-methylbutane (on dry ice) immediately after decapitation and dissection, transfers the tissue to another receptacle, and stores the samples at −80°C until required for analysis. For the determination of biogenic amines, amino acids, and psychotropic drugs, brain tissue samples are homogenized with five volumes of ice-cold HPLC grade water. Depending on the assay, brain tissue homogenates are manipulated with specific solvents to achieve optimum isolation of the analytes of interest. For example, the determination of biogenic amines requires the addition of 10% of 1 M $HClO_4$ containing ascorbic acid to a portion of brain homogenate, whereas an assay for amino acid determinations requires the addition of methanol to another sample of brain tissue homogenate, and a separate portion of homogenate is usually reserved for drug analysis.

The second step in analysis involves converting the sample into a form that is compatible with the method that will be used for identification of the compound. The compound(s) are usually extracted from the matrix followed by a sample purification process and possible derivatization to enhance sensitivity of detection and/or selectivity for analysis. The degree of purification required is dependent on the selected analytical method and on the tolerance of the specific type of detection system prone to contamination. For example, the determination of psychotropic drugs in plasma and serum by LC/MS/MS (Wood and Morris, 2002), involves a simple protein precipitation step with acetonitrile for a satisfactory purification process. Preparation of the analyte for HPLC analysis usually involves either an organic solvent extraction at a pH at which the analyte is greater than 99% unionized or protein precipitation using inorganic salts such as phosphates, or organic solvents such as acetonitrile or methanol. The analyte in the solvent extract or the supernatant with the protein precipitate removed is determined by a suitable analytical method usually following a purification and preconcentration step depending on the concentrations present and/or instrument sensitivity desired. In quantitative measurements, an internal standard possessing structural similarity to the compounds of interest is added to the matrix prior to pH adjustment. This is done to compensate for variable recovery and for variations observed in detector responses.

Many psychotropic compounds are lipophilic strong bases and accordingly, at high pH values, the free base can be extracted into an organic solvent (such as butyl acetate, hexane, heptane, diethyl ether, or ethyl acetate) from plasma, serum, or tissue homogenate or supernatant. The compound can be reextracted from the organic solvent into an aqueous acidic phase (e.g., hydrochloric acid solution). This allows for the separation of the analyte(s) from endogenous plasma constituents, and is known as an "acid wash or clean up" procedure. The acidic aqueous layer may be basified again and the analyte reextracted into an organic layer. The organic layer may be taken to dryness for concentration of the analyte; the residue is then taken up in a solvent suitable for HPLC analysis. The overall recovery for this three-step extraction process is often of the order of 60–80%. The sample volume required depends on the absolute concentration present and the selected method of analysis.

Disposable, prepackaged solid phase extraction (SPE) cartridges have introduced a wide range of phases that use polarity, hydrophobicity, or ionization as trapping mechanisms. SPE offers several benefits over liquid–liquid extraction such as less solvent consumption, smaller sample volumes, and increased sample throughput. It is relatively easy to use and is often incorporated into fully automatic sample preparation systems, but the cost of the various cartridges available must be taken into consideration. Other benefits include a variety of functionalities with different modes of sorbent interactions available to the analyst. The sample matrix can be an organic solvent or aqueous. The analyte is released from the cartridge by altering the polarity or the pH (Smith, 2003). Libraries of applications are found on manufacturers' Web sites. Use of SPE cartridges has enabled online extraction and quantification of drugs such as the antipsychotic quetiapine by HPLC (Hasselstrom and Linnet, 2003).

6.2 Chirality

Compounds with chiral centers of asymmetry are found in biological systems such as sugars and amino acids as well as in many marketed drugs. The presence of a chiral center (usually a carbon atom with four different substituents attached to it) in a molecule means that the molecule can exist as two nonsuperimposable mirror images (enantiomers). Many drugs with chiral centers are marketed as a racemate (equal mixture of two enantiomers). Drug enantiomers may display large differences in activity in a biological system and are often not differentiated by conventional chromatography. Chiral chromatography has become a valuable tool for the determination of enantiomers and for isolation of pure enantiomers in preparative work in industry.

Chiral resolution by HPLC can by divided into three categories: (1) a direct resolution using a chiral stationary phase (CSP); (2) addition of a chiral agent to the mobile phase, which reacts with the enantiomeric analytes (chiral mobile phase additive method (CMPA)); (3) an indirect method that utilizes a precolumn diastereomer formation with a chiral derivatization reagent (Mišl'anová and Hutta, 2003).

The major advantage of using CSPs for chiral analysis is that no racemization occurs during analysis. Separation occurs due to the stability differences of the transient diastereomeric complexes formed between the chiral selector in the stationary phase and each enantiomer in the chromatographic system. Some disadvantages of this method of analysis include the high price of most chiral columns and lower efficiencies of columns than those in conventional HPLC (Toyo'oka, 2002; Mišl'anová and Hutta, 2003). Column choice may also be a difficult decision. Resolution of a pair of enantiomers is not easy to predict even with knowledge of the macrostructure of the chiral stationary phase. Commercial databases, such as CHIRBASE, compiling literature on enantioseparation data, are available (Maier et al., 2001).

There are several types of chiral selectors used for CSP analysis. Protein-based or affinity-phase [e.g., ovomucoid (OVM), α1-acid glycoprotein, ovoglycoprotein (OVG), avidin (AVD), flavoprotein (FLA)] are compatible with an aqueous mobile phase and exhibit enantioselectivity for a wide range of compounds. Most protein-based CSPs have a silica-based stationary phase, which limits the eluent to a pH of between 3 and 8 (Haginaka, 2001). Chiral drugs containing one or more basic nitrogen atoms and one or more hydrogen acceptor or hydrogen donor groups can be resolved using a Cellobiohydrolase I (CBH I) column. Polysaccharide derivatives or helical-phase (cellulose esters, cellulose carbamates, amylase carbamates) are the most widely used (Toyo'oka, 2002; Mišl'anová and Hutta, 2003). Macrocyclic antibiotics (i.e., glycopeptides, ansamycins, polypeptides, and aminoglycosides) offer far greater selectivity because they have a number of stereogenic centers and functional groups, allowing them to have multiple interactions with chiral molecules (Ward and Farris, 2001). Chirobiotic V, a glycopeptide column, has been used to quantitate the enantiomers of citalopram and its demethylated metabolites (Kosel et al., 1998). Other types discriminate enantiomers by formation of inclusion complexes in chiral cavities. These phases include molecularly imprinted polymers, cyclodextrins, and crown-ether. Proton donor and proton acceptor type stationary phases make up another category that includes Pirkle type and low-molecular weight CSPs (Liu and Liu, 2002).

Manipulation of mobile phase and temperature parameters can have some unusual effects on chiral separations. Variation of temperature and mobile phase composition has been reported to reverse the elution order on protein phases and polysaccharide phases (Persson and Andersson, 2001).

Analysis using a CMPA is usually resolved on a nonchiral column. A transient diastereomeric complex is formed between the enantiomer and the chiral component in the mobile phase, similar to the complexes formed with chiral stationary phases. A review by Liu and Liu (2002) cites several papers where addition of CPMAs has been used in analyzing amphetamine-related compounds. Some CPMAs include amino acid enantiomers, metal ions, proteins, and cyclodextrins. Advantages of this method of analysis include the use of less expensive columns and more flexibility in the optimization of chiral separation (Mišl'anová and Hutta, 2003).

Formation of diastereomers by tagging with a chiral derivatization reagent is suitable for trace analysis in biological matrices because of the option of coupling a highly sensitive reagent with a high molar absorptivity or fluorescence. This is a major advantage of this type of analysis. Problems may include optical purity and stability of the reagent, and the possibility of racemization during the tagging process. Derivatizing reagents for both UV and fluorescence are discussed in a detailed review by Toyo'oka (2002).

6.3 Endogenous Compounds

6.3.1 Biogenic Amines

The use of HPLC to analyze biogenic amines and their acid metabolites is well documented. HPLC assays for classical biogenic amines such as norepinephrine (NE), epinephrine (E), dopamine (DA), and 5-hydroxytryptamine (5-HT, serotonin) and their acid metabolites are based on several physicochemical properties that include a catechol moiety (aryl 1,2-dihydroxy), basicity, easily oxidized nature, and/or native fluorescence characteristics (Anderson, 1985). Based on these characteristics, various types of detector systems can be employed to assay low concentrations of these analytes in various matrices such as plasma, urine, cerebrospinal fluid (CSF), tissue, and dialysate.

Electrochemical detectors coupled to HPLC systems employing reverse-phase chromatography with an ion-pair reagent offer a wide range of versatility for the analysis of biogenic amines by permitting manipulation of various parameters to enhance separation and sensitivity. A change in the molarity of the ion-pair reagent, such as sodium octyl sulfate, will increase the retention time for amines but have little effect on acid metabolites. Raggi and colleagues (1999) assayed E, NE, and DA in human plasma using a C8 column and coulometric end-point detection. An SPE procedure provided the sample cleanup step (human plasma is usually cleaned up for biogenic amines by absorption on alumina or boric acid gel, or extraction with organic solvents). This rapid SPE method achieved recoveries of 92 to 98%.

Numerous assays are also available in the literature for analysis of biogenic amines and their acid metabolites in brain tissue. For example, Chi and colleagues (1999) developed a rapid and sensitive assay for analyzing NE, DA, 5-HT, 5-hydroxyindole-3-acetic acid (5-HIAA), and homovanillic acid (HVA) in rat brain. The assay used a C18 column (150×4.6 mm) coupled to an amperometric electrochemical detector. The mobile phase consisted of a phosphate buffer (pH 4.75) and octane sulphonic acid as an ion-pair reagent in acetonitrile. The sensitivity of the analytes reported was 3–8 pg on column.

Microdialysis presents its own challenges as a matrix for analysis. Often the analyst is limited by sample volume, and the stability of biogenic amines in solution, particularly at a basic pH, can be questionable. Addition of a small amount of ascorbic acid or another antioxidant may be necessary before storage. Parent and colleagues (2001) analyzed NE, DA, 5-HT, 5-HIAA, HVA and 3,4-dihydroxyphenylacetic acid (DOPAC) in rat microdialysate, employing an amperometric detector combined with a RP8 column. The mobile phase consisted of a phosphate buffer (pH 2.9) with sodium octyl sulphate as the ion-pair reagent in acetonitrile. Parsons and colleagues (1998) described a microbore analysis of DA and cocaine out of a single 5 μl sample of dialysate. It utilized an electrochemical detector with two glassy carbon electrodes in tandem with a UV detector set at an absorbance wavelength of 225 nm.

An assay for NE, E, L-DOPA, DA, 3-nitrotyrosine, *m*-,*o*-, and *p*-tyrosine compared an amperometric detector with a CoulArray detector. A CoulArray detector has the sensitivity of a coulometric detector applied to eight different electrodes to give an "array" of applied voltages. A C18 column with a mobile phase consisting of an acetate buffer (pH 4.75) and sodium citrate in methanol was used. The assay was applied to plasma samples after a molecular mass cutoff cleanup procedure. The CoulArray method was approximately 10× more sensitive than the amperometric method, but both assays gave sufficient sensitivity for analysis (Kumarathasan and Vincent, 2003).

Another application utilizing the coulometric array detector is the simultaneous determination of biogenic amines, kynurenine, and indole derivatives of tryptophan. The method employed a C18 column with a phosphate–acetate mobile phase (pH 4.1) containing methanol and sodium octyl sulphonate (Vaarman et al., 2002).

Fluorescence detection, alone or with the aid of derivatizing reagents to enhance detector responses and improve the chromatographic resolution, has also been used for the determination of biogenic amines. Lakshmana and Trichur (1997) used native fluorescence to analyze NE, DA, and 5HT in rat brain utilizing an isocratic separation on an ODS C18 column. The detection limits reported were 100–250 pg on column.

Derivatizing reagents such as 9-fluorenylmethyloxycarbonyl chloride (FMOC-Cl) are commonly used for the determination of both primary and secondary amines in biological fluids. Chan and colleagues (2000) derivatized NE, E, and DA together with the metabolites normetanephrine (NM) and metanephrine (MN) using FMOC-Cl as the derivatizing reagent. A C8 column was used with a linear gradient. Excitation and emission wavelengths used were 263 and 313 nm respectively. Urine samples were analyzed by direct derivatization followed by a chloroform cleanup step. The lowest level of quantification reported was 1,250 fmol.

Fluorescence detection with precolumn derivatization has also been applied to microdialysate samples. Kehr et al. (2001) derivatized NE and 5HT with benzylamine. The excitation and emission wavelengths used were 345 and 480 nm respectively, and limits of detection reported were 40 atmol/10 μl for NE and 10 atmol/10 μl for 5-HT.

6.3.2 Amino Acids

Amino acids are organic compounds containing both amino and carboxyl groups. Since these compounds do not possess adequate chromophores or fluorophores, the majority of HPLC applications are based on precolumn derivatization procedures. Some of these reagents include 1-dimethyl aminonaphthalenesulfonyl chloride (DANSYL), phenylisothiocyanate (PITC), 6-aminoquinoquinolyl-*N*-hydroxysuccinimidyl carbarmate (AQC), *ortho*-phthalaldehyde (OPA), and naphthalene-2,3-dicarboxaldehyde (NDA). PITC and DANSYL form stable derivatives but they both generate fluorescent by-products that may interfere in the assay. OPA has been used extensively for the determination of primary amines. It neither fluoresces nor is it electroactive, but the derivatives formed have fluorescence and electrochemical characteristics. A major disadvantage of its use is the instability of the derivative that is formed. Autosamplers that allow programming for the addition of a reagent and a time delay before injection onto the analytical column have overcome this disadvantage. NDA derivatives exhibit more stability than OPA derivatives. They also possess fluorescence and electrochemical characteristics (Shah et al., 2002).

Our laboratory has applied the OPA derivatization procedure to tissue, microdialysates, plasma, cells harvested from cell culture, and CSF. The matrix is treated appropriately for protein precipitation and cleanup, usually by homogenizing in ice cold methanol and centrifuging at 12,000 g for 4 min. OPA reacts with primary amines in an alkaline medium using 2-mercaptoethanol as the reducing agent, to form highly fluorescent thioalkylsubstituted isoindoles (Parent et al., 2001). A phosphate buffer (pH 6.2), containing methanol, acetonitrile, and tetrahydrofuran as organic modifiers, is run as a gradient through a C18 column using fluorescence detection with an excitation of 260 nm and emission wavelength of 455 nm. Piepponen and Skujins (2001) analyzed glutamate (GLU), glycine (GLY), taurine (TAU), and γ-aminobutyric acid (GABA) in microdialysate in a similar manner.

NDA derivatization has also been automated for analysis of amino acids in brain tissue and microdialysates (Shah et al., 1999). NDA reacts with primary amines in the presence of cyanide to form a highly stable N-substituted 1-cyanobenz[*f*]isoindole (CBI) derivative. Addition of a nucleophile, such as cyanide, hydrogen sulphite, isothiocyanate, or 2-mercaptoethanol, is essential for the formation of the derivative.

Derivatives formed having a substituent other than a cyano group on the indole ring have decreased stability and reduced fluorescence response.

Recent progress in chiral analytical chemistry has enabled the analyses of D-amino acids in mammalian tissue. Considerable interest has been generated in *D*-serine (*D*-Ser) and *D*-aspartate (*D*-Asp), specifically. *D*-Ser is suggested to have a role in N-methyl- D-aspartate (NMDA) receptor-mediated transmission, and *D*-Asp is suggested to function as a regulator of hormonal secretion (Hamase et al., 2002).

OPA in combination with chiral thiols is one method used to determine amino acid enantiomers. A highly fluorescent diastereomeric isoindole is formed and can be separated on a reverse-phase column. Some of these chiral thiols include *N*-acetyl-L-cysteine (NAC), *N*-*tert*-butyloxy-carbonyl- *L*-cysteine (Boc-*L*-Cys), *N*-isobutyryl- *L*-cysteine (IBLC), and *N*-isobutyryl- *D* -cysteine (IBDC). Replacing OPA–IBLC with OPA–IBDC causes a reversal in the elution order of the derivatives of D- and L-amino acids on an ODS column (Hamase et al., 2002). Nimura and colleagues (2003) developed a novel, optically active thiol compound, *N*-(*tert*-butylthiocarbamoyl)- *L*-cysteine ethyl ester (BTCC). This reagent was applied to the measurement of *D*-Asp with a detection limit of approximately 1 pmol, even in the presence of large quantities of *L*-ASP.

CSPs and chiral mobile phase additives have also been used in the separation of amino acid enantiomers. Another technique that should be mentioned is an analysis system employing column-switching. *D*- and *L*- amino acids are first isolated as the racemic mixture by reverse-phase HPLC. The isolated fractions are introduced to a second column (a CSP or a mobile phase containing a chiral selector) for separation of enantiomers. Long et al. (2001) applied this technique to the determination of *D*- and *L*-Asp in cell culture medium, within cells and in rat blood.

6.3.3 Acetylcholine and Choline

Analysis of the quaternary amines acetylcholine (ACh) and choline (Ch) presents several challenges. These compounds are neither electroactive nor UV-absorbing. Relatively higher concentrations of Ch in comparison to ACh may contribute to difficulties in chromatographic separation of these highly polar analytes. Proper treatment of tissue samples to prevent postmortem changes of ACh and Ch is essential. Shahed and colleagues (1996) compared two methods of tissue fixation in rat and mice brain samples. ACh concentrations obtained after microwave fixation of the brains were significantly greater than those obtained by freeze fixation. Ch concentrations were comparable between both methods.

Analysis of ACh in blood and CSF poses two obstacles—low concentrations (<20 pmol/ml) and very fast hydrolysis by esterase (Tsai, 2000). An acetylcholinesterase inhibitor such as physostigmine may be added to artificial CSF used for the collection of microdialysates to help overcome this problem (Kato et al., 1996).

The most widely used analysis technique employs HPLC/EC coupled to a postcolumn enzyme reactor. ACh and Ch are separated on a reverse-phase column. Detection is addressed by the use of an immobilized enzyme reactor (IMER) consisting of the enzymes acetylcholinesterase and choline oxidase covalently bonded to a solid matrix and packed into a postcolumn enzyme reactor. Acetylcholinesterase hydrolyzes ACh to Ch, which is oxidized by choline oxidase to betaine and hydrogen peroxide; hydrogen peroxide is the species that is detected electrochemically (Tsai, 2000). A precolumn IMER containing choline oxidase may be installed to remove Ch if interference with the measurement of ACh becomes a problem. The hydrogen peroxide generated is not retained on the column but will be detected electrochemically, broadening the solvent front. A precolumn IMER containing a combination of choline oxidase and catalase or peroxidase and choline oxidase reduces this response (Tsai, 2000; Kato et al., 1996).

Several different types of electrodes have been used in this analysis. Conventionally, a platinum (Pt) electrode has been employed in ACh and Ch analysis. Yasumatsu and colleagues (1998) used this method to measure ACh and Ch in tissue extracts and microdialysates, achieving mean concentrations of 0.06 pmol/10 μl and 0.64 pmol/10 μl of preoptic or anterior hypothalamus dialysate respectively. Rakovska and colleagues (2003) also analyzed microdialysates with a similar procedure, achieving limits of detection of 500 fmol for ACh and 250 fmol for Ch.

Frölich and colleagues (1998) analyzed ACh in human CSF by different methods, which included: thermospray/mass spectroscopy, HPLC/mass spectroscopy, HPLC–EC Pt electrode and gas chromatography/mass spectroscopy (GC/MS). An SPE extraction was used for cleanup and concentration. Samples were run with and without the IMER to rule out any interference by physostigmine, a cholinesterase inhibitor, in the HPLC–EC assay. HPLC–EC and GC–MS gave data correlations with similar sensitivities, but the HPLC–EC values were 39% lower. Analysis using thermospray/mass spectroscopy and HPLC/ mass spectroscopy did not provide adequate sensitivity and the data obtained were inconsistent.

A horseradish peroxidase-osmium redox polymer-modified glassy carbon electrode (HRP-GCE) has also been applied to this analysis to improve sensitivity and reduce problems with faradic interference. Kato and colleagues (1996) employed this electrode in measurement of basal ACh in microdialysates using a precolumn enzyme reactor. This system was three to five times more sensitive than a conventional Pt electrode. ACh in rat hippocampus dialysate was quantitated at 9 ± 5 fmol/15 μl ($n = 8$). ACh was analyzed in PC12 cells in a similar assay by Kim and colleagues (2004). No precolumn enzyme reactor was employed.

Osborne and Yamamoto (1998) compared disposable, film carbon electrodes (PFCEs) and glassy carbon electrodes that were both modified with cast-coated Osmium-polyvinylpyrridine-wired horseradish peroxidase gel polymer (Os-gel-HRP). Sensitivities for ACh were 16 and 10 fmol/10 μl respectively.

6.3.4 Polyamines

Polyamines such as putrescine (PUT), spermine (SP), spermidine (SPD), and cadaverine (CAD) do not absorb in the ultraviolet region nor do they have native fluorescence. Dansyl chloride, fluorescamine, *o*-phthaldialdehyde (OPA)-2-mercaptoethanol or OPA-ethanethiol have been used as derivatizing reagents for fluorometric determination of polyamines. UV/VIS detection requires derivatization reactions with reagents such as benzoyl chloride, *p*-toluenesulfonic chloride (tosyl chloride), 2,4-dinitrofluorobenzene, 4-fluoro-3-nitrobenzotriflouride, and 4-dimethylaminoazobenzene-4′-sulfonyl chloride (dansyl chloride) (Teti et al., 2002; Molins-Legua et al., 1999). Generally, the fluorometric reagents only react with the primary amino groups. The acid chlorides form derivatives with both primary and secondary amino groups (including imidazole nitrogen) phenolic hydroxyls and some alcohols. Acid chloride reaction products have the advantage of being more stable than flourescamine or OPA derivatives (Teti et al., 2002).

Venza and colleagues (2001) applied fluorescence detection using OPA-2-mercaptoethanol to analyze SP, SPD, and PUT in saliva samples using reverse-phase chromatography. Detection limits reported were 0.04, 0.05, and 0.06 nmol/ml for SP, SPD, and PUT respectively.

Aboul-Enein and Al-Duraibi (1998) employed dansyl chloride in a fluorescence assay for PUT, SP, SPD, and their acetylated derivatives by ion-pair reverse-phase chromatography. This assay could be applied to the separation of free and acetylated polyamines in biological samples. Dansyl chloride has also been used as the fluorescence reagent in the determination of polyamines in urine by Molins-Legua and colleagues (1999). Derivatization was carried out within the C18 cartridges that were used during the SPE extraction procedure. Recoveries were 80–95% for all four polyamines analyzed and the limit of detection was 10 ng/ml.

A fluorescence assay based on an intramolecular-forming derivatization with a pyrene labeling reagent, 4-(1-pyrene) butyric acid *N*-hydroxysuccinimide ester (PSE), was applied to polyamines by Nohta and colleagues (2000). Polyamines were converted to the corresponding dipyrene- to tetrapyrene-labeled derivatives by reaction with PSE for 20 min at 100°C. One excited pyrene can form an excited-state complex (intramolecular excimer) with the other ground-state pyrene in the molecule. A longer wavelength (450–500 nm) is emitted by the excimer than by the pyrene monomer, which emits normal fluorescence (350–400 nm). Detection limits reported were 1 (PUT), 1 (CAD), 5 (SPD), and 8 (SP) fmol on column. This method could be applied to biological investigations of polyamines. A similar assay was used for the determination of histamine in urine by Yoshitake et al. (2003).

Another example of precolumn derivatization employed benzoyl chloride to derivatize polyamines and acetylpolyamines; UV detection was used (Taibi et al., 2000).

6.3.5 Steroids

Steroids are biologically active, relatively low molecular weight substances that are naturally synthesized from cholesterol. Steroids containing 21 carbon atoms are known as *pregnanes* whereas those containing 18 or 19 carbon atoms are known as *estranes* and *andronanes* respectively. The individual metabolites are characterized by the presence or absence of certain functional groups. The most common functional groups include hydroxy, ketones, and aldehydes and are located in positions C-3, C-5, C-11, C-17, C-18, C-20, or C-21. The term *neuroactive steroids* refers to steroid hormones that are active on neuronal tissue. They are either synthesized endogenously in the brain or by peripheral endocrine glands. A common feature for the neuroactive steroids is that they rapidly alter the excitability of neurons by binding to membrane-bound receptors such as those for inhibitory and/or excitatory neurotransmitters (e.g., 5-HT, NE, GABA, glutamate, and DA).

The steroids are highly lipophilic although they are found to be conjugated to hydrophilic ligands such as sulphates, glucuronide derivatives, or bound to plasma proteins (Nozaki, 2001).

Early assays for steroids involved liquid–liquid extraction (LLE) techniques, which are still widely used. However, the popularity of LLE has declined as more efficient techniques such as solid-phase extraction (SPE) have been introduced. In LLE techniques, the polarity of the steroids and their ability to bind to proteins are important aspects when selecting various solvents for their extraction from biological matrices. LLE is a nonselective procedure and nonsteroidal lipids are likely to be coextracted and interfere with the analysis.

Precautions should be taken when selecting various vessels of differing materials for the extraction of these steroids from their matrices. For example, the compatibility of some solvents with plastic extraction tubes may present some problems. If glass vessels are to be selected, it is recommended that all glassware is silanized prior to use since several steroids bind to glass surfaces.

The LLE of the pregnane steroids is usually carried out using either diethyl ether or ethyl acetate and the extracts are then evaporated to dryness. This step is followed by a reconstitution step into various solvents depending on the choice of analytical identification (Yamada et al., 2000). Some investigators have incorporated a protein precipitation strategy utilizing acetonitrile prior to LLE for the identification of alphaxalone, pregnenolone, and 3α-hydroxy-5β-pregnan-20-one from plasma (Visser et al., 2000).

The extraction of steroids from tissues normally requires a mixture of solvents (e.g., methanol, isopropanol, hexane, and chloroform). Studies have also shown that pregnanes have been successfully extracted from brain tissue when the brain samples are allowed to stand in 95% aqueous ethanol for 7 days at 4°C (Wang et al., 1997; Bixo et al., 1984).

The steroids 17β-hydroxyandrost-4-en-3-one, 3α-hydroxy-5a-pregnan-20-one, progesterone, pregnenolone, 3β-hydroxy-5α-pregnan-20-one, 3α-hydroxyandrost-5-en-17-one, together with some of their related metabolites have been isolated from plasma, urine, bovine serum, human CSF and brain tissue utilizing C18 SPE columns (Draisci et al., 2000; Kim et al., 2000; Vallee et al., 2000). Investigators have also extracted a number of steroids with a combination of C18 and ion-exchange sorbent columns (Que et al., 2000).

HPLC is the most versatile separation technique for neuroactive steroids as it is generally a nondestructive technique not normally requiring derivatization processes; however, the consumption and cost of mobile phase solvents is considerably higher than those in other techniques. Despite a wide variety of bonded phases available, reverse-phase HPLC (RP-HPLC) using octadecyl silica (ODS) or C18 columns is the most commonly used and is the principal separation mode for small nonvolatile compounds such as the neuroactive steroids, although there have been several instances where normal-phase HPLC (NP-HPLC) has also been used. In general, the optimization for the separation process on a RP column with either methanol or acetonitrile mixed with buffers in various proportions is the initial choice of mobile phase selection. However, if separating a series of steroids and metabolites, a more complex mixture of eluents is sometimes required. Investigators have also used more specialized "mixed-mode" columns that contain both polar and nonpolar bonded silica functional groups such as 50% cyano-propyl (CN) with C18 material. Normal phase chromatography has also been utilized. A study by Pearson and colleagues (2000) has shown separation profiles for progesterone, pregnenolone, and some of its neuroactive metabolites, namely

5α-pregnan-3,20-dione, 5β-pregnan-3,20-dione, and 3β-hydroxy-5α-pregnan-20-one using a 5 × 100 mm silica column with 0.18% ethanol in methylene chloride as the mobile phase. The analytes were extracted from plasma with hexane and toluene. Following evaporation to dryness all the samples were reconstituted in methylene dichloride prior to injection onto the HPLC analytical column. A 1-h reequilibration time for the column was required before subsequent samples were injected. Fractions were collected and assayed by a radioimmunoassay. The separation time for all five steroids was 30 min. A study by AbuRuz and colleagues (2003) has shown a simple method for the simultaneous determination of prednisolone and cortisol using hydrophilic/lipophilic (polar/nonpolar) or mixed mode SPE cartridges from plasma and urine. The chromatographic separation was performed on a silica column with UV/VIS end-point detection set at 240 nm. The authors reported limits of detection and quantitation for prednisolone in plasma and urine of approximately 5–7 and 8–11 ng/ml respectively; for cortisol, these values were 4–6 and 7–10 ng/ml in plasma and urine respectively. The extraction recovery for both compounds averaged 90%. Kuronen and colleagues (1998) reported a multisteroid (24 steroids) screening method in serum utilizing RP-HPLC with UV/VIS diode-array end-point detection operated at a wavelength 205 nm. The analytical columns used were C18 bonded phase materials operated at a temperature of 22°C. The mobile phase consisted of two separate solvents of water and acetonitrile run in gradient mode. Serum samples were preheated at 40°C for 10 min with phosphoric acid prior to extraction on SPE C18-bonded silica column cartridges. The eluents of the SPE procedures were evaporated to dryness and analytes reconstituted into ethanol prior to injection onto the RP analytical columns.

A prevailing problem with neuroactive steroid analysis by HPLC is the lack of sufficient chromaphores or fluorophores within their chemical structures to allow suitable spectrophotometric end-point detection such as with UV/VIS or fluorescence with adequate sensitivity. The multitude of structural isomers of the metabolites also decreases the applicability of RP-HPLC since the chromatographic profiles become very complex with "co-eluting peaks." Due to these inherent problems, it is often necessary to derivatize this group of compounds prior to chromatographic separation and suitable end-point detection to allow their direct determination at physiological concentrations.

Studies have shown that precolumn derivatization of certain neuroactive steroids and their metabolites to create fluorescent esters is easily achieved by reacting with 9-anthroyl nitrile at the 21 hydroxyl group position. A study by Shibata and colleagues (1998) has analyzed glucocorticoids, namely cortisol, cortisone, 18-hydroxycortisol, 18-hydroxycortisone, and 18-oxocortisol, in plasma and urine, using 9-anthroyl nitrile as the derivatizing reagent in the presence of a mixture of catalysts (triethylamine and quinuclidine) to form fluorescent esters (➋ *Figure 2-1*).

These compounds were analyzed by using normal phase silica columns operated at ambient temperature. Mobile phase constituents included diethylene dioxide/ethyl acetate/chloroform/hexane–pyridine

◘ Figure 2-1
Fluorescent steroid derivatives formed by the reaction with 9-anthroyl nitrile

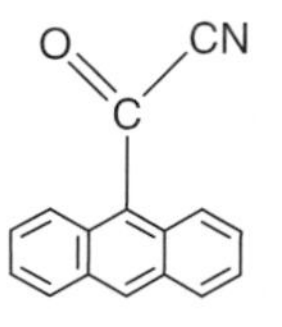

+

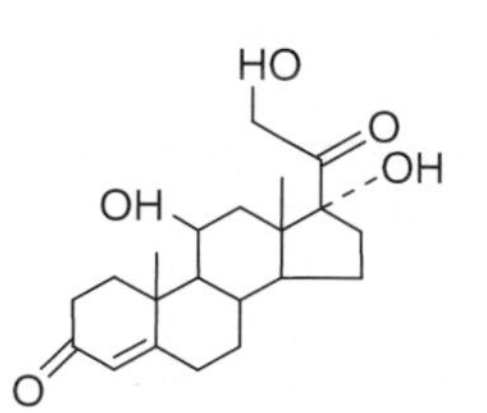

→

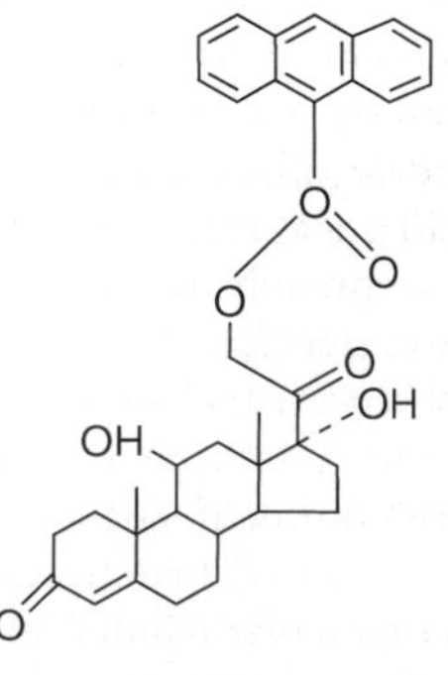

9-anthroyl nitrile hydroxycortisol 9-anthroyl nitrile derivative

mixtures. The analysis time for complete separation of these steroids was approximately 100 min (Shibata et al., 1998). Another example using anthroyl nitrile as the derivatizing reagent in the presence of quinuclidine was reported by Kurosawa and associates (1995) on the determination of 18-hydroxycortisol, 18-hydroxycortisone, and 18-oxocortisol.

Others have reported the use of fluorescent dansyl derivatives for the identification of 17α-oestradiol from serum samples (➋ *Figure 2-2*) (Nozaki et al., 1988). The derivatization of the keto functional group of particular steroids (ketosteroids) can also be achieved by hydrazone formation (➋ *Figure 2-3*); the most

Figure 2-2
Fluorescent steroid derivatives formed by the reaction with Dansyl chloride

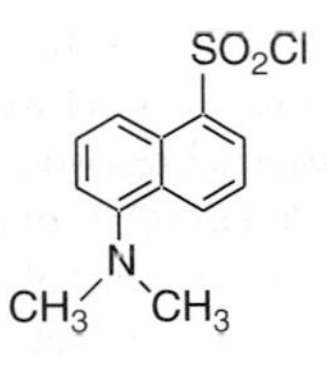

+

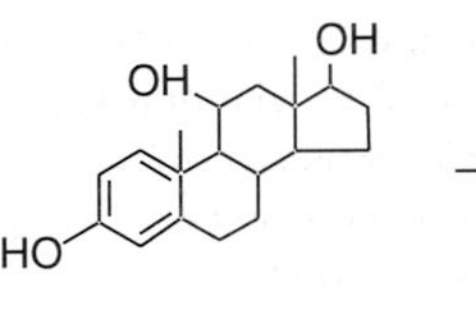

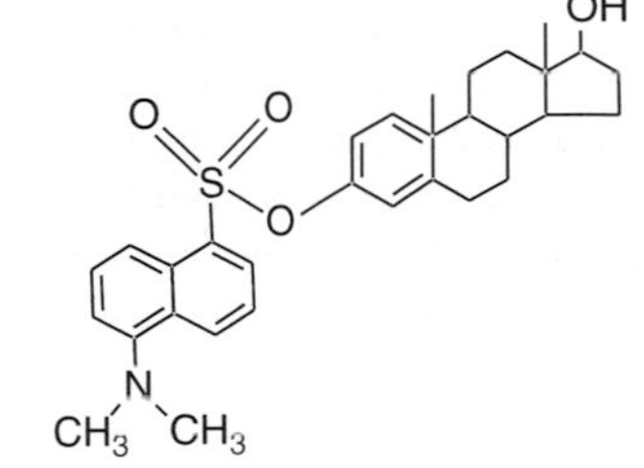

Dansyl chloride | oestradiol | Dansyl derivative

Figure 2-3
Fluorescent steroid derivatives formed by the reaction with a hydrazine reagent

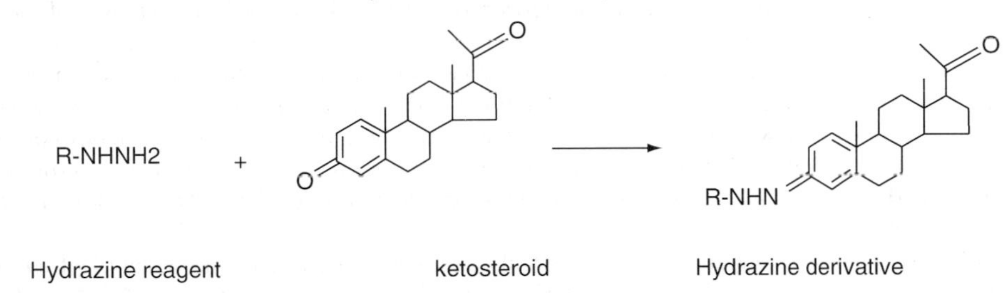

commonly used reagent is dansylhydrazine or DNSH. Visser and colleagues (2000) have used DNSH for the determination of alphaxalone and pregnanolone in plasma. The isolation of these compounds from their plasma matrices included protein precipitation with acetonitrile followed by the derivatization procedure and extraction into dichloromethane. The separation of these steroids was performed using RP-HPLC with fluorescence detection. The mobile phase consisted of an acetate buffer with acetonitrile. The total run time for the complete separation of these steroids was 35 min. It should also be noted that different catalysts have been used to influence the rate and yield of these hydrazine derivatives. These include hydrochloric acid, acetic acid, trichloroacetic acid, and trifluoroacetic acid.

6.4 Exogenous Compounds

6.4.1 Antidepressants

The determination of these compounds in biological fluids is crucial for several reasons such as investigating overdosing (toxicological monitoring), conducting pharmacokinetic and bioavailability studies, and measuring compliance.

6.4.1.1 Tricyclic antidepressants Despite the numerous publications over the past 30 years on the determination of the TCAs (Tricyclic Antidepressants) by HPLC to establish possible therapeutic windows, both therapeutic drug monitoring and pharmacokinetic calculations have revealed there is considerable variation (10- to 50-fold) in plasma concentrations between individuals with these drugs. The plasma concentrations are usually in the range of 50–300 ng/ml.

The routine determination of TCAs has often utilized RP-HPLC with some form of spectrophotometric or, in some cases, electrochemical detection. These systems include UV/VIS, fluorescence, or amperometric/coulometric end-point detection. The TCAs are usually lipophilic strong bases (possess high pKa values) and therefore isolated as free bases by alteration of the biological matrix to pH 12 followed by extraction into an organic solvent.

In LLE it is common to employ a mixture of organic solvents such as 1–10% isoamyl alcohol in heptane or hexane to enhance the recovery and to minimize emulsion formation.

Studies have shown that most of the tricyclic and tetracyclic antidepressants can be determined by RP-HPLC with UV/VIS detection. Presented here are some examples of HPLC determination of TCAs.

Tanaka and colleagues (1997) have developed a sensitive assay for the simultaneous determination of amitryptiline (AMI), imipramine (IMI), nortryptyline (NT), amoxapine, clomipramine (CMI), desipramine (DMI), doxepin, maprotiline, and mianserin in human biological samples using RP-HPLC with UV/VIS detection. Nyanda and associates (2000) have also developed a simple HPLC assay for the determination of commonly used TCAs in plasma. The extraction recoveries reported were 89–108% with a lower limit of quantitation (LLOQ) of 15 ng/ml. The authors reported inter-assay coefficients of variation ranging from 1.1 to 13% depending on the compound investigated. Queroz and colleagues (1995) have also reported simultaneous determinations of IMI, DMI, AMI, and NT in human plasma. Detection was achieved with UV/VIS monitor set at 254 nm.

The determination of TCAs and metabolites using RP-HPLC combined with SPE extraction technologies has also been reported by several investigators. Studies have shown that extraction of TCAs and associated metabolites from their plasma matrices can be achieved using CN, C2, or C18 bonded phase cartridges using 1 ml of plasma. The chromatographic determination was usually performed using either a cyano-propyl or C18 analytical column with UV/VIS detection. The LLOQs reported were in the range of 5–15 ng/ml. The extraction recoveries ranged from 65 to 110% (Weigmann et al., 1998; Olesen et al., 2000; Palego et al., 2000; Pirola et al., 2002). Recently, further applications using smaller sample volumes and fluorescence detection have also been reported (Maris et al., 1999; Shams et al., 2004).

6.4.1.2 Selective serotonin reuptake inhibitors (SSRIs) and other newer antidepressants Due to their widely differing structures in comparison to TCAs, several methods for their determination in biological matrices have been developed for each compound individually. Presented here are examples of HPLC determinations for several of these drugs.

Duverneuil and coworkers (2003) have developed a method for the determination of 11 of the most commonly prescribed "non-tricyclic" antidepressants and some of their metabolites; these include paroxetine, fluoxetine, norfluoxetine, sertraline, citalopram, fluvoxamine mirtazapine, venlafaxine, and O-desmethylvenlafaxine. The method involves an LLE procedure followed by an HPLC separation with photodiode-array UV detection at three different wavelengths (220, 240, and 290 nm). The total run time was 18 min. The extraction recoveries were calculated to be in the range of 74–109% and the lower limit of detection (LLOD) reported was 2.5–5 ng/ml. A method published by Tournel and associates (2001) also reported the simultaneous determination of several newer antidepressants by RP-HPLC with UV detection. The compounds were isolated from human serum using an LLE process. The LLOQ ranged from 15–50 ng/ml depending on the analyte of interest. The total run time for all compounds eluted was approximately 20 min.

Further applications for the determination of the newer antidepressants have employed precolumn derivatization, which included a reaction with dansyl chloride or 4-(N-chloroformylmethyl-N-methyl) amino-7-nitro-2,1,3,-benzoxadiazole (NBD-COCL) followed by separation on ODS C18 analytical columns maintained at either 35°C or 70°C using either isocratic or gradient elution with fluorescence end-point detection. The compounds were isolated from human plasma or serum by LLE or SPE techniques using volumes ranging from 100 μl to 1 ml. The extraction recoveries and LLOQs achieved were greater than 90% and 5–10 ng/ml respectively (Lucca et al., 2000; Waschgler et al., 2002; Guo et al., 2003).

6.4.2 Anxiolytics

6.4.2.1 Benzodiazepines The benzodiazepines (BZDs) are characterized by the presence of the phenyl ring fused to a partially saturated seven-membered ring with nitrogen at positions 1 and 4. Several examples of BZDs include: diazepam, alprazolam, lorazepam, flurazepam, temazopam, midazolam, flunitrazepam, medazolam, and oxazepam.

The techniques for isolation of the BZDs from their biological matrices are similar to those for several other psychotropic compounds requiring chromatographic separation. Solvents used to extract BZDs include diethyl ether, toluene, dichloromethane, butyl chloride, toluene, and ethyl acetate or various combinations of these solvents including diethyl ether/propan-2-ol (70:30, v/v), hexane/dichloromethane (50:50, v/v), butylchloride/ethylacetate (20:80, v/v), and dichloromethane/diethyl ether (40:60). There is little to distinguish between these solvents with regard to recoveries, which are in excess of 60% for most BZDs. The extractions are usually conducted under alkaline conditions, with either sodium carbonate, sodium phosphate, sodium borate/tetraborate, or Trizma buffer (pH 9–10) used to adjust the pH value of the sample matrix. An acid back extraction has been used to clean up initial extracts, but this does not offer any particular advantage over the use of organic solvents alone. SPE approaches to BZDs generally use octadecylsilane C18 bonded phase cartridges, but C2 bonded columns have also been used successfully. Studies by Casas and colleagues (1993) investigated the degree of extractability using several different SPE cartridges and concluded that C2 cartridges provided the best combination of optimum recovery and cleaner extracts from urine compared to C8 or C18, phenyl and cyclohexyl phases, whereas CN provided little retention due to its relatively polar nature. Drug screening methods using fully automated SPE systems with Bond–Elut Certify cartridge columns for plasma and whole blood have also been reported. The authors reported recoveries of greater than 80% for various BZDs including ketazolam, lorazepam, diazepam, oxazepam, and flunitazepam (Chen et al., 1993; Huang et al., 1996).

Some applications also involved direct injection techniques in which BZDs are preferentially absorbed onto a precolumn and are back-flushed onto the analytical column using column switching techniques (Lauber et al., 1994; Iwase et al., 1994).

Several investigators have reported the determination of numerous BZDs by RP-HPLC with UV absorbance detection (Berhault et al., 1996; Tanaka et al., 1998; Wilhelm et al., 2001).

6.4.2.2 Non-benzodiazepines Buspirone, gepirone, and ipsapirone are non-BZD anxiolytics and structurally belong to the azapirone group of compounds. They have in common a major metabolite namely, 1-(2 pyrimidinyl) piperazine (1-PP). Investigators have reported methods for the determination of all these compounds and their common metabolite (1-PP) by RP-HPLC. Studies by Odontiadis and Franklin (1996) have shown the simultaneous determination of buspirone and 1-PP using RP-HPLC with coulometric electrochemical detection. The extraction from human plasma was performed with SPE using ion-exchange cartridges containing an acetic acid functional group (carboxymethyl or CBA). The analytical column used to separate these compounds together with various other psychotropic compounds was a "mixed-mode" (CN/C18: 5 × 250 mm). The LLOQ values for buspirone and 1-PP were 0.5 and 2 ng/ml respectively. Others have also reported RP-HPLC assays with coulometric detection for azapirone type compounds. Ary and associates (1998) used coulometric detection for the determination of buspirone. The chromatography was performed using a Supelcosil ABZ plus C18 material. Buspirone was extracted using Bond–Elut C18 SPE cartridges. The LLOD reported was 100 pg/ml in human plasma.

Studies by Farina and associates (1992) reported the analysis of buspirone and related compounds using a C18 analytical column with an elution solvent containing a phosphate buffer with an ion-pairing reagent (sodium dodecyl sulphate or SDS) and acetonitrile run in gradient mode. Detection was measured using fluorescence with excitation and emission wavelengths of 237 and 374 nm respectively. Bianchi and Caccia

(1988) described the simultaneous determination of buspirone, geprione, and ipsapirone together with 1-PP in plasma and brain tissue by gradient elution RP-HPLC with UV/VIS end-point detection. The analytes were extracted using LLE techniques.

6.4.3 Antipsychotics

6.4.3.1 Phenothiazines The phenothiazines (PTZs) undergo extensive metabolism. Metabolic routes include S-oxidation, aromatic hydroxylation, N-dealkylation, N-oxidation, and a combination of these processes. Chlorpromazine, for example, possesses 168 possible metabolites, a large proportion of which are pharmacologically active compounds. The development of an HPLC assay capable of resolving a large number of these metabolites is virtually impossible and assays that permit the simultaneous determination of the parent compound and a selected number of active metabolites must suffice. The PTZ group of compounds includes chlorpromazine, thioridazine, fluphenazine, and perphenazine.

Other first generation (atypical) antipsychotics include thioxanthenes, haloperidol, pimozide, and loxapine.

Several HPLC methods have been reported for the determination of PTZs by HPLC. Recently, Pistos and Stewart (2003) reported a method for the simultaneous determination of promethazine, promazine, chlorpromazine, prochlorpromazine, ethopromazine, and trifluoroperazine using a C18 Hisep analytical column maintained at 23°C with UV/VIS absorption at 254 nm. The isolation of the analytes from their plasma matrices was performed by filtering the plasma (0.22 μm filters) and injecting 20 μl of the supernate directly onto the analytical column. The extraction recoveries averaged 79% for all the analytes of interest. The investigators reported that the method requires no sample pretreatment when using a Hisep analytical column and possesses the necessary sensitivity for pharmacokinetic and toxicological studies. A study by Larsimont and associates (1998) reported a method for the determination of promazine in human plasma. The RP-HPLC assay with coulometric detection involved a single step LLE using a pentane/propanol mixture and separated on a Spherisorb CN analytical column. The detector was fitted with a guard cell operating at 0.9V preceding a high- sensitivity analytical cell with detector electrodes 1 and 2 operated at 0.4 and 0.75V respectively. The extraction recovery for the analytes averaged 72%, with an LLOQ of 0.25 ng/ml. Recently, other RP-HPLC methods for the determination of promethazine, chlorprothixene, levomepromazine, fluphenazine, perphenazine, and associated metabolites using UV/VIS coupled to amperometric and/or coulometric end-point detection have been reported. The isolation of the analytes was performed using either LLE or SPE techniques. The extraction recoveries ranged from 68 to 92% depending on the analytes of interest. The analytical column used for the separation of these compounds was a Nucleosil CN or C18 bonded phase materials. The LLOD ranged from 0.1 to 1 ng/ml using sample volumes of 0.2–1 ml (Bagli et al., 1994; Foglia et al., 1995; Luo et al., 1997; Vanapalli et al., 2001). An excellent review of the determination of the PTZ group of compounds by HPLC can be referred to by Hefnawy (2002). Applications for the determination of amoxapine, loxapine, haloperidol, and associated metabolites by RP-HPLC can also be found in the literature (Cheung et al., 1991; Hue et al., 1998; Walter et al., 1998; Seno et al., 2000; Angelo and Petersen, 2001; Arinobu et al., 2002).

6.4.2.2 Atypical antipsychotics The second generation or so-called atypical antipsychotics have chemical, pharmacological, and clinical properties that are different from those of the classical antipsychotics/neuroleptics. The most commonly used atypicals include clozapine, olanzapine, risperidone, and quetiapine.

Recently, three papers have reported the determination of risperidone and its active metabolite 9-hydroxyrisperidone using LLE and SPE technologies. The analytical columns used to separate these compounds were C4 or C18 bonded phases of 3 μm or 5 μm particle sizes with UV/VIS detection. Mobile phases consisted of phosphate buffers (pH 3–4) in acetonitrile. The sample volumes used ranged from 200 μl to 1 ml, with extraction recoveries averaging 90%. The limits of quantitation ranged from 0.5 to 10 ng/ml in human plasma (Nagasaki et al., 1999; Avenso et al., 2000; Titier et al., 2002). A study by Titier showed the simultaneous determination of clozapine, olanzapine, haloperidol, risperidone, and its active metabolites by RP-HPLC in human plasma. The assay involved LLE with a hexane/isoamyl alcohol mixture using sample volumes of 500 μl. The analytes were separated on a Symmetry C8 analytical column and monitored with a UV/VIS detector at various wavelengths, depending on the analyte of interest. The mobile phase was run in gradient mode at a flow rate of 1.5 ml/min. The extraction recoveries ranged from 60 to 92% with LLOQ of 5 ng/ml for all the analytes (Titier et al., 2003). A method used for the simultaneous determination of clozapine and olanzapine together with their desmethyl metabolites has been described by Weigmann and colleagues (2001). The method involves an RP-HPLC method using a Hypersil CN (cyanopropyl) SPE column for eliminating interfering plasma constituents prior to the chromatographic separation on a Hypersil ODS C18 analytical column. The analytes were monitored at a wavelength of 254 nm. The LLOQ was 10–20 ng/ml. Another application for the determination of olanzapine using RP-HPLC with coulometric detection has been demonstrated by Bao and Potts (2001). They have shown that olanzapine can be measured in rat brain tissue using LLE techniques from brain homogenates. The extraction involved the addition of sodium carbonate to brain tissue and an organic solvent mixture of dichloromethane in cyclohexane. The separation of olanzapine together with its internal standard was achieved using a YMC Basic HPLC column. The operating potentials for the guard cell and analytical cell for detectors 1 and 2 were −0.3V −0.2V and +0.2V respectively. The extraction recoveries were greater than 82% with an LLOQ of 0.5 ng/ml of brain tissue homogenate. An excellent review of the analysis of atypical antipsychotics has been written by Raggi (2002). Further applications found in the literature include papers by Kollroser and Schober (2002), Liu et al. (2001), Berna et al. (2002), and Dusci et al. (2002).

6.4.4 Stimulants and Alkaloids

Mancinelli and colleagues (1999) reported the simultaneous determination of 3-4-methylenedioxymethamphetamine (MDMA), N-ethyl-3,4-methylenedioxyamphetamine (MDEA), methylenedioxyamphetamine (MDA), and N-methyl-1-(1,3-benzdioxol-5-yl)-2-butamine (MBDB) by RP-HPLC with fluorometric detection. The method required a 100 μl sample volume of serum, urine, or saliva; separation was achieved on a LiChrospher 100RP-18 analytical column using an acetonitrile/water mobile phase (pH 11.4) set at a flow rate of 1 ml/min. The LLOQ and LLOD were 50 and 10 ng/ml respectively.

Santagati and associates (2002) reported a method for the determination of amphetamine and one of its metabolites, 4-hydroxynorephedrine by RP-HPLC with precolumn derivatization and amperometric electrochemical detection. The derivatization was performed with 2,5-dihydroxybenzaldehyde as the electroactive reagent. The compounds were separated on a Hypersil ODS RP-18. The detector oxidation was set at +0.6 volts.

Several other methods have been published using RP-HPLC for the determination amphetamines and related derivatives. Studies have shown the determination of amphetamine and related derivatives in plasma, urine, and hair by RP-HPLC with precolumn derivatization and either UV/VIS or fluorescence detection. Various methods are employed by SPE technologies using C18 cartridges for sample cleanup prior to derivatization. The derivatized compounds were separated on analytical columns of various C18 bonded phase materials. The methods generally used water/acetonitrile mobile phases operated in gradient mode. All studies reported extraction recoveries of 85–102% for all the analytes, with LLOQs ranging from 5 to 60 ng/ml (Tedeschi et al., 1993; Falco et al., 1996; Hernandez et al., 1997; Al-Dirbashi et al., 1997; Al-Dirbashi et al., 2000; Soares et al., 2001).

Several methods have been reported for the determination of alkaloids from biological matrices by RP-HPLC with UV/VIS, diode-array, fluorescence, or coulometric electrochemical detection or tandem arrangements of these selected detectors. Studies have reported the simultaneous determination of morphine, codeine, and their glucuronidated metabolites isolated from plasma or urine matrices using SPE technologies usually with nonpolar bonded phases and in some instances weak and strong cation exchange sorbents. Assays by Rotshteyn and Weingarten (1996) and Ary and Rona (2001), for example, have shown the simultaneous determination of morphine and its glucuronidated metabolites from plasma by HPLC with fluorescence and electrochemical detection in tandem. The analytes were extracted with C2 or C18 bonded phase SPE cartridges. Electrochemical detection was used for the identification of morphine and fluorescence and/or UV for its glucuronides. The LLOQs ranged from 0.5 to 10 ng/ml, depending on the analytes of interest. Further applications for codeine and other opioids together with their respective metabolites also used RP-HPLC with multiple detector configurations (He et al., 1998; Sun et al., 2000; Dams et al., 2002).

7 Conclusions

The determination of endogenous compounds and drugs in biological matrices has always presented a formidable challenge as one has to consider various factors before attempting to develop a suitable HPLC assay. These include the physicochemical properties of the compound such as the pKa value, solubility, volatility, particular functional groups (e.g., possessing chromophores, fluorophores, or electroactive characteristics), potential metabolites, and the required sensitivity and specificity. All these aspects will determine the type of extraction processes, analytical column selection, and suitable detector systems to be used as part of the HPLC apparatus.

Acknowledgments

The authors are grateful to the Beebensee Schizophrenia Research Unit (Dr. J. Odontiadis), the Canada Research Chair Program (Dr. J. Odontiadis), and the Canadian Institutes for Health Research (CIHR) (Ms. G. Rauw) for their financial support.

References

Aboul-Enein HY, Al-Duraibi IA. 1998. Separation of several free polyamines and their acetylated derivatives by ion-pair reversed-phase high performance liquid chromatography. Biomed Chromatogr 12: 291-293.

Abu Ruz S, Millership J, Heaney L, McElnay J. 2003. Simple liquid chromatography method for the rapid simultaneous determination of prednisolone and cortisol in plasma and urine using hydrophilic lipophilic balanced solid phase extraction cartridges. J Chromatogr B Biomed Sci Appl 798(2): 193-201.

Al-Dirbashi O, Kuroda N, Akiyama S, Nakashima K. 1997. High-performance liquid chromatography of methamphetamine and its related compounds in human urine following derivatization with fluorescein isothiocyanate. J Chromatogr B Biomed Sci Appl 695(2): 251-258.

Al-Dirbashi O, Wada M, Kuroda N, Takahashi M, Nakashima K. 2000. Achiral and chiral quantification of methamphetamine and amphetamine in human urine by semi-micro column high-performance liquid chromatography and fluorescence detection. J Forensic Sci 45(3): 708-714.

Anderson G. 1985. Liquid chromatographic analysis of monoamines and their metabolites. Neuromethods 2. Amines and their Metabolites. Boulton AA, Baker GB, Baker JM, editors. New Jersey: Humana Press; pp. 347-358.

Angelo HR, Petersen A. 2001. Therapeutic drug monitoring of haloperidol, perphenazine, and zuclopenthixol in serum by a fully automated sequential solid phase extraction followed by high-performance liquid chromatography. Ther Drug Monit 23(2): 157-162.

Arinobu T, Hattori H, Iwai M, Ishii A, Kumazawa T, et al. 2002. Liquid chromatographic-mass spectrometric determination of haloperidol and its metabolites in human plasma and urine. J Chromatogr B Analyt Technol Biomed Life Sci 776(1): 107-113.

Ary K, Rona K. 2001. LC determination of morphine and morphine glucuronides in human plasma by coulometric and UV detection. J Pharm Biomed Anal 26(2): 179-187.

Ary K, Rona K, Ondi S, Gachalyi B. 1998. High-performance liquid chromatographic method with coulometric detection for the determination of buspirone in human plasma by means of a column-switching technique. J Chromatogr A 797(1–2): 221-226.

Avenso A, Facciola G, Salemi M, Spina E. 2000. Determination of risperidone and its major metabolite 9-hydroxyrisperidone in human plasma by reversed-phase liquid chromatography with ultraviolet detection. J Chromatogr B Biomed Sci Appl 746(2): 173-181.

Bagli M, Rao ML, Hoflich G. 1994. Quantification of chlorprothixene, levomepromazine and promethazine in human serum using high-performance liquid chromatography with coulometric electrochemical detection. J Chromatogr B Biomed Appl 657(1):141-148.

Bao J, Potts BD. 2001. Quantitative determination of olanzapine in rat brain tissue by high-performance liquid chromatography with electrochemical detection. J Chromatogr B Biomed Sci Appl 752(1): 61-67.

Berna M, Ackermann B, Ruterbories K, Glass S. 2002. Determination of olanzapine in human blood by liquid chromatography–tandem mass spectrometry. J Chromatogr B Analyt Technol Biomed Life Sci 767(1): 163-168.

Berthault F, Kintz P, Mangin P. 1996. Simultaneous high-performance liquid chromatographic analysis of flunitrazepam and four metabolites in serum. J Chromatogr B Biomed Appl 685(2): 383-387.

Bianchi G, Caccia S. 1988. Simultaneous determination of buspirone, gepirone, ipsapirone and their common metabolite 1-(2-pyrimidinyl)piperazine in rat plasma and brain by high-performance liquid chromatography. J Chromatogr 431(2): 477-480.

Bixo M, Backstrom T, Winblad B. 1984. Progesterone distribution in the brain of the PMSG treated female rat. Acta Physiol Scand 122(3): 355-359

Casas M, Berrueta LA, Gallo B, Vicente F. 1993. Solid-phase extraction of 1,4-benzodiazepines from biological fluids. J Pharm Biomed Anal 11(4–5): 277-284.

Chan ECY, Wee PY, Ho PY, Ho PC. 2000. High-performance liquid chromatographic assay for catecholamines and metanephrines using fluorimetric detection with pre-column 9-fluorenylmethyloxycarbonyl chloride derivatization. J Chromatogr B 749: 179

Chen XH, Franke JP, Ensing K, Wijsbeek J, De Zeeuw RA. 1993. Pitfalls and solutions in the development of a fully automated solid-phase extraction method for drug screening purposes in plasma and whole blood. J Anal Toxicol 17 (7): 421-426.

Cheung SW, Tang SW, Remington G. 1991. Simultaneous quantitation of loxapine, amoxapine and their 7- and 8-hydroxy metabolites in plasma by high-performance liquid chromatography. J Chromatogr 564(1): 213-221.

Chi JD, Odontiadis J, Franklin M. 1999. Simultaneous determination of catecholamines in rat brain tissue by high-performance liquid chromatography. J Chromatogr B 731: 361

Dams R, Benijts T, Lambert WE, De Leenheer AP. 2002. Simultaneous determination of in total 17 opium alkaloids and opioids in blood and urine by fast liquid chromatography-diode-array detection-fluorescence detection, after solid-phase extraction. J Chromatogr B Analyt Technol Biomed Life Sci 773(1): 53-61.

Draisci R, Pallaeschi L, Ferretti E, Lucentini L, Cammarata P. 2000. Quantitation of anabolic hormones and their metabolites in bovine serum and urine by liquid chromatography–tandem mass spectrometry. J Chromatogr A 870(1–2): 511-522.

Dusci LJ, Hackett P, Fellows LM, Ilett KF. 2002. Determination of olanzapine in plasma by high-performance liquid chromatography using ultraviolet absorbance detection. J Chromatogr B Analyt Technol Biomed Life Sci 773(2): 191-197.

Duverneuil C, Grandmaison GL, Mazancourt P, Alvarez JC. 2003. A high-performance liquid chromatography method with photodiode-array UV detection for therapeutic drug monitoring of the nontricyclic antidepressant drugs. Ther Drug Mon 25(5): 565-573.

Ernst DJ, Ernst C. 2003. Proper handling and storage of blood specimens. Home Health Nurse 21: 266

Falco P, Cabeza A, Legua C, Kohlmann M. 1996. Amphetamine and methamphetamine determination in urine by reversed-phase high-performance liquid chromatography with simultaneous sample clean-up and derivatization with naphthoquinone 4-sulphonate on solid-phase cartridges. J Chromatogr B Biomed Sci Appl 687(1): 239-246.

Farina A, Doldo A, Quaglia MG. 1992. Analysis of new serotonergic anxiolytics by liquid chromatography. J Pharm Biomed Anal 10(10–12): 889-893.

Foglia JP, Sorisio D, Kirshner MA, Mulsant BH, Perel JM. 1995. Quantitative determination of perphenazine and its metabolites in plasma by high-performance liquid chromatography and coulometric detection. J Chromatogr B Biomed Appl 668(2): 291-297.

Frölich L, Dirr A, Götz ME, Gsell W, Reichmann H, et al. 1998. Acetylcholine in human CSF: methodological considerations and levels in dementia of Alzheimer type. J Neural Transm 105: 961

Guo X, Fukushima T, Li F, Imai K. 2003. Determination of fluoxetine and norfluoxetine in rat plasma by HPLC with pre-column derivatization and fluorescence detection. Bio Med Chromatogr 17(1): 1-5.

Haginaka J. 2001. Protein-based chiral stationary phases for high performance liquid chromatography enantioseparations. J Chromatogr A 906: 253-273.

Hamase K, Morikawa A, Zaitsu K. 2002. D-amino acids in mammals and their diagnostic value. J Chromatogr B 781: 73-91.

Hasselstrom J, Linnet K. 2003. Fully automated on-line quetiapine in human serum by solid phase extraction and liquid chromatography. J Chromatogr B 798: 9

He H, Shay SD, Caraco Y, Wood M, Wood AJ. 1998. Simultaneous determination of codeine and its seven metabolites in plasma and urine by high-performance liquid chromatography with ultraviolet and electrochemical detection. J Chromatogr B Biomed Sci Appl 708(1–2): 185-193.

Hefnawy MM. 2002. Analysis of certain tranquilizers in biological fluids. J Pharm Biomed Anal 27(5): 661-678.

Hernandez R, Falco P, Cabeza A. 1997. Liquid chromatographic analysis of amphetamine and related compounds in urine using solid-phase extraction and 3,5-dinitrobenzoyl chloride for derivatization. J Chromatogr Sci 35(4): 169-175.

Huang ZP, Chen XH, Wijsbeek J, Franke JP, De Zeeuw RA. 1996. An enzymic digestion and solid-phase extraction procedure for the screening for acidic, neutral, and basic drugs in liver using gas chromatography for analysis. J Anal Toxicol 20(4): 248-254.

Hue B, Palomba B, Giacardy-Paty M, Bottai T, Alric R, et al. 1998. Concurrent high-performance liquid chromatographic measurement of loxapine and amoxapine and of their hydroxylated metabolites in plasma. Ther Drug Monit 20(3): 335-339.

Iwase H, Gondo K, Koike T, Ono I. 1994. Novel precolumn deproteinization method using a hydroxyapatite cartridge for the determination of theophylline and diazepam in human plasma by high-performance liquid chromatography with ultraviolet detection. J Chromatogr B Biomed Appl 655(1): 73-81. Erratum in: J Chromatogr B Biomed Appl 660(2): 422.

Kato T, Liu JK, Yamamoto K, Osborne PG, Niwa O. 1996. Detection of basal acetylcholine release in the microdialysis of rat frontal cortex by high- performance liquid chromatography using a horseradish peroxidase-osmium redox polymer electrode with pre-enzyme reactor. J Chromatogr B 682: 162-166.

Kehr J, Yoshitake T, Wang FH, Wynick D, Holmberg K, et al. 2001. Microdialysis in freely moving mice: Determination of acetylcholine, serotonin and noradrenaline release in galanin transgenic mice. J Neuro Sci Method 109: 71-80.

Kim YS, Zhang H, Kim HY. 2000. Profiling neurosteroids in cerebrospinal fluids and plasma by gas chromatography/electron capture negative chemical ionization mass spectrometry. Anal Biochem 277(2): 187-195.

Kim D-K, Natarajan N, Prabhakar NR, Kumar GK. 2004. Facilitation of dopamine and acetylcholine release by intermittent hypoxia in PC12 cells: involvement of calcium and reactive oxygen species. J Appl Physiol 96: 1206-1215.

Kollroser M, Schober C. 2002. Direct-injection high performance liquid chromatography ion trap mass spectrometry for the quantitative determination of olanzapine, clozapine and N-desmethylclozapine in human plasma. Rapid Commun Mass Spectrom 16(13): 1266-1272.

Kosel M, Eap CB, Amey M, Baumann P. 1998. Analysis of citalopram and its demethylated metabolites using chiral liquid chromatography. J Chromatogr B 719: 234-238.

Kumarathasan P, Vincent R. 2003. New approach to simultaneous analysis of catecholamines and tyrosines in biological fluids. J Chromatogr A 987: 349-358.

Kuronen P, Volin P, Laitalainen T. 1998. Reversed-phase high-performance liquid chromatographic screening method for serum steroids using retention index and diode-array detection. J Chromatogr B 718(2): 211-224.

Kurosawa S, Yoshimura T, Kurosawa H, Chiba K. 1995. Simultaneous determination of 18-oxygenated corticosteroids by high performance liquid chromatography with fluorescence detection. J Liq Chromatogr 18: 2383-2396.

LaCourse W. 2000. Column liquid chromatography: equipment and instrumentation. Anal Chem 72: 37R-51R.

LaCourse W. 2002. Column liquid chromatography: equipment and instrumentation. Anal Chem 74: 2813-2832.

Lakshmana MK, Trichur TR. 1997. An isocratic assay for norepinephrine, dopamine and 5-hydroxytryptamine using their native fluorescence by high-performance liquid chromatography with fluorescence detection in discrete brain areas of rat. Anal Biochem 246: 166-170.

Larsimont V, Meins J, Buschges HF, Blume H. 1998. Validated high-performance liquid chromatographic assay for the determination of promazine in human plasma. J Chromatogr B Biomed Sci Appl 719(1–2): 222-226.

Lauber R, Mosimann M, Buhrer M, Zbinden AM. 1994. Automated determination of midazolam in human plasma by high-performance liquid chromatography using column switching. J Chromatogr B Biomed Appl 654(1): 69-75.

LeGatt D. 1988. High performance liquid chromatography analysis. Neuromethods 10. Analysis of Psychiatric Drugs. Boulton AA, Baker GB, Coutts RT, editors. Clifton, NJ: Humana Press; pp. 155-202.

Lenox RH, Kant GJ, Meyerhoff JL. 1982. Rapid enzyme inactivation. Handbook of Neurochemistry 2nd edn, Vol. 2: Experimental Neurochemistry. Lajtha A, editor. New York: Plenum; pp. 77-100.

Liu YY, Troostwijk JAE, Guchelaar HJ. 2001. Simultaneous determination of clozapine, norclozapine and clozapine-N-oxide in human plasma by high-performance liquid chromatography with ultraviolet detection. Biomed Chromatogr 15(4): 280-286.

Liu J-T, Liu RH. 2002. Enantiomeric composition of abused amine drugs: chromatographic methods of analysis and data interpretation. J Biochem Biophys Methods 54: 115-146.

Long Z, Nimura N, Adachi M, Sekine M, Hanai T, et al. 2001. Determination of D- and L-aspartate in cell culturing medium, within cells of MPT1 cell line and in rat blood by a column-switching high-performance liquid chromatographic method. J Chromatogr B 761: 99-106.

Lucca A, Gentilini G, Lopez-Silva S, Soldarin A. 2000. Simultaneous determination of human plasma levels of four selective serotonin reuptake inhibitors by high-performance liquid chromatography. Ther Drug Monit 22(3): 271-276.

Luo JP, Hubbard JW, Midha KK. 1997. Sensitive method for the simultaneous measurement of fluphenazine decanoate and fluphenazine in plasma by high-performance liquid chromatography with coulometric detection. J Chromatogr B Biomed Sci Appl 688(2): 303-308.

Maier NM, Pilar F, Lindner, W. 2001. Separation of enantiomers: needs, challenges, perspectives. J Chromatogr A 906: 3-33.

Mancinelli R, Gentili S, Guiducci M, Macchia T. 1999. Simple and reliable high-performance liquid chromatography fluorimetric procedure for the determination of amphetamine-derived designer drugs. J Chromatogr B Biomed Sci Appl 735(2): 243-253.

Maris FA, Dingler E, Niehues S. 1999. High-performance liquid chromatographic assay with fluorescence detection for the routine monitoring of the antidepressant mirtazapine and its demethyl metabolite in human plasma. J Chromatogr B Biomed Sci Appl 721(2): 309-316.

Mei H, Hseih Y, Nardo C, Xu X, Wang S, et al. 2003. Investigation of matrix effects in bioanalytical high performance liquid chromatography/tandem, mass spectrometrometic assay: application to drug discovery. Rapid Commun Mass Spectrom 17: 97-103.

Mišl'anová C, Hutta M. 2003. Role of biological matrices during the analysis of chiral drugs by liquid chromatography. J Chromatogr B 797: 91-109.

Molins-Legua C, Campínc-Falcó P, Sevillano-Cabeza A, Pedrón-Pons M. 1999. Urine polyamines determination using dansyl chloride derivatization in solid-phase extraction cartridges and HPLC. Analyst 124: 477-482.

Nagasaki T, Ohkubo T, Sugawars K, Yasui N, Furukori H, et al. 1999. Determination of risperidone and 9-hydroxyrisperidone in human plasma by high-performance liquid chromatography: Application to therapeutic drug monitoring in Japanese patients with schizophrenia. J Pharm Biomed Anal 19(3–4): 595-601.

Nimura N, Fujiwara T, Watanabe A, Sekine M, Furuchi T, et al. 2003. A novel chiral thiol reagent for automated precolumn derivatization and high-performance liquid chromatographic enantioseparation of amino acids and its application to the aspartate racemase assay. Anal Biochem 315: 262-269.

Nohta H, Satozono H, Koiso K, Yoshida H, Ishida J, et al. 2000. Highly selective fluorometric determination of polyamines based on intramolecular excimer-forming derivatization with a pyrene-labeling reagent. Anal Chem 72: 4199-4204.

Nozaki O. 2001. Steroid analysis for medical diagnosis. J Chromatogr A 935: 267-278.

Nozaki O, Ohba Y, Imai K. 1988. Determination of serum estradiol by normal-phase high-performance liquid chromatography with peroxyoxalate chemiluminescence detection. Anal Chim Acta 205: 255-260.

Nyanda AM, Nunes MG, Ramesh A. 2000. A simple high-performance liquid chromatography method for the quantitation of tricyclic antidepressant drugs in human plasma or serum. J Toxicol Clin Toxicol 38(6): 631-636.

Odontiadis J, Franklin M. 1996. Simultaneous quantitation of buspirone and its major metabolite 1-(2-pyrimidinyl)piperazine in human plasma by high-performance liquid chromatography with coulometric detection. J Pharm Biomed Anal 14(3): 347-351.

Olesen OV, Plougmann P, Linnet K. 2000. Determination of nortriptyline in human serum by fully automated solid-phase extraction and on-line high-performance liquid chromatography in the presence of antipsychotic drugs. J Chromatogr B 746(2): 233-239.

Operation and maintenance of the ZQ with Masslynx v. 4.0, Vol. 2, Waters, (2002) Appendix E.

Osborne PG, Yamamoto K. 1998. Disposable, enzymatically modified printed film carbon electrodes for use in the high-performance liquid chromatographic-electrochemical detection of glucose or hydrogen peroxide from immobilized enzyme reactors. J Chrom B 707: 3-8.

Palego L, Marazziti D, Biondi L, Giannaccini G, Sarno N. 2000. Simultaneous plasma level analysis of clomipramine, N-desmethylclomipramine, and fluvoxamine by reversed-phase liquid chromatography. Ther Drug Monit 22(2): 190-194.

Parent M, Bush D, Rauw G, Master S, Vaccarino F, et al. 2001. Analysis of amino acids and catecholamines, 5-hydroytryptamine and their metabolites in brain areas in the rat using *in vivo* microdialysis. Methods 23: 11-20.

Parsons LH, Kerr TM, Weiss F. 1998. Simple microbore high-performance liquid chromatographic method for the determination of dopamine and cocaine from a single in vivo brain microdialysis sample. J Chromatogr B 709: 35-45.

Pearson BE, Beverley E, Allison CM. 2000. Determination of progesterone and some of its neuroactive ring A-reduced metabolites in human serum. J Steroid Biochem Mol Bio 74 (3): 137-142.

Persson B-A, Andersson S. 2001. Unusual effects of separation conditions on chiral separations. J Chromatogr A 906: 195-203.

Piepponen TP, Skujins A. 2001. Rapid and sensitive step gradient assays of glutamate, glycine, taurine and γ-aminobutyric acid by high-performance liquid chromatography-fluorescence detection with *o*-phthalaldehyde-mercaptoethanol derivatization with emphasis on microdialysis samples. J Chromatogr B 757: 277-283.

Pirola R, Mundo E, Bellodi L, Bareggi SR. 2002. Simultaneous determination of clomipramine and its desmethyl and hydroxy metabolites in plasma of patients by high-performance liquid chromatography after solid-phase extraction. J Chromatogr B 772(2): 205-210.

Pistos C, Stewart JT. 2003. Direct injection HPLC method for the determination of selected phenothiazines in plasma using a Hisep column. Biomed Chromatogr 17(7): 465-470.

Que AH, Palm A, Baker AG, Novotny MV. 2000. Steroid profiles determined by capillary electrochromatography, laser-induced fluorescence detection and electrospray-mass spectrometry. J Chromatogr A 887(1–2): 379-391.

Queroz RH, Lanchote VL, Bonato PS, de Carvaldo D. 1995. Simultaneous HPLC analysis of tricyclic antidepressants and metabolites in plasma samples. Pharm Acta Helv 70 (2):181-186.

Raggi MA. 2002. Therapeutic drug monitoring: chemical-clinical correlations of atypical antipsychotic drugs. Curr Med Chem 9(14): 1397-1349.

Raggi MA, Sabbioni C, Casamenti G, Gerra G, Calonghi N, et al. 1999. Determination of catecholamines in human plasma by high-performance liquid chromatography with electrochemical detection. J Chrom B 730: 201-211.

Rakovska A, Javitt D, Raichev P, Ang R, Balla A, et al. 2003. Physiological release of striatal acetylcholine (in vivo): effect of somatostatin on dopaminergic-cholinergic interaction. Brain Res Bull 61: 529-536.

Rotshteyn Y, Weingarten B. 1996. A highly sensitive assay for the simultaneous determination of morphine, morphine-3-glucuronide, and morphine-6-glucuronide in human plasma by high-performance liquid chromatography with electrochemical and fluorescence detection. Ther Drug Monit 18(2): 179-188.

Santagati NA, Ferrara G, Marrazzo A, Ronsisvalle G. 2002. Simultaneous determination of amphetamine and one of its metabolites by HPLC with electrochemical detection. J Pharm Biomed Anal 30(2): 247-255.

Seno H, Hattori H, Ishii A, Kumazawa T, Watanabe-Suzuki K, et al. 2000. Analyses of butyrophenones and their analogues in whole blood by high-performance liquid chromatography-electrospray tandem mass spectrometry. J Chromatogr B Biomed Sci Appl 746(1): 3-9.

Shah AJ, de Biasi V, Taylor SG, Roberts C, Hemmati P, et al. 1999. Development of a protocol for the automated analysis of amino acids in brain tissue samples and microdialysates. J Chromatogr B 735: 133-140.

Shah AJ, Crespi F, Heidbreder C. 2002. Amino acid neurotransmitters: Separation approaches and diagnostic value. J Chromatogr B 781: 151-163.

Shahed AR, Werchan PM, Stavinoha WB. 1996. Differences in acetylcholine but not choline in brain tissue fixed by freeze fixation or microwave heating. Methods Find Exp Clin Pharmacol 18: 349-351.

Shams M, Hiemke C, Hartter S. 2004. Therapeutic drug monitoring of the antidepressant mirtazapine and its N-demethylated metabolite in human serum. Ther Drug Monit 26(1): 78-84.

Shibata N, Hayakawa T, Takada K, Hoshino N, Minouchi T, et al. 1998. Simultaneous determination of glucocorticoids in plasma or urine by high-performance liquid chromatography with precolumn fluorimetric derivatization by 9-anthroyl nitrile. J Chromatogr B 706(2): 191-199.

Skoog D, Holler FJ, Nieman TA. 1998. Chromatography. Principles of Instrumental Analysis. Philadelphia: Harcourt Brace College Publishers; pp. 735-736.

Smith R. 2003. Before the Injection-modern methods of sample preparation for separation techniques. J Chromatogr A 1000: 3

Soares ME, Carvalho F, Bastos ML. 2001. Determination of amphetamine and its metabolite p-hydroxyamphetamine in rat urine by reversed-phase high-performance liquid chromatography after dabsyl derivatization. Biomed Chromatogr 15(7): 452-426.

Sun L, Hall G, Lau CE. 2000. High-performance liquid chromatographic determination of cocaine and its metabolites in serum microsamples with fluorimetric detection and its application to pharmacokinetics in rats. J Chromatogr B Biomed Sci Appl 745(2): 315-323.

Taibi G, Schiavo MC, Gueli MC, Calanni Rindina P, Muratore R, et al. 2000. Rapid and simultaneous high-performance liquid chromatography assay of polyamines and monoacetylpolyamines in biological specimens. J Chromatogr B 745: 431-437.

Tanaka E, Terada M, Nakamura T, Misawa S, Wakasugi C. 1997. Forensic analysis of eleven cyclic antidepressants in human biological samples using a new reversed-phase chromatographic column of 2 micron porous microspherical silica gel. J Chromatogr B Biomed Sci Appl 692(2): 405-412.

Tanaka E, Terada M, Misawa S, Wakasugi C. 1998. Erratum to "Simultaneous determination of twelve benzodiazepines in human serum using a new reversed-phase chromatographic column on a 2um porous microspherical silica gel." J Chromatogr B Biomed Appl 709(2): 324. [J Chromatogr B 682 (1996) 173].

Tedeschi L, Frison G, Castagna F, Giorgetti R, Ferrara SD. 1993. Simultaneous identification of amphetamine and its derivatives in urine using HPLC-UV. Int J Legal Med 105 (5): 265-269.

Teti D, Visalli M, McNair H. 2002. Analysis of polyamines as markers of (patho) physiological conditions. J Chromatogr B 781: 107-149.

Titier K, Deridet E, Cardone E, Abouelfath A, Moore N. 2002. Simplified high-performance liquid chromatographic method for determination of risperidone and 9-hydroxyrisperidone in plasma after overdose. J Chromatogr B Analyt Technol Biomed Life Sci 772(2): 373-378.

Titier K, Bouchet S, Pehourcq F, Moore N, Molimard M. 2003. High-performance liquid chromatographic method with diode array detection to identify and quantify atypical antipsychotics and haloperidol in plasma after overdose. J Chromatogr B Analyt Technol Biomed Life Sci 788(1): 179-185.

Tournel G, Houdret N, Hedouin V, Deveaux M, Gosset D, Lhermitte M. 2001. High-performance liquid chromatographic method to screen and quantitate seven selective serotonin reuptake inhibitors in human serum. J Chromatogr B Biomed Sci Appl 761(2): 147-158.

Toyo'oka T. 2002. Resolution of chiral drugs by liquid chromatography based upon diastereomer formation with chiral derivatization reagents. J Biochem Biophys Methods 54: 25-56.

Tsai T-H. 2000. Separation methods used in the determination of choline and acetylcholine. J Chromatogr B 747: 111-122.

Vaarman A, Kask A, Mäeorg U. 2002. Novel and sensitive high-performance liquid chromatographic methods based on electrochemical coulometric array detection for simultaneous determination of catecholamines, kynurenine and indole derivatives of tryptophan. J Chrom B 769: 145-153.

Vallee M, Rivers JD, Koob GF, Purdy RH, Fitzgerald RL. 2000. Quantification of neurosteroids in rat plasma and brain following swim stress and allopregnanolone administration using negative chemical ionization gas chromatography/mass spectrometry. Anal Biochem 287(1): 153-166.

Vanapalli SR, Kambhampati SP, Putcha L, Bourne DW. 2001. A liquid chromatographic method for the simultaneous determination of promethazine and three of its metabolites in plasma using electrochemical and UV detectors. J Chromatgr Sci 39(2): 70.

Venza M, Visalli M, Cicciu D, Teti D. 2001. Determination of polyamines in human saliva by high performance liquid chromatography with fluorescence detection. J Chromatogr B 757: 111-117.

Visser SAG, Smulders CJGM, Gladdines WWFT, Irth H. 2000. High-performance liquid chromatography of the neuroactive steroids alphaxalone and pregnanolone in plasma using dansyl hydrazine as fluorescent label: application to a pharmacokinetic-pharmacodynamic study in rats. J Chromatogr B 745: 357-363.

Walter S, Bauer S, Roots I, Brockmoller J. 1998. Quantification of the antipsychotics flupentixol and haloperidol in human serum by high-performance liquid chromatography with ultraviolet detection. J Chromatogr B Biomed Sci Appl 720 (1–2): 231-237.

Wang MD, Wahlstrom G, Backstrom T. 1997. The regional brain distribution of the neurosteroids pregnenolone and pregnenolone sulfate following intravenous infusion. J Steroid Biochem Mol Biol 62(4): 299-306.

Ward TJ, Farris AB III. 2001. Chiral separations using macrocyclic antibiotics: a review. J Chromatogr A 906: 73-89.

Waschgler R, Hubmann MR, Conca A, Moll W, Konig P. 2002. Simultaneous quantification of citalopram, clozapine, fluoxetine, norfluoxetine, maprotiline, desmethylmaprotiline and trazodone in human serum by HPLC analysis. Int J Clin Pharmacol Ther 40(12): 554-559.

Weigmann H, Hartter S, Hiemke C. 1998. Automated determination of clomipramine and its major metabolites in human and rat serum by high-performance liquid chromatography with on-line column-switching. J Chromatogr B Biomed Sci App 710(1–2): 227-233.

Weigmann H, Hartter S, Maehrlein S, Kiefer W, Kramer G, et al. 2001. Simultaneous determination of olanzapine, clozapine and demethylated metabolites in serum by on-line column-switching high-performance liquid chromatography. J Chromatogr B Biomed Sci Appl 759(1): 63-71.

Wilhelm M, Battista HJ, Obendorf D. 2001. HPLC with simultaneous UV and reductive electrochemical detection at the hanging mercury drop electrode: a highly sensitive and selective tool for the determination of benzodiazepines in forensic samples. J Anal Toxicol 25(4): 250-257.

Wood M, Morris M. 2002. Simultaneous quantification of psychotherapeutic drugs in human plasma by tandem mass spectrometry. Proc Ann Meet American Society for Mass Spectrometry.

Yamada H, Kuwahara Y, Takamastu Y, Hayase T. 2000. A new sensitive determination method of estradiol in plasma using peroxyoxalate ester chemiluminescence combined with an HPLC system. Biomed Chromatogr 14(5): 333-337.

Yasumatsu M, Yazawa T, Otokawa M, Kuwasawa K, Hasegawa H, et al. 1998. Monoamines, amino acids and acetylcholine in the preoptic area and anterior hypothalamus of rats: measurements of tissue extracts and in vivo microdialysates. Comp Biochem Physiol A Mol Integr Physiol 121: 13-23.

Yoshitake T, Ichinose F, Yoshida H, Todoroki K, Kehr J, et al. 2003. A sensitive and selective determination method of histamine by HPLC with intramolecular excimer-forming derivatization and fluorescence detection. Biomed Chromatogr 17: 509-516.

3 Cytochrome P450 Reactions in the Human Brain

R. L. Haining

Abstract: The activation of a carbon atom through hydroxylation is one of the most useful and fundamental enzymatic processes in the body. Such oxidations are often carried out by the enzyme cytochrome P450. First noted as a hepatic heme-containing enzyme, we now know that 'cytochrome P450' constitutes a superfamily in humans, containing dozens of individual but related enzymes. Because of its heavy presence in liver, cytochrome P450 is often thought of as evolved primarily to rid the body of unwanted toxins. However, due to the utility of the enzymatic reaction, P450 enzymes are not limited to liver but are instead employed ubiquitously throughout the body with very different outcomes. In the liver, toxins are made soluble for elimination through urinary and biliary pathways. In other tissues including brain, P450 enzymes play crucial roles in *biosynthetic* pathways involving natural, hydrophobic substrates. This chapter summarizes much of the information available regarding the presence and function of cytochrome P450s in human brain. While performing research and attempting to make order of the many isoforms, I was struck by the ability to place every known human brain P450 into three major chemical pathways: those involving a) cholesterol, b) arachadonic acid, and c) retinoic acid. That is, a majority of P450 enzymes implicated in human brain catalyze reactions within 7 metabolic steps of cholesterol itself, while the remainders, with the exception of one, catalyze reactions metabolically related to arachadonic acid. Each of these pathways leads to key signaling molecules, thus it would appear that there is tremendous overlap between xenobiotic and biochemical pathways as catalyzed by cytochrome P450 in brain. My hope is that the information contained herein will prove useful for others to better visualize, understand, and further study the global implications of P450-catalzyed reactions in the human brain.

List of Abbreviations: AA, arachidonic acid; AD, Alzheimer's Disease; BBB, blood-brain barrier; CHAPS, 3-[(3-Cholamidopropyl)dimethylammonio]-1-propanesulfonate; CYP, cytochrome P450; DEAE, diethylaminoethyl; DHEA, dehydroepiandrosterone; EET, epoxyeicosatrienoic acid; ER, endoplasmic reticulum; GAPDH, glyceraldehyde-3-phosphate dehydrogenase; HETE, hydroxyeicosatrienoic acid; HPLC, high performance liquid chromatography; ICC, immunocytochemistry; IHC, immunohistochemistry; IL, Interleukin; LTB4, leukotriene B4; MALDI-TOF, matrix-assisted laser desorption ionization time-of-flight; MW, molecular weight; OHase, hydroxylase; TNF, tumor necrosis factor; PREG, pregnenolone; PB, phenobarbital; PVDF, polyvinyl difluoride; RA, retinoic acid; RIA, Radio-Immunological Assay; RT-PCR, Reverse Transcription Polymerase Chain Reaction; RT-RT-PCR, Real-time RT-PCR; SDS-PAGE, sodium dodecyl sulfate polyacrylamide gel electrophoresis; WB, Western blot

1 Introduction

1.1 Extrahepatic Metabolism

The study of cytochrome P450-catalyzed drug metabolism has a long and colorful history, with over four decades worth of untangling the various functions of this enzyme family from the human liver. As such, there are many pieces of dogma that have perhaps inadvertently been cemented into the literature and in the minds of many researchers and physicians alike. One example is the idea that drug metabolism in humans is largely confined to the liver. While it is true that the majority of drugs circulating throughout the bloodstream are indeed subject to metabolism by hepatic cytochrome P450s, it is now very clear that xenobiotic metabolism can also occur in many extrahepatic tissues. First-pass metabolism, for example, was once considered to be caused by the passage of absorbed chemical entities from the gut and into the blood, where it must pass through the liver before reaching general circulation. It was believed that the loss of parent drug to this first-pass effect was due to metabolism in the liver; however, we now know that the intestine itself contributes substantially to this effect. Many drugs distribute and accumulate in tissues far removed from the blood stream. Such drugs must then be metabolized to more polar compounds near their site of accumulation to be removed efficiently. In the central nervous system, the blood–brain barrier (BBB) serves to efficiently block most polar xenobiotic compounds while sequestering others. Codeine, which is efficiently transported across the BBB, while its analgesic potential must be activated by P450-catalyzed conversion to morphine to unleash its analgesic potential. Therefore, drug metabolism should be viewed as a whole-body phenomenon and not merely a hepatic phenomenon. At least 38 of the known human p450 genes have a demonstrated presence in the brain. Many of them are dedicated to endogenous pathways and presumably participate very little in drug metabolism. However, many of them have the ability to carry out both endogenous and xenobiotic reactions, making the knowledge of their function in human brain necessary to understand the full effects of drug toxins on neurochemical pathways.

1.2 Multiplicity and Nomenclature

We now know that there are at least 18 separate gene families comprising 57 expressed cytochrome P450 genes and a further 59 pseudogenes (Nebert and Russell, 2002). To further add to the complexity, it has been shown recently that several P450 mRNA species contain exons encoded by these pseudogenes through alternate splicing pathways, which can be highly tissue specific. Examples now exist in which a single P450 gene may be alternately spliced for distinct localization in the endoplasmic reticulum, mitochondrial membrane, or both. Thus, we are still unclear about the true number of human cytochrome P450s until we solve the human proteome puzzle.

Standardized rules have been adopted for organizing and naming P450 isoforms and alleles, providing clarity to investigators in the field. Briefly, cytochrome P450's are sorted into families (1, 2, 3, and so on.) based on an amino acid identity of approximately 40% or greater. Subfamilies are recognized when they contain over 55% identity (2D or 3A for example) and individual genes may have up to 97% identity (2D6 or 3A4). Individual isoforms are therefore named "*CYP*," which represents cytochrome P450, followed by the family, subfamily, and gene designation, e.g., *CYP2D6* is a human isoform of great interest to neurochemistry and neuropharmacology. Further complexity of P450s is introduced by the existence of polymorphisms in most, if not all, drug metabolizing P450s, a topic that has been discussed in detail by me and others recently.

The complexity of the P450 system can be a daunting prospect for novices and P450 experts alike. With the advent of the computer era, a number of resources are now available at our fingertips to make the task of appreciating them more manageable. David Nelson maintains an internet web site dedicated to organizing the taxonomy of P450 throughout the kingdoms (http://drnelson.utmem.edu/CytochromeP450.html). In addition, he lists the historical name changes that have occurred before the adoption of standardized nomenclature, helping make the task of reading the earlier literature possible. Kirill N. Degtyarenko maintains another extremely useful web site at http://www.icgeb.org/~p450srv/, which includes links to much of the P450 literature, the reactions catalyzed by individual isoforms, nomenclature, and a host of other information. Magnus Ingelman-Sundberg, Ann Daly, and Daniel Nebert maintain the home page of the Human Cytochrome P450 (*CYP*) Allele Nomenclature Committee at http://www.imm.ki.se/CYPalleles/, which keeps track of the ever expanding list of *minor* P450 allelic variants that can nonetheless cause *major* drug interactions. David Flockhart maintains another useful web site at http://medicine.iupui.edu/flockhart/, which tabulates the P450 routes of drug elimination to help gauge the risk of drug–drug interactions.

The multiplicity of P450 raises the immediate question of why these enzymes have evolved in such a complicated fashion in extrahepatic tissues. In the liver, being the body's designated organ of detoxification, this diversity makes sense—it has been argued to be the result of "plant–animal warfare" that occurred throughout the evolution as animals sought new sources of food. In addition, the ingestion of plant alkaloids and other chemicals foreign to animals may have played a role in the evolution of cytochrome P450 in the brain. However, many, if not all, of the extrahepatic P450 enzymes are capable of metabolizing endogenous chemicals as well as xenobiotics. Clearly then, P450s would have evolved *in the presence of* such endogenous molecules, such that Mother Nature could not have designed the enzymes merely to rid the body of unwanted chemicals. Consider that the ability to oxidize carbon is one of the most difficult organic chemistry reactions (beside free radical mechanisms), so that the evolution of P450 catalysis must largely have occurred to exploit the potential of this reaction. Indeed, P450s are found ubiquitously across kingdoms in Nature, highlighting the usefulness of the metabolic pathways catalyzed by these enzymes.

1.3 Chapter Scope

The purpose of this chapter is to collect and summarize our state of knowledge regarding the location, function, and methods used to study human brain P450s. Although I have attempted to be thorough, a quick PubMed search on the term "P450" will convince even those with the greatest mental capacity that the P450 field is enormous. No single review or book chapter could do justice to it and I apologize in advance to those whose contributions appear to have been overlooked. In addition, many studies have, of course, been done on animals or with animal P450s, most notably those from rats, mice, dogs, guinea pigs, hamsters, zebra fish, and frogs. For the purposes of clarity, brevity, and sanity, I have largely limited discussion and references to those specific isoforms for which there is evidence of expression in the human brain. In many instances, human and animal P450s have very similar structures, substrate specificities, and expression patterns. Rat 2D4, for example, has a very similar substrate preference as human 2D6. However, the substrate specificity of the major mouse brain *CYP2D* species, *CYP2D22*, is very different from that of human *CYP2D6* (Yu and Haining, 2006). A great deal of caution must be exercised when extrapolating animal data to the human condition. In the cases where sufficient evidence exists for relevance to human brain P450, I have tried to include such data. However, I have made no attempt to be thorough with the animal brain P450 literature. I hope and believe that the readers of this chapter will be most interested in the human condition, while recognizing the limitations of studying human P450 in vivo and thus the need for animal models. It is my hope to convince readers of this chapter that there is no substitute for the study of the actual human proteins. I have also not attempted to provide recipes and protocols or a review of the psychopharmacology literature, rather I hope that the discussion of the tools, and links to papers that describe the methods others have used, may spark ideas and facilitate the tremendous work we still have ahead of us before we can answer the question "What are drug metabolizing enzymes doing in the brain?"

2 Methods Used to Study Brain Cytochrome P450

2.1 Measurement of Characteristic Enzyme Activity

2.1.1 Enrichment/Purification of P450 from Brain

The measurement of P450 enzyme activities in tissue preparations high in P450 content is fraught with enough difficulties because of the inherent complexity of the gene superfamily and the extensive substrate overlap. The brain has the added complexity of being easily 100-fold less abundant in P450 content overall, with large regional variations due to tissue- and/or cell-specific P450 isoform expression. Simple whole tissue homogenates from the brain have been found to be unsuitable for most P450 studies. Either simply not enough enzyme is present or the signal is blocked by endogenous background noise, either way making P450 content often unquantifiable by traditional spectroscopic means.

Perhaps the most obvious technique, given P450s' hepatic history, is to make a microsomal preparation of the brain tissue. In this method, the tissue is typically first homogenized with a Teflon pestle and glass mortar in a buffered solution with detergent and protease inhibitors. Significant problems with foaming can arise owing to the high lipid content, thus care must be taken at this step. The tissue homogenate is then subjected to low-speed centrifugation ($3000 \times g$) to pellet nuclei and large cell debris. Centrifugation at $10{,}000 \times g$ then pellets the mitochondrial fraction from the supernatant. Finally, high-speed centrifugation ($100{,}000 \times g$) is used to pellet the microsomal fraction, leaving the soluble cytosolic fraction behind. For proteomic analysis, the tissue should first be homogenized in a suitable detergent, preferably a zwitterionic one such as CHAPS, which allows solubilization of cytosolic and membrane proteins in a single step. This detergent was shown to be superior for the 2D-gel proteomic analysis of rat

Table 3-1
Commercial availability of P450s and related material

Isoform	P450s	Antibodies	cDNA probe	Assays
CYP19 (aromatase)	B, U	U		
CYP11B2 (Ald. Synth)		U		
CYP11A1 (P450scc)	U	R		
CYP8A1 (PGI2 synth)			O	
CYP4F3	B			
CYP4F2	B	R		
CYP4A11	B, R	R, U		
CYP1A1	B	B, I, R, O, U	O, U	
CYP1A2	B, I, R, O, U	B, R, O, U	O	B, I
CYP1B1	B	B		
CYP2A6	B, R	B, I, R, U		
CYP2B6	B, I	B, I, R, U		B, I
CYP2C8	B, R	B, I, R, U		
CYP2C9	B, I, R, U	B, I, R, U	O, U	B, I
CYP2C18	B	I		
CYP2C19	B, I, R, O, U	B, I, R, U		B, I
CYP2D6	B, I, R, O, U	B, R, O, U	O	B, I
CYP2E1	B, I, R, O, U	B, R, O, U	O, U	I
CYP3A4	B, I, R, O, U	B, R, O, U	O, U	B, I
CYP3A5	B, I, U	B, R, U		I
CYP3A7	B			
P450 Reductase	B, I, R, O, U	B, R, U		
Cytochrome b5	B, I, R, O			

B—BD-Gentest, www.bdbiosciences.com; I—Invitrogen/Panvera, www.invitrogen.com; R—Research Diagnostics, Inc., www.researchd.com; O—Oxford Biomedical, www.oxfordbiomed.com; U—U.S. Biological, www.usbio.net.

brain (Carboni et al., 2002). Two-dimensional gel electrophoresis can then be used to isolate individual proteins for immunoidentification.

The Ravindranath laboratory is perhaps the only one to report a purification or partial purification of multiple P450 isoforms from human brain autopsy tissue. They used sequential chromatography on octylamino-Sepharose 4B, DEAE-Sephacel, and DEAE-cellulose to isolate 4 distinct populations of P450 (Bhamre et al., 1993). Ikeda et al., used immunoaffinity purification of an unknown antigen that, by circumstance, isolated *CYP39* from ciliary processes of bovine ocular tissue (Ikeda et al., 2003). Given the wealth of commercially available antibody preparations (*Table 3-1*), a similar technique should prove highly useful in the determination of brain-specific isoform variation; however, to this author's knowledge it has not been employed to date. Our laboratory has tested a polyclonal anti-*CYP2D6* antibody raised in rabbit against highly purified recombinant *CYP2D6* (Yu et al., 2001) to immunoprecipitate solubilized human brain protein. Homogenates were first subjected to $CaCl_2$ precipitation (Bhagwat et al., 1995b) followed by resuspension and incubation with antibody. Protein-G-agarose beads were then used to adhere to the antibody and make the protein collectable by centrifugation. Analysis of the immunoprecipitated protein by SDS-PAGE followed by Western blot with a commercially monoclonal anti-*CYP2D6* antibody (BD-Gentest) revealed two distinct bands in the vicinity of 55 kDa, which differ by Ca. 3 kDa (not shown). We have begun the process of analyzing these proteins by MALDI – TOF mass spectrometry and were very interested to learn of the findings of Pai et al. regarding a *CYP2D6/7* variant expressed in the human brain (Pai et al., 2004). Following enrichment, it is now a standard technique to transfer protein to PVDF membrane for direct *N*-terminal sequencing, as in the example provided by Wu (Wu et al., 1999) and we intend to perform this on our human 2D proteins as well.

2.1.2 Enzyme Assays

The high lipid and cholesterol content of brain makes the finding of cholesterol or arachidonic acid (AA) metabolites formed in vitro by P450 using standard HPLC methods extremely difficult, necessitating radioactivity or other means to enhance the signal. Another complication in the brain is the finding that some very important isoforms can exist in more than one location inside the cell, being targeted both to the endoplasmic reticulum (as assumed from their microsomal fractionation) and to the inner mitochondrial membrane, each with distinct enzyme activities and redox partner preferences (Bhagwat et al., 1995a; Addya et al., 1997; Robin et al., 2001).

Most notably characterized among the brain P450s that catalyze endogenous reactions are *CYP19* (aromatase), *CYP11A1* (P450scc or cholesterol side-chain cleavage enzyme), and *CYP7B1* (oxysterol 7α-hydroxylase) owing to a great interest in their role in neurosteroid hormone biosynthesis. It is now very clear that several enzymes involved in cholesterol biosynthetic or degradation pathways, as well as the steroidogenic acute regulatory protein (King et al., 2002), are present and active in the brain tissue (➋ *Table 3-2*, ➋ *Figure 3-1*). Aromatase activity, for example, was assayed in human fetal tissue by determining the rate of incorporation of tritium from [1β ^{3}H]androstenedione into water. These investigators found low but measurable aromatase activity in human fetal brain tissue (Doody and Carr, 1989). Using a similar assay, experiments with the cerebral cortex and subcortical white matter specimens from children and adults revealed significantly higher aromatase activity in the cerebral cortex than in subcortical white matter, but no sex or age differences were found (Steckelbroeck et al., 1999). Dehydroepiandrosterone (DHEA) 7α-hydroxylase (*CYP7B1*) activity was also found in these tissues with this method (Steckelbroeck et al., 2002). Warner et al. reported the use of radiolabeled 5α-androstane-3β,17β-diol to measure P450 (*CYP7B*)-mediated reactions in rat brain tissue (Warner et al., 1991). The products were separated by TLC and/or HPLC followed by autoradiography or scintillation counting (Sundin et al., 1987).

Akwa et al. (1992) showed, using a deuterated substrate and a mass spectrometry-based assay, that DHEA and pregnenolone (PREG) are converted by rat brain microsomes into the respective 7α-hydroxylated derivatives, indicating the presence of *CYP7B1* (Akwa et al., 1992). Brain extracts from humans also contain *CYP7B1* as judged by the release of tritium from [7α-^{3}H]pregnenolone and conversion of DHEA to a product with a chromatographic mobility indistinguishable from 7α-hydroxy DHEA (Rose et al., 1997; Yau et al., 2003). The activity of *CYP11A1* was measured in extracts of amphibian brain by radio-immunoassay (RIA) and shown to produce both PREG and its sulfate ester (Takase et al., 1999). Using radiolabeled cholesterol, Brown et al. (2000) were able to measure the formation of PREG in human neuronal cell culture. They found that oligodendrocytes are capable of synthesizing PREG, but neither human astrocytes nor neurons are able to make radiolabeled PREG (Brown et al., 2000).

Kawato et al. (2002) reported the use of a unique Radio Immuno Assay (RIA) technique in which they were able to measure steroidogenic P450 enzyme activity directly in the rat brain slices. These workers measured the conversion of [7-^{3}H]-pregnenolone to [7-^{3}H]-DHEA (CYP17) and [7-^{3}H]-DHEA to [^{3}H] 17β-estradiol (CYP19) during a 5-h incubation, then extracted metabolites with methanol/acetic acid followed by HPLC separation, identification, and quantification (Kawato et al., 2002). However, this method has not been widely adopted as of now. *CYP46* in human brain catalyzes the 24-hydroxylation of cholesterol, thought to be a major pathway for the removal of cholesterol from the brain (Bjorkhem et al., 1999). Lund et al. were able to measure 24-hydroxylase from transfected cells carrying the cloned human *CYP46* gene (Lund et al., 1999). This group then followed up with a *CYP46*-knockout mouse to show that the synthesis of new cholesterol was reduced by approximately 40% in the brains of these mice (Lund et al., 2003). Kehl et al. (2002) provide one of the best descriptions of a radioactive AA assay in a freely accessible online journal, a technique very useful for "fingerprinting" P450 isoforms based on their metabolite profile. Briefly, these investigators started with ^{14}C-labeled AA (Amersham, Arlington Heights, IL); the reactions were terminated by the addition of formic acid followed by ethyl acetate extraction and HPLC on a C18 reverse phase column. Detection was then achieved with an online radioactive flow detector (Kehl et al., 2002).

Certain prototypic drug substrates have been used to characterize enzyme activity in the human brain tissue. Amitriptyline, for example, was shown to be demethylated to nortriptyline by both rat and human

Table 3-2

Cytochromes P450 with primarily endogenous function in human brain (Amruthesh et al., 1992; Degtyarenko and Fabian, 1996; Nelson, 1998; Chang and Kam, 1999; Omiecinski et al., 1999; Nebert and Russell, 2002; Stoffel-Wagner, 2003)

P450 reactions	Highlights of P450 expression in brain
CYP51 Lanosterol 14-demethlyase	• *CYP51* found in rat brain microsomal fraction (Aoyama et al., 1996a). • Human *CYP51* is 93% and ~40% identical to rat and fungus (Aoyama et al., 1996b; Stromstedt et al., 1996). • *CYP51* is essential for the viability of mammals (Aoyama et al., 1996b; Rozman et al., 1996). • 0.286 per *GAPDH* mRNA in adult human brain by RT-RT-PCR (Nishimura et al., 2003).
CYP46 Cholesterol 24–OHase; homeostasis	• Human *CYP46* in brain microsomal fraction by RNA and protein blots (Lund et al., 1999). • Found in neurons of human brain by IHC; found in glial cells of AD but not control brain; *CYP46* distribution differs in brains of control and AD patients (Bogdanovic et al., 2001). • 24-OH is a major pathway for cholesterol turnover in mouse and human brain (Lund et al., 2003). • *CYP46* mRNA markedly insensitive to transcriptional regulation (Ohyama et al., 2006). • 0.162 per *GAPDH* mRNA in adult human brain by RT-RT-PCR (Nishimura et al., 2003).
CYP39A1 Oxysterol 7α-OHase; bile synthesis	• Hepatic *CYP39A1* mRNA not induced by cholesterol or bile acids, unlike other sterol 7alpha-hydroxylases; protein found in hepatic microsomal fraction (Li-Hawkins et al., 2000). • Found in ocular tissues; immunoaffinity purified MW 44 kDa (Ikeda et al., 2003). • 0.0032 per *GAPDH* mRNA in adult human brain by RT-RT-PCR (Nishimura et al., 2003).
CYP27 Sterol 27–OHase, VitD 25-OHase; bile	• Mitochondrial inner membrane of pig and rabbit liver (Axen et al., 1994). • Recombinant *CYP27* requires adrenodoxin and adrenodoxin reductase (Pikuleva et al., 1997). • Mutations in sterol 27-hydroxylase in cerebrotendinous xanthomatosis (Miki et al., 1986; Cali et al., 1991; Wakamatsu et al., 1999; Sugama et al., 2001; Guyant-Marechal et al., 2005; Lorincz et al., 2005). • 0.0158 per *GAPDH* mRNA in adult human brain by RT-RT-PCR (Nishimura et al., 2003).
CYP27B1 25-hydroxy VitD-1α-OHase	• In mouse cerebellum and cerebral cortex by IHC and WB (Zehnder et al., 2001). • Gene amplification for 25-hydroxyvitamin D(3) 1,alpha-hydroxylase and nonfunctional mRNA splice variants in human glioblastoma multiforme (Maas et al., 2001; Diesel et al., 2004, 2005). • 0.0006 per *GAPDH* mRNA in adult human brain by RT-RT-PCR (Nishimura et al., 2003).
CYP24 VitD 24, 25-OHase	• Mitochondrial localization in rat yolk sac (Danan et al., 1982). • *CYP24* mRNA upregulated by calcitriol in glioblastoma multiforme (Diesel et al., 2005). • *CYP24* mRNA subject to alternative splicing in human myelomonocytic cell line (Ren et al., 2005). • 0.0017 per *GAPDH* mRNA in adult human brain by RT-RT-PCR (Nishimura et al., 2003).

Table 3-2 (continued)

P450 reactions	Highlights of P450 expression in brain
CYP21A Sterol 21-OHase	• *CYP21* in tractus reticulothalamicus and other ascending fibers of rat brain (Iwahashi et al., 1993). • Alternatively mRNA splicing in rat adrenal (Zhou et al., 1997b). • *CYP21* mRNA expressed in human amygdala, caudate nucleus, cerebellum, corpus callosum, hippocampus, spinal cord, and thalamus (Yu et al., 2002). • *CYP21* mRNA detected in fetal brain by RT-RT-PCR (Pezzi et al., 2003).
CYP19 (aromatase) Steroid 19–OHase; estrogen and estrone synthesis	• Protein identified by IHC in the ventral pallidum, cerebral cortex, the amygdaloid area, the nucleus of the diagonal band, others of rat brain (Shinoda et al., 1989). • Less activity in human fetal brain than other tissues assayed by $[^3H]$ release from 1-androstenedione into water (Doody and Carr, 1989). • Found by IHC in amygdaloid structures, supraoptic nucleus, reticular thalamic nucleus, olfactory tract and piriform cortex, less so in the paraventricular and arcuate nuclei and hippocampus of rat (Sanghera et al., 1991). • Aromatase immunoreactivity in rat, human glioblastoma (Yague et al., 2004). • Human fetal brain and intestine had more *CYP19* mRNA than other tissues (Price et al., 1992). • IHC shows CYP19 in hypothalamus and limbic system of rat brain (Tsuruo et al., 1994). • IHC shows CYP19 in neuronal perikarya of quail, rat, monkey, human (Naftolin et al., 1996). • *CYP19* mRNA of adult Rhesus monkey is high in the bed nucleus of the stria terminalis $>$ medial preoptic/anterior hypothalamus $>$ amygdala; intermediate in the medial basal hypothalamus $>$ lateral preoptic/anterior hypothalamus; and low in the septum $>$ lateral-dorsal-medial hypothalamus; undetectable in cingulate and parietal cortex, hippocampus, and cerebellum (Abdelgadir et al., 1997). • Protein in neurons of rat medial amygdaloid nucleus by IHC. Enzyme activity in fetal rat brain competitively inhibited by dopamine and norepinephrine (Jimbo et al., 1998). • Higher aromatase activity in human cerebral cortex than subcortical white matter; no sex or age differences (Steckelbroeck et al., 1999). • Enzyme activity regulated by androgens in monkey brain reproductive areas (Roselli and Resko, 2001). • Duplicate *CYP19* genes in Zebra fish for specific expression in ovary and brain (Chiang et al., 2001). • Mammary glands of *CYP19*-humanized male mice undergo development resembling terminally differentiated female glands (Li et al., 2002). • Alternate promoters for tissue-specific regulation of *CYP19* expression (Kamat et al., 2002). • Mouse gene has three promoters to direct expression in ovary, testis, and brain (Golovine et al., 2003). • 0.0019 per *GAPDH* mRNA in adult human brain by RT-RT-PCR (Nishimura et al., 2003).
CYP17 Sterol 17β-OHase; Pregnenolone 17β-OHase; Pregnenolone→ 17β-OH- pregnenolone→ dehydroepiandrosterone (DHEA)	• *CYP17* not detectable in the brains of rat or guinea-pig by IHC (Le Goascogne et al., 1991). • Found in microsomal fraction of rat and mouse embryonic but not adult CNS (Compagnone et al., 1995). • *CYP17* mRNA undetectable in hamster brain (Cloutier et al., 1997). • Rat astrocytes and neurons express *CYP17*, synthesize DHEA from pregnenolone. Astrocytes metabolize DHEA to sex steroid hormones (Zwain and Yen, 1999). • *CYP17* expressed regionally in male and female avian brain (Matsunaga et al., 2001).

Table 3-2 (continued)

P450 reactions	Highlights of P450 expression in brain
	• *CYP17* mRNA expressed in human amygdala, caudate nucleus, cerebellum, corpus callosum, hippocampus, spinal cord, and thalamus (Yu et al., 2002). • *CYP17* mRNA below detection in whole adult human brain by RT-RT-PCR (Nishimura et al., 2003). • *CYP17* mRNA not found in adult mouse brain by RT-PCR (Choudhary et al., 2003).
CYP11A1 P450scc; Cholesterol side-chain cleavage enzyme→ Pregnenolone	• Found in mitochondrial inner membrane of rat (Costa et al., 1994). • *CYP11A1* mRNA in adult rat brain, primary glial cultures, *C6* glioma cells (Mellon and Deschepper, 1993; Zhang et al., 1995). • Protein in somata and dendrites of Purkinje cells in adult rat cerebella by IHC (Ukena et al., 1998). • *CYP11A1* in amphibian brain makes pregnenolone and its sulfate ester (Takase et al., 1999). • Human oligodendrocytes make pregnenolone, astrocytes, neurons do not (Brown et al., 2000). • *CYP11A1* in pyramidal and granular neurons of rat brain by IHC (Kimoto et al., 2001). • Transcriptionally regulated in embryonic rodent nervous system (Hammer et al., 2004). • *CYP11A1* mRNA in human amygdala, caudate nucleus, cerebellum, corpus callosum, hippocampus, spinal cord, and thalamus (Yu et al., 2002). • *CYP11A1* mRNA below quantitation limits in whole human brain (Nishimura et al., 2003).
CYP11B1 11β-OHase; sterol 7-OHase	• *CYP11B1* mRNA in whole rat brain and hypothalamus by RT-PCR; protein in cerebellum, especially Purkinje cells and hippocampus by IHC (MacKenzie et al., 2000). • *CYP11B1* mRNA in human amygdala, caudate nucleus, corpus callosum, spinal cord, and thalamus (Yu et al., 2002). • *CYP11B1* mRNA below quantitation limits in whole human brain (Nishimura et al., 2003).
CYP11B2 (aldosterone synthase) Sterol 11-OHase; 11-deoxycorticosterone →aldosterone 11-deoxycortisol→ cortisol	• Found in mitochondrial fraction of liver; mRNA not detected in rat brain (Mellon, 1993). • *CYP11B2* mRNA in rat hypothalamus, hippocampus, amygdala, cerebrum, and cerebellum by RT-PCR and Southern blot (Gomez-Sanchez et al., 1997). • Protein in rat brain by IHC (MacKenzie et al., 2000). • mRNA in human caudate nucleus, corpus callosum, spinal cord, and thalamus (Yu et al., 2002). • *CYP11B* mRNA in frog telencephalon, diencephalon, midbrain, and cerebellum by RT-PCR. In situ hybridization shows *CYP11B* mRNA in cells of pallium mediale in telencephalon, nucleus preopticus in diencephalon, stratum griseum superficiale tecti in midbrain, and Purkinje cells in cerebellum (Takase et al., 2002).
CYP8B1 Sterol 12α–OHase; Bile synthesis	• Northern blot shows only hepatic expression in rabbit (Eggertsen et al., 1996). • Found in rabbit liver microsomal fraction (Andersson et al., 1998). • *CYP8B* was expressed specifically in rat liver (Ishida et al., 1999). • 0.0120 *CYP8B1* mRNA/*GAPDH* mRNA in human brain by RT-RT-PCR (Nishimura et al., 2003).
CYP8A1 Prostacyclin Synthase Prostaglandin H2→ Prostacyclin (Prostaglandin I2)	• Low but present in rat brain by in situ hybridization (Tone et al., 1997). • *CYP8A1* in neurons of bovine, rat and human brain, especially Purkinje cells of cerebellum and cortical neurons, but not glial cells by IHC (Mehl et al., 1999). • *CYP8A1* mRNA and protein in blood vessels of human brain, pyramidal cells of cortex and hippocampus, and Purkinje cells of cerebellum (Siegle et al., 2000). • 0.000161 per *GAPDH* mRNA in whole adult human brain by RT-RT-PCR (Nishimura et al., 2003).

Table 3-2 (continued)

P450 reactions	Highlights of P450 expression in brain
CYP7B1 Oxysterol 7α -OHase (DHEA, pregnenolone, 25-hydroxy cholesterol 7α-OHase) 24-hydroxycholesterol 7-OHase	• Rat brain microsomes 7-α-hydroxylate DHEA and pregnenolone (Akwa et al., 1992). • Mouse brain metabolizes [7α-^{3}H] pregnenolone and makes 7α-OH-DHEA (Rose et al., 1997). • *CYP7B1* mRNA found in human hippocampus by Northern blot (Wu et al., 1999). • Enzyme activity higher in human cerebral neocortex than subcortical white matter; no sex differences (Steckelbroeck et al., 2002). • Enzyme activity in sheep, marmoset, and human brain extracts. *CYP7B* mRNA in human hippocampus, cerebellum, cortex; less in dentate neurons from Alzheimer's diseased subjects by in situ hybridization (Yau et al., 2003). • IHC shows expression in neurons only of human hippocampus (Trap et al., 2005). • 0.0375 per *GAPDH* mRNA in adult human brain by RT-RT-PCR (Nishimura et al., 2003). • 7α-hydroxy-DHEA synthesis in frontal cortex, hippocampus, amygdala, cerebellum, and striatum of Alzheimer's patients and controls (Weill-Engerer et al., 2003).
CYP5A1 Thromboxane synthase Prostaglandin H2→ Thromboxane A2	• TXA2 and PGI2 imbalance underlies ischemic cerebrovascular disease (Uyama et al., 1985). • Increase in TXA2 synthesis at 5 min and 1 week of reperfusion suggests role in acute events and later stages of neurological dysfunction (Shohami et al., 1987). • TXA2 synthetase inhibitor decreases vasospasm and reduces neurological deterioration after subarachnoid hemorrhage (Tokiyoshi et al., 1991). • Hypothermia improves imbalances of TXA2 and PGI2 after traumatic brain injury in humans (Aibiki et al., 2000). • 0.0146 per *GAPDH* mRNA in adult human brain by RT-RT-PCR (Nishimura et al., 2003).
CYP4F3 LTB4 ω-hydroxylase	• Multiple *CYP4F* isoforms in rat brain (Kawashima and Strobel, 1995). • Tissue-specific *CYP4F3* isoforms regulated by alternate promoter usage and mutually exclusive exon splicing, which changes substrate specificity (Christmas et al., 1999, 2001). • *CYP4F3* in humans restricted to polymorphonuclear leukocytes (Kikuta et al., 2002). • 0.0074 per *GAPDH* mRNA in adult human brain by RT-RT-PCR (Nishimura et al., 2003).
CYP4F2 FA ω-OHase, LTB4	• 20-HETE and acute cerebral blood flow drop after brain trauma in rats (Kehl et al., 2002). • 0.0008 per *GAPDH* mRNA in adult human brain by RT-RT-PCR (Nishimura et al., 2003).
CYP4A11 FA ω-OHase	• *CYP4A* mRNA detected by RT-PCR but not Northern blot in rat brain. WBs detect a *CYP4A*-specific protein (Stromstedt et al., 1994). • Cultured rat neurons, astrocytes, and oligodendrocytes all contained omega-oxidation activity. WBs detect *CYP4A* protein (Alexander et al., 1998). • *CYP4A* protein and mRNA localized to cerebral arteriolar muscle of rat by IHC, in situ hybridization and RT-PCR (Gebremedhin et al., 2000). • 20-HETE affects cerebral blood flow after subarachnoid hemorrhage in rats (Kehl et al., 2002; Miyata et al., 2005). • 0.0139 per *GAPDH* mRNA in adult human brain by RT-RT-PCR (Nishimura et al., 2003).
CYP2J2 AA→ EET's	• Enzyme activity in rat brain parenchymal tissue and perivascular astrocytes (Alkayed et al., 1996). • Mouse *CYP2J9* expressed primarily in brain (Qu et al., 2001). • 0.0128 per *GAPDH* mRNA in adult human brain by RT-RT-PCR (Nishimura et al., 2003).

Table 3-2 (continued)

P450 reactions	Highlights of P450 expression in brain
CYP2U1 AA ω-1 OHase	• *CYP2U1* mRNA expression in human thymus greater than heart and brain but similar in rats. WB shows protein in rat brain limbic, cortex, and thymus (Karlgren et al., 2004). • *CYP2U1* mRNA most abundant in thymus and the cerebellum (Chuang et al., 2004).
CYP26A1 (P450RAI-1) RA→4-hydroxy-RA/4-oxo-RA	• RT-PCR shows *CYP26* mRNA in human olfactory bulb, temporal cortex, and hippocampus; less in parietal cortex, medulla/pons region, and putamen (Ray et al., 1997). • Human *CYP26* metabolizes RA to 4-OH-RA, 4-oxo-RA, and 18-OH-RA (White et al., 1997). • RT-RT-PCR and Northern blot show mRNA expression in fetal, adult brain (Trofimova-Griffin and Juchau, 1998). • *CYP26A1* mRNA in human cerebellum, less in cerebral cortex, medulla, occipital pole, frontal lobe, and temporal lobe (White et al., 2000). • 0.0053 per *GAPDH* mRNA in adult human brain by RT-RT-PCR (Nishimura et al., 2003).

microsomes. Inhibition studies using ketoconazole, furafylline, sulfaphenazole, omeprazole, and quinidine suggested that *CYP3A4* was responsible for carrying out this reaction (Voirol et al., 2000). Dextromethorphan was metabolized to dextrorphan in rat brain microsomes, and this metabolism was inhibited by quinidine and by a polyclonal antibody against *CYP2D6*. However, only the addition of exogenous reductase allowed the measurement of this activity in human brain microsomes (Voirol et al., 2000). Tyndale et al. also measured dextromethorphan turnover by rat brain membranes and noted a significant loss of activity depending on the buffer used for homogenization and on undergoing freeze/thaw cycling (Tyndale et al., 1999).

2.2 Hybridization Techniques

2.2.1 Nucleic Acid Hybridization

Nucleic acid hybridization techniques suffer from the high homology between many P450 mRNA species; thus, the use of a probe from a given isoform in another species, organ, or tissue may lead to spurious results. With the sequencing of the human genome and the identification of all P450-related coding regions, it is now possible however to design very specific probes to detect the presence of individual mRNA segments. Beginning with a commercial or self-prepared cDNA library, Southern blotting can be used to identify nearly any stretch of unique sequence already known or predicted to occur. In this technique, cDNA is first separated by gel electrophoresis, denatured by raising the pH, and transferred to nitrocellulose. Radiolabeled RNA or oligonucleotides can then be used to identify sequences of interest. Ray et al. provide a description of the methods they used to identify *CYP26* in a cDNA library enriched for minor transcripts by subtractive hybridization (Ray et al., 1997). Alternatively, mRNA from brain regions or cell culture may be directly separated by electrophoresis and then hybridized with an antisense mRNA or oligonucleotide probe sequences to specifically detect the presence of P450 mRNA via Northern blot. The isolation of mRNA, once an onerous task, has become much simpler in recent years owing to the availability of commercial isolation kits; therefore, this topic will not be addressed in the interest of brevity.

2.2.2 Antibody Hybridization

Western blot and immunohistochemical detection are the two most widely used techniques to confirm the expression of a given P450 isoform in isolated brain regions or thin slices. Antibodies detect actual protein,

Figure 3-1
Cytochrome P450 pathways in human brain: cholesterol, steroid hormones, vitamin D, and bile

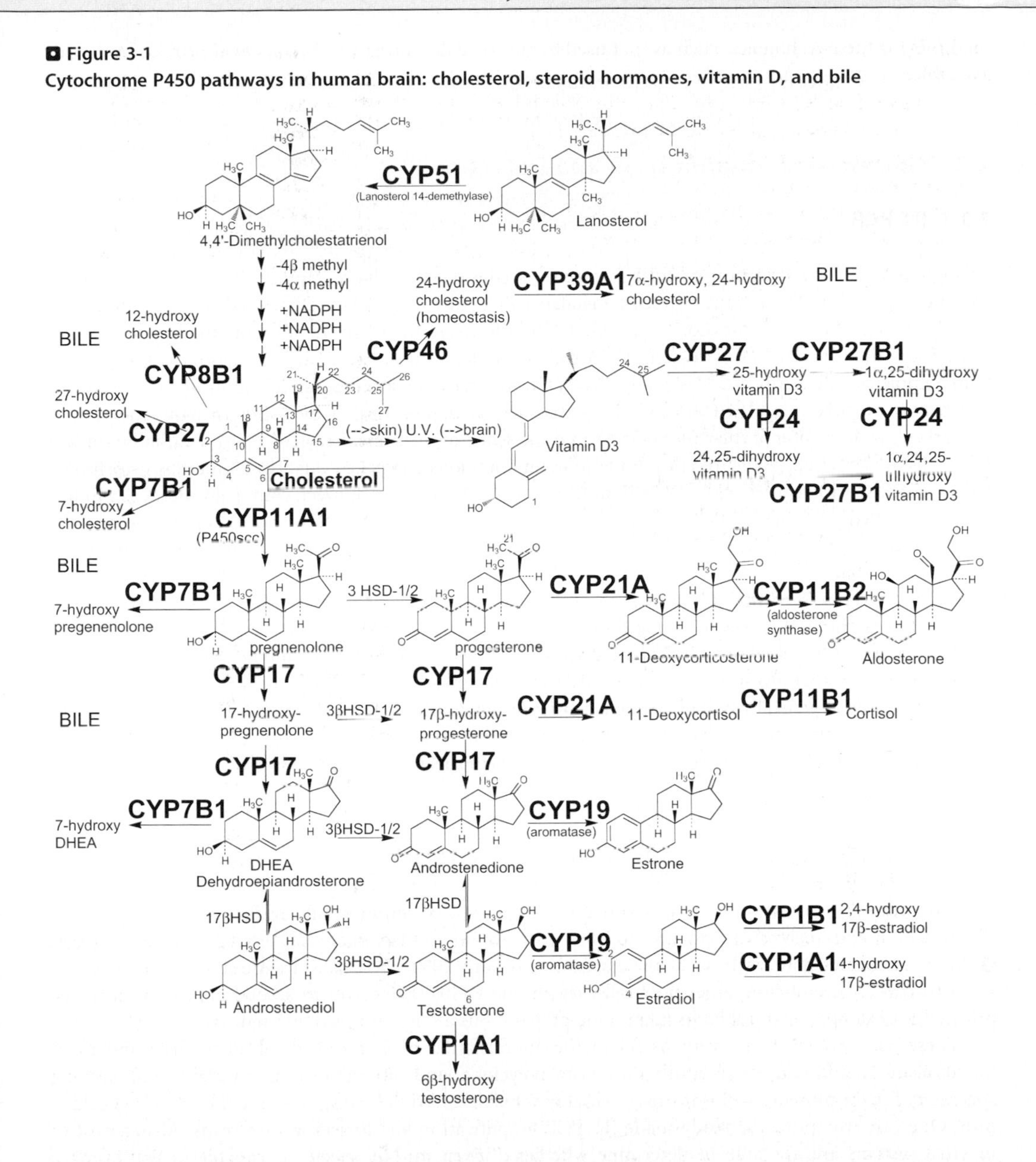

and given that not all mRNA species are translated, may be the only way to be sure of protein expression. However, the evolutionary conservation of overall P450 protein structure causes significant problems with antibody cross-reactivity between isoforms. Indeed, many related P450s are nearly identical in amino acid sequences. Thus, polyclonal antibodies are often unsuitable for a given isoform unless they are back-adsorbed against abundant P450s and other nonspecific proteins. Specific monoclonal antibodies, many of which are now commercially available, and polyclonal antibodies raised against unique peptides have provided more reliable data. Still, brain slices have proven extremely difficult to penetrate with such large molecules. Detergents and the use of several freeze–thaw cycles are techniques that have been used to allow better penetration by anti-P450 antibodies. The solubilization steps required for Western blot largely avoid this problem; however, this also denatures the protein and may alter antibody specificity. Several published methods for these techniques, such as that used by the Tyndale laboratory (Miksys et al., 2002), are readily available.

2.3 Nucleic Acid Amplification and Detection

2.3.1 RT-PCR

Though indirect, the detection of mRNA species is considered to be a fairly strong evidence for P450 enzyme expression. The advent of readily available PCR machines in the 1990s led to an explosion of knowledge regarding the expression of specific P450 isoforms in the brain. In addition, the human genome project and the concomitant increase in DNA technology have made the design of oligonucleotide primers quite simple. RT-PCR kits are now readily available commercially for the amplification of PCR products, which may be either visualized directly on agarose gels or probed further by Southern blot as for cDNA (as earlier). A discussion of their merits is beyond the scope of the present review. Using a variation on RT-PCR, Nishimura et al. (2003) recently undertook an ambitious project to quantify numerous cytochrome P450 mRNA species in a series of cDNA libraries from Clontech (www.clontech.com; Palo Alto, CA). In this technique called real-time RT-PCR or RT-RT-PCR, the amplification of mRNA species is monitored in real time by measuring the accumulation of fluor-labeled PCR products. These authors compared the amount of mRNA to a cellular marker mRNA, in this case that encoding glyceraldehyde phosphate dehydrogenase (GAPDH), and report their results as a ratio of P450 mRNA to GAPDH mRNA (➋ *Tables 3-2* and ➋ *3-3*). A single human brain donor was used and mRNA was quantified from whole brain, thus no information is available regarding regional or subcellular localization. However, their work would seem to corroborate the findings of many other investigators, in that mRNA species already believed to be present from the literature indeed appeared at generally higher frequency. Interestingly, a number of P40 mRNA species not known to be expressed in brain tissue are implicated even now, (Nishimura et al., 2003) highlighting how far we have to go before we understand the role of individual P450s in specific brain processes.

2.4 Freely and Commercially Available Tools for the Study of P450 in Brain

In addition to those mentioned already, the number and type of commercially available tools to aid in the identification and analysis of cytochrome P450 in the brain and elsewhere has exploded in recent years. ➋ *Table 3-1* collates some of the available materials as of early 2004, and this will undoubtedly expand in the years to come. Several companies now sell cytochrome P450s themselves in several expression formats, purified for example or supplied in membrane preparations along with NADPH reductase.

These have proven to be very useful in the confirmation of isoforms involved in drug and toxin metabolism. In addition, many antibodies, both polyclonal and monoclonal, are available with varying specificity for the human P450 isoforms—a highly valuable and time-saving resource for immunolocalization. One can even purchase ready made RT-PCR amplification kits for certain isoforms. Also, a host of in vitro systems are available to determine whether a given mRNA species is capable of expressing a functional enzyme. Other valuable resources are the cDNAs and cDNA libraries available, which have been derived from specific brain regions. Before one attempts to clone their own cDNA or make their own library, a quick scan of the publicly available resources is essential. Programs like the Clone Ranger offered through Invitrogen's web site (➋ *Table 3-1*) allow one to enter any unique DNA sequence of interest and scan thousands of premade and often pretested full-length cDNAs for the particular gene of interest from numerous tissue sources.

In addition to commercial sources, the knowledgeable P450 investigator needs to be aware of the freely available resources on the Internet. In addition to the web sites listed in the introduction, the human and mouse genomes are now simpler to scan for genes of interest. Many DNA design and analysis utilities are also not hard to find. For protein analysis, there is at least one automated modeling program that can take

Table 3-3

Cytochromes P450 with xenobiotic function in human brain (Amruthesh et al., 1992; Degtyarenko and Fabian, 1996; Nelson, 1998; Chang and Kam, 1999; Omiecinski et al., 1999; Nebert and Russell, 2002; Stoffel-Wagner, 2003)

P450	P450 expression in brain
CYP1A1	• *CYP1A1* mRNA in cerebellum, frontal, occipital, pons, red nucleus, substantia nigra regions of human brain by RT-PCR. *CYP1A1* protein in neurons of the substantia nigra, red nucleus, pons, median raphe, locus ceruleus, inferior vestibular nucleus, dorsal motor nucleus of the vagus, thalamus and some but not all astrocytes by IHC (Farin and Omiecinski, 1993). • Targeted to mitochondrial and microsomal fractions in hepatic tissue; by IHC, WB (Bhagwat et al., 1995a; Addya et al., 1997). • *CYP1A1* mRNA found in human brain by RT-PCR (Yun et al., 1998). • WB shows *CYP1A1*induction in arachnoid, dura mater, choroid plexus, pineal gland, and pituitary of rat brain by β-naphthoflavone (Morse et al., 1998). • No *CYP1A1* protein or activity in purified rat brain mitochondria (Morse et al., 1998). • *CYP1A1*mRNA in 22 glioblastomas and 4 anaplastic astrocytomas by RT-PCR (Kirches et al., 1999). • *CYP1A1* mRNA induced in rat olfactory bulb, striatum-caudate, hypothalamus, hippocampus, cortex, cerebellum, substantia nigra by tetrachlorodibenzo-*p*-dioxin (Huang et al., 2000; Korkalainen et al., 2005). • Lipopolysaccharide downregulates *CYP1A* activity in the rat brain (Monshouwer et al., 2000; Nicholson and Renton, 2001). • *CYP1A1* mRNA mainly in mitochondria of the human brain (Zhang et al., 2001). • Hematin increased EROD-activity tenfold in cytosol of mouse olfactory bulb; cytosolic *CYP1A1* protein in *COS-1* cells increased with heme synthesis inhibitor (Meyer et al., 2002). • IHC reveals *CYP1A1* induction in endothelial cells in the choroid plexus, veins in the leptomeninges, and cerebral veins of AhR agonist-pretreated mice (Granberg et al., 2003). • Protein expression downregulated in malignant medulloblastoma (Wu et al., 2005). • Developmental and tissue-specific regulation in transgenic mice (Galijatovic et al., 2004). • Novel splice variant found in the human brain (Chinta et al., 2005b). • Neonatal (Desaulniers et al., 2005) or prenatal (Johri et al., 2006) exposure to AHR agonist imprints adult brain mRNA expression in rat. • Found mRNA in rat olfactory mucosa, not liver (Minn et al., 2005). • Catalyzes 6-hydroxlation of melatonin in vitro (Ma et al., 2005). • CNS inflammation induced by LPS downregulates 1A1 activity in rat; loss prevented by beta adrenergic receptor agonist (Abdulla and Renton, 2005). • 0.0004 *CYP1A1* mRNA/*GAPDH* mRNA in whole human brain by RT-RT-PCR (Nishimura et al., 2003).
CYP1A2	• *CYP1A2* mRNA detected at low levels in the cerebellum, frontal, occipital, pons, red nucleus, and substantia nigra regions of the human brain (Farin and Omiecinski, 1993). • WB shows little *CYP1A2* protein in microsomes from cortex, cerebellum, brainstem, thalamus, hippocampus, and striatum of rat (Morse et al., 1998). • No *CYP1A2* protein nor EROD activity detected in purified rat brain mitochondria (Morse et al., 1998). • Found mRNA in rat olfactory mucosa and liver (Minn et al., 2005). • Prenatal exposure to deltamethrin imprints adult brain mRNA expression in rat (Johri et al., 2006). • RT-PCR for *CYP1A2* is negative in a series of 22 human glioblastomas and 4 anaplastic astrocytomas (Kirches et al., 1999). • CNS inflammation induced by LPS downregulates 1A2 activity in rat; loss prevented by beta adrenergic receptor agonist (Abdulla and Renton, 2005). • Catalyzes 6-hydroxlation of melatonin in vitro (Ma et al., 2005).

Table 3-3 (continued)

P450	P450 expression in brain
CYP1B1	• IHC negative for *CYP1B1* in normal tissue, positive in brain tumor (Murray et al., 1997). • Northern blot reveals *CYP1B1* mRNA in human CNS, highest in putamen; RT-PCR confirms *CYP1B1* in human astrocytoma cells (Rieder et al., 1998). • *CYP1B1* mRNA in normal human brain (McFadyen et al., 1999). • 7,12-dimethylbenz(a)anthracene induces *CYP1B1* protein seen by WB in astrocytoma cell line; *CYP1B1* in blood–brain interface of temporal lobe in human brain by IHC (Rieder et al., 2000). • No *CYP1B1* mRNA detected in fetal brain cDNA library (Jansson et al., 2001). • Human *CYP1B1* transgene expressed in mouse brain under Tet promoter (Hwang et al., 2001). • In situ hybridization and IHC reveal *CYP1B1* mRNA and protein in brain parenchymal and stromal tissue; *CYP1B1* protein was nuclear (Muskhelishvili et al., 2001). • IHC shows constitutive and inducible *CYP1B1* in artery smooth muscle cells, leptomeninges, cerebral arteries/arterioles, and ependymal cells (Granberg et al., 2003). • *CYP1B1* mRNA in male and female Rhesus monkey frontal cortex, hippocampus, thalamus, and amygdala; *CYP1B1* protein colocalized with mRNA in female brains but restricted to hippocampal pyramidal neurons in male brains (Scallet et al., 2005). • Catalyzes 6-hydroxlation of melatonin in vitro and in transgenic mouse brain (Ma et al., 2005). • 0.0017 *CYP1B1* mRNA/*GAPDH* mRNA in whole human brain by RT-RT-PCR (Nishimura et al., 2003).
CYP2A6/7	• 0.0087 P450 mRNA/*GAPDH* mRNA in whole human brain by RT-RT-PCR (Nishimura et al., 2003). • *CYP2A6/7* allele (*CYP2A6*12*) is a hybrid allele; 5′ regulatory region and exons 1–2 are of *CYP2A7* origin and exons 3–9 are of *CYP2A6* origin, resulting in 10 amino acid substitutions and reduced enzyme activity compared with the *CYP2A6*1* allele (Oscarson et al., 2002).
CYP2A6	• 0.0073 *CYP2A6* mRNA/*GAPDH* mRNA in whole human brain by RT-RT-PCR (Nishimura et al., 2003).
CYP2A7	• 0.0006 P450 mRNA/*GAPDH* mRNA in whole human brain by RT-RT-PCR (Nishimura et al., 2003). • *CYP2A7* "pseudogene" cDNA generated a protein of molecular mass 49 kDa of unknown function in *COS-7* cells (Ding et al., 1995).
CYP2A13	• *CYP2A13* mRNA was detected in human brain by RT-PCR (Su et al., 2000).
CYP2B6	• PB inducible P450 2B1/2B2 O-dealkylates 7-pentoxyresorufin in rat brain preparation (Parmar et al., 1998). • Nicotine induces several enzyme activities including *CYP2B1/2B2* in rat brain (Zevin and Benowitz, 1999). • Immunoblot shows *CYP2B6* expressed in human brain (Gervot et al., 1999). • Phenobarbital, *trans*-stilbene oxide, diallyl sulfide, diphenylhydantoin, amitriptyline induce *CYP2B1* and *CYP2B2* mRNA in rat striatum and cerebellum (Schilter et al., 2000). • *CYP2B* in rat brain extracts confirmed by immunoblotting (Albores et al., 2001). • *CYP2B* in PB treated rat neuronal cells, especially the reticular neurons in midbrain (Upadhya et al., 2002). • IHC shows *CYP2B6* in neurons and astrocytes, higher in smokers and alcoholics especially in cerebellar Purkinje cells and hippocampal pyramidal neurons, also caudate nucleus and putamen (Miksys et al., 2003). • *CYP2B6* protein in frontal cortical pyramidal cells, cerebellar Purkinje cells and neurons in the substantia nigra of African Green monkey induced by chronic nicotine (Lee et al., 2006). • 0.0016 P450 mRNA/*GAPDH* mRNA in whole human brain by RT-RT-PCR (Nishimura et al., 2003).
CYP2C8	• *CYP2C8* mRNA detected in human brain by RT-PCR (McFadyen et al., 1998). • *CYP2C8* mRNA detected in human brain by RT-PCR (Klose et al., 1999). • 0.0021 P450 mRNA/*GAPDH* mRNA in whole human brain by RT-RT-PCR (Nishimura et al., 2003).
CYP2C9	• *CYP2C9* mRNA not detected in brain by RT-PCR (Klose et al., 1999). • *CYP2C9* mRNA detected by RT-PCR in each of ten human brain tumors (Knupfer et al., 1999). • 0.0006 *CYP2C9* mRNA/*GAPDH* mRNA in whole human brain by RT-RT-PCR (Nishimura et al., 2003).

Table 3-3 (continued)

P450	P450 expression in brain
CYP2C18	• *CYP2C18* mRNA detected in human brain by RT-PCR (Klose et al., 1999).
	• 0.0015 P450 mRNA/*GAPDH* mRNA in whole human brain by RT-RT-PCR (Nishimura et al., 2003).
CYP2C19	• *CYP2C19* mRNA not detected in human brain by RT-PCR (Klose et al., 1999).
	• 0.000014 P450 mRNA/*GAPDH* mRNA in whole human brain by RT-RT-PCR (Nishimura et al., 2003).
CYP2D6 P450db1 Debrisoquine OHase Sparteine OHase	• MPTP and anti rat-liver 2D antisera inhibit bufuralol 1′-hydroxylase in rat brain tissue. Immunoblot shows protein in rat and human brain microsomes (Fonne-Pfister et al., 1987).
	• [3H]GBR-12935 binds dopamine transporter and canine neuronal P450IID1 (Niznik et al., 1990).
	• *CYP2D1* activity found unevenly throughout canine brain (Tyndale et al., 1991).
	• *CYP2D6* deficients show less tonic pain tolerance than extensive metabolizers (Sindrup et al., 1993b).
	• [3H]GBR-12935 binds to nondopaminergic piperazine acceptor sites in canine brain, one of which is *CYP2D1*. No regional differences in seen in binding (Allard et al., 1994).
	• Dark Agouti rats, which do not express *CYP2D1* mRNA in the liver, express as much 2D4 mRNA in the brain as Wistar rats (Wyss et al., 1995).
	• Testosterone induces *CYP2D* mRNA the absence of estrogen in female rat brain by RT-PCR. *CYP2D* mRNA appears throughout rat brain (Bergh, 1996).
	• RT-PCR, Southern blot, sequencing and gene-specific PCR characterize alternatively spliced forms of *CYP2D* mRNA in human breast tissue (Huang et al., 1997).
	• Testosterone increases expression of *CYP2D* in intact female rat brain; progesterone decreases; coadministration gives intermediate effect (Baum and Strobel, 1997).
	• MPTP N-demethylation is catalyzed by *CYP2D6*. In situ hybridization shows *CYP2D6* localized in pigmented neurons of human substantia nigra (Gilham et al., 1997).
	• RT-PCR for *CYP2D6* negative in 22 glioblastomas and 4 anaplastic astrocytomas (Kirches et al., 1999).
	• Alternative *CYP2D6* splicing also occurs in the human brain (Woo et al., 1999).
	• Quinidine and anti-*CYP2D6* antibody inhibit DXM metabolism in rat brain microsomes; DXM metabolism shows significant regional variation (Tyndale et al., 1999).
	• DXM metabolism in rat brain microsomes is inhibited by quinidine and anti-*CYP2D6* antibody; human brain required exogenous reductase (Voirol et al., 2000).
	• *CYP2D* protein is present in rat brain mitochondrial and microsomal membranes (Miksys et al., 2000).
	• Neuronal and glial cells of human brain show *CYP2D6* mRNA in neocortex, caudate nucleus, putamen, globus pallidus, hippocampus, hypothalamus, thalamus, substantia nigra, and cerebellum by in situ hybridization. IHC shows *CYP2D6* protein in pyramidal cells of the cortex, pyramidal cells of the hippocampus, and Purkinje cells of the cerebellum but absent in glial cells (Siegle et al., 2001).
	• RT-PCR, Southern blot, slot blot, immunoblot, and ICC reveal correlation of mRNA and *CYP2D6* protein across 13 brain regions; higher expression is seen in brains from alcoholics versus nonalcoholics. In hippocampus this was localized in *CA1–3* pyramidal cells and dentate gyrus granular neurons. In cerebellum this was localized in Purkinje cells and their dendrites (Miksys et al., 2002).
	• *CYP2D6* mRNA and protein in human brain regions by RT-PCR, Northern blot, and Immunoblot, in situ hybridization and IHC shows constitutive *CYP2D6* mRNA expression in neurons of cerebral cortex, Purkinje and granule cell layers of cerebellum, reticular neurons of midbrain and pyramidal neurons of *CA1, CA2,* and *CA3* subfields of hippocampus. Protein found in cortex, cerebellum, midbrain, striatum, and thalamus of human brain. IHC shows *CYP2D6* in dendrites of Purkinje and cortical neurons and neuronal soma (Chinta et al., 2002).

Table 3-3 (continued)

P450	P450 expression in brain
	• Alternate splicing of *CYP2D1* mRNA specific to rat brain (Chinta et al., 2005b). • 0.0022 *CYP2D6* mRNA/*GAPDH* mRNA in whole human brain by RT-RT-PCR (Nishimura et al., 2003). • MDMA and MDA produce serotonergic nerve damage in rats, but not when injected directly into the brain. HLMs make two glutathione regioisomers from MDA, both of which depend on *CYP2D* activity. The 5-(glutathion-S-yl)-alpha-MeDA adduct is behaviorally active in the rat (Easton et al., 2003).
CYP2E1	• *CYP2E1* mRNA detected in cerebellum, frontal cortex, occipital cortex, pons, red nucleus, and substantia nigra of human brain by RT-PCR (Farin and Omiecinski, 1993). • ICC of astrocyte cell culture shows *CYP2E1* over cytoplasm and processes but more pronounced over nuclear membrane; immunogold labeling agrees. Ethanol causes an increase in *CYP2E1* as determined by ICC and Dot blot (Montoliu et al., 1995). • *CYP2E1* mRNA and protein found in rat hippocampus, mainly microsomal fraction. *CYP2E1* is induced by ethanol in cortical glial cultures (Tindberg and Ingelman-Sundberg, 1996). • Northern blots and RNase protection assay show *CYP2E1* mRNA increases as a function of gestational age in human prenatal whole brain (Boutelet-Bochan et al., 1997). • *CYP2E1* mRNA is confirmed in rat brain by RT-PCR (Yoo et al., 1997). • *CYP2E1*, associated with free radical production, is selectively localized in nigral dopamine-containing cells (Jenner, 1998). • Nicotine induces several P450s in animal brains (review; Zevin and Benowitz, 1999). • RT-PCR for *CYP2E1* is positive in 22 glioblastomas and 4 anaplastic astrocytomas (Kirches et al., 1999). • Human brain *CYP2E1* increases at about gestational day 50; then a fairly constant level is maintained through at least day 113 (Brzezinski et al., 1999). • *CYP2E1* mRNA seen in neurons of the cerebral cortex, Purkinje and granule cells of cerebellum, granule cells of dentate gyrus, and pyramidal neurons of *CA1, CA2*, and *CA3* subfields of hippocampus in rat and human brain by in situ hybridization; ethanol-induced brain expression in rat (Upadhya et al., 2000). • Rat liver shows mitochondrial and microsomal *CYP2E* (Robin et al., 2001). • IHC shows ethanol induction of *CYP2E1* in olfactory bulbs, frontal cortex, hippocampus, and cerebellum of male rats; nicotine induces in olfactory bulbs, frontal cortex, olfactory tubercle, cerebellum, and brainstem. More *CYP2E1* found in brains from human alcoholics and alcoholic smokers in granular cells of the dentate gyrus, pyramidal cells of *CA2* and *CA3* hippocampus, and cerebellar Purkinje cells. More *CYP2E1* in frontal cortices of alcoholic smokers versus all nonsmokers. *CYP2E1* in cultured human neuroblastoma cells induced by nicotine (Howard et al., 2003). • Chronic nicotine induces *CYP2E1* expression in cortical pyramidal neurons and cerebellar Purkinje cells; no change seen in temporal cortex, hippocampus, putamen, or thalamus of monkey brain (Joshi and Tyndale, 2005, 2006). • Prenatal exposure to deltamethrin imprints adult brain mRNA expression in rat (Johri et al., 2006). • 0.0189 *CYP2E1* mRNA/*GAPDH* mRNA in whole human brain by RT-RT-PCR (Nishimura et al., 2003).
CYP3A3/4	• Immunoblot shows low or negligible *CYP3A* levels in human brain tissues (Yang et al., 1994). • 0.0073 *CYP3A3/4* mRNA/*GAPDH* mRNA in whole human brain by RT-RT-PCR (Nishimura et al., 2003).
CYP3A4 Major adult P450	• RT-PCR shows more *CYP3A* mRNA in pons versus cerebellum, frontal, occipital, and red nucleus but none in substantia nigra region of human brain (Farin and Omiecinski, 1993). • Immunoblot shows low or negligible *CYP3A* levels in human brain tissues (Yang et al., 1994).

Table 3-3 (continued)

P450	P450 expression in brain
	• *CYP3A* mRNA in normal human brain demonstrated by RT-PCR (McFadyen et al., 1998). • *CYP3A* in tumor cells, neurons in normal brain; astrocytes were negative by IHC (Kirches et al., 1999). • RT-PCR of ten brain tumors reveals none positive for *CYP3A4* (Knupfer et al., 1999). • Phenobarbital induces *CYP3A1* mRNA in rat striatum and cerebellum (Schilter et al., 2000). • Rat and human brain microsomes demethylate amitriptyline. Chemical and antibody inhibition suggests *CYP3A4* is responsible (Voirol et al., 2000). • More α-hydroxy alprazolam is formed from alprazolam in human and rat brain than liver microsomes by constitutive *CYP3A* (Pai et al., 2002). • 0.0054 *CYP3A4* mRNA/*GAPDH* mRNA in whole human brain by RT-RT-PCR (Nishimura et al., 2003).
CYP3A5	• Immunoblot indicates low or negligible *CYP3A* in human brain (Yang et al., 1994). • *CYP3A5* mRNA identified by RT-PCR in human pituitary gland and localized by IHC to growth hormone containing cells of the anterior pituitary (Murray et al., 1995). • RT-PCR of ten brain tumors reveals three positive for *CYP3A5* (Knupfer et al., 1999). • 0.0005 *CYP3A5* mRNA/*GAPDH* mRNA in whole human brain by RT-RT-PCR (Nishimura et al., 2003).
CYP3A7 Major fetal P450	• Immunoblot indicates low or negligible *CYP3A* in human brain (Yang et al., 1994). • Human *CYP3A7* mRNA is expressed in brains of adult transgenic mice (Li et al., 1996). • 0.0060 *CYP3A7* mRNA/*GAPDH* mRNA in whole human brain by RT-RT-PCR (Nishimura et al., 2003).
NADPH Reductase	• Reductase in catecholamine-containing structures of rat, monkey brain by IHC (Haglund et al., 1984). • Rat brain DT-diaphorase enzyme hypothesized to be cytochrome P450 reductase (Kemp et al., 1988). • High enzyme activity found in immature rat brain and neuronal cultures (Ghersi-Egea et al., 1989). • IHC of human brain medulla localizes reductase to neuronal cell body (Ravindranath et al., 1990). • Reductase purified from rat brain same as liver by SDS-PAGE and WB (Bergh and Strobel, 1992). • NADPH-cytochrome P450 reductase was purified from rat brain microsomes. • Immunoblot shows constitutive expression in rat and human brain (Anandatheerthavarada et al., 1992). • High NADPH cytochrome P450 reductase activity is found in immature rat brain and neuronal cultures, but very little cytochrome P450, thus NADPH cytochrome P450 reductase may be involved in cytochrome P450 independent pathways (Ghersi-Egea et al., 1993). • Eight human brain tumors contained DT-diaphorase, NADH cytochrome b5 reductase and NADPH cytochrome P450 reductase as assessed by enzyme activity and WB (Rampling et al., 1994). • RT-PCR shows reductase mRNA throughout rat brain (Bergh and Strobel, 1996). • P450 reductase-dependent metabolism is active in both neuronal and glial cells; and ubiquitously present among major brain structures of rat (Teissier et al., 1998). • Amyloid peptides induce cytochrome P450 reductase in neuroblastoma cells. IHC shows strong reductase surrounding amyloid deposits of transgenic mice. Reductase expression pattern resembled neuritic and oxidative stress markers, while nontransgenic mice showed none. IHC of brains from Alzheimer's patients showed reductase in the cytoplasm of cortical neurons versus none in controls (Pappolla et al., 2001).

any amino acid sequence, compare it with the known protein crystal data bank, choose a suitable template (if one exists), and give a series of models complete with statistical analyses (Guex and Peitsch, 1997). Today's computers are easily capable of handling the graphics required to view and manipulate the models to analyze the possible role of individual amino acids. Many of the obstacles that have hindered progress in the past have now become a thing of the past. As such, there has been and will continue to be an explosion in our understanding of P450 in the brain.

3 Summary of Brain P450s with Primarily Endogenous Substrates

3.1 Cholesterol, Neurosteroid Hormones, Vitamin D3, and Bile

3.1.1 *CYP51*

Despite the complicated nature of ➋ *Figure 3-1*, it becomes simpler when one realizes that all the cytochrome P450 pathways shown are within seven metabolic steps of cholesterol itself. Few of these enzymes have known xenobiotic substrates. The most abundant P450 mRNA of all the isoforms in the human brain, as estimated by Nishimura et al. at 0.286 P450 mRNA per *GAPDH* mRNA, was *CYP51* (Nishimura et al., 2003). This should perhaps not be surprising, since it is one of the only two P450s now known to span kingdoms (Nebert and Russell, 2002). This isoform catalyzes the 14-demethylation of lanosterol, a crucial step toward the synthesis of cholesterol (➋ *Table 3-2*; ➋ *Figure 3-1*). Inhibition of this enzyme may underlie the mechanism of action of azole antifungals (Kelly et al., 2001). The K_i for azalanstat (RS-21607, (2S,4S)-*cis*-2[1H-imidazol-1-yl)methyl]-2-[2-(4-chlorophenyl)ethyl]-4-[[(4-aminophenyl)thio] methyl]-1,3-dioxolane), for example, was determined to be 840 pM (Swinney et al., 1994).

3.1.2 *CYP46*

The next most abundant P450 mRNA in the human brain appears to be *CYP46*, which encodes for cholesterol 24-hydroxylase. The *CYP46* enzyme, like *CYP51*, is crucial in the homeostasis of cholesterol in the brain. The 24-hydroxy cholesterol metabolite, unlike cholesterol itself, is readily excreted from the brain and is believed to constitute the major route of recirculation from the brain to the liver (Bjorkhem et al., 1999). In addition, the *CYP46* reaction is the first step in an elimination pathway for excess brain cholesterol. However, it has also been shown to exhibit a broad substrate specificity, both endogenous and xenobiotic, and thus may wear more than one hat (Mast et al., 2003). Recently, polymorphism within the human *CYP46* gene was implicated as a risk factor for Alzheimer's disease (Golanska et al., 2005). However, at this time, no clear association has emerged (Juhasz et al., 2005; Tedde et al., 2005). The expression of *CYP46* mRNA appears markedly insensitive to transcriptional regulators but sensitive to oxidative stress (Ohyama et al., 2006).

3.1.3 *CYP39A1*

This isoform was found as the result of a mouse deficient in the oxysterol 7a-hydroxylase gene (*CYP7B1* in humans), which nonetheless did not accumulate 24-hydroxy cholesterol from the *CYP46* pathway, suggesting the existence of another 7α-hydroxylase (Li-Hawkins et al., 2000). However, very little evidence exists for its expression in human brain. Nishimura et al. quantified *CYP39A1* mRNA by RT-RT-PCR and reported some 100-fold less than they reported for *CYP51* mRNA (Nishimura et al., 2003). There is no other evidence for the expression of this isoform in the brain as of this writing and this is believed to be the first report of its kind. The isoform has been shown to be sexually dimorphic in the liver (Li-Hawkins et al., 2000).

3.1.4 *CYP27*

Sterol 27-hydroxylases are known to be crucial to cholesterol homeostasis (Pikuleva et al., 1997) and bile synthesis (Chiang, 1998) in the liver. *CYP27* is a mitochondrial isoform that requires adrenodoxin and adrenodoxin reductase to exhibit activity (Pikuleva et al., 1997). The only direct evidence for expression of this isoform again comes from the single brain examined by Nishimura et al. (Nishimura et al., 2003). The defects in the *CYP27* gene, however, are known to cause cerebrotendinous xanthomatosis, a serious progressive neurological disease, (Miki et al., 1986; Cali et al., 1991; Wakamatsu et al., 1999; Sugama et al., 2001) however, this effect may be accounted for by disruptions in the hepatic bile synthesis.

3.1.5 *CYP27B1*

CYP27B1 is a 27-hydroxy vitamin D3 1α-hydroxylase that has been localized in the human and mouse brain by immunohistochemistry, Northern blot, and Western blot (Zehnder et al., 2001). Eyles et al. recently noted strong immunohistochemical staining for *CYP27B1* in the human hypothalamus and in the large neurons of the substantia nigra and propose that the widespread distribution of 1α-OHase and the vitamin D-receptor suggest that vitamin D may have autocrine/paracrine properties in the human brain (Eyles et al., 2005). In addition to 27-hydroxy vitamin D3, *CYP27B1* will also 1α-hydroxylate 24, 25-dihydroxy vitamin D3 (Axen et al., 1994). Found at 500-fold lower quantity than *CYP51* (Nishimura et al., 2003), this isoform was recently shown by PCR of tissue-specific cDNA libraries and RT-PCR of tissue samples to be induced by gene amplification in several malignant glioblastomas. In addition to a full-length coding mRNA, a number of truncated splice variants were also found (Maas et al., 2001). Of a total of 16 splice variants of *CYP27B1* identified in glioblastoma multiforme, none were found to encode a functional enzyme (Diesel et al., 2005).

3.1.6 *CYP24*

The only evidence found for the presence of *CYP24* in the normal human brain is again that provided by Nishimura et al. (2003). Recently however, Diesel et al. noted that the addition of calcitriol led to an elevated expression of *CYP24* (and *CYP27B1*) mRNA in glioblastoma multiforme cell lines (Diesel et al., 2005). Induced by 1,25-hydroxyvitamin D3 (Matkovits and Christakos, 1995), this enzyme catalyzes the hydroxylation of 25-hydroxy and 1,25-dihydroxy vitamin D3 at the 24th position to form the corresponding *di*-or *tri*-hydroxy metabolites. Using tritiated substrates, Danan et al. found that the associated enzyme activity was prevalent in the mitochondrial fraction of rat yolk sac, but not rat fetal brain tissue (Danan et al., 1982). Recently, *CYP24* mRNA was found to be subject to alternative splicing in human myelomonocytic cell lines (Ren et al., 2005), raising the possibility of tissue-specific alternate splicing seen in other P450s.

3.1.7 *CYP21A*

The *CYP21A* gene encodes a 21-hydroxylase, which converts progesterone to 11-deoxycorticosterone on the synthesis pathway toward aldosterone (following hydroxylation by *CYP11B2*). It also catalyzes the hydroxylation of 17β-hydroxy progesterone to 11-deoxy cortisol. In adult rat brain, *CYP21* was found to be mainly localized in the tractus reticulothalamicus and other ascending fibers, therefore, these authors suggested that deoxycorticosterone or its derivatives could be involved in the regulation of consciousness and the induction of anesthesia (Iwahashi et al., 1993). *CYP21* was found to be subject to splice variation in rat, as Zhou et al. found two distinct mRNA species in kidney, aorta, liver, cerebellum, hypothalamus, brain stem, heart, and cerebrum, but not the hippocampus as determined by Southern blot. The cDNA from only one of these transcripts appeared to result in functional protein (Zhou et al., 1997b). Recently, Yu et al. were able to detect *CYP21* mRNA in all the human brain regions that they examined by RT-PCR (Yu et al., 2002). *CYP21* mRNA has also been shown to be present in human fetal brain (Pezzi et al., 2003).

3.1.8 *CYP19*

CYP19, otherwise known as aromatase, catalyzes the aromatization of androstenedione to estrone and testosterone to estradiol, and thus plays a crucial role in the synthesis and regulation of steroid hormones. As such, a great deal of effort has gone into the localization of this enzyme activity in brain tissue. Doody and Carr were perhaps the first to note low-level aromatase activity in the human brain tissue. These investigators followed the incorporation of tritium into water from 1-[^{3}H]androstenedione and found that fetal brain had significant aromatase activity, though 120-fold lower than that found in the placenta and sevenfold lower than in the liver (Doody and Carr, 1989). Price et al. confirmed the presence of message a few years later by using a competitive PCR technique, showing that fetal liver had by far the highest levels of mRNA, followed by the brain, and intestine (Price et al., 1992). More recently, *CYP19* protein was localized by ICC to the hypothalamus and limbic regions in the brains of several species, including monkey and human. The protein was localized throughout the axons and dendrites of affected cells (Naftolin et al., 1996). Yague et al. recently found immunohistochemical evidence for *CYP19* expression in human and rat glioblastomas (Yague et al., 2004).

Roselli and Resko noted that *CYP19* induction was sexually dimorphic, being stimulated by physiological doses of testosterone to a greater extent in males than in female rats (Roselli and Resko, 1997). Another group prepared microsomes from numerous biopsy specimens of human temporal lobe, and then measured aromatase activity via tritium incorporation into water with the androstenedione assay. These workers found higher aromatase activity in cerebral cortex than in subcortical white matter, but no sex or age differences. Perhaps in premonition, they also noted that the tritium release from brain microsomes was incompletely inhibited by the aromatase-specific inhibitor atamestane (Steckelbroeck et al., 1999). Chiang et al. soon reported that in zebra fish, two distinct *CYP19* genes are present, one which encodes for brain aromatase and the other for that found in ovaries (Chiang et al., 2001). Later, the human gene was found to be regulated by mRNA splice variation in a tissue-specific fashion using alternate promoters (Kamat et al., 2002), while the mouse gene was reported to use three alternate promoters for expression in ovaries, testes, and brain (Golovine et al., 2003). Thus, the regulation of *CYP19* appears to fall into the expanding subset of P450s known to be expressed in alternate forms. In addition to steroid regulation, Sun et al. noted that the regulatory region of the *CYP19* gene contained a retinoid responsive element (Sun et al., 1998).

Sex steroid hormones may play in important role in neuroprotection. Accordingly, aromatase has been shown to be upregulated in the brain following traumatic CNS injury (Garcia-Ovejero et al., 2005), and polymorphism within the *CYP19* gene has been associated with the risk of developing Alzheimer's disease in humans (Iivonen et al., 2004). Ishunina et al. noted decreased aromatase immunoreactivity in the hypothalamus, but not in the basal forebrain nuclei of tissues with Alzheimer's disease (Ishunina et al., 2005). These results will undoubtedly lead to many more exciting discoveries with potential therapeutic value.

3.1.9 *CYP17*

Conflicting evidence exists regarding the expression of this isoform, the 17b-hydroxylase, in the brain. *CYP17* catalyzes the 17-hydroxylation of PREG and the further two-carbon elimination of this product to form DHEA. It also carries out the analogous reactions with progesterone to create androstenedione. Recently, Yu et al. were able to detect *CYP17* mRNA in several regions of the human brain (amygdala, caudate nucleus, cerebellum, corpus callosum, hippocampus, spinal cord, and thalamus) using a sensitive RT-PCR assay (Yu et al., 2002). Before this, mRNA and/or protein have been reported in embryonic rodent brain microsomes (Compagnone et al., 1995), neonatal rat astrocytes and neurons (Zwain and Yen, 1999), and adult male avian brain (Matsunaga et al., 2001), but not adult rodent brains (Le Goascogne et al., 1991; Compagnone et al., 1995; Cloutier et al., 1997). Nishimura and colleagues, however, report that they were unable to detect *CYP17* mRNA in whole human brain (Nishimura et al., 2003), leaving the Yu report (Yu et al., 2002) the only study to date to claim *CYP17* in the human brain. This distinction is crucial, however, given the necessary role of this enzyme in the completion of the neurosteroid hormone biosynthetic pathways. Interestingly *CYP17* may be induced by intracellular free radicals (Brown et al., 2000).

3.1.10 *CYP11A1*

This inner mitochondrial membrane isoform, historically, known as the cholesterol side-chain cleavage enzyme, plays a crucial role in the conversion of cholesterol into PREG. Numerous laboratories have reported the presence of mRNA, protein, and/or enzyme activity in the rat brain (Mellon and Deschepper, 1993; Costa et al., 1994; Zhang et al., 1995; Ukena et al., 1998; Kimoto et al., 2001). However, not until the year 2000 did Brown et al. report on the actual synthesis of PREG from a radiolabeled precursor in the human brain cell lines. They found that oligodendrocytes, but not astrocytes or neurons, were able to synthesize PREG from cholesterol (Brown et al., 2000). Transcriptional regulation of P450scc gene expression was reported in the embryonic rodent nervous system (Hammer et al., 2004). Yu et al. showed, using RT-PCR, the presence of *CYP11A* mRNA in human amygdala, caudate nucleus, cerebellum, corpus callosum, hippocampus, spinal cord, and thalamus (Yu et al., 2002). Nishimura et al. however, were apparently unable to detect this mRNA isoform in whole human brain (Nishimura et al., 2003), leaving some doubt regarding its presence.

3.1.11 *CYP11B1* and *CYP11B2*

The terminal stages of cortisol and aldosterone production in the human adrenal gland are catalyzed by the enzymes 11beta-hydroxylase and aldosterone synthase, which are encoded by the *CYP11B1* and *CYP11B2* genes, respectively (MacKenzie et al., 2000). The only evidence for the presence of *CYP11B1* mRNA in the human brain is, as in the case of *CYP11A1*, that provided by Yu et al. (2002), who report expression in several brain regions. However again, Nishimura et al. were unable to confirm this independently (Nishimura et al., 2003), though data from the rat (MacKenzie et al., 2000) would appear to support the findings of Yu et al. Most of the available evidence appears to support the expression of *CYP11B2*, also known as aldosterone synthase, in multiple brain regions (➋ *Table 3-2*). Thus, it appears highly likely that the synthesis of aldosterone occurs locally in the human brain.

3.1.12 *CYP8B1* and *CYP7B1*

This gene encodes for a cholesterol 12-hydroxylase enzyme activity in humans and is most likely involved in the biosynthesis of bile. Nishimura et al. are so far the only group to report the detection of *CYP8B1* mRNA in the human brain. All the available rodent data indicate that this pathway is liver-specific, however. Most interestingly, this gene appears to be very ancient as evidenced by the fact that it has no introns (Gafvels et al., 1999). *CYP7B1*, a cholesterol, DHEA, and PREG 7α-hydroxylase (Rose et al., 1997; Wu et al., 1999; Martin et al., 2001), are involved in the clearance of these neuroactive compounds through bile. All the available evidence for the expression of this isoform in animal and human brain is in agreement, and thus its presence seems to be certain. In addition, *CYP7B1* appears to be inducible by IL-1β and TNFα (Payne et al., 1995), as well as azalanstat in hamster (Burton et al., 1995). *CYP7B1* was shown to be sexually dimorphic, with more expression observed in males than in females (Stapleton et al., 1995; Rose et al., 1997; Li-Hawkins et al., 2000). Recently, Trap et al. developed a *CYP7B1*-specific antibody following cDNA immunization of a mouse. Immunohistochemical detection of P4507B1 in slices of human hippocampus using this antibody indicated expression in neurons only (Trap et al., 2005).

3.2 Arachidonic Acid/Leukotriene/Prostaglandin/Thromboxane

A striking feature of the collection of human brain cytochrome P450s discussed in the present review is the obvious major classes they fall into. With the exception of *CYP26A1*, which appears to be alone in the retinoic acid metabolic pathway, all the other brain P450s, including those normally considered "xenobiotic-metabolizing," are either involved in the metabolism (or synthesis, in the case of *CYP51*, the most universal of P450s) of cholesterol or the metabolism of AA. Whether their primary role lies in endogenous synthetic or elimination pathways, or in the protection against xenobiotics, many of these isoforms can, and probably do, both. The previous section dealt with the role of brain cytochrome P450s in brain cholesterol pathways. Once and still thought to be a minor contributor to the overall body synthesis of steroid hormones, the localized synthesis in the brain turns out to play a commanding role in hepatic and extrahepatic metabolism nonetheless.

Infants require a substantial supplementation of AA, which is normally supplied through breast milk. Almost 10% of the membrane phospholipid content of breast fed infants was found to be AA in one study (Koletzko et al., 1996). A crucial factor of the developing infant brain is the amount and type of polyunsaturated fatty acids they receive from their diet. That is, the ratio of dietary n-3 fatty acids (those in which the unsaturation begins 3 carbons from the terminal carbon) to *n*-6 fatty acids can be optimized to

Figure 3-2
Cytochrome P450 pathways in human brain: arachidonic acid cascade

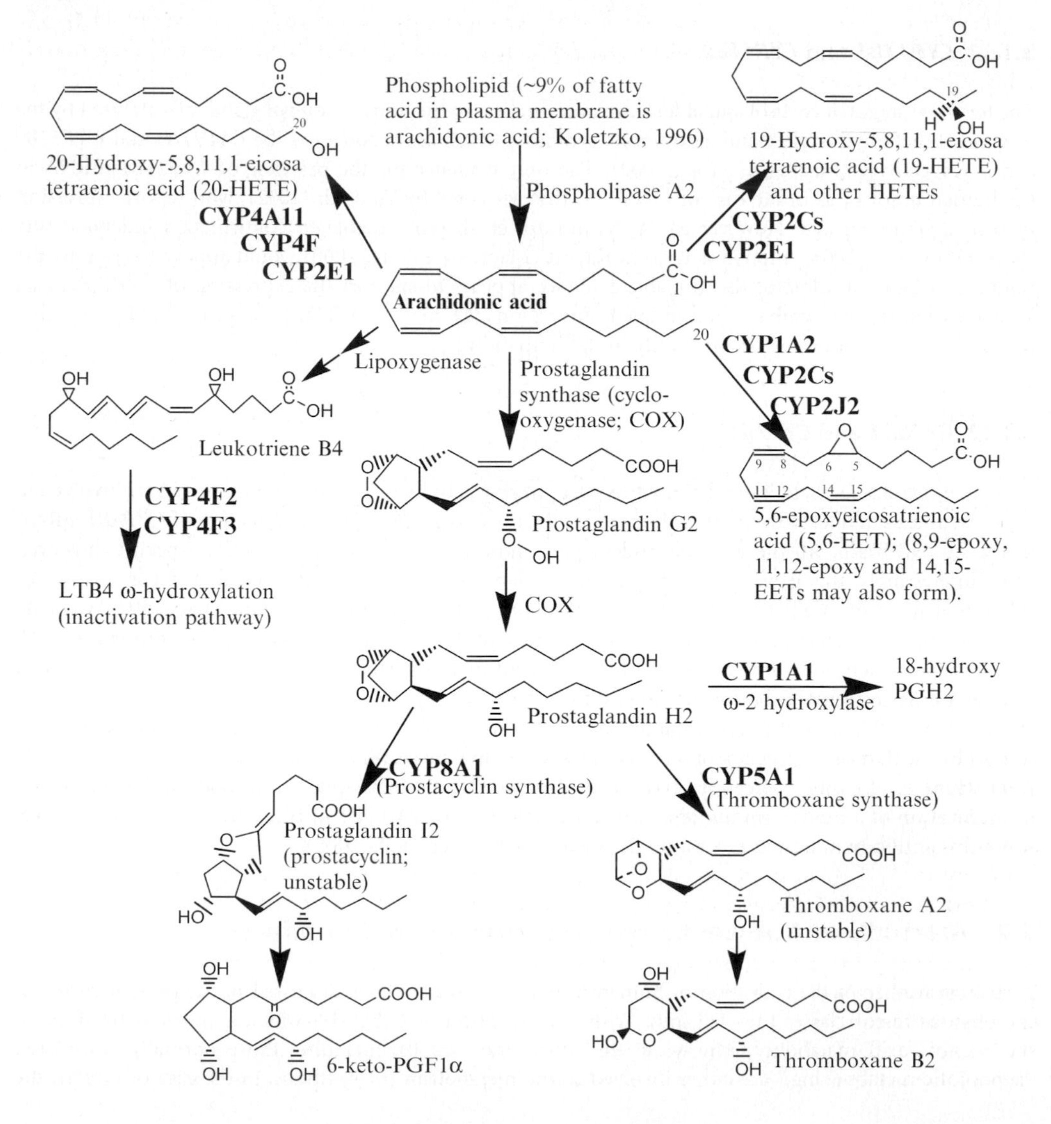

meet the demands of the developing human brain (Jumpsen et al., 1997). AA, being an *n*-6 fatty acid that cannot be synthesized de novo but derives through the metabolism of linoleic acid, is tightly controlled. Its role in the brain cannot be understated, as AA metabolites ultimately control vascular tone and therefore the blood flow to the brain. Such regulation must thus occur in a highly regulated fashion. Therefore, it is not surprising to find P450s in the vascular endothelium.

The AA cascade begins with its release from membrane phospholipid through the action of phospholipase A2. A number of vasoactive compounds arise from the direct epoxidation or hydroxylation of AA. One major signaling branch point that involves cytochrome P450s occurs a step further downstream (➲ *Figure 3-2*). The conversion of AA into prostaglandin G2 or prostaglandin E2 (the peroxy-ring opened derivative of G2) is catalyzed by cyclooxygenase. Subsequent reduction of the peroxy moiety to the hydroxyl results in the production of prostaglandin H2. At this branch point, prostacyclin H2 can be acted on by either of the two enzymes: prostacyclin synthase, otherwise known as *CYP8A1*, or thromboxane A2 synthase (*CYP5A1*). Both the resulting bicyclic metabolites are highly active and also quickly hydrolyzed to the corresponding monocyclic compound, substantially altering their properties in the process. Thus, these messengers are ideally suited as short acting messengers for their crucial role in turning on or off the supply of glucose and oxygen to active regions of the brain. An imbalance in the ratio of these two metabolites is associated with ischemic cerebrovascular disease in humans, highlighting the crucial nature of this branch point (Uyama et al., 1985).

3.2.1 *CYP8A1* (Prostacyclin Synthase)

The role of prostacyclin synthesis in the brain has been known for many years, but the role of individual P450 isoforms in the localized synthesis of HETEs, ETEs, prostaglandins, and thromboxanes only began to be appreciated since the late twentieth century. Tone et al. used in situ hybridization to show that the comparatively low levels of prostacyclin synthase in rat brain were found primarily localized to smooth muscle cells such as those of the arteries (Tone et al., 1997). Mehl et al. showed with immunohistochemistry that prostacyclin (PGI2)-synthase was present in the neuronal cells of bovine, rat, and human brain, most abundantly in Purkinje cells of the cerebellum and cortical neurons, but not in glial cells (Mehl et al., 1999). The next year, Siegle et al. performed a thorough study and found prostacyclin synthase and its mRNA in blood vessels throughout the brain. They also found expression in neuronal and glial cells such as microglia and oligodendrocytes, the strongest expression being in large principal neurons including pyramidal cells of the cortex, pyramidal cells of the hippocampus, and Purkinje cells of the cerebellum (Siegle et al., 2000). Despite the clear role for this enzyme in the brain, Nishimura et al. (2003) reported only very low but measurable *CYP8A1* mRNA in whole human brain.

3.2.2 *CYP5A1* (Thromboxane A2 Synthase)

The other P450 involved in the vascular tone regulation equation discussed above is *CYP5A1*, the enzyme that catalyzes the synthesis of thromboxane A2, the vasoconstriction component (➲ *Figure 3-2*). In addition, like its counterpart described above, the disruption of a balance between the thromboxane synthesis and prostacyclin synthesis was recognized early on to underlie ischemia in human trauma patients (Shohami et al., 1987). The ability of the brain tissue to synthesize eicosanoids has been known for a long time. Bishai and Coceani painstakingly separated neuronal and glial cell fractions from the rat brain and proved that they differed only slightly in their ability to produce products from the cyclooxygenase pathway of the AA cascade (Bishai and Coceani, 1992). However, only following the undertaking of the human genome sequencing project and the standardization of the superfamily nomenclature was the tangle of P450s and the genes encoding them unraveled, so that the study of distinct yet highly homologous isoforms can now be accurately distinguished. What seems clear given the abundant literature, beyond the scope of the current review, is that *CYP5A1* is universally present throughout the human brain owing to its central role in the movement of blood. Recently, Chevalier et al. screened a population of 200 individuals and found several allelic variants of the gene, which have the potential to underlie a number of cardiovascular diseases. However, the effects of these changes are unknown presently (Chevalier et al., 2001). A specific thromboxane synthase inhibitor, sodium (E)-3-[p-(1H-imidazol-1-ylmethyl)phenyl]-2-propenoate], has been used experimentally in Japan under the name "Cataclot" to protect against the cerebral vasospasm, which occurs after subarachnoid hemorrhage (Tokiyoshi et al., 1991). However, it carries with it the risk of severe internal bleeding. Another thromboxane synthase inhibitor known as ozagrel hydrochloride (OKY-046; (E)-3-[4-(1-imidazolylmethyl)phenyl]-2-propenoic acid hydrochloride monohydrate) has been used experimentally for about 20 years and is now available as an antiasthmatic agent in Japan (Uyama et al., 1985). Ketoconazole has long been used as an inhibitor of thromboxane synthase, and thromoboxane synthesis was shown to be upregulated by leukotriene B4 in cultured bovine aortic endothelial cells (Dunham et al., 1984).

3.2.3 *CYP4F3* and *CYP4F2*

Another major branch point from AA involving cytochrome P450 is following the action of lipoxygenase and subsequent synthesis of leukotriene B4, an inflammatory lipid mediator. Leukotriene B4 is one of the weapons employed by polymorphonuclear leukocytes (PMNs) during the body's response to bacterial infections. Their invasion into the brain and production of oxygen free radicals following infection or brain damage underlie much of the additional damage resulting from the inflammation itself. LTB4 was also shown to upregulate the synthesis of thromboxane in primary cultures of bovine aortic endothelial cells (Dunham et al., 1984). Both *CYP4F3* and *CYP4F2* are ω-hydroxylases of LTB4, a step that inactivates its primary inflammatory properties; thus, these enzymes play a crucial role in the development of inflammatory diseases. LTB4 hydroxylase was identified as a novel cytochrome P450 in 1993. It appears to be primarily restricted to PMNs themselves and exhibits very high affinity for the substrate LTB4. *CYP4F2* on the other hand is more broad in its distribution and substrate preference, acting as a more general ω-hydroxylase in the efficient synthesis of 20-hydroxyeicosatrieneoic acid (20-HETE) from AA (Powell et al., 1998). Interestingly, *CYP4F2* was also shown recently to efficiently ω-hydroxylate tocopherol (vitamin E); thus, it has been proposed to regulate this crucial pathway (Sontag and Parker, 2002). However, the role of this in the brain is not clear at this point. Inhibitors of ω-hydroxylation that have been used experimentally in animals include 17-octadecynoic acid (17-ODYA) and *N*-hydroxy-*N'*-(4-butyl-2methylphenyl) formamidine (HET0016; Kehl et al., 2002). A further related isoform, *CYP4F3B*, has recently been identified in humans and found to be more broad in its tissue distribution (Kikuta et al., 1998, 2002). However, direct evidence for the existence of and localization of *CYP4F* isoforms in human brain remains scarce.

3.2.4 *CYP4A11*

CYP4A11 is undoubtedly a ω-hydroxylase; however, there appears to be some discrepancy in the literature as to whether it is able to metabolize AA or not. Several P450 isoforms are now known to be capable of carrying out this reaction. Therefore, the true capability for an enzyme to carry out this reaction *in vivo* depends on its regional and subcellular localization and how frequently it is exposed to free AA. Imaoka et al. (1993) expressed a cDNA encoding *CYP4A11* in a baculovirus-mediated insect cell system and noted that although the enzyme was an efficient lauric acid w-hydroxylase, it did not hydroxylate AA. Jerry Lasker's laboratory on the other hand went to great lengths to purify a number of individual human P450 isoforms from human liver microsomes to characterize their enzymatic activity toward AA. *CYP4A11* and *CYP4F2* were both found to efficiently make 20-HETE from AA; however, the K_m(AA) was measured to be tenfold higher in *CYP4A11*, indicating that *CYP4F2* is more likely to be the major AA ω-hydroxylase (Powell et al., 1998). However, again in 2000 it was reported that recombinant *CYP4A11* was inactive toward AA, further confusing the situation (Kawashima et al., 2000). Lasker again countered with a report indicating that anti-*CYP4A11*, not just anti-*CYP4F2* antibodies, substantially inhibited 20-HETE formation (33 and 66%, respectively) in human kidney microsomes (Lasker et al., 2000). More recently, Kehl and coworkers examined the turnover of ^{14}C-AA from commercially purchased recombinant human P450 isoforms, including *CYP4F2* and *CYP4A11*. They found that both isoforms did carry out this reaction, and that the compound HET0016 was a potent ω-hydroxylase inhibitor of both *CYP4F2* and *CYP4A11* with IC (50)s of 125 and 42 nM, respectively (Kehl et al., 2002). Perhaps this explains the question once and for all, unless *CYP4A11* turns out to have multiple identities, as in the case of *CYP1A1*, *CYP19*, and other P450s.

In 1994, the Gustafsson laboratory reported that CYP4A family mRNA species were detectable in the rat brain by RT-PCR and protein by Western blot. However, they were unable to detect message by Northern blot (Stromstedt et al., 1994). Alexander et al. examined the ω-oxidation activity in general of rat brain microsomes and found that the rates were on the same order as that found in liver microsomes. Individually cultured rat neurons, astrocytes, and oligodendrocytes were all capable of carrying out this reaction. Western blot then confirmed a protein recognizable by rat liver *CYP4A* in rat brain (Alexander et al., 1998). Gebremedhin et al. used in situ hybridization, immunohistochemistry, and RT-PCR to localize *CYP4A* protein and mRNA to arteriolar muscle of rat cerebrum, lending more credence to the hypothesis that ω-hydroxylation of AA by *CYP4A* is also an important modulator of vascular tone and blood flow in the brain (Gebremedhin et al., 2000). Miyata et al. later found that inhibition of 20-HETE synthesis opposed the cerebral vasospasm following subarachnoid hemorrhage and reduced the infarct size in ischemic models of stroke (Miyata et al., 2005).

3.2.5 *CYP2J2*

One of the most recent additions to the human brain P450s, *CYP2J2*, was apparently first isolated from a human cDNA library by Wu and coworkers in 1996. Northern blot hybridization of regional cDNA libraries revealed that this mRNA was expressed primarily in the heart. In addition, these workers expressed the cDNA in vitro and found that the recombinant enzyme was an efficient AA epoxygenase (Wu et al., 1996). Epoxidation has long been known to occur in brain tissue, however, and the individual isoforms and their localized purposes are only beginning to be understood. A mouse analog of human *CYP2J2*, *CYP2J9*, was recently shown to be primarily localized in the brain by Qu et al., but this enzyme appears to be primarily an AA ω-1 hydroxylase rather than an epoxidse (Qu et al., 2001). In addition, true to the reputation of *CYP2* enzymes, *CYP2J2* was shown to metabolize xenobiotics in addition to its supposed endogenous role. Hashizume showed that metabolism by human intestinal microsomes of the antihistamine ebastine was substantially inhibited by an anti-*CYP2J* antibody (Hashizume et al., 2002). Concurrently, Matsumoto et al. determined that astemizole was substantially affected by first-pass metabolism through intestinal *CYP2J2* (Matsumoto et al., 2002). Judging by the amount of mRNA found by Nishimura et al. in whole human brain, we have undoubtedly not heard the last of this story.

3.2.6 *CYP2U1*

Finally, the newest member of the brain P450s was identified only in 2004 by Chuang et al. and independently by the Ingelman-Sundberg laboratory. Chuang et al. used Northern blot of regional tissues and RT-PCR to demonstrate that *CYP2U1* transcripts were displayed most prominently in the human thymus and the brain. Expression of the recombinant enzyme revealed it to be almost exclusively an ω- and ω-1 hydroxylase of AA (Chuang et al., 2004). In agreement, the Ingelman-Sundberg laboratory found that the *CYP2U1* mRNA was most prominent in the human thymus, followed by the brain and heart, whereas in rat the analogous transcript was approximately equal in the brain and thymus. At the protein level, *CYP2U1* was found much more abundantly in the rat brain, especially in the cortex and limbic region (Karlgren et al., 2004).

3.3 All-*Trans*-Retinoic Acid Metabolism: *CYP26A1*

Apparently alone among brain P450s, being not involved in either the cholesterol-related or the AA-related pathways, *CYP26A1* was first identified as a retinoic acid (RA) inducible P450 in zebra fish (White et al., 1996), where it was shown to be essential for determining territories of hindbrain and spinal cord (Emoto et al., 2005). Ray et al. reported the first mouse and human *CYP26* genes following their screening of a mouse cDNA library, which had been enriched by subtractive hybridization for genes upregulated during in vitro differentiation from mouse embryonic stem cells. These researchers identified *CYP26* as a new member of the P450 family and suggested that it plays a role in retinoic acid hydroxylation (➲ *Figure 3-3*) on the basis of analogy to the zebra fish homolog found by White et al. the previous year (Ray et al., 1997). *CYP26* mRNA was shown to be present in the developing mouse brain and liver as well as embryo. Subsequently, the brain from a human heart attack victim was collected within 2 h of death and the regions dissected for mRNA isolation. RT-PCR was then used to identify *CYP26* mRNA in human olfactory bulb, temporal cortex, and hippocampus, with lesser amounts in the parietal cortex, medulla/pons region, and putamen. Under the same conditions (33 PCR cycles), the frontal cortex, caudate, cerebellum, thalamus, and spinal cord did not generate an observable PCR product. In addition, these workers found that acute administration of retinoic acid in adult mice increased the levels of *CYP26* mRNA in the liver, but not in the brain (Ray et al., 1997).

At about the same time, White et al. also cloned the human *CYP26* gene, but also showed that the expressed gene product indeed accepted RA as a substrate and catalyzed the synthesis of 4-OH-RA, 4-oxo-RA, and 18-OH-RA (White et al., 1997). Simultaneously, a third group reported the discovery of mouse *CYP26* and postulated that its role was in the inactivation of RA based on the hyposensitivity to RA observed on over-expression of the *CYP26* gene. They localized the expression of the gene in developing mouse embryos to the posterior neural plate and neural crest cells for cranial ganglia (Fujii et al., 1997). The next year, the laboratory of Mont Juchau reported that the highest levels of transcription in adult human tissue were found in the liver, heart, pituitary gland, adrenal gland, placenta, and regions of the brain. In

Figure 3-3
Cytochrome P450 pathways in human brain: retinoic acid

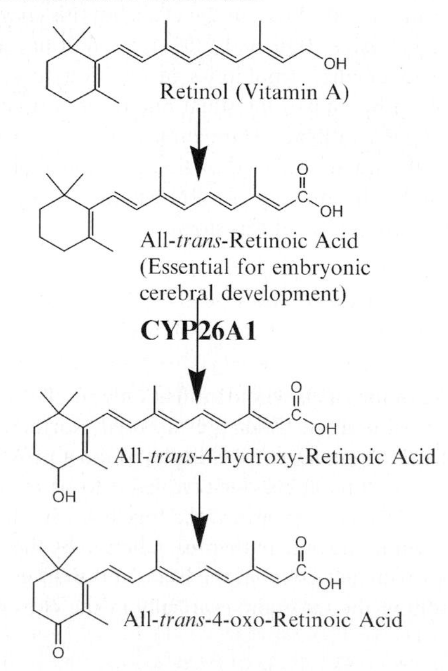

fetal tissue, mRNA was most abundant in the brain with levels comparable to mRNA levels in adult tissues (Trofimova-Griffin and Juchau, 1998). Shortly after this, White and coworkers again reported their findings regarding the discovery and localization of a second form of human *CYP26*, preferentially localized to the human brain tissue. They found that levels of transcript encoding this isoform were highest in the cerebellum and pons, but were also detectable in the cerebral cortex, medulla, occipital pole, frontal lobe, and the temporal lobe (White et al., 2000).

Note that all the published works referenced in the *CYP26A* discussion above, with the exception of the Trofimova-Griffin paper, appear in freely accessible journals online and include very good descriptions of the methodologies used, including links to referenced works. Note that although the reactions catalyzed by human *CYP26* are generally considered to be an inactivation of all-*trans*-retinoic acid, the retinoid ligand 4-*oxo*-retinoic acid is in fact a highly active modulator of positional specification in its own right (Pijnappel et al., 1993).

4 Human Xenobiotic Metabolizing P450s in the Brain

Not enough information exists regarding *CYP1A2* or any of the *CYP2A* isoforms to make any sound conclusions regarding their localization or function in the human brain. Therefore, although included in ➜ *Table 3-3*, they are omitted from the discussion below.

4.1 *CYP1A1*

CYP1A1 is able to catalyze several endogenous reaction pathways, notably prostaglandin ω-2 hydroxylation, testosterone 6β-hydroxylation, and 17β-estradiol 4-hydroxylation (Theron et al., 1985; Omiecinski et al., 1999). There is now a convincing accumulation of evidence for the expression of *CYP1A1* in animal and human brain tissue. Particularly interesting in the case of *CYP1A1* is the dual nature of this isoform in the brain. The story really begins to unravel in the 1990s. Ghersi-Egea et al. noted in 1993 that the P450 content of the mitochondrial fraction of the human brain appeared to be much higher than the microsomal content, but that the P450-dependent enzyme activities appeared to be microsomal in origin. The Ravindranath laboratory, one of the more prolific in the field of brain P450 and associated activities, noted in 1995 that the 1A1, 2B1/2B2, and 2E1 isoforms appeared to be inducible in the mitochondrial fraction of rat brain preparations as determined by cross reactivity with antisera against purified hepatic enzymes. Interestingly, although the mitochondrial enzyme activities resembled microsomal activities, they were not completely inhibited by anti-P450-reductase antibodies as were those in the microsomal fraction (Bhagwat et al., 1995a). Addya et al. later reported that a mitochondrial form of *CYP1A1* found in rat liver was in fact an *N*-terminally truncated form of the microsomal enzyme. The *CYP1A1* amino acid sequence contains an internal mitochondrial targeting signal consisting of positive charged residues immediately following the microsomal targeted *N*-terminal membrane spanning alpha-helix (Addya et al., 1997). The same group also reported that the enzyme activities of multiple mitochondrial forms were inhibited by antimicrosomal P450 isoforms and by antibodies against adrenodoxin (Anandatheerthavarada et al., 1997). They then reported again in 2001 on the specific amino acids in the alternate form of *CYP1A1*, which were responsible for the novel electron donor preference (Anandatheerthavarada et al., 2001). The data of Zhang et al. would appear to support the *CYP1A1* phenomenon in the human brain, in that they found *CYP1A1* mRNA and enzyme activity localized in the mitochondria (Zhang et al., 2001).

In rodents, *CYP1A1* appears to be highly inducible in the endothelial cells of the choroid plexus, in veins in the leptomeninges, and in cerebral veins of mice pretreated with beta-naphthoflavone or other aryl-hydrocarbon receptor agonists (Granberg et al., 2003). Minn et al. found *CYP1A1* mRNA in rat olfactory mucosa but not liver (Minn et al., 2005). Using humanized transgenic mice carrying *CYP1A1*, Galijatovic et al. showed in 2004 that the regulation of *CYP1A1* occurred in a developmental and tissue-specific manner (Galijatovic et al., 2004). Abdulla and Renton found that CNS inflammation induced by LPS results in the downregulation of both *CYP1A1* and 1A2 activities in rat. Furthermore, stimulation of beta-adrenergic receptors modulated this induction through a cAMP-mediated pathway (Abdulla and Renton, 2005). A most interesting report in 2005 found that neonatal rodents exposed to aryl hydrocarbon receptor agonists show long-term imprinting effects on *CYP1A1* expression in several different tissues, including the brain (Desaulniers et al., 2005). Similarly, Johri found that exposure to low doses of deltamethrin prenatally was sufficient to imprint the expression of *CYP1A1, 1A2,* and *2E1,* among others, in the adult rat brain (Johri et al., 2006).

The Ravindranath laboratory recently confirmed the presence of *CYP1A1* mRNA and protein regionally in the human brain and also reported a novel splice variant, which was expressed in the brain, but not in the liver, for the same individual (Chinta et al., 2005a). Wu et al. recently performed an immunohistochemical examination on 11 pairs of human medulloblastoma and noncancerous cerebellar tissues for the presence of *CYP1A1, CYP1B1,* and the aryl hydrocarbon receptor (AhR). Interestingly, they found that the expression of both *CYP1A1* and *CYP1B1* was downregulated in malignant tissues, and that this regulation was AhR-independent (Wu et al., 2005). Ma et al. found that recombinant human *CYP1A1,* in addition to *CYP1B1*and *CYP1A2,* was capable of hydroxylating the pineal hormone melatonin in vitro, raising the interesting possibility of a role for this enzyme in sleep induction and light sensitivity (Ma et al., 2005).

4.2 *CYP1B1*

This isoform can metabolize estradiol (Theron et al., 1985; Jansson et al., 2001), and thus like *CYP1A1,* its presence in the brain could have a profound impact on neurosteroid hormone homeostasis. Using an antipeptide antibody specific for *CYP1B1,* Murray et al. were able to show an immunoreactive protein in several human tumors. However, they did not detect the protein in normal human tissue (Murray et al., 1997). Rieder et al. used a radiolabeled *CYP1B1* probe to identify regional expression of mRNA, particularly in the putamen (Rieder et al., 1998). They later went on to use Western blot and immunohistochemistry to determine that *CYP1B1* was located at the blood–brain interface areas of the temporal lobe, suggesting a role in the metabolism and exclusion of xenobiotics from the brain. They also showed that the same protein appeared to be induced by 7,12-dimethylbenz(a)anthracene in an astrocytoma cell line (Rieder et al., 2000). Muskhelishvili et al. used in situ hybridization and immunohistochemical localization to examine the levels of *CYP1B1* in several normal human tissues. Among their conclusions were that *CYP1B1* mRNA and protein are expressed in parenchymal and stromal tissue from the human brain. Interestingly, they report *CYP1B1* immunostaining in the nuclear compartment of cells (Muskhelishvili et al., 2001). A transgenic mouse carrying the human *CYP1B1* gene has recently been created and shown to express the human *CYP1B1* enzyme in mouse brain in a dioxin-inducible manner by immunohistochemical staining techniques (Hwang et al., 2001).

Ideally, the expression pattern in the transgenic mouse model will turn out to mimic that in the human brain and thus enable an examination of its in vitro role in the metabolism of endogenous and foreign chemicals in the brain. As noted above for *CYP1A1,* Ma et al. recently noted that recombinant human *CYP1B1* is capable of the 6-hydroxylation of melatonin in vitro. Furthermore, examination of the catalytic activity of brain homogenates from transgenic mice carrying the human *CYP1B1* gene indicates that this isoform is capable of contributing to this reaction in vivo (Ma et al., 2005). In 2005, Scallet et al. reported the localization of *CYP1B1* mRNA and protein regionally in the brain of a Rhesus monkey. They found widespread distribution of *CYP1B1* mRNA in both male and female monkey frontal cortex, hippocampus, thalamus, and amygdala. Although *CYP1B1* protein is colocalized with its mRNA in the female brains, it was largely restricted to hippocampal pyramidal neurons of the male brains, suggestive of sex-dependent differences in estradiol metabolism in the monkey brain (see ➋ *Figure 3-1;* Scallet et al., 2005).

4.3 *CYP2B6*

In rat brain, the analogous *CYP2B* isoforms are well known to be inducible by phenobarbitol and nicotine (Parmar et al., 1998; Zevin and Benowitz, 1999; Schilter et al., 2000; Albores et al., 2001; Upadhya et al.,

2002). The Tyndale laboratory recently reported *CYP2B6* protein in several specific cell types of the brain of African Green monkeys, including frontal cortical pyramidal cells, cerebellar Purkinje cells, and neurons in the substantia nigra and found that, like in rodents, protein was induced by chronic nicotine exposure (Lee et al., 2006). At least three studies have specifically addressed the localization of *CYP2B6* in the human brain by Immunoblot (Gervot et al., 1999), RT-RT-PCR (Nishimura et al., 2003), or immunohistochemistry (Miksys et al., 2003). The latter study is by far the most comprehensive, locating *CYP2B6* to individual brain cell types (cerebellar Purkinje cells and hippocampal pyramidal neurons). These workers noted that the level of this enzyme in human brain was highly elevated in the brains of smokers and alcoholics, confirming the conserved nature of this inducibility across species. Thus, the evidence for a xenobiotic metabolizing role of this enzyme in human brain is rather compelling. Endogenous compounds may also be metabolized by *CYP2B6*, as it 16-hydroxylates testosterone for example (Omiecinski et al., 1999).

4.4 *CYP2C*

The evidence for the individual expression of *CYP2C8*, *2C9*, *2C18*, and *2C19* isoforms in human brain is very limited at this time. All the evidence available so far appears to arise from the RT-PCR analysis of mRNA species (➋ *Table 3-3*), largely because this was the only reliable method of distinguishing these P450s in humans before specific antibodies were widely available. Murine *CYP2C* isoforms have all been shown to metabolize AA (Luo et al., 1998), and recombinant human *CYP2Cs* appear to be very similar in this regard. *CYP2C8*, for example, makes EETs (Zeldin et al., 1996) and HETEs, as does *CYP2C9* but with different metabolite profiles (Bylund et al., 1998). The endogenous neurochemicals 5-hydroxytryptamine and adrenaline are *CYP2C* inhibitors in vitro (Gervasini et al., 2001). Thus, it is tempting to speculate that human *CYP2C* isoforms are involved in the regulation of vascular tone and blood flow in brain micro-vessels. This raises some intriguing hypotheses regarding the pharmacology of nonsteroidal antiinflammatory drugs (NSAIDs), many of which were shown to be metabolized by *CYP2C9*, and the endogenous role of this subfamily of P450.

4.5 *CYP2D6*

Because of its preference for monoamine-like molecules and its role in the metabolism of many psychoactive compounds, *CYP2D6* is perhaps the most interesting of the xenobiotic metabolizing enzymes to be implicated in the human brain. As such, a great deal of information is now available regarding its presence there (➋ *Table 3-3*).

4.5.1 Evidence for the Expression of *CYP2D6* in the Human Brain

Evidence for the expression of *CYP2D6* in the human brain has steadily accumulated over the last decade. In the beginning, most of the data were circumstantial and subject to much debate. *CYP2D6* is known to metabolize dozens of drugs of clinical significance whose mechanisms of action are central. In addition, several known *CYP2D6* substrates and inhibitors are psychoactive drugs of abuse. The fates of codeine, hydrocodone, oxycodone, dextromethorphan, amphetamine, para-methoxyamphetamine, 3,4-methylene-dioxyamphetamine, and 3,4-methylenedioxy methamphetamine (PMA, MDA and MDMA, respectively) are all largely affected by this polymorphic P450 (Sellers et al., 1997). *CYP2D6* is believed to metabolize the conversion of codeine to morphine, leading to the potentiation of codeine's analgesic effect (Dayer et al., 1988; Desmeules et al., 1991). *CYP2D6* is known to catalyze the detoxification pathway for 1-methyl-4-phenyl-1,2,3,6-tetrahydropyridine (MPTP), a proneurotoxin formed as a contaminant during the synthesis of meperidine or "street heroin" (Fonne-Pfister et al., 1987; Gilham et al., 1997). MPTP causes Parkinsonian symptoms in susceptible individuals, raising great interest in the possibility that *CYP2D6* may have an endogenous crucial protective function against the effects of aging (Coleman et al., 1996). Furthermore,

(-)-deprenyl, whose mechanism of action is considered to involve inhibition of monoamine oxidase B (MAO-B), is metabolized, at least in part, by *CYP2D6*, and the selective serotonin reuptake inhibitors fluoxetine and paroxetine are also potent *CYP2D6* inhibitors (Brosen and Skjelbo, 1991; Brosen et al., 1993). Interestingly, *CYP2D6* is believed to metabolize the conversion of codeine to morphine, leading to the potentiation of codeine's analgesic effect at the site of action (Dayer et al., 1988; Desmeules et al., 1991). Kornhuber et al. recently examined the regional distribution of levomepromazine and its desmethyl metabolite (formation catalyzed by *CYP2D6*) in the human brain tissue. Interestingly, they found the highest levels of residual parent drug in the basal ganglia, a structure reported to be relatively deficient in *CYP2D6* (Kornhuber et al., 2006). Cross inhibition by multiple *CYP2D6* psychoactive drugs has long made tissue-localized 2D6 activity in human brain an attractive postulate. Accordingly, *CYP2D6* polymorphism is suggested to affect moods (Llerena et al., 1993; Martinez et al., 1997) and pain tolerance (Sindrup et al., 1993a). However, the existence of null metabolizers makes the presence of functional *CYP2D6* apparently nonessential, such that one might expect any role in this capacity to be modulation and not a primary metabolic pathway.

Recently, strong evidence in support of *CYP2D6* protein and mRNA in specific cell types in the human brain was reported. Neuronal cells, as well as glial cells, showed labeling for mRNA in brain regions such as the neocortex, caudate nucleus, putamen, globus pallidus, hippocampus, hypothalamus, thalamus, substantia nigra, and cerebellum. In contrast, *CYP2D6* protein was primarily localized in large principal neurons such as pyramidal cells of the cortex, pyramidal cells of the hippocampus, and Purkinje cells of the cerebellum. In glial cells, *CYP2D6* protein was absent (Siegle et al., 2001). Another report revealed a slightly different pattern of expression and found that *CYP2D6* appeared to be upregulated in the brains from alcoholic patients (Miksys et al., 2002). When *2D6* protein was localized subcellularly in human neurons, the signal was detected not only in neuronal soma but also in dendrites of Purkinje and cortical neurons (Chinta et al., 2002). One recent report places 2D6 on the cytosolic face of the endoplasmic reticulum, unlike most P450s whose active sites face the luminal side (Zhou et al., 1997a). Another study in which the *CYP2D6* was expressed in yeast and COS7 cells found that in addition to its unusual orientation in the endoplasmic reticulum, 2D6 was partially N-glycosylated and was located at the plasma membrane (Loeper et al., 1998). This fact implies a different protein sorting mechanism than is found in hepatic tissue for other P450s and argues in favor of the possibility that an alternate isoform of 2D6 could be expressed in the brain. One concern that Voirol et al. (2000) point out is that to measure *CYP2D6* enzyme activity in human brain preparations, it was necessary to add exogenous reductase. However, it is well known that peroxides such as cumene hydroperoxide and even hydrogen peroxide itself can serve as cofactors in P450-mediated reactions, and thus there may be no need for a P450 reductase enzyme under certain circumstances (Chan et al., 2001). The important thing to be noted is that the metabolism of tyramine by MAO-B is known to produce a H_2O_2 molecule, which causes oxidative damage to mitochondrial DNA of the dopaminergic neuron (Hauptmann et al., 1996).

4.5.2 Alternate Splicing

A largely unexplored possible source of variation in human *CYP2D6* protein apart from point mutations in allelic variants is through alternate splicing of a pre-mRNA species. A recent analysis of the P450 superfamily reveals slice sites with "negative information content," indicating the need to invoke novel splicing regulatory mechanisms to compensate for the weak splicing signals (Rogan et al., 2003). Huang et al. (1997) reported the presence of six distinct *CYP2D* mRNA species in human breast tissue. Using gene-specific oligonucleotide probes, they were able to trace four of these transcripts to expression from the so-called *CYP2D7P* pseudogene, rather than *2D6* itself. Two transcripts appeared to arise from the *2D6* gene, one of which was the expected wild type or known hepatic version of the enzyme (Kimura et al., 1989). A second transcript was detected, but was approximately 141 nucleotide bases shorter in length than expected for full-length protein. On closer examination, it was found that this message was completely lacking exon 6 of *CYP2D6*. Woo and coworkers (Woo et al., 1999) found a similar pattern of splice variation when they examined 94 different brain samples for *2D6* mRNA. In this case, only some 48%

of the samples contained any message corresponding to the known full-length hepatic transcript. Interestingly, the most abundant transcript, found in 87% of their samples, was the same shortened clone as that found by Huang et al. The deletion of the 141 base fragment results in a frame mutation with a high likelihood of being translated. Little is known about the function, if any, of the predominant neuronal splice variant arising from the *2D6/2D7* genes (Huang et al., 1997; Woo et al., 1999). Apart from the previously mentioned dually targeted P450s, precedence also exists with neuronal enzymes involved in catecholamine synthesis, such as tyrosine hydroxylase for example, in which a single gene can result in 4 distinct protein isoforms (O'Malley et al., 1987) via mRNA splice variation. Therefore, a possibility exists that a distinct form of the enzyme, produced as a result of alternate splicing of *CYP2D6* pre-mRNA, is expressed in the human brain. Given the dual targeting and alternate electron donor preference noted with *CYP1A1*, it is very possible that such a protein would not exhibit the same substrate specificity as its microsomal counterpart. Deserving mention here is the fact that the well-known "artifactual" mutation in the original *CYP2D6* cDNA at amino acid position 374 is in fact encoded for by the analogous exon of *CYP2D7*.

At first glance, any predicted protein expressed from this message would appear to be nonfunctional. After all, exon 6 of the genomic *CYP2D6* locus encodes for the I-helix of this enzyme according to P450 homology models (Lewis et al., 1997; Modi et al., 1997). In typical cytochrome P450s, designed to metabolize hydrophobic substrates from out of the membrane, the I-helix serves as the backbone of the active site and includes amino acid residues involved in the transfer of protons from the solvent into the active site. However, consider the possibility that a hemoprotein not entirely unlike *CYP2D6* could be designed to metabolize water-soluble substrates instead of lipophilic ones. Indeed the preference of the hepatic form for amine-containing substrates, which are more soluble than their hydrocarbon counterparts, distinguishes this enzyme family. Such an isoform may have no need for an I-helix, if indeed protons were readily available from bulk solvent. Interestingly, submission of the predicted amino acid sequence for automated protein modeling (Combet et al., 2002) resulted in a defensible model based on the rabbit *CYP2C5* crystal structure (not shown). Notably, of all the enzymes known to be involved in catecholamine biosynthesis whose structure is known, each has mixed α/β protein folds and open or solvent-exposed active sites (Erlandsen et al., 2000). Contrast this with the prevailing theories of hepatic cytochrome P450s, which are thought to recruit substrates through a membrane-accessible face (Krainev et al., 1991; Nakayama et al., 2001). Furthermore, in a recent report by Chinta et al., Western blot analysis of *CYP2D* protein in human brain tissue reveals a protein of MW~52KDa, rather than ~56KDa as predicted from the amino acid sequence. Interestingly, the protein appeared to migrate slightly slower when analyzed from midbrain, striatum, and thalamus regions than it did from the cortex and cerebellum, indicating possibly multiple cell-specific isoforms (Chinta et al., 2002). This group also reported alternate splicing of a rat analog of human *CYP2D6*, *CYP2D1*, specifically in the rat brain (Chinta et al., 2005b). News also arrived from the same laboratory of a functional *CYP2D7* hybrid expressed in the human brain, which, most interestingly, appeared to metabolize codeine more efficiently than hepatic *CYP2D6* (Pai et al., 2004). However, this report has been met with some skepticism, as no evidence for functional CYP2D7 transcripts was observed in Asian, Caucasian, or African American individuals when an attempt was made to confirm the Ravindranath result independently (Gaedigk et al., 2005).

4.5.3 Possible Endogenous Substrates of *CYP2D6*

The fairly unique substrate specificity of *CYP2D6* among P450s has led us and others to speculate on the possible endogenous substrates for this enzyme in the human brain. In addition to the neurotransmitters discussed later, it is also very possible that *CYP2D6* could be involved in the synthesis or degradation of endogenous neuropeptide opiates. *CYP2D6* exhibits progesterone 2β-, 6β-, 16α-, and 21-hydroxylase activities (Hiroi et al., 2001; Kishimoto et al., 2004). Hiroi et al. were also among the first to report that the conversion of tyramine into dopamine noted in human liver microsomes was catalyzed mostly by the *CYP2D6* isoform. The K_m values of *CYP2D6*, expressed in yeast, for *p*-tyramine and *m*-tyramine were

190.1 ± 19.5 μM and 58.2 ± 13.8 μM, respectively (Hiroi et al., 1998). These values agree well with kinetic analyses from the present author's laboratory showing that the wild-type *CYP2D6.1* isoform catalyzed the formation of dopamine from *m*-tyramine with a K_m of 75 ± 10 μM and a V_{max} of 8.2 ± 0.4 pmol/pmol P450/min and from *p*-tyramine with a K_m of 152 ± 52 μM and a V_{max} of 7.6 ± 1.2 pmol/pmol P450/min. Pharmacogenetic variants of *CYP2D6* exhibited an impaired ability to catalyze this reaction (Haining and Yu, 2003). Unlike dextromethorphan, which formed a distinct Type I binding spectrum with the *CYP2D6.1* isoform, *p*-tyramine and *m*-tyramine formed a peculiar reverse Type I binding spectra (Haining and Yu, 2003). This suggests that *p*-tyramine and *m*-tyramine have a similar binding orientation to each other in the active site of *CYP2D6*, but different from dextromethorphan. It is known that most of the substrates of *CYP2D6* contain a basic nitrogen atom, which is located about 5 or 7Å away from the site of oxidation with a flat hydrophobic area nearby (Strobl et al., 1993; de Groot et al., 1997). However, several substrates with a distance of approximately 10Å between the basic nitrogen atom and the site of oxidation are also known. More detailed pharmacophore and homology models for *CYP2D6* have been developed to predict the interactions between *CYP2D6* enzyme and it substrates (de Groot et al., 1999; Ekins et al., 1999).

Excess tyramine in the diet can cause a hypertensive crisis in individuals taking monoamine oxidase (MAO) inhibitors. This phenomenon is widely known as the "cheese effect" because of the unusually high concentrations of tyramine found in aged cheeses and air-cured meats. Irreversible MAO-A inhibition and concomitant reduction in the peripheral deamination of tyramine by MAO-A leads to the well-known increase in levels and release of catecholamines and subsequent elevation in blood pressure. A primary source of tyramine in vivo is that produced from tyrosine by aromatic amino acid decarboxylase (➋ *Figure 3-4*). Interestingly, the concentration of tyramine in brain has been shown to change directly with dietary protein content. Tyramine concentrations rise as much as two- to threefold between 0% and 10% dietary protein content. This increase produces a clear stimulation of the rate of dopamine synthesis, notably in the hypothalamus, where a significant amount of *CYP2D6* isoform is observed. In addition, there is evidence that differences in tyramine and dopamine levels as regulated by allelic variants may lead to the differences in appetite motivation among *CYP2D6* polymorphic individuals (Fernstrom and Fernstrom, 2001). Tyramine is also known to enhance the impulse propagation mediated release of catecholamines in the rat brain (Knoll et al., 1999). Furthermore, *CYP2D1* appears to colocalize with the dopamine transporter in canine brain, further strengthening the hypothesis of a role for *2D6* in dopamine synthesis in the brain (Niznik et al., 1990). Tryptamine, another possible endogenous substrate candidate for *CYP2D6*, also formed a reverse Type I binding spectrum with *CYP2D6*. Although it has been reported that *CYP2D6* was capable of carrying out the deamination of tryptamine to tryptophol (Martinez et al., 1997), Dr. Aiming Yu showed while working in Frank Gonzalez' laboratory that, in fact, MAO-B is responsible for selective conversion of tryptamine to tryptophol (via indole-3-acetaldehyde and further reduction by aldehyde reductase) and that *CYP2D6* does not participate in this process (Yu et al., 2003).

4.5.4 *CYP2D6*-Humanized Transgenic Mouse

The complete wild type human *CYP2D6* gene, including its hepatic regulatory sequence, but lacking the upstream pseudogenes, was microinjected by Gonzalez and coworkers into a fertilized FVB/N mouse egg to create *CYP2D6* transgenic mice. *CYP2D6*-specific protein expression was detected in the liver, kidney, and intestine from only the *CYP2D6*-humanized mice. Debrisoquine clearance of this drug was markedly higher (~sixfold) and its half-life significantly reduced (twofold) in humanized mice when compared with wild-type animals (Corchero et al., 2001). In our hands, incubation of *Tg2D6* mouse liver microsomes with exogenous cytochrome b_5, MPTP, and NADPH resulted in ~sixfold greater PTP production than that in parental FVB/N mice based on overall P450 content. Though mice have at least three related *CYP2D* isoforms, data from the present author's laboratory indicate that the closest mouse relative to human *CYP2D6*, which is *CYP2D22* at 75% amino acid identity, does not share the same substrate specificity but in fact more closely resembles that of *CYP3A4* (Yu and Haining, 2006). *CYP2D22* enzyme as predicted from

Figure 3-4
Neurotransmitter biosynthesis pathways: possible involvement of CYP2D6?

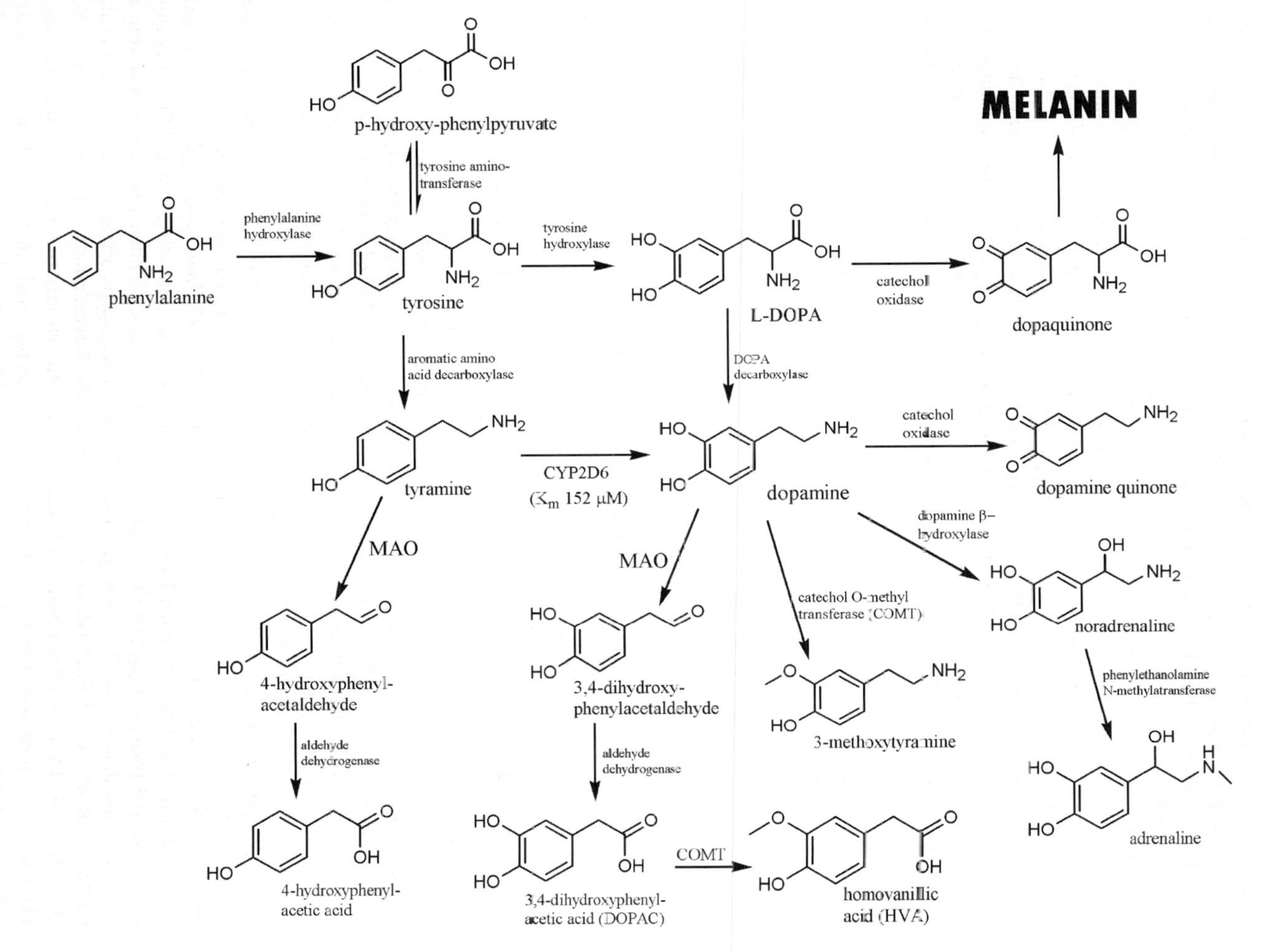

the cDNA has an identical *N*-terminal amino acid sequence as that of the partially purified *CYP2D* (Blume et al., 2000). Of the mouse *CYP2D* isoforms (*CYP2D9, 2D22,* and *2D26*), only *CYP2D22* mRNA appeared to be present in mouse brain (Choudhary et al., 2003). Incubation of purified, recombinant *CYP2D22* with rat cytochrome P450 reductase, cytochrome b5, lipid, NADPH, and MPTP resulted in no measurable production of PTP when analyzed by HPLC with fluorescence detection. Production of PTP by *CYP2D6* on the other hand proceeded with apparent biphasic kinetics (Korzekwa et al., 1998), the "low K_m" component estimated as 10.3 ± 6.8 μM and the "high K_m" component estimated as 106 ± 19 μM with a V_{max} of 912 ± 74 pmole (nmol P450)$^{-1}$min^{-1}.

Our data are consistent with an earlier published report, in which a *CYP2D* enzyme partially purified from mouse liver microsomes did not exhibit debrisoquine 4-hydroxylase activity (Masubuchi et al., 1994). Thus, the mouse would appear to be a good model for the *CYP2D6* poor metabolizer phenotype, and we were very excited about the prospect of *2D6* being expressed in the brains of these mice. However, although we are able to specifically detect the presence of *CYP2D6* mRNA (as distinguished from *CYP2D22* mRNA) in the mouse brain, we have been unable to locate the protein by Western blot or immunohistochemistry. Dr. Aiming Yu also was unable to locate immunoreactive protein in whole-brain homogenates using a commercially available monoclonal antibody (BD-Gentest) directed against human *2D6* (personal communication). The Tyndale laboratory recently reported their more thorough investigation, which came to the same conclusion as we have (Miksys et al., 2005). Though disappointing, this nonetheless provides fodder for the notion that a tissue-specific mRNA splicing phenomenon involving the *CYP2D7* pseudogene or some other signal is necessary for expression in the brain.

4.6 *CYP2E1*

Hepatic *CYP2E1* is notorious for its role in the metabolism of and induction by ethanol. Thus, it is not surprising that numerous experiments have been focused at determining the presence and inducibility of this isoform in animal and human brain. Numerous pieces of evidence now appear to confirm that human *CYP2E1* is indeed present and inducible in the brain. Farin and Omiecinskiet al. detected *CYP2E1* mRNA in several human brain regions using RT-PCR (Farin and Omiecinski, 1993). Montoliu went on to show that *CYP2E1* was expressed and was induced by ethanol in primary rat astrocyte cell culture. Interestingly, the protein was localized subcellularly to the nuclear membrane and the cytoplasm. They also noted a concomitant increase in reactive oxygen species as a result of the upregulation of *CYP2E1*, suggesting a role for this enzyme in ethanol-induced cell damage (Montoliu et al., 1995). Vasilou et al. used a transgenic mouse model to show that the levels of *CYP2E1* expression in mouse brain contribute to ethanol sensitivity (Vasiliou et al., 2006). Tindberg and Ingelman-Sundberg found *CYP2E1* in the microsomal fraction of rat hippocampus. They also found that acute ethanol administration resulted in a marked induction of *CYP2E1*-like enzyme activity in cortical glial cell culture (Tindberg and Ingelman-Sundberg, 1996). Yoo et al. confirmed the presence of *CYP2E1* mRNA in rat brain (Yoo et al., 1997).

In 1997, the laboratory of Mont Juchau reported the presence of *CYP2E1* in human fetal brain tissue and noted that the levels of this transcript increased with gestational age and suggested a role for this enzyme in fetal alcohol syndrome (Boutelet-Bochan et al., 1997; Brzezinski et al., 1999). Jenner et al. suggested that *CYP2E1* is selectively localized in nigral dopamine-containing cells and thus may contribute to nigrostriatal toxicity in chemically induced Parkinson's disease (Jenner, 1998). The next year, Kirches et al. reported that out of 22 human glioblastomas and 4 anaplastic astrocytomas all tested positive for *CYP2E1* mRNA by RT-PCR (Kirches et al., 1999). The Ravindranath laboratory studied the levels of *CYP2E1* mRNA by Northern blot and fluorescence in situ hybridization in rat and human brain tissue. These workers report a constitutive presence of *2E1* in several neuronal cell types and the inducibility of this isoform by ethanol in rat brain (Upadhya et al., 2000). The Tyndale laboratory recently showed that in monkey brain, *CYP2E1* is upregulated in a cell-specific manner by chronic nicotine exposure (Joshi and Tyndale, 2005), but that this pattern is not retained in an acute exposure model (Joshi and Tyndale, 2006).

Recently, Robin et al. isolated *CYP2E1* from both the microsomal and mitochondrial fractions of rat liver, suggesting that this enzyme may now join the others known to be dually targeted within the cell. However, unlike the *CYP1A1* mitochondrial form, the *CYP2E1* mitochondrial form did not appear to be *N*-terminally truncated (Robin et al., 2001). More recently, Howard et al. confirmed the ethanol and nicotine inducibility of rat brain *CYP2E1* in a very convincing study. They went further to show the presence of *CYP2E1* in several specific human brain regions and a consistent increase in the *2E1* content of brains from alcoholics and smokers. Furthermore, they noted a significant induction by nicotine in the level of *CYP2E1* immunostaining in neuroblastoma cell culture (Howard et al., 2003). In addition to its probable role in the metabolism of xenobiotics in the brain, the constitutive low-level presence suggests that there may be an endogenous function as well. *CYP2E1* is known to act as a fatty acid hydroxylase, and thus it may fall into the AA cascade (➋ *Figure 3-2*).

4.7 CYP3A

Evidence for the expression of one or more *CYP3A* isoforms (*3A3, 3A4, 3A5,* or *3A7*) in the human brain is accumulating (➋ *Table 3-3*), but has so far been somewhat inconsistent and no clear consensus has emerged on which isoform is expressed. RT-PCR has implicated the presence of *CYP3A* mRNA in several normal human brain regions (Farin and Omiecinski, 1993; McFadyen et al., 1998; Nishimura et al., 2003). However, Knupfer et al. were unable to detect *CYP3A4* by RT-PCR in any of the 10 brain tumors they tested. They did, however, find *CYP3A5* mRNA in 3 of the tumors (Knupfer et al., 1999). Kirches et al. on the other hand noted a very strong signal for CYP3A by RT-PCR in 22 glioblastomas and 4 anaplastic astrocytomas, while antibodies to human liver *CYP3A4* stained only a subfraction of tumor cells of each type and some neurons in normal brain areas, whereas astrocytes were negative (Kirches et al., 1999). Several groups have now attempted to measure *CYP3A*-like enzyme activity in brain tissues. Voirol et al. reported a few years ago the measurement of amitriptyline demethylation activity in rat and human brain microsomes. They found that this activity that was *CYP3A*-mediated as determined by chemical and antibody inhibition (Voirol et al., 2000). More recently, Pai et al. measured alprazolam hydroxylation in rat and human liver and brain preparations and found that the enzyme activity was higher in the brain preparations than liver. Furthermore, inhibition studies implicated *CYP3A* in the brain reaction, as was determined previously for the liver catalyzed reaction (Pai et al., 2002). On balance, it appears that the evidence for *CYP3A* expression in human brain is mounting, but it is still not clear about the individual isoforms involved.

The function of *CYP3A* in human brain is largely unknown. *CYP3A* is known to be inducible in the liver and has the capability of metabolizing a wide assortment of compounds, both xenobiotic and endogenous. Rosenbrock examined phenytoin-inducible P450-activity in rat brain microsomes and noted the formation of testosterone metabolites that appeared to be *CYP3A* mediated (Rosenbrock et al., 1999). Schilter et al. demonstrated markedly induced levels of *CYP2B* and *CYP3A1* mRNA in the striatum and cerebellum of mice following induction with phenobarbital, diphenylhydantoin, amitriptyline, *trans*-stilbene oxide, or diallyl sulfide (Schilter et al., 2000). *CYP3A7* is the major human P450 expressed in fetal liver (Yang et al., 1994), but little is known about the levels of this isoform in fetal brain. Recently, a transgenic mouse has been shown to incorporate the human *CYP3A7* gene, which appears to be transcribed in the brain (Li et al., 1996). *CYP3A4* is known to metabolize sterols (Furster and Wikvall, 1999; Bodin et al., 2002; Handschin et al., 2002), and given the now established role of neurosteroid hormone biosynthesis in the brain, it seems likely that *CYP3A* isoforms in the brain will affect these cholesterol-related pathways (➋ *Figure 3-1*), at least to some extent. Bylund et al. studied AA metabolism by *CYP3A4* and found that it produced 13-, 10-, and 7-hydroxyeicosanoids (Bylund et al., 1998). The other important endogenous chemicals, including adrenaline, serotonin, and 5-hydroxytryptophol, have been shown to be potent inhibitors of *CYP3A* enzyme activity, exhibiting K_i values of 42.3, 26.4, and 43 μM, respectively (Martinez et al., 2000). Chen et al. noted that human fetal liver *CYP3A7* is capable of carrying out retinoic acid 4-hydroxylation (Chen et al., 2000), a finding that has important implications for the role of this P450 in neuronal development, even more so if it turns out to be abundantly expressed in the human brain.

4.8 NADPH Cytochrome P450 Reductase

Though not a P450, the involvement of NADPH cytochrome P450 reductase (reductase) in the activation of numerous P450s in the human brain gives us an understanding of its crucial role. Ghersi-Egea et al. noted that the levels of cytochrome P450 and the level of reductase did not correlate well in the human brain, suggesting a role for reductase other than its namesake (Ghersi-Egea et al., 1989, 1993). Kemp et al. suggested that rat brain reductase may in fact be related or identical to an NADPH-dependent diaphorase, an enzyme thought to be involved in the protection from semiquinone free radical damage in neurons due to its ability to transfer two electrons simultaneously (Kemp et al., 1988). However, Rampling et al. studied the hypoxic environment of human tumors and were able to distinguish the expression of three separate reductases by Western blot and enzyme activity, namely DT-diaphorase, NADH cytochrome b5 reductase, and NADPH cytochrome P-450 reductase (Rampling et al., 1994). A valid question that arises is whether the presence of cytochrome P450 reductase is even necessary for many P450-mediated reactions or not. Note the *CYP1A1* story here in which the mitochondrial form of the enzyme exhibits a distinct electron donor preference (adrenodoxin reductase versus NADPH cytochrome P450 reductase) and metabolic profile. In addition, consider the known ability of some P450s to obtain electrons from peroxides or lipid peroxidation products. This raises the interesting possibility that mitochondrial P450s could be preferentially activated when the mitochondria is under oxidative stress.

5 Concluding Remarks

From the evidence accumulated herein, it is apparent that cytochrome P450s are much more than hepatic drug- and xenobiotic-metabolizing enzymes. In human brain, they play crucial roles in key signaling pathways and have tremendous implications for human health and drug safety. Inadvertent inhibition of these isoforms by drug entities, not predicted by current hepatic screening methods, undoubtedly underlies many clinical side effects and drug interactions. The pharmaceutical industry would be well advised to pay heed to such possible interactions, and indeed, potential drug targets, as new drugs are developed for CNS disorders in order to attain maximum safety and efficacy.

References

Abdelgadir SE, Roselli CE, Choate JV, Resko JA. 1997. Distribution of aromatase cytochrome P450 messenger ribonucleic acid in adult rhesus monkey brains. Biol Reprod 57: 772-777.

Abdulla D, Renton KW. 2005. Beta-adrenergic receptor modulation of the LPS-mediated depression in *CYP1A* activity in astrocytes. Biochem Pharmacol 69: 741-750.

Addya S, Anandatheerthavarada HK, Biswas G, Bhagwat SV, Mullick J, et al. 1997. Targeting of NH_2-terminal-processed microsomal protein to mitochondria: a novel pathway for the biogenesis of hepatic mitochondrial P450MT2. J Cell Biol 139: 589-599.

Aibiki M, Maekawa S, Yokono S. 2000. Moderate hypothermia improves imbalances of thromboxane A2 and prostaglandin I2 production after traumatic brain injury in humans. Crit Care Med 28: 3902-3906.

Akwa Y, Morfin RF, Robel P, Baulieu EE. 1992. Neurosteroid metabolism. 7 alpha-Hydroxylation of dehydroepiandrosterone and pregnenolone by rat brain microsomes. Biochem J 288(Pt 3): 959-964.

Albores A, Ortega-Mantilla G, Sierra-Santoyo A, Cebrian ME, Munoz-Sanchez JL, et al. 2001. Cytochrome P450 2B (*CYP2B*)-mediated activation of methyl-parathion in rat brain extracts. Toxicol Lett 124: 1-10.

Alexander JJ, Snyder A, Tonsgard JH. 1998. Omega-oxidation of monocarboxylic acids in rat brain. Neurochem Res 23: 227-233.

Alkayed NJ, Narayanan J, Gebremedhin D, Medhora M, Roman RJ, et al. 1996. Molecular characterization of an arachidonic acid epoxygenase in rat brain astrocytes. Stroke 27: 971-979.

Allard P, Marcusson JO, Ross SB. 1994. [3H]GBR-12935 binding to cytochrome P450 in the human brain. J Neurochem 62: 342-348.

Amruthesh SC, Falck JR, Ellis EF. 1992. Brain synthesis and cerebrovascular action of epoxygenase metabolites of arachidonic acid. J Neurochem 58: 503-510.

Anandatheerthavarada HK, Boyd MR, Ravindranath V. 1992. Characterization of a phenobarbital-inducible cytochrome P-450, NADPH-cytochrome P-450 reductase and reconstituted cytochrome P-450 mono-oxygenase system from rat brain. Evidence for constitutive presence in rat and human brain. Biochem J 288(Pt 2): 483-488.

Anandatheerthavarada HK, Addya S, Dwivedi RS, Biswas G, Mullick J, et al. 1997. Localization of multiple forms of inducible cytochromes P450 in rat liver mitochondria: immunological characteristics and patterns of xenobiotic substrate metabolism. Arch Biochem Biophys 339: 136-150.

Anandatheerthavarada HK, Amuthan G, Biswas G, Robin MA, Murali R, et al. 2001. Evolutionarily divergent electron donor proteins interact with P450MT2 through the same helical domain but different contact points. EMBO J 20: 2394-2403.

Andersson U, Eggertsen G, Bjorkhem I. 1998. Rabbit liver contains one major sterol 12alpha-hydroxylase with broad substrate specificity. Biochim Biophys Acta 1389: 150-154.

Aoyama Y, Horiuchi T, Yoshida Y. 1996a. Lanosterol 14-demethylase activity expressed in rat brain microsomes. J Biochem (Tokyo) 120: 982-986.

Aoyama Y, Noshiro M, Gotoh O, Imaoka S, Funae Y, et al. 1996b. Sterol 14-demethylase P450 (P45014DM*) is one of the most ancient and conserved P450 species. J Biochem (Tokyo) 119: 926-933.

Axen E, Postlind H, Sjoberg H, Wikvall K. 1994. Liver mitochondrial cytochrome P450 *CYP27* and recombinant-expressed human *CYP27* catalyze 1 alpha-hydroxylation of 25-hydroxyvitamin D3. Proc Natl Acad Sci USA 91: 10014-10018.

Baum LO, Strobel HW. 1997. Regulation of expression of cytochrome P-450 2D mRNA in rat brain with steroid hormones. Brain Res 765: 67-73.

Bergh AF, Strobel HW. 1992. Reconstitution of the brain mixed function oxidase system: purification of NADPH-cytochrome P450 reductase and partial purification of cytochrome P450 from whole rat brain. J Neurochem 59: 575-581.

Bergh AF, Strobel HW. 1996. Anatomical distribution of NADPH-cytochrome P450 reductase and cytochrome P4502D forms in rat brain: effects of xenobiotics and sex steroids. Mol Cell Biochem 162: 31-41.

Bhagwat SV, Boyd MR, Ravindranath V. 1995a. Brain mitochondrial cytochromes P450: xenobiotic metabolism, presence of multiple forms and their selective inducibility. Arch Biochem Biophys 320: 73-83.

Bhagwat SV, Boyd MR, Ravindranath V. 1995b. Rat brain cytochrome P450. Reassessment of monooxygenase activities and cytochrome P450 levels. Drug Metab Dispos 23: 651-654.

Bhamre S, Anandatheerathavarada HK, Shankar SK, Boyd MR, Ravindranath V. 1993. Purification of multiple forms of cytochrome P450 from a human brain and reconstitution of catalytic activities. Arch Biochem Biophys 301: 251-255.

Bishai I, Coceani F. 1992. Eicosanoid formation in the rat cerebral cortex. Contribution of neurons and glia. Mol Chem Neuropathol 17: 219-238.

Bjorkhem I, Diczfalusy U, Lutjohann D. 1999. Removal of cholesterol from extrahepatic sources by oxidative mechanisms. Curr Opin Lipidol 10: 161-165.

Blume N, Leonard J, Xu ZJ, Watanabe O, Remotti H, et al. 2000. Characterization of *Cyp2d22*, a novel cytochrome P450 expressed in mouse mammary cells. Arch Biochem Biophys 381: 191-204.

Bodin K, Andersson U, Rystedt E, Ellis E, Norlin M, et al. 2002. Metabolism of 4 beta -hydroxycholesterol in humans. J Biol Chem 277: 31534-31540.

Bogdanovic N, Bretillon L, Lund EG, Diczfalusy U, Lannfelt L, et al. 2001. On the turnover of brain cholesterol in patients with Alzheimer's disease. Abnormal induction of the cholesterol-catabolic enzyme *CYP46* in glial cells. Neurosci Lett 314: 45-48.

Boutelet-Bochan H, Huang Y, Juchau MR. 1997. Expression of *CYP2E1* during embryogenesis and fetogenesis in human cephalic tissues: implications for the fetal alcohol syndrome. Biochem Biophys Res Commun 238: 443-447.

Brosen K, Skjelbo E. 1991. Fluoxetine and norfluoxetine are potent inhibitors of P450IID6–the source of the sparteine/debrisoquine oxidation polymorphism. Br J Clin Pharmacol 32: 136-137.

Brosen K, Hansen JG, Nielsen KK, Sindrup SH, Gram LF. 1993. Inhibition by paroxetine of desipramine metabolism in extensive but not in poor metabolizers of sparteine. Eur J Clin Pharmacol 44: 349-355.

Brown RC, Cascio C, Papadopoulos V. 2000. Pathways of neurosteroid biosynthesis in cell lines from human brain: regulation of dehydroepiandrosterone formation by oxidative stress and beta-amyloid peptide. J Neurochem 74: 847-859.

Brzezinski MR, Boutelet-Bochan H, Person RE, Fantel AG, Juchau MR. 1999. Catalytic activity and quantitation of cytochrome P-450 2E1 in prenatal human brain. J Pharmacol Exp Ther 289: 1648-1653.

Burton PM, Swinney DC, Heller R, Dunlap B, Chiou M, et al. 1995. Azalanstat (RS-21607), a lanosterol 14 alpha-demethylase inhibitor with cholesterol-lowering activity. Biochem Pharmacol 50: 529-544.

Bylund J, Kunz T, Valmsen K, Oliw EH. 1998. Cytochromes P450 with bisallylic hydroxylation activity on arachidonic and linoleic acids studied with human recombinant enzymes and with human and rat liver microsomes. J Pharmacol Exp Ther 284: 51-60.

Cali JJ, Hsieh CL, Francke U, Russell DW. 1991. Mutations in the bile acid biosynthetic enzyme sterol 27-hydroxylase underlie cerebrotendinous xanthomatosis. J Biol Chem 266: 7779-7783.

Carboni L, Piubelli C, Righetti PG, Jansson B, Domenici E. 2002. Proteomic analysis of rat brain tissue: comparison of protocols for two-dimensional gel electrophoresis analysis based on different solubilizing agents. Electrophoresis 23: 4132-4141.

Chan TS, Moridani M, Siraki A, Scobie H, Beard K, et al. 2001. Hydrogen peroxide supports hepatocyte P450 catalysed xenobiotic/drug metabolic activation to form cytotoxic reactive intermediates. Adv Exp Med Biol 500: 233-236.

Chang GW, Kam PC. 1999. The physiological and pharmacological roles of cytochrome P450 isoenzymes. Anaesthesia 54: 42-50.

Chen H, Fantel AG, Juchau MR. 2000. Catalysis of the 4-hydroxylation of retinoic acids by *CYP3A7* in human fetal hepatic tissues. Drug Metab Dispos 28: 1051-1057.

Chevalier D, Lo-Guidice JM, Sergent E, Allorge D, Debuysere H, et al. 2001. Identification of genetic variants in the human thromboxane synthase gene (*CYP5A1*). Mutat Res 432: 61-67.

Chiang EF, Yan YL, Guiguen Y, Postlethwait J, Chung B. 2001. Two *Cyp19* (P450 aromatase) genes on duplicated zebrafish chromosomes are expressed in ovary or brain. Mol Biol Evol 18: 542-550.

Chiang JYL. 1998. Regulation of bile acid synthesis. Front Biosci 3: D176-193.

Chinta SJ, Pai HV, Ravindranath V. 2005b. Presence of splice variant forms of cytochrome P4502D1 in rat brain but not in liver. Brain Res Mol Brain Res 135: 81-92.

Chinta SJ, Kommaddi RP, Turman CM, Strobel HW, Ravindranath V. 2005a. Constitutive expression and localization of cytochrome P-450 1A1 in rat and human brain: presence of a splice variant form in human brain. J Neurochem 93: 724-736.

Chinta SJ, Pai HV, Upadhya SC, Boyd MR, Ravindranath V. 2002. Constitutive expression and localization of the major drug metabolizing enzyme, cytochrome P4502D in human brain. Brain Res Mol Brain Res 103: 49-61.

Choudhary D, Jansson I, Schenkman JB, Sarfarazi M, Stoilov I. 2003. Comparative expression profiling of 40 mouse cytochrome P450 genes in embryonic and adult tissues. Arch Biochem Biophys 414: 91-100.

Christmas P, Ursino SR, Fox JW, Soberman RJ. 1999. Expression of the *CYP4F3* gene. tissue-specific splicing and alternative promoters generate high and low K(m) forms of leukotriene B(4) omega-hydroxylase. J Biol Chem 274: 21191-21199.

Christmas P, Jones JP, Patten CJ, Rock DA, Zheng Y, et al. 2001. Alternative splicing determines the function of *CYP4F3* by switching substrate specificity. J Biol Chem 276: 38166-38172.

Chuang SS, Helvig C, Taimi M, Ramshaw HA, Collop AH, et al. 2004. *CYP2U1*, a novel human thymus- and brain-specific cytochrome P450, catalyzes omega- and (omega-1)-hydroxylation of fatty acids. J Biol Chem 279: 6305-6314.

Cloutier M, Fleury A, Courtemanche J, Ducharme L, Mason JI, et al. 1997. Characterization of the adrenal cytochrome P450C17 in the hamster, a small animal model for the study of adrenal dehydroepiandrosterone biosynthesis. DNA Cell Biol 16: 357-368.

Coleman T, Ellis SW, Martin IJ, Lennard MS, Tucker GT. 1996. 1-Methyl-4-phenyl-1,2,3,6-tetrahydropyridine (MPTP) is N-demethylated by cytochromes P450 2D6, 1A2 and 3A4-implications for susceptibility to Parkinson's disease. J Pharmacol Exp Ther 277: 685-690.

Combet C, Jambon M, Deleage G, Geourjon C. 2002. Geno3D: automatic comparative molecular modelling of protein. Bioinformatics 18: 213-214.

Compagnone NA, Bulfone A, Rubenstein JL, Mellon SH. 1995. Steroidogenic enzyme P450c17 is expressed in the embryonic central nervous system. Endocrinology 136: 5212-5223.

Corchero J, Granvil CP, Akiyama TE, Hayhurst GP, Pimprale S, et al. 2001. The *CYP2D6* humanized mouse: effect of the human *CYP2D6* transgene and HNF4alpha on the disposition of debrisoquine in the mouse. Mol Pharmacol 60: 1260-1267.

Costa E, Auta J, Guidotti A, Korneyev A, Romeo E. 1994. The pharmacology of neurosteroidogenesis. J Steroid Biochem Mol Biol 49: 385-389.

Danan JL, Delorme AC, Mathieu H. 1982. Presence of 25-hydroxyvitamin D3 and 1,25-dihydroxyvitamin D3 24-hydroxylase in vitamin D target cells of rat yolk sac. J Biol Chem 257: 10715-10721.

Dayer P, Desmeules J, Leemann T, Striberni R. 1988. Bioactivation of the narcotic drug codeine in human liver is mediated by the polymorphic monooxygenase catalyzing debrisoquine 4-hydroxylation (cytochrome P-450 dbl/bufI). Biochem Biophys Res Commun 152: 411-416.

de Groot MJ, Ackland MJ, Horne VA, Alex AA, Jones BC. 1999. Novel approach to predicting P450-mediated drug metabolism: development of a combined protein and pharmacophore model for *CYP2D6*. J Med Chem 42: 1515-1524.

de Groot MJ, Bijloo GJ, Martens BJ, van Acker FA, Vermeulen NP. 1997. A refined substrate model for human cytochrome P450 2D6. Chem Res Toxicol 10: 41-48.

Degtyarenko KN, Fabian P. 1996. The directory of P450-containing systems on WorldWide Web. Comput Appl Biosci 12: 237-240.

Desaulniers D, Xiao GH, Leingartner K, Chu I, Musicki B, et al. 2005. Comparisons of brain, uterus, and liver mRNA expression for cytochrome P450s, DNA methyltransferase-1, and catechol-o-methyltransferase in prepubertal female

Sprague-Dawley rats exposed to a mixture of aryl hydrocarbon receptor agonists. Toxicol Sci 86: 175-184.

Desmeules J, Gascon MP, Dayer P, Magistris M. 1991. Impact of environmental and genetic factors on codeine analgesia. Eur J Clin Pharmacol 41: 23-26.

Diesel B, Radermacher J, Bureik M, Bernhardt R, Seifert M, et al. 2005. Vitamin D(3) metabolism in human glioblastoma multiforme: functionality of *CYP27B1* splice variants, metabolism of calcidiol, and effect of calcitriol. Clin Cancer Res 11: 5370-5380.

Diesel B, Seifert M, Radermacher J, Fischer U, Tilgen W, et al. 2004. Towards a complete picture of splice variants of the gene for 25-hydroxyvitamin D31alpha-hydroxylase in brain and skin cancer. J Steroid Biochem Mol Biol 89–90: 527-532.

Ding S, Lake BG, Friedberg T, Wolf CR. 1995. Expression and alternative splicing of the cytochrome P-450 *CYP2A7*. Biochem J 306(Pt 1): 161-166.

Doody KJ, Carr BR. 1989. Aromatase in human fetal tissues. Am J Obstet Gynecol 161: 1694-1697.

Dunham B, Shepro D, Hechtman HB. 1984. Leukotriene induction of TxB2 in cultured bovine aortic endothelial cells. Inflammation 8: 313-321.

Easton N, Fry J, O'Shea E, Watkins A, Kingston S, et al. 2003. Synthesis, in vitro formation, and behavioural effects of glutathione regioisomers of alpha-methyldopamine with relevance to MDA and MDMA (ecstasy). Brain Res 987: 144-154.

Eggertsen G, Olin M, Andersson U, Ishida H, Kubota S, et al. 1996. Molecular cloning and expression of rabbit sterol 12alpha-hydroxylase. J Biol Chem 271: 32269-32275.

Ekins S, Bravi G, Binkley S, Gillespie JS, Ring BJ, et al. 1999. Three and four dimensional-quantitative structure activity relationship (3D/4D-QSAR) analyses of *CYP2D6* inhibitors. Pharmacogenetics 9: 477-489.

Emoto Y, Wada H, Okamoto H, Kudo A, Imai Y. 2005. Retinoic acid-metabolizing enzyme *Cyp26a1* is essential for determining territories of hindbrain and spinal cord in zebrafish. Dev Biol 278: 415-427.

Erlandsen H, Abola EE, Stevens RC. 2000. Combining structural genomics and enzymology: completing the picture in metabolic pathways and enzyme active sites. Curr Opin Struct Biol 10: 719-730.

Eyles DW, Smith S, Kinobe R, Hewison M, McGrath JJ. 2005. Distribution of the vitamin D receptor and 1 alpha-hydroxylase in human brain. J Chem Neuroanat 29: 21-30.

Farin FM, Omiecinski CJ. 1993. Regiospecific expression of cytochrome P-450s and microsomal epoxide hydrolase in human brain tissue. J Toxicol Environ Health 40: 317-335.

Fernstrom JD, Fernstrom MH. 2001. Diet, monoamine neurotransmitters and appetite control. Nestle Nutr Workshop Ser Clin Perform Programme 5: 117-131.

Fonne-Pfister R, Bargetzi MJ, Meyer UA. 1987. MPTP, the neurotoxin inducing Parkinson's disease, is a potent competitive inhibitor of human and rat cytochrome P450 isozymes (P450bufI, P450db1) catalyzing debrisoquine 4-hydroxylation. Biochem Biophys Res Commun 148: 1144-1150.

Fujii H, Sato T, Kaneko S, Gotoh O, Fujii-Kuriyama Y, et al. 1997. Metabolic inactivation of retinoic acid by a novel P450 differentially expressed in developing mouse embryos. EMBO J 16: 4163-4173.

Furster C, Wikvall K. 1999. Identification of *CYP3A4* as the major enzyme responsible for 25-hydroxylation of 5beta-cholestane-3alpha,7alpha,12alpha-triol in human liver microsomes. Biochim Biophys Acta 1437: 46-52.

Gaedigk A, Gaedigk R, Leeder JS. 2005. *CYP2D7* splice variants in human liver and brain: does *CYP2D7* encode functional protein? Biochem Biophys Res Commun 336: 1241-1250.

Gafvels M, Olin M, Chowdhary BP, Raudsepp T, Andersson U. et al. 1999. Structure and chromosomal assignment of the sterol 12alpha-hydroxylase gene (*CYP8B1*) in human and mouse: eukaryotic cytochrome P-450 gene devoid of introns. Genomics 56. 184-196.

Galijatovic A, Beaton D, Nguyen N, Chen S, Bonzo J, et al. 2004. The human *CYP1A1* gene is regulated in a developmental and tissue-specific fashion in transgenic mice. J Biol Chem 279: 23969-23976.

Garcia-Ovejero D, Azcoitia I, Doncarlos LL, Melcangi RC, Garcia-Segura LM. 2005. Glia-neuron crosstalk in the neuroprotective mechanisms of sex steroid hormones. Brain Res Brain Res Rev 48: 273-286.

Gebremedhin D, Lange AR, Lowry TF, Taheri MR, Birks EK, et al. 2000. Production of 20-HETE and its role in autoregulation of cerebral blood flow. Circ Res 87: 60-65.

Gervasini G, Martinez C, Agundez JA, Garcia-Gamito FJ, Benitez J. 2001. Inhibition of cytochrome P450 2C9 activity in vitro by 5-hydroxytryptamine and adrenaline. Pharmacogenetics 11: 29-37.

Gervot L, Rochat B, Gautier JC, Bohnenstengel F, Kroemer H, et al. 1999. Human *CYP2B6*: expression, inducibility and catalytic activities. Pharmacogenetics 9: 295-306.

Ghersi-Egea JF, Minn A, Daval JL, Jayyosi Z, Arnould V, et al. 1989. NADPH:cytochrome P-450(c) reductase: biochemical characterization in rat brain and cultured neurons and evolution of activity during development. Neurochem Res 14: 883-887.

Ghersi-Egea JF, Perrin R, Leininger-Muller B, Grassiot MC, Jeandel C, et al. 1993. Subcellular localization of cytochrome P450, and activities of several enzymes responsible for drug metabolism in the human brain. Biochem Pharmacol 45: 647-658.

Gilham DE, Cairns W, Paine MJ, Modi S, Poulsom R, et al. 1997. Metabolism of MPTP by cytochrome P4502D6 and the demonstration of 2D6 mRNA in human foetal and adult brain by in situ hybridization. Xenobiotica 27: 111-125.

Golanska E, Hulas-Bigoszewska K, Wojcik I, Rieske P, Styczynska M, et al. 2005. *CYP46*: a risk factor for Alzheimer's disease or a coincidence? Neurosci Lett 383: 105-108.

Golovine K, Schwerin M, Vanselow J. 2003. Three different promoters control expression of the aromatase cytochrome p450 gene (*cyp19*) in mouse gonads and brain. Biol Reprod 68: 978-984.

Gomez-Sanchez CE, Zhou MY, Cozza EN, Morita H, Foecking MF, et al. 1997. Aldosterone biosynthesis in the rat brain. Endocrinology 138: 3369-3373.

Granberg L, Ostergren A, Brandt I, Brittebo EB. 2003. *CYP1A1* and *CYP1B1* in blood-brain interfaces: *CYP1A1*-dependent bioactivation of 7,12-dimethylbenz(a)anthracene in endothelial cells. Drug Metab Dispos 31: 259-265.

Guex N, Peitsch MC. 1997. SWISS-MODEL and the Swiss-PdbViewer: an environment for comparative protein modeling. Electrophoresis 18: 2714-2723.

Guyant-Marechal L, Verrips A, Girard C, Wevers RA, Zijlstra F, et al. 2005. Unusual cerebrotendinous xanthomatosis with fronto-temporal dementia phenotype. Am J Med Genet A 139: 114-117.

Haglund L, Kohler C, Haaparanta T, Goldstein M, Gustafsson JA. 1984. Presence of NADPH-cytochrome P450 reductase in central catecholaminergic neurones. Nature 307: 259-262.

Haining R, Yu A. 2003. Cytochrome P450 Pharmacogenetics. Cytochrome p450 and Drug Metabolism, Lee J, Editor. FontisMedia; pp. 343-387.

Hammer F, Compagnone NA, Vigne JL, Bair SR, Mellon SH. 2004. Transcriptional regulation of P450scc gene expression in the embryonic rodent nervous system. Endocrinology 145: 901-912.

Handschin C, Podvinec M, Amherd R, Looser R, Ourlin JC, et al. 2002. Cholesterol and bile acids regulate xenosensor signaling in drug-mediated induction of cytochromes P450. J Biol Chem 277: 29561-29567.

Hashizume T, Imaoka S, Mise M, Terauchi Y, Fujii T, et al. 2002. Involvement of *CYP2J2* and *CYP4F12* in the metabolism of ebastine in human intestinal microsomes. J Pharmacol Exp Ther 300: 298-304.

Hauptmann N, Grimsby J, Shih JC, Cadenas E. 1996. The metabolism of tyramine by monoamine oxidase A/B causes oxidative damage to mitochondrial DNA. Arch Biochem Biophys 335: 295-304.

Hiroi T, Imaoka S, Funae Y. 1998. Dopamine formation from tyramine by *CYP2D6*. Biochem Biophys Res Commun 249: 838-843.

Hiroi T, Kishimoto W, Chow T, Imaoka S, Igarashi T, et al. 2001. Progesterone oxidation by cytochrome P450 2D isoforms in the brain. Endocrinology 142: 3901-3908.

Howard LA, Miksys S, Hoffmann E, Mash D, Tyndale RF. 2003. Brain *CYP2E1* is induced by nicotine and ethanol in rat and is higher in smokers and alcoholics. Br J Pharmacol 138: 1376-1386.

Huang P, Rannug A, Ahlbom E, Hakansson H, Ceccatelli S. 2000. Effect of 2,3,7,8-tetrachlorodibenzo-p-dioxin on the expression of cytochrome P450 1A1, the aryl hydrocarbon receptor, and the aryl hydrocarbon receptor nuclear translocator in rat brain and pituitary. Toxicol Appl Pharmacol 169: 159-167.

Huang Z, Fasco MJ, Kaminsky LS. 1997. Alternative splicing of *CYP2D* mRNA in human breast tissue. Arch Biochem Biophys 343: 101-108.

Hwang DY, Chae KR, Shin DH, Hwang JH, Lim CH, et al. 2001. Xenobiotic response in humanized double transgenic mice expressing tetracycline-controlled transactivator and human *CYP1B1*. Arch Biochem Biophys 395: 32-40.

Iivonen S, Corder E, Lehtovirta M, Helisalmi S, Mannermaa A, et al. 2004. Polymorphisms in the CYP19 gene confer increased risk for Alzheimer disease. Neurology 62: 1170-1176.

Ikeda H, Ueda M, Ikeda M, Kobayashi H, Honda Y. 2003. Oxysterol 7alpha-hydroxylase (*CYP39A1*) in the ciliary nonpigmented epithelium of bovine eye. Lab Invest 83: 349-355.

Imaoka S, Ogawa H, Kimura S, Gonzalez FJ. 1993. Complete cDNA sequence and cDNA-directed expression of *CYP4A11*, a fatty acid omega-hydroxylase expressed in human kidney. DNA Cell Biol 12: 893-899.

Ishida H, Kuruta Y, Gotoh O, Yamashita C, Yoshida Y, et al. 1999. Structure, evolution, and liver-specific expression of sterol 12alpha-hydroxylase P450 (*CYP8B*). J Biochem (Tokyo) 126: 19-25.

Ishunina TA, van Beurden D, van der Meulen G, Unmehopa UA, Hol EM, et al. 2005. Diminished aromatase immunoreactivity in the hypothalamus, but not in the basal forebrain nuclei in Alzheimer's disease. Neurobiol Aging 26: 173-194.

Iwahashi K, Kawai Y, Suwaki H, Hosokawa K, Ichikawa Y. 1993. A localization study of the cytochrome P-450(21)-linked monooxygenase system in adult rat brain. J Steroid Biochem Mol Biol 44: 163-169.

Jansson I, Stoilov I, Sarfarazi M, Schenkman JB. 2001. Effect of two mutations of human *CYP1B1*, *G61E* and *R469W*, on stability and endogenous steroid substrate metabolism. Pharmacogenetics 11: 793-801.

Jenner P. 1998. Oxidative mechanisms in nigral cell death in Parkinson's disease. Mov Disord 13(Suppl 1): 24-34.

Jimbo M, Okubo K, Toma Y, Shimizu Y, Saito H, et al. 1998. Inhibitory effects of catecholamines and maternal stress on aromatase activity in the fetal rat brain. J Obstet Gynaecol Res 24: 291-297.

Johri A, Dhawan A, Lakhan Singh R, Parmar D. 2006. Effect of prenatal exposure of deltamethrin on the ontogeny of xenobiotic metabolizing cytochrome P450s in the brain and liver of offsprings. Toxicol Appl Pharmacol 214: 279-289.

Joshi M, Tyndale RF. 2006. Regional and cellular distribution of *CYP2E1* in monkey brain and its induction by chronic nicotine. Neuropharmacology 50: 568-575.

Joshi M, Tyndale RF. 2006. Induction and Recovery Time Course of Rat Brain *CYP2E1* following Nicotine treatment. Drug Metab Dispos 34: 647-652.

Juhasz A, Rimanoczy A, Boda K, Vincze G, Szlavik G, et al. 2005. *CYP46* T/C polymorphism is not associated with Alzheimer's dementia in a population from Hungary. Neurochem Res 30: 943-948.

Jumpsen J, Lien EL, Goh YK, Clandinin MT. 1997. Small changes of dietary (n-6) and (n-3)/fatty acid content ration alter phosphatidylethanolamine and phosphatidylcholine fatty acid composition during development of neuronal and glial cells in rats. J Nutr 127: 724-731.

Kamat A, Hinshelwood MM, Murry BA, Mendelson CR. 2002. Mechanisms in tissue-specific regulation of estrogen biosynthesis in humans. Trends Endocrinol Metab 13: 122-128.

Karlgren M, Backlund M, Johansson I, Oscarson M, Ingelman-Sundberg M. 2004. Characterization and tissue distribution of a novel human cytochrome P450-CYP2U1. Biochem Biophys Res Commun 315: 679-685.

Kawashima H, Strobel HW. 1995. cDNA cloning of three new forms of rat brain cytochrome P450 belonging to the *CYP4F* subfamily. Biochem Biophys Res Commun 217: 1137-1144.

Kawashima H, Naganuma T, Kusunose E, Kono T, Yasumoto R, et al. 2000. Human fatty acid omega-hydroxylase, *CYP4A11*: determination of complete genomic sequence and characterization of purified recombinant protein. Arch Biochem Biophys 378: 333-339.

Kawato S, Hojo Y, Kimoto T. 2002. Histological and metabolism analysis of P450 expression in the brain. Methods Enzymol 357: 241-249.

Kehl F, Cambj-Sapunar L, Maier KG, Miyata N, Kametani S, et al. 2002. 20-HETE contributes to the acute fall in cerebral blood flow after subarachnoid hemorrhage in the rat. Am J Physiol Heart Circ Physiol 282: H1556-1565.

Kelly SL, Lamb DC, Cannieux M, Greetham D, Jackson CJ, et al. 2001. An old activity in the cytochrome P450 superfamily (*CYP51*) and a new story of drugs and resistance. Biochem Soc Trans 29: 122-128.

Kemp MC, Kuonen DR, Sutton A, Roberts PJ. 1988. Rat brain NADPH-dependent diaphorase. A possible relationship to cytochrome P450 reductase. Biochem Pharmacol 37: 3063-3070.

Kikuta Y, Kusunose E, Kusunose M. 2002. Prostaglandin and leukotriene omega-hydroxylases. Prostaglandins Other Lipid Mediat 68–69: 345-362.

Kikuta Y, Kato M, Yamashita Y, Miyauchi Y, Tanaka K, et al. 1998. Human leukotriene B4 omega-hydroxylase (*CYP4F3*) gene: molecular cloning and chromosomal localization. DNA Cell Biol 17: 221-230.

Kimoto T, Tsurugizawa T, Ohta Y, Makino J, Tamura H, et al. 2001. Neurosteroid synthesis by cytochrome p450-containing systems localized in the rat brain hippocampal neurons: N-methyl-D-aspartate and calcium-dependent synthesis. Endocrinology 142: 3578-3589.

Kimura S, Umeno M, Skoda RC, Meyer UA, Gonzalez FJ. 1989. The human debrisoquine 4-hydroxylase (*CYP2D*) locus: sequence and identification of the polymorphic *CYP2D6* gene, a related gene, and a pseudogene. Am J Hum Genet 45: 889-904.

King SR, Manna PR, Ishii T, Syapin PJ, Ginsberg SD, et al. 2002. An essential component in steroid synthesis, the steroidogenic acute regulatory protein, is expressed in discrete regions of the brain. J Neurosci 22: 10613-10620.

Kirches E, Scherlach C, von Bossanyi P, Schneider T, Szibor R, et al. 1999. MGMT- and P450 3A-inhibitors do not sensitize glioblastoma cell cultures against nitrosoureas. Clin Neuropathol 18: 1-8.

Kishimoto W, Hiroi T, Shiraishi M, Osada M, Imaoka S, et al. 2004. Cytochrome P450 2D catalyze steroid 21-hydroxylation in the brain. Endocrinology 145: 699-705.

Klose TS, Blaisdell JA, Goldstein JA. 1999. Gene structure of CYP2C8 and extrahepatic distribution of the human *CYP2Cs*. J Biochem Mol Toxicol 13: 289-295.

Knoll J, Yoneda F, Knoll B, Ohde H, Miklya I. 1999. (-)1-(Benzofuran-2-yl)-2-propylaminopentane, [(-)BPAP], a selective enhancer of the impulse propagation mediated release of catecholamines and serotonin in the brain. Br J Pharmacol 128: 1723-1732.

Knupfer H, Knupfer MM, Hotfilder M, Preiss R. 1999. P450-expression in brain tumors. Oncol Res 11: 523-528.

Koletzko B, Decsi T, Demmelmair H. 1996. Arachidonic acid supply and metabolism in human infants born at full term. Lipids 31: 79-83.

Korkalainen M, Linden J, Tuomisto J, Pohjanvirta R. 2005. Effect of TCDD on mRNA expression of genes encoding bHLH/PAS proteins in rat hypothalamus. Toxicology 208: 1-11.

Kornhuber J, Weigmann H, Rohrich J, Wiltfang J, Bleich S, et al. 2006. Region specific distribution of levome promazine in the human brain. J Neural Transm 113: 387-397.

Korzekwa KR, Krishnamachary N, Shou M, Ogai A, Parise RA, et al. 1998. Evaluation of atypical cytochrome P450 kinetics with two-substrate models: evidence that multiple substrates can simultaneously bind to cytochrome P450 active sites. Biochemistry 37: 4137-4147.

Krainev AG, Weiner LM, Kondrashin SK, Kanaeva IP, Bachmanova GI. 1991. Substrate access channel geometry of soluble and membrane-bound cytochromes P450 as studied by interactions with type II substrate analogues. Arch Biochem Biophys 288: 17-21.

Lasker JM, Chen WB, Wolf I, Bloswick BP, Wilson PD, et al. 2000. Formation of 20-hydroxyeicosatetraenoic acid, a vasoactive and natriuretic eicosanoid, in human kidney. Role of *Cyp4F2* and *Cyp4A11*. J Biol Chem 275: 4118-4126.

Le Goascogne C, Sananes N, Gouezou M, Takemori S, Kominami S, et al. 1991. Immunoreactive cytochrome P-450(17 alpha) in rat and guinea-pig gonads, adrenal glands and brain. J Reprod Fertil 93: 609-622.

Lee AM, Miksys S, Palmour R, Tyndale RF. 2006. *CYP2B6* is expressed in African Green monkey brain and is induced by chronic nicotine treatment. Neuropharmacology 50: 441-450.

Lewis DF, Eddershaw PJ, Goldfarb PS, Tarbit MH. 1997. Molecular modelling of cytochrome P4502D6 (*CYP2D6*) based on an alignment with *CYP102*: structural studies on specific *CYP2D6* substrate metabolism. Xenobiotica 27: 319-339.

Li X, Warri A, Makela S, Ahonen T, Streng T, et al. 2002. Mammary gland development in transgenic male mice expressing human P450 aromatase. Endocrinology 143: 4074-4083.

Li Y, Yokoi T, Kitamura R, Sasaki M, Gunji M, et al. 1996. Establishment of transgenic mice carrying human fetus-specific *CYP3A7*. Arch Biochem Biophys 329: 235-240.

Li-Hawkins J, Lund EG, Bronson AD, Russell DW. 2000. Expression cloning of an oxysterol 7alpha-hydroxylase selective for 24-hydroxycholesterol. J Biol Chem 275: 16543-16549.

Llerena A, Edman G, Cobaleda J, Ben'itez J, Schalling D, et al. 1993. Relationship between personality and debrisoquine hydroxylation capacity. Suggestion of an endogenous neuroactive substrate or product of the cytochrome P4502D6. Acta Psychiatrica Scandinavica 87: 23-28.

Loeper J, Le Berre A, Pompon D. 1998. Topology inversion of *CYP2D6* in the endoplasmic reticulum is not required for plasma membrane transport. Mol Pharmacol 53: 408-414.

Lorincz MT, Rainier S, Thomas D, Fink JK. 2005. Cerebrotendinous xanthomatosis: possible higher prevalence than previously recognized. Arch Neurol 62: 1459-1463.

Lund EG, Guileyardo JM, Russell DW. 1999. cDNA cloning of cholesterol 24-hydroxylase, a mediator of cholesterol homeostasis in the brain. Proc Natl Acad Sci USA 96: 7238-7243.

Lund EG, Xie C, Kotti T, Turley SD, Dietschy JM, et al. 2003. Knockout of the cholesterol 24-hydroxylase gene in mice reveals a brain-specific mechanism of cholesterol turnover. J Biol Chem 278: 22980-22988.

Luo G, Zeldin DC, Blaisdell JA, Hodgson E, Goldstein JA. 1998. Cloning and expression of murine *CYP2Cs* and their ability to metabolize arachidonic acid. Arch Biochem Biophys 357: 45-57.

Ma X, Idle JR, Krausz KW, Gonzalez FJ. 2005. Metabolism of melatonin by human cytochromes P450. Drug Metab Dispos 33: 489-494.

Maas RM, Reus K, Diesel B, Steudel WI, Feiden W, et al. 2001. Amplification and expression of splice variants of the gene encoding the P450 cytochrome 25-hydroxyvitamin D(3) 1, alpha-hydroxylase (*CYP 27B1*) in human malignant glioma. Clin Cancer Res 7: 868-875.

MacKenzie SM, Clark CJ, Fraser R, Gomez-Sanchez CE, Connell JM, et al. 2000. Expression of 11beta-hydroxylase and aldosterone synthase genes in the rat brain. J Mol Endocrinol 24: 321-328.

Martin C, Bean R, Rose K, Habib F, Seckl J. 2001. *cyp7b1* catalyses the 7alpha-hydroxylation of dehydroepiandrosterone and 25-hydroxycholesterol in rat prostate. Biochem J 355: 509-515.

Martinez C, Agundez JA, Gervasini G, Martin R, Benitez J. 1997. Tryptamine: a possible endogenous substrate for *CYP2D6*. Pharmacogenetics 7: 85-93.

Martinez C, Gervasini G, Agundez JA, Carrillo JA, Ramos SI, et al. 2000. Modulation of midazolam 1-hydroxylation activity in vitro by neurotransmitters and precursors. Eur J Clin Pharmacol 56: 145-151.

Mast N, Norcross R, Andersson U, Shou M, Nakayama K, et al. 2003. Broad substrate specificity of human cytochrome P450 46A1 which initiates cholesterol degradation in the brain. Biochemistry 42: 14284-14292.

Masubuchi Y, Hosokawa S, Horie T, Suzuki T, Ohmori S, et al. 1994. Cytochrome P450 isozymes involved in propranolol metabolism in human liver microsomes. The role of CYP2D6 as ring-hydroxylase and *CYP1A2* as N-desisopropylase. Drug Metab Dispos 22: 909-915.

Matkovits T, Christakos S. 1995. Variable in vivo regulation of rat vitamin D-dependent genes (osteopontin, Ca, Mg-adenosine triphosphatase, and 25-hydroxyvitamin D3 24-hydroxylase): implications for differing mechanisms of regulation and involvement of multiple factors. Endocrinology 136: 3971-3982.

Matsumoto S, Hirama T, Matsubara T, Nagata K, Yamazoe Y. 2002. Involvement of *CYP2J2* on the intestinal first-pass metabolism of antihistamine drug, astemizole. Drug Metab Dispos 30: 1240-1245.

Matsunaga M, Ukena K, Tsutsui K. 2001. Expression and localization of cytochrome P450 17 alpha-hydroxylase/c17,20-lyase in the avian brain. Brain Res 899: 112-122.

McFadyen MC, Melvin WT, Murray GI. 1998. Regional distribution of individual forms of cytochrome P450 mRNA in normal adult human brain. Biochem Pharmacol 55: 825-830.

McFadyen MC, Murray GI, Melvin WT. 1999. Cytochrome P450 *CYP 1B1* mRNA in normal human brain. Mol Pathol 52: 164

Mehl M, Bidmon HJ, Hilbig H, Zilles K, Dringen R, et al. 1999. Prostacyclin synthase is localized in rat, bovine and human neuronal brain cells. Neurosci Lett 271: 187-190.

Mellon SH, Deschepper CF. 1993. Neurosteroid biosynthesis: genes for adrenal steroidogenic enzymes are expressed in the brain. Brain Res 629: 283-292.

Meyer RP, Podvinec M, Meyer UA. 2002. Cytochrome P450 *CYP1A1* accumulates in the cytosol of kidney and brain and is activated by heme. Mol Pharmacol 62: 1061-1067.

Miki H, Takeuchi H, Yamada A, Nishioka M, Matsuzawa Y, et al. 1986. Quantitative analysis of the mitochondrial cytochrome P-450-linked monooxygenase system: NADPH-hepatoredoxin reductase, hepatoredoxin, and cytochrome P-450s27 in livers of patients with cerebrotendinous xanthomatosis. Clin Chim Acta 160: 255-263.

Miksys S, Lerman C, Shields PG, Mash DC, Tyndale RF. 2003. Smoking, alcoholism and genetic polymorphisms alter *CYP2B6* levels in human brain. Neuropharmacology 45: 122-132.

Miksys S, Rao Y, Hoffmann E, Mash DC, Tyndale RF. 2002. Regional and cellular expression of *CYP2D6* in human brain: higher levels in alcoholics. J Neurochem 82: 1376-1387.

Miksys S, Rao Y, Sellers EM, Kwan M, Mendis D, et al. 2000. Regional and cellular distribution of *CYP2D* subfamily members in rat brain. Xenobiotica 30: 547-564.

Miksys SL, Cheung C, Gonzalez FJ, Tyndale RF. 2005. Human *CYP2D6* and mouse *CYP2DS*: organ distribution in a humanized mouse model. Drug Metab Dispos 33: 1495-1502.

Minn AL, Pelczar H, Denizot C, Martinet M, Heydel JM, et al. 2005. Characterization of microsomal cytochrome P450-dependent monooxygenases in the rat olfactory mucosa. Drug Metab Dispos 33: 1229-1237.

Miyata N, Seki T, Tanaka Y, Omura T, Taniguchi K, et al. 2005. Beneficial effects of a new 20-hydroxyeicosatetraenoic acid synthesis inhibitor, TS-011 [N-(3-chloro-4-morpholin-4-yl) phenyl-N′-hydroxyimido formamide], on hemorrhagic and ischemic stroke. J Pharmacol Exp Ther 314: 77-85.

Modi S, Gilham DE, Sutcliffe MJ, Lian LY, Primrose WU, et al. 1997. 1-methyl-4-phenyl-1,2,3,6-tetrahydropyridine as a substrate of cytochrome P450 *2D6*: allosteric effects of NADPH-cytochrome P450 reductase. Biochemistry 36: 4461-4470.

Monshouwer M, Agnello D, Ghezzi P, Villa P. 2000. Decrease in brain cytochrome P450 enzyme activities during infection and inflammation of the central nervous system. Neuroimmunomodulation 8: 142-147.

Montoliu C, Sancho-Tello M, Azorin I, Burgal M, Valles S, et al. 1995. Ethanol increases cytochrome P4502E1 and induces oxidative stress in astrocytes. J Neurochem 65: 2561-2570.

Morse DC, Stein AP, Thomas PE, Lowndes HE. 1998. Distribution and induction of cytochrome P450 1A1 and 1A2 in rat brain. Toxicol Appl Pharmacol 152: 232-239.

Murray GI, Pritchard S, Melvin WT, Burke MD. 1995. Cytochrome P450 *CYP3A5* in the human anterior pituitary gland. FEBS Lett 364: 79-82.

Murray GI, Taylor MC, McFadyen MC, McKay JA, Greenlee WF, et al. 1997. Tumor-specific expression of cytochrome P450 *CYP1B1*. Cancer Res 57: 3026-3031.

Muskhelishvili L, Thompson PA, Kusewitt DF, Wang C, Kadlubar FF. 2001. In situ hybridization and immunohistochemical analysis of cytochrome P450 1B1 expression in human normal tissues. J Histochem Cytochem 49: 229-236.

Naftolin F, Horvath TL, Jakab RL, Leranth C, Harada N, et al. 1996. Aromatase immunoreactivity in axon terminals of the vertebrate brain. An immunocytochemical study on quail, rat, monkey and human tissues. Neuroendocrinology 63: 149-155.

Nakayama K, Puchkaev A, Pikuleva IA. 2001. Membrane binding and substrate access merge in cytochrome P450 *7A1*, a key enzyme in degradation of cholesterol. J Biol Chem 276: 31459-31465.

Nebert DW, Russell DW. 2002. Clinical importance of the cytochromes P450. Lancet 360: 1155-1162.

Nelson DR. 1998. Cytochrome P450 nomenclature. Methods Mol Biol 107: 15-24.

Nicholson TE, Renton KW. 2001. Role of cytokines in the lipopolysaccharide-evoked depression of cytochrome P450 in the brain and liver. Biochem Pharmacol 62: 1709-1717.

Nishimura M, Yaguti H, Yoshitsugu H, Naito S, Satoh T. 2003. Tissue distribution of mRNA expression of human cytochrome P450 isoforms assessed by high-sensitivity real-time reverse transcription PCR. Yakugaku Zasshi 123: 369-375.

Niznik HB, Tyndale RF, Sallee FR, Gonzalez FJ, Hardwick JP, et al. 1990. The dopamine transporter and cytochrome P45OIID1 (debrisoquine 4-hydroxylase) in brain: resolution and identification of two distinct [3H]GBR-12935 binding proteins. Arch Biochem Biophys 276: 424-432.

Ohyama Y, Meaney S, Heverin M, Ekstrom L, Brafman A, et al. 2006. Studies on the transcriptional regulation of cholesterol 24-hydroxylase (*CYP46A1*): marked insensitivity toward different regulatory axes. J Biol Chem 281: 3810-3820.

O'Malley KL, Anhalt MJ, Martin BM, Kelsoe JR, Winfield SL, et al. 1987. Isolation and characterization of the human tyrosine hydroxylase gene: identification of 5′ alternative splice sites responsible for multiple mRNAs. Biochemistry 26: 6910-6914.

Omiecinski CJ, Remmel RP, Hosagrahara VP. 1999. Concise review of the cytochrome P450s and their roles in toxicology. Toxicol Sci 48: 151-156.

Oscarson M, McLellan RA, Asp V, Ledesma M, Ruiz ML, et al. 2002. Characterization of a novel *CYP2A7/CYP2A6* hybrid allele (*CYP2A6*12*) that causes reduced *CYP2A6* activity. Hum Mutat 20: 275-283.

Pai HV, Upadhya SC, Chinta SJ, Hegde SN, Ravindranath V. 2002. Differential metabolism of alprazolam by liver and brain cytochrome (P4503A) to pharmacologically active metabolite. Pharmacogenomics J 2: 243-258.

Pai HV, Kommaddi RP, Chinta SJ, Mori T, Boyd MR, et al. 2004. A frame-shift mutation and alternate splicing in human brain generates a functional form of the pseudogene, cytochrome P4502D7 that demethylates codeine to morphine. J Biol Chem 279: 27383-27389.

Pappolla MA, Omar RA, Chyan YJ, Ghiso J, Hsiao K, et al. 2001. Induction of NADPH cytochrome P450 reductase by the Alzheimer beta-protein. Amyloid as a "foreign body." J Neurochem 78: 121-128.

Parmar D, Dhawan A, Seth PK. 1998. Evidence for O-dealkylation of 7-pentoxyresorufin by cytochrome P450 *2B1/2B2* isoenzymes in brain. Mol Cell Biochem 189: 201-205.

Payne DW, Shackleton C, Toms H, Ben-Shlomo I, Kol S, et al. 1995. A novel nonhepatic hydroxycholesterol 7 alpha-hydroxylase that is markedly stimulated by interleukin-1 beta. Characterization in the immature rat ovary. J Biol Chem 270: 18888-18896.

Pezzi V, Mathis JM, Rainey WE, Carr BR. 2003. Profiling transcript levels for steroidogenic enzymes in fetal tissues. J Steroid Biochem Mol Biol 87: 181-189.

Pijnappel WW, Hendriks HF, Folkers GE, Brink van den CE, Dekker EJ, et al. 1993. The retinoid ligand 4-oxo-retinoic acid is a highly active modulator of positional specification. Nature 366: 340-344.

Pikuleva IA, Bjorkhem I, Waterman MR. 1997. Expression, purification, and enzymatic properties of recombinant human cytochrome P450c27 (*CYP27*). Arch Biochem Biophys 343: 123-130.

Powell PK, Wolf I, Jin R, Lasker JM. 1998. Metabolism of arachidonic acid to 20-hydroxy-5,8,11,14-eicosatetraenoic acid by P450 enzymes in human liver: involvement of *CYP4F2* and *CYP4A11*. J Pharmacol Exp Ther 285: 1327-1336.

Price T, Aitken J, Simpson ER. 1992. Relative expression of aromatase cytochrome P450 in human fetal tissues as determined by competitive polymerase chain reaction amplification. J Clin Endocrinol Metab 74: 879-883.

Qu W, Bradbury JA, Tsao CC, Maronpot R, Harry GJ, et al. 2001. Cytochrome P450 *CYP2J9*, a new mouse arachidonic acid omega-1 hydroxylase predominantly expressed in brain. J Biol Chem 276: 25467-25479.

Rampling R, Cruickshank G, Lewis AD, Fitzsimmons SA, Workman P. 1994. Direct measurement of pO2 distribution and bioreductive enzymes in human malignant brain tumors. Int J Radiat Oncol Biol Phys 29: 427-431.

Ravindranath V, Anandatheerthavarada HK, Shankar SK. 1990. NADPH cytochrome P-450 reductase in rat, mouse and human brain. Biochem Pharmacol 39: 1013-1018.

Ray WJ, Bain G, Yao M, Gottlieb DI. 1997. *CYP26*, a novel mammalian cytochrome P450, is induced by retinoic acid and defines a new family. J Biol Chem 272: 18702-18708.

Ren S, Nguyen L, Wu S, Encinas C, Adams JS, et al. 2005. Alternative splicing of vitamin D-24-hydroxylase: a novel mechanism for the regulation of extrarenal 1,25-dihydroxyvitamin D synthesis. J Biol Chem 280: 20604-20611.

Rieder CR, Ramsden DB, Williams AC. 1998. Cytochrome P450 1B1 mRNA in the human central nervous system. Mol Pathol 51: 138-142.

Rieder CR, Parsons RB, Fitch NJ, Williams AC, Ramsden DB. 2000. Human brain cytochrome P450 1B1: immunohistochemical localization in human temporal lobe and induction by dimethylbenz(a)anthracene in astrocytoma cell line (MOG-G-CCM). Neurosci Lett 278: 177-180.

Robin MA, Anandatheerthavarada HK, Fang JK, Cudic M, Otvos L, et al. 2001. Mitochondrial targeted cytochrome P450 2E1 (P450 MT5) contains an intact N terminus and requires mitochondrial specific electron transfer proteins for activity. J Biol Chem 276: 24680-24689.

Rogan PK, Svojanovsky S, Leeder JS. 2003. Information theory-based analysis of *CYP2C19*, *CYP2D6* and *CYP3A5* splicing mutations. Pharmacogenetics 13: 207-218.

Rose KA, Stapleton G, Dott K, Kieny MP, Best R, et al. 1997. Cyp7b, a novel brain cytochrome P450, catalyzes the synthesis of neurosteroids 7alpha-hydroxydehydroepiandrosterone and 7alpha-hydroxypregnenolone. Proc Natl Acad Sci USA 94: 4925-4930.

Roselli CE, Resko JA. 1997. Sex differences in androgen-regulated expression of cytochrome P450 aromatase in the rat brain. J Steroid Biochem Mol Biol 61: 365-374.

Roselli CE, Resko JA. 2001. Cytochrome P450 aromatase (*CYP19*) in the non-human primate brain: distribution, regulation, and functional significance. J Steroid Biochem Mol Biol 79: 247-253.

Rosenbrock H, Hagemeyer CE, Singec I, Knoth R, Volk B. 1999. Testosterone metabolism in rat brain is differentially enhanced by phenytoin-inducible cytochrome P450 isoforms. J Neuroendocrinol 11: 597-604.

Rozman D, Stromstedt M, Waterman MR. 1996. The three human cytochrome P450 lanosterol 14 alpha-demethylase (*CYP51*) genes reside on chromosomes 3, 7, and 13: structure of the two retrotransposed pseudogenes, association with a line-1 element, and evolution of the human *CYP51* family. Arch Biochem Biophys 333: 466-474.

Sanghera MK, Simpson ER, McPhaul MJ, Kozlowski G, Conley AJ, et al. 1991. Immunocytochemical distribution of aromatase cytochrome P450 in the rat brain using peptide-generated polyclonal antibodies. Endocrinology 129: 2834-2844.

Scallet AC, Muskhelishvili L, Slikker W Jr, Kadlubar FF. 2005. Sex differences in cytochrome P450 1B1, an estrogen-metabolizing enzyme, in the rhesus monkey telencephalon. J Chem Neuroanat 29: 71-80.

Schilter B, Andersen MR, Acharya C, Omiecinski CJ. 2000. Activation of cytochrome P450 gene expression in the rat brain by phenobarbital-like inducers. J Pharmacol Exp Ther 294: 916-922.

Sellers EM, Otton SV, Tyndale RF. 1997. The potential role of the cytochrome P-450 2D6 pharmacogenetic polymorphism in drug abuse. NIDA Res Monogr 173: 6-26.

Shinoda K, Yagi H, Fujita H, Osawa Y, Shiotani Y. 1989. Screening of aromatase-containing neurons in ratforebrain: an immunohistochemical study with antibody against human placental antigen X-P2 (hPAX-P2). J Comp Neurol 290: 502-515.

Shohami E, Jacobs TP, Hallenbeck JM, Feuerstein G. 1987. Increased thromboxane A2 and 5-HETE production following spinal cord ischemia in the rabbit. Prostaglandins Leukot Med 28: 169-181.

Siegle I, Fritz P, Eckhardt K, Zanger UM, Eichelbaum M. 2001. Cellular localization and regional distribution of *CYP2D6* mRNA and protein expression in human brain. Pharmacogenetics 11: 237-245.

Siegle I, Klein T, Zou MH, Fritz P, Komhoff M. 2000. Distribution and cellular localization of prostacyclin synthase in human brain. J Histochem Cytochem 48: 631-641.

Sindrup SH, Poulsen L, Brosen K, Arendt N-L, Gram LF. 1993a. Are poor metabolisers of sparteine/debrisoquine less pain tolerant than extensive metabolisers? Pain 53: 335-339.

Sindrup SH, Poulsen L, Brosen K, Arendt-Nielsen L, Gram LF. 1993b. Are poor metabolisers of sparteine/debrisoquine less pain tolerant than extensive metabolisers? Pain 53: 335-339.

Sontag TJ, Parker RS. 2002. Cytochrome P450 omega-hydroxylase pathway of tocopherol catabolism. Novel mechanism of regulation of vitamin E status. J Biol Chem 277: 25290-25296.

Stapleton G, Steel M, Richardson M, Mason JO, Rose KA, et al. 1995. A novel cytochrome P450 expressed primarily in brain. J Biol Chem 270: 29739-29745.

Steckelbroeck S, Heidrich DD, Stoffel-Wagner B, Hans VH, Schramm J, et al. 1999. Characterization of aromatase cytochrome P450 activity in the human temporal lobe. J Clin Endocrinol Metab 84: 2795-2801.

Steckelbroeck S, Watzka M, Lutjohann D, Makiola P, Nassen A, et al. 2002. Characterization of the dehydroepiandrosterone (DHEA) metabolism via oxysterol 7alpha-hydroxylase and 17-ketosteroid reductase activity in the human brain. J Neurochem 83: 713-726.

Stoffel-Wagner B. 2003. Neurosteroid biosynthesis in the human brain and its clinical implications. Ann N Y Acad Sci 1007: 64-78.

Strobl GR, von Kruedener S, Stockigt J, Guengerich FP, Wolff T. 1993. Development of a pharmacophore for inhibition of human liver cytochrome P-450 *2D6*: molecular modeling and inhibition studies. J Med Chem 36: 1136-1145.

Stromstedt M, Rozman D, Waterman MR. 1996. The ubiquitously expressed human *CYP51* encodes lanosterol 14 alpha-demethylase, a cytochrome P450 whose expression is regulated by oxysterols. Arch Biochem Biophys 329: 73-81.

Stromstedt M, Warner M, Gustafsson JA. 1994. Cytochrome P450s of the *4A* subfamily in the brain. J Neurochem 63: 671-676.

Su T, Bao Z, Zhang QY, Smith TJ, Hong JY, et al. 2000. Human cytochrome P450 *CYP2A13*: predominant expression in the respiratory tract and its high efficiency metabolic activation of a tobacco-specific carcinogen, 4-(methylnitrosamino)-1-(3-pyridyl)-1-butanone. Cancer Res 60: 5074-5079.

Sugama S, Kimura A, Chen W, Kubota S, Seyama Y, et al. 2001. Frontal lobe dementia with abnormal cholesterol metabolism and heterozygous mutation in sterol 27-hydroxylase gene (*CYP27*). J Inherit Metab Dis 24: 379-392.

Sun T, Zhao Y, Mangelsdorf DJ, Simpson ER. 1998. Characterization of a region upstream of exon I.1 of the human *CYP19* (aromatase) gene that mediates regulation by retinoids in human choriocarcinoma cells. Endocrinology 139: 1684-1691.

Sundin M, Warner M, Haaparanta T, Gustafsson JA. 1987. Isolation and catalytic activity of cytochrome P-450 from ventral prostate of control rats. J Biol Chem 262: 12293-12297.

Swinney DC, So OY, Watson DM, Berry PW, Webb AS, et al. 1994. Selective inhibition of mammalian lanosterol 14 alpha-demethylase by RS-21607 in vitro and in vivo. Biochemistry 33: 4702-4713.

Takase M, Ukena K, Tsutsui K. 2002. Expression and localization of cytochrome P450 (11beta,aldo) mRNA in the frog brain. Brain Res 950: 288-296.

Takase M, Ukena K, Yamazaki T, Kominami S, Tsutsui K. 1999. Pregnenolone, pregnenolone sulfate, and cytochrome P450 side-chain cleavage enzyme in the amphibian brain and their seasonal changes. Endocrinology 140: 1936-1944.

Tedde A, Rotondi M, Cellini E, Bagnoli S, Muratore L, 2006. Lack of association between the *CYP46* gene polymorphism and Italian late-onset sporadic Alzheimer's disease. Neurobiol Aging 27: 773.e1-773.e3

Teissier E, Fennrich S, Strazielle N, Daval JL, Ray D, et al. 1998. Drug metabolism in in vitro organotypic and cellular models of mammalian central nervous system: activities of membrane-bound epoxide hydrolase and NADPH-cytochrome P-450 (c) reductase. Neurotoxicology 19: 347-355.

Theron CN, Russell VA, Taljaard JJ. 1985. Evidence that estradiol-2/4-hydroxylase activities in rat hypothalamus and hippocampus differ qualitatively and involve multiple forms of P-450: ontogenetic and inhibition studies. J Steroid Biochem 23: 919-927.

Tindberg N, Ingelman-Sundberg M. 1996. Expression, catalytic activity, and inducibility of cytochrome P450 *2E1* (*CYP2E1*) in the rat central nervous system. J Neurochem 67: 2066-2073.

Tokiyoshi K, Ohnishi T, Nii Y. 1991. Efficacy and toxicity of thromboxane synthetase inhibitor for cerebral vasospasm after subarachnoid hemorrhage. Surg Neurol 36: 112-118.

Tone Y, Inoue H, Hara S, Yokoyama C, Hatae T, et al. 1997. The regional distribution and cellular localization of mRNA encoding rat prostacyclin synthase. Eur J Cell Biol 72: 268-277.

Trap C, Nato F, Chalbot S, Kim SB, Lafaye P, et al. 2005. Immunohistochemical detection of the human cytochrome P4507B1: production of a monoclonal antibody after cDNA immunization. J Neuroimmunol 159: 41-47.

Trofimova-Griffin ME, Juchau MR. 1998. Expression of cytochrome P450RAI (*CYP26*) in human fetal hepatic and cephalic tissues. Biochem Biophys Res Commun 252: 487-491.

Tsuruo Y, Ishimura K, Fujita H, Osawa Y. 1994. Immunocytochemical localization of aromatase-containing neurons in the rat brain during pre- and postnatal development. Cell Tissue Res 278: 29-39.

Tyndale RF, Li Y, Li NY, Messina E, Miksys S, et al. 1999. Characterization of cytochrome P-450 *2D1* activity in rat brain: high-affinity kinetics for dextromethorphan. Drug Metab Dispos 27: 924-930.

Tyndale RF, Sunahara R, Inaba T, Kalow W, Gonzalez FJ, et al. 1991. Neuronal cytochrome P450IID1 (debrisoquine/sparteine-type): potent inhibition of activity by (-)-cocaine and nucleotide sequence identity to human hepatic P450 gene *CYP2D6*. Mol Pharmacol 40: 63-68.

Ukena K, Usui M, Kohchi C, Tsutsui K. 1998. Cytochrome P450 side-chain cleavage enzyme in the cerebellar Purkinje neuron and its neonatal change in rats. Endocrinology 139: 137-147.

Upadhya SC, Chinta SJ, Pai HV, Boyd MR, Ravindranath V. 2002. Toxicological consequences of differential regulation of cytochrome p450 isoforms in rat brain regions by phenobarbital. Arch Biochem Biophys 399: 56-65.

Upadhya SC, Tirumalai PS, Boyd MR, Mori T, Ravindranath V. 2000. Cytochrome P4502E (*CYP2E*) in brain: constitutive expression, induction by ethanol and localization by fluorescence in situ hybridization. Arch Biochem Biophys 373: 23-34.

Uyama O, Nagatsuka K, Nakabayashi S, Isaka Y, Yoneda S, et al. 1985. The effect of a thromboxane synthetase inhibitor, OKY-046, on urinary excretion of immunoreactive thromboxane B2 and 6-keto-prostaglandin F1 alpha in patients with ischemic cerebrovascular disease. Stroke 16: 241-244.

Vasiliou V, Ziegler TL, Bludeau P, Petersen DR, Gonzalez FJ, et al. 2006. *CYP2E1* and catalase influence ethanol sensitivity in the central nervous system. Pharmacogenet Genomics 16: 51-58.

Voirol P, Jonzier-Perey M, Porchet F, Reymond MJ, Janzer RC, et al. 2000. Cytochrome P-450 activities in human and rat brain microsomes. Brain Res 855: 235-243.

Wakamatsu N, Hayashi M, Kawai H, Kondo H, Gotoda Y, et al. 1999. Mutations producing premature termination of translation and an amino acid substitution in the sterol 27-hydroxylase gene cause cerebrotendinous xanthomatosis associated with parkinsonism. J Neurol Neurosurg Psychiatry 67: 195-198.

Warner M, Ahlgren R, Zaphiropoulos PG, Hayashi S, Gustafsson JA. 1991. Identification and localization of cytochromes P450 expressed in brain. Methods Enzymol 206: 631-640.

Weill-Engerer S, David JP, Sazdovitch V, Liere P, Schumacher M, et al. 2003. In vitro metabolism of dehydroepiandrosterone (DHEA) to 7alpha-hydroxy-DHEA and Delta5-androstene-3beta,17beta-diol in specific regions of the aging brain from Alzheimer's and non-demented patients. Brain Res 969: 117-125.

White JA, Beckett-Jones B, Guo YD, Dilworth FJ, Bonasoro J, et al. 1997. cDNA cloning of human retinoic acid-metabolizing enzyme (hP450RAI) identifies a novel family of cytochromes P450. J Biol Chem 272: 18538-18541.

White JA, Guo YD, Baetz K, Beckett-Jones B, Bonasoro J, et al. 1996. Identification of the retinoic acid-inducible all-trans-retinoic acid 4-hydroxylase. J Biol Chem 271: 29922-29927.

White JA, Ramshaw H, Taimi M, Stangle W, Zhang A, et al. 2000. Identification of the human cytochrome P450, P450RAI-2, which is predominantly expressed in the adult cerebellum and is responsible for all-trans-retinoic acid metabolism. Proc Natl Acad Sci USA 97: 6403-6408.

Woo SI, Hansen LA, Yu X, Mallory M, Masliah E. 1999. Alternative splicing patterns of *CYP2D* genes in human brain and neurodegenerative disorders. Neurology 53: 1570-1572.

Wu ML, Li H, Wu DC, Wang XW, Chen XY, et al. 2005. *CYP1A1* and *CYP1B1* expressions in medulloblastoma cells are AhR-independent and have no direct link with resveratrol-induced differentiation and apoptosis. Neurosci Lett 384: 33-37.

Wu S, Moomaw CR, Tomer KB, Falck JR, Zeldin DC. 1996. Molecular cloning and expression of *CYP2J2*, a human cytochrome P450 arachidonic acid epoxygenase highly expressed in heart. J Biol Chem 271: 3460-3468.

Wu Z, Martin KO, Javitt NB, Chiang JY. 1999. Structure and functions of human oxysterol 7alpha-hydroxylase cDNAs and gene *CYP7B1*. J Lipid Res 40: 2195-2203.

Wyss A, Gustafsson JA, Warner M. 1995. Cytochromes P450 of the *2D* subfamily in rat brain. Mol Pharmacol 47: 1148-1155.

Yague JG, Lavaque E, Carretero J, Azcoitia I, Garcia-Segura LM. 2004. Aromatase, the enzyme responsible for estrogen biosynthesis, is expressed by human and rat glioblastomas. Neurosci Lett 368: 279-284.

Yang HY, Lee QP, Rettie AE, Juchau MR. 1994. Functional cytochrome P4503A isoforms in human embryonic tissues: expression during organogenesis. Mol Pharmacol 46: 922-928.

Yau JL, Rasmuson S, Andrew R, Graham M, Noble J, et al. 2003. Dehydroepiandrosterone 7-hydroxylase *CYP7B*: predominant expression in primate hippocampus and reduced expression in Alzheimer's disease. Neuroscience 121: 307-314.

Yoo M, Ryu HM, Shin SW, Yun CH, Lee SC, et al. 1997. Identification of cytochrome P450 2E1 in rat brain. Biochem Biophys Res Commun 231: 254-256.

Yu A, Dong H, Lang D, Haining RL. 2001. Characterization of dextromethorphan O- and N-demethylation catalyzed by highly purified recombinant human cyp2d6. Drug Metab Dispos 29: 1362-1365.

Yu AM, Granvil CP, Haining RL, Krausz KW, Corchero J, et al. 2003. The relative contribution of monoamine oxidase and cytochrome p450 isozymes to the metabolic deamination of the trace amine tryptamine. J Pharmacol Exp Ther 304: 539-546.

Yu L, Romero DG, Gomez-Sanchez CE, Gomez-Sanchez EP. 2002. Steroidogenic enzyme gene expression in the human brain. Mol Cell Endocrinol 190: 9-17.

Yu AM, Haining RL. 2006. Expression, Purification and characterization of mouse CYP2d22. Drug Metab Dispos 34: 1167-1174

Yun CH, Park HJ, Kim SJ, Kim HK. 1998. Identification of cytochrome P450 *1A1* in human brain. Biochem Biophys Res Commun 243: 808-810.

Zehnder D, Bland R, Williams MC, McNinch RW, Howieison M. 2001. Extrarenal expression of 25-hydroxyvitamin d(3)-1 alpha-hydroxylase. J Clin Endocrinol Metab 86: 888-894.

Zeldin DC, Moomaw CR, Jesse N, Tomer KB, Beetham J, et al. 1996. Biochemical characterization of the human liver cytochrome P450 arachidonic acid epoxygenase pathway. Arch Biochem Biophys 330: 87-96.

Zevin S, Benowitz NL. 1999. Drug interactions with tobacco smoking. An update. Clin Pharmacokinet 36: 425-438.

Zhang P, Rodriguez H, Mellon SH. 1995. Transcriptional regulation of P450scc gene expression in neural and steroidogenic cells: implications for regulation of neurosteroidogenesis. Mol Endocrinol 9: 1571-1582.

Zhang SH, Li Q, Yao SL, Zeng BX. 2001. Subcellular expression of *UGT1A6* and *CYP1A1* responsible for propofol metabolism in human brain. Acta Pharmacol Sin 22: 1013-1017.

Zhou L, Erickson RR, Hardwick JP, Park SS, Wrighton SA, et al. 1997a. Catalysis of the cysteine conjugation and protein binding of acetaminophen by microsomes from a human lymphoblast line transfected with the cDNAs of various forms of human cytochrome P450. J Pharmacol Exp Ther 281: 785-790.

Zhou MY, del Carmen Vila M, Gomez-Sanchez EP, Gomez-Sanchez CE. 1997b. Cloning of two alternatively spliced 21-hydroxylase CDNAs from rat adrenal. J Steroid Biochem Mol Biol 62: 277-286.

Zwain IH, Yen SS. 1999. Dehydroepiandrosterone: biosynthesis and metabolism in the brain. Endocrinology 140: 880-887.

4 Practical Enzymology

Quantifying Enzyme Activity and the Effects of Drugs Thereupon

A. Holt

Abstract: Neuroscientists may wish to quantify an enzyme activity for one of many reasons. In order to do so, the researcher must be able to set up an assay appropriately, and this requires some understanding of the kinetic behavior of the enzyme toward the substrate used. Furthermore, such an understanding is vital if the inhibitory effects of a drug are to be assessed appropriately. This chapter outlines key principles that must be adhered to, and describes basic approaches by which rather complex kinetic data might be obtained, in order that enzyme kinetics and inhibitor kinetics might be studied successfully by the nonexpert.

1 Why Enzymes?

Changes in enzyme activities can occur as a consequence of drug administration and as a consequence of, or as a causative factor in, a variety of pathophysiological conditions. Effecting a change in one or more of these activities might be beneficial in alleviating disease symptoms, whereas activity of a particular enzyme may represent a direct or indirect surrogate marker of disease progression and of the success of the chosen treatment. More than 50% of novel potential drug targets predicted from mining the human genome are enzymes (Terstappen and Reggiani, 2001) and it is inevitable that expertise in obtaining and interpreting enzymatic data will assume an increasing degree of importance in the ensuing drug discovery race. Thus, there are several reasons why the academic or industrial neuroscientist would benefit from an increased understanding of enzyme behavior and of the mechanisms by which drugs can alter this behavior. Yet, perhaps as a result of the widespread promotion to undergraduates of tools such as the Lineweaver–Burk plot and of the perception that enzymology and enzyme kinetics are not sufficiently cutting edge to attract grant funding, a general malaise is now prevalent among occasional users of the discipline. Indeed, most now appear to rank enzymology in terms of experimental complexity somewhere between the Lowry protein assay and preparation of physiological saline solution. But should this be a matter for concern? After all, are not enzymes simple beasts, faithfully and unvaryingly complying with Michaelis–Menten kinetics, and with inhibitors either competitive or noncompetitive in their actions? If the answers to these questions were "yes," this would indeed be a short chapter. However, it is not, and the literature is littered with pseudokinetics derived from poorly designed and misinterpreted experiments. The following pages attempt to address this issue and assist the neuroscientist in avoiding the most common errors and pitfalls. For more detailed coverage of these and related topics, the reader is referred to *Enzyme kinetics. Behavior and analysis of rapid equilibrium and steady-state enzyme systems* by Segel (1993) and *Fundamentals of Enzymology* by Price and Stevens (1999). The former text provides an in-depth coverage of models, equations, and mathematics of enzyme behavior whereas the latter deals more with the practicalities of studying enzymes and relates kinetic data to enzyme structure and function.

2 What Enzymes Do

2.1 Structure and Function

Enzymes are proteins, which catalyze, or speed up, chemical reactions by reducing the activation energy required for the reaction to proceed. Binding of one or more substrates to the active (catalytic) site of the enzyme places crucial chemical groups on the substrate(s) and enzyme at optimal distances and angles relative to one another, and for a prolonged period of time, in contrast to the situation where reactants are free in solution and where optimal alignment of reactants occurs both rarely and fleetingly. Some distortion of intra-substrate bonds may also occur, increasing the likelihood of reaction. Most enzymatic reactions proceed via recognized organic reaction mechanisms, such as acid–base catalysis, or nucleophilic/electrophilic displacements, with the "reactants" comprising structural components from both the enzyme and the substrate(s). During the course of these reactions, the enzyme may also undergo some degree of structural alteration, perhaps donating or accepting a chemical group, but regeneration of active enzyme must then occur to allow continued participation of the enzyme in subsequent reactions.

For many enzyme reactions, additional factors besides the enzyme protein and substrate are necessary for the reaction to proceed. *Coenzymes* are usually large organic molecules that bind to the enzyme protein and participate in the chemical reaction with substrate, often as carriers of a chemical group. These coenzymes may dissociate from the enzyme protein and undergo regeneration at a separate location, perhaps under the control of a second enzyme, before associating once again with the enzyme of interest. The use of the term *prosthetic groups* to refer to such cofactors is usually limited to more tightly bound molecules, which do not dissociate but instead are regenerated in situ. Many enzymes also contain one or more *activators* bound to the protein, which are often metal ions. These metals may play a structural role, ligating the charged amino acids in the protein, and may also participate in the enzymatic reaction itself. Also, a metal ion may be involved in posttranslational modification of the protein backbone to confer activity upon an otherwise inactive enzyme (Klinman, 2003). Activators are designated as *essential* if their absence renders the enzyme protein catalytically inactive, and *nonessential* if their presence simply increases an existing activity of the enzyme.

Enzymes may exist as simple monomers, or as homo- or heterodimers, or as multimers. In multimeric enzymes, each component monomer may possess a catalytic site; alternatively, the catalytic site may be located at the interface between two or more monomers, or only one monomer of a heteromultimer may possess an active site. It is not uncommon in dimeric or multimeric enzymes containing two or more active sites for some degree of cooperativity to exist between the sites, with respect to the substrate binding or substrate turnover number (Monod et al., 1965).

2.2 What Is Enzyme Activity?

Enzyme activity refers to the rate at which a particular enzyme catalyzes the conversion of a particular substrate (or substrates) to one or more products under a given set of conditions. Usually, activity refers to the contribution of many enzyme molecules (often expressed simply as "activity per mg of protein" or similar) but, in its simplest form, activity refers to the contribution of a single enzyme molecule. The *turnover number* of an enzyme–substrate combination refers to the number of substrate molecules metabolized in unit time (usually a period of 1 s) under a given set of conditions (see later). These definitions appear, at first glance, to be largely self-explanatory. However, many factors contribute to the final activity of an enzyme, and these must be considered during any assessment of such activity.

The rate at which a particular substrate is metabolized depends firstly upon the turnover number of the enzyme for that substrate, which itself is a measure of how rapidly a single chemical reaction takes place within a single enzyme molecule. This rate is affected by factors such as temperature, pH, and ionic strength. Also, enzyme protein should be in an optimally active form in so much as the protein present has undergone any necessary posttranslational modification and any essential and nonessential activators and coenzymes are present at concentrations sufficiently high to allow each enzyme molecule to exhibit optimal activity. If any of these factors are altered, the result can be an apparent change in "enzyme activity" in the absence of any change in the concentration of protein. It should thus be apparent that the common practice of assessing changes in "enzyme activity" by using electrophoretic approaches to measure altered protein expression, or worse still, content of DNA or RNA coding for an enzyme protein, is in many cases entirely inappropriate. Such techniques yield exactly the information that they are designed to yield, but confirm nothing with regard to changes in enzyme activity.

All other things being equal, the rate of turnover of a particular substrate is directly proportional to the number (or the concentration) of enzyme molecules present. This is an important property to confirm during, for example, the development of a novel assay system.

Perhaps the most important variable influencing enzyme activity is the concentration of substrate present. An enzyme is "most active" when a substrate is present at a *saturating concentration*, that is, a concentration at which an enzyme molecule remains unoccupied for the shortest time that is physically possible following vacation of the active site by the reaction product. At subsaturating substrate concentrations, the length of time for which the active site remains unoccupied following product release is longer than this shortest possible time. It should be apparent, then, that at subsaturating substrate concentrations, the rate of product formation resulting from the action of a single enzyme molecule, or from the combined action of many enzyme molecules, will be lower than the maximum possible rate that occurs at saturating substrate concentrations. The turnover number for a particular enzyme–substrate combination refers to the number of molecules of substrate metabolized in unit time by a single enzyme molecule when all substrates are present at saturating concentrations. Similarly, V_{max} refers to the rate of product formation resulting from the combined action of many enzyme molecules, when all substrates are present at saturating concentrations.

The relationship between substrate concentration ($[S]$) and reaction velocity (v, equivalent to the degree of binding of substrate to the active site) is, in the absence of cooperativity, usually hyperbolic in nature, with binding behavior complying with the law of mass action. However, the equation describing the hyperbolic relationship between v and $[S]$ can be simple or complex, depending on the enzyme, the identity of the substrate, and the reaction conditions. Quantitative analyses of these v versus $[S]$ relationships are referred to as *enzyme kinetics.*

2.3 What Are the Consequences of Changing Enzyme Activity?

Enzymes convert substrate(s) to product(s). Both substrates and products can possess biological activity, and in regulating the concentrations of substrates and products, enzymes are responsible in part for regulating the extent and duration of biological responses to these bioactive compounds.

Substrates are often the product of an upstream enzymatic reaction, whereas products often act as substrates for a downstream enzyme:

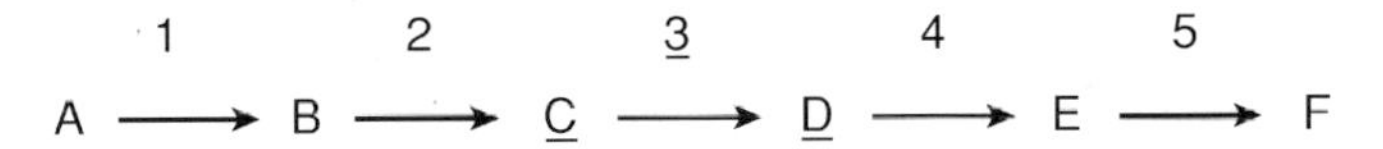

In the above scheme, enzyme 3 converts substrate C to product D. What would be the consequences of an increase in the activity of enzyme 3? The short answer is that it is impossible to say, unless the nature of the increase in activity is known. For example, the proposed effect might result from an increase in the affinity of the enzyme for substrate C in the presence of a nonessential activator of enzyme 3, which increases substrate affinity. If substrate C is present at a subsaturating concentration with respect to enzyme 3, then the rate of formation of product D would likely increase to some degree. However, if C is present at a saturating concentration, there would be no change in the rate of formation of D. Alternatively, the proposed effect may result from an increase in availability of enzyme protein. If this protein is fully active with respect to cofactor, activator availability, etc., then there would be an initial increase in the rate of formation of product D. However, this would result directly in a decrease in [C]. If enzyme 2 is rate limiting within this pathway, and if [C] is subsaturating with respect to enzyme 3, the decrease in [C] would result in a decrease in the rate of formation of product D by enzyme 3. Thus, the increase in the rate of formation of D due to the increased enzyme concentration could be offset to a substantial degree by a reduction in reaction velocity due to the decreased substrate concentration. On the other hand, if enzyme 2 is not rate limiting, an increased rate of formation of product D might be maintained. However, if [D] is subsaturating as a substrate for enzyme 4, an increased rate of formation of D might result simply in a consequential increase in the rate of formation of product E. If increased availability of enzyme 3 protein is not matched by an increased availability of cofactors, or if the subsequent increase in turnover of substrate C leads to depletion of an essential cofactor, then the increase in enzyme protein may prove to be largely inconsequential with respect to altering substrate and product concentrations. Finally, as elevated levels of substrates for, or products of a particular enzyme can, in some cases, lead to the phenomena known as substrate inhibition or product inhibition of that enzyme, or of an enzyme immediately upstream or downstream from the enzyme of interest, the ability to predict physiological outcomes can be further complicated.

So what would be the consequences of an increase in the activity of enzyme 3? The long answer is that even if the nature of the increase in enzyme activity is known, it is often difficult to predict effects on concentrations of substrates and products, and thus on their biological effects, with any certainty unless something is known regarding the behavior of enzymes upstream and downstream from the enzyme of interest. However, in general, an increase in the activity of an enzyme is likely to cause some degree of reduction in the concentration of one or more upstream compounds, particularly of those that are the products of an upstream rate-limiting step. It could also cause some degree of increase in the concentration of one or more downstream compounds, particularly of those that are substrates for a downstream rate-limiting step. The converse generally holds true for a reduction in activity of an enzyme. It is for these reasons that ex vivo analyses of drug effects on enzyme activities are often accompanied by analyses of tissue concentrations of substrates and products.

The situation in a test tube is usually much more straightforward; enzyme activity is reflected as a decrease in substrate concentration and an increase in product concentration, and measurement of the initial rate of change of concentration of either – the reaction velocity – constitutes a direct measurement of enzyme activity.

3 Measuring Reaction Velocity

3.1 What Question Are You Addressing?

There are several reasons as to why someone might wish to measure enzyme activity. Sometimes the reasons are entirely qualitative. *Is a particular enzyme activity present in a particular tissue, or not? What is the subcellular location of the enzyme?* However, in most cases, some quantitative information is required. *How well does an enzyme metabolize a particular substrate? Is drug A an inhibitor of enzyme B? What are the in vivo effects of drug A on enzyme activity? What is the mechanism by which drug A inhibits enzyme B? Is drug X more potent than drug Y as an inhibitor? Do changes in enzyme activity parallel disease progression or therapeutic effect of a drug treatment?* In order to address quantitative problems, it is invariably necessary, at least in preliminary experiments, to examine enzyme kinetics. Such studies involve examinations of initial rates of substrate metabolism in the presence of different concentrations of substrate, and, if necessary, in the absence or presence of inhibitor(s) of interest.

3.2 What To Measure?

As the enzyme itself is usually the focus of interest, information on the behavior of that enzyme can be obtained by incubating the enzyme with a suitable substrate under appropriate conditions. A suitable substrate in this context is one which can be quantified by an available detection system (often absorbance or fluorescence spectroscopy, radiometry or electrochemistry), or one which yields a product that is similarly detectable. In addition, if separation of substrate from product is necessary before quantification (for example, in radioisotopic assays), this should be readily achievable. It is preferable, although not always possible, to measure the appearance of product, rather than the disappearance of substrate, because a zero baseline is theoretically possible in the former case, improving sensitivity and resolution. Even if a product (or substrate) is not directly amenable to an available detection method, it may be possible to derivatize the product with a chemical species to form a detectable adduct, or to subject a product to a second enzymatic step (known as a coupled assay, discussed further later) to yield a detectable product. But, regardless of whether substrate, product, or an adduct of either is measured, the parameter we are interested in determining is *the initial rate of change of concentration*, which is determined from the initial slope of a concentration versus time plot.

If measurements are made in a system containing a purified or a cloned and expressed enzyme, then assays are relatively straightforward if the species being measured is chemically stable under the reaction conditions. However, if tissue homogenates are used as a source of enzyme, the potential exists for the product of interest to be further metabolized by one or more contaminating enzymes, or even for the substrate to be subjected to an alternative metabolic process. This can be particularly problematic when the substrate added to measure enzyme activity is also an endogenous substrate for the enzyme of interest. In such cases, it may be necessary to "trap" the desired product, either chemically or enzymatically, as soon as it is released, or to inhibit any enzymes responsible for metabolism and removal of the desired product or for alternative metabolism of the substrate. For example, many enzyme reactions generate hydrogen peroxide, which allows measurement of enzyme activity through coupling to peroxidase. Contamination of tissue samples with catalase can result in a substantial underestimation of enzyme activity unless an inhibitor of catalase is included (Margoliash et al., 1960).

Several things may be done if the researcher has difficulty in detecting an enzyme activity of interest in a homogenate, or elsewhere. A more sensitive assay technique may be used, if one is available. The concentration of enzyme may be increased, as the rate of product formation is directly proportional to [E]. The incubation time for enzyme with substrate can be increased, although the *caveats* discussed in ➋ Section 3.3.2 must be borne in mind. The reaction volume may also be increased, while maintaining concentrations of reactants constant; this approach is particularly useful if product is separated and detected by chromatography, or if a column is used to separate radiolabeled substrate from product, because the increased amount of product formed in unit time will result in enhanced signal size.

It may also improve the likelihood of detecting enzyme activity if substrate concentration is increased. This approach is useful only if the initial substrate concentration is below K_M (the Michaelis constant; see ➋ Section 4.1.2) because enzyme activity can only be doubled from that at K_M even if the substrate concentration is increased to infinitely high levels. Furthermore, in many assay systems, blank readings are proportional to substrate concentration, and there is little to be gained by a tenfold increase in $[S]$ that results in a 70% increase in v and a 1,000% increase in the blank signal.

3.3 Continuous and Discontinuous Approaches

Enzyme reactions can be measured continuously or discontinuously, with each approach having both advantages and drawbacks.

3.3.1 Continuous Assays

In a continuous assay system, the appearance of product (or disappearance of substrate) is monitored in real time during the period of incubation of substrate with enzyme. This approach may only be used when the product possesses chemical characteristics sufficiently different from the substrate to preclude any masking or augmentation of the measured signal from the species of interest by a contaminating signal from the other. For example, if both substrate and product are fluorescent, one, or both of λ_{ex} or λ_{em} for the product should be quite different from those for the substrate. Continuous assay methods are most commonly used with absorbing or fluorogenic species, with output from the measuring device (often a platereader or cuvette spectrometer) taking the form of a signal versus time plot. The signal would typically be an absorbance or fluorescence reading, which (usually) relates directly to the concentration of the species of interest (John, 1993).

Most spectrometric instrumentation is now accompanied by a software package that allows calculation from data of what is often (and incorrectly) referred to as V_{max}. In fact, the calculation done by the software is a linear regression yielding the slope corresponding to the initial rate of product formation, and this is usually observed as a tangent to an early part of the data curve (➋ *Figure 4-1*). Although this rate is usually the most rapid (or "maximum") attainable with respect to the measured data, it is not a V_{max} rate unless substrate is present at a saturating concentration – and in kinetic studies, this is rarely the case.

The major advantage associated with continuous assays is that the initial rate of product formation can be determined with complete confidence, and any unusual behavior of the enzyme would be immediately apparent. The major disadvantage is a question of throughput; an instrument such as a platereader would remain dedicated to the reading of a single plate for the duration of the enzyme–substrate incubation period, compared with an equivalent discontinuous assay where an entire plate may be measured in a

Figure 4-1
Typical output from an absorbance spectrophotometer (dotted line) showing increasing absorbance with time as a result of formation of a colored product. The solid line is a tangent to the initial slope of the (dotted) curve, and the slope of the tangent represents the initial reaction velocity (*v*) in this cuvette

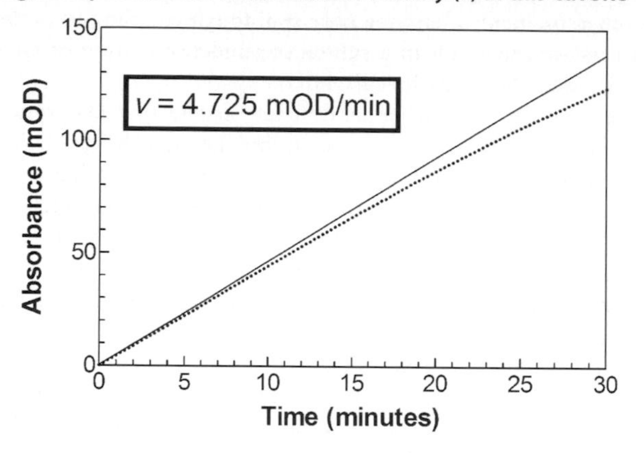

matter of seconds. Nevertheless, the advantages certainly outweigh the disadvantages, and if a choice exists between a continuous or discontinuous approach to assess enzyme kinetic behavior, a continuous route should be chosen unless there is good reason to do otherwise.

3.3.2 Discontinuous Assays

In a discontinuous assay, the enzyme is incubated with substrate for a period of time, the reaction is stopped (sometimes by addition of strong acid or base, perhaps along with a derivatizing agent of some kind), product may then be separated from unreacted substrate, and a single measurement reflecting product (or sometimes substrate) concentration is made. It is necessary to use a discontinuous approach when there is a requirement to separate product from substrate before making a measurement. For example, radiometric assays generate a radioactive product from a radioactive substrate, and the means by which "product" is measured (i.e., determination of radioactivity associated with the product) cannot differentiate between substrate and product in this regard. Therefore, usually either a solvent extraction step or a chromatography procedure is first used to separate product before counting of radioactivity associated with the product. Similarly, an absorbent or fluorogenic substrate may possess spectral characteristics too similar to those of the product of interest to allow differentiation, and separation prior to quantification of product is necessary. Many liquid- and gas-chromatography-based assays are designed to circumvent this problem, with detection and quantification of product in the eluate achieved through absorbance, fluorescence, electrochemical, or mass spectrometric techniques.

The major advantage associated with the discontinuous approach is that only a single measurement is made, facilitating data analysis. In addition, for spectrometer- and platereader-based assays, many more samples can be measured in unit time, compared with the equivalent continuous assay system.

The major disadvantage associated with the discontinuous approach is that only a single measurement is made, "facilitating" data analysis. The discontinuous approach is perfectly acceptable if all the necessary preliminary experiments have been completed. But, as is evident too often in the literature, they have not been done, and because the data output from such experiments does not permit visualization of the enzyme's behavior throughout the incubation period, the researcher remains blissfully unaware that any problem exists.

Figure 4-2a shows theoretical output from a continuous platereader assay of enzyme activity at five substrate concentrations between 1 and 15 μM. It is clearly apparent that the rate of product formation

Figure 4-2

(a) Output from an absorbance platereader of enzymatic turnover of a substrate (K_M 3 μM) at five different subsaturating concentrations. (b) Michaelis–Menten plots of data shown in *Figure 4-2a* **with *v* values determined from tangents to continuous data (correct) and from theoretical endpoint readings after 10 min and 40 min of incubation. The marked error in the curve determined from 40 min endpoint data is very apparent, with the impact most evident on examination of derived kinetic constants (*Table 4-1*)**

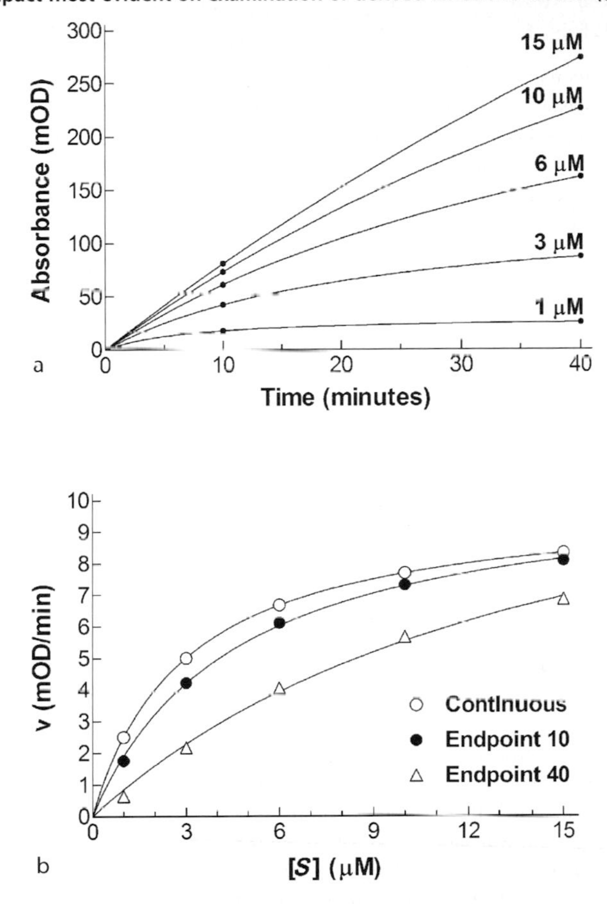

decreases more rapidly in wells containing lower concentrations of substrate, presumably in this case due to substrate depletion. *Figure 4-2a* shows the Michaelis–Menten hyperbolic plots (v versus $[S]$) constructed using initial velocity values obtained from initial tangents to continuous data and from discontinuous (endpoint) readings following a 10- or 40-min incubation period. *Table 4-1* lists the kinetic constants and their associated errors obtained from these hyperbolae. Although fits to discontinuous data show (deceivingly) acceptable goodness-of-fit (R^2) values, the degrees of error in the kinetic constants relative to the correct values, obtained from tangents to the continuous data, are quite substantial, particularly in the endpoint data obtained after 40 min.

The reason for these substantial errors is quite simply that in a discontinuous assay, the researcher *assumes* that product formation remains (approximately) linear *for the duration of the incubation period.* Although this may hold true for high substrate concentrations (in this example), it is clear from continuous data that deviation from linearity is substantial at lower substrate concentrations. Accordingly, what *must* be done prior to use of a discontinuous assay protocol is a timecourse assessment, in which the concentration of product formed from both low *and high* concentrations of substrate is determined at various time

Table 4-1
Kinetic constants determined from nonlinear regression of data in Figure 4-2b

	$K_M \pm$ S.E.M (μM)	K_M expressed as % continuous	$V_{max} \pm$ S.E.M. (mOD/min)	V_{max} expressed as % continuous	R^2
Continuous	2.99 ± 0.01	100	9.99 ± 0.01	100	1.000
Endpoint 10 min	4.60 ± 0.26	154	10.6 ± 0.22	107	0.999
Endpoint 40 min	15.6 ± 2.8	523	14.2 ± 1.5	142	0.996

Initial velocities of product formation were obtained from absorbance versus time data shown in Figure 4-2a, either by determination of slopes of tangents to early data points (Continuous) or from theoretical single endpoint readings made after 10 min (Endpoint 10 min) or 40 min (Endpoint 40 min) of incubation with enzyme. Rectangular hyperbolae were fitted with the Michaelis–Menten equation (Figure 4-2b) and K_M, V_{max} and goodness-of-fit values were determined. Kinetic constants obtained from endpoint readings have been expressed as a percentage of the correct values obtained directly from the continuous data

points to determine the duration of pseudolinearity of product formation, and thus a suitable incubation time. High concentrations of substrate should be included because linearity might be lost as a result of product inhibition of the enzyme, or of depletion of a coupling or derivatizing reagent during the reaction. Note also that unless substrate is present at a saturating concentration (which is rarely the case), the rate of product formation is *never* linear. A useful rule-of-thumb is that if substrate depletion does not exceed 10% of total substrate present during incubation with enzyme, an endpoint reading expressed *per* unit time provides an acceptable approximation for initial reaction rate. However, for the reasons outlined earlier, preliminary timecourse studies should still be done.

In the earlier example, how might the assay have been done successfully in a discontinuous manner? Based on the lowest concentration of substrate used, depletion of substrate would have exceeded 10% in less than 2 min following the addition of enzyme. When very short incubation times are involved, factors such as time taken for sample temperature equilibration and time taken to pipette reactants into multiple wells become a source of increasing error, and incubation times of at least 5 to 10 min are thus preferable. In order to prolong the incubation time while minimizing the degree of substrate depletion, several things could be done. The assay could be run at a suboptimal pH, or at a lower temperature. However, although velocities would certainly be reduced as a consequence of taking such steps, it is possible that they would result, at least in part, from subtle structural changes within the enzyme (and perhaps the substrate) and the enzyme's behavior under these altered conditions may not represent a true reflection of events occurring under conditions of "normal" pH and temperature. The most appropriate step, then, would be *to reduce the concentration of enzyme present*, as velocity is directly proportional to enzyme concentration. Concentrations of substrate, cofactors, and other reactants should not be altered as the affinities of these species for the enzyme would not change because of the reduction in enzyme concentration. If sensitivity becomes an issue, the *volume* of the reaction could be increased, maintaining the *concentrations* of substrate and cofactors constant while maintaining the *amount* of enzyme constant (thereby reducing the concentration of enzyme). Sensitivity may be improved as the *amount* of product corresponding to 10% of the substrate is increased with increased volume, even though the *change in concentration* of that product remains consistent with the lower volume assay. This can lead to increased peak height and volume in chromatographic analyses and increased absorbance readings obtained from a platereader (due to the increased pathlength) but not from a spectrophotometer (where pathlength is determined by cuvette width).

Discontinuous approaches are used through necessity in radiometric assays (Oldham, 1993) and in liquid- and gas-chromatography-based protocols (Syed, 1993). Many commercially available kits designed to measure a particular enzyme activity combine absorbance and fluorescence platereader-based technologies with assay protocols that could be done continuously. However, since many users of these kits have little experience in enzymology, assay instructions usually outline a simplified procedure whereby a discontinuous (endpoint) reading is instead taken, negating any apparent necessity to calculate a slope, and thus an initial rate. Although some such kits do yield useful data if instructions are followed meticulously, the commercial pressure to demonstrate "improved sensitivity" compared with competitors' products means that others suggest incubation times beyond what would be acceptable kinetically. Accordingly, if it becomes necessary to make use of a commercial kit, read the assay continuously if possible, and take the time to obtain your own preliminary timecourse data if it is not.

3.3.3 Coupled Assays

A coupled assay is one in which a product of enzyme activity is further metabolized by a second added enzyme to yield a new product, which can be quantified easily. A coupled assay is usually used when the primary products of the enzyme of interest are not readily amenable to direct quantification. Usually the coupling enzyme is added to the assay, although it is possible to make use of an enzyme already present in a homogenate. In those cases where the coupling enzyme exhibits acceptable activity under the pH and buffer conditions employed to assay the enzyme of interest, coupled assays can often be done continuously. Furthermore, it is possible to use more than one coupling enzyme; in this regard, many reactions can ultimately be coupled to peroxidase, which directly generates chromogens or fluorophores in the presence of commercially available substrates, such as Amplex Red. For example, glutamine aminotransferase generates pyruvate and methionine from alanine and α-keto-γ-methiolbutyrate. Neither pyruvate nor methionine lends itself to easy direct quantification in a tissue homogenate environment. However, inclusion of commercially available pyruvate oxidase would generate hydrogen peroxide, and in the presence of peroxidase and a suitable cosubstrate, the transaminase activity may be readily amenable to continuous measurement with a double-coupled assay in a platereader.

If a coupled assay is used, it is vital that coupling enzymes are present in excess. In other words, the rate of the measured reaction must be directly proportional to the concentration of the (initial) enzyme of interest, and any lag period in generation of the chromophore or fluorophore should be minimized. Consideration should also be given to contaminating enzymes in tissue homogenates that may deplete those products used as substrates by subsequent coupling enzymes. If this becomes a problem, then a more pure enzyme preparation may need to be used, or an inhibitor of the competing enzyme might also be included. Alternatively, increasing the concentration of the coupling enzyme can minimize the impact of a reaction product being lost to an alternative pathway. Finally, if a coupled assay system is being used to assess effects of enzyme inhibitors, it is necessary to ensure that the inhibitors are acting to inhibit the enzyme of interest, and not one of the coupling enzymes present.

4 Kinetic Experiments

In examinations of effects of drugs on enzyme activities, we are usually interested in determining the degree to which activity is altered by the drug, as well as whether the change in activity results from a change in substrate affinity, a change in turnover number, or a change in both parameters. Furthermore, as enzymes regulate concentrations of bioreactive molecules, many researchers have a particular interest in enzyme activity as it relates to one particular substrate or product of interest. However, most enzymes bind two (bireactant) or more (multireactant) substrates, and generate two or more products. Yet, biochemistry textbooks usually limit discussion of enzyme kinetics to relatively uncommon unireactant systems, in which a single substrate is metabolized. Kinetic approaches to examine unireactant systems may be suitable for examinations of multisubstrate systems – or they may not. Major complicating factors in this regard include the fact that binding of one substrate to an enzyme may alter the binding characteristics of subsequent substrates in a concentration-dependent manner, or that binding of two or more substrates may only occur in a particular sequence. The following sections will attempt to simplify and summarize acceptable approaches for determining kinetic constants.

4.1 Determining Kinetic Constants in Unireactant Systems

4.1.1 Background

In a unireactant system, an enzyme (E) reacts rapidly with a single substrate molecule (S) to form an enzyme–substrate complex (ES), with subsequent breakdown of the complex yielding product (P), as well as free enzyme:

$$E + S \underset{k_{-1}}{\overset{k_1}{\rightleftharpoons}} ES \xrightarrow{k_p} E + P \tag{1}$$

If it is assumed that ES is formed at the same rate at which it breaks down to E + P (a "steady-state" assumption), and that the concentration of S is much larger than the concentration of E, then the change in reaction velocity, v, relative to changes in $[S]$, is described by the Michaelis–Menten equation:

$$\frac{v}{V_{max}} = \frac{[S]}{(K_M + [S])} \tag{2}$$

In other words, the velocity expressed as a proportion of the maximum possible velocity (V_{max}) is dependent both upon the substrate concentration, $[S]$, and on a constant, K_M, referred to as the Michaelis constant.

Velocity data may be plotted in any one of a number of ways to illustrate the relation between v and $[S]$, and plotting procedures are detailed later in this chapter. However, the Michaelis–Menten equation is an equation for a rectangular hyperbola, and plotting v versus $[S]$ yields a hyperbolic curve, in the absence of cooperative or other unusual behaviors (*Figure 4-3*).

Figure 4-3
A hyperbolic Michaelis–Menten plot for a simple unireactant enzyme system

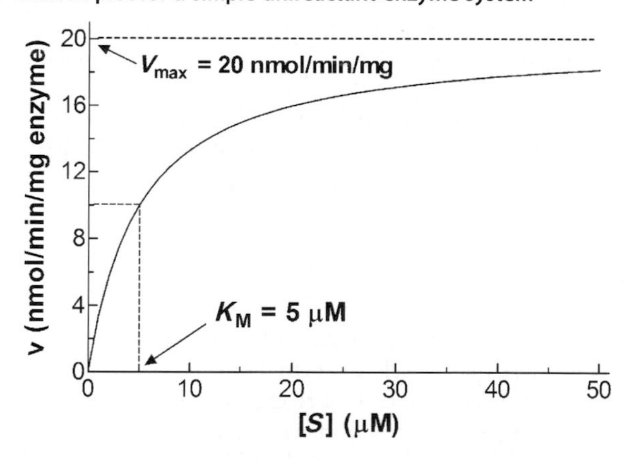

4.1.2 What Is K_M and Why Do We Want to Determine Its Value?

The K_M value for a particular enzyme–substrate combination under a given set of reaction conditions is the concentration of substrate at which the enzyme is working at exactly 50% of V_{max}. From Eq. 1, it is tempting to assume that K_M represents a dissociation constant for the ES complex. However, this is only true if k_p (also referred to as k_2 or k_{cat}) is very small relative to k_{-1}. When k_p is large relative to k_{-1}, K_M is essentially a kinetic constant for the formation of product. Thus, K_M is a dynamic or pseudoequilibrium constant, related to k_1, k_{-1}, and k_p (Segel, 1993):

$$K_M = \frac{k_{-1} + k_p}{k_1} \quad (3)$$

There are several reasons why it is useful to determine K_M (Segel, 1993):

If K_M for an endogenous substrate is known, then it establishes an approximate value for the endogenous concentration of that substrate. This is because it would not make sense kinetically or metabolically for a substrate to be present at concentrations significantly below or above its K_M value. In the former case, the majority of enzyme protein molecules, synthesized by the cell at significant metabolic cost, would be redundant, and in the latter case, the cell would be unable to respond immediately to fluctuations in substrate concentration because enzymes would approach saturation and v would thus vary only marginally with fluctuations in $[S]$.

Under given reaction conditions, K_M is a constant for a particular enzyme–substrate combination. Thus, enzymes from different tissues, or from different species, or from tissue from a single species at different stages of development, can be compared easily to assess whether the proteins have subtle structural and mechanistic differences, or indeed, whether entirely different enzymes might be responsible for catalyzing the same reaction.

Some endogenous ligands may alter a K_M in an allosteric manner; for example, K_M determined in vitro, which seems unreasonably high based on other evidence might suggest the presence of an endogenous allosteric enzyme activator.

It is necessary to determine K_M before any assessment of the effects of potential inhibitors (or activators) that may act through altering K_M.

4.1.3 Experimental Details

It has already been stated that a suitable quantitative assay technique must be available to measure the reaction of interest and it is assumed that the experimenter has determined optimal reaction conditions for the enzyme of interest. All kinetic assay techniques assume that v is a variable and that $[S]$ is known; as such, preparation of substrate must be meticulous in terms of ensuring that concentrations are correct, and this in turn will rely upon factors such as good weighing and pipetting techniques with calibrated instruments capable of precise, accurate, and sufficiently sensitive measurement.

If an approximate K_M value for the enzyme–substrate combination of interest is known, a full-scale kinetic assay may be done immediately. However, often an approximate value is not known and it is necessary first to do a "range finding" or "suck and see" preliminary assay. For such an assay, a concentrated substrate solution is prepared and tenfold serial dilutions of the substrate are made so that a range of substrate concentrations is available within which the experimenter is confident the K_M value lies. Initial velocities are determined at each substrate concentration, and data may be plotted either hyperbolically (as v versus $[S]$) or with $[S]$ values expressed as $\log_{10}$ values. In the latter case, a sigmoidal curve is fitted to data with a three parameter logistic equation (➋ Eq. 4):

$$v = \text{Bottom} + \frac{(V_{max} - \text{Bottom})}{1 + 10^{(\log K_M - \log[S])}} \quad (4)$$

The "Bottom" of the curve should ideally equal zero, if appropriate blank assays have been included; in other words, v should equal zero when $[S]$ is zero.

Results from an appropriate range-finding assay done before obtaining data shown in ➋ *Figure 4-3* are presented in ➋ *Figure 4-4*; since such a curve will span a region between 10% and 90% of V_{max} over an 81-fold substrate concentration range, 10-fold substrate concentration separations in the range-finding assay will inevitably yield at least one point on the steep part of the curve, facilitating estimation of K_M.

If an approximate K_M value is known, a kinetic assay may be done to obtain an accurate determination of K_M and V_{max}. This is achieved simply by determining values for v (initial velocities) for a range of appropriate substrate concentrations.

Figure 4-4

Range finding v versus $[S]$ data fitted to Eq. 4 yields a sigmoidal curve. The estimated K_M value, the concentration of substrate at which $v = 10$ μmol/h/mg, is 5 μM in this example. At least one, and usually two data points will lie on the steep portion of a sigmoidal curve if each $[S]$ is tenfold higher than the last

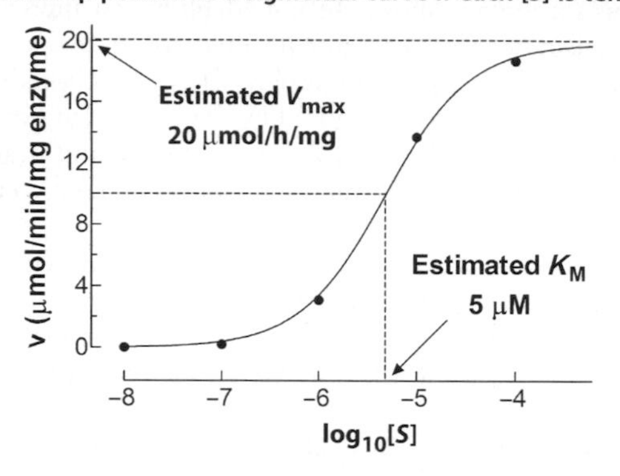

Because K_M represents the substrate concentration at which reaction velocity is half of V_{max}, it is not possible to determine K_M with any great certainty in the absence of a reasonable estimate for V_{max}. In other words, although the substrate concentration range must surround the K_M value, it must also extend beyond K_M to a sufficient degree that the enzyme is also measured working close to V_{max}. A range of $1/2 \times K_M$ to $10 \times K_M$ is usually sufficient (Henderson, 1993), with concentrations spaced more closely at low values of $[S]$. An absolute minimum of five substrate concentrations should be included, not including the blank assay (equivalent to zero substrate), while twelve concentrations are almost always more than adequate (Henderson, 1993). Furthermore, it is always advisable to make several determinations at a single value of $[S]$ (four determinations should really be considered the minimum) and more robust curve fits are usually obtained from fewer points with a greater number of replicates than from more points of fewer replicates. So, in a platereader-based assay, for example, if 36 wells are available for a particular kinetic assay, a more representative curve would be obtained from quadruplicate determinations at nine concentrations than from triplicate determinations at twelve concentrations.

There is also some merit in choosing substrate concentrations that result in incremental increases in v of similar magnitude. *Table 4-2* indicates suitable substrate concentrations, expressed as fractions or multiples of the K_M value, leading to consistent increases in v, and this could be used as a template when designing a kinetic assay.

Table 4-2

Suggested substrate concentrations (relative to K_M) yielding consistent incremental increases in v

$[S]$ ($\times K_M$)	(0)	0.25	0.4	0.55	0.8	1.1	1.5	2.1	3.2	5	10
V (% V_{max})	(0)	20	29	35	44	52	60	68	76	83	91

Figure 4-5 illustrates the relationship between $[S]$ and v when substrate concentrations are chosen based on the data shown in *Table 4-2*.

4.1.4 Plotting Raw Data

Historically, data have been transformed to facilitate plotting on linear plots such as Lineweaver–Burk ($1/v$ versus $1/[S]$), Hanes–Woolf ($[S]/v$ versus $[S]$), or Eadie–Hofstee ($v/[S]$ versus v). However, with the present availability of affordable nonlinear regression and graphing software packages such as GraphPad Prism,

Figure 4-5
A hyperbolic plot illustrating the (approximately) equal incremental increases in *v* when [*S*] is increased with respect to K_M as indicated in Table 4-2

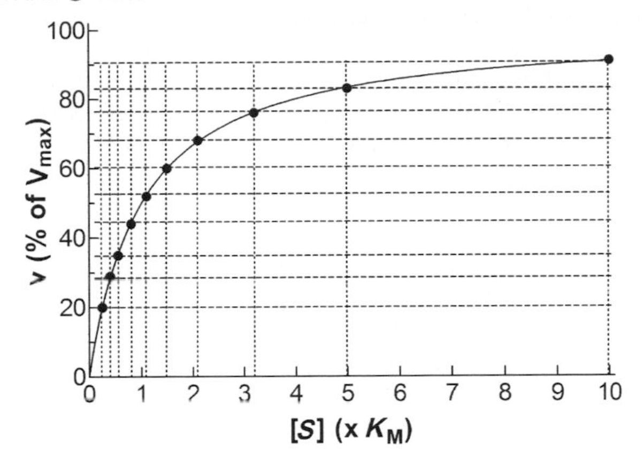

there are no reasons whatsoever to support the continued use of such linear approaches to determine kinetic constants. Indeed, the Lineweaver–Burk and Eadie–Hofstee plots introduce unnecessary errors as a result of the transformations done. Accordingly, there are two approaches, outlined here, which can be recommended for different reasons to obtain kinetic constants. That said, there remains justification for presenting data in the form of a linear plot – particularly on a Lineweaver–Burk plot – simply for display purposes, as diverse inhibitor mechanisms are often readily identifiable from such plots.

The direct linear plot (*Figure 4-6*) requires no transformation of data, and this approach is arguably the most appropriate for determining K_M and V_{max} values. Individual values of [*S*] are plotted as their negatives on the *x*-axis and measured values of *v* are plotted on the *y*-axis. Corresponding pairs of (−[*S*], 0)

Figure 4-6
A direct linear plot made from seven pairs of (*v*, [*S*]) data. The dotted lines mark the lowest and highest points of intersection. Clearly, a graph showing 21 horizontal and 21 vertical dotted lines, equivalent to the number of intersections from seven data pairs (see text), would be cluttered and difficult to interpret, and these lines are not shown. Rather, the hatched lines indicate K_M and V_{max} values obtained from nonlinear regression of the same data; not surprisingly, these lie close to the median intersection points that would be obtained from a full direct linear analysis

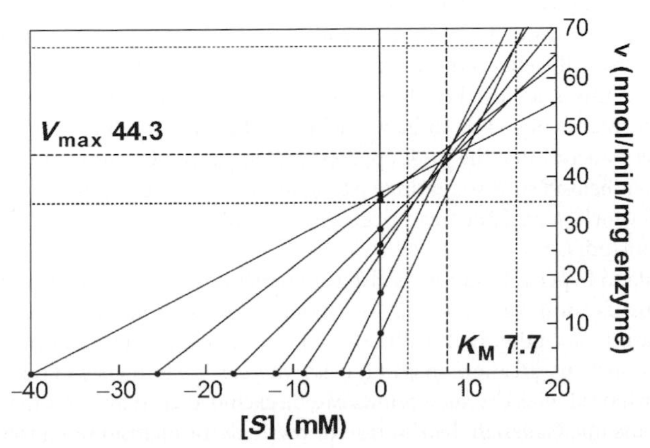

and $(0, v)$ values are then joined by a straight line, and each line extrapolated to the right of the y-axis. In the perfect experiment, all extrapolated lines will intersect at a single point, with coordinates (K_M, V_{max}). However, in reality, experimental error dictates that a series of intersections will be obtained; for n determinations of v, there will be $[n(n-1)/2]$ intersections. Thus, for the data shown in ➲ *Figure 4-6*, where v was determined at only seven substrate concentrations, there will be $[7(6)/2]$ or 21 intersections. Each of these intersections is then extrapolated vertically to the x-axis, and horizontally to the y-axis (not shown in ➲ *Figure 4-6*). Of all the points at which these vertical lines intersect the x-axis, the best K_M value is the *median* value (in the case of ➲ *Figure 4-6*, the value of the 11th of 21 intersection points). Similarly, the best V_{max} value is the *median* intersection point with the y-axis of all horizontal lines drawn. If there were an even number of intersections with the axes, for example, 20 intersections, the best value for the kinetic constant of interest would lie midway between the 10th and 11th intersections.

It should be immediately apparent that plotting kinetic data in this manner quickly results in cluttered graphs, and it becomes rather difficult to locate appropriate intersection points among the resulting jumble of lines. In ➲ *Figure 4-6*, only the lowest and highest "V_{max}" and "K_M" values have been indicated (dotted lines), whereas the hatched lines show values for V_{max} and K_M determined by nonlinear (hyperbolic) regression of the raw data. Not surprisingly, the values obtained by nonlinear regression lie in the same area in which median intersection values would be expected to lie. One can imagine that determining v at ten or more substrate concentrations would result in a graph too cluttered to be practical.

Further drawbacks associated with the direct linear plot include the fact that this analysis does not readily lend itself to standard computerized graphing methods (for example, use of GraphPad Prism), although specialized software is available (Henderson, 1993). Of course, one of the major advantages of the direct linear plot is the ability to obtain kinetic constants by eye, without the need for a computer. However, for presentation purposes, the use of graphing software is still desirable. Furthermore, any behavior more complicated than simple, single substrate kinetics – for example, turnover in the presence of an inhibitor, or multisubstrate kinetics – cannot readily be shown on a direct linear plot. This is in contrast with the flexibility afforded by nonlinear regression approaches.

Fitting velocity data directly to a hyperbolic curve has several advantages over linear methods, transformed or otherwise. The major advantages are that no transformation of data is necessary, curves are fitted easily with currently available graphing software, and variations in behavior from a simple Michaelis–Menten one-substrate equation usually result in an equation which still describes a hyperbola, thus requiring no change in the analytical approach.

There are, of course, potential drawbacks to this procedure. In the literature, curve fitting has generally been achieved through unweighted least-squares nonlinear regression analysis. An unweighted approach is perfectly acceptable if random errors in measured replicate values of v are normally distributed, and if there is no systematic variation between standard deviations going from low to high values of v. This said, it would likely prove time-consuming and perhaps expensive to run tests for normality and distribution of errors. Nevertheless, evidence increasingly supports the likelihood that in a great many cases, measurement errors are more likely to increase at higher values of v (Storer et al., 1975; Nimmo and Mabood, 1979). Thus, in order that all determined values of v have a similar influence on goodness-of-fit, it may be worthwhile at least to assess the effects of incorporating a relative weighting scheme (weighting by $1/Y^2$) on the goodness-of-fit, and this is achieved easily with software packages such as GraphPad Prism. However, if altering weighting schemes is observed to have a marked effect on goodness-of-fit, the responsibility lies with the experimenter to determine the most appropriate weighting scheme to use. The relative merits and practical considerations behind such weighting systems are discussed in a useful publication (Motulsky and Christopoulos, 2003) which is available free-of-charge as an online resource (www.graphpad.com/manuals/prism4/RegressionBook.pdf).

Fitting kinetic data to hyperbolic curves is also very sensitive to the presence of outlying points; if there is an experimental justification for removing such points (for example, the experimenter is aware of an error made with regard to an individual reading) then one is justified in excluding the point in question. Even in the absence of such justification, methods exist whereby the influence of outliers on goodness-of-fit can be minimized, or indeed whereby such points can be excluded entirely from the analysis. Curve-fitting methods that do not assume Gaussian distribution of residuals, or inclusion of a weighting function (such as that used in the Tukey Biweight approach), are appropriate means by which this might be achieved (Motulsky and Christopoulos, 2003).

If kinetic plots are constructed from a sufficient number of points composed of an appropriate number of replicate values, and if both the assay system used and the experimenter's technique are consistent, robust, accurate, and precise (including meticulous preparation of substrate concentrations), it is likely that variations in best-fit K_M and V_{max} values (and ranges) with different weighting procedures will be relatively small, and perhaps negligible. The extreme example of this would be the "perfect" experiment, in which even plotting data on a Lineweaver–Burk graph would yield kinetic constants identical to the "true" values. The graphing approach favored by the author is to determine kinetic constants from hyperbolic fits without weighting of points, whereas Lineweaver–Burk plots are often used *for display purposes only.* Indeed, the "best-fit" straight line on these plots may have intercepts and slopes constrained to values calculated from K_M and V_{max} values obtained from nonlinear regression of hyperbolae. Thus, for example, when the slope of a Lineweaver–Burk plot (which equals K_M/V_{max}) must be determined to allow a secondary slope replot to be made (see Section 5.2.3 later), the "slope" is actually calculated from K_M and V_{max} values determined from a hyperbola; the Lineweaver–Burk plot itself is *never* used to determine kinetic constants from velocity data.

4.1.5 Circumstances Under Which Enzyme Behavior Is Not Fully Described by the Michaelis–Menten Equation

Under many circumstances, the behavior of a simple unireactant enzyme system cannot be described by the Michaelis–Menten equation, although a v versus $[S]$ plot is still hyperbolic and can be described by a modified version of the equation. For example, as will be discussed later, when enzyme activity is measured in the presence of a competitive inhibitor, hyperbolic curve fitting with the Michaelis–Menten equation yields a perfectly acceptable hyperbola, but with a value for K_M which is apparently different from that in the control curve (*Figure 4-7*). Of course, neither the affinity of the substrate for the active site nor the turnover number for that substrate is actually altered by the presence of a competitive

Figure 4-7
Hyperbolic curve fits to control enzymatic data and to data obtained in the presence of a competitive inhibitor. Curve fitting to the Michaelis–Menten equation results in two different values for K_M. However, K_M does not, in actuality, change, and the value in the presence of inhibitor (15 μM) is an apparent value. Fitting with the correct equation, that for turnover in the presence of a competitive inhibitor (Eq. 5), results in plots identical in appearance to those obtained with the Michaelis–Menten equation. However, nonlinear regression now reveals that K_M remains constant at 5 μM and that $[I]/K_i = 2.5$; with knowledge of $[I]$, calculation of K_i is straightforward

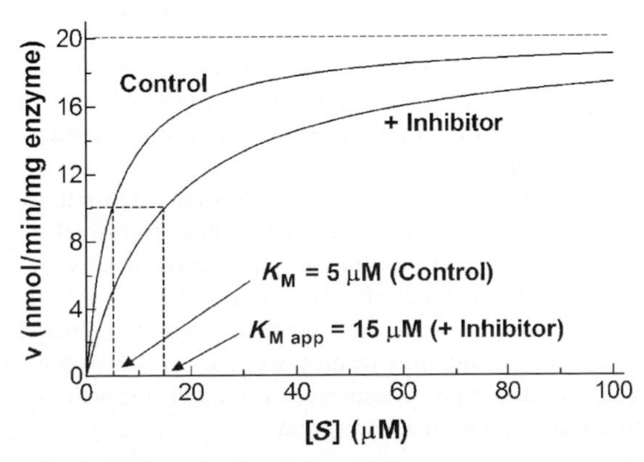

inhibitor, and the new K_M value obtained is not a real value, but an apparent value determined in part by the concentration of the competitive inhibitor present, and its affinity for the active site. The reaction scheme for the interaction of a simple full competitive inhibitor with a unireactant enzyme system can be drawn as follows:

$$\begin{array}{l} E + S \overset{Ks}{\longleftrightarrow} ES \overset{k_p}{\longrightarrow} E + P \\ + \\ I \\ \updownarrow K_i \\ EI \end{array}$$

From such a scheme, it is possible to derive an equation describing the relationship between v and $[S]$ in the presence of a competitive inhibitor, I, with a dissociation constant of K_i for the enzyme (➋ Eq. 5):

$$v = \frac{V_{max}[S]}{K_M\left(1 + \frac{[I]}{K_i}\right) + [S]} \tag{5}$$

It is apparent that ➋ Eq. 5 is a variation of the Michaelis–Menten equation. The inhibitor data shown in ➋ *Figure 4-7* can instead be fitted to ➋ Eq. 5, holding K_M (and V_{max}) constant to their control values (5 μM and 20 nmol/min/mg, respectively). The curve obtained is identical to that fitted with the Michaelis–Menten equation (➋ *Figure 4-7*), but nonlinear regression now yields the information that K_i of the inhibitor equals 40% of the concentration at which it was included in the assay to obtain the best-fit curve. In other words, if the concentration of inhibitor present in the experiment shown in ➋ *Figure 4-7* was 25 μM, the K_i for the inhibitor is 10 μM.

Although inhibition kinetics will be discussed in more detail later (➋ Section 5), the reason for providing this example is to illustrate that it will often be necessary to use a variation of the Michaelis–Menten equation. If the reason for doing so is to obtain inhibitor dissociation constants, as in the earlier example, then most basic enzymology texts list equations, such as ➋ Eq. 5, appropriate for simple inhibitors that act through recognized mechanisms. For example, equations are typically supplied describing competitive and noncompetitive inhibition, and sometimes uncompetitive inhibition and the behavior of Lineweaver–Burk plots in the presence of these simple inhibitors could be considered as being relatively common knowledge. The problems begin when the behavior of an enzyme–inhibitor system is not described adequately by one of these common mechanisms. This occurs more often when substrate binding displays evidence of cooperativity, or when inhibitors bind to one or more allosteric sites distinct from the active site. What should be done when such a situation is encountered?

4.1.6 Addressing More Complex Behavior

If unireactant enzyme behavior is not described adequately by existing equations, a reaction scheme should be constructed which seems to describe the behavior based on available experimental data. In doing so, Occam's razor should be applied and the simplest possible explanations should be first discounted before more complex explanations are proposed.

For example, experimental data might reveal that a novel enzyme inhibitor causes a concentration-dependent increase in K_M, with no effect on V_{max}, and with Lineweaver–Burk plots indicative of competitive inhibition. However, even at very high inhibitor concentrations and very low substrate concentrations, it is observed that the degree of inhibition levels off when some 60% of activity still remains. Furthermore, it has been confirmed that only one enzyme is present, and all appropriate "blank" rates have been accounted for. It is clear that full competitive inhibition cannot account for such observations because complete inhibition can be attained at infinitely high concentrations of a full competitive inhibitor. Thus, it is likely that the inhibitor binds to the enzyme at an allosteric site.

The simplest reaction scheme describing these observations might be drawn as follows:

$$
\begin{array}{ccc}
E + S \xleftrightarrow{Ks} & ES \xrightarrow{k_p} E + P \\
+ & + \\
I & I \\
\updownarrow K_i & \updownarrow \alpha K_i \\
EI + S \xrightarrow{\alpha Ks} & ESI \xrightarrow{k_p} EI + P
\end{array}
$$

Thereafter, a reference text such as "Enzyme Kinetics" (Segel, 1993) should be consulted to determine whether or not the proposed mechanism has been described and characterized previously. For the example given, it would be found that the proposed mechanism corresponds to a system referred to as partial competitive inhibition, and an equation is provided which can be applied to the experimental data. If the data can be fitted successfully by applying the equation through nonlinear regression, the proposed mechanism would be supported; further secondary graphing approaches to confirm the mechanism are also provided in texts such as "Enzyme Kinetics," and values could be obtained for the various associated constants. If the data cannot be fitted successfully, the proposed reaction scheme should be revisited and altered appropriately, and the whole process repeated.

It is possible, of course, that a devised reaction scheme may be novel and that no existing equation can be identified. In such a case, it will be necessary to derive an equation to describe the proposed mechanism, and thereafter to attempt once again to fit the experimental data by nonlinear regression. If this is deemed necessary, there are two approaches that might be taken.

4.1.7 Rapid Equilibrium and Steady-State Approaches to Derive Equations

The rapid equilibrium approach for deriving rate equations (Segel, 1993) is the simplest approach available. This approach assumes that only the early components of a reaction are at equilibrium:

$$E + S \underset{k_{-1}}{\overset{k_1}{\rightleftharpoons}} ES \xrightarrow{k_p} E + P$$

In order for an equilibrium to exist between E + S and ES, the rate constant k_p would have to be much smaller than k_{-1} However, for the majority of enzyme activities, this assumption is unlikely to hold true. Nevertheless, the rapid equilibrium approach remains a most useful tool since equations thereby derived often have the same form as those derived by more correct steady-state approaches (see later), and although steady-state analyses of very complex systems (such as those displaying cooperative behavior) are almost impossibly complicated, rapid equilibrium assumptions facilitate relatively straightforward derivations of equations in such cases.

A description of the approach taken to derive a rapid equilibrium equation is beyond the scope of this chapter. However, a step-by-step guide has been published (Segel, 1993) with which an equation describing even a rather complex system can be generated in a very short time. Equations can then be programmed into software such as GraphPad Prism and curve fitting becomes a rapid and straightforward process.

The steady-state approach for deriving rate equations makes no assumptions regarding the relative magnitudes of k_p and k_1. The ES complex need not be in equilibrium with E and S, but the assumption is made that soon after initiation of an enzyme reaction, the concentration of ES would reach a near constant or steady-state level. Early theories devised by Briggs and Haldane in 1925 are mathematically taxing and do not lend themselves easily to use by the nonexpert. An improved approach was devised by Cleland and it is his system which has largely been adopted by enzymologists (Cleland, 1963a, b, c; Segel, 1993). However, as the number of substrates and products, and thus the number of enzyme species, increases, the mathematics associated with Cleland's approach becomes increasingly complex. To facilitate steady-state kinetic analyses of more complex systems, King and Altman have devised a method based on Cleland's approach which can be applied to systems of moderate complexity by nonexperts (King and Altman, 1956; Segel, 1993). Once again, a description of these methodologies is beyond the scope of this chapter, and the reader is directed to the step-by-step approach outlined by Segel if it becomes necessary to derive an equation under steady-state conditions (Segel, 1993).

It is possible to test the applicability of the rapid equilibrium or steady-state mechanisms to a particular enzyme reaction, if the concentration of enzyme is known (Price and Stevens, 1999). If

$$v = k_p[ES] \tag{6}$$

then

$$v = k_p \frac{[E][S]}{K_M + [S]} \tag{7}$$

From Eq. 3, if $k_p < k_{-1}$, then $K_M = k_p/k_1$. Therefore, k_p/K_M (more usually written as k_{cat}/K_M) is equal to k_1, the rate constant for the association of substrate with enzyme. When the magnitude of k_1 is determined only by rates of diffusion, and thus is as fast as is physically possible, its value lies in the region of 10^9 $M^{-1}s^{-1}$. Thus, if k_{cat}/K_M has a value higher than around 10^7 $M^{-1}s^{-1}$ then steady-state conditions are present. However, equilibrium conditions apply if $k_p < k_{-1}$, because $K_M = k_{-1}/k_1$, and k_{cat}/K_M would have a value much lower than 10^9 $M^{-1}s^{-1}$.

4.2 Determining Kinetic Constants in Bireactant and Multireactant Systems

Procedures necessary to determine correctly kinetic constants for reactions involving more than one substrate or product can be a great deal more complex than those described earlier for single substrate reactions. For most neurobiologists, the requirement is simply to obtain a "correct" K_M and V_{max} before undertaking subsequent studies, with the minimum of fuss, or to determine the mechanism and/or potency of an inhibitor of the enzyme of interest. There are ways in which relatively straightforward experiments can be done to yield meaningful kinetic constants in such reaction systems. Nevertheless, it is useful to have a general understanding of what factors influence kinetic constants in these more complex systems so that the experimenter more fully appreciates the limitations of simplified approaches.

4.2.1 Reaction Nomenclature

Involvement of one, two, three, or four substrates or products is signified by combining two of the terms *Uni, Bi, Ter,* and *Quad,* respectively. Therefore, a reaction using two substrates and yielding one product is referred to as a "Bi–Uni" reaction.

In a *sequential* reaction, all the substrates must be bound to the enzyme before any release of product can occur. Sequential systems can be either *ordered* or *random.* In an ordered sequential reaction, substrates must bind to the enzyme in a particular order, whereas in a random sequential system, substrates may bind to the enzyme in any order. In reaction schemes, substrates are usually abbreviated as A, B, C, and D in the order that they bind to the enzyme, whereas products are abbreviated as P, Q, R, and S in the order that they leave the enzyme. Sequential binding of substrates is a consequence of their orientation within the enzyme active site.

In a *nonsequential* reaction, following binding of some, but not all, substrates, release of one or more products occurs prior to binding of remaining substrates and release of remaining products. *Ping Pong* reactions are examples of nonsequential kinetics. For example, the enzyme monoamine oxidase catalyzes a Ping Pong reaction, binding amine substrate and then releasing ammonia anaerobically. Thereafter, binding of molecular oxygen as a substrate is followed by the release of hydrogen peroxide as a product. Naming Ping Pong reactions using the terminology outlined earlier can be made more useful by breaking down the binding order into component parts. For example, a Ping Pong Bi-Ter reaction (two substrates, three products) might be written in longer form as a Uni Uni Uni Bi Ping Pong reaction if binding of one substrate is followed by the release of one product, then binding of a second substrate and release of two more products. In contrast, an ordered Bi-Ter reaction would require that both substrates bind before all three products are released.

4.2.2 Equations for Multireactant Systems

Multireactant kinetics are generally more complex than unireactant kinetics, not just because more species are involved but also because each substrate has its own affinity for the enzyme which may be influenced by the presence of other substrates bound to the protein. The possibility of ordered reactions or nonsequential kinetic behavior adds further to the potential complexity. The cytochromes P_{450} exemplify such a diverse range of behaviors, and thus the diversity of kinetic approaches required, in a single enzyme superfamily (Shou et al., 2001). Segel has provided an extensive choice of suitable equations to describe a great many multireactant systems, along with details of how these systems can be identified graphically (Segel, 1993). These equations take into account the effects of different concentrations of all substrates, each with their own affinity for the active site. If a suitable equation cannot be found, it is nevertheless possible to write one, in which case the approach of King and Altman (see ➔ Section 4.1.7) is perhaps the most user-friendly (King and Altman, 1956; Segel, 1993).

More often than not, the neuroscientist will be interested in one particular substrate (or product) of a multisubstrate enzyme. Furthermore, an inhibitor of interest will often bind to the binding site for only one of several substrates in such a system. For example, GABA aminotransferase (EC 2.6.1.19) catalyzes a Ping Pong Bi-Bi reaction with 4-aminobutanoate (GABA) and 2-oxoglutarate as substrates, and succinate semialdehyde and glutamate as products. The compound 4-amino-hex-5-enoic acid (vigabatrin) is a substrate analog of GABA which acts as a potent inhibitor at the GABA-binding site (Lippert et al., 1977). Thus, any kinetic examination of the effects of this inhibitor might focus on interactions of the inhibitor with the GABA-binding site, and would be largely unconcerned with interactions between the enzyme and 2-oxoglutarate. Is it necessary to apply a multireactant equation to this system in order to assess the inhibitory potency and mechanism of vigabatrin?

Thankfully, such multisubstrate systems can usually be simplified quite markedly, facilitating the use of unireactant equations and graphing procedures. In a multisubstrate system, the "true" K_M and V_{max} values for a particular substrate are those determined when all other substrates are present at saturating concentrations, and when concentrations of products, which might inhibit enzyme activity, are negligible. Accordingly, in the example given earlier, it would be acceptable to assess the kinetic behavior of the enzyme at a range of concentrations of GABA in the presence of a fixed saturating concentration of 2-oxoglutarate, and then to repeat these analyses in the presence of the inhibitor of interest, which is known to interact with the enzyme at the GABA-binding site. In the absence of complicating effects due to substrate cooperativity, such a system could be analyzed appropriately through the application of unireactant kinetic principles.

5 Measuring Enzyme Inhibition

Many drugs are effective as a result of inhibiting one or more enzymes. Enzyme inhibitors reduce the rate of formation of product. In a closed system such as an enzyme assay in a test tube, they do not alter the amount of product that is ultimately generated; rather, it takes longer in the presence of an inhibitor to generate a given amount of product.

Drugs can inhibit enzymes through several general mechanisms. The mechanism by which inhibition occurs dictates the approach by which the inhibition should be quantified. Inhibitors may act to alter K_M, or V_{max}, or both; in this context, "alter" may mean that the true values of K_M and V_{max} really have been altered, or it may mean that these values have *apparently* changed in the presence of the inhibitor but that, in actual fact, they have not. In either case, it is vital that before any inhibitor has been introduced to a system, a robust and reproducible assay to measure K_M and V_{max} is in place, and good estimates for these kinetic constants have been obtained.

5.1 Reversible, Irreversible, and Tight-Binding Inhibitors

5.1.1 Definitions and Importance

Binding of a *reversible inhibitor* to an enzyme is rapidly reversible and thus bound and unbound enzymes are in equilibrium. Binding of the inhibitor can be to the active site, or to a cofactor, or to some other site on the protein leading to allosteric inhibition of enzyme activity. The degree of inhibition caused by a reversible inhibitor is not time-dependent; the "final" level of inhibition is reached almost instantaneously, on addition of inhibitor to an enzyme or enzyme–substrate mixture.

Binding of an *irreversible inhibitor* is not reversible and while there may be an initial reversible stage, the enzyme is ultimately unable to turn over substrate, and synthesis of new enzyme protein may be necessary to regain activity. In such cases, binding is usually covalent. Again, binding of the inhibitor can be to the active site, or to a cofactor, or to some other site on the protein leading to allosteric inhibition of enzyme activity. If binding is to a cofactor which itself binds reversibly to the enzyme protein, enzyme activity may be regenerated through exchange of the cofactor–inhibitor complex with an unbound cofactor from bulk solvent. Such a situation rather complicates assessment of inhibition. The degree of inhibition caused by an irreversible inhibitor may be time-dependent, in as much as it may be necessary to preincubate the enzyme with the inhibitor for a period of time to achieve the maximum attainable degree of inhibition.

Some inhibitors bind tightly to the enzyme and may remain bound for some time, but ultimately dissociate from the enzyme. These are termed *tight-binding inhibitors.* If the initial association between enzyme and a tight-binding inhibitor exhibits time-dependence, the term *slow tight-binding inhibitor* is used. Binding of tight-binding inhibitors can, as with other types of inhibitors, occur at several possible places on an enzyme molecule. Analysis of tight-binding inhibition is complicated by the fact that an appreciable reduction in the concentration of unbound inhibitor may occur following addition of the inhibitor to the enzyme. If appreciable inhibition occurs at a concentration of the inhibitor similar to the concentration of enzyme present, tight-binding inhibition should be suspected.

In some cases, an inhibitor can bind to more than one site on an enzyme protein, with inhibition resulting from binding at multiple sites. Binding affinities at the two (or more) sites may be different, and mechanisms of inhibition may be different; for example, high-affinity inhibition might occur through an allosteric site and lower affinity inhibition through the active site. Analysis of such systems is complex and may require a combination of several of the approaches outlined later.

Prior to any kinetic analysis of inhibitory behavior, it is necessary to determine whether inhibition is reversible, irreversible, or tight binding. This is because the approaches used to quantify inhibitor potencies are different in each case. Furthermore, use of an inappropriate procedure often leads to an incorrect interpretation of results. For example, the literature is littered with inhibitor data showing "noncompetitive" or "mixed" inhibitory effects of a great many compounds on a great many enzymes. The important implication from such results is that the inhibitor of interest binds to a regulatory site, rather than to the active site of these enzymes. However, inhibitors that bind tightly or irreversibly *to the active site* will often yield plots suggesting noncompetitive or mixed inhibition when the experimental and graphing procedures used were those suitable only for reversible inhibitors.

5.1.2 Assessing Reversibility of Inhibition

Reversibility of inhibition can be assessed by dialyzing the inhibited enzyme versus buffer, or by passing the inhibited enzyme down a desalting column, or by a dilution procedure. In each case, an uninhibited (control) sample of enzyme should be treated in parallel with the inhibited enzyme sample, as control enzyme activity often changes to some degree as a result of the experimental procedure.

In a dialysis procedure, enzyme should be preincubated with a concentration of inhibitor sufficient to reduce activity by at least 90%. A solution containing the inhibited enzyme is then placed inside a small bag fashioned from dialysis membrane with a molecular weight cutoff value substantially lower than the mass of the enzyme. The sealed dialysis bag is then placed in a beaker containing a large volume of buffer, and is left for a period of time in order for equilibration of unbound inhibitor to occur between the inside and the outside of the dialysis bag. Stirring the external buffer facilitates diffusion of inhibitor, as does dialyzing at room temperature, or at 37°C, if thermostability or proteolytic degradation are not matters for concern; otherwise, dialysis is most often done in a cold room, at 4°C. Replacement of the dialysis buffer is also advisable in the interest of economy; dialysis of a 1 mL volume of enzyme versus two consecutive volumes of 100 mL of buffer (resulting in a 10,000-fold dilution of inhibitor) is equivalent to a single dialysis versus a volume of 10 L. A total dilution factor of at least 1,000 should be incorporated into the design of a dialysis experiment when buffer volumes are being considered.

At the end of the dialysis procedure, the bag is blotted dry and enzyme solution is removed prior to assessment of activity. It may be necessary to include cofactor in the assay mixture if it is possible that a dissociable cofactor was lost from the enzyme during dialysis. As the volume containing enzyme inside the dialysis bag changes to some degree during the dialysis procedure, it is usually necessary to correct measured enzyme activity to reflect this change in volume, and a correction based on protein concentration is often done in this regard. It is normal for activity thus measured to be expressed as a fraction of that in a parallel (dialyzed) control experiment.

A (rapidly) reversible inhibitor will permit rapid and complete recovery of enzyme activity by dialysis. However, irreversible inhibitors are not removed by this procedure. Recovery from tight-binding inhibition is usually slow; it is not uncommon for several dialysis bags containing enzyme to be prepared and for activity in each to be determined at various time points following the commencement of dialysis. The "off-rate" of these inhibitors is generally more rapid at higher temperatures.

Desalting columns provide a more rapid means by which inhibitor reversibility might be assessed. Disposable size-exclusion (desalting) columns can be purchased from companies such as Bio-Rad. When an enzyme–inhibitor mixture is passed down a desalting column, unbound inhibitor is able to enter small pores in the resin from which the enzyme is excluded, because of its size. This has the effect of diluting unbound inhibitor in the microenvironment of each enzyme molecule, resulting in dissociation of a reversible inhibitor in an attempt to reach a new point of equilibrium, and is analogous to the effects of dialysis. Irreversible inhibitors will, of course, remain bound to the enzyme. Once again, a control enzyme should be "desalted" in a parallel experiment. If instructions for using the column are followed carefully, enzyme (with or without bound inhibitor) will appear in an initial small elution volume, and it is this enzyme fraction that should be assessed for activity. Unbound inhibitor will wash through the column shortly thereafter, if further buffer is applied to the top of the column, and if this is allowed to drain into the fraction containing the enzyme, a reversible inhibitor will simply reassociate with free enzyme. Transit time of enzymes through such columns is very rapid, with the result that tight-binding inhibitors with a slow "off-rate" may remain bound to the enzyme, and this procedure does not readily allow differentiation between these and irreversible inhibitors.

A rapid dilution procedure is routinely used in the author's laboratory to assess reversibility, and it is particularly enlightening if enzyme activity is then determined in a continuous spectrophotometric assay. A microplate assay is set up, in triplicate, as outlined in ➋ *Table 4-3*, later. It is assumed, for the purposes of this example, that the K_M for substrate is 10 μM and that the IC_{50} for the inhibitor in the presence of 10 μM substrate is 200 μM. The concentration of enzyme used in control wells should be at least tenfold greater than the minimum concentration necessary to catalyze a quantifiable increase in product concentration over the duration of the incubation of substrate with enzyme.

Table 4-3
Assay setup for a dilution experiment to assess inhibitor reversibility

Assay group	Initial volume	Enzyme	[Inhibitor]	[Substrate]
(1) Control	220 μL	1 Unit	0	10 μM
(2) Diluted control	180 μL	–	–	10 μM
(3) Inhibitor 300 μM	220 μL	1 Unit	450 μM	10 μM
(4) Diluted inhibitor 300 μM	180 μL	–	–	10 μM
(5) Inhibitor 30 μM	200 μL	0.1 Unit	45 μM	10 μM

Assays are set up containing enzyme and inhibitor (or water for controls) in the initial volumes shown. Inhibitor concentrations listed are initial concentrations. Enzyme and inhibitor should be preincubated together if time-dependent inhibition is suspected. The assay is started by transfer of 20 μL of the contents of each well in groups 1 and 3 to a corresponding well in groups 2 and 4, respectively. Thereafter, 100 μL of substrate at an initial concentration of 30 μM (3 × K_M) is added to all wells. The reaction wells then contain components at the concentrations listed in *Table 4-4*.

Table 4-4

Final reaction conditions for a dilution experiment to assess inhibitor reversibility

Assay group	Final volume	Enzyme	[Inhibitor]	[Substrate]
(1) Control	300 μL	0.9 Unit	0	10 μM
(2) Diluted control	300 μL	0.1 Unit	0	10 μM
(3) Inhibitor 300 μM	300 μL	0.9 Unit	300 μM	10 μM
(4) Diluted inhibitor 300 μM	300 μL	0.1 Unit	30 μM	10 μM
(5) Inhibitor 30 μM	300 μL	0.1 Unit	30 μM	10 μM

The plate can then be read continuously (whenever possible) in a microplate reader. *Figure 4-8* shows results which would be expected if the inhibitor was either fully reversible or irreversible. Activities are expressed as a percentage of that in the Control (group 1) wells. It should be apparent that if the inhibitor in question is fully reversible, activity measured for group 4 should be equal to the level of activity measured in group 5. However, if inhibition is irreversible, the ratio between activities measured in group 4 to that in group 3 should equal the ratio of activity in group 2 to that in the control group, group 1. In this example, that ratio is approximately 11%.

If an inhibitor has a slow "off-rate", this may be observed in a continuous assay as a slow recovery in enzyme activity, with an initial rate equivalent to that expected with a reversible inhibitor increasing slowly until the rate is equivalent to that expected with a reversible inhibitor. In the earlier example, the rate measured in group 4 would thus be expected to increase slowly almost ninefold, as inhibitor dissociates from the enzyme.

Irreversible inhibitors may be distinguished graphically from reversible noncompetitive inhibitors by plotting V_{max} versus $[E]_t$, where $[E]_t$ represents the total units of enzyme activity in the assay. For a noncompetitive inhibitor, the slope of the curve in the presence of inhibitor will be less than that of the control plot, and the plot will pass through the origin. If the inhibitor is instead irreversible, the slope of the curve in the presence of inhibitor will be identical to that of the control data, and the line will intersect the horizontal $[E]_t$ axis at a point equivalent to the concentration of enzyme irreversibly inactivated (Segel, 1993).

5.2 Identifying Mechanisms and Quantifying Effects of Reversible Inhibitors

5.2.1 Experimental Approach

When reversibility experiments have determined that the effects of an inhibitor of interest are rapidly reversible, it is a relatively straightforward process to obtain kinetic data yielding valuable information regarding the mechanism and potency of inhibition.

It is a worthwhile exercise first to construct a preliminary "inhibitor plot." Holding the substrate concentration constant at a value around the K_M value, effects of a range of concentrations of inhibitor are assessed and plotted on a semilogarithmic graph, in the same way that a preliminary estimate for K_M was

Figure 4-8
Typical results obtained from a dilution experiment to assess reversibility of inhibition, setup according to *Table 4-3* and *Table 4-4*. Group numbers and descriptions are identical to those used in *Table 4-3* and *Table 4-4*. Two potential outcomes are shown for the test group (Group 4). An outcome similar to that shown by the heavily hatched bar, in which the measured velocity is similar to that in Group 5, indicates that inhibition is rapidly reversible. An outcome similar to that shown by the stippled bar, in which the measured velocity is much lower than that in Group 3, indicates irreversible inhibition. In particular, the ratio of column heights 4 to 3 should be similar to the ratio of column heights 2 to 1 if inhibition is irreversible (data are expressed as a percentage of the control rate in column 1, and plotting column 1 is thus unnecessary). A measured value for Group 4 somewhere between the two (extreme) options shown may indicate slow reversibility of inhibition

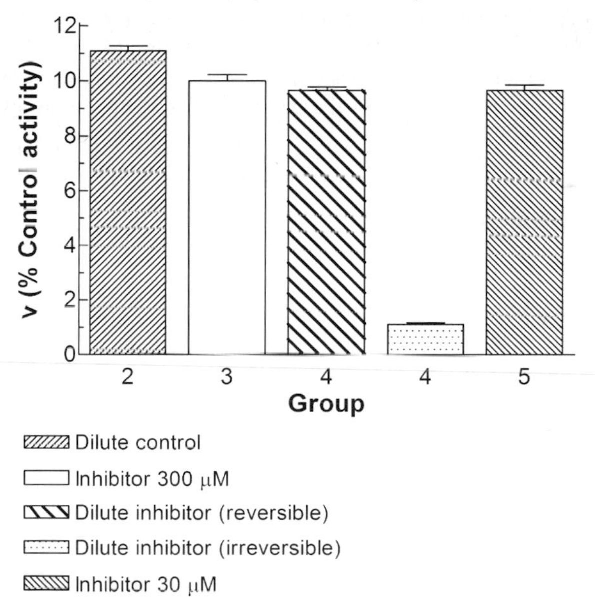

obtained in *Figure 4-4*. A sigmoidal fit yields an estimate for the IC_{50} value, knowledge of which facilitates more in-depth examinations of inhibitor kinetics.

Thereafter, K_M and V_{max} values for substrate turnover are determined in the absence (controls) and presence of several concentrations of the inhibitor of interest. It is recommended that substrate turnover in the presence of at least four concentrations of inhibitor are examined, at concentrations between $1/3 \times IC_{50}$ and $4 \times IC_{50}$. Velocity data are then plotted versus substrate concentration, yielding a control plot and plots at each of the concentrations of inhibitor assessed. Hyperbolic curves are then fitted to data with the Michaelis–Menten equation, or with whichever variation of the Michaelis–Menten equation was found to describe control enzyme behavior most appropriately (see Section 4.1.4 *et seq.*). In this way, a pattern of changes in K_M and V_{max}, or both, should become apparent with changing inhibitor concentration.

While it is a standard practice to use the same range of substrate concentrations (between $1/4 \times K_M$ and $10 \times K_M$) for the control group and at each concentration of inhibitor used, this approach may be inappropriate if the inhibitor markedly increases K_M. For example, if the highest concentration of an inhibitor examined causes an apparent fivefold increase in the K_M value, the substrate concentrations will then extend across a range between $0.05 \times K_M$ and $2 \times K_M$. This is not an appropriate range yielding data from which V_{max}, and thus K_M, might be determined with any confidence. For this reason, it may be worthwhile to include two or three substrate concentrations in the assay, which extend markedly beyond $10 \times K_M$, with velocity data at these points used only if necessary to facilitate construction of appropriate hyperbolae in the presence of inhibitor.

5.2.2 Interpreting Kinetic Plots

Although nonlinear regression of hyperbolic data is arguably the most appropriate means by which K_M and V_{max} values can be determined, and thus by which the concentration-dependent effects of an inhibitor on these parameters can be measured, any pattern of inhibitory behavior is not immediately obvious on visual examination of hyperbolic plots. However, plotting data on Lineweaver–Burk plots does lead to distinct patterns, often allowing immediate diagnosis of an inhibitor mechanism. It should be remembered, of course, that the use of Lineweaver–Burk plots to obtain kinetic data is strongly discouraged, as discussed in Section 4.1.4.

When data in the presence of an enzyme inhibitor are presented in the form of a Lineweaver–Burk plot, a series of straight lines should be obtained. The slopes of these lines may or may not change, and the lines may or may not intersect at a common point. The relationships between slopes, intersection points, and inhibitor mechanisms are outlined later. Further information regarding these mechanisms, including velocity equations describing data obtained in the presence of inhibitors with diverse mechanisms, can be found in (Segel, 1993).

5.2.3 Categories of Inhibition

Reversible inhibition is usually either *full* (linear) or *partial* (hyperbolic). The reason for use of the terms linear and hyperbolic to differentiate between the two mechanisms will become apparent later in this section.

In the presence of a full inhibitor, enzyme activity can be driven to zero. In other words, when all the enzyme proteins are bound by inhibitor (and thus exist as EI), the enzyme cannot yield product.

In the presence of a partial inhibitor, an enzyme is still capable of yielding product even when all the enzyme proteins are bound by inhibitor.

Kinetics of full and partial inhibitors are generally competitive, noncompetitive, uncompetitive, or mixed. The definitions of these behaviors are discussed later. *Table 4-5* lists effects on control K_M and V_{max} values usually observed in the presence of these classes of inhibitors.

Table 4-5
Effects of reversible inhibitors on kinetic constants for substrate turnover

Inhibitor class	Effect on K_M	Effect on V_{max}
Full competitive	↑	≈
Full noncompetitive	≈	↓
Full uncompetitive	↓	↓
Full mixed	↑	↓
Partial competitive	↑	≈
Partial noncompetitive	≈	↓
Partial uncompetitive	↓	↓
Partial mixed	↑ or ↓	↑ or ↓

Key: ↑ = increase, ↓ = decrease, ≈ = no change

When plotted on double reciprocal axes, inhibitor data for full inhibitory mechanisms cannot be distinguished easily from those for partial inhibitory mechanisms. However, with suitable data, careful inspection of Lineweaver–Burk plots may reveal subtle differences; these become clear in secondary plots (replots) of slopes or intercepts, as shown later. The use of K_S rather than K_M (later) reflects the convention employed by Segel (1993); as has been discussed earlier, this dissociation constant provides a good indication of the value of K_M if rapid equilibrium conditions exist.

All the examples discussed here, with their associated graphs, are based on inclusion of a theoretical inhibitor with a K_i of 3 μM. Graphical examples are shown for $[I] = 0$ (control), 1, 3, and 10 μM ($1/3 \times K_i$, $1 \times K_i$ and $3.3 \times K_i$, respectively) and, in some cases, for an infinitely high concentration of inhibitor.

Figure 4-9a shows a reaction scheme for interactions of enzyme and substrate with a full competitive inhibitor. The inhibitor interacts with the same site as does the substrate, resulting in an apparent increase

Figure 4-9

Full and partial competitive inhibitory mechanisms. (a) Reaction scheme for full competitive inhibition indicates binding of substrate and inhibitor to a common site. (b) Lineweaver–Burk plot for full competitive inhibition reveals a common intercept with the 1/*v* axis and an increase in slope to infinity at infinitely high inhibitor concentrations. In this example, $K_i = 3$ μM. (c) Replot of Lineweaver–Burk slopes from (b) is linear, confirming a full inhibitory mechanism. (d) Reaction scheme for partial competitive inhibition indicates binding of substrate and inhibitor to two mutually exclusive sites. The presence of inhibitor affects the affinity of enzyme for substrate and the presence of substrate affects the affinity of enzyme for inhibitor, both by a factor α. (e) Lineweaver–Burk plot for partial competitive inhibition reveals a common intercept with the 1/*v* axis and an increase in slope to a finite value at infinitely high inhibitor concentrations. In this example, $K_i = 3$ μM and $\alpha = 4$. (f) Replot of Lineweaver–Burk slopes from (e) is hyperbolic, confirming a partial inhibitory mechanism

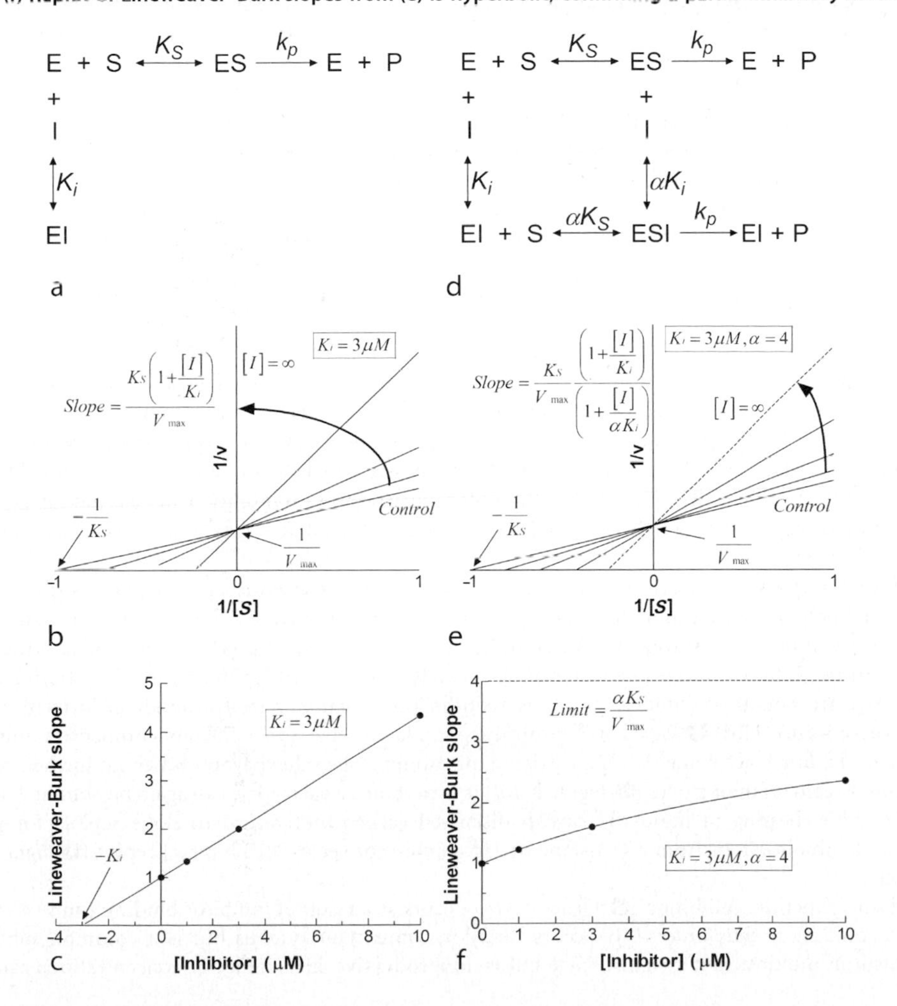

in the K_S value. A Lineweaver–Burk plot (➤ *Figure 4-9b*) reveals a common intersection point on the $1/v$ axis for data obtained at different inhibitor concentrations. It can be seen that as inhibitor concentration increases toward infinity, the slope of the Lineweaver–Burk plot increases toward infinity. Thus, a replot of the slopes versus inhibitor concentrations (➤ *Figure 4-9c*) generates a straight line, which intersects the $[I]$ axis at a value equal to $-K_i$.

In contrast, ➤ *Figure 4-9d* shows a reaction scheme outlining partial (hyperbolic) competitive inhibition. What is both immediately apparent and vitally important is the fact that the inhibitor binds not to the substrate-binding site but to a second allosteric site. Accordingly, there can be no true competition between substrate and inhibitor. However, the affinity of the substrate for the active site is reduced when inhibitor is bound to the inhibitory site, and vice-versa, although V_{max} is unchanged because both ES and ESI can yield product with equal ease. Thus, the behavior has all the kinetic attributes of competitive inhibition, and a Lineweaver–Burk plot (➤ *Figure 4-9e*) is similar to that for a full competitive inhibitor in that there is a common intercept on the $1/v$ axis. More careful examination of the slopes on a Lineweaver–Burk plot indicates that there exists a maximum slope value that does not increase at increasing inhibitor concentrations. This is because binding of inhibitor to the allosteric site displays typical saturable (hyperbolic) binding behavior, with the maximum effect on substrate affinity occurring when inhibitor binds all allosteric sites. The saturable nature of this effect is evident in ➤ *Figure 4-9f*, which reveals that a replot of slopes versus inhibitor concentrations is hyperbolic. A value for α, the degree to which K_S is increased in the presence of a saturating concentration of inhibitor, can be determined from the plateau of the hyperbolic replot. Note, however, that as the slope of a control plot is greater than zero, the hyperbolic curve does not pass through the origin but instead is displaced upward on the y-axis by a distance equal to the slope of the control plot. To calculate α in this way, a nonlinear regression program can be used to fit a hyperbolic curve to slope data *if data are expressed as a change in slope (or Δ slope) relative to control*, in other words, so that the Δ slope value for the control (zero inhibitor) curve equals zero. However, if this is done, the plateau of the hyperbola will equal $(\alpha K_S/V_{max})-(K_S/V_{max})$.

In order to determine a K_i value for a partial inhibitor, competitive or otherwise, $1/\Delta$ slope or $1/\Delta$ y-intercept versus 1/[inhibitor] data can be plotted (➤ *Figure 4-13*), with the resulting straight line yielding values for K_i, α, and/or β (Segel, 1993).

➤ *Figure 4-10a* shows a reaction scheme for interactions of enzyme and substrate with a full noncompetitive inhibitor. The inhibitor interacts with a site distinct from the active site, and the ESI complex is incapable of yielding product. It is thus possible, at saturating concentrations of inhibitor, to drive all enzymes to a nonproductive form, and so activity can be completely inhibited. Furthermore, the affinity of the inhibitor for the saturable allosteric inhibitory site remains independent of substrate concentration. A Lineweaver–Burk plot (➤ *Figure 4-10b*) reveals a common intersection point on the $1/[S]$ axis for the data obtained at different inhibitor concentrations. It can be seen that as inhibitor concentration increases toward infinity, the slope of the Lineweaver–Burk plot increases toward infinity. Thus, a replot of the slopes versus inhibitor concentrations (➤ *Figure 4-10c*) generates a straight line, which intersects the $[I]$ axis at a value equal to $-K_i$.

➤ *Figure 4-10d* shows a reaction scheme for interactions of enzyme and substrate with a partial (hyperbolic) noncompetitive inhibitor. The mechanism is similar to that of a full noncompetitive inhibitor, but product can be released from the ESI complex. However, release of product from ESI is slower than from ES, and k_p is thus reduced by a factor, β. As a result, at infinitely high inhibitor concentrations when all enzymes are bound by inhibitor, V_{max} is reduced but reaction velocity cannot be driven to zero. A Lineweaver–Burk plot (➤ *Figure 4-10e*) appears similar to that for a full noncompetitive inhibitor, with the major difference being that there exists a maximum slope value beyond which no further increase can occur. A replot of these slopes (➤ *Figure 4-10f*) is hyperbolic; a value for β can again be obtained directly from this plot (bearing in mind the *caveats* discussed earlier with regard to slope replots for partial competitive inhibitors), or from a $1/\Delta$ slope or $1/\Delta$ y-intercept versus 1/[inhibitor] replot (➤ *Figure 4-13*; see later).

Full uncompetitive inhibition (➤ *Figure 4-11a*) occurs as a result of inhibitor binding (only) to the ES complex; binding is thus *ordered*. It occurs rarely in unireactant systems but is a common inhibitory mechanism in multireactant systems. Since ESI is nonproductive, high inhibitor concentrations can drive

Figure 4-10

Full and partial noncompetitive inhibitory mechanisms. (a) Reaction scheme for full noncompetitive inhibition indicates binding of substrate and inhibitor to two mutually exclusive sites. The presence of inhibitor prevents release of product. (b) Lineweaver–Burk plot for full noncompetitive inhibition reveals a common intercept with the 1/[*S*] axis and an increase in slope to infinity at infinitely high inhibitor concentrations. In this example, $K_i = 3$ μM. (c) Replot of Lineweaver–Burk slopes from (b) is linear, confirming a full inhibitory mechanism. (d) Reaction scheme for partial noncompetitive inhibition indicates binding of substrate and inhibitor to two mutually exclusive sites. The presence of inhibitor alters (reduces) the rate of release of product by a factor β. (e) Lineweaver–Burk plot for partial noncompetitive inhibition reveals a common intercept with the 1/[*S*] axis and an increase in slope to a finite value at infinitely high inhibitor concentrations. In this example, $K_i = 3$ μM and $\beta = 0.5$. (f) Replot of Lineweaver–Burk slopes from (e) is hyperbolic, confirming a partial inhibitory mechanism

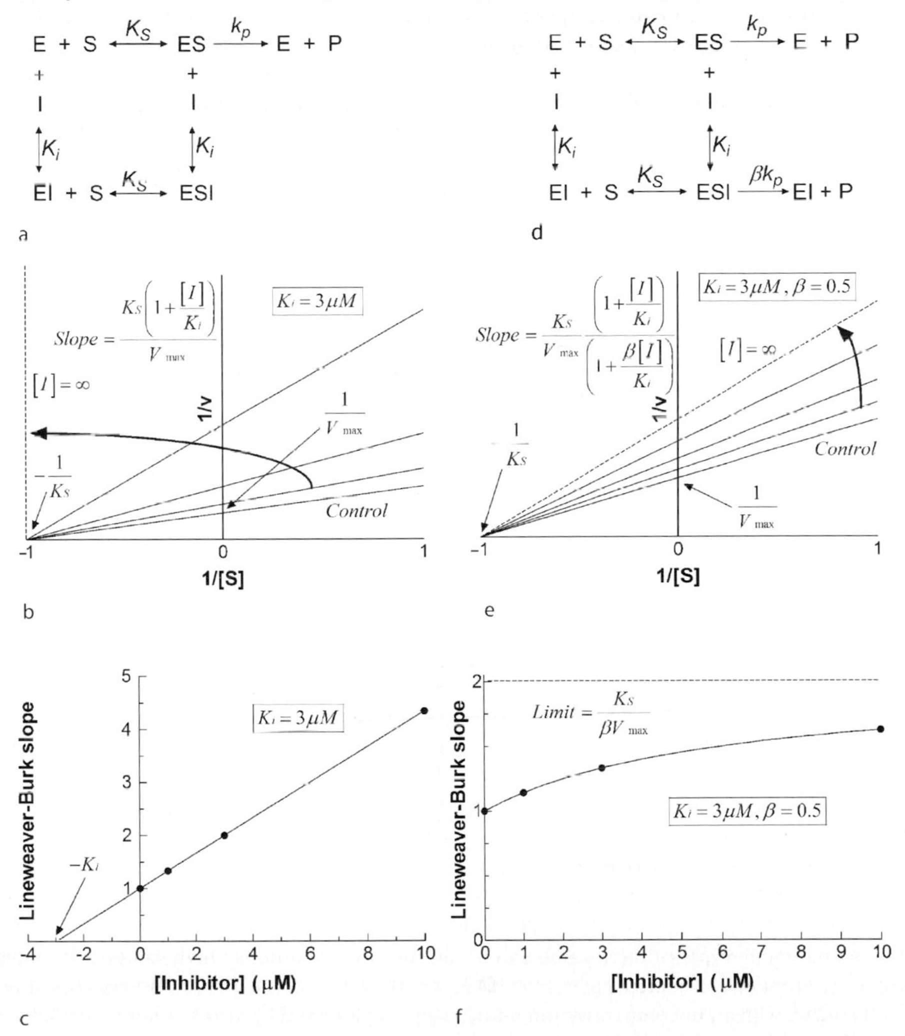

the reaction velocity to zero, and V_{max} is reduced in the presence of an uncompetitive inhibitor. However, in contrast with noncompetitive inhibition, K_M is also reduced by a similar degree. This is because binding of *I* to ES results in a depletion of ES, thereby moving the equilibrium between E, S, and ES to the right. This is apparent in a Lineweaver–Burk plot (*Figure 4-11b*) as a parallel shift of the control plot in the presence of

Figure 4-11
Full and partial uncompetitive inhibitory mechanisms. (a) Reaction scheme for full uncompetitive inhibition indicates ordered binding of substrate and inhibitor to two mutually exclusive sites. The presence of inhibitor prevents release of product. (b) Lineweaver–Burk plot for full uncompetitive inhibition reveals a series of parallel lines and an increase in the 1/*v* axis intercept to infinity at infinitely high inhibitor concentrations. In this example, $K_i = 3\ \mu M$. (c) Replot of Lineweaver–Burk slopes from (b) is linear, confirming a full inhibitory mechanism. (d) Reaction scheme for partial uncompetitive inhibition indicates random binding of substrate and inhibitor to two mutually exclusive sites. The presence of inhibitor alters the rate of release of product (by a factor β) and the affinity of enzyme for substrate (by a factor α) to an identical degree, while the presence of substrate alters the affinity of enzyme for inhibitor by α. (e) Lineweaver–Burk plot for partial uncompetitive inhibition reveals a series of parallel lines and an increase in the 1/*v* axis intercept to a finite value at infinitely high inhibitor concentrations. In this example, $K_i = 3\ \mu M$ and $\alpha = \beta = 0.5$. (f) Replot of Lineweaver–Burk slopes from (e) is hyperbolic, confirming a partial inhibitory mechanism

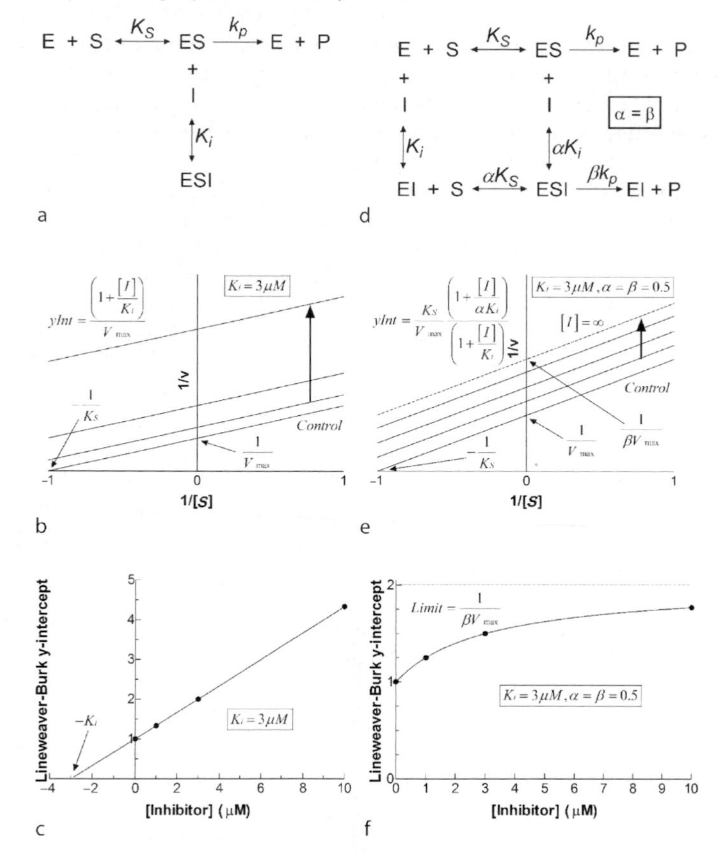

inhibitor. As the *y*-intercepts of these parallel lines can increase to infinitely high values at infinitely high inhibitor concentrations, a *y*-intercept replot (*Figure 4-11c*) is a straight line intersecting the *x*-axis at $-K_i$. Of course, with an uncompetitive inhibitor, a slope replot would generate a horizontal line yielding no information with regard to inhibitory potency.

Partial uncompetitive inhibition does not resemble full uncompetitive inhibition in terms of having an ordered mechanism, but it instead represents a very specific form of partial mixed inhibition (discussed later). However, it is sometimes referred to as partial uncompetitive inhibition due to the parallel displacement of Lineweaver–Burk plots in the presence of inhibitor, and it is thus related to full uncompetitive inhibition in the same way that partial competitive inhibition is related to full competitive inhibition.

Binding of a partial uncompetitive inhibitor (➜ *Figure 4-11d*) occurs at a site distinct from the active site. As a result of inhibitor binding, affinity of the substrate for the active site is increased, and thus K_S is reduced by a factor, α ($0 < \alpha < 1$). However, although the ESI complex is able to yield product, the rate of product formation, k_p, is reduced by a factor, β, compared with control ($0 < \beta < 1$). Furthermore, in the specific case of partial uncompetitive inhibition, α is equal to β. As a result, Lineweaver–Burk plots are displaced in a parallel manner with respect to control (➜ *Figure 4-11e*), the degree of inhibition reaching a maximum when the *y*-intercept is equal to $1/\beta V_{max}$. A *y*-intercept replot (➜ *Figure 4-11f*) is hyperbolic, and K_i may again be obtained from a $1/\Delta$ *y*-intercept versus 1/[inhibitor] replot (➜ *Figure 4-13*).

Mixed inhibition, which affects both K_M and V_{max}, may occur through several mechanisms; the most common are discussed later.

Full mixed inhibition (or linear mixed inhibition) occurs through a reaction mechanism (➜ *Figure 4-12a*) similar to that for noncompetitive inhibition. However, with a mixed inhibitor, the affinity for substrate of the EI complex is altered compared with that of free enzyme, and K_S is thus changed by a factor, α. However, as ESI is nonproductive, V_{max} is reduced in the presence of inhibitor, and reaction velocity can be driven to zero at infinitely high concentrations of inhibitor. A Lineweaver–Burk plot (➜ *Figure 4-12b*) reveals a common intersection above the *x*-axis (if affinity is reduced) or below the *x*-axis (if affinity is increased). In the example shown, α equals 2 and affinity is thus reduced. As V_{max} can be driven to zero, the slope of a plot at an infinitely high concentration of inhibitor approaches infinity. Thus, a slope replot (➜ *Figure 4-12c*) is a straight line, intersecting the *x*-axis at $-K_i$.

In partial (hyperbolic) mixed inhibition (➜ *Figure 4-12d*), binding of inhibitor to a site distinct from the active site results in altered affinity of enzyme for substrate (by a factor, α) as well as a change (by a factor, β) in the rate at which product can be released from ESI. The effects of a partial mixed inhibitor on a Lineweaver–Burk plot depend upon the actual values, and on the relative values, of α and β. Once again, inhibitor plots can intersect the control plot above or below, but not on, the *x*-axis, and to the left or to the right of, but not on, the *y*-axis. Because V_{max} cannot be driven to zero, a maximum Lineweaver–Burk slope is reached at infinitely high inhibitor concentrations beyond which no further increase occurs.

In the example Lineweaver–Burk plot shown in ➜ *Figure 4-12e*, both affinity for substrate and k_p are reduced, and plots intersect to the left of the *y*-axis and above the *x*-axis. However, if both α and β were less than one, an intercept in the lower left quadrant would be possible, although if β was much greater than α, the intercept might instead lie in the upper right hand quadrant (Segel, 1993). The complexity of the relationships between α and β demands that each example of partial mixed inhibition be examined thoroughly, and when the position of a common intersection point has been determined, it is recommended that the experimenter consults (Segel, 1993) or a similar text in order that the data are interpreted correctly thereafter.

As with other cases of partial inhibition, a slope replot of Lineweaver–Burk data (➜ *Figure 4-12f*) is hyperbolic, and values for K_i, α, and β can, as before, be obtained from a $1/\Delta$ slope or $1/\Delta$ *y*-intercept versus 1/[inhibitor] replot (➜ *Figure 4-13*).

Equations describing all the above inhibitory behaviors have been outlined in detail by Segel (1993).

The relatively straightforward graphing procedures described earlier can be used to determine estimates for K_i, α and β for most reversible inhibitors. The K_i value is the concentration of inhibitor at which 50% of enzymes are bound in the absence of substrate. The value α indicates the maximum degree to which K_S may change in the presence of a saturating concentration of inhibitor, as well as the maximum degree to which K_i may change in the presence of a saturating concentration of substrate. It is necessary for both dissociation constants to be altered to a similar degree as the overall equilibrium constant for the formation of ESI must be the same, regardless of the reaction pathway taken (Segel, 1993). The value β indicates the maximum degree to which k_p may change in the presence of a saturating concentration of inhibitor. If k_p is not very small relative to k_{-1} then any inhibitor that affects k_p will also alter K_S, and inhibition will be mixed. It stands to reason, therefore, that in any system in which an inhibitor exhibits pure noncompetitive kinetics, rapid equilibrium conditions must exist (Segel, 1993).

The ability of a reversible inhibitor to alter substrate affinity on binding to the enzyme, and vice versa, as well as the ability of the inhibitor to alter k_p, generally occur because of *allosteric* effects. As the allosteric effects of a molecule are related to its size, shape, and charge distribution, it might be expected that different

Figure 4-12

Full and partial mixed inhibitory mechanisms. (a) Reaction scheme for full mixed inhibition indicates binding of substrate and inhibitor to two mutually exclusive sites. The presence of inhibitor prevents release of product and alters the affinity of enzyme for substrate to the same degree (α) that the affinity of enzyme for inhibitor is altered by the presence of substrate. (b) Lineweaver–Burk plot for full mixed inhibition reveals a common intercept at a point which does not lie on either axis. In this example, $K_i = 3\ \mu M$ and $\alpha = 2$. (c) Replot of Lineweaver–Burk slopes from (b) is linear, confirming a full inhibitory mechanism. (d) Reaction scheme for partial mixed inhibition indicates binding of substrate and inhibitor to two mutually exclusive sites. The presence of inhibitor alters the rate of release of product by a factor β, and the affinity of enzyme for substrate by a factor α, while the presence of substrate alters the affinity of enzyme for inhibitor by α. (e) Lineweaver–Burk plot for partial mixed inhibition reveals a common intercept at a point that does not lie on either axis. In this example, $K_i = 3\ \mu M$, $\alpha = 4$ and $\beta = 0.5$. (f) Replot of Lineweaver–Burk slopes from (e) is hyperbolic, confirming a partial inhibitory mechanism

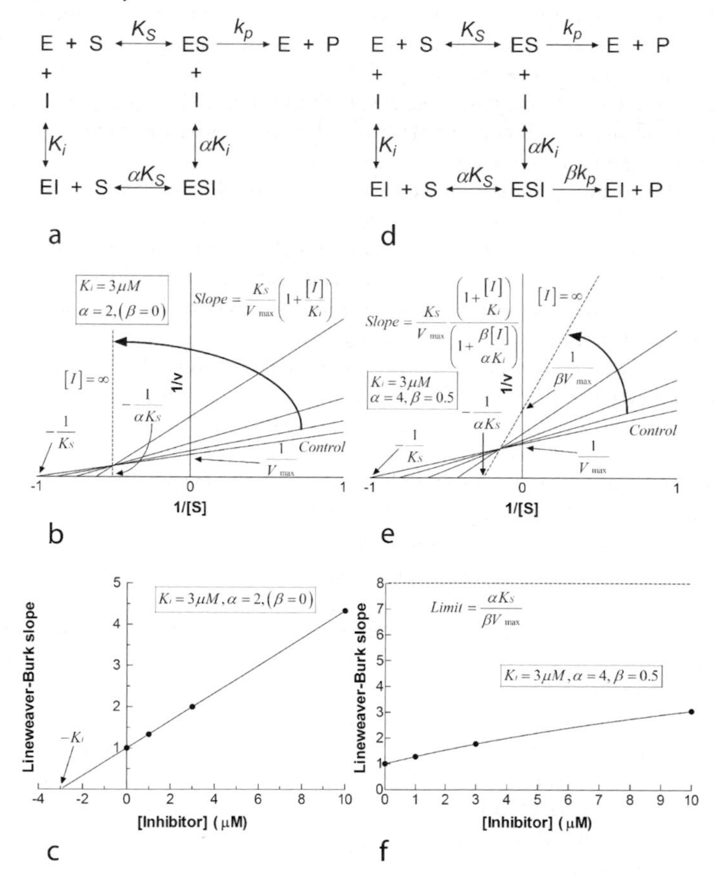

inhibitors will have different effects on K_S and k_p values for a single substrate. However, sometimes overlooked is the equally obvious deduction that α and β values for a single inhibitor would be expected to be different for different substrate molecules. Thus, while K_i should not change, since this value reflects inhibitor affinity in the absence of substrate, values of α and β are associated with particular enzyme–substrate combinations, and are not unique to an individual inhibitor.

With knowledge of K_i, α and β for a reversible inhibitor, along with K_M and V_{max} values for the substrate of interest, it is possible to fit hyperbolic curves to multiple data sets by applying the appropriate inhibitor equation. It is a relatively straightforward task to enter such equations into graphing programs such as

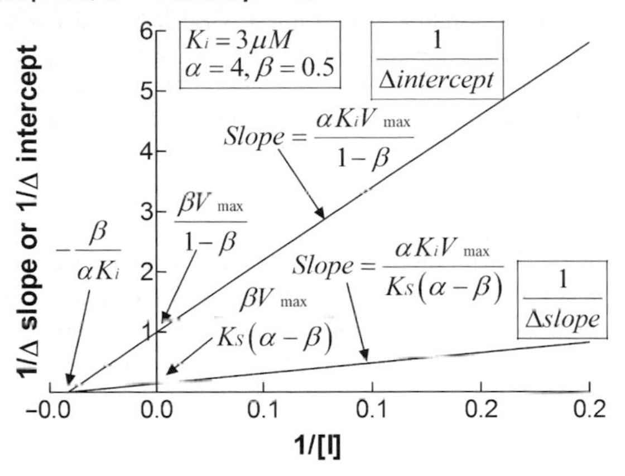

Figure 4-13
Secondary (reciprocal) replots for hyperbolic inhibition systems. When data from *Figure 4-9f*–*Figure 4-12f* are plotted on reciprocal axes, a straight line is obtained. Values for one or both α and β may equal 1, depending upon the inhibitor mechanism being examined. The meaning of the axes intersection points and slope of the straight line depend upon whether the primary replot data (e.g., *Figure 4-9f*) show slope versus [*I*] (shallow line) or *x*-axis intercept versus [*I*] (steep line) data. The plots shown are those for an inhibitor with $K_i = 3$ μM and, when appropriate, $\alpha = 4$ and/or $\beta = 0.5$

GraphPad Prism, then holding K_M and V_{max} constants during nonlinear regression analyses in order to obtain best-fit values for K_i, α and β. In this regard, the latest versions of Prism allow "global curve fitting" whereby simultaneous analyses of all data sets facilitate the determination of global best-fit values for constants (Holt et al., 2004).

5.2.4 A Note on Enzyme Activation

It is not a requirement that binding of an allosteric modulator to an enzyme must result in inhibition of activity; indeed, in some mixed inhibition systems described earlier, both α and β have values between 0 and 1. If an increase in substrate affinity outweighs a decrease in k_p at lower substrate concentrations, an increase in enzyme activity may occur, relative to control values, and the use of the term "inhibitor" to classify such a compound is open to debate.

It is perfectly possible for some substrate–modulator combinations to result in an increase in substrate affinity, an increase in the rate of product formation, or both. The same analytical approaches may be used to study such compounds as have been described earlier to assess inhibitory mechanisms and potencies. However, with an allosteric activator, the dissociation constant might better be termed K_a, and values for α and β are more likely to be less than one, and greater than one, respectively. As is the case for inhibition, allosteric enzyme activation would be expected to exhibit substrate dependence (Holt et al., 2004).

5.3 Tight-Binding, Slow-Binding, and Slow, Tight-Binding Inhibitors

The equations discussed thus far are based on the assumption that there is no significant change in the concentration of unbound inhibitor as a result of formation of an EI complex. However, if an inhibitor demonstrates extremely high affinity for an enzyme, concentrations of inhibitor used to achieve partial inhibition will be sufficiently low such that significant depletion of unbound inhibitor occurs because of binding to enzyme. Consequently, graphical methods described earlier through which K_i might be calculated are inappropriate. Instead, Dixon has devised a method that allows determination of K_i values while, at the same time, confirming depletion of a tight-binding inhibitor (Dixon, 1972; Segel, 1993). This approach may be applied easily to tight-binding competitive inhibitors, or to tight-binding noncompetitive inhibitors.

At a single nonsaturating substrate concentration, v is determined in a control assay and in the presence of several concentrations of a tight-binding competitive inhibitor ($[I]_t$). The resulting plot of v versus $[I]_t$ (*Figure 4-14a*) is a concave plot. If straight lines are then drawn between the initial control velocity, v_0, and a series of points on the curve where v equals $v_0/2$, $v_0/3$, $v_0/4$, etc., these lines will yield a series of intercepts with the x-axis, with each intersection point separated by a distance equal to $K_{i\ app}$. If this experiment is repeated at several substrate concentrations, a value for $K_{i\ app}$ can be determined at each substrate concentration. Thereafter, a replot of $K_{i\ app}$ versus $[S]$ (*Figure 4-14b*) yields a straight line that intersects the y-axis at a value equal to K_i for the tight-binding competitive inhibitor.

Figure 4-14

Plots devised by Dixon to determine K_i for tight-binding inhibitors. (a) A primary plot of v versus total inhibitor present ($[I]_t$) yields a concave line. In this example, $[S] = 3 \times K_M$ and thus $v = 67\%$ of V_{max}. Straight lines drawn from v_0 (when $[I]_t = 0$) through points corresponding to $v_0/2$, $v_0/3$, etc. intersect with the x-axis at points separated by a distance $K_{i\ app}$, when inhibition is competitive. When inhibition is noncompetitive, intersection points are separated by a distance equivalent to K_i. The positions of lines for $n = 1$ and $n = 0$ can then be deduced and the total enzyme concentration, $[E]_t$, can be determined from the distance between the origin and the intersection point of the $n = 0$ line on the x-axis. If inhibition is competitive, this experiment is repeated at several different substrate concentrations such that a value for $K_{i\ app}$ is obtained at each substrate concentration. (b) Values for $K_{i\ app}$ are replotted versus $[S]$, and the y-intercept yields a value for K_i. If inhibition is noncompetitive, this replot is not necessary (see text)

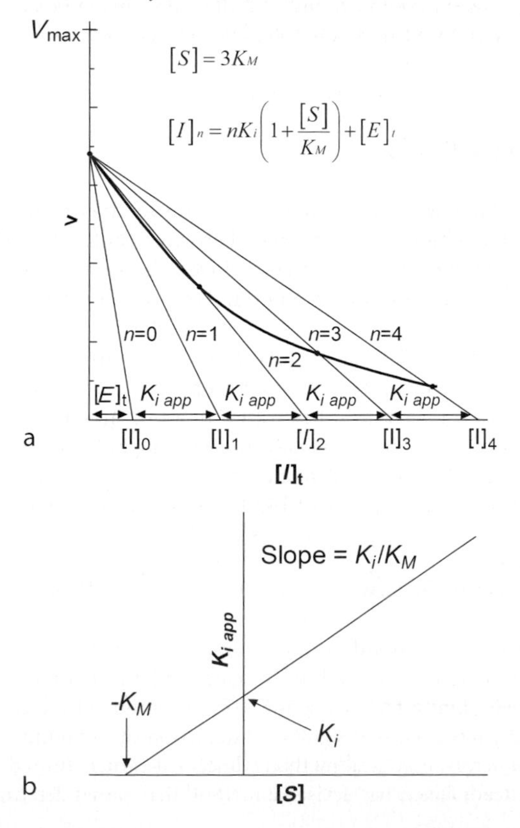

When the tight-binding inhibitor is instead a noncompetitive inhibitor, an experiment done in an identical manner yields a plot similar to that shown in ➋ *Figure 4-14a*, except that *x*-axis intercepts are now each separated by a distance equal to K_i, and a replot is not necessary.

Henderson has also described a graphical method which can distinguish between, and quantify the effects of, tight-binding inhibitors acting through competitive, noncompetitive, and uncompetitive mechanisms (Henderson, 1972). The practicalities of this method are discussed by Segel (1993).

Some inhibitors interact very slowly with the enzyme protein, and onset of inhibition thus exhibits time-dependence. These inhibitors are generally referred to as *slow-binding* inhibitors, and as *slow tight-binding* inhibitors if the potency of inhibition is extremely high. Analysis of these inhibitory mechanisms is complex because binding and dissociation rate constants may be determined in addition to K_i values. Indeed, a complete analysis may require extensive use of specialized computer software, and the complexities of such analyses preclude their discussion in this chapter. However, the reader is directed to several publications from Morrison's laboratory if a slow-binding mechanism is suspected for an inhibitor of interest (Morrison, 1982; Morrison and Stone, 1985; Sculley and Morrison, 1986; Morrison and Walsh, 1988).

5.4 Irreversible Inhibitors

Irreversible enzyme inhibition can occur through a variety of mechanisms. From the point of view of the neuroscientist, an irreversible inhibitor is one which inhibits an enzyme for an extended period of time. Often, a covalent bond is formed and recovery of enzyme activity may require synthesis of new enzyme protein. However, some covalent bonds are slowly reversible, while other enzymes may recover slowly because of metabolism of inhibitor. Formation of a covalent bond is not necessary for inhibition to be considered irreversible, as a reversible tight-binding inhibitor may dissociate from an enzyme so slowly that inhibition is, to all intents and purposes, irreversible. Thus, irreversible inhibitors can usually be categorized as belonging to one of four groups, based on mechanism. One of these groups, slow, tight-binding inhibitors, has already been mentioned (see ➋ Section 5.3). The others are discussed briefly later.

5.4.1 Transition State Analogs

Transition state analogs are compounds that exhibit some structural similarity to a transition state intermediate formed during conversion of substrate to product. These intermediates are generally bound more tightly to the enzyme than are substrates and products, and an inhibitor with similar properties is thus likely to bind tightly to the enzyme and dissociate very slowly from its binding site (Silverman, 1995). Because inhibition is reversible, the effects of the drug may be analyzed as described for tight-binding inhibitors (earlier). However, more often than not, a simple IC_{50} value will be determined. A range of concentrations of the inhibitor is preincubated with enzyme before the addition of a single concentration of substrate (generally at a concentration above K_M) and remaining activity is measured. Data are expressed as percentage of activity in a control assay versus $\log_{10}$[inhibitor]. A sigmoidal curve fitted to the data by nonlinear regression yields an IC_{50}, the concentration of inhibitor reducing control activity by 50% (assuming that complete inhibition is attainable). These values are not tremendously useful, inasmuch as they are not necessarily constant and may vary with substrate concentration and with enzyme concentration. Nevertheless, the potencies of several inhibitors may be compared quite rapidly by generation of IC_{50} values if experimental conditions are similar in each case.

5.4.2 Affinity Labeling Agents

Such drugs interact very rapidly and reversibly with the enzyme. Once bound, a time-dependent reaction of an enzyme nucleophile with bound inhibitor occurs, resulting in formation of a covalent bond. Although it is possible to determine the inactivation rate for the covalent step, once again, the neuroscientist is likely to be more interested in an estimate of potency, and the determination of an IC_{50} value is often sufficient.

5.4.3 "Mechanism-Based" (Suicide or k_{cat}) Inhibitors

All inhibitors act through a mechanism of some sort, and the term "mechanism-based inhibitor" is thus not particularly revealing. These inhibitors bind, reversibly and competitively, with the enzyme active site, and they may then be metabolized to completion and leave as a product. However, during the metabolic process, which involves generation of a transition state intermediate, a proportion of enzymes will interact irreversibly with the transition state intermediate, and inhibition is permanent. In effect, the enzyme commits "suicide" by converting an otherwise harmless drug molecule into an irreversible inhibitor. Such a mechanism can impart high degrees of specificity and potency upon drugs that act in this manner.

$$E + I \underset{k_{off}}{\overset{k_{on}}{\rightleftharpoons}} EI \xrightarrow{k_2} EI' \xrightarrow{k_4} EI''$$

$$EI' \xrightarrow{k_3} E + I'$$

In the earlier scheme, I′ represents a product formed by metabolism of the inhibitor by the enzyme. This product may be released into bulk solvent, or may interact (often covalently) with a suitably reactive component of the enzyme within the active site. This irreversibly inactivated enzyme complex is shown as EI″. There are two kinetic constants that can be obtained from relatively straightforward experiments with a suicide inhibitor. The K_I value is an equilibrium constant for the initial reversible step, and all the rate constants from the above scheme contribute to its value. The rate of irreversible inactivation of enzyme at a saturating concentration of the suicide inhibitor is given by k_{inact}, to which only k_2, k_3, and k_4 contribute (Silverman, 1995). At infinitely high concentrations of the inhibitor, the half-life for inactivation is equal to $\ln 2/k_{inact}$.

From the above scheme, it is clearly not necessary that every conversion of EI to E′ results in irreversible enzyme inactivation. The *partition ratio* for an inhibitor is the ratio of product released to enzyme inactivated. Mathematically, the partition ratio is equal to k_3/k_4, with a ratio of zero indicating that every turnover of inhibitor molecule results in enzyme inactivation. The partition ratio does not depend on the concentration of the inactivator, but rather on the reactivity of I′, its proximity to a suitable target on the enzyme with which it might react, and its rate of diffusion away from the active site.

If suicide inhibition is suspected, it must be confirmed that generation of the active inhibitor species is a process that depends upon enzyme activity. Furthermore, irreversible interaction of I′ with the enzyme must occur within the active site; diffusion of I′ from the active site, and subsequent interaction with a distal site on the enzyme, might be thought of as generation of an affinity labeling agent. Inclusion of an electrophile-trapping reagent, such as a thiol, in the reaction mixture, should facilitate confirmation of any involvement of a reactive species that diffuses from the active site prior to inactivating the enzyme.

The procedure described by Kitz and Wilson is perhaps the most appropriate by which the effects of a suicide inhibitor might be quantified (Kitz and Wilson, 1962; Silverman, 1995). The procedure has been described in detail by Silverman, and a brief outline is provided here (Silverman, 1995).

At least five reaction tubes are prepared containing a small volume of concentrated enzyme, as well as any cofactors and other buffer components necessary to facilitate enzyme activity. At $t = 0$, a small volume of inhibitor is added to each reaction tube, with each tube receiving a different concentration of inhibitor. Immediately, a small volume aliquant of the reactant mixture from each tube is removed and diluted immediately into a large volume of reaction buffer containing a high concentration of substrate. Initial reaction velocity is determined for each. Thereafter, at predetermined time points, further aliquants are withdrawn from the enzyme–inhibitor mixture and are diluted into a large volume of substrate. This should be repeated until initial velocity data for at least five time points have been determined. For a set of readings at a single time point, each initial rate is expressed as a percentage of that in a corresponding control assay, to which solvent, rather than inhibitor, was added. Plotting logarithms of rates versus preincubation times yields a series of (pseudo first order) straight lines (➋ *Figure 4-15a*). The appearance of curved lines indicates that the reaction may be more complex than expected (Silverman, 1995). The slope of each straight line corresponds to an inactivation rate, $-k$. A reciprocal replot ($1/k$ versus 1/[Inhibitor]; ➋ *Figure 4-15b*) yields a straight line, which intersects the x-axis at $-1/K_I$, and the y-axis at $1/k_{inact}$. If the straight line intersects with the origin, k_{inact} is fast relative to the formation of EI, and it then becomes necessary to alter reaction conditions (temperature, pH) to slow k_{inact} and permit saturation kinetics, in order that K_I and k_{inact} might be determined graphically.

Figure 4-15
Kitz and Wilson graphing procedure for analysis of suicide inhibitor potency. (a) Time-dependence of onset of inhibition is determined at several inhibitor concentrations by plotting activity remaining after preincubation for varying times. Each inhibitor concentration thus yields a separate straight line; a control plot containing no inhibitor is also included. A semilogarithmic scale results in straight line plots, each with a slope equal to $-k$; if a linear y-axis scale is used, plots are exponential and k can be determined directly by nonlinear regression of data. (b) A reciprocal replot ($1/k$ versus $1/[I]$) yields a straight line with x- and y-intercepts providing values for the reciprocals of K_I and k_{inact}, respectively

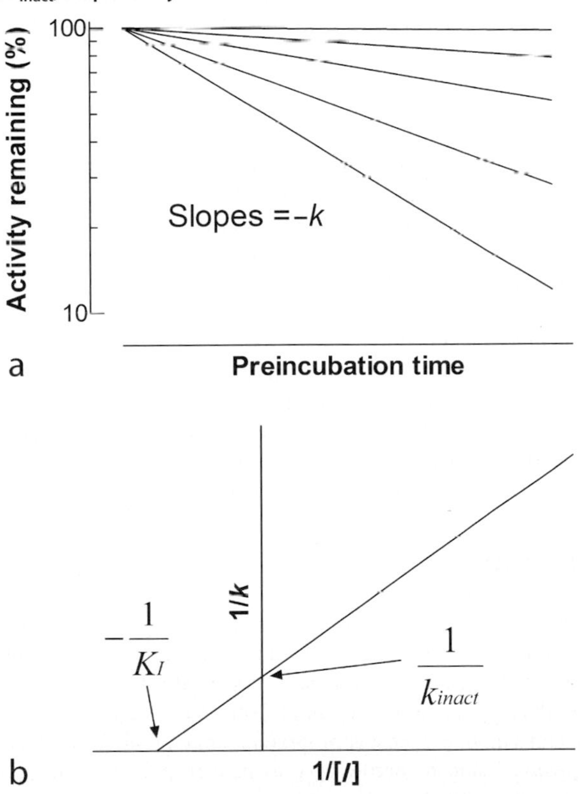

5.5 Inhibition In Vivo and Ex Vivo

Often, demonstration of potency and selectivity in vitro is assumed to extrapolate to effects of inhibitors administered in vivo. Unfortunately, such assumptions are often invalid, and selectivity and potency must also be assessed in vivo or ex vivo. There are several reasons for such discrepancies between in vitro and in vivo data.

The pharmacokinetics of the inhibitor will determine its concentration–time profile at the site of action. Thus, the lipophilicity and degree of ionization of the molecule, and its susceptibility to metabolism by hepatic cytochromes P_{450} and by other enzymes, will determine the extent to which the drug crosses membrane barriers into different tissue types, as well as the duration of time that the drug remains in the target tissue at or above a minimally effective concentration. Furthermore, hepatic metabolism may result in the presence in the plasma of oxidized drug metabolites that maintain inhibitory potency, but lose selectivity with respect, for example, to enzyme subtypes. In addition, pharmacokinetics may introduce "tissue selectivity" to enzyme inhibition, where none existed in vitro; for example, peripheral DOPA decarboxylase inhibitors are selective over decarboxylase in the brain only because they fail to access the brain to a significant degree.

When examining the effects of inhibitors in vivo, the most common approach is to administer the inhibitor to an animal and, after an appropriate time, kill the animal and measure remaining enzyme activity in a tissue homogenate ex vivo. While such an analysis is perfectly acceptable when the inhibitor binds tightly or irreversibly to the enzyme, problems occur when inhibition is reversible, regardless of the mechanism. The act of homogenizing a tissue in buffer serves to dilute the reversible inhibitor present in the tissue, such that a new degree of inhibition is reached, and this is exacerbated when an aliquant of homogenate is then diluted further as a result of addition of substrate and other reaction components. As a result, an inhibitor could be present in a tissue at a concentration sufficient to inhibit activity almost entirely, yet no inhibition may be observed when the enzyme is assayed ex vivo. Furthermore, the inhibitor may be competing in vivo with endogenous substrates for the enzyme, further complicating any attempts at interpretation and extrapolation of data obtained ex vivo.

A method has been described by which the effects of reversible competitive monoamine oxidase inhibitors might be estimated successfully ex vivo (Green, 1984). This method relies upon the ability of a reversible competitive inhibitor (perhaps administered chronically) to protect against the effects of an irreversible inhibitor (administered as a single dose) that binds to the enzyme active site. As inhibition by an irreversible inhibitor can be measured quite easily ex vivo, the degree of irreversible inhibition in an animal coadministered a reversible competitive inhibitor would be less than that in a control animal that received only the irreversible inhibitor. The difference would provide an estimate of the degree to which enzyme was bound (protected) by the reversible inhibitor in vivo.

It may also be useful to examine the in vivo effects of a reversible inhibitor by measuring tissue concentrations of endogenous substrates and metabolites of the enzyme of interest, or perhaps by examining enzyme-mediated disappearance of a xenobiotic substrate (or, preferably, appearance of a metabolite), administered to the animal as a "tracer". If the latter route is chosen, it is important to ensure that disappearance of the tracer is indeed a result of the action of the enzyme of interest.

6 Concluding Remarks

The most important message this chapter attempts to convey is that it is not possible to understand the mechanism and quantify the effects of an inhibitor unless the kinetic behavior of the enzyme of interest toward the substrate used is fully understood. One or two relatively straightforward and carefully done experiments are often all that is necessary to accomplish the latter objective; and drug effects on enzyme activity might then be quantified in the most appropriate manner. Only by understanding the enzyme and inhibitor mechanisms involved, and by becoming proficient in plotting and interpreting kinetic graphs correctly, may the researcher begin to recognize unusual or unexpected behavior and thus avoid publication of erroneous or misleading data and conclusions.

Acknowledgments

Work in the author's laboratory is supported by the Canadian Institutes of Health Research (CIHR), the Faculty of Medicine and Dentistry of the University of Alberta, the Northern Alberta Clinical Trial and Research Centre, and BioTie Therapies Corp. (Turku, Finland).

References

Cleland WW. 1963a. The kinetics of enzyme-catalyzed reactions with two or more substrates or products. I. Nomenclature and rate equations. Biochim Biophys Acta 67: 104.

Cleland WW. 1963b. The kinetics of enzyme-catalyzed reactions with two or more substrates or products. II. Inhibition: nomenclature and theory. Biochim Biophys Acta 67: 173.

Cleland WW. 1963c. The kinetics of enzyme-catalyzed reactions with two or more substrates or products. III. Prediction of initial velocity and inhibition patterns by inspection. Biochim Biophys Acta 67: 188.

Dixon M. 1972. The graphical determination of K_M and K_I. Biochem J 129: 197.

Green AL. 1984. Assessment of the potency of reversible MAO inhibitors in vivo. Monoamine oxidase and disease. Tipton KF, et al. editors. London: Academic; pp. 73-81.

Henderson PJ. 1972. A linear equation that describes the steady-state kinetics of enzymes and subcellular particles interacting with tightly bound inhibitors. Biochem J 127: 321.

Henderson PJF. 1993. Statistical analysis of enzyme kinetic data. Enzyme assays. A practical approach. Eisenthal R, Danson MJ, editors. Oxford: Oxford University Press; pp. 277-316.

Holt A, Wieland B, Baker GB. 2004. Allosteric modulation of semicarbazide-sensitive amine oxidase activities in vitro by imidazoline receptor ligands. Br J Pharmacol 143: 495.

John RA. 1993. Photometric assays. Enzyme assays. A practical approach. Eisenthal R, Danson MJ, editors. Oxford: Oxford University Press; pp. 59-92.

King EL, Altman C. 1956. A schematic method of deriving the rate laws for enzyme-catalyzed reactions. J Phys Chem 60: 1375.

Kitz R, Wilson IB. 1962. Esters of methanesulfonic acid as irreversible inhibitors of acetylcholinesterase. J Biol Chem 237: 3245.

Klinman JP. 2003. The multi-functional topa-quinone copper amine oxidases. Biochim Biophys Acta 1647: 131.

Lippert B, Metcalf BW, Jung MJ, Casara P. 1977. 4-amino-hex-5-enoic acid, a selective catalytic inhibitor of 4-aminobutyric-acid aminotransferase in mammalian brain. Eur J Biochem 74: 441.

Margoliash E, Novogrodsky A, Schejter A. 1960. Irreversible reaction of 3-amino1:2:4-triazole and related inhibitors with the protein of catalase. Biochem J 74: 339.

Monod J, Wyman J, Changeux J-P. 1965. On the nature of allosteric transitions: a plausible model. J Mol Biol 12: 88.

Morrison JF. 1982. The slow-binding and slow, tight-binding inhibition of enzyme-catalysed reactions. Trends Biochem Sci 7: 102.

Morrison JF, Stone SR. 1985. Approaches to the study and analysis of the inhibition of enzymes by slow- and tight-binding inhibitors. Comments Mol Cell Biophys 2: 347.

Morrison JF, Walsh CT. 1988. The behavior and significance of slow-binding enzyme inhibitors. Adv Enzymol Relat Areas Mol Biol 61: 201.

Motulsky HJ, Christopoulos A. 2003. Fitting models to biological data using linear and nonlinear regression. A practical guide to curve fitting. San Diego: GraphPad Software Inc.

Nimmo IA, Mabood SF. 1979. The nature of the random experimental error encountered when acetylcholine hydrolase and alcohol dehydrogenase are assayed. Anal Biochem 94: 265.

Oldham KG. 1993. Radiometric assays. Enzyme assays. A practical approach. Eisenthal R, Danson MJ, editors. Oxford: Oxford University Press; pp. 93-122.

Price NC, Stevens L. 1999. Fundamentals of Enzymology. New York: Oxford University Press.

Sculley MJ, Morrison JF. 1986. The determination of kinetic constants governing the slow, tight-binding inhibition of enzyme-catalysed reactions. Biochim Biophys Acta 874: 44.

Segel IH. 1993. Enzyme kinetics. Behavior and analysis of rapid equilibrium and steady-state enzyme systems. New York: Wiley.

Shou M, Lin Y, Lu P, Tang C, Mei Q, et al. 2001. Enzyme kinetics of cytochrome P450-mediated reactions. Curr Drug Metab 2: 17.

Silverman RB. 1995. Mechanism-based enzyme inactivators. Methods Enzymol 249: 240.

Storer AC, Darlison MG, Cornish-Bowden A. 1975. The nature of experimental error in enzyme kinetic measurements. Biochem J 151: 361.

Syed SEH. 1993. High performance liquid chromatographic assays. Enzyme assays. A practical approach. Eisenthal R, Danson MJ, editors. Oxford: Oxford University Press; pp. 123-166.

Terstappen GC, Reggiani A. 2001. In silico research in drug discovery. Trends Pharmacol Sci 22: 23.

5 Fluorescence Measurements of Receptor–Ligand Interactions

S. M. J. Dunn

Abstract: Fluorescence techniques are widely used to study the interaction of ligands with their receptors. If the binding of a ligand is accompanied by a suitable change in fluorescence, this can be monitored to provide information on not only the properties of the equilibrium complex but also on the kinetics of the interaction i.e., the conformational changes that lead to the equilibrium state. Examples of suitable fluorescence signals that can be exploited include changes in protein intrinsic fluorescence, the use of fluorescent ligands or the use of extrinsic fluorescent reporter groups, either reversibly or covalently bound to the receptor protein. In this chapter, we provide a brief overview of the theory of fluorescence and the methods that can be used to monitor receptor-ligand interactions. We also discuss some of the fluorescence techniques that have been developed to study the binding of ligands to the nicotinic acetylcholine receptor, one of the best characterized neurotransmitter receptors.

List of Abbreviations: ACh, acetylcholine; GABA, γ–aminobutyric acid; IANBD, 4-[N-[2-(iodoacetoxy)-ethyl]-N-methylamino]-7-nitrobenz-2-oxa-1,3-diazole; 5-IAS, 5-(iodoacetamido)salicylic acid; nAChR, nicotinic acetylcholine receptor

1 Introduction

As discussed elsewhere in this volume, the use of radiolabelled derivatives of natural ligands and drugs has had a huge impact on our understanding of receptors and their binding sites. However, there are also a number of biophysical approaches for studying receptor–ligand interactions that can afford many advantages over radioligand binding studies. These methods include a variety of spectroscopic approaches such as absorbance spectrophotometry, spectrofluorimetry, circular dichroism, electron spin resonance and nuclear magnetic resonance. This review focuses on the use of fluorescence techniques to monitor the binding of ligands to their receptors. We will discuss how fluorescence can be used to provide information on the affinities and, in some cases, also the stoichiometries of receptor–ligand interactions. The use of fluorescence techniques to monitor the kinetics of the protein conformational changes that are induced by ligand binding and to probe the locations of binding sites within a receptor protein will be briefly described. In the latter part of the chapter, we will use the example of one of the best characterized neurotransmitter receptors, the nicotinic acetylcholine receptor (nAChR) from *Torpedo* electric organ, to illustrate how fluorescence approaches have been used to further our knowledge of the structure and function of this protein.

2 Simple Theory of Fluorescence

The discussion that follows is a simplified description of the theory of fluorescence. Readers who are interested in more detailed information are referred to comprehensive reviews (e.g., Freifelder, 1982; Lakowicz, 1983; Harris and Bashford, 1987).

2.1 Excitation and Emission Spectra

Fluorescence occurs when a molecule absorbs a photon and emits this energy in the form of light at a longer wavelength. All organic molecules absorb energy at certain wavelengths in the electromagnetic spectrum but only a small fraction of these show significant fluorescence at room temperature. Thus fluorescence measurements can be used to detect a population of molecules that may be present as only a minor component of a complex mixture. As discussed below, this confers both *specificity* and *sensitivity* in fluorescence assays.

For biochemical studies, the ultraviolet/visible region of the spectrum (200 – 900 nm) is by far the most important. As illustrated in the simplified diagram in *Figure 5-1*, the absorption of a photon by a molecule rapidly (within about 10^{-15} s) promotes the transition of an electron from a ground state (S_0) to a higher energy state, shown here as S_1 (note that higher energy states, S_2, S_3, etc. may also exist). Within each energy

Figure 5-1
The principles of fluorescence showing energy levels in the ground state (S_0) and the first excited state (S_1). Various vibrational energy levels exist within each state (illustrated here as 0, 1 and 2)

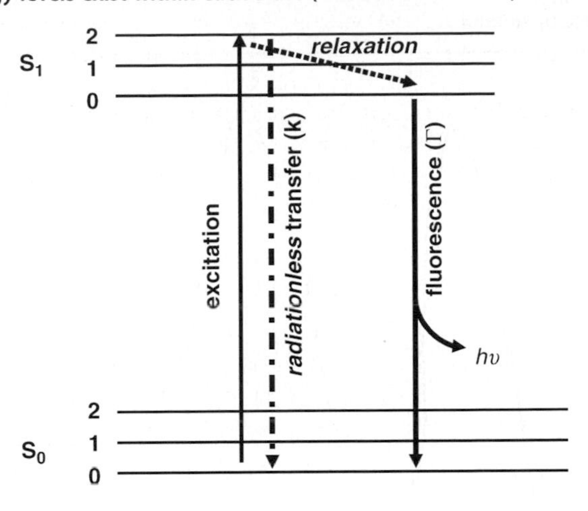

level, there are a number of different vibrational levels as originally proposed by Jablonski in 1935. Some of these states are depicted here as 0, 1 and 2. In most cases, the energy that had been absorbed is lost by very rapid intramolecular processes or by collisional interactions with other molecules (often solvent molecules) in the vicinity. Under these circumstances, the energy is lost primarily as heat and no fluorescence is observed. Fluorescence occurs when the return of the electron to the ground state is accompanied by the emission of a photon, i.e., at least part of the light energy absorbed is re-emitted as light.

Figure 5-1 shows an example of a fluorescent molecule that, upon irradiation, is excited to a higher vibrational level within S_1. Some of the energy of the excited state is lost by thermal equilibration (e.g., collisional quenching) to a lower energy state within S1. Subsequently, the electronic transition back from the lower energy state of S_1 to S_0 is accompanied by either a radiationless transition (k, where k is the rate constant of this transition) or by the emission of light (occurring with a rate constant of Γ). In the latter case, this is fluorescence (or phosphorescence if the emission is long-lived, a phenomenon that usually occurs only at low temperature). In all biochemical experiments that are carried out in solution, the consequence of the loss of some of the absorbed energy in the excited state through various quenching mechanisms is that the energy emitted is less than the energy absorbed. This leads to a fundamental property of biochemical fluorescence experiments, i.e., *the wavelength at which a molecule fluoresces (emits) is longer than the wavelength at which it is excited (absorbs).* This phenomenon, which is commonly referred to as the "Stokes' shift" (see Lakowicz, 1983), is illustrated in the excitation and emission spectra of the aromatic amino acid, tryptophan, in *Figure 5-2.* In aqueous solution, tryptophan absorbs ultraviolet light with an excitation maximum at around 280 nm and it exhibits broad fluorescence emission with a maximum at around 350 nm. Superimposed on this figure is the ultraviolet absorbance spectrum of tryptophan; as can be seen, the *fluorescence excitation spectrum overlaps that of the absorbance spectrum.* Also illustrated in this figure is another common feature of fluorescence, i.e., that *the emission spectrum is often the mirror image of the excitation spectrum.* Various physical and chemical phenomena can, however, complicate these observations and these have been extensively reviewed.

2.2 The Importance of Environment on Fluorescence

When tryptophan is dissolved in water, it shows the fluorescence characteristics illustrated in *Figure 5-2.* However, one especially useful property of tryptophan fluorescence is that its emission spectrum is highly sensitive to the polarity of its environment. In less polar solvents (alcohols, alkanes, etc.), the emission

Figure 5-2
Fluorescence of an aqueous solution (10 μM) of tryptophan. Superimposed on the figure is its absorbance spectrum (dashed line). Tryptophan fluorescence shows an excitation maximum at about 280 nm and broad fluorescence emission at 350–360 nm

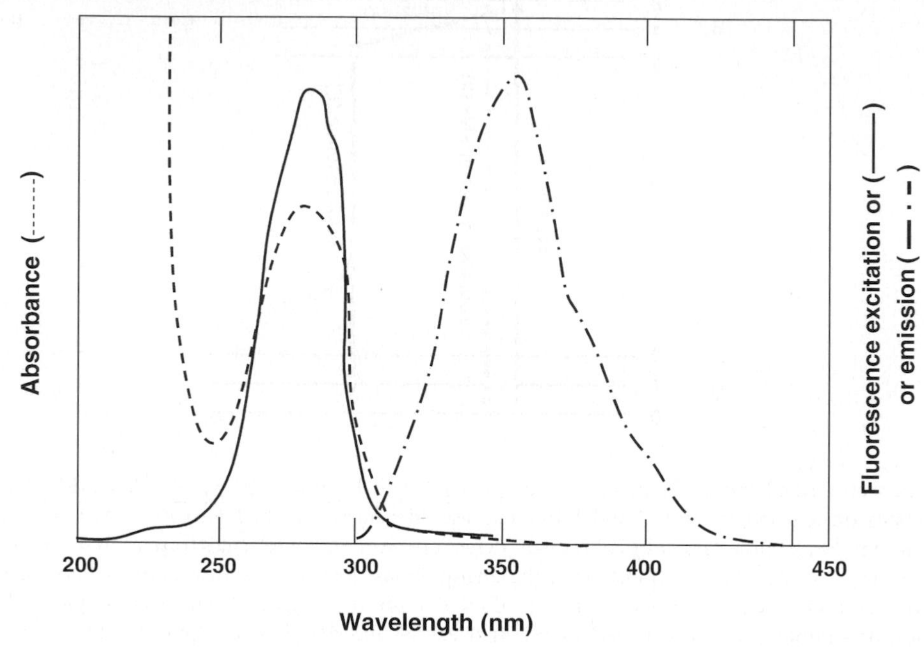

maximum (λ_{max}) of tryptophan is shifted to lower wavelengths (a "blue shift") and the magnitude of its fluorescence emission is increased. Although there are exceptions, this is a property displayed by most fluorophores. Fluorescence is also temperature sensitive and, for most fluorophores in aqueous solution, it is higher at low temperatures due to the reduced rate of collisional deactivation processes occurring during the lifetime of the excited state. As discussed below, tryptophan residues are the major contributors to protein fluorescence accounting for approximately 90% of the total emission. Changes in tryptophan fluorescence that accompany protein conformational changes are monitored frequently in studies of ligand binding. By studying the effects of ligand binding on the λ_{max} of protein fluorescence and its magnitude, we can make certain predictions about the types of conformational transitions that have occurred. When the binding of a ligand to its receptor protein is accompanied by an increase in protein fluorescence and a blue shift, this suggests that tryptophan residues may have moved into a more hydrophobic environment such as an interior crevice within the protein. The opposite effect, i.e., a decrease in fluorescence and a shift to longer wavelengths (a "red shift") suggests movement towards a more polar environment, perhaps to the exterior surface of the protein thus allowing for greater exposure to the solvent. Other environmental factors that affect fluorescence include pH, ionic strength, the presence of quenching molecules (such as iodide or cesium ions) and the proximity of neighboring chemical groups to the fluorophore. While these factors introduce considerable complexity and variability into fluorescence measurements, all can be exploited usefully in studies of protein–ligand interactions.

2.3 Fluorescence Resonance Energy Transfer

One common approach in the study of ligand binding to their receptors is to monitor fluorescence changes not directly but via "energy transfer". The principles of this method are illustrated in *Figure 5-3*.

Figure 5-3

Fluorescence energy transfer. Panel (a) shows the fluorescence characteristics of a hypothetical protein A (excitation and emission maxima at 290 nm and 350 nm, respectively). In (b), Ligand B fluoresces with an emission maximum at 450 nm when excited at 350 nm. Panel (c) shows that the formation of the protein–ligand complex can be monitored using an excitation wavelength of 290 nm and recording the decrease in protein fluorescence at 350 nm or the increase in ligand fluorescence at 450 nm

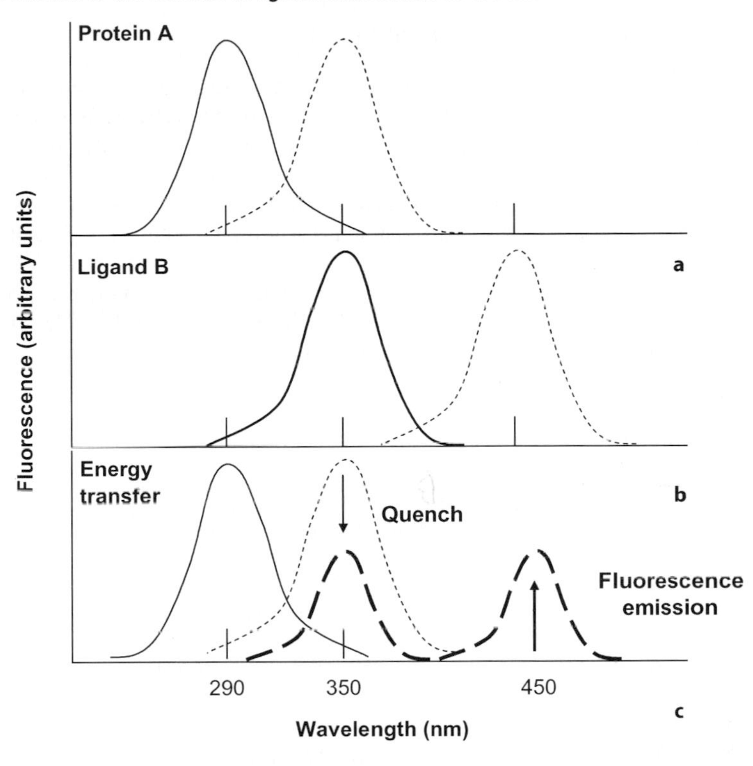

Simplistically, energy transfer requires two (or more) fluorophores with overlapping spectral characteristics, one of which acts as a fluorescence "donor" and the other as an "acceptor". In *Figure 5-3*, Protein A is fluorescent (due to the presence of aromatic amino acids such as tryptophan; see *Figure 5-2*) and it shows typical spectral properties with an excitation maximum at about 290 nm and broad fluorescence emission at 330–350 nm. Ligand B is also fluorescent and it absorbs light at 340 nm and fluoresces at about 450 nm. Examples of ligands with such fluorescence characteristics are the common enzyme cofactors, NADH and NADPH. Due to spectral overlap, when the ligand binds to the protein, it can absorb the fluorescence that is emitted by the protein. Using an excitation wavelength of 290 nm, formation of the complex thus results in a quench in Protein A fluorescence at 350 nm and the appearance of a new fluorescent peak at 450 nm (due to Ligand B emission), i.e., the energy has been transferred. As discussed below, energy transfer can occur only over relatively short distances (less than about 70Å between donor and acceptor molecules). These spectral and distance restraints provide two major advantages in biochemical studies. Firstly, monitoring of fluorescence via energy transfer provides for increased specificity and reduces some of the artifacts that are invariably encountered in spectroscopic studies. Secondly, information on the efficiency of energy transfer between donor and acceptor molecules can be used to define distances between binding sites or other structural features on a receptor protein (see below).

2.4 Quantification of Fluorescence

Quantification of fluorescence usually involves studies of the quantum yield (Q) and/or the fluorescence lifetime (τ). In simple terms, the quantum yield is a measure of fluorescence efficiency and this is defined by the ratio of the number of emitted photons to the number absorbed. In the terminology of *Figure 5-1*, this is given by:

$$Q = \Gamma/(\Gamma + k) \tag{1}$$

In biochemical studies, absolute values of Q are rarely measured (such quantitation is notoriously difficult). However, relative values can be estimated by comparing the fluorescence emission of the molecule in question with that of a standard fluorophore such as quinine bisulfate. The greater the ratio, the higher is the fluorescence of the molecule. As noted above, most molecules do not fluoresce (i.e., $Q = 0$); this is usually because all of the absorbed energy is dissipated by nonradiationless decay mechanisms (k). Highly fluorescent molecules may have a quantum yield that approaches (but cannot exceed) unity since most of the absorbed energy is emitted in the form of light.

The lifetime of the excited state (τ) is a measure of the average time that a molecule spends in this state before returning to the ground state. For most molecules, this lifetime is of the order of 10 ns. For the hypothetical fluorophore illustrated in *Figure 5-1*, this is described by:

$$\tau = 1/(\Gamma + k) \tag{2}$$

In the absence of nonradiative processes ($k = 0$), the lifetime (the "intrinsic lifetime") becomes:

$$\tau_0 = 1/\Gamma \tag{3}$$

which leads to a simple relationship between lifetime and quantum yield:

$$Q = \tau/\tau_0 \tag{4}$$

The above show that both the quantum yield and fluorescence lifetime can be modified by any factor that affects the relative contributions of the nonradiative (k) and radiative (Γ) decay processes. As described in Section 2.2, these factors include environmental factors such as solvent polarity, ionization and temperature.

As described briefly above, energy transfer fluorescence is often used to monitor binding processes. The efficiency of transfer decreases as the sixth power of the distance between the donor and acceptor fluorophores:

$$\text{Efficiency} = \frac{R_0^6}{R_0^6 + r^6} \tag{5}$$

where r is the distance between the fluorescent pairs and R_0 is a constant that depends on the spectral characteristics of the particular donor–acceptor interaction (Förster, 1948). Measurement of the magnitude of energy transfer can thus be used as a "spectroscopic ruler" (Stryer, 1978) to measure the distance between fluorescence donors and acceptors such as between a bound ligand and some fluorescent moiety in the receptor molecule. An example of how this has been applied in the study of the nAChR is the measurement of energy transfer between membrane-partitioned fluorescent probes and a bound fluorescent agonist, dansyl-C6-choline (see below) which revealed that the agonist binding sites are located approximately 30 Å above the membrane surface (Valenzuela et al., 1994).

3 Instrumentation for Studying Fluorescence

Instruments for studying UV-visible absorbance, i.e., spectrophotometers are available in many biomedical laboratories and most investigators are familiar with the basic components:

1. A light source, which is usually tungsten for measurements in the visible range (350–800 nm) or deuterium for ultraviolet emission (200–350 nm).
2. A monochromator to select the wavelength for absorbance (or in simpler models, an optical filter to allow only certain wavelengths to pass through to the sample).
3. A sample holder to insert the solution of interest.
4. A photodetector to detect the amount of light that has been transmitted through the sample (or the amount of light that has been absorbed).
5. Some mechanism for recording the output, e.g., a meter, chart recorder, printer, computer, etc.).

The design of a research spectrofluorimeter is essentially the same except for two important differences (see *Figure 5-4*). Firstly the detector is positioned to measure light emission at right angles to the incident beam. This change in detection angle minimizes the contribution from the incident light and selects for detection of only emitted light (this accounts for the greater *sensitivity* of fluorescence experiments). Secondly, a second monochromator (or optical filter) is used to select for the emission characteristics of the fluorophore. This capitalizes on the differences in the excitation and emission properties of fluorescent molecules thus providing for *selectivity.* Most modern spectrofluorimeters also use a xenon light source rather than the tungsten/deuterium sources commonly used in spectrophotometers. Although xenon lamps are less stable than the other sources and suffer from the disadvantage of "spiking" they provide a higher intensity of light over the UV/visible range and are thus more useful for measurements of poorly fluorescent compounds. An important consideration in fluorescence experiments is that the magnitude of the observed signal depends on the instrument being used and thus, unlike absorbance measurements, fluorescence is normally described in relative rather than absolute terms.

Figure 5-4
The design of a research spectrofluorimeter showing the basic components

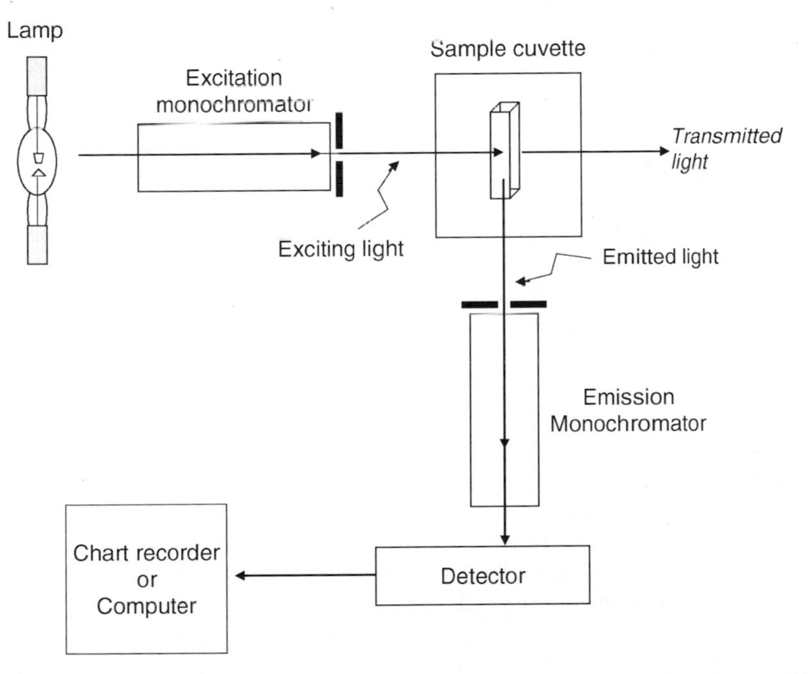

4 Choice of Fluorophore

The success of a fluorescence experiment to study the binding of a ligand to its receptor protein is critically dependent upon the choice of the fluorescent signal that will be monitored. Identification of a suitable probe is the first stumbling block in these experiments and this has often proved to be an insurmountable problem. Techniques must be developed and optimized for each receptor–ligand interaction of interest. As described in Section 2.2, changes in fluorescence arise from changes in the environment of the fluorophore (polarity, ionic strength, pH, etc.). All receptor proteins and many receptor ligands are flexible in solution and at least one (and often both) is likely to change conformation upon formation of the receptor–ligand complex. If either (or both) is fluorescent this can provide a suitable fluorescence signal. However, the magnitude of the fluorescence change may be small and there can be many confounding factors. A ligand that may seem to have ideal fluorescence properties in aqueous solution may not be so accommodating when it encounters its binding site in the interior of a protein. These experimental predicaments often necessitate the development of indirect methods such as the use of extrinsic fluorescent probes to act as "reporter" groups. Below is a general description of the most widely used techniques.

4.1 Protein Intrinsic Fluorescence

Many proteins are fluorescent due to their complement of aromatic amino acids (tryptophan, tyrosine, phenylalanine) which, depending on their environment, absorb light at 270–290 nm and emit at 320–340 nm. As noted above, tryptophan tends to dominate the protein fluorescence spectrum and is thus the most often utilized. Phenylalanine has a very low quantum yield (Q, see Section 2.4) and, although tyrosine is highly fluorescent in solution, its fluorescence is usually much weaker in proteins due to intramolecular quenching arising either from its ionisation and/or its proximity to quenching factors that are intrinsic to the protein (amino groups, carboxyl groups, tryptophan, etc.). Thus, while changes in protein fluorescence provide the most direct measurement of receptor–ligand interactions, nature is not always kind to the experimenter.

4.2 Fluorescent Ligands

A huge number of fluorescent probes are available from commercial sources (see Haugland, 1996). Some receptor ligands are naturally fluorescent and some can be derivatized to make them fluorescent. In the latter case, it is important to verify that the derivatization *per se* does not adversely affect the binding properties of the ligand. In these studies, the goal is that the properties of the fluorescent analogue closely mimic those of the natural ligand or drug under investigation.

4.3 Extrinsic Probes

When neither the protein nor the ligand provides a suitable fluorescence signal to monitor their interaction, the next most common technique is to use an extrinsic probe as a reporter fluorophore. As discussed below with reference to the nAChR, these probes can be reversibly bound or covalently attached to the receptor protein. An ideal reporter group is one that has no effect on the properties of the receptor, i.e., the fluorophore itself does not perturb the system under investigation. The aim in these studies is to develop an approach in which the fluorescence of the probe changes in response to a changing environment and thus acts as a faithful monitor of the receptor–ligand interaction of interest.

4.4 Allosteric Ligands

Many natural compounds and drugs do not directly activate or inhibit a receptor process; rather they bind to allosteric sites on the receptor protein to modulate the effects of the endogenous ligand. Examples of these are local anesthetics acting on the nAChR or benzodiazepine site ligands modulating the $GABA_A$ receptor. On their own, these drugs do not activate the receptor but they influence the effects of the endogenous neurotransmitters on receptor-mediated conductance changes. Such fluorescent allosteric ligands can provide information not only on their own binding to the receptor but also on their allosteric interactions with other binding sites and the effects of these on protein conformation.

5 Equilibrium Studies of Receptor–Ligand Interactions Monitored by Fluorescence

5.1 Introduction

The principle of equilibrium fluorescence measurements is identical to that of radiolabelled ligand binding experiments, i.e., the ligand (L) binds to its receptor (R) and the formation of the complex (RL) is measured:

$$R + L \rightleftharpoons RL$$

In the case of radioligand binding, free and bound ligand must be physically separated and either the depletion in the concentration of free radioligand or the increase in the concentration of the radioactive complex is quantified. In fluorescence experiments, the bound and free ligand do not need to be separated but there must be a specific and measurable change in fluorescence as a consequence of binding. As discussed below, the lack of requirement to physically separate bound and free ligand affords a major advantage to fluorescence studies, i.e., sites of low affinity (rapidly dissociating) can be measured in fluorescence experiments whereas they are intransigent to most studies using radiolabelled ligands.

In order to illustrate the principles of equilibrium fluorescence assays, we will use the hypothetical Protein A/ Ligand B interaction shown in ➲ *Figure 5-3*. A preparation of Protein A is placed in a fluorescence cuvette (quartz in this instance since the excitation wavelength, 290 nm, is in the ultraviolet range). Protein A may be in solution (if it is a soluble protein) or in suspension (if it is a membrane-bound protein). Using a typical 1 cm^2 cuvette, the minimum volume required is about 2 ml for most instruments but smaller sample volumes are possible using "microcuvettes", i.e., those with shorter excitation and/or emission pathlengths. Excitation and emission spectra are recorded to determine suitable excitation and emission wavelengths for titration. In the case of protein fluorescence quenching studies, these would be approximately 280 and 350 nm, respectively. Using these wavelengths, the changes in fluorescence of Protein A in response to the addition of small aliquots (1–5 μl) of ligand B are recorded. The binding of ligand B to Protein A causes a quench in the protein intrinsic fluorescence (➲ *Figure 5-5a*). These data can be used to obtain information on the equilibrium affinity of the complex (K_d) in a manner that is analogous to radioligand binding experiments.

As shown in ➲ *Figure 5-5b* the binding of Ligand B to Protein A, also gives rise to an increase in Ligand B fluorescence when it is excited indirectly from the receptor protein. This "energy transfer fluorescence" can be measured in a similar manner to that described above using the same excitation wavelength (280 nm) but now monitoring the ligand fluorescence at 430 nm. ➲ *Figure 5-5c* shows the third way in which the same receptor–ligand interaction can be monitored by fluorescence techniques. In this case, changes in the fluorescence of Ligand B is monitored directly rather than via energy transfer from receptor protein. Since ligand B is fluorescent, its fluorescence may change upon complex formation. Thus, using excitation and emission wavelengths of 340 and 430 nm (for the ligand) we can see a concentration-dependent increase in fluorescence that is due to the changes in the environment of the ligand itself. ➲ *Figure 5-5c* also shows a control titration in which a parallel titration of buffer by the ligand is used to correct for the natural fluorescence of the ligand. Here the advantages and disadvantages of fluorescence experiments come to the fore: the exquisite sensitivity of such experiments requires individual assessment of the suitable controls. In the examples shown in ➲ *Figure 5-5*, the three methods of measurement give the same measure of affinity (apparent K_d = 100 nM).

Figure 5-5
Fluorescence titration experiments. In each case a solution of Protein A (2 ml) was titrated with aliquots (2μl) of a concentrated solution of Ligand B to give the indicated concentrations of ligand. The changes in fluorescence that accompanied complex formation were measured using the appropriate wavelengths for (a) protein quenching, (b) energy transfer or (c) enhancement of ligand fluorescence. The solid lines show the observed fluorescence change and the dashed lines show control titrations that were carried out in parallel in order to determine any nonspecific effects (see text for details). After subtraction of nonspecific effects, the data in each case show saturable binding of Ligand B to Protein A with an apparent K_d of 1 μM

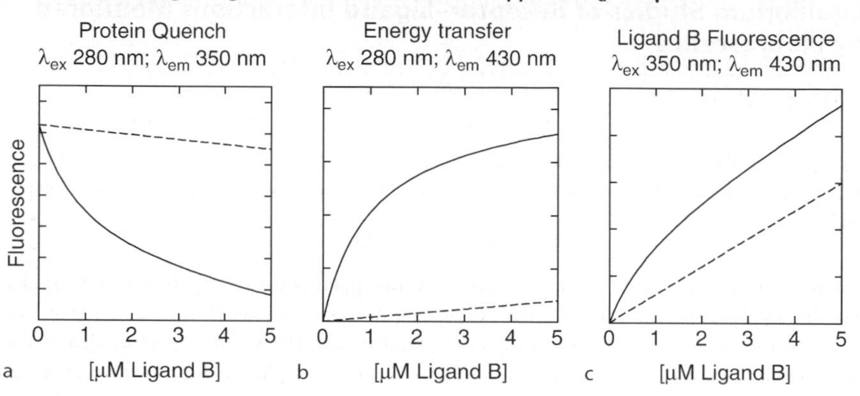

5.2 Practical Considerations in Equilibrium Fluorescence Studies

- If either the protein or ligand is light sensitive then some photolysis ("photodecomposition") may occur upon exposure to light. Under these circumstances, it is essential to shut off the excitation light source between measurements (usually by closing the excitation shutter that is provided in most instruments). The slit-width on the excitation side can also be reduced in order to minimize overall exposure (this can often be compensated by an increase in the slit-width on the emission side to increase the magnitude of emitted fluorescence).
- The binding of ligands and the protein conformational changes that are induced by binding may be very slow and this requires some knowledge of the kinetics of the interaction. This can be readily estimated by making fluorescence measurements at different times after ligand addition to ensure that an equilibrium fluorescence level has indeed been reached.
- As in studies of radioligand binding, there are often nonspecific effects in fluorescence measurements. In quantitative studies, it is therefore important to include a control sample that is titrated in parallel to allow quantification of the specific effects. In the case of the fluorescence quenching data shown in *Figure 5-5a*, this may be titration of a standard tryptophan solution which starts with approximately the same fluorescence as the protein in question (this corrects mainly for dilution effects) or titration of a parallel sample in which specific binding sites have been blocked by an excess of an inhibiting ligand (although this may prove difficult if the inhibiting ligand also affects the protein fluorescence).

5.3 Analysis of Equilibrium Fluorescence Titration Data

5.3.1 Binding Affinity

The mathematical treatment of each of the fluorescence curves shown in *Figure 5-5* is identical. First nonspecific effects must be subtracted so that in each case we have a relationship between the concentration of ligand and the change in fluorescence. The binding equation is formally identical to the terminology for radioactive ligand binding described elsewhere in this volume. For the simplest of binding interactions, i.e., a reversible bimolecular association/unimolecular dissociation reaction:

$$R + L \underset{k_{-1}}{\overset{k_1}{\rightleftharpoons}} RL$$

the concentration of the complex, [RL] is given by:

$$[RL] = \frac{[R_0][L]}{K_d + [L]} \tag{6}$$

where $[R_0]$ and [L] are the concentrations of total receptor sites and *free* ligand, respectively. The dissociation constant, K_d denotes the concentration of ligand which occupies 50% of the receptor binding sites ($[RL] = 0.5[R_0]$), see C). In fluorescence studies, we do not usually directly measure the concentration of the RL complex; rather we measure some change in fluorescence, denoted by Fl. The above equation should therefore be modified:

$$Fl = \frac{F_{\max}[L]}{K_d + [L]} \tag{7}$$

where Fl is the measured change in fluorescence (after subtracting for background effects) and F_{max} is the maximum fluorescence change when all available sites are saturated. The relationship between the measured fluorescence and [L] is thus hyperbolic. Here it is important to consider an important aspect of such fluorescence titrations. In radioligand binding studies, both the bound and free ligand concentrations can be directly measured. However, in fluorescence studies, we do not usually have a direct measure of the *free* ligand concentration, although we do know how much total ligand ($[L_0]$) has been added. Thus, if there is a significant depletion of free ligand as a consequence of binding to the receptor, this introduces a significant error into estimates of the affinity when the fluorescence signal is related directly to the added ligand concentration. Depletion due to ligand binding can however, be corrected for by using nonlinear regression curve fitting techniques and the following equation:

$$Fl = F_{RL}[RL] = 0.5F_{RL}\{(L_0 + K_d + R_0) - \sqrt{[(L_0 + R_0 + K_d)^2 - 4R_0L_0]}\} \tag{8}$$

where F_{RL} is the fluorescence change per unit concentration of the RL complex.

5.3.2 Binding Site Stoichiometry

While equilibrium fluorescence titrations are very useful for measuring the affinities of ligand for their receptors, measurements of the concentration of receptor binding sites are often problematic. This arises from the difficulty of quantitating the absolute value of fluorescence signal changes. As noted above, fluorescence is often described in “arbitrary units”. The maximum fluorescence change when all the binding sites are occupied cannot be related readily to the number of sites in the assay. Under certain circumstances, however, fluorescence can provide a very accurate measurement of receptor concentration. Such circumstances arise when the binding of a ligand is of such high affinity that virtually all added ligand will bind to the protein of interest until all sites are occupied. At this point, a clear “break point” in the titration will be observed (see ➲ *Figure* 5-6). We have, for example, used such “stoichiometric titrations” to quantify the concentration of high affinity agonist binding sites in fluorescently labeled *Torpedo* nAChR preparations (Dunn and Raftery, 1993).

6 Kinetic Studies: General Theory and Practice

A major advantage of fluorescence studies of receptor–ligand complex formation over radiolabelled ligand binding techniques is that the binding reaction can be monitored in “real time”. As discussed below, this is especially important in the case of receptor proteins that undergo conformational changes upon ligand

Figure 5-6
Fluorescence measurements of binding site stoichiometries. The solid line shows a theoretical curve in which a receptor solution ($R_0 = 1\ \mu M$) is titrated with a ligand which binds with a K_D of 10 nM. At low ligand concentrations, virtually all ligand will bind and there is a measurable breakpoint when all receptor sites are saturated (1 μM). The dashed line shows theoretical data for the same interaction but using a K_D of 1 μM. Thus an estimate of the density of receptor sites (R_0) can be obtained only for a high affinity ligand binding to a high concentration of receptor

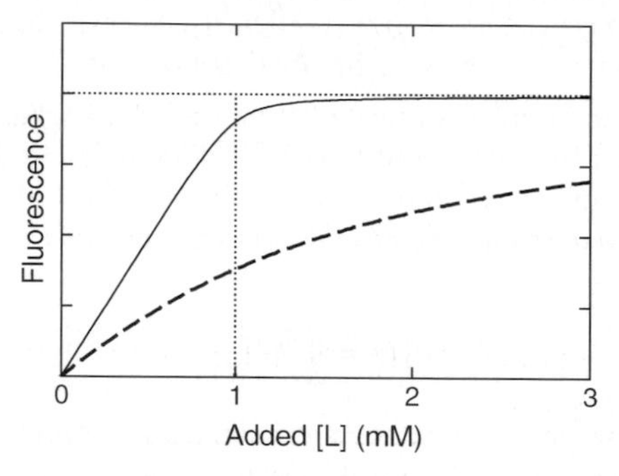

Figure 5-7
The principles of the stopped-flow technique showing the principle components of a research instrument

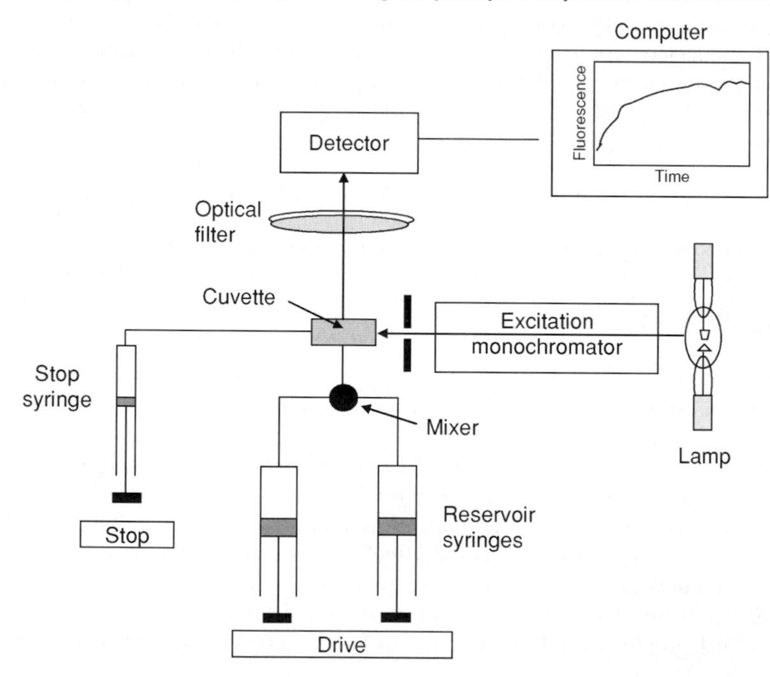

binding. The time resolution of radiolabelled ligand binding assays is limited by the necessity of using some physical technique to separate bound from free ligand. In common filtration experiments using membrane-bound receptors, for example, the radioligand is incubated with the receptor preparation of interest until equilibrium is reached. The membrane-bound complex is then collected by filtration through a glass fiber filter that allows the free ligand to pass through while the bound ligand is trapped by the filter. The separation techniques take time; in manual filtration assays, even the most dexterous of experimenters cannot complete the task in less than 2–3 s. Automated techniques can reduce this time to the order of 10 ms but still the experiments are arduous with each time point requiring a separate experiment. Fluorescence experiments allow for continuous monitoring of the conformational changes that are mediated by ligand binding provided, of course, that there is a suitable fluorescence signal to monitor these transitions. *Figure 5-7* illustrates the principles of the widely used stopped-flow technique. Briefly the ligand and receptor of interest are loaded into separate "drive" syringes. Under electronic control, equal volumes of the two components are mixed (via a specially designed mixing chamber) and the reaction mixture flows through a cuvette and into a third "stop" syringe which is oriented such that the plunger moves back upon filling to stop the flow and activate the recording procedures (within about 1 ms). Changes in the fluorescence of the reactant mixture in the cuvette can thus be continually monitored from almost the time of mixing (stop of flow) until the time that equilibrium is reached.

7 Ligand–Receptor Interactions: The Example of the Nicotinic Acetylcholine Receptor

The use of fluorescence to study receptor–ligand interactions can be illustrated by studies of ligand binding to the nAChR. This receptor is one of the best characterized neurotransmitter receptors due mainly to its abundance in the electric organ of *Torpedo* fish from which it was first purified and characterized. The *Torpedo* receptor is very similar to the muscle-type (peripheral) receptor that is localized to the mammalian neuromuscular junction. It is also the prototype of a family of homologous receptors that includes the neuronal nAChRs, $GABA_A$, glycine and serotonin type 3 ($5HT_3$) receptors. Each receptor in the family is a pentameric complex of homologous subunits that assemble to form a central ion pore (see *Figure 5-8*). In the case of the *Torpedo* receptor, this pore is cation (Na^+) selective and the receptor has a defined subunit structure with two identical α subunits in addition to a β, γ and δ subunit. The *Torpedo* receptor, like most

Figure 5-8
(a) The *Torpedo* nicotinic acetylcholine receptor is a pentamer of homologous subunits that assemble to form a central ion pore. (b) Cartoon depicting some of the conformations of the receptor that may exist in the absence or presence of bound ACh (•). Transitions between the different conformations are dictated by the concentration of and the time of exposure to the neurotransmitter

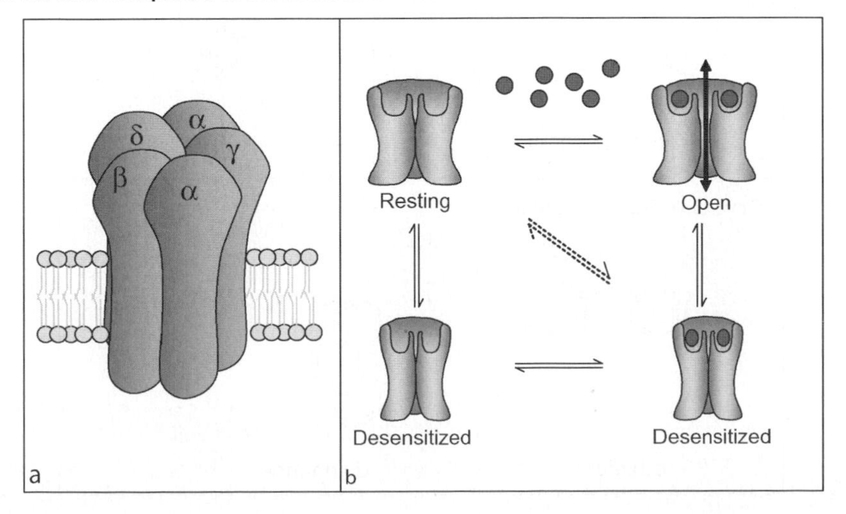

other members of the family has two major functional characteristics; activation and desensitization. Activation (channel opening) occurs very rapidly (within milliseconds) of neurotransmitter release (in this case acetylcholine, ACh). Upon prolonged exposure to the agonist, the receptor desensitizes, i.e., the channel closes and the receptor enters a refractory, nonconducting state. Another commonality among members of this receptor family is that activation is an intrinsically low affinity event (ACh activation of the *Torpedo* nAChR has an EC_{50} of about 50 μM) whereas the affinity for the neurotransmitter in the equilibrium, desensitized state is high (K_D of about 30 nM for [^{3}H]ACh binding to the *Torpedo* nAChR). Thus the nAChR exists in a number of different conformational states and transitions between these states depends on the time of exposure to agonist. A major goal in the study of this receptor family is to provide a quantitative description of the conformational changes that are involved in receptor activation and desensitization. In the case of the *Torpedo* nAChR, spectrofluorimetric techniques have proved to be of particular importance in this pursuit.

Figure 5-9 shows some of the fluorescence techniques that have been used to monitor ligand binding to the *Torpedo* nAChR. These include (A) fluorescent agonists, such as dansyl-C6-choline (Heidmann and Changeux, 1979) and NBD-5-acylcholine (Prinz and Maelicke, 1983), (B) fluorescent reporter groups such as 5-IAS (Dunn et al., 1980) and IANBD (Dunn and Raftery, 1982) which can be covalently incorporated into the receptor protein by virtue of their ability to react with available sulfhydryl goups and (C) reversible

Figure 5-9
Structure of fluorescent probes that have been used to study ligand binding to the *Torpedo* nicotinic acetylcholine receptor

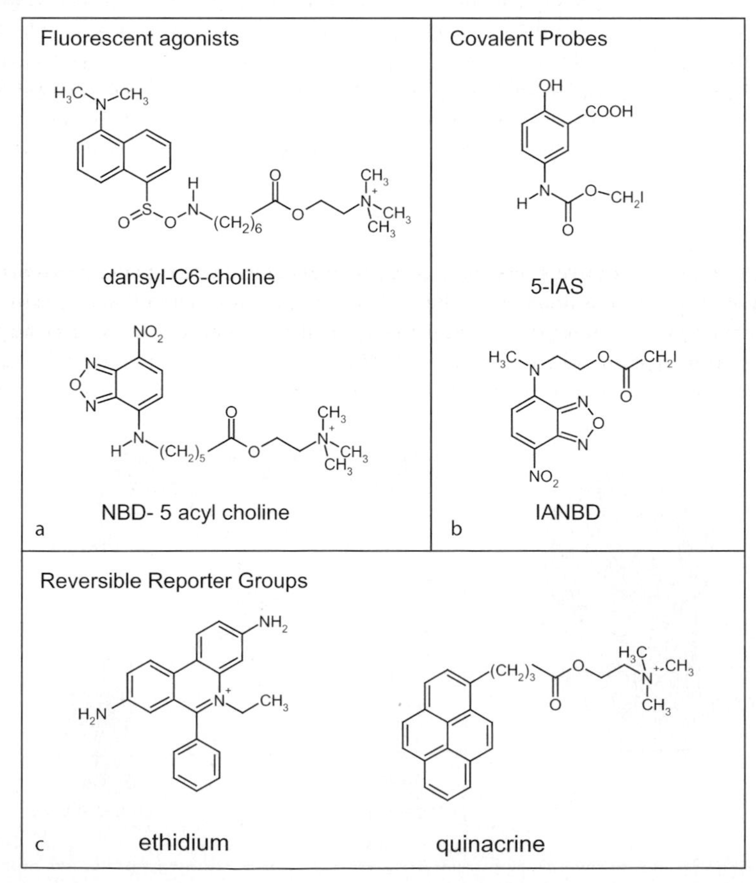

reporter groups such as ethidium (Quast et al., 1979) and quinacrine (Grűnhagen et al., 1977) which also act as noncompetitive blockers of the nAChR.

In general the kinetics of agonist binding are complex and the multiple phases in the fluorescence changes observed over different timescales have been analyzed in terms of multiple changes in the conformation of the receptor–ligand complex. Most investigators have interpreted the observations as being consistent with a modified version of the two-state model originally proposed by Katz and Thesleff (1957):

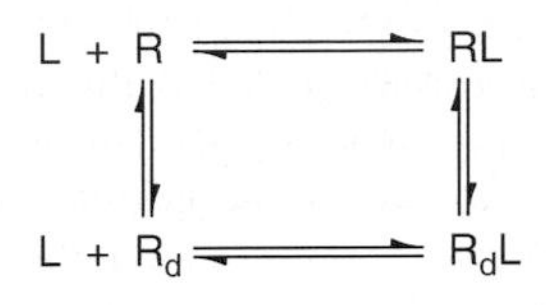

in which R and R_d represent the resting and desensitized states of the receptor. The results of our studies using covalently attached fluorophores to monitor the kinetics of agonist binding to the membrane-bound nAChR have led to the proposal of an alternative model in which receptor activation and inactivation are mediated by distinct classes of sites of low and high affinity, respectively. The fluorescence changes accompanying the binding of the agonist, carbamylcholine to receptors that had been labelled by reaction with 5-IAS (see Figure 5-9) occurred in three readily distinguishable phases occurring over distinct timescales of seconds to minutes. None of the observed conformational changes were sufficiently fast to account for the rapidity of channel activation and the apparent K_d value of even the resting state for ACh (approximately 2 μM) did not correlate with the measured EC_{50} value for ACh-mediated conductance changes (approximately 100 μM). In equilibrium fluorescence titration experiments, the apparent K_d for ACh binding was very similar to the equilibrium affinity measured in radiolabelled ACh binding studies (approximately 30 nM). We, therefore, suggested that the conformational changes that were being reported by the bound fluorophore were more likely to reflect changes that were associated with receptor desensitisation. In contrast, when agonist induced conformational changes were monitored using another covalently bound fluorophore, NBD (see Figure 5-9), the fluorescence change occurred in a single fast

Figure 5-10

The interaction of ACh with the *Torpedo* nAChR. The data shown compare the equilibrium binding parameters obtained from fluorescence studies using covalently attached fluorescent probes and those obtained from radiolabelled [^{3}H]ACh binding studies or functional measurements of cation flux. These data support a model in which the *Torpedo* nAChR carries sites of different affinities for ACh. We have previously suggested that occupancy of the lower affinity sites leads to channel activation whereas the higher affinity sites may play a role in desensitization processes

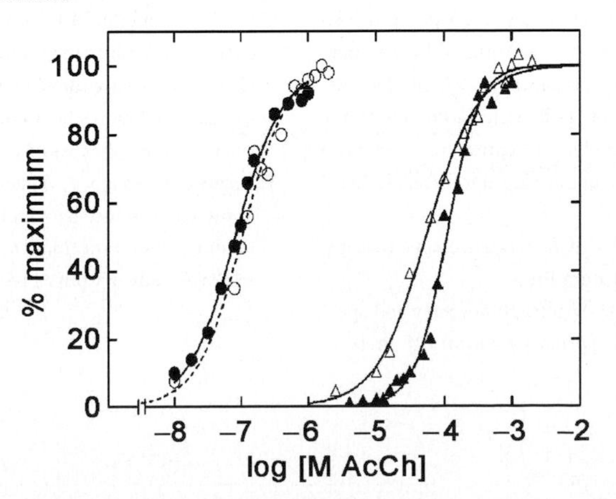

phase that approached a rate of 600 s^{-1} and the equilibrium affinity for ACh measured in fluorescence titration experiments was approximately 100 and in close agreement with the EC_{50} for ion flux responses. Some of these results are shown in Figure 5-10.

8 Conclusions

The above discussion provides only a brief overview of how fluorescence techniques can be used to study the interactions of ligands with their receptors. We have focused on the quantitation of the binding parameters and compared the data with that which may be obtained with those from radiolabelled ligand binding studies. The number of applications of fluorescence in the study of neurochemistry and molecular biology is ever increasing. Outside the scope of this review is, for example, the use of fluorescence microscopy to monitor cell surface expression and targeting of receptors or the use of fluorescence probes to monitor ion transport into and out of cells.

Most of the above studies have been made possible because of the high density of nAChRs in the *Torpedo* electroplax membrane. However, with recent technological developments, it may be hoped that similar approaches will prove useful when applied to receptors of much lower abundance.

References

Dunn SMJ, Raftery MA. 1982. Activation and desensitization of the *Torpedo* acetylcholine receptor: Evidence for separate binding sites. Proc Natl Acad Sci USA 79: 6757-6761.

Dunn SMJ, Raftery MA. 1993. Cholinergic binding sites on the pentameric acetylcholine receptor of *Torpedo californica*. Biochemistry 32: 8608-8615.

Dunn SMJ, Blanchard SG, Raftery MA. 1980. Kinetics of carbamoylcholine binding to membrane-bound acetylcholine receptor monitored by fluorescence changes of a covalently bound probe. Biochemistry 19: 5645-5652.

Förster T. 1948. Intermolecular energy migration and fluorescence. Surface density. Ann Phys 2: 55-75.

Förster T. 1959. Tenth Spiers Memorial Lecture. Transfer mechanisms of electronic excitation. Discussions Faraday Soc 27: 7-17.

Freifelder D. 1982. *Physical biochemistry: applications to biochemistry and molecular biology.* New York: W. H. Freeman.

Grünhagen HH, Iwatsubo M, Changeux JP. 1977. Fast kinetic studies on the interaction of cholinergic agonists with the membrane-bound acetylcholine receptor from Torpedo marmorata as revealed by quinacrine fluorescence. Eur J Biochem 80: 225-242.

Lakowicz JR. 1983. Principles of fluorescence spectroscopy. New York and London: Plenum Press.

Harris DA, Bashford CL. 1987. Spectrophotometry and spectrofluorimetry: a practical approach. Oxford: IRL Press.

Haugland RP. 1996. Handbook of fluorescent probes and research chemicals, sixth edition. Eugene, OR: Molecular Probes, Inc.

Heidmann T, Changeux J.-P. 1979. Fast kinetic studies on the interaction of a fluorescent agonist with the membrane-bound acetylcholine receptor from *Torpedo marmorata.* Eur J Biochem 94: 255-279.

Jablonski A. 1935. Űber den mechanismus des photoluminesczenz von farbstoff-phosphoren. Z Phys 93: 38-46.

Katz B, Thesleff S. 1957. A study of the 'desensitisation' produced by acetylcholine at the motor enplate. J Physiol 138: 63-80.

Prinz H, Maelicke A. 1983. Interaction of cholinergic ligands with the purified acetylcholine receptor protein. II. Kinetic studies. J Biol Chem 258: 10273-10282.

Quast U, Schimerlik M I, Raftery MA. 1979. Ligand-induced changes in membrane-bound receptor observed by ethidium fluorescence. Biochemistry 18: 1891-1901.

Stryer L. 1978. Fluorescence energy transfer as a spectroscopic ruler. Ann Rev Biochem 47: 819-846.

Valenzuela F, Weign P, Yguerabide J, Johnson DA. 1994. Transverse distance beween the membrane and the agonist binding sites on the *Torpedo* acetylcholine receptor: a fluorescence study. Biophys J 66: 674-682.

6 The Analyses of Neurotransmitters, Other Neuroactive Substances, and Their Metabolites Using Mass Spectrometry

B. D. Sloley · G. Rauw · R. T. Coutts

Abstract: The precise analysis of neurotransmitters and other neuroactive substances requires sensitive and selective techniques capable of isolating specific signals originating from compounds of interest from a large background of other biological interferences. Mass spectrometry is capable of identifying mass signals that are characteristic of compounds of identified or predicted interest. By combining mass spectrometry with a separation procedure (gas chromatography, liquid chromatography or others) it is possible to develop an accurate quantitative technique that eliminates almost all potentially interfering signals. This chapter summarizes the ways in which mass spectrometry is used to analyze neuroactive compounds and how it is combined with separation methods to provide accurate, quantitative estimates of the concentrations of these compounds in biological or other preparations. The advantages and limitations of particular methods are discussed and an extensive list of references is provided.

List of Abbreviations: APCI, Atmospheric Pressure Chemical Ionization; API-electrospray, Atmospheric Pressure Ionization Electrospray; APPI, Atmospheric Pressure Photoionization; CE, Capillary Electrophoresis; CI, Chemical Ionization; CID, Collision Induced Fragmentation; DC, Direct Current; ECNI, Electron Capture Negative Ion Chemical Ionization; EI, Electron Ionization; ESI, Electrospray Ionization; FAB, Fast Atom Bombardment; GC, Gas Chromatography; HIC, Hydrophobic Interaction Chromatography; HILIC, Hydrophilic Interaction Chromatography; HPLC, High Performance Liquid Chromatography; LC, Liquid Chromatography; MALDI, Matrix-Assisted Laser Desorption/Ionization; MS, Mass Spectrometry; MS/MS, Tandem Mass Spectrometry; m/z, Mass to Charge; PCI, Positive Chemical Ionization; PFTBA, Perfluorotributylamine; RF, Radio Frequency; SIM, Single Ion Mode; TOF, Time of Flight; UV, Ultraviolet; VIS, Visible

1 Introduction

Mass spectrometric methods for the identification and structural elucidation of endogenous neuroactive substances and pharmacologically active exogenous chemicals as well as the metabolites of these materials have been utilized for more than four decades. Mass spectrometry (MS) is a very powerful analytical technique, but until recently, it has been of limited use in routine analyses due to its considerable expense. The recent advent of reasonably priced mass spectrometers now permits many more researchers to use mass spectrometry, not only for chemical identification, but also for routine rapid, high throughput, highly specific quantitative analyses. Initially, mass spectrometers were not interfaced with separation instruments, and chemicals were analyzed by direct methods and this required that samples be relatively pure. Later, mass spectrometers were interfaced with gas chromatography (GC), which permitted the separation of mixtures of compounds and the interpretation of signals derived from chromatographically isolated compounds as they passed through the spectrometer. The limitations of gas chromatography necessitated that the compounds of interest were stable and volatile. Analysis of nonvolatile materials required that they be derivatized to increase their volatility. Furthermore, biological samples are usually contained within an aqueous matrix, and aqueous samples can cause harm to a GC system; thus, tedious extraction procedures may have to be applied (Rittenbach and Baker, 2006). The advent of new and ingenious interfaces now permits mass spectrometers to be interfaced with liquid chromatography (LC) and capillary electrophoresis (CE) systems, which allow for the analysis of nonvolatile materials in aqueous solutions. As new, more robust, and less expensive mass spectrometers have entered the market, there has been an exponential growth in mass spectral analytical applications. In particular, the advent of atmospheric pressure ionization-electrospray (API-electrospray) and atmospheric pressure chemical ionization (APCI) systems, which can process material from LC and CE separation systems and can be interfaced with a variety of mass spectrometers, has greatly increased the diversity of analytes that can be examined. These systems are able to measure both more polar and higher molecular weight chemicals than GC/MS could do previously. The use of single quadrupole mass spectrometers connected to GC or LC systems is very popular for routine quantitative analysis. The useful polarity and molecular weight ranges for GC/MS, APCI/MS, and API-electrospray/MS are illustrated in *Figure 6-1.*

Figure 6-1
Relative applicability of GC/MS and LC/MS techniques

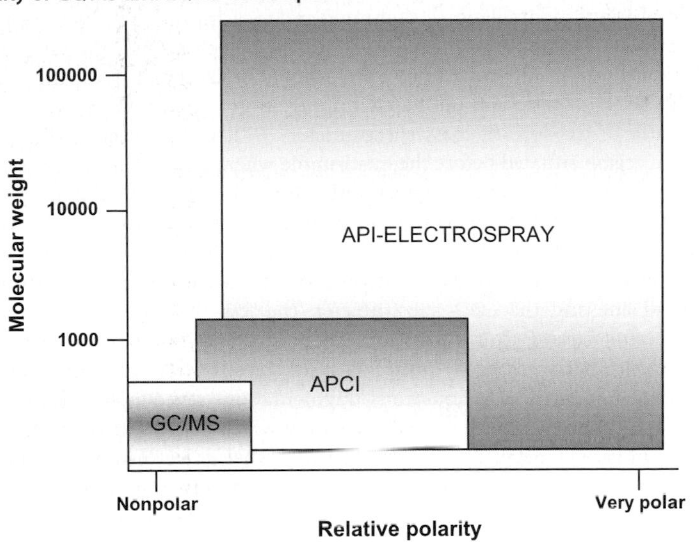

This chapter has two aims. First, techniques that use mass spectrometry for the identification and structural elucidation of neuroactive chemicals are reviewed. Second, and with greater emphasis, this chapter attempts to describe the routine use and pitfalls encountered when using mass spectrometry—especially GC/MS and LC/MS—for rapid, specific, quantitative analyses of neuroactive compounds and metabolites in a variety of biological matrices.

2 Basic Principles

The fundamental principles of mass spectrometry are relatively simple. The initial difficulty is in introducing a large enough amount of identical ions derived from a particular molecule into the mass spectrometer in order to produce a signal sufficiently intense to provide meaningful data. Once a molecule enters a mass spectrometer, it can be manipulated to provide the desired information such as high-resolution weight estimates, isotope ratios, fragmentation patterns, and quantitative data.

The introduction of a sample into the mass spectrometer can be accomplished in a number of ways, all of which basically involve the production of dispersed and vaporized ions derived from the chemical of interest. Various sample introduction techniques provide a basis for connecting mass spectrometers to diverse instrumental separation methods, thereby increasing the range of analytes that can be evaluated by mass spectrometers. A number of interfaces are described in great detail in later sections. At present, a simple method of direct sample introduction is discussed. This technique involves the use of a temperature-controlled probe in conjunction with an electron ionization source. With this method, the sample probe is placed in the vacuum near an electron source at the entrance of the mass spectrometer. The sample mixture is heated and individual chemicals volatilize at different temperatures. A portion of the vapor produced will enter the ionization chamber, where exposure to electron bombardment will occur. This will lead to the formation of charged molecules when vaporized molecules collide with electrons emitted by a hot tungsten or rhenium wire. This provides positively charged ions because of the loss of an electron from the parent molecule. An electrical potential is generated across the ionization chamber, pushing any charged molecules into the entrance of the mass spectrometer as soon as they are generated. Usually, the sample is closely bracketed by the cathode and the mass spectrometer entrance whereas the anode resides at the back of the machine, thus providing a charge differential that will push any positively charged ion down the length of the mass spectrometer. Once the ion passes through the mass spectrometer entrance, it encounters a high vacuum where interactions with atmospheric and sample molecules are minimized. As the ions continue through the vacuum, various manipulations (skimming and focusing) occur, which are aimed at providing a sample that is traveling in a uniform direction toward the detector. Once the sample is being directed uniformly, it enters an electric field generated by a quadrupole where molecules change direction according to their mass-to-charge (m/z) ratio and ultimately impinge on the detector, which provides a signal that indicates the m/e ratio of the ion. In mass spectrometers with collision-induced fragmentation (CID) capabilities, there is a region situated before the quadrupole where inert molecules, argon for example, are introduced into the vacuum to provide opportunities for collisions with the sample ions. Collisions between the inert molecules and the analyte result in fragmentation of the sample (parent) ions and a resultant spectrum of fragment ions of varying masses. These fragments are separated by the quadrupole and will produce information that permits elucidation of the analyte's chemical structure. In general, the purer the compound entering the mass spectrometer, the lesser chance there is that contaminating chemicals will produce interfering signals and the cleaner will be the resultant spectrum.

Tandem mass spectrometry (MS/MS) utilizes a mass spectrometer configuration that permits the selection of specific ions of interest. Spontaneous fragmentation- or collision-induced fragmentation of these selected ions can produce smaller fragments masses of which permit extensive structural elucidation. MS/MS is described in greater detail later in this chapter. Alternative methods for the introduction of relatively clean samples—including nonvolatile, unstable compounds, and negatively charged ions found in complex biological matrices—into the mass spectrometer are more sophisticated and often quite ingenious. These methods are discussed in the sections where separation methods combined with MS are described.

3 Mass Spectrometric Instrumentation Used in Infusion or Direct Analytical Methods for Chemical Identification and Structure Elucidation

Direct methods are usually performed either to identify chemicals or to elucidate structures by means of interpretation of the fragmentation pattern rather than to provide routine quantitative data for already characterized chemicals. As a result, high-resolution direct analytical methods often require more expensive mass spectrometers that have high mass resolution capabilities and MS/MS configurations. As these spectrometers are expensive, it is not cost-effective to have them attached to separation systems (GC, LC, or CE) where chromatographic runs often exceed 30 min. Rather, the purification of the compounds of interest is first performed elsewhere and the purified sample is introduced directly into the mass spectrometer as a bolus. In this manner, the desired information is often rapidly obtained (within 2 min). This usually requires that the sample is relatively pure and free of contaminants, especially structural isomers and nonvolatile salts, to provide artifact-free data. However, certain introduction techniques can, by their nature, provide some degree of partial purification of the sample as it enters the mass spectrometer. For example, when using a temperature-controlled probe, gradual heating of the sample in the sample chamber of an electron ionization or chemical ionization mass spectrometer will permit selective analysis of compounds of different vapor pressures in a mixture.

Alternatively, direct methods (syringe infusion, flow injection) can be used as a preliminary step in determining the optimal MS detector conditions for particular molecules. This is especially useful when the MS is attached to a liquid chromatograph, in which the fluid entering the MS will vary with gradient elutions (varied solvent and salt compositions) or with sample types (varied sample matrices, extraction solvents, included salts, etc.), which can affect the predominant ion type and sensitivity, or produce other matrix effects such as ion suppression or extraneous signals.

3.1 Electron Ionization (EI) Mass Spectrometry

EI/MS, formerly electron-impact ionization, methods were among the earliest developed. EI/MS is performed in a vacuum and the ions produced are positively charged ($M^{+\bullet}$) due to the loss of an electron from the molecule of interest. Interpretation of mass spectra is somewhat different from other softer ionization techniques, which produce positively charged ions by the addition of protons (MH^+) or other positively charged ions (MNa^+, MK^+, MNH_4^+). EI usually leads to the fragmentation of the molecular ion. Consequently, the molecular ion is often not observed. This is a particular problem with conjugated biomolecules (glucuronides and other sugar conjugates, sulfates and other biological metabolites), which fragment easily at the conjugation site. As a result, information on an unconjugated molecule can be collected, but little information is obtained on the type of parent conjugate present.

This technique is primarily used for the high-resolution analysis of smaller molecules and it often provides excellent fragment spectra. For this reason, it is used for preliminary structural elucidation of synthetic compounds, including potential neuropharmaceuticals, or for structural confirmation of drugs destined for biological experiments.

Very few, if any, recent biomedical publications describe the use of ion-impact mass spectrometry without the use of GC or some other separation method because most biological samples are chemically complex. The production of clean and useful EI mass spectrometric signals requires the substance of interest to be very pure, and thus direct EI experiments are usually confined to preliminary studies of highly purified biomolecules or to studies on the metabolism of pure materials. Two publications that describe direct EI methods applicable to biochemical analysis and neuropharmaceutical studies are those of Costa et al. (1992) and Karminski-Zamola et al. (1995).

3.2 Chemical Ionization (CI) Mass Spectrometry

CI/MS produces low-energy ions and, as a result, a spectrum with limited fragmentation is obtained in which the molecular species (MH^+) is usually readily observed. Thus, CI/MS is complementary to EI/MS in that it can provide more information regarding the molecular ion. This is particularly useful in studies of drug conjugates, where the route of metabolism can be determined as a signal from the intact conjugate rather than the daughter drug is more likely to be obtained. With CI/MS the polarity of the accelerating electrical field can be reversed, which permits the analysis of negative ions formed by intact compounds with acidic groups or electronegative elements.

As with EI/MS, most biomedical applications involve interfacing CI/MS with a chromatographic instrument, usually GC. Again, as with EI/MS, CI/MS linked to a high-resolution mass spectrometer is particularly useful in elucidating the exact mass and structure of pure materials and is especially useful for the rapid structural elucidation of synthetic compounds. Thus, references to direct CI techniques are more often found in journals that are more concerned with structural and synthetic chemistry than in those dedicated to medical or biological applications. Initially, CI/MS was used for structural identification of a number of neuroactive materials including biogenic amines (Milne et al., 1973), sphingolipids (Markey and Wenger, 1974), and alkaloids (Fales et al., 1970), and by using stable isotopes as internal standards direct, rapid, quantitative mass spectrometric procedures were developed for some trace substances. These techniques eventually incorporated GC instruments (Hashimoto and Miyazaki, 1979; Durden and Boulton, 1988).

3.3 Direct Injection Atmospheric Pressure Ionization Electrospray (API-electrospray; ESI) MS

Direct injection API-Electrospray MS is capable of analyzing much larger and less volatile substances than either EI/MS or CI/MS. As a result, this method is often used to provide structural information on peptides, proteins, and polymers derived from both natural and synthetic processes; it is also useful in the analysis of many natural compounds including molecules such as saponins and flavonol glycosides, derived from plants. When using direct injection API-electrospray, partial purification and LC preparation are performed elsewhere and a collected fraction is dissolved in an appropriate solvent and injected as a bolus into the mass spectrometer (flow or direct injection or syringe infusion). This has an advantage, as the mass spectrometers, especially high-resolution instruments with MS/MS capability, are not tied up with long LC separations. Structural information and even identification can be obtained within 1 or 2 min. Automated high-throughput systems coupled to microtiter plates have been described elsewhere (Felten et al., 2001). This process assumes that the fraction under investigation is relatively pure and the amounts of salts, which can cause matrix effects and ion suppression, are held to a minimum. Early in the evolution of API-electrospray mass spectrometers, the sample was oriented directly into the mass spectrometer. This had the disadvantage that all materials including salts from the mobile phase as well as the sample entered the system. This often resulted in a rapid buildup of salt at the entry to the mass spectrometer and this area had to be cleaned frequently. Furthermore, samples often contained high concentrations of salts resulting from ion exchange separations or other purification techniques in which the final sample was not sufficiently desalted. These salts often caused ion suppression or provided confusing artifacts and the mass signals obtained were less than optimal. This problem was solved to a substantial degree by the advent of an orthogonal arrangement for the introduction of sample [see Henry (1999) for a review] in which the material passes in a fine mist at right angles to the aperture of the MS. The electrical potential is at a right angle to the sample spray and directed down the length of the MS. Thus, only charged molecules are directed into the aperture. The use of volatile salts such as ammonium formate and ammonium acetate and volatile acids and bases such as acetic acid, trifluoroacetic acid, and ammonium hydroxide in the preparation of samples greatly reduces the amount of maintenance required to keep the MS operating efficiently.

Although the usual upper limit of the API-electrospray spectrometer is only 1,500 to 2,000 da, it is capable of providing molecular weight estimates for much larger molecules. This is because most peptides and other polymers (nucleosides and polyethylene glycols, for example) are multiply charged. The detector will often record mass-to-charge signals for molecules with 1 to 100 protons or more. This provides a lattice of signals derived from mass-to-charge ratios, and by using a simple algorithm a molecular weight can be calculated. As can be imagined, signals derived from several different multiply charged molecules contained within a mixture can become extremely confusing and as a consequence, the purer the analyte directly introduced into the mass spectrometer the better and more useful is the resulting signal.

A further application of direct injection API-electrospray to the structural elucidation of peptides is to partially fragment the peptide. In this way, one can often determine the partial or complete amino acid sequence of the peptide of interest (Ramstrom et al., 2003). This is particularly useful in rapidly establishing amino acid sequences of synthetic peptides in which the confirmation of a sequence can be done in a few moments rather than going through a complex amino acid analysis derived from a sequential hydrolysis process.

Several useful publications describing early direct injection API-electrospray methods applicable to neurochemical studies include studies on acetylcholinesterase phosphonylation products (Barak et al., 1997); glycerophosphocholine lipids (Harrison and Murphy, 1996); receptor antibody characterization (Lewis et al., 1994); neuropeptide Y analogs (Beck-Sickinger et al., 1994); and large biomolecules (Loo et al., 1992).

3.4 Direct Injection Atmospheric Pressure Chemical Ionization (APCI) MS

Chemicals that can be analyzed by direct injection APCI/MS must have some degree of volatility. APCI/MS is complementary to direct injection API-electrospray/MS. It can detect polar, volatile compounds that are either not volatile enough for CI/MS or EI/MS or are too volatile for optimal API-electrospray/MS analysis. The greatest advantage that direct injection APCI/MS has over other ionization methods is that it can actively generate ions from neutral analytes. Thus, use of direct injection APCI/MS makes analysis of low- to medium-polarity analytes, which must be dissolved in partially aqueous solutions, more accessible to mass spectrometry. One disadvantage of direct injection APCI is lower reproducibility than other ionization methods. As with other direct injection methods, the purer the sample the more informative is the mass spectrum obtained. Direct injection APCI is usually used for preliminary method development to optimize detector conditions for the analysis of molecules of molecular weight less than 1,500 and is often used for relatively nonpolar molecules in which direct injection ESI-electrospray provides rather weak signals. The detector conditions derived from direct injection APCI experiments can then be combined with a chromatographic separation procedure to provide an optimized analytical procedure. There are few published medical or biological applications dedicated to direct injection APCI analyses.

3.5 Fast Atom Bombardment (FAB) Ionization

FAB is a particle desorption technique that involves the bombardment of a solid analyte-matrix by a fast particle beam, usually a neutral inert gas (typically Ar or Xe) at bombardment energies of 4–10 keV. The matrix is a small organic species such as glycerol or 3-nitrobenzyl alcohol, which is used to maintain a homogeneous surface for bombardment. The particle beam is focused at the analyte-matrix surface, where it transfers much of its energy to the immediate surroundings. This produces momentary collisions and some chemical species are ejected off the surface as positive and negative ions. The ions are then accelerated into the mass analyzer where their *m/z* ratios are determined. FAB is a comparatively soft ionization technique and is well suited to the analysis of relatively polar, low volatility species including peptides. FAB is not particularly amenable to interfacing with chromatographic systems and thus some degree of purification of the analyte is performed prior to its introduction into the FAB instrument.

3.6 Matrix-Assisted Laser Desorption/Ionization (MALDI) MS

In this ionization process, the analyte is combined with a matrix compound such as α-cyano-4-hydroxycinnamic acid or 2,5-dihydroxybenzoic acid in a molar ratio of approximately 1:1,000 and placed on a metallic plate. The matrix compound is selected to absorb strongly at the laser wavelength used such that when exposed to the laser beam rapid heating occurs and the sample is vaporized without extensive fragmentation. The analyte ions are electrostatically propelled into the mass analyzer. As with API-electrospray and APCI/MS, multiply charged ions are produced. Identification of the multiply charged ions associated with a particular molecule permits the analysis of compounds with molecular weights larger than the dynamic range of the detector. Recent work has investigated the use of matrix additives or tags to enhance MALDI signal intensities much as derivatization methods are used to enhance volatility in EI/MS, CI/MS, and GC methods (Pashkova et al., 2004) or to protect labile compounds from photodegradation (Low et al., 2004). MALDI/MS is most often performed in high-resolution mass spectrometers including time-of-flight spectrometers (MALDI/TOF) to permit high-resolution analysis of large molecules. Recently, an interface between liquid delivery systems and MALDI/MS has been developed (Daniel et al., 2004) and one can soon expect many LC/MALDI/MS techniques to appear in the literature. MALDI/MS is primarily used for examining the primary structure of peptides, proteins, and oligonucleotides (Che et al., 2001; Blondal et al., 2003; Stuhler and Meyer, 2004; Poljak et al., 2004) and can be used to determine sequence alterations and aid in determining structural alterations caused by the binding of chemicals such as substrates and inhibitors, to enzymes and other proteins. One exciting aspect of MALDI/MS is that samples are not confined merely to dried preparations of synthetic products or purified extracts. Actual pieces of biological tissues can be impregnated with appropriate matrices and deposited on the sampling surface. In the case of nervous tissue, especially that of invertebrates, which possess large anatomically identified neurons, the laser can be directed to selected structures or cells and anatomical distributions of particular peptides, proteins, and other chemicals in the tissue can be determined (Rubakhin et al., 2003; Fournier et al., 2003).

Several additional useful publications of MALDI/MS methods applicable to neurotransmitter analysis and neuropharmaceutical studies include those of molecular profiling in Parkinson's disease (Pierson et al., 2004); spatial profiling of ganglia (Kruse and Sweedler, 2003); binding of drugs to enzymes (Guo et al., 2004); and oligonucleotide sequencing (Hong et al., 2004).

3.7 Quadrupole Ion Traps

Quadrupole ion traps are very useful instruments that can be interfaced between various ionization sources and the mass detector. These instruments store ions of a selected mass-to-charge (*m/z*) ratio within their field. Thus, many ions from a sample may enter the ion trap but only one *m*/*z* is retained and stored. The mass analyzer is turned off while the trap accumulates the selected ions and the unwanted molecules are directed to waste. The desired ion is allowed to accumulate, then the mass analyzer is activated, and a negative charge is applied to one end cap of the ion trap to propel the selected ion into the mass analyzer. This provides a relatively pure and amplified signal, which permits an enhanced accuracy in the estimation of the molecular weight of the parent ion and its fragments. Ion traps are particularly useful in providing accurate molecular weights of peptides, proteins, and oligonucleotides and for analyzing compounds and metabolites found in very low concentrations in nervous tissue. Several articles that describe applications for ion traps that measure endogenous neurochemicals include the analysis of the human pituitary proteome (Zhan and Desiderio, 2003), phospholipids (Kakela et al., 2003), neuropeptide FF-related peptides (Burlet-Schiltz et al., 2002), neurosteroids (Mitamura et al., 2000), dynorphin metabolites (Prokai and Zharikova, 1998), and phosphorylated proteins (Yamauchi et al., 1998). Ion traps have also been applied to the measurement of exogenous chemicals such as new central nervous system agents (Prokai et al., 2001) and cocaine in nervous tissue (Hernandez et al., 1994).

3.8 Tandem Mass Spectrometry (MS/MS)

MS/MS is an extremely powerful tool using instruments that employ a complex configuration of multiple quadrupoles. These instruments are often composed of three quadrupoles in a series (triple quad) where particular ions are selected after they have passed through the first quadrupole, and fragmented in the adjoining collision chamber (second quadrupole), then sent through to the third quadrupole. This will result in a mass spectrum derived from one selected ion. If the operator sequentially selects ions of interest or uses further quadrupoles and collision chambers one can, in theory, fragment a selected ion down to its smallest stable components, providing exhaustive details of its chemical structure.

As one usually wishes to fragment a number of different selected ions derived from a single molecule over an extended period of time, the sample is usually introduced as a relatively pure and concentrated material using syringe infusion or flow injection at a low flow rate. The continuous introduction of the analyte allows the experimenter to select different ions or fragmentation conditions over time so that numerous different fragmentation experiments can be performed on a single sample, resulting in structural information derived from an exhaustive breakdown of different structural elements. This is also possible when the system is connected to a chromatographic system. In this configuration, the flow from the chromatograph is often slowed down so that the peak of interest has a longer residency time in the mass spectrometer rather than passing through the detector in a matter of seconds. This permits sufficient time for a number of experiments using fragmentation of different selected ions to be performed.

The literature contains numerous references to the use of MS/MS in the determination of new neuropeptides in identified cells of invertebrates (Bulau et al., 2004, for a recent example) and this technique is now being applied to in situ analysis of vertebrate tissues (Fournier et al., 2003). MS/MS is also used for studies of neuropeptide processing (Nilsson et al., 2001), pharmacokinetics of synthetic peptides (Mock et al., 2002), nonpeptide drug metabolism (Kamel et al., 2003), identification of peptides purified by immunoaffinity (Suresh Babu et al., 2004), and MALDI/MS/MS techniques adaptable to brain dialysis (Bogan and Agnes, 2004).

4 Mass Spectrometric Methods Linked to Chromatographic Separation Techniques for Routine Quantitative Analyses

Recently, comparatively inexpensive, very reliable, and stable single quadrupole mass spectrometers have entered the market. These spectrometers can be coupled to GC, LC, and CE separation methods simply by modifying the sampling interfaces. Although these detectors are more expensive than most conventional detectors including the versatile electron capture and diode array absorbance detectors used for GC and LC respectively, the reduction in sample preparation effort and their increased specificity can often rapidly offset the initial additional cost outlay. For example, many amine neurotransmitters, including serotonin, are analyzed either by GC following a time-consuming extraction and derivatization process or by LC using an electrochemical detector. Both methods require extensive and specific chromatographic separations to isolate the serotonin from other amine neurotransmitters and their metabolites, which can interfere with the serotonin signal. These chromatographic separations often take more than 25 min because longer retaining compounds such as tryptophan must be eluted from the column prior to a subsequent injection. In contrast, if one uses the same tissue preparation commonly used in the LC-electrochemical detection method (i.e., precipitation and centrifugal removal of insolubles in about 5 volumes of 0.2N perchloric acid) and API-electrospray, one can use single ion mode (SIM) to selectively detect only the ions associated with serotonin (usually MH^+, MNH_4^+, MNa^+), thereby eliminating almost all other interfering signals. Consequently, complete resolution of all compounds is not required and a rapid chromatographic separation using a small cartridge column (2.1 mm × 30 mm, 3.5 μM particle size) can permit the measurement of serotonin with a complete run time of less than 8 min. Under these conditions, there is often enough resolution to monitor other amines including dopamine and several characteristic amine metabolites. Additional information can be gleaned by programming the detector to monitor a range of molecular weights (SCAN mode) for a set portion of time and subsequently extracting ions of potential interest from the data. The amount of information gleaned from a simple preparation using a rapid separation and a minimum of mobile phase can be extensive almost to the point of being overwhelming.

4.1 Gas Chromatography Linked to Mass Spectrometry

A gas chromatograph with a capillary column coupled to a mass spectrometer is an ideal analytical partnership. Effluent from the column has an elevated temperature and the molecules of interest are in a vapor state and ready to enter the ion source. This eliminates the need for desolvation that is required in high-performance liquid chromatography (HPLC)–MS.

4.1.1 GC/MS Interface

Gas chromatography is discussed in detail in another chapter in this book, but there are special considerations that must be taken into account when using it as an inlet for an MS. Scrupulous attention must be paid to keeping the system clean. Any handling of inlet supplies, ferrules, columns, etc. requires the use of gloves because fingerprint oils are a source of contamination. The GC carrier gas, usually helium, should be at least 99.99% and a gas purifier trap is suggested. Low bleed septa and columns should be used. For the inlet, 10% graphite, 90% vespel ferrules are recommended and 100% vespel ferrules are recommended for the MS interface. Graphite ferrules can easily be overtightened and can affect chromatography, causing peak tailing. Vespel ferrules will shrink when they are heated so they can be preconditioned by keeping them in the GC oven or they can be tightened after the oven has cycled a few times. Choosing an appropriate inlet liner can be difficult because so many different liners are available. Choice is based on the mode of injection (i.e., split or splitless) and the characteristics of the analyte and solvent. Most methods employed to analyze neurochemicals use a splitless mode of injection and a single, tapered deactivated glass liner. Deactivated glass wool may be added to keep nonvolatiles from getting to the column. Too much glass wool or glass wool that has been significantly broken can greatly decrease sensitivity.

4.1.2 GC/MS Configurations

There are five major components of a mass spectrometer coupled to a GC. They are, vacuum system, ion source, mass filter, detector, and data system.

For proper operation, the ion source, mass filter, and detector must all be operated under vacuum (approximately 10^{-5} torr). This vacuum makes it possible to move ions from the ion source to the detector without any collision with other ions and molecules. This is referred to as a mean free path and is defined as the average distance an ion must travel before it strikes something. A high mean free path ensures predictable and reproducible fragmentation, high sensitivity, and reliable mass analysis. Generally, two pumps are required to achieve an effective vacuum. The foreline pump is also referred to as a roughing pump and produces an initial level of vacuum. The second pump is a high vacuum pump and can be either a diffusion pump or a turbomolecular pump. Typically, CI/MS requires a higher vacuum and a turbomolecular pump is used.

Several types of ion sources may be employed in GC/MS. These include EI and CI, which may be positive (PCI) or electron capture negative ion chemical ionization (ECNI). FAB, field ionization, and plasma desorption MS are now less frequently used. EI requires the most energy and generates the most fragments. Plasma desorption is the softest of the ionization methods; it requires the least amount of energy and generates little or no fragmentation. FAB, field ionization, and plasma desorption are methods of direct injection that require a source that is under a vacuum. Most applications to neurochemistry and drug analysis use EI and CI sources.

EI is probably one of the most common ionization techniques employed. A beam of electrons emitting from a tungsten filament at a voltage of 70 eV bombards the analyte molecule, a single electron is expelled producing a positively charged molecular radical ion (M^+). Fragmentation may be reduced by lowering the electron energy but the number of molecular ions produced is also reduced (Honour, 2003). A voltage of 70 eV is used because this energy is well in excess of the energy required to ionize and fragment molecules. Ionization efficiency curves plateau around 70 eV (Gross, 2004). Mass spectral libraries have been created based on the unique fragmentation of molecules at a voltage of 70 eV. This technique, coupled with a quadrupole detector, generates very stable and reproducible spectra commonly called "classical spectra," which can be used in the identification of unknowns (Prest, 1999).

Chemical ionization MS involves a much lower energy than EIMS so it is a much gentler ionization technique and results in less fragmentation. For chemical ionization to occur, a reagent gas is added to the ion source. This gas can be methane, isobutane, or ammonia. Methane is the most common and the easiest to work with. During PCI, reagent ions (RH^+) are produced when electrons from the filament ionize the reagent gas. Addition of the reagent gas to the source causes an increase in pressure, which enables a variety of ionization mechanisms. Proton transfer ($M + RH^+ \rightarrow (M+H)^+ + R$), and adduct formation ($(M + RH^+) \rightarrow (R+M+H)^+$), are the most common PCI reactions (Prest, 1999). Another ionization mechanism is hydride abstraction: ($R^+ + M \rightarrow (M\text{–}H)^+ + RH$).

ECNI/MS operates in a similar manner to an electron capture detector (ECD). To perform ECNI, analyzer voltage polarities are switched to select negative ions. The reagent gas in this technique is referred to as a buffer gas. The electrons emitted from the filament are thermalized or slowed down by collisions with this gas within the source. This allows the electrons to be captured by electrophilic analyte molecules to form M^- or other negatively charged fragments (Prest, 1999). Dissociative electron capture is similar to an electron capture CI/MS, but the sample molecule dissociates (fragments) yielding a negative ion (Leis et al., 2004).

Both PCI and ECNI/MS are very sensitive to the presence of contaminating water or oxygen molecules, which can greatly decrease sensitivity as well as contaminate the source itself.

Following ionization and fragmentation, the ions produced are separated based on their m/z ratio in the mass filter (analyzer). There are several types of analyzers used in GC/MS, including time-of-flight (TOF), magnetic sector, and radiofrequency, which include both quadrupole and ion trap.

In a TOF mass analyzer, ions of different m/z ratios are determined by the time they take to travel through a field-free path of known length between the source and the detector. The ions are accelerated by application of an electrical potential. The velocity of an ion is inversely proportional to its mass; therefore, each m/z has a characteristic TOF (de Hoffman and Stroobant, 2001).

The quadrupole analyzer uses oscillating electrical fields to separate ions based on their stable trajectories. The analyzer consists of four parallel poles placed between the ion source and the detector in such a manner that the path of the ion beam travels through the middle. The rods have DC voltages applied to opposite rods to carry the same charge, yielding one set of positive rods and one set of negative rods. All four rods have an oscillating radio frequency (RF) applied to them (Honour, 2003). If the ion mass is too low, the ion travels toward the positive rods; if it is too high, it travels toward the negative rods. Only an ion of a particular mass will have stable oscillations and exit the end of the quadrupole and be detected.

The ion trap is a similar analyzer. There are two end cap electrodes that are at ground potential. An electrostatic field is generated by a donut-shaped hyperbolic electrode within the cap, which maintains ions in a stable trajectory. Changing electrode voltages ejects ions of a particular mass from the trap into the detector (Honour, 2003).

Ions are accelerated to a high velocity after leaving the ion source before entering the magnetic sector analyzer. The ions pass through a magnetic field that is perpendicular to the direction of ion motion. This deflects ions according to their size and momentum. Ions with different *m/z* ratios will have different flight paths according to the radius of the arc in which they travel. Ions can be focused on the detector by varying the magnetic field (Honour, 2003).

This is only a brief overview of the analyzers used in GC/MS. Very detailed descriptions of all of these types can be found in de Hoffman and Stroobant (2001) and Gross (2004).

A detector counts the ions leaving the mass analyzer, and a signal is generated. This detector is typically an electron multiplier or electron horn. The signal generated is usually captured by a computer, which is then used to generate the appropriate data and run quantification software.

An important step in setting up the GC/MS for analysis is a correct tuning of the instrument. Tuning establishes a relationship between RF and *m/z* by calibration using a reference material such as perfluorotributylamine (PFTBA). This optimizes the instrument response by adjusting voltages applied to the ion acceleration lenses of the ion source as well as some of the components of the ion optical system (Gross, 2004). Each type of ionization source has a respective tune procedure, which can be invaluable when trouble shooting the instrument.

4.1.3 GC/MS Sample Handling

Handling of samples for analysis by chromatographic methods is discussed elsewhere in this book in the chapters on gas chromatography by Rittenbach and Baker and in the chapter on high-performance liquid chromatography by Odontiadis and Rauw. Preparing samples for GC/MS may include a derivatization step to improve sensitivity, particularly in ECNI, as well as volatility. An excellent review of derivatization for ECNI is given by Leis et al. (2004). Segura et al. (1998) present a broader review of derivatizing agents for mass spectrometry, especially for drugs of abuse.

4.1.4 Selected GC/MS Methods Applicable to Neurochemical Studies

Gas chromatography was first coupled with a mass spectrometer in 1956 and GC/MS became commercially available in 1957 (de Hoffman and Stroobant, 2001). Examples of some of those applications pertinent to neurochemistry are listed in ➲ *Table 6-1*.

4.2 Liquid Chromatography Linked to Mass Spectrometry (LC/MS)

One of the simplest separation methods that can be interfaced with mass spectrometry to provide quantitative analyses is HPLC. The two predominating interfaces between liquid chromatographs and mass spectrometers are APCI and API-electrospray. These interfaces are described earlier in ➲ Section 3. For simplicity, robustness, and ease of use, it is difficult to better these configurations. Previously, HPLC has been coupled to UV absorbance, UV/VIS diode array absorbance, fluorescence, electrochemical, refractive index, or evaporative spray light scattering detectors. The difficulty with all of these detectors is that the analyzed chemicals must have properties that allow their detection by the selected detector and, as these properties are often common to a number of analytes within the sample, chromatographic separations must totally resolve the signal of the chemical of interest from any interfering signals. This is often difficult to do in complex biological matrices. On the other hand, LC/MS relies on a property that is possessed by all

Table 6-1
Selected references of GC-based methods for the analyses of endogenous and exogenous chemicals of neurobiological interest

Chemical group	Derivatization	Ionization	GC reference
Endogenous Neurochemicals			
Amino acids	Pentafluorobenzyl Chloroformate	ECNI	Simpson et al. (1996)
Amino acids (review)	Trifluoroacetic anhydride Pentafluoracetic anhydride Heptafluorobutyryl anhydride		Shah et al. (2002)
Amino acids (chiral analysis)	Esterification with deuterium chloride followed by derivatization with trifluoroacetic anhydride		Nkoihara and Gerhardt (2001)
Biogenic amines Antidepressants Antipsychotics (review)	Acetic anhydride		Baker et al. (1994)
Dopamine, norsalsolinol, and salsolinol		ECNI	Watson et al. (1990)
Dopaminergic metabolites	Trimethylsilyl derivatives	EI	Musshoff et al. (2003) Loutelier-Bourhis et al. (2004)
Putrescine	Pentafluoropropionic Anhydride		Noto et al. (1987)
Neurosteroids	Carboxymethoxylamine Pentafluorobenzyl bromide Bis(trimethylsilyl) trifluoroacetamide	ECNI	Kim et al. (2000)
Neurosteroids (in this book)			Purdy et al. (2006)
Exogenous Chemicals			
Amphetamine, methamphetamine, cocaine and tetrhydrocannabinol		EI	Yonamine et al. (2003)
Amphetamine, methamphetamine enantiomers	S-(-)-heptafluorobutylprolyl chloride	ECNI	Peters et al. (2002)
Amphetamine-type stimulants and related drugs	Heptafluorobutyric anhydride	EI	Kankaanpää et al. (2004)
Benzodiazepines		EI	Inoue et al. (2000)
Antidepressants, Neuroleptics	Acetic anhydride with pyridine	EI	Bickeboeller-Friedrich and Maurer (2001)
Dexamethasone		ECNI	Hidalgo et al. (2003)
Drugs of abuse (review)			Segura et al. (1998)
Forensic toxicology, doping control and biomonitoring (review)			Maurer (2002)

chemicals, a molecular mass. If a chemical can adopt a charge under the chromatographic conditions used, it can be detected and quantified. With the use of chromatography, there is little chance that mass signals from one molecule will overlap with identical signals from another molecule because relatively few chemicals in a biological cocktail have identical molecular weights, the exception being optical and structural isomers. In fact, if one examines a biochemical database such as BioPath (www.mol-net.de/databases/biopath.html), it appears that in a particular biological tissue there are rarely more than six compounds and often less than three compounds with overlapping signals for each molecular weight even at the low resolution obtained with less accurate mass spectrometers. A caveat may be that plants often produce several structural isomers of compounds, but in our experience, these positional isomers are usually separated by the initial chromatographic step.

The use of existing LC/MS instruments requires that the investigator appreciate certain limitations associated with the linking of separation techniques to the mass spectrometer interface to optimize signal production and to maintain operating efficiency. These limitations are described in the following sections.

4.2.1 Liquid Chromatography Columns

Liquid chromatography columns that have minimal bleed of the stationary phase are necessary for LC/MS. Columns that are not compatible with extreme ranges of pH should not be used and care should be taken with LC/MS compatible columns that the pH range of the mobile phases used never exceeds that recommended by the manufacturer. The most common separation modes associated with LC/MS are reverse phase (C_4, C_8, C_{18}, phenyl-hexyl), normal phase (NH_2, CN), hydrophobic interaction (HIC), and hydrophilic interaction (HILIC). These modes are compatible with organic solvent gradients. They also use relatively low salt concentrations and these salts include volatiles such as ammonium formate and ammonium acetate. Gel permeation columns can also be used provided the column does not bleed and the mobile phase contains low salt concentrations. It is wise to avoid normal phase separations using silica and ion exchange systems, which use high salt concentrations as they cause rapid buildup of unwanted materials at the interface of the LC system with the mass spectrometer and there can be problems with ion suppression and extraneous noise. Most manufacturers of LC columns now have products specifically designed to be compatible with LC/MS.

Column size is another important consideration. For equipment designed for most routine laboratory HPLC situations the relative sensitivity of API-electrospray instruments is better at low flow rates (0.2–0.8 mL/min) whereas the relative sensitivity of APCI instruments is enhanced at high flow rates (0.5–2 mL/min). As a result, small columns are appropriate for API-electrospray/MS and, if only one or two compounds of interest are found in a particular sample, high-resolution separations are not necessary. For API-electrospray analysis of complex samples, 150 mm × 4.1 mm I.D., 3 μm columns (flow 0.5–1.0 mL/min) are usually sufficient. For drug quantification involving analysis of single or low numbers of compounds, small columns such as 30 mm × 2.1 mm I.D., 3.5 μm columns (flow rate 0.2–0.4 mL/min) provide sufficient separation and a saving in both column cost and solvent utilization. The reduced injection volume required for the small columns often results in better resolution and increased sensitivity.

4.2.2 Liquid Chromatography Mobile Phases Applicable to Mass Spectrometry

The mobile phases used to provide separations that interface cleanly with the MS are of great importance. Both isocratic and gradient elution can be used. High purity (HPLC grade) water, acetonitrile, and C_1 to C_4 alcohols are compatible with API-electrospray and APCI. Less polar solvents such as hexane, cyclohexane, toluene, and ethyl acetate are also compatible with APCI. In general, it is advisable to always have an organic solvent present in the mobile phase to reduce surface tension, which enhances the formation of smaller, more uniform droplets and also aids vaporization and ionization and hence provides greater sensitivity.

The addition of buffering salts to the mobile phase often improves chromatographic separation, provides a stable pH during separation, and reduces problems associated with column disturbances produced by highly variable samples. These salts are usually volatile (examples are ammonium formate, ammonium acetate, and *t*-ethylammonium hydroxide) and the concentrations used are usually less than 10 mM. With the advent of orthogonal interfaces for ESI and APCI, the absolute requirement for volatile salts has disappeared. However, the prolonged use of nonvolatile salts is not recommended as the accumulation of salts in the spray chamber of the MS reduces sensitivity and increases maintenance requirements.

In order to produce the ions necessary for analysis, the pH of the mobile phase often has to be modified. Volatile organic acids (formic acid, acetic acid, and trifluoroacetic acid) and volatile bases (ammonium hydroxide) are used to provide this modification. With the analysis of basic compounds, a lower pH mobile phase (pH 2–4) is used and the ions examined are positively charged (positive mode). For the analysis of acidic compounds, a higher pH mobile phase (pH 7–10) is used and the ions produced are negatively charged (negative mode). Many compounds are charged at relatively neutral pH (5–7) and the pH adjustment of the mobile phase is not necessary to obtain ion signals for these compounds. Column manufacturers usually provide the appropriate pH range for mobile phases. This information is especially useful when basic mobile phases are being used. The use of surfactants and inorganic acids is usually avoided in LC/MS applications.

4.2.3 Isocratic and Gradient Elutions

Both isocratic and gradient elution methods are applicable to LC/MS. Gradient elution is particularly effective for the analysis of complex samples (especially when the composition of organic solvent is altered over time). Usually, the aqueous and organic mobile phases used in gradient elution have salts added to them such that the salt concentration and the pH remain relatively constant throughout the analysis. The appropriate use of a gradient (i.e., loading a slightly nonpolar sample onto a column that has been equilibrated with relatively polar initial mobile phase conditions) can produce a degree of sample concentration at the head of the column, which can ultimately provide sharper peaks and enhanced sensitivity. Gradient elution is also appropriate for the analysis of samples containing compounds of significantly different polarities, and the inclusion of a column wash, analogous to a bakeout period used in GC, toward the end of the procedure ensures that unwanted materials are not retained on the column only to appear in later analyses. Possible disadvantages of this technique are that gradient elution requires a reequilibration period at the end of each run and that the use of high concentrations of organic solvent during the wash procedure can produce some column bleed.

Isocratic elution is very useful for high-throughput analyses of less complex mixtures. With such analyses, it is usual that only one or two compounds are being analyzed and it has been previously established that no interfering compounds are being retained on column that could produce extraneous signals in later analyses. In such circumstances, isocratic elution removes the need for longer columns. A single isocratic pump can be used; preparation of more than one mobile phase is not needed and run times can often be reduced to less than 3 min because reequilibration is unnecessary.

4.2.4 Mass Spectrometric Interfaces for Routine Quantitative Analysis

The two most common LC/MS interfaces used for routine quantitative analyses are APCI and API-electrospray. The principles of these techniques in direct infusion analyses have been described earlier (see Sections 3.3 and 3.4). As API-electrospray has a broader application profile, its use is more widespread than APCI. Other configurations including EI, atmospheric pressure photoionization (APPI), and thermospray interfaces with liquid chromatographs are available but are less commonly used for high throughput or routine analysis.

4.2.4.1 Atmospheric Pressure Chemical Ionization (APCI)/MS APCI/MS is used to analyze compounds of intermediate molecular weight (100–1,500 da) and intermediate polarity and is particularly useful for the analysis of biochemicals such as triacylglycerides, carotenoids, and lipids (Byrdwell, 2001). For volatile, nonpolar compounds of low molecular weight, GC/MS is preferred to APCI/MS whereas API-electrospray/MS provides better results for larger, more polar materials. The selection of APCI/MS over GC/MS or API-electrospray/MS depends on the compounds to be analyzed. Many LC/MS instruments can be easily switched between APCI/MS and API-electrospray/MS so that it can be rapidly determined which ionization process is more suitable to a given chemical. Additional manipulations such as pre and postcolumn derivatization reactions (Nagy et al., 2004; Peters et al., 2004) or coulometric oxidation (Diehl et al., 2001) can make the chemicals of interest more amenable to detection by APCI.

The process whereby APCI produces ionized chemicals involves two stages, flash vaporization and gas-phase chemical ionization. Flash vaporization occurs when the fluid eluting from the LC apparatus is forced through a small aperture (capillary probe) to provide it with a high linear velocity. The eluant passes through a region where a high-temperature heater and a nebulizer rapidly vaporize the fluid stream, producing a chemical stream that is then passed through a region where it is exposed to electrons from a corona discharge electrode. This results in the production of primary ions. A high temperature is present at the heater–fluid interface, but analytes are not usually thermally degraded. As long as solvent is present, the highest temperature that the analyte is exposed to is the boiling point of the solvent. This generally reduces the possibility of sample decomposition, although some thermally labile materials may be susceptible to degradation. Gas-phase ionization occurs when ions generated by the corona discharge, designated primary ions, react with the mobile phase molecules to produce stable secondary or reagent ions. Molecules from the sample introduced into the mobile phase react with the reagent ions and typically become protonated or deprotonated. In positive mode APCI, the protonated molecules are detected, whereas in negative mode APCI deprotonated ions are detected. The relationship between the proton affinity of the sample molecules and the solvent vapors determines how much of the analyte will become protonated or deprotonated. Most mobile phase solvents (water, methanol, acetonitrile) have low proton affinities and are good choices for use in APCI analyses. Positive ions associated with APCI include those resulting from proton transfer (MH^+ ions), adduct attachment ions, e.g., MNH_4^+ ions, and charge exchange (M^+) ions. Negative mode signals can result from proton abstraction forming $(M–H)^-$, adduct attachment with a negative ion providing MCl^- for example, charge exchange from another negatively charged ion resulting in the formation of M^-, and from electron capture producing M^- ions.

Factors that affect sensitivity in APCI analyses include rate of solvent flow, analyte concentration, matrix constituents, nitrogen flow rate in the nebulizer, and probe temperature. With most quantitative APCI applications, a flow rate of 0.2 to 2.0 mL/min is optimal. This flow rate is higher than that used for API-electrospray as the high linear velocity of flow and the high temperature used promote more efficient evaporation. Saturation effects are a common difficulty with APCI applications, and the linear dynamic range of these detectors is only about 2–3 orders of magnitude. One indication of sample saturation is the formation of multimers of the analyte produced by incomplete separation of analyte molecules during formation of the gas-phase ions. This results in the formation of complexes such as M_2H^+, M_2Na^+, and M_3H^+. Such signals are indicative of oversaturated samples. Matrix constituents can compete for ionization and are commonly responsible for the suppression of the observed signal. Nebulization must be consistent for accurate and sensitive analyses, and thus the nitrogen flow through the nebulizer must be sufficient to promote the formation of small droplets. Poor nebulization will reduce sensitivity and increase variability. Capillary probe temperature also affects nebulization, and the temperature must be high enough to promote efficient nebulization and solvent evaporation, yet low enough to prevent thermal degradation of the analytes. Balancing these conditions effectively will optimize sensitivity and the quantitative accuracy of the analysis.

A list of APCI applications to a number of chemicals of neurochemical interest is provided in ➲ *Table 6-2*.

4.2.4.2 Atmospheric Pressure Ionization (API)-Eelectrospray/MS For high-throughput routine analyses, API-electrospray/MS is the most often used LC/MS interface. Its ability to analyze molecules with a wide range of polarities and molecular weights ranging from the smallest fragments to those with a mass more than 100,000 da means that it is one of the most versatile analytical techniques available.

Accurate quantitative analysis using API-electrospray/MS requires the investigator to have some understanding of the physical and chemical properties involved in the production of ions as many factors will affect the absolute quantities of ions produced and thus the magnitude of the signal. Adjustments of a number of parameters will affect aspects of the signal detected, including the types of ions observed, the number of ions present, and the linear dynamic range of the detector response.

API-electrospray ionization involves three stages. First, there is the formation of charged droplets. Once the droplets are formed, solvent evaporation and droplet fission occur. Droplet fission is due to an increase in charge repulsion at the surface of the droplet as the solvent evaporates. Once the droplets become small enough (<10 nm), it is believed that charge repulsion produces ion evaporation from the surface of the droplet. Thus, ions are transferred from the solution to the gas phase. Factors affecting the production of the desired ions include analyte concentration, flow rate, matrix content, and analyte surface activity. In

Table 6-2

Selected references of APCI-based methods for the analyses of endogenous and exogenous chemicals of neurobiological interest

Chemical group	Specific chemicals	APCI reference
Endogenous		
Neurochemicals		
Carboxylic acids,	Acetone, propionaldehyde with derivatization	Nagy et al. (2004)
aldehydes, and ketones	Carboxylic acids, aldehydes, and ketones with postcolumn derivatization	Peters et al. (2004)
Amino Acids	L-arginine and metabolites	Huang et al. (2004)
	Underivatized amino acids (comparison of APCI with other methods)	Kwon and Moini (2001)
Sphingomyelins and	Shingomyelins and ceramides	Karlsson et al. (1998)
Ceramides	Various ceramide species	Pettus et al. (2004), Couch et al. (1997)
Coenzyme Q	Coenzyme Q	Hansen et al. (2004)
Oxosteroids	Numerous derivatized oxosteroids	Higashi et al. (2003)
Neurosteroids	Neurosteroids with derivatization	Mitamura and Shimada (2001), Purdy et al. (2006)
Catecholamines and Metanephrines	Epinephrine, norepinethrine, dopamine, normetanephrine, metanephrines	Chan and Ho (2000)
Phenylethylamines	1-phenylethylamine and related chemicals	Bogusz et al. (2000)
Serotonin	Serotonin, related indoles	Danaceau et al. (2003)
Anandamide related chemicals	Anandamide and its analogs	Koga et al. (1995), Koga et al. (1997)
	2-arachidonylglycerol	Huang et al. (1999)
Exogenous Chemicals		
Neuroleptics	Risperidone and its metabolites	Moody et al. (2004)
	Numerous neuroleptics including clozapine, flupenthiol, haloperidol	Kratzsch et al. (2003)
	Olanzapine	Bogusz et al. (1999)
Benzodiazepines	23 benzodiazepines	Kratzsch et al. (2004)
	11 benzodiazepines and their metabolites	Miki et al. (2002)
	Flunitrazepam and its metabolites	Kollroser and Schober (2002)
Antidepressants	Fluoxetine	Shen et al. (2002)
	Pramipexole	Lau et al. (1996)
	Tricyclic antidepressants	Kagan et al. (2004)
Amphetamines	MDA, MDMA, MDE, amphetamine, methamphetamine	Cristoni et al. (2004), Nordgren and Beck (2003)
Opiates	Various opiates and metabolites	Scheidweiler and Huestis (2004)
	Methadone and metabolites	Dams et al. (2003)
	Morphine, codeine, and metabolites	Bogusz et al. (1997)
Cocaine	Cocaine and metabolites	Scheidweiler and Huestis (2004), Lin et al. (2003)
Alkaloids	Ephedra alkaloids and caffeine	Jacob et al. (2004)
	Diterpinoid alkaloids	Wada et al. (2000)
	Norditerpenoid alkaloids	Gardner et al. (1999)
Barbiturates	Barbital, allobarbital, phenobarbital, butalbital	Jones et al. (2003)
Hallucinogens	N,N-dimethyltryptamine, O-methyl-bufotenine	Barker et al. (2001)
	Psilocybin	Bogusz (2000)
Cannabinoids	Numerous cannabinoids	Backstrom et al. (1997)
Nicotine	Nicotine and cotinine	Xu et al. (1996)
Caffeine	Caffeine	Gardinali and Zhao (2002)

most API-electrospray applications, the linear dynamic range of the detector is only about 2–3 orders of magnitude, with signal saturation occurring at higher concentrations. At high analyte concentrations, signals are produced by incomplete separation of analyte molecules during the formation of the gas phase. This results in the formation of ions such as M_2H^+, M_2Na^+, and M_3H^+. Such signals are indicative of oversaturated samples. In order to perform quantitative work, it is preferable that the amount of analyte introduced to the system falls within the linear range of the detector. This is not always possible or practical. As a result, most of the software associated with API-electrospray detectors can fit standard curves to curvilinear data and thus provide an extension of the dynamic range of the detector. Curvilinear standard curves should be used with caution and measurement of samples at concentrations approaching signal saturation should be evaluated with care. The flow rate of the mobile phase eluting from the liquid chromatograph also has a profound effect on the API-electrospray response. As droplet size increases with flow rate, an increased flow will result in a decreased signal response. Furthermore, signal stability is reduced as droplet uniformity decreases, and the rate of accumulation of undesired salts at the mass spectrometer entrance is more rapid and increased maintenance is mandatory. With most quantitative API-electrospray applications, a flow rate of 100–500 μL/min is optimal. Alternative configurations such as microspray and nanospray also have their optimal flow rates. Signal response for the analyte of interest decreases in the presence of other electrolytes. This is termed a matrix effect and most often results in ion suppression, as there is competition between the analyte and electrolyte ions for the charges available and for occupation of the droplet surface. Such effects are most often seen with syringe infusion, flow injection, or very rapid analyses in which chromatographic separation of the analyte from the solvent is insufficient. In general, matrix effects can be reduced if the analyte is more basic than the matrix (positive mode), if it is more acidic than the matrix (negative mode), or it is sufficiently polar to form stable adducts. A reduction in matrix artifacts can also be accomplished by ensuring a sufficient chromatographic separation of the analyte from any electrolytes and by establishing standard curves using dilutions in materials similar to those in which the sample has been prepared. Analyte surface activity is another important factor affecting analyte response. Ions that favor the surface of the droplet are more likely to form gas-phase ions. This property is a two-edged sword. Some very sensitive methods have been developed for compounds with high surface activities such as tertiary and quaternary amines. However, because of competition for conversion to gas-phase ions the presence of compounds such as surfactants, which are often found in tissue preparations, dramatically suppresses the response to the desired analytes. Many biological and pharmaceutical preparations contain, and certain chromatographic separations utilizes, surfactants and other similar materials including Triton, fatty acids, lauryl sulfate, polyethylene glycols, saponins, and ion-pairing agents such as sodium dodecyl sulfate. It is best to avoid these compounds, but if it is not possible, the analyte of interest must be chromatographically resolved from the surfactant.

The API-electrospray interface with chromatographic equipment is not perfect and conditions in the spray chamber can alter over time. It is essential, especially when performing analyses involving complex and rather concentrated samples or a large number of extended separations to use intermittent quality control checks. Over time (usually greater than 24 h), the spray chamber can accumulate materials that affect vaporization and ionization and thus sensitivity. When using a diode array absorbance detector inline with an API-electrospray MS detector, it is often observed that mass spectral signals decline over time whereas the absorbance signals remain undiminished.

A list of API-electrospray applications to a number of chemicals of neurochemical interest is provided in *Table 6-3*.

4.3 Capillary Electrophoresis Linked to Mass Spectrometry (CE/MS)

CE is considered to be a separation method that compliments LC. Many compounds including amino acids, peptides, and proteins (Moini, 2004) can be separated with LC and with CE, but CE has some advantages over LC in the separation of various organic materials including oligosaccharides (Zamfir et al., 2002; Li et al., 2004a), oligonucleotides (von Brocke et al., 2003), and neurotransmitter metabolites including serotonin conjugates (Stuart et al., 2003), which are difficult to retain and separate on conventional reverse

Table 6-3
Selected references of API-electrospray based methods for the analyses of endogenous and exogenous chemicals of neurobiological interest

Chemical group	Specific chemicals	API-electrospray reference
Endogenous Neurochemicals		
Amino Acids	Amino acids with derivatization	Liu et al. (2004)
Nucleosides and deoxynucleosides	Inosine, guanosine, adenosine	Zhu et al. (2001)
	Purine and pyrimidine bases, ribonucleosides	Dobolyi et al. (1998)
Acetylcholine	Acetylcholine and choline	Dunphy and Burinsky (2003)
	Acetylcholine	Hows et al. (2002)
Sphingomyelins and Ceramides	Ceramides	Camera et al. (2004), Colsch et al. (2004),
	Sphingosine, sphinganine	Lieser et al. (2003)
Neurosteroids	Sulfated neurosteroids	Griffiths et al. (1999)
	Ketonic neurosteroids	Liu et al. (2003)
Amines	Histamine, tryptamine, agmatine with derivatization	Song et al. (2004)
	Phenethylamine, tryptamine and others	Vorce and Sklerov (2004)
Catecholamines and Metanephrines	Catecholamines and metabolites	Tornkvist et al. (2004), Lang et al. (2004)
	Dopamine, norepinephrine, serotonin	Hows et al. (2004)
	6-Hydroxydopamine	Liao et al. (2003)
	Dopamine, 6-hydroxydopamine	Hao et al. (2002)
	L-Dopa, dopamine	Li et al. (2000)
	Norepinephrine	Neubecker et al. (1998)
Serotonin	Serotonin and metabolites	Semak et al. (2004)
Anandamide-related chemicals	Endocannabinoids with Ag^{+} binding	Kingsley and Marnett (2003)
	acylethanolamides	Giuffrida et al. (2000)
Neuropeptides	Galanin	Norberg et al. (2004)
	Somatostatin	Zhou et al. (2003)
	RFamide-related	Ukena et al. (2002)
	Dynorphin metabolism	Prokai et al. (1998), Sandin et al. (1997)
	Substance P, bradykinins	D'Agostino et al. (1997)
	Bradykinnin metabolite BK1-5	Murphey et al. (2001)
	Enkephalins	Lorenz et al. (1999), Marquez et al. (1997)
	Neurotensin	Andren and Caprioli (1999)
	Neuropeptide Y	Racaityte et al. (2000)
	Neuropeptide FF-related	Bonnard et al. (2001)
	Numerous pituitary peptides	Desiderio (1999)
	Peptide fragments	Skold et al. (2002)
Exogenous Chemicals		
Neuroleptics	Pimpamperone	Muller et al. (2000)
	Various neuroleptics and metabolites	Josefsson et al. (2003)
	Various neuroleptics	Gutteck and Rentsch (2003)
	Risperidone, 9-hydroxyrisperidone	Aravagiri and Marder (2000)
	Chlorpromazine, trifluoperazine, flupenthixol, risperidone	McClean et al. (2000)
	Fluspirilene	Swart et al. (1998)

Table 6-3 (continued)

Chemical group	Specific chemicals	API-electrospray reference
	Haloperidol and metabolites	Iwahashi et al. (2001)
	Quetiapine	Li et al. (2004b)
	Phenothiazine derivatives	Seno et al. (1999)
Benzodiazepines	Midazolam	Quintela et al. (2004)
	Clobazam	Proenca et al. (2004)
	Alrazolam, estazolam, midazolam, and metabolites	Toyo'oka et al. (2003)
	Numerous benzodiazepines	Miki et al. (2002)
Antidepressants	Fluoxetine and metabolites	Souverain et al. (2003)
	Maprotiline, citalopram, and metabolites	Muller et al. (2000)
	Various antidepressants	Gutteck and Rentsch (2003)
β-agonists	Clenbuterol, salbutamol, cimaterol	Lau et al. (2004)
	Salbutamol	Schmeer et al. (1997)
Serotonin agonists	Buspirone and others	Kowalski et al. (2003)
Amphetamines	Numerous amphetamines	Stanaszek and Piekoszewski (2004)
Opiates	Opiates, amphetamines	Mortier et al. (2002)
	Opiates	Cailleux et al. (1999)
Cocaine	Cocaine	Fuh et al. (2001), Cailleux et al. (1999)
	Cocaine metabolites	Lin et al. (2003)
Alkaloids	Indole alkaloids	McClean et al. (2002), Forsstrom et al. (2001)
Barbiturates	Phenobarbital, butalbital, pentobarbital, thiopental	Spell et al. 1998
LSD related	LSD	Bodin and Svensson (2001)
	LSD, iso-LSD	Canezin et al. (2001)
	LSD, 2-oxo-3-hydroxy-LSD	Sklerov et al. (2000)
Phencyclidine	Phencyclidine	Schneider et al. (1998)
Cannabinoids	Cannabinoids	Maralikova and Weinmann (2004), Tai and Welch (2000), Breindahl and Andreasen (1999)
Antimigraines	Sumatriptan, naratriptan, zolmitriptan, rizatriptan	Vishwanathan et al. (2000)
Nicotine	Nicotine	Taylor et al. (2004)
	Nicotine, cotinine	Chetiyanukornkul et al. (2004)
	Nicotine, cotinine, 3-hydroxycotinine	Tuomi et al. (1999)
Caffeine	Caffeine	Tuomi et al. (1999)
	Caffeine and theobromine	Thevis et al. (2004)
	Caffeine and metabolites	Schneider et al. (2003)
Synthetic peptides	Cysteinyldopaenkephalins	Rosei et al. (2000)
	Neuropeptide FF antagonist	Prokai et al. (2000)
Wide ranging screening assays for numerous compounds	Diverse drugs	Kronstrand et al. (2004), Gergov et al. (2003)

phase HPLC columns due to their very polar nature. The most common configuration involves connecting the outlet of the CE system to an API-electrospray source. This configuration is accurately illustrated by Smith et al. (1988).

Several additional useful publications demonstrating practical applications of CE/MS methods for neurotransmitter analysis and neuropharmaceutical studies are those of Larsson and Lutz (2000) (neuropeptides including substance P); Hettiarachchi et al. (2001) (synthetic opioid peptides); Varesio et al. (2002) (amyloid-beta peptide); Zamfir and Peter-Katalinic (2004) (gangliosides); Peterson et al. (2002) (catecholamines and metanephrines); Cherkaoui and Veuthey (2002) (fluoxetine); and Smyth and Brooks (2004) (various lower molecular weight molecules including benzodiazepines, steroids, and cannabinols).

5 Conclusions

The use of mass spectrometry for the analysis of neuroactive substances, neuropharmaceuticals, and their metabolites is a rapidly expanding field. Mass spectrometry has become increasingly affordable and versatile in the last decade and its application has rapidly expanded from analysis of small volatile molecules to nonvolatile, unstable, and often very large molecules, including peptides, proteins, oligosaccharides, and oligonucleotides, as new interfaces that provide access of diverse chemicals to mass spectrometers become available. Relatively, simple single quadrupole mass spectrometers are now used for routine quantitative measurements. These mass spectrometers are rapidly becoming the detectors of choice for analytical laboratories due to their versatility, which permits the selective quantitative analysis of designated molecules in complex biological specimens. These selective analyses can often be accomplished rapidly with minimal cleanup using a gradient separation on a small column.

Acknowledgments

The authors are grateful to the Canadian Institutes for Health Research (CIHR) (Ms. G. Rauw) for their financial support. Dr. Sloley wishes to thank Dr. Y.K. Tam for permission to use Novokin Biotech Inc. facilities for production of this manuscript. Dr. Soheir Tawfik is thanked for technical assistance.

References

Andren PE, Caprioli RM. 1999. Determination of extracellular release of neurotensin in discrete rat brain regions utilizing in vivo microdialysis/electrospray mass spectrometry. Brain Res 845: 123.

Aravagiri M, Marder SR. 2000. Simultaneous determination of risperidone and 9-hydroxyrisperidone in plasma by liquid chromatography/electrospray tandem mass spectrometry. J Mass Spectrom 35: 718.

Backstrom B, Cole MD, Carrott MJ, Jones DC, Davidson, G, et al. 1997. A preliminary study of the analysis of Cannabis by supercritical fluid chromatography with atmospheric pressure chemical ionization mass spectroscopic detection. Sci Justice 37: 91.

Baker GB, Coutts RT, Holt A. 1994. Derivatization with acetic anhydride, under both aqueous and anhydrous conditions: application to the analysis of biogenic amines and psychiatric drugs by gas chromatography and mass spectrometry. J Pharmacol Toxicol Methods 31: 141.

Barak R, Ordentlich A, Barak D, Fischer M, Benschop HP, et al. 1997. Direct determination of the chemical composition of acetylcholinesterase phosphonylation products utilizing electrospray-ionization mass spectrometry. FEBS Lett 407: 347.

Barker SA, Littlefield-Chabaud MA, David C. 2001. Distribution of the hallucinogens N,N-dimethyltryptamine and 5-methoxy-N,N-dimethyltryptamine in rat brain following intraperitoneal injection: application of a new solid-phase extraction LC-APCI-MS-MS-isotope dilution method. J Chromatogr B Biomed Appl 751: 37.

Beck-Sickinger AG, Wieland HA, Wittenben H, Willim KD, Rudolf K, et al. 1994. Complete L-alanine scan of neuropeptide Y reveals ligands binding to Y1 and Y2 receptors with distinguished conformations. Eur J Biochem 225: 947.

Bickeboeller-Friedrich J, Maurer H. 2001. Screening for the detection of new antidepressants, neuroleptics, hypnotics and their metabolites in urine by GC-MS developed by rat liver microsomes. Ther Drug Monit 23: 61.

Blondal T, Waage BG, Smarason SV, Jonsson F, Fjalldal SB, et al. 2003. A novel MALDI-TOF based methodology for genotyping single nucleotide polymorphisms. Nucleic Acids Res 15: e155.

Bodin K, Svensson JO. 2001. Determination of LSD in urine with high-performance liquid chromatography—mass spectrometry. Ther Drug Monit 23: 389.

Bogan MJ, Agnes GR. 2004. Wall-less sample preparation of micron-sized sample spots for femtomole detection limits of proteins from liquid based UV-MALDI matrices. J Am Soc Mass Spectrom 15: 486.

Bogusz MJ. 2000. Liquid chromatography-mass spectrometry as a routine method in forensic sciences: a proof of maturity. J Chromatogr B Biomed Sci Appl 748: 3.

Bogusz MJ, Kruger KD, Maier RD. 2000. Analysis of underivatized amphetamines and related phenylethylamines with high-performance liquid chromatography-atmospheric pressure chemical ionization mass spectrometry. J Anal Toxicol 24: 77.

Bogusz MJ, Maier RD, Erkens M, Driessen S. 1997. Determination of morphine and its 3- and 6-glucuronides, codeine, codeine-glucuronide and 6-monoacetylmorphine in body fluids by liquid chromatography atmospheric pressure chemical ionization mass spectrometry. J Chromatogr B Biomed Sci Appl 703: 115.

Bogusz MJ, Kruger KD, Maier RD, Erkwoh R, Tuchtenhagen F. 1999. Monitoring of olanzapine in serum by liquid chromatography-atmospheric pressure chemical ionization mass spectrometry. J Chromatogr B Biomed Sci Appl 732: 257.

Bonnard E, Burlet-Schltitz O, Frances B, Mazarguil H, Monsarrat B, et al. 2001. Identification of neuropeptide FF-related peptides in rodent spinal cord. Peptides 22: 1085.

Breindahl T, Andreasen K. 1999. Determination of 11-nor-delta9-tetrahydrocannabinol-9-carboxylic acid in urine using high-performance liquid chromatography and electrospray ionization mass spectrometry. J Chromatogr B Biomed Sci Appl 732: 155.

von Brocke A, Freudemann T, Bayer E. 2003. Performance of capillary gel electrophoretic analysis of oligonucleotides coupled on-line with electrospray mass spectrometry. J Chromatogr A 991: 129.

Bulau P, Meisen I, Schmitz T, Keller R, Peter-Katalinic J. 2004. Identification of neuropeptides from the sinus gland of the crayfish *Orconectes limosus* using nanoscale on-line liquid chromatography tandem mass spectrometry. Mol Cell Proteomics 3: 558.

Burlet-Schiltz O, Marzarguil H, Sol JC, Chaynes P, Monsarrat B, et al. 2002. Identification of neuropeptide FF-related peptides in human cerebrospinal fluid by mass spectrometry. FEBS Lett 532: 313.

Byrdwell WC. 2001. Atmospheric pressure chemical ionization mass spectrometry for analysis of lipids. Lipids 36: 327.

Cailleux A, Le Bouil A, Auger B, Bonsergent G, Turcant A, et al. 1999. Determination of opiates and cocaine and its metabolites in biological fluids by high-performance liquid chromatography with electrospray tandem mass spectrometry. J Anal Toxicol 23: 620.

Camera E, Picardo M, Presutti C, Catarcini P, Fanali S. 2004. Separation and characterization of sphingoceramides by high-performance liquid chromatography-electrospray ionization mass spectrometry. J Sep Sci 27: 971.

Canezin J, Cailleux A, Turcant A, Le Bouil A, Harry P, et al. 2001. Determination of LSD and its metabolites in human biological fluids by high-performance liquid chromatography with electrospray tandem mass spectrometry. J Chromatogr B Biomed Sci Appl 765: 15.

Chan EC, Ho PC. 2000. High-performance liquid chromatography/atmospheric pressure chemical ionization mass spectrometric method for the analysis of catecholamines and metanephrines in human urine. Rapid Commun Mass Spectrom 14: 1959.

Che FY, Yan L, Li H, Mzhavia N, Devi LA, et al. 2001. Identification of peptides from brain and pituitary of Cpe (fat)/Cpe(fat) mice. Proc Natl Acad Sci USA 98: 9971.

Cherkaoui S, Veuthey JL. 2002. Nonaqueous capillary electrophoresis-electrospray-mass spectrometry for the analysis of fluoxetine and its related compounds. Electrophoresis 23: 442.

Chetiyanukornkul T, Toriba A, Kizu R, Kimura K, Hayakawa K. 2004. Hair analysis of nicotine and cotinine for evaluating tobacco smoke exposure by liquid chromatography-mass spectrometry. Biomed Chromatogr 18: 655.

Colsch B, Afonso C, Popa I, Portoukalian J, Fournier F, Tabet JC, Baumann N. 2004. Characterization of the ceramide moieties of sphingoglycolipids from mouse brain by ESI-MS/MS: identification of ceramides containing sphingadienine. J Lipid Res 45: 281.

Costa C, Bertazzo A, Allegri G, Toffano G, Curcuruto O, et al. 1992. Melanin biosynthesis from dopamine. II. A mass spectrometric and collisional spectroscopic investigation. Pigment Cell Res 5: 122.

Couch LH, Churchwell MI, Doerge DR, Tolleson WH, Howard PC. 1997. Identification of ceramides in human cells using liquid chromatography with detection by atmospheric pressure chemical ionization-mass spectrometry. Rapid Commun Mass Spectrom 11: 504.

Cristoni S, Bernardi LR, Gerthoux P, Gonella E, Mocarelli P. 2004. Surface-activated chemical ionization ion trap mass spectrometry in the analysis of amphetamines in diluted urine samples. Rapid Commun Mass Spectrom 18: 1847.

D'Agostino PA, Hancock JR, Provost LR. 1997. Analysis of bioactive peptides by liquid chromatography-high-resolution electrospray mass spectrometry. J Chromatogr A 767: 77.

Dams R, Murphy CM, Choo RE, Lambert WE, De Leenheer AP, et al. 2003. LC-atmospheric pressure chemical ionization-MS/MS analysis of multiple illicit drugs, methadone, and their metabolites in oral fluid following protein precipitation. Anal Chem 75: 798.

Danaceau JP, Anderson GM, McMahon WM, Crouch DJ. 2003. A liquid chromatographic-tandem mass spectrometric method for the analysis of serotonin and related indoles in human whole blood. J Anal Toxicol 27: 440.

Daniel JM, Ehala S, Friess SD, Zenobi R. 2004. On-line atmospheric pressure matrix-assisted desorption/ionization mass spectrometry. Analyst 129: 574.

Desiderio DM. 1999. Mass spectrometric analysis of neuropeptidergic systems in the human pituitary and cerebrospinal fluid. J Chrmoatogr B Biomed Sci Appl 731: 3.

Diehl G, Liesener A, Karst U. 2001. Liquid chromatography with post-column electrochemical treatment and mass spectrometric detection of non-polar compounds. Analyst 126: 288.

Dobolyi A, Reichart A, Szikra T, Szilagy N, Kekesi AK, et al. 1998. Analysis of purine and pyrimidine bases, nucleosides and deoxynucleosides in brain microsamples (microdialysates and micropunches) and cerebrospinal fluid. Neurochem Int 32: 247.

Dunphy R, Burinsky DJ. 2003. Detection of choline and acetylcholine in a pharmaceutical preparation using high-performance liquid chromatography/electrospray ionization mass spectrometry. J Pharm Biomed Anal 31: 901.

Durden DA, Boulton AA. 1988. Analysis of tryptamine at the femtomole level in tissue using negative ion chemical ionization gas chromatography-mass spectrometry. J Chromatogr 440: 253.

Fales HM, Lloyd HA, Milne GW. 1970. Chemical ionization of complex molecules. II. Alkaloids. J Am Chem Soc 92: 1590.

Felten C, Foret F, Minarik M, Goetzinger W, Karger BL. 2001. Automated high throughput infusion ESI-MS with direct coupling to a microtiter plate. Anal Chem 73: 1449.

Forsstrom T, Tuominen J, Karkkainen J. 2001. Determination of potentially hallucinogenic N-dimethylated indoleamines in human urine by HPLC/ESI-MS-MS. Scand J Clin Lab Invest 61: 547.

Fournier I, Day R, Salzet M. 2003. Direct analysis of neuropeptides by in situ MALDI-TOF mass spectrometry in the rat brain. Neuro Endocrinol Lett 24: 9.

Fuh MR, Tai YL, Pan WH. 2001. Determination of free-form of cocaine in rat brain by liquid chromatography-electrospray mass spectrometry with in vivo microdialysis. J Chromatogr B Biomed Sci Appl 752: 107.

Gardinali PR, Zhao X. 2002. Trace determination of caffeine in surface water samples by liquid chromatography–atmospheric pressure chemical ionization–mass spectrometry (LC-APCI-MS). Environ Int 28: 521.

Gardner DR, Panter KE, Pfister JA, Knight AP. 1999. Analysis of toxic norditerpenoid alkaloids in Delphinium species by electrospray, atmospheric pressure chemical ionization, and sequential tandem mass spectrometry. J Agric Food Chem 47: 5049.

Gergov M, Ojanperä I, Vuori E. 2003. Simultaneous screening for 238 drugs in blood by liquid chromatography-ionspray tandem mass spectrometry with multiple-reaction monitoring. J Chromatogr B 795: 41.

Giuffrida A, Rodriguez de Fonseca F, Piomelli D. 2000. Quantification of bioactive acylethanolamides in rat plasma by electrospray mass spectrometry. Anal Biochem 280: 87.

Griffiths WJ, Liu S, Yang Y, Purdy RH, Sjovall J. 1999. Nano-electrospray tandem mass spectrometry for the analysis of neurosteroid sulphates. Rapid Commun Mass Spectrom 13: 1595.

Gross JH. 2004. Mass spectrometry. Germany: Springer.

Guo Z, Wagner CR, Hanna PE. 2004. Mass spectrometric investigation of the mechanism of inactivation of hamster arylamine N-acetyltransferase 1 by N-hydroxy-2-acetylaminofluorene. Chem Res Toxicol 17: 275.

Gutteck U, Rentsch KM. 2003. Therapeutic drug monitoring of 13 antidepressant and five neuroleptic drugs in serum with liquid chromatography-electrospray ionization mass spectrometry. Clin Lab Med 41: 1571.

Hansen G, Christensen P, Tuchsen E, Lund T. 2004. Sensitive and selective analysis of coenzyme Q10 in human serum by negative APCI LC-MS. Analyst 129: 45.

Hao C, March RE, Croley TR, Chen S, Legault MG, et al. 2002. Study of the neurotransmitter dopamine and the neurotoxin 6-hydroxydopamine by electrospray ionization coupled with tandem mass spectrometry. Rapid Commun Mass Spectrom 16: 591.

Harrison KA, Murphy RC. 1996. Direct mass spectrometric analysis of ozonides: application to unsaturated glycerophosphocholine lipids. Anal Chem 68: 3224.

Hashimoto Y, Miyazaki H. 1979. Simultaneous determination of endogenous norepinephrine and dopamine-beta-hyroxylase activity in biological materials by chemical ionization mass fragmentography. J Chromatogr 168: 59.

Henry CM. 1999. Electrospray in flight: orthogonal acceleration brings the advantages of time of flight to electrospray. Anal Chem 71: 197A.

Hernandez A, Andollo W, Hearn WL. 1994. Analysis of cocaine and metabolites in brain using solid phase extraction and full-scanning gas chromatography/ion trap mass spectrometry. Forensic Sci Int 13: 149.

Hettiarachchi K, Ridge S, Thomas DW, Olson L, Obi CR, et al. 2001. Characterization and analysis of biphalin: an opioid peptide with a palindromic sequence. J Pept Res 57: 151.

Hidalgo OH, Lopez MJ, Carazo EA, Larrea MSA, Reuvres TBA. 2003. Determination of dexamethasone in urine by gas chromatography with negative chemical ionization. J Chromatogr B 788: 137.

Higashi T, Takido N, Shimada K. 2003. Detection and characterization of 20-oxosteroids in rat brains using LC-electron capture APCI-MS after derivatization with 2-nitro-4-trifluoromethylphenylhydrazine. Analyst 128: 130.

de Hoffman E, Stroobant V. 2001. Mass spectrometry principles and applications, 2nd edn. West Sussex, England: Wiley.

Hong SP, Kim NK, Hwang SG, Chung HJ, Kim S, et al. 2004. Detection of hepatitis B virus YMDD variants using mass spectrometric analysis of oligonucleotide fragments. J Hepatol 40: 837.

Honour JW. 2003. Benchtop mass spectrometry in clinical biochemistry. Ann Clin Biochem 40: 628.

Hows ME, Lacroix L, Heidbreder C, Organ AJ, Shah AJ. 2004. High-performance liquid chromatography/tandem mass spectrometric assay for the simultaneous measurement of dopamine, norepinephrine, 5-hydroxytryptamine and cocaine in biological samples. J Neurosci Methods 138: 123.

Hows ME, Organ AJ, Murray S, Dawson LA, Foxton R, et al. 2002. High-performance liquid chromatography/tandem mass spectrometry assay for the rapid high sensitivity measurement of basal acetylcholine from microdialysates. J Neurosci Methods 121: 33.

Huang LF, Guo FQ, Liang YZ, Li BY, Cheng BM. 2004. Simultaneous determination of L-arginine and its mono- and demethylated metabolites in human plasma by high-performance liquid chromatography-mass spectrometry. Anal Bioanal Chem 380: 643.

Huang SM, Strangman NM, Walker JM. 1999. Liquid chromatographic-mass spectrometric measurement of the endogenous cannabinoid 2-arachidonylglcerol in the spinal cord and peripheral nervous system. Zhongguo Yao Li Xue Bao 20: 1098.

Inoue H, Maeno Y, Iwasa M, Matoba R, Nagas M. 2000. Screening and determination of benzodiazepines in whole blood using solid-phase extraction and gas chromatography/mass spectrometry. Forensic Sci Int 113: 367.

Iwahashi K, Anemo K, Nakamura K, Fukunishi I, Igarashi K. 2001. Analysis of the metabolism of haloperidol and its neurotoxic pyridinium metabolite in patients with drug-induced parkinsonism. Neuropsychobiology 44: 126.

Jacob P III, Haller CA, Duan M, Yu L, Peng M, et al. 2004. Determination of ephedra alkaloid and caffeine concentrations in dietary supplements and biological fluids. J Anal Toxicol 28: 152.

Jones JJ, Kidwell H, Games DE. 2003. Application of atmospheric pressure chemical ionization mass spectrometry in the analysis of barbiturates by high-speed analytical countercurrent chromatography. Rapid Commun Mass Spectrom 17: 1565.

Josefsson M, Kronstrand R, Andersson J, Roman M. 2003. Evaluation of electrospray ionization liquid chromatography-tandem mass spectrometry for rational determination of a number of neuroleptics and their major metabolites in human body fluids and tissues. J Chromatogr B Analyt Technol Biomed Life Sci 789: 151.

Kagan M, Chlenov M, Kraml CM. 2004. Normal-phase high-performance liquid chromatographic separations using ethoxynonafluorobutane as hexane alternative. II. Liquid chromatography-atmospheric pressure chemical ionization-mass spectrometry applications with methanol gradients. J Chromatogr A 1033: 321.

Kakela R, Somerharju P, Tyynela J. 2003. Analysis of phospholipids molecular species in brains from patients with infantile and juvenile neuronal-ceroid lipofuscinosis using liquid chromatography-electrospray ionization mass spectrometry. J Neurochem 84: 1051.

Kamel AM, Zandi KS, Massefski WW. 2003. Identification of the degradation product of ezlopitant, a non-peptidic substance P antagonist receptor, by hydrogen deuterium exchange, electrospray ionization tandem mass spectrometry (ESI/MS/MM) and nuclear magnetic resonance (NMR) spectroscopy. J Pharm Biomed Anal 31: 1211.

Kankaanpää A, Gunnar T, Ariniemi K, Lillsunde P, Mykänen S, et al. 2004. Single-step procedure for gas chromatography-mass spectrometry screening and quantitative determination of amphetamine-type stimulants and related drugs in blood, serum, oral fluids and urine samples. J Chromatogr B 810: 57.

Karlsson AA, Michelsen P, Odham G. 1998. Molecular species of sphingomyelin: determination by high-performance liquid chromatography/mass spectrometry with electrospray and high-performance liquid chromatography/tandem mass spectrometry with atmospheric pressure chemical ionization. J Mass Spectrom 33: 1192.

Karminski-Zamola G, Dogan J, Boykin DW, Bajic M. 1995. Mass spectral fragmentation patterns of some new benzo [b]thiophene- and thieno[2,3-b] thiopene-2,5-dicarbonyldichlroide and –dicarbonyldianilides and anilidoquinolones. Rapid Commun Mass Spectrom 9: 282.

Kim Y-S, Zhang H, Kim H-Y. 2000. Profiling neurosteroids in cerebrospinal fluids and plasma by gas chromatography/electron capture negative chemical ionization mass spectrometry. Anal Biochem 277: 187.

Kingsley PJ, Marnett LJ. 2003. Analysis of endocannabinoids by Ag^+ coordination tandem mass spectrometry. Anal Biochem 314: 8.

Koga D, Santa T, Fukushima T, Homma H, Imai K. 1997. Liquid chromatographic-atmospheric pressure chemical ionization mass spectrometric determination of anandamide and its analogs in rat brain and peripheral tissues. J Chromatogr B Biomed Sci Appl 690: 7.

Koga D, Santa T, Hagiwara K, Imai K, Takizawa H, et al. 1995. High-performance liquid chromatography and fluorometric detection of arachidonylethanolamide and its analogues, derivatized with 4-(N-chloroformylmethyl-N-methyl) amino-7-N,N-dimethylaminosulphonyl-2,1,3-benzoxadiazole (DBD-COCl). Biomed Chromatogr 9: 56.

Kollroser M, Schober C. 2002. Simultaneous analysis of flunitrazepam and its major metabolites in human plasma by high performance liquid chromatography tandem mass spectrometry. J Pharm Biomed Anal 28: 1173.

Kowalski P, Suder P, Kowalska T, Silberring J, Duszynska B, et al. 2003. Electrospray mass spectrometric studies of non-covalent complexes busperone hydrochloride and other serotonin 5-HT(1A) receptor ligands containing arylpiperazine moieties. Rapid Commun Mass Spectrom 17: 2139.

Kratzsch C, Peters FT, Kraemer T, Weber AA, Maurer HH. 2003. Screening, library-assisted identification and validated quantification of fifteen neuroleptics and three of their metabolites in plasma by liquid chromatography/mass spectrometry with atmospheric pressure chemical ionization. J Mass Spectrom 38: 283.

Kratzsch C, Tenberken O, Peters FT, Weber AA, Kraemer T, et al. 2004. Screening, library-assisted identification and validated quantification of 23 benzodiazepines, flumazenil, zalepone, zolpidem and zopiclone in plasma by liquid chromatography/mass spectrometry with atmospheric pressure chemical ionization. J Mass Spectrom 39: 856.

Kronstrand R, Nystrom I, Strandberg J, Druid H. 2004. Screening for drugs of abuse in hair with ion spray LC-MS-MS. Forensic Sci Int 145: 183.

Kruse R, Sweedler JV. 2003. Spatial profiling invertebrate ganglia using MALDI MS. J Am Soc Mass Spectrom 14: 752.

Kwon JY, Moini M. 2001. Analysis of underivatized amino acid mixtures using high performance liquid chromatography/dual oscillating nebulizer atmospheric pressure microwave induced plasma ionization-mass spectrometry. J Am Soc Mass Spectrom 12: 117.

Larsson M, Lutz ES. 2000. Transient isotachophoresis for sensitivity enhancement in capillary electroporesis-mass spectrometry for peptide analysis. Electrophoresis 21: 2859.

Lau JH, Khoo CS, Murby JE. 2004. Determination of clenbuterol, salbutamol, and cimaterol in bovine retina by electrospray ionization-liquid chromatography-tandem mass spectrometry. J AOAC Int 87: 31.

Lau YY, Selenka JM, Hanson GD, Talaat R, Ichhpurani N. 1996. Determination of pramipexole (U-98,528) in human plasma by high-performance liquid chromatography with atmospheric pressure chemical ionization tandem mass spectrometry. J Chromatogr B Biomed Appl 683: 209.

Leis HJ, Fauler G, Rechberger GN, Windischhofer W. 2004. Electron-capture mass spectrometry: a powerful tool in biomedical trace level analysis. Curr Med Chem 11: 1585.

Lewis DA, Guzzetta AW, Hancock WS, Costello M. 1994. Characterization of humanized anti-TAC, an antibody directed against the interleukin 2 receptor, using electrospray ionization mass spectrometry by direct infusion, LC/MS, and MS/MS. Anal Chem 66: 585.

Li J, Purves RW, Richards JC. 2004a. Coupling capillary electrophoresis and high-field asymmetric waveform ion mobility spectrometry mass spectrometry for the analysis of complex lipopolysaccharides. Anal Chem 76: 4676.

Li KY, Cheng ZN, Li X, Bai XL, Zhang BK, et al. 2004b. Simultaneous determination of quetiapine and three metabolites in human plasma by high-performance liquid chromatography-electrospray ionization mass spectrometry. Acta Pharmacol Sin 25: 110.

Li W, Rossi DT, Fountain ST. 2000. Development and validation of a semi-automated method for L-DOPA and dopamine in rat plasma using electrospray LC/MS/MS. J Pharm Biomed Anal 24: 325.

Liao PC, Kuo YM, Chang YC, Lin C, Cherng CF, et al. 2003. Striatal formation of 6-hydroxydopamine in mice treated with pargyline, pyrogallol and methamphetamine. J Neural Transm 110: 487.

Lieser B, Liebisch G, Drobnik W, Schmitz G. 2003. Quantification of sphingosine and shinganine from crude lipid extracts by HPLC electrospray ionization tandem mass spectrometry. J Lipid Res 44: 2209.

Lin SN, Walsh SL, Moody DE, Foltz RL. 2003. Detection and time course of cocaine N-oxide and other cocaine metabolites in human plasma by liquid chromatography/tandem mass spectrometry. Anal Chem 75: 4335.

Liu S, Sjovall J, Griffiths WJ. 2003. Neurosteroids in rat brain: extraction, isolation, and analysis by nanoscale liquid chromatography-electrospray mass spectrometry. Anal Chem 75: 5835.

Liu Z, Minkler PE, Lin D, Sayre LM. 2004. Derivatization of amino acids with N,N-dimethyl-2,4-dinitro-5-fluorobenzylamine for liquid chromatography/electrospray ionization mass spectrometry. Rapid Commun Mass Spectrom 18: 1059.

Loo JA, Quinn JP, Ryu SI, Henry KD, Senko MW, et al. 1992. High resolution tandem mass spectrometry of large biomolecules. Proc Natl Acad Sci USA 89: 286.

Lorenz SA, Moy MA, Dolan AR, Wood TD. 1999. Electrospray ionization fourier transform mass spectrometry quantification of enkephalin using an internal standard. Rapid Commun Mass Spectrom 13: 2098.

Loutelier-Bourhis C, Legros H, Bonnet JJ, Costentin J, Lange CM. 2004. Gas chromatography/mass spectrometric identification of dopaminergic metabolites in striata of rats treated with L-DOPA. Rapid Commun Mass Spectrom 18: 571.

Low W, Kang J, Di Gruccio M, Kirby D, Perrin M, et al. 2004. MALDI-MS analysis of peptides modified with photolabile arylazido groups. J Am Soc Mass Spectrom 15: 1156.

Maralikova B, Weinmann W. 2004 Simultaneous determination of Delta9-tetrahydrocannabinol, 11-hydroxy-Delta9-tetrahydrocannabinol and 11-nor-9-carboxy-Delta9-tetrahydrocannnabinol in human plasma by high-performance liquid chromatography/tandem mass spectrometry. J Mass Spectrom 39: 526.

Markey SP, Wenger, DA. 1974. Mass spectra of complex molecules. I. Chemical ionization of sphingolipids. Chem Phys Lipids 12: 182.

Marquez CD, Weintraub ST, Smith PC. 1997. Quantitative analysis of two opiod peptides in plasma by liquid chromatography-electrospray ionization tandem mass spectrometry. J Chromatogr Biomed Sci Appl 694: 21.

Maurer HH. 2002. Role of gas chromatography-mass spectrometry with negative ion chemical ionization in clinical and forensic toxicology, doping control, and biomonitoring. Ther Drug Monit 24: 247.

McClean S, O'Kane EJ, Smyth WF. 2000. Electrospray ionization-mass spectrometric characterization of selected anti-psychotic drugs and their detection and determination in human hair samples by liquid chromatography-tandem mass spectrometry. J Chromatogr B Biomed Sci Appl 740: 141.

McClean S, Robinson RC, Shaw C, Smyth WF. 2002. Characterization and determination of indole alkaloids in frog-skin secretions by electrospray ionization trap mass spectrometry. Rapid Commun Mass Spectrom 16: 346.

Miki A, Tatsuno M, Katagi M, Nishikawa M, Tsuchihashi H. 2002. Simultaneous determination of eleven benzodiazepine hypnotics and eleven relevant metabolites in urine by column-switching liquid chromatography-mass spectrometry. J Anal Toxicol 26: 87.

Milne GW, Fales HM, Colburn RW. 1973. Chemical ionization mass spectrometry of complex molecules: biogenic amine. Anal Chem 45: 1952.

Mitamura K, Shimada K. 2001. Derivatization in liquid chromatography/mass spectrometric analysis of neurosteroids. Se Pu 19: 508.

Mitamura K, Yatera M, Shimada K. 2000. Studies on neurosteroids. Part XIII. Characterization of catechol estrogens in rat brains using liquid chromatography-mass spectrometry-mass spectrometry. Analyst 125: 811.

Mock S, Shen X, Tamvakopoulos C. 2002. Determination of melanotan-II in rat plasma by liquid chromatography/tandem mass spectrometry: determination of pharmacokinetic parameters in rat following intravenous administration. Rapid Commun Mass Spectrom 16: 2142.

Moini M. 2004. Capillary electrophoresis-electrospray ionization mass spectrometry of amino acids, peptides and proteins. Methods Mol Biol 276: 253.

Moody DE, Laycock JD, Huang W, Foltz RL. 2004. A high-performance liquid chromatographic-atmospheric pressure chemical ionization-tandem mass spectrometric method for determination of risperidone and 9-hydroxyrisperidone in human plasma. J Anal Toxicol 28: 494.

Mortier KA, Maudens KE, Lambert WE, Clauwaert KM, Van Bocxlaer JF, et al. 2002. Simultaneous, quantitative determination of opiates, amphetamines, cocaine and benzoylcognine in oral fluid by liquid chromatography quadrupole-time-of-flight mass spectrometry. J Chromatogr B Analyt Technol Biomed Life Sci 779: 321.

Muller C, Vogt S, Goerke R, Kordon A, Weinmann W. 2000. Identification of selected psychopharmaceuticals and their metabolites in hair by LC/ESI-CID/MS and LC/MS/MS. Forensic Sci Int 113: 415.

Murphey LJ, Hachey DL, Vaughan DE, Brown NJ, Morrow JD. 2001. Quantification of BK1-5, the stable bradykinin plasma metabolite in humans, by a highly accurate liquid chromatographic tandem mass spectrometric assay. Anal Biochem 292: 87.

Musshoff F, Lachenmeier DW, Kroener L, Schmidt P, Dettmeyer R, et al. 2003. Simultaneous gas chromatographic-mass spectrometric determination of dopamine, norsalsolinol and salsolinol enantiomers in brain samples of a large human collective. Cell Mol Biol 49: 837.

Nagy K, Pollreisz F, Takats Z, Veky K. 2004. Atmospheric pressure chemical ionization mass spectrometry of aldehydes in biological matrices. Rapid Commun Mass Spectrom 18: 2473.

Neubecker TA, Coombs MA, Quijano M, O'Neill TP, Cruze CA, et al. 1998. Rapid and selective method for norepinephrine in rat urine using reversed-phase ion-pair high-performance liquid chromatography-tandem mass spectrometry. J Chromatogr B Biomed Sci Appl 718: 225.

Nilsson CL, Brinkmalm A, Minthon L, Blennow K, Ekman R. 2001. Processing of neuropeptide Y, galanin, and somatostatin in the cerebrospinal fluid of patients with Alzheimer's disease and frontotemporal dementia. Peptides 22: 2105.

Nokihara K, Gerhardt J. 2001. Development of an improved automated gas-chromatographic chiral analysis system: application to non-natural amino acids and natural protein hydrolysates. Chirality 13: 431.

Norberg A, Griffths WJ, Hjelmqvist L, Jornvall H, Rokaeus A. 2004. Identification of variant forms of the neuroendocrine peptide galanin. Rapid Commun Mass Spectrom 18: 1583.

Nordgren HK, Beck O. 2003. Direct screening of urine for MDMA and MDA by liquid chromatography-tandem mass spectrometry. J Anal Toxicol 27: 15.

Noto T, Hasegawa T, Kamimura H, Nakao J, Hashimoto H, Nakajima T. 1987. Determination of putrescine in brain tissue using gas chromatography-mass spectrometry. Anal Biochem 160: 371-375.

Odontiadis J, Rauw G. 2006. High performance liquid chromatography. Handbook of Neurochemistry and Molecular Biology, Vol. 18 General Techniques. Baker GB, Dunn SMJ, Holt A, editors. New York: Springer, in press.

Pashkova A, Moskovets E, Karger BL. 2004. Coumarin tags for improved analysis of peptides by MALDI-TOF MS and MS/MS. 1. Enhancement in MALDI MS signal intensities. Anal Chem 76: 4550.

Peters FT, Kraemer T, Maurer HH. 2002. Drug testing in blood: validated negative-ion chemical ionization gas chromatographic-mass spectrometric assay for determination of amphetamine and methamphetamine enantiomers and its application to toxicology cases. Clin Chem 48: 1472.

Peters R, Hellenbrand J, Mengerink Y, Wal Van der S. 2004. On-line determination of carboxylic acids, aldehydes and ketones by high-performance liquid chromatography-diode array detection-atmospheric pressure chemical ionization mass spectrometry after derivatization with 2-nitrophenylhydrazine. J Chromatogr A 1031: 35.

Peterson ZD, Collins DC, Bowerbank CR, Lee ML, Graves SW. 2002. Determination of catecholamines and metanephrines in urine by capillary electrophoresis-electrospray ionization-time-of-flight mass spectrometry. J Chromatogr B Analyt Technol Biomed Life Sci 776: 221.

Pettus BJ, Baes M, Busman M, Hannun YA, Van Veldhoven PP. 2004. Mass spectrometric analysis of ceramide perturbations in brain and fibroblasts of mice and human patients with peroxisomal disorders. Rapid Commun Mass Spectrom 18: 1569.

Pierson J, Norris JL, Aerni HR, Svenningsson P, Caprioli RM, et al. 2004. Molecular profiling of experimental Parkinson's disease: direct analysis of peptides and proteins on brain tissue sections by MALDI mass spectrometry. J Proteome Res 3: 289.

Poljak A, McLean CA, Sachdev P, Brodaty H, Smythe GA. 2004. Quantification of hemorphins in Alzheimer's disease brains. J Neurosci Res 75: 704.

Prest HF. 1999. Ionization methods in gas phase mass spectrometry: operating modes of 5973*Network* series MSDs. Agilent technologies, Application Note (23), 5698–7957E, www.agilent.com/chem.

Proenca P, Teixeira H, Pinheiro J, Marques EP, Vieira DN. 2004. Forensic intoxication with clobazam: HPLC/DAD/MSD analysis. Forensic Sci Int 143: 205.

Prokai L, Kim HS, Zharikova A, Roboz J, Ma L, et al. 1998. Electrospray ionization mass spectrometric and liquid chromatographic-mass spectrometric studies on the metabolism of synthetic dynorphin A peptides in brain tissue in vitro and in vivo. J Chromatogr A 800: 59.

Prokai L, Zharikova AD. 1998. Identification of synaptic metabolites of dynorphin A (1–8) by electrospray ionization tandem mass spectrometry. Rapid Commun Mass Spectrom 12: 1796.

Prokai L, Zharikova AD, Janaky T, Prokai-Tatrai K. 2000. Exploratory pharmacokinetics and brain distribution study of a neuropeptide FF antagonist by liquid chromatography/atmospheric pressure ionization tandem mass spectrometry. Rapid Commun Mass Spectrom 14: 2412.

Prokai L, Zharikova A, Janaky T, Li X, Braddy AC, et al. 2001. Integration of mass spectrometry into early-phase discovery and development of central nervous system agents. J Mass Spectrom 36: 1211.

Purdy RH, Fitzgerald RL, Everhart ET, Mellon SH, Alomary AA, Parsons LH. 2006. The analysis of neuroactive steroids by mass spectrometry. Handbook of Neurochemistry and Molecular Biology, Vol. 18 General Techniques. Baker GB, Dunn SMJ, Holt A, editors. New York: Springer, in press.

Quintela O, Cruz A, Concheiro M, Castro AD, Lopez-Rivadulla M. 2004. A sensitive, rapid and specific determination of midazolam in human plasma and saliva by liquid chromatography/electrospray mass spectrometry. Rapid Commun Mass Spectrom 18: 2976.

Racaityte K, Lutz ESM, Unger KK, Lubda D, Boos KS. 2000. Analysis of neuropeptide Y and its metabolites by high-performance liquid chromatography-electrospray ionization mass spectrometry and integrated sample clean-up with a novel restricted-access sulphonic acid cation exchanger. J Chromatogr A 890: 135.

Ramstrom M, Hagman C, Tsybin YO, Markides KE, Hakansson P, et al. 2003. A novel mass spectrometric approach to the analysis of hormonal peptides in extracts of mouse pancreatic islets. Eur J Biochem 270: 3146.

Rittenbach K, Baker GB. 2006. Gas chromatography. Handbook of Neurochemistry and Molecular Biology, Vol. 18 General Techniques. Baker GB, Dunn SMJ, Holt A, editors. New York: Springer, in press.

Rosei MA, Coccia R, Foppoli C, Blarzino C, Cini C, et al. 2000. Cysteinyldopaenkephalins: synthesis, characterization and binding to bovine brain opioid receptors. Biochem Biophys Acta 1478: 19.

Rubakhin SS, Greenough WT, Sweedler JV. 2003. Spatial profiling with MALDI MS: distribution of neuropeptides within single neurons. Anal Chem 75: 5374.

Sandin J, Tan-No K, Kasakov L, Nylander I, Winter A, et al. 1997. Differential metabolism of dynorphins in substantia nigra, striatum, and hippocampus. Peptides 18: 949.

Scheidweiler KB, Huestis MA. 2004. Simultaneous quantification of opiates, cocaine, and metabolites in hair by LC-APCI-MS/MS. Anal Chem 76: 4358.

Schmeer K, Sauter T, Schmid J. 1997. Rapid screening of salbutamol in plasma by column-switching high-performance liquid chromatography-electrospray mass spectrometry. J Chromatogr A 777: 67.

Schneider S, Kuffer P, Wennig R. 1998. Determination of lysergide (LSD) and phencyclidine in biosamples. J Chromatogr B Biomed Sci Appl 713: 189.

Schneider H, Ma L, Glatt H. 2003. Extractionless method for the determination of urinary caffeine metabolites using high-performance liquid chromatography coupled with tandem mass spectrometry. J Chromatogr B Analyt Technol Biomed Life Sci 789: 227.

Segura J, Ventura R, Jurado C. 1998. Derivatization procedures for gas chromatographic determination of xenobiotics in biological samples, with special attention to drugs of abuse and doping agents. J Chromatogr B 713: 61.

Semak I, Korik E, Naumova M, Wortsman J, Slominski A. 2004. Serotonin metabolism in rat skin: characterization by liquid chromatography-mass spectrometry. Arch Biochem Biophys 421: 61.

Seno H, Hattori H, Ishoo A, Kumazawa T, Watanabe-Suzuki K, et al. 1999. High performance liquid chromatography/electrospray tandem mass spectrometry for phenothiazines with heavy side chains in whole blood. Rapid Commun Mass Spectrom 13: 2394.

Shah AJ, Crespi F, Heidbreder C. 2002. Amino acid neurotransmitters: separation approaches and diagnostic value. J Chromatogr B 781: 51.

Shen Z, Wang S, Bakhtiar R. 2002. Enantiometric separation and quantification of fluoxetine (Prozac) in human plasma by liquid chromatography/tandem mass spectrometry using liquid-liquid extraction in 96-well plate format. Rapid Commun Mass Spectrom 16: 332.

Simpson JT, Torok DS, Girard JE, Markey SP. 1996. Analysis of amino acids in biological fluids by pentafluorobenzyl chloroformate derivatization and detection by electron capture negative chemical ionization mass spectrometry. Anal Biochem 233: 58.

Sklerov JH, Magluilo J Jr, Shannon KK, Smith ML. 2000. Liquid chromatography-electrospray ionization mass spectrometry for the detection of lysergide and a major metabolite, 2-oxo-3-hydroxy-LSD, in urine and blood. J Anal Toxicol 24: 543.

Skold K, Svensson M, Kaplan A, Bjorkesten L, Astrom J, et al. 2002. A neuroproteomic approach to targeting neuropeptides in the brain. Proteomics 2: 447.

Smith RD, Olivares JA, Nguyen NT, Udseth HR. 1988. Capillary zone electrophoresis-mass spectrometry using an electrospray ionization interface. Anal Chem 60: 436.

Smyth WF, Brooks P. 2004. A critical evaluation of high performance liquid chromatography ionization-mass spectrometry and capillary electrophoresis-electrospray-mass spectrometry for the detection and determination of small molecules of significance in clinical and forensic science. Electrophoresis 25: 1413.

Song Y, Quan Z, Evans JL, Byrd EA, Liu YM. 2004. Enhancing capillary liquid chromatography/tandem mass spectrometry of biogenic amines by pre-column derivatization with 7-fluoro-4-nitrobenzoxadiazole. Rapid Commun Mass Spectrom 18: 989.

Souverain S, Mottaz M, Cherkaoui S, Veuthey JL. 2003. Rapid analysis of fluoxetine and its metabolite in plasma by LC-MS with column-switching approach. Anal Bioanal Chem 377: 880.

Spell JC, Srinivasan K, Stewart JT, Bartlett MG. 1998. Supercritical fluid extraction and negative ion electrospray liquid chromatography tandem mass spectrometry analysis of phenobarbital, butalbital, pentobarbatal and thiopentoal in human serum. Rapid Commun Mass Spectrom 12: 890.

Stanaszek R, Piekoszewski W. 2004. Simultaneous determination of eight underivatized amphetamines in hair by high-performance liquid chromatography-atmospheric pressure chemical ionization mass spectrometry (HPLC-APCI-MS). J Anal Toxicol 28: 77.

Stuart JN, Zhang X, Jakubowski JA, Romanova EV, Sweedler JV. 2003. Serotonin catabolism depends upon location of release: characterization of sulfated and gamma-glutamylated serotonin metabolites in *Aplysia californica*. J Neurochem 84: 1358.

Stuhler K, Meyer HE. 2004. MALDI: more than peptide mass fingerprints. Curr Opin Mol Ther 6: 239.

Suresh Babu CV, Lee J, Lho DS, Yoo YS. 2004. Analysis of substance P in rat brain by means of immunoaffinity capture and matrix-assisted laser desorption/ionization time-of-flight mass-spectrometry. J Chromatogr B Analyt Technol Biomed Life Sci 807: 307.

Swart KJ, Sutherland FC, van Essen GH, Hundt HK, Hundt AF. 1998. Determination of fluspirilene in human plasma by liquid chromatography-tandem mass spectrometry with electrospray ionization. J Chromatogr A 828: 219.

Tai SS, Welch MJ. 2000. Determination of 11-nor-delta9-tetrahydrocannabinol-9-carboxylic acid in a urine-based standard reference material by isotope-dilution liquid chromatography-mass spectrometry with electrospray ionization. J Anal Toxicol 24: 385.

Taylor PJ, Forrest KK, Landsberg PG, Mitchell C, Pillans PI. 2004. The measurement of nicotine in human plasma by high-performance liquid chromatography-electrospray-tandem mass spectrometry. Ther Drug Monit 26: 563.

Thevis M, Opfermann G, Krug O, Schanzer W. 2004. Electrospray ionization mass spectrometric characteri zation and quantitation of xanthine derivatives using isotopically labeled analogues: an application for equine doping control analysis. Rapid Commun Mass Spectrom 18: 1553.

Tornkvist A, Sjoberg PJ, Markides KE, Bergquist J. 2004. Analysis of catecholamines and related substances using porous graphite carbon as separation media in liquid chromatography tandem mass spectrometry. J Chromatogr B Analyt Technol Biomed Life Sci 801: 323.

Toyo'oka T, Kumaki Y, Kanbori M, Kato M, Nakahara Y. 2003. Determination of hypnotic benzodiazepines (alprazolam, estrazolam, and midazolam) and their metabolites in rat hair and plasma by reversed-phase liquid-chromatography with electrospray ionization mass spectrometry. J Pharm Biomed Anal 30: 1773.

Tuomi T, Johnsson T, Reijula K. 1999. Analysis of nicotine, 3-hydroxycotinine, cotenine, and caffeine in urine of passive smokers by HPLC-tandem mass spectrometry. Clin Chem 45: 2164.

Ukena K, Iwakoshi E, Minakata H, Tsutsui K. 2002. A novel rat hypothalamic RFamide-related peptide identified by immunoaffinity chromatography and mass spectrometry. FEBS Lett 512: 255.

Varesio E, Rudaz S, Krause KH, Veuthey JL. 2002. Nanoscale liquid chromatography and capillary electrophoresis coupled to electrospray mass spectrometry for the detection of amyloid-beta peptide related to Alzheimer's disease. J Chromatogr A 974: 135.

Vishwanathan K, Bartlett MG, Stewart JT. 2000. Determination of antimigraine compounds rizatriptan, zolmitriptan, naratriptan and sumatriptan in human serum by liquid chromatography/electrospray tandem mass spectrometry. Rapid Commun Mass Spectrom 14: 168.

Vorce SP, Sklerov JH. 2004. A general screening and confirmation approach to the analysis of designer tryptamines and phenylethylamines in blood and urine using GC-EI-MS and HPLC-electrospray-MS. J Anal Toxicol 28: 407.

Wada K, Mori T, Kawahara N. 2000. Stereochemistry of nor-diterpenoid alkaloids by liquid chromatography/atmospheric pressure chemical ionization mass spectrometry. J Mass Spectrom 35: 432.

Watson DG, Midgeley JM, Chen RN, Huang W, Bain GM, et al. 1990. Analysis of biogenic amines and their metabolites in biological tissues and fluids by gas-chromatography-negative ion chemical ionization mass spectrometry (GC-NICIMS). J Pharm Biomed Anal 8: 899.

Xu AS, Peng LL, Havel JA, Petersen ME, Fiene JA, et al. 1996. Determination of nicotine and cotinine in human plasma by liquid chromatography-tandem mass spectrometry with atmospheric-pressure chemical ionization interface. J Chromatogr B Biomed Appl 682: 249.

Yamauchi E, Kiyonami R, Kanai M, Taniguchi H. 1998. The C-terminal conserved domain of MARCKS is phosphorylated in vivo by praline-directed protein kinase. Application of ion trap mass spectrometry to the determination of protein phosphorylation sites. J Biol Chem 273: 4367.

Yonamine M, Tawil N, Moreau RL, Silva OA. 2003. Solid-phase micro-extraction-gas chromatography-mass spectrometry and headspace-gas chromatography of tetrahydrocannabinol, amphetamine, methamphetamine, cocaine and ethanol in saliva samples. J Chromatogr B Anal Technol Biomed Life Sci 789: 73.

Zamfir A, Peter-Katalinic J. 2004. Capillary electrophoresis-mass spectrometry for glycoscreening in biomedical research. Electrophoresis 25: 1949.

Zamfir A, Seidler DG, Kresse H, Peter-Katalinic J. 2002. Structural characterization of chondroitin/dermatin sulfate oligosaccharides from bovine aorta by capillary electrophoresis and electrospray ionization quadrupole time-of-flight tandem mass spectrometry. Rapid Commun Mass Spectrom 16: 2015.

Zhan X, Desiderio DM. 2003. Heterogeneity analysis of the human pituitary proteome. Clin Chem 49: 1740.

Zhou HH, Ma RL, Sheng LS, Xiang BR, An DK. 2003. Determination of first-order structure of somatostatin by electrospray ionization mass spectrometry. Yao Xue Xue Bao 38: 617 (article in Chinese).

Zhu Y, Wong PS, Zhou Q, Sotoyama H, Kissinger PT. 2001. Identification and determination of nucleosides in rat brain microdialysates by liquid chromatography/electrospray tandem mass spectrometry. J Pharm Biomed Anal 26: 967.

7 The Analysis of Neuroactive Steroids by Mass Spectrometry

R. H. Purdy · R. L. Fitzgerald · E. T. Everhart · S. H. Mellon · A. A. Alomary · L. H. Parsons

Abstract: The effects of neuroactive steroids on the nervous system have increasingly become a rewarding topic for neuoscientific investigations. The use of radioimmunoassay in the past has sometimes provided invalid results, such as those reported for pregnenolone sulfate. Therefore, a variety of mass spectroscopic methods have been developed, which incorporate unambiguous criteria of identification. However, since concentrations of these neuroactive steroids in the plasma and central nervous system are normally very low, it is necessary to employ ultrasensitive and specific methodology to measure these compounds in select regions of the brain. The most sensitive method of gas chromatography/mass spectrometry reported thus far is electron capture-negative chemical ionization/mass spectrometry performed with selective ion monitoring. This allows several related neuroactive steroids and their deuterated internal standards to be measured in the same chromatogram. Such methods and further refinements are under continual development.

List of Abbreviations: API/MS, atmospheric pressure chemical ionization/mass spectrometry; DHEA, dehydroepiandrosterone; DHEAS, dehydroepiandrosterone sulfate; ESI, electrospray ionization; FAB, fast atom bombardment; FT-ICR, Fourier transform ion cyclotron resonance; GC, gas chromatography; GC/MS, gas chromatography/mass spectroscopy; GC/EC-NCI/MS, gas chromatography/electron capture-negative chemical ionization/mass spectrometry; HFBA, heptafluorobutyrate; HPLC, high performance liquid chromatography; ID, isotope dilution; LC/MS, liquid chromatography/mass spectroscopy; MS, mass spectrometry; MS/MS, tandem mass mass spectrometry; m/z, mass/charge ratio; PREG, pregnenolone; PREGS, pregnenolone sulfate; RIA, radioimmunoassay; SIM, selective ion monitoring; S/N, signal/noise ratio; TH PROG, tetrahydroprogesterone

1 Introduction

Since the earliest applications of mass spectrometry to the structural characterization of organic compounds, steroids have been extensively utilized in the development of mass spectral analysis. Steroids biosynthesized from cholesterol in the central and peripheral nervous systems are now commonly referred to as *neurosteroids*. This term implies that the greatest concentration of these compounds is in nervous tissues, rather than in peripheral blood. Thus, it was reported by Baulieu and Robel and their colleagues in the 1980s (Corpéchot et al., 1981, 1983) by means of radioimmunoassay (RIA) that certain steroids—for example, dehydroepiandrosterone (DHEA), pregnenolone (PREG), and their sulfates and lipoidal esters—exist in higher concentrations in central nervous system tissue than in blood. This early finding was the first presumed evidence for the synthesis of neurosteroids in the brain.

A variety of endogenous and synthetic steroids have now been demonstrated to alter neuronal activity. Paul and Purdy (1992) proposed that this group of neurosteroids be named *neuroactive steroids*, a nomenclature defining their activity. Further complicating this terminology, some authors more generally define neurosteroids both as neuroactive compounds produced de novo in the nervous system and as those steroids derived from circulating precursors, which are metabolized to neuroactive compounds in the nervous system.

Many methods have been used to quantify steroidal compounds. These include RIA, gas chromatography–mass spectrometry (GC/MS), high-performance liquid chromatography (HPLC), and liquid chromatography–mass spectrometry (LC/MS). Although these techniques are successful in the analysis of steroids, it has been difficult to achieve quantitative analysis of small samples of neurosteroids because of their low concentrations in nervous tissues. Highly specific analytical methods are required to analyze small quantities of neurosteroids and their sulfates. Only with extremely sensitive methods of analysis is it possible to discover whether neurosteroids are synthesized in nervous tissues in quantities sufficient to affect neuronal activity, and whether these neurosteroids are distributed uniformly in brain.

The most suitable and effective method to date that is capable of analyzing several neurosteroids simultaneously is gas chromatography/electron capture–negative chemical ionization/mass spectrometry (GC/EC–NCI/MS) with selected ion monitoring (SIM). This technique allows a focus on a few specific ions bearing structural information.

In this chapter, we review various methods that are used for analyzing neuroactive steroids, and discuss the advantages and disadvantages of each technique. We focus on relatively recent methods of mass spectrometry employed for the analysis of neuroactive steroids and their sulfates. In addition, we examine some promising methods that enable the simplification of sample pretreatment procedures for the measurement of low levels of neuroactive steroids in small samples. Lowering detection limits will facilitate a better understanding of the physiological function of neuroactive steroids, and a clearer comprehension of the mechanisms by which these steroids regulate brain function. Furthermore, increased efficiency of low-level analysis will increase the throughput, which is important for research purposes as well as clinical analysis. A cogent comparison of various procedures for the analysis of steroids and their derivatives by MS is presented in the recent monograph of Siuzdak (2003), which is also recommended for investigators interested in the practical aspects of biomolecular analysis by MS.

2 Requirements for Identification of Neuroactive Steroids

During the initial identification of dehydroepiandrosterone sulfate (DHEAS) (Corpéchot et al., 1981) and pregnenolone sulfate (PREGS) (Corpéchot et al., 1983) in the rat brain, a conjugated steroid fraction from brain extracts was prepared by chromatography on a column of Sephadex LH-20, and termed the "sulfate fraction." This was free from unconjugated steroids, steroidal esters of fatty acids (lipoidal steroids), and steroidal glucosiduronates. This "sulfate fraction" was then hydrolyzed by solvolysis in ethyl acetate for 12–16 h at 37°C and the hydrolyzed products were purified on a column of Lipidex 5000. The purified steroids were converted to their trimethylsilyl ethers and characterized by GC/MS as DHEA and PREG. On the basis that the levels of DHEA and PREG separately measured from the hydrolyzed "sulfate fraction" from brain extracts by RIA were markedly elevated compared to corresponding levels in blood, and were found in extracts of brain tissue from rats previously adrenalectomized and orchiectomized, DHEAS and PREGS were described as *neurosteroids.* A considerable body of electrophysiological, pharmacological, and physiological work has been subsequently carried out on these two presumed neurosteroid sulfates (e.g., Vallée et al., 2003). Meanwhile, Prasad et al. (1994) demonstrated that when extracts of rat brain were heated and treated with triethylamine or with various reducing agents such as ferrous sulfate, larger amounts of DHEA and PREG could be measured by GC/MS as compared to simple extraction without such treatments. They suggested that there might be steroidal hydroperoxides or peroxides in brain that have not yet been characterized. Two laboratories have independently cast grave doubts on the existence of significant amounts of DHEAS and PREGS in the adult rat brain following the reports of Shimada et al. (1998, 2002). Using a nonexchangeable internal standard of [3β,11,11-$^{2}H_3$]-allopregnanolone sulfate and a tracer amount of [1,2,6,7-$^{3}H_4$]DHEA sulfate, Liu et al. (2003) were unable to find detectable amounts of DHEAS, PREGS, or other pregnanolone sulfates in nonhydrolyzed extracts of the adult Sprague–Dawley male and female rats using liquid chromatography-micro-electrospray mass spectrometry (LC/micro-ESI-MS/MS; see ➲ section 9). Shimada's group (Mitamura et al., 1999) also found only low levels of PREGS (0.53 ± 0.28 ng/g) in nonhydrolyzed extracts of the adult Wistar rat using a unique derivatization procedure followed by LC/micro-ESI-MS/MS. This compares to the value of about 20-ng/g brain of PREGS originally reported using solvolysis and measurement by RIA (Corpéchot et al., 1983). Subsequently, the laboratories who originally reported the isolation of DHEAS and PREGS from the rat brain reinvestigated the matter, using a different method for partial purification of these sulfates from the rodent brain prior to MS. They found <1 pmol/g DHEAS or PREGS in rat or mouse brain (Liere et al., 2004). At present, there are no reports of other PREG-containing compounds that could account for this 20-ng/g brain level. Thus, the nature of the majority of PREG containing compound or compounds in the "sulfate fraction" from extracts of the rodent brain remains a mystery. It would require low microgram amounts of such a compound to be identified (after purification) using the most sensitive microprocedure of high-resolution proton magnetic resonance instrumentation.

In our laboratories, we have abandoned using RIA for the measurement of neuroactive steroids except where the RIA procedure has been strictly validated by prior MS identification for each experimental protocol. Immunoassays that measure low concentrations of steroids in complex biological matrices are highly variable (Herold and Fitzgerald, 2003). The calculation is based on two measurements, before and after addition of the antisera, compared to a standard curve. The immunoassay provides no criteria for the characterization of the ligand as compared to any MS procedure.

3 Derivatization of Neuroactive Steroids

Neuroactive steroids are comparatively polar compounds that, in most cases, are derivatized by a suitable reagent before being analyzed by GC/MS. It is essential to add functional groups that enhance ionization efficiency, decrease polarity, and increase volatility of the steroid, thus making it more easily detected by GC/MS. Suitable derivatization of neuroactive steroids also results in enhanced sensitivity and specificity of mass spectra.

Most commonly, silyl ether derivatives are employed for GC/MS analyses of steroids. Silyl derivatives can be prepared from alcoholic, phenolic, and carboxyl groups. For example, Diallo et al. (2004) have reported a GC/MS procedure for measuring DHEA and its C7-oxygenated metabolites in human plasma as their mono- or di-silyl ethers. Enolizable ketones also form silyl derivatives, but yields are typically not as good as those for alcoholic or phenolic groups. For hydroxyl groups, silylation is usually employed. For carbonyl and carboxyl groups, other derivatives are frequently used. Oximes and methoximes are most commonly used for carbonyl groups.

After formation of an oxime derivative with hydroxylamine hydrochloride, it can be further derivatized with a silyl reagent. Dehennin and Scholler (1973) demonstrated that the C3 carbonyl group with a Δ^4-conjugated double bond, as in progesterone or testosterone, forms almost exclusively the 3,5-dienol ether with heptafluorobutyric anhydride when the reaction is performed in acetone (e.g., Lière et al., 2000). The use of polyfluorinated derivatives and NCI GC/MS produces greatly enhanced sensitivity. Four common polyfluorinated derivatives of PREG are shown in ➲ *Figure 7-1*. Because PREG has a single hydroxyl group (3β), a monoheptafluorobutyrate is formed (➲ *Figure 7-1a*) using heptafluorobutyric anhydride or heptafluorobutyrylimidazole. For androstane-3,17-diols or pregnane-3,20-diols, diheptaflourobutyrates are formed. The ketone group of PREG can also be reacted to form the pentafluorinated oxime derivative shown in ➲ *Figure 7-1b* (Vallée et al., 2000) in one step, or the similar, but somewhat larger, pentafluorinated derivative shown in ➲ *Figure 7-1c* (Kim et al., 2000), in two steps. Using the procedure developed by Uzunova et al. (1998)—the separation of heptafluorobutyrate (HFBA) esters of allopregnanolone and its isomers—only three peaks were obtained from the four isomers. Using a similar GC/MS analysis of allopregnanolone and its isomers in the SIM mode, Hill et al. (2000) found that the sensitivities of the assay were rather variable; 0.16, 0.034, 0.61, and 0.673 pg for the 3β,5β, 3α,5α, 3α,5β, and 3β,5α isomers, respectively. Strohle et al. (2002) were also able to obtain baseline separation of these four isomers extracted from human plasma. The separation of the HFBA derivatives of nine endogenous neuroactive steroids found in human plasma has been accomplished, as shown in ➲ *Figure 7-2* (Everhart et al., unpublished). After extraction with dichloromethane/isooctane, the analytes were converted to HFBA derivatives, and extracted into a pentane/octane mixture, followed by evaporation of the pentane. For increased sensitivity and reproducibility, all glassware employed in the extraction and derivatization procedures were silanized before use.

It is important to note that, with certain oximes, *syn-* and *anti*-isomers are formed because of the derivatization, relative to the single pair of electrons on the oxime. This can bring about multiple chromatographic peaks, as Fitzgerald and Herold (1996) observed with testosterone. When oximes are formed, the alcoholic groups still need to be derivatized with a silylating reagent (➲ *Figure 7-1b* and ➲ *7-1c*) in order to be analyzed by GC/MS.

Numerous workers have demonstrated the applicability of electrospray ionization mass spectrometry (ESI/MS) for the detection and analysis of biomolecules with highly electronegative groups (reviewed by Wood et al., 2003, and for neutral steroids by Higashi and Shimada, 2004). The sensitivity of detection of neurosteroids can also be enhanced by derivatization when they are analyzed by nano-electrospray/mass spectrometry procedures. Neurosteroid sulfates can be easily prepared in a single-step reaction in pyridine with the *N,N*-dimethylformamide complex of sulfur trioxide (Chatman et al., 1999). Another elegant

Figure 7-1
Structures of four derivatives of pregnenolone employed in GC/MS: in (a) the 3-heptafluorobutyrate, in (b) the 3-trimethylsilyl ether 20-pentafluorobenzyloxime, in (c) the 3-trimethylsilyl ether 20-pentafluorobenzylcarboxymethoxime, and in (d) the 20-oxime

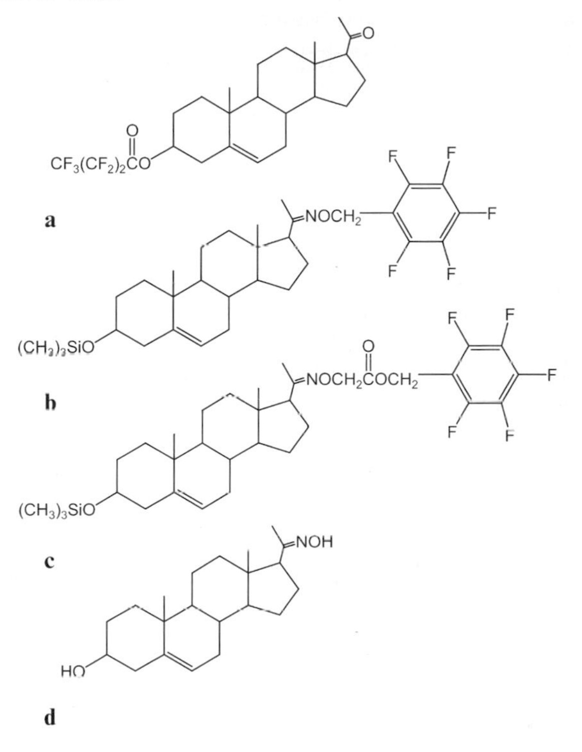

example was presented by Liu et al. (2000) when they formed *mono-* or *bis*-oxime derivatives from *mono-* or *di*-ketosteroids with hydroxylamine hydrochloride, as illustrated by PREG oxime shown in *Figure 7-1d.* This procedure provided derivatives with detection limits by nano-ESI/MS that were approximately 20 times lower than detection limits of the underivatized neurosteroid. In this study, the oximes were found to give profuse positive ions by ES when sprayed from a suitable solvent. Both deuterated neurosteroid sulfates and neurosteroid oximes can easily be prepared in stable crystalline form for use as internal standards.

Derivatization can also lead to the formation of characteristic fragments, and to the shifting of main fragment ions to higher masses with lower matrix background. In choosing which derivatives to use, one must consider several aspects of MS instrumentation; for example, the ionization mode, the resolution of the mass spectrometer, and the potential increase in selectivity and/or sensitivity, which can be obtained by using tandem mass spectrometry (MS/MS). In some cases, when structurally related compounds are not separated in the underivatized form, they may be resolved as derivatives. Shackleton et al. (1997) have utilized the formation of water-soluble hydrazones of a variety of testosterone esters, whose general structure is shown in *Figure 7-3,* for the analysis by electrospray MS of these anabolic lipoidal steroids in human plasma. This reaction, employing the Girard reagent trimethylamino-acetohydrazide hydrochloride, makes the classical use of Girard's Reagent T to form water-soluble derivatives of oxosteroids for isolation from complex biological mixtures.

4 Columns for Separation of Neuroactive Steroid Derivatives

In selecting columns, the general rule is that columns with a polar stationary phase are used to separate polar compounds, whereas columns with nonpolar stationary phases are used to separate nonpolar

Figure 7-2
Selected ion monitoring GC/MS trace of the HFBA derivatives of nine neuroactive steroids found in human plasma. The separations were accomplished on a Restek Rtx-200 MS column. Relative absorbance was measured at negative CI for m/z values of 474 (pregnanolones), 492 (pregnenolone), 664 (androstanediols), and 706 (pregnanediolones). (Everhart et al., unpublished)

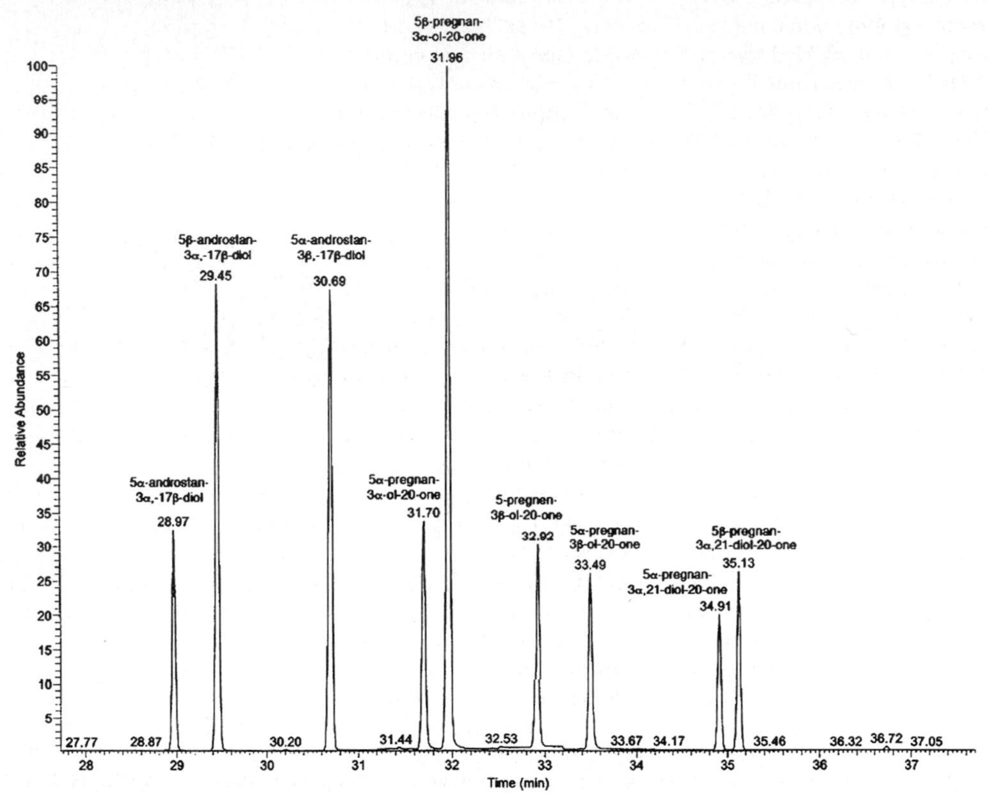

Figure 7-3
General structure of a fatty acid ester (R) of testosterone Girard hydrazone

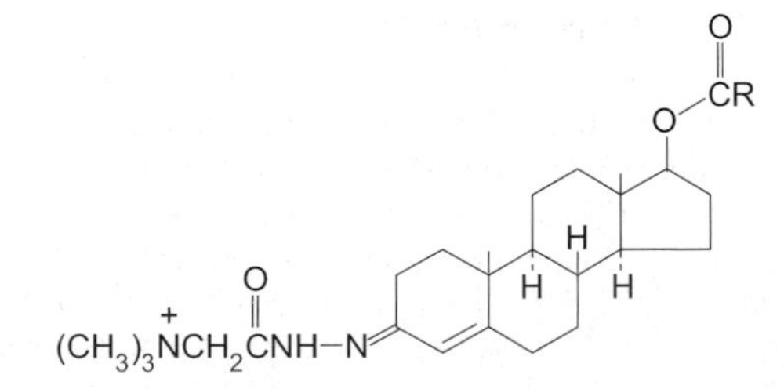

compounds. Separation of nonpolar substances on nonpolar columns is a nonselective process, with compounds usually eluted in the order of increasing boiling point. A variety of capillary columns are used in the analysis of steroids by GC/MS. These include DB-1 (dimethyl polysiloxane), DB-5 (5% phenylmethyl polysiloxane), and DB-17 (50% phenylmethyl polysiloxane). A capillary column with a split/splitless injector is usually used because of its advantages over packed columns in terms of efficiency, reduction in analysis time, and resolving power.

As previously mentioned, the presence of polar functional groups on some neuroactive steroids requires that derivatization be performed to decrease polarity and increase volatility for separation by GC. This additional sample handling results in artifact formation in some cases, representing an undesirable aspect of steroid analysis by GC. An alternative approach is to employ HPLC for neurosteroid separation. Several different types of separations can be performed based on analyte adsorption (normal- and reversed-phase chromatography), ionic interactions between the analyte and the stationary phase (ion-exchange chromatography), and physical size of the analyte (size-exclusion chromatography). The separation of steroids by HPLC is most often performed using adsorption-based techniques. Although other separation approaches have been described (Pearson Murphy and Allison, 2000; Lière et al., 2000), reversed-phase HPLC (RP-HPLC) is most commonly employed for neuroactive steroid separation (Wei et al., 1990; Ma and Kim, 1997; Kuronen et al., 1998; Rule and Henion, 1999). In this technique, the stationary phase is nonpolar (typically a silica bead backbone on which carbon chains are bonded). Varying the organic strength of an aqueous mobile phase controls the selectivity of the separation. Many advances have recently been made in the types of stationary phases available for RP-HPLC, and accordingly, there is a broad range of column types that are suitable for steroid separations. In general, highly selective separations can be achieved using small-pore stationary phases (e.g., 100 Å) that offer exceptionally high amounts of surface area on which the molecules can interact. Stationary phases made with Lichrosorb, an irregular porous packing material manufactured in Germany by E. Merck, have been successfully employed in the separation of neuroactive steroids (Lière et al., 2000; Weill-Engerer et al., 2003). Secondary interactions between steroids and most RP stationary phases (e.g., silanol interactions) are typically not problematic, and thus it is not necessary to use highly deactivated or shielded stationary phases.

The recent development of mixed-mode stationary phases may provide new approaches for the separation of neuroactive steroids from biological matrices. These stationary phases typically embed a charged functional group within the nonpolar carbon chains typically employed in RP separations, and as such separations based on both RP and ion-exchange characteristics can be designed. Examples of this column type are the Primesep line of columns from SIELC Technologies. This separation approach may be quite useful for the separation of steroid sulfates from complex mixtures.

In general, the use of microbore or capillary columns will improve the sensitivity of neuroactive steroid analyses by LC-MS. The use of much slower mobile phase flow rates with these columns (nanoliters to microliters per minute) allows for more efficient ionization of the column eluent by either electrospray ionization or chemical ionization. Using gradient analyses, relatively large sample injections can be made with the steroid content concentrated at the head of the column with the mobile phase containing a modest organic concentration. The steroids are subsequently recovered by gradually increasing the organic solvent (gradient elution).

5 Isotopic Dilution in Quantitative GC/MS

It is critical when performing quantitative GC/MS procedures that appropriate internal standards are employed to account for variations in extraction efficiency, derivatization, injection volume, and matrix effects. For isotope dilution (ID) GC/MS analyses, it is crucial to select an appropriate internal standard. Ideally, the internal standard should have the same physical and chemical properties as the analyte of interest, but will be separated by mass. The best internal standards are nonradioactive stable isotopic analogs of the compounds of interest, differing by at least 3, and preferably by 4 or 5, atomic mass units. The only property that distinguishes the analyte from the internal standard in ID is a very small difference in mass, which is readily discerned by the mass spectrometer. Isotopic dilution procedures are among the most accurate and precise quantitative methods available to analytical chemists. It cannot be emphasized too strongly that internal standards of the same basic structure compensate for matrix effects in MS. Therefore, in the ID method, there is an absolute reference (i.e., the response factors of the analyte and the internal standard are considered to be identical; Pickup and McPherson, 1976).

In the ID method, a known exact weight of the internal standard (typically a deuterium-labeled analog of the compound to be analyzed) is added to the biological sample to be analyzed. The internal standard is then subjected to exactly the same conditions as the analyte of interest, thus compensating for any loss of analyte throughout the entire analytical process. The peak of the labeled standard is shifted to a different position in the spectrum according to the number and the nature of the atoms added in the labeling procedure. When subjected to MS analysis, the amount of the analyte is calculated relative to signal intensity of the internal standard. The ratio of the two signal intensities is used to calculate their relative proportion. For this procedure, it is essential that the internal standard have the isotopes incorporated in stable positions, and that the isotope is contained in the fragments monitored.

Numerous deuterated neuroactive steroids are now available from commercial sources (Cambridge Isotope Laboratories Inc, Andover, MA; C/D/N/Isotopes Inc., Pointe-Claire, Quebec, Canada; Steraloids Inc., Newport, RI). The most useful of these deuterated steroids are those containing two or more nonexchangeable deuteriums, such as the [3α, 7,7-$^2H_3$]-analogs of DHEA sulfate (*Figure 7-4c*) and PREG sulfate (*Figure 7-4d*). With steroids containing exchangeable deuterium, for example [17,21,21,21-$^2H_4$]-analogs of PREG or TH PROG, it is recommended that stock solutions be stored in mono-deuteroethanol or 95% mono-deuterated ethanol and 5% ^2H_2O, rather than ethanol or methanol, since the steroid can lose deuterium over time through exchange with the solvent or its impurities.

Figure 7-4
Chemical structures of (a) DHEA sulfate, (b) PREG sulfate, (c) [3α,7,7,$^2H_3$]-DHEA sulfate and (d) [3α,7,7,$^2H_3$]-PREG sulfate

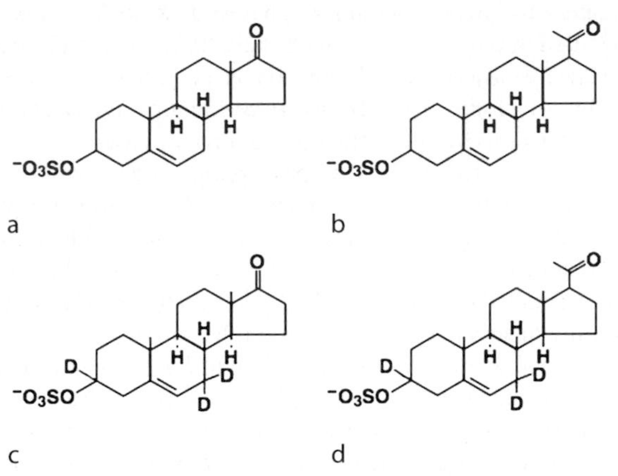

Deuterated steroids are also valuable for use in metabolic studies. For example, the isolation of di-deuterotestosterone in urine after oral administration in men of [16,16-$^2H_2$]DHEA demonstrated that DHEA had potential as an anabolic steroid (Dehennin et al., 1998). Similarly, after treatment of rats with [1-$^2H_2$]ethanol, mono-deuterotestosterone was isolated from plasma (Alomary et al., 2003).

6 Modes of Data Acquisition in MS

In all ionization techniques and chromatographic methods, there are three possible modes of acquiring data: scanning, SIM, and selected-reaction monitoring. In the scan mode, a range of mass to charge (m/z) (e.g., from 50 to 700 m/z) is analyzed. Generally, an initial analysis of compounds is made in the scan mode to identify all the ions of interest. In the SIM mode, monitoring is only performed on the ions selected to be characteristic of the analyte of interest. Thus, based on the full-scan spectra, ions are selected for SIM analysis. SIM can be especially useful if the aim of the analysis is to detect target compounds with known spectral characteristics and with maximum sensitivity. Good ions for SIM analysis typically are at high m/z and contain a unique fragment of the molecule of interest. For example, it is not good practice to monitor the

73-m/z ion for silylated compounds, as it is characteristic of the derivatizing reagent, and therefore not specific for any particular analyte. Because of the increase in dwell time associated with the ions of interest, SIM provides improved sensitivity compared with the scan mode. SIM offers 100 to 1000 times greater sensitivity than that obtained in the full-scan mode. However, if there is no need for high sensitivity, SIM is not always preferred, because in SIM the full range of ions is not recorded. If information different from that originally sought is required later, it will not be available, and the analysis will have to be repeated. Examples of SIM chromatograms are shown in *Figures 7-2* and *7-5*.

Figure 7-5

The GC/MS analysis by selected ion monitoring of neuroactive steroids extracted from the cortex of the male rat brain. (a) The derivatives of DHEA and testosterone were quantified using the m/z 535 ion and the trideutero internal standards at m/z 538. (b) The pregnenolone derivative was quantified by using the m/z 563 ion and its trideutero internal standard at m/z 566. (c) The derivatives of allopregnanolone and epiallopregnanolone were monitored at m/z 407 ion and their tetradeutero-internal standards at m/z 411. Taken from Vallée et al. (2000) with permission

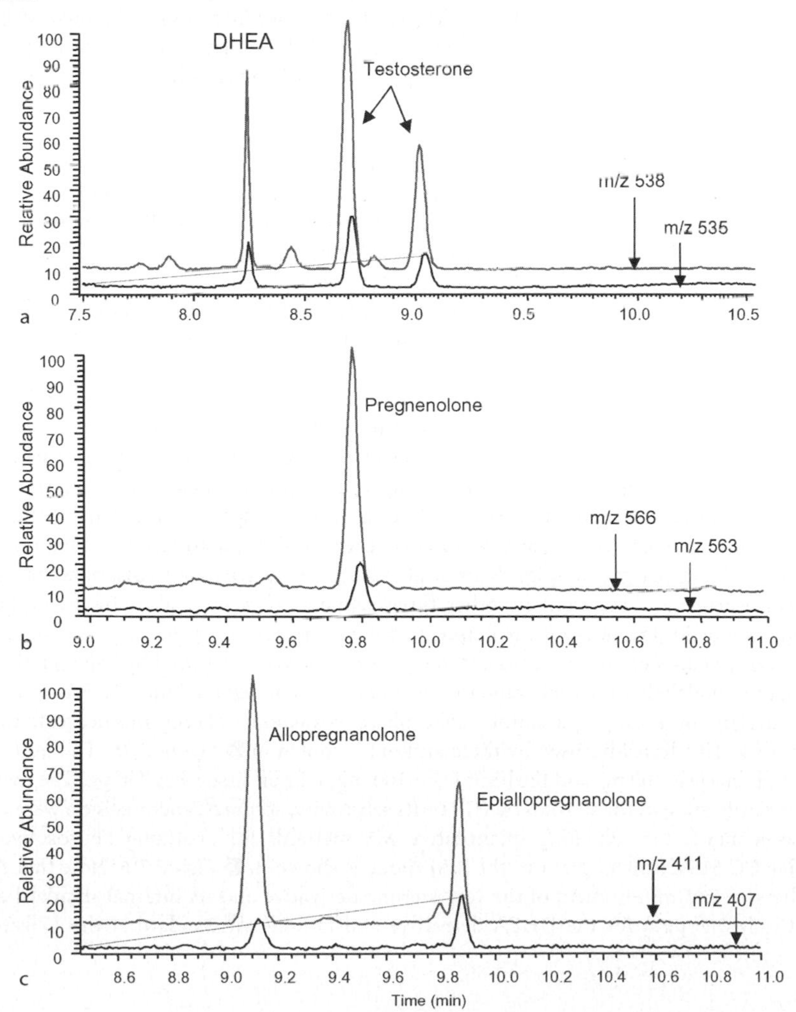

7 Analysis of Unconjugated Neuroactive Steroids by GC/MS

Many groups have chosen to include an HPLC purification step before GC/MS analysis (Cheney et al., 1995; Uzunov et al., 1998; Shimada et al., 1998; Lière et al., 2000). Alternatively, multiple clean-up steps have been employed. These pre- and post-steps were labor-intensive, and had the further drawback of failing to use the appropriate internal standard for each steroid under analysis. Presumably, this failure was because of the nonavailability of all the required deuterated internal standards. It is important to note that deuterated internal standards prepared by exchange labeling procedures cannot be used in procedures requiring prior purification by HPLC, because such deuterated internal standards are partially separated from the nondeuterated steroids of interest during most HPLC procedures. This is caused by the introduction of the more acidic deuterium atoms adjacent to enolizable carbonyl groups. It is often difficult to broaden the retention window to collect a wider fraction without introducing additional impurities.

Our review of the literature indicates that the manner in which the limit of detection of steroid is expressed can vary widely. The amount of steroid that can be detected with any given method depends on several variables. Some authors define their limit of detection as the smallest amount of steroid that can be detected using standards free of a biological matrix. Other authors express the limit of detection as the smallest amount of steroid that can reliably be detected from a specific amount of a biological matrix (e.g., 250 pg of allopregnanolone from 100 mg of brain tissue). We believe that the best way to express sensitivity is the latter method. Not only is good sensitivity required, but also it is crucial that the strength of the assay be demonstrated with quality control samples. These are typically run at three concentrations spanning the range of interest for each analytical batch.

Cheney et al. (1995) analyzed steroids by coupling an HPLC purification step with GC/MS. The steroids were initially characterized by their HPLC retention times compared with the retention times of tritium-labeled recovery standards. Next, the neurosteroids were characterized by their GC retention times. Finally, they were identified by their unique fragmentation spectra following derivatization with heptafluorobutyric anhydride or methoxyamine hydrochloride. For structural identification, the mass spectra were compared to appropriate reference standards. This approach is highly specific, and its sensitivity is increased by the use of SIM. The detection limit for measuring allopregnanolone achieved in the 1995 study was 0.63 pmol (0.2 ng) starting from ~100–300 mg of brain tissue.

Kim et al. (2000) used a GC/MS method to identify and quantify neurosteroids in rat and human plasma and cerebrospinal fluid (CSF). Their procedure was notable for its relatively simple sample-preparation scheme; only a C18 solid-phase extraction column was used for initial purification of neurosteroids. In most cases, deuterium-labeled internal standards were used for the analyte of interest. Derivatization was accomplished using the method of Hubbard et al. (1994), whose product with PREG is shown in ➋ *Figure 7-1c.* This method allowed allopregnanolone, pregnanolone, testosterone, and androsterone to be quantified in CSF and plasma samples. Kim et al. (2000) easily separated epiallopregnanolone (which they termed iso-pregnanolone) from its other three isomers. They were also able to partially separate pregnanolone (about 80% resolved) from allopregnanolone and epipregnanolone. A similar GC/MS procedure for analyzing neurosteroids in rat plasma and brain tissue was developed by Vallée et al. (2000). They employed deuterium-labeled internal standards to quantify allopregnanolone, DHEA, testosterone, and epiallopregnanolone by means of a simple, solid-phase extraction isolating the neurosteroids from the biological matrix. The derivative used by these authors is shown in ➋ *Figure 7-1b.* The limit of detection from 1 ml of plasma was 100 pg, and the limit from 100 mg of brain tissue was 250 pg. The method did not require prior purification of the steroids by HPLC. Its sensitivity, accuracy, and precision were validated and can serve as a model for validating quantitative MS methods for analyzing neuroactive steroids. A representative GC/MS chromatogram in the SIM mode is shown in ➋ *Figure 7-6.* Note that there are two peaks for the *syn*- and *anti*-isomers of the testosterone derivative and its internal standard at the 3-keto groups, but only one peak for the DHEA derivative and its internal standard at the 17-keto groups, as expected.

Figure 7-6
Analysis of a 50-pg sample of neuroactive steroids by reverse phase HPLC coupled with ESI/MS using the negative ion mode of SIM. (Parsons et al., unpublished)

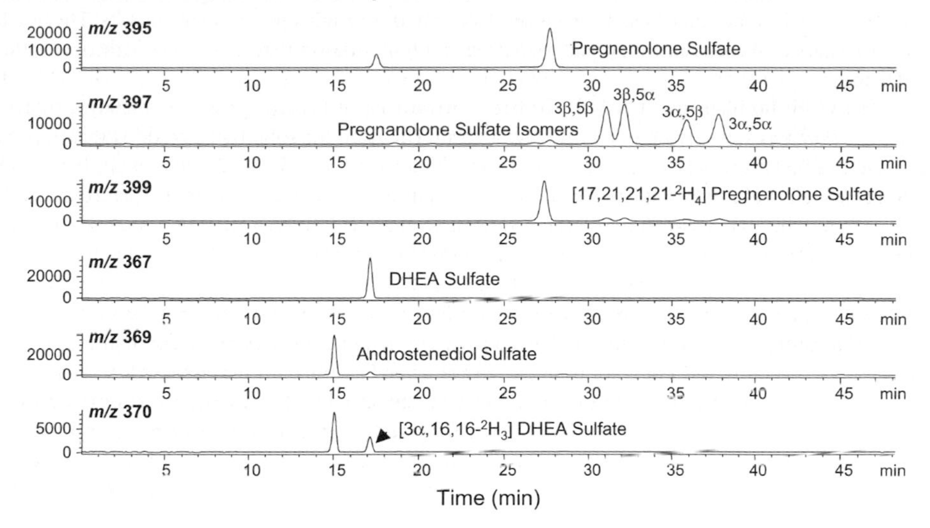

8 Analysis of Neuroactive Steroids by MS/MS

Mass spectrometers are commonly combined with separation devices such as GC and LC for analysis of neurosteroids. MS/MS is a similar technique using a second mass spectrometer for the first-stage separation device. MS/MS instrumentation is also connected with LC or GC to form the multihyphenated techniques of LC/MS/MS and GC/MS/MS. These multidimensional analytical systems can be useful in the identification of compounds in complex mixtures and in the determination of structures of unknown substances.

The MS/MS technique makes use of the first m/z separation to select a single (precursor) mass that is characteristic of a given analyte in a mixture. The mass-selected ions are then bombarded with a neutral gas in a process called collisional activation or collision-induced dissociation. This causes the ion to fall apart, producing fragment (product) ions. The product ions are then separated by the second mass spectrometer according to m/z. The resulting MS/MS spectrum consists entirely of product ions from the selected precursor. Chemical background and other mixture components are decreased or absent. The chemical noise may decrease more than that of the signal ions, so the signal-to-noise ratio increases, resulting in greater sensitivity.

9 Analysis of Neuroactive Steroid Sulfates by MS

Although GC/MS is a suitable method for analyzing steroids, it is difficult to analyze polar, nonvolatile steroids, such as sulfated steroids, by GC/MS directly without derivatization. GC/MS is usually used for the indirect analysis of sulfated steroids, by hydrolyzing conjugates to the parent alcohols, and then using a suitable reagent to derivatize those alcohols before analysis. Meng et al. (1996) described a method of solvolysis of steroid sulfates in tetrahydrofuran with dilute trifluoroacetic acid as a catalyst. They used a labeled internal standard to monitor the efficiency of solvolysis. However, solvents such as tetrahydrofuran or ether often contain peroxides, whose presence reduces the effective recovery of the parent neurosteroids during solvolysis.

Direct analysis of sulfated steroids by MS has become common, especially since the advent of soft ionization techniques (Bowers and Sanaullah, 1996; Murry et al., 1996). The gentle process of proton transfer leading to high intensity of intact molecular ions that is characteristic of such soft ionization techniques as ESI and chemical ionization can produce pseudo-molecular ions. Electron impact ionization (EI) is often used to generate ions for mass analysis of molecules that can be vaporized without decomposition. In EI, ionization occurs when electrons are accelerated through a potential of 70 volts. This is a highly energetic, or "hard," process, and may lead to extensive fragmentation that leaves very little or no trace of a molecular ion.

Once fast atom bombardment (FAB) had been introduced, it became possible to analyze intact polar biomolecules (such as sulfated steroids) directly, without the need for solvolysis and derivatization. Several studies have used FAB to analyze steroid sulfates (Shackleton and Straub, 1982; Liehr et al., 1982; Dumasia et al., 1983; Gaskell et al., 1983; Kingston et al., 1985). In addition, secondary ion mass spectrometry (SIMS) has been used for the analysis of steroid sulfates (Shimada et al., 1998). FAB and SIMS analyses both require extensive purification of samples, which can result in limited sensitivity. As FAB is not a particularly "soft" ionization technique, a softer method of ionization (e.g., ESI) is expected to provide a better possibility of forming intact molecular ions of sulfated neurosteroids with less fragmentation. Furthermore, ESI has the advantage of a minimal amount of matrix (solvent)-associated chemical noise, which makes it possible to analyze low levels of steroid sulfates. For example, Bowers and Sanaullah (1996) were able to analyze steroid glucosiduronates and sulfates from a subnanogram level directly by HPLC ESI/MS. Reverse-phase HPLC was capable of resolving all the isomers studied. In this study, detection limits of 3–25 pg injected on a packed capillary column were achieved. However, no biological analyses were reported. Later, Shimada et al., (1998) analyzed PREGS in rat brain by ESI, with SIM of the $[M\text{-}H]^-$ ion at m/z 395. The detection limit achieved by this group using this technique was 1 ng/injection. Griffiths et al. (1999) analyzed steroid sulfates in the rat brain by nano-ES/MS. They were able to analyze deuterated neurosteroid sulfates by nano-ES-MS/MS at a level of 50 pg/mg of brain tissue. This study demonstrated that complete structural information could be obtained from 1 ng (3 pmol) of steroid sulfate, whereas fragment ions of the sulfate esters can be obtained from only 3 pg (10 fmol) of sample.

Chatman et al. (1999) also used nano-ESI for analyzing neurosteroids. In nano-ESI, samples are sprayed from pulled capillaries that are usually coated with metal. This allows analysis of very small (e.g., ~1 μl) sample volumes, on the order of 10–40 nl/min. The result is decreased sample consumption, more stable spray from a wide pH range and from high buffer concentrations, and less contamination of the instrument. In this study, Chatman et al. (1999) developed a method of extracting unconjugated steroids and steroid sulfates from biofluids. Their new method of extraction allowed them to isolate the unconjugated steroids and their sulfate esters separately, in a two-step procedure using diethyl ether/hexane (90:10, v/v) initially, and then chloroform/2-butanol (50:50, v/v) in the second step, to extract steroid sulfates. Precursor ion scanning (one of the scan modes used to carry out MS/MS) performed with a triple-quadrupole mass spectrometer was used to quantify the steroids. (Precursor ion scanning also reduces interference from chemical noise, and thus improves the signal:noise ratio.) Chatman et al. (1999) also used deuterated steroids as internal standards in the quantification. The limit of detection for steroid sulfates from the biological matrix was 200 amol/μl (~80 fg/μl), with only about 1 μl of sample being injected. Endogenous levels of the unconjugated and sulfated steroids were detected and quantified from physiological samples. The positive mode of nano-ESI was employed for analysis of the unconjugated steroids, whereas negative nano-ESI was used to analyze the steroid sulfates. Anionic species of steroid sulfates were detected in negative nano-ESI at concentrations lower than the protonated $[M + H]^+$ or $[M - H_2O + H]^+$ species of the unconjugated steroids in the positive mode.

Liu et al. (2001) developed a method of extracting unconjugated steroids and steroid sulfates that was based on capillary liquid chromatography (LC)/micro-ESI mass spectrometry. In this study, rat brain was extracted with ethanol and passed through a lipophilic cation and anion exchanger. The sulfated neurosteroids were eluted from the anion exchanger. They converted oxosteroids in the unconjugated fraction to their oximes, which were then sorbed on a bed of the cation exchanger and eluted. The fractions containing the oximes and the sulfates, respectively, were concentrated to 100 μl by micro-solid-phase extraction. Capillary LC/micro-ES/MS was used to analyze the neurosteroid sulfate fraction. Deuterated neurosteroids were used as internal standards. For example, $[^2H_4]$ PREGS was used in the analysis of PREGS, which is the most abundant neurosteroid sulfate in the rat brain. A detection limit of 1 pg on-column was achieved (S/N 3:1) for PREGS. More recently, these authors have reported that they were unable to detect either PREGS or DHEAS in purified extracts of the adult male and female Sprague–Dawley rat brain (Liu et al., 2003; see section 2). In our laboratory, we have been able to separate DHEAS, androstenediol sulfate, PREGS, and all four of the pregnanolone sulfate isomers using an RP-HPLC coupled with ESI-MS. A chromatogram of the standards is shown in *Figure 7-5.* In our laboratory we have developed an assay for DHEAS, androstenediol sulfate, PREGS, and all four of the pregnanolone sulfate isomers using a reverse phase capillary HPLC coupled with μESI-MS. This method provides full chromatographic resolution of each analyte as shown in *Figure 7-6,* and through the use of selected ion monitoring (SIM), the limit of quantitation of each analyte is approximately 5 pg on-column. Using this method and a solid phase extraction procedure, we have found very low levels of DHEAS (0.8 ± 0.2 pg/mg; 110 ± 35 total pg/sample) and PREGS (0.7 ± 0.1 pg/mg; 100 ± 20 total pg/sample) in brain tissue from neonatal rats (tissue harvested on embryonic day 20). Such embryonic brain tissue was reported by Caldeira et al. (2004) to contain about 5 ng/g DHEA and 45 ng/g PREGS by RIA using commercially-available antisera after partial purification of the "sulfate fraction" by the method of Liere et al. (2000).

Kobayashi et al. (1993) used APCI/MS to study steroid fragmentation. Their study showed that APCI/MS could be applied to determine the molecular weight of polar nonvolatile thermolabile steroids without derivatization. The steroids used in this study were divided into two groups according to their mass spectral profiles, Group A having a carbonyl group at C3 together with a double bond at C4-5, and Group B having a hydroxyl group at C3. In Group A, the predominant peak observed corresponded to the protonated molecular ion $[M+H]^+$. The fragment ion corresponding to the elimination of CH_2OH, $COCH_3$, and/or $COCH_2OH$ from the steroid skeleton appeared as a base peak in some steroids of Group A. In Group B, because of the loss of water, predominantly $[M+H\text{-}H_2O)]^+$ and/or $[M+H\text{-}2H_2O)]^+$ ions were observed. The protonated molecular ions were also prevalent in Group B.

Watson et al. (1985) performed analysis of conjugated steroids by coupling LC to MS by means of a thermospray (Th)-LC/MS interface. In their studies, negative ion Th-mass spectra were recorded for several steroid conjugates. This interface vaporizes the sample before it passes through a small orifice at the tip of a heated stainless steel tube in the ion source. Ionization occurs at the same time, usually promoted by a volatile buffer, such as ammonium acetate.

One of the most sensitive techniques for analyzing biological samples involves the coupling of ESI with Fourier transform ion cyclotron resonance (FT-ICR) MS. ESI is considered to be the "softest" ionization technique available (Whitehouse et al., 1985). It is capable of producing intact molecular ions, which leads to enhancement of sensitivity. ICR is a nondestructive detector that allows detection of multiple ions leading to improved sensitivity (Marshall et al., 1985). There have been several studies that have focused on the ability of FT-ICR MS to improve detection limits. For example, Valaskovic et al. (1996) used ESI coupled with FT-ICR MS to acquire mass spectra from attomole quantities of analyte. Later studies obtained mass spectra from zeptomole ranges by using ESI in tandem with FT-ICR MS (Belov et al., 1999). For this reason, the pairing of ESI with FT-ICR is expected to be the most sensitive and rapid method for analyzing neurosteroid sulfates in the future.

10 Conclusions

Concentrations of neuroactive steroids that are present in brain and plasma are normally very low. Therefore, several prepurification steps are generally required before analyzing these samples. It is necessary to improve detection limits and to introduce new methods of analyzing these samples directly from solution, without these labor-intensive and time-consuming prepurification procedures. Now, most neurosteroid analyses are performed by RIA or by GC/MS. The most sensitive GC/MS reported so far is GC/EC-NCI/MS, wherein MS in performed in the SIM mode. Neurosteroid sulfates can be directly analyzed by MS, without derivatization, by using soft ionization methods such as FAB and ESI. These methods are currently undergoing further refinement and development.

Acknowledgments

This is publication number 16768-NP from The Scripps Research Institute. The research was supported by National Institutes of Health Grants AA06420 and AA07456 from the National Institute on Alcohol Abuse and Alcoholism. We are grateful for the assistance of Elizabeth Gordon and Michael Arends in the preparation of this manuscript.

References

Alomary AA, Vallée M, O'Dell LE, Koob GF, Purdy RH, et al. 2003. Acutely administered ethanol participates in testosterone synthesis and increases testosterone in rat brain. Alcohol Clin Exp Res 27: 38-43.

Belov ME, Gorshkov MV, Anderson GA, Udseth HR, Smith RD. 1999. On improving the performance of a FT-ICR mass spectrometer with an external accumulation device, Proceedings of the 47th ASMS Conference on Mass Spectrometry and Allied Topics, pp. 767–768.

Bowers LD, Sanaullah. 1996. Direct measurement of steroid sulfate and glucuronide conjugates with high-performance liquid chromatography-mass spectrometry. J Chromatogr B 687: 61-68.

Caldeira JC, Wu Y, Mameli M, Purdy RH, Li P-K, Akwa Y, Savage DD, Engen JR, Valenzuela CF. 2004. Fetal alcohol exposure alters neurosteroid levels in the developing rat brain. J Neurochem 90: 1530-1539.

Chatman K, Hollenbeck T, Hagey L, Vallée M, Purdy RH, et al. 1999. Nanoelectrospray mass spectrometry and precursor ion monitoring for quantitative steroid analysis and attomole sensitivity. Anal Chem 71: 2358-2363.

Cheney DL, Uzunov D, Costa E, Guidotti A. 1995. Gas chromatography–mass fragmentometric quantification of 3α-hydroxy-5α-pregnan-20-one (allopregnanolone) and its precursor in blood and brain of adrenalectomized and castrated rats. J Neurosci 16: 4641-4650.

Corpéchot C, Robel P, Axelson M, Sjövall J, Baulieu E-E. 1981. Characterization and measurement of dehydroepiandrosterone sulfate in rat brain. Proc Natl Acad Sci USA 78: 4704-4704.

Corpéchot C, Synguelakis M, Talha S, Axelson M, Sjövall J, et al. 1983. Pregnenolone and its sulfate ester in the rat brain. Brain Res 270: 119-125.

Dehennin L, Scholler R. 1973. Dienol heptafluorobutyrates as derivatives for gas liquid chromatography of steroidal Δ^4-3-ketones determination of the structure of the isomeric dienol esters. Tetrahedron 29: 1591-1594.

Dehennin L, Ferry M, Lafarge P, Pérès G, Lafarge J-P. 1998. Oral administration of dehydroepiandrosterone to healthy men: alteration of the urinary androgen profile and consequences for the detection of abuse in sport by gas chromatography–mass spectrometry. Steroids 63: 80-87.

Diallo S, Lecanu L, Greeson J, Papadopoulos V. 2004. A capillary gas chromatography/mass spectrophotometric method for the quantification of hydroxysteroids in human plasma. Anal Biochem 324: 123-130.

Dumasia MC, Houghton E, Bradley CV, Williams DH. 1983. Studies related to the metabolism of anabolic steroids in the horse: the metabolism of 1-dehydrosterone and the use of fast atom bombardment mass spectrometry in the identification of steroid conjugates. Biomed Mass Spectrom 10: 434-440.

Fitzgerald RL, Herold DA. 1996. Serum total testosterone: immunoassay compared with negative chemical ionization gas chromatography–mass spectrometry. Clin Chem 42: 749-755.

Gaskell SJ, Brownsey BG, Brooks PW, Green BN. 1983. Fast atom bombardment mass spectrometry of steroid sulphates: qualitative and quantitative analyses. Biomed Mass Spectrom 10: 215-219.

Griffiths WJ, Liu S, Yang Y, Purdy RH, Sjövall J. 1999. Nanoelectrospray tandem mass spectrometry for the analysis of neurosteroid sulphates. Rapid Commun Mass Spectrom 13: 1595-1610.

Herold DA, Fitzgerald RL. 2003. Immunoassays for testosterone in women better than a Guess? Clin Chem 49: 1250-1251.

Higashi T, Shimada K. 2004. Derivatization of neutral steroids to enhance their detection characteristics in liquid chromatography–mass spectrometry. Anal Bioanal Chem 378: 875-882.

Hill M, Parízek A, Bicíková M, Havlíková H, Klak J, et al. 2000. Neuroactive steroids, their precursors, and polar conjugates during parturition and postpartum in maternal and umbilical blood: 1. identification and simultaneous determination of pregnanolone isomers. J Steroid Biochem Mol Biol 75: 237-244.

Hubbard WC, Bickel C, Schleimer RP. 1994. Simultaneous quantitation of endogenous levels of cortisone and cortisol in human nasal and bronchoalveolar lavage fluids and plasma via gas chromatography–negative ion chemical ionization mass spectrometry. Anal Biochem 221: 109-117.

Kim YS, Zhang H, Kim HY. 2000. Profiling neurosteroids in cerebrospinal fluids and plasma by gas chromatography/

electron capture negative chemical ionization mass spectrometry. Anal Biochem 277: 187-195.

Kingston EE, Beynon JH, Newton RP, Liehr JG. 1985. The differentiation of isomeric biological compounds using collision-induced dissociation of two ions generated by fast atom bombardment. Biomed Mass Spectrom 12: 525-534.

Kobayashi Y, Saiki K, Watanabe F. 1993. Characteristics of mass fragmentation of steroids by atmospheric pressure chemical ionization/mass spectrometry. Biol Pharm Bull 16: 1175-1178.

Kuronen P, Volin P, Laitalaninen T. 1998. Reversed-phase high-performance liquid chromatographic screening method for serum steroids using retention index and diode-array detection. J Chromatog B 718: 211-224.

Liehr JG, Beckner CF, Baltore AM, Caprioli RM. 1982. Fast atom bombardment mass spectrometry of estrogen glucuronides and sulfates. Steroids 39: 599-605.

Lière P, Akwa Y, Engerer SW, Eychenne B, Pianos A, et al. 2000. Validation of an analytical procedure to measure trace amounts of neurosteroids in brain tissue by gas chromatography-mass spectrometry. J Chromatog B 739: 301-312.

Liere P, Pianos A, Eychenne B, Cambourg A, Liu S, Griffiths W, Schumacher M, Sjovall J, and Baulieu EE. 2004. J Lipid Res 45: 2287-2302.

Liu S, Sjövall J, Griffiths WJ. 2000. Analysis of oxosteroids by nano-electrospray mass spectrometry of their oximes. Rapid Commun Mass Spectrom 14: 390-400.

Liu S, Sjövall J, Griffiths WJ. 2001. Analysis of neurosteroids in brain by nanoscale capillary liquid chromatography/micro-electrospray mass spectrometry. J Am Soc Mass Spectrom 14: 390-400.

Liu S, Sjövall J, Griffiths WJ. 2003. Neurosteroids in rat brain: extraction, isolation, and analysis by nanoscale liquid chromatography – electrospray mass spectrometry. Anal Chem 75: 5835-5846.

MaY-C, Kim H-Y. 1997. Determination of steroids by liquid chromatography/mass spectrometry. J Am Soc Mass Spectrom 8: 1010-1020.

Marshall AG, Wang TCL, Ricca TL. 1985. Tailored excitation for Fourier transform ion cyclotron resonance mass spectrometry. J Am Chem Soc 107: 7893-7897.

Meng LJ, Griffiths WJ, Sjövall J. 1996. The identification of novel steroid *N*-acetyl-glucosaminides in the urine of pregnant women. J Steroid Biochem Mol Biol 58: 137-142.

Mitamura K, Yatera M, Shimada K. 1999. Quantitative determination of pregnenolone 3-sulfate in rat brains using liquid chromatography/electrospray ionization–mass spectrometry. Anal Sci 15: 951-955.

Murry S, Rendell NB, Taylor GW. 1996. Microbore high-performance liquid chromatography–electrospray ionization mass spectrometry of steroid sulphates. J Chromatogr A 738: 191-199.

Paul SM, Purdy RH. 1992. Neuroactive steroids. FASEB J 6: 2311-2322.

Pearson Murphy BE, Allison CM. 2000. Determination of progesterone and some of its neuroactive ring A-reduced metabolites in human serum. J Steroid Biochem Mol Biol 74: 137-142.

Pickup JF, McPherson K. 1976. Theoretical considerations in stable isotope dilution mass spectrometry for organic analysis. Anal Chem 48: 1885-1890.

Prasad VVK, Vegesna SR, Welch M, Lieberman S. 1994. Precursors of the neurosteroids. Proc Natl Acad Sci USA 91: 3220-3223.

Rule G, Henion J. 1999. High-throughput sample preparation and analysis using 96-well membrane solid-phase extraction and liquid chromatography–tandem mass spectrometry for the determination of steroids in human urine. J Am Soc Mass Spectrom 10: 1322-1327.

Shackleton CH, Straub KM. 1982. Direct analysis of steroid conjugates: the use of secondary ion mass spectrometry. Steroids 40: 35-51.

Shackleton CHL, Chuang H, Kim J, de la Torre X, Segura J. 1997. Electrospray mass spectrometry of testosterone esters: potential for use in doping control. Steroids 62: 523-529.

Shimada K, Higashi T, Mitamura K. 2002. Development of analyses of biological steroids using chromatography – special reference to Vitamin D compounds and neurosteroids. Chromatography 24: 1-6.

Shimada K, Mukai Y, Yago KJ. 1998. Studies of neurosteroids. VII. Characterization of pregnenolone, its sulfate and dehydroepiandrosterone in rat brains using liquid chromatography/mass spectrometry. J Liq Chrom Rel Technol 21: 765-775.

Siuzdak G. 2003. The expanding role of mass spectrometry in biotechnology. MCC Press; San Diego: pp. 182-189.

Strohle A, Romeo E, di Michele F, Pasini A, Yassouridis A, et al. 2002. GABA(A) receptor-modulating neuroactive steroid composition in patients with panic disorder. Am J Psychiatry 159: 145-147.

Uzunova V, Sheline Y, David JM, Rasmusson A, Uzunov DP, et al. 1998. Increase in the cerebrospinal fluid content of neurosteroids in patients with unipolar major depression who are receiving fluoxetine or fluvoxamine. Proc Natl Acad Sci USA 95: 3239-3244.

Valaskovic M, Kelleher NK, McLafferty FW. 1996. Attomole protein characterization by capillary electrophoresis–mass spectrometry. Science 273: 1199-1202.

Vallée M, Purdy RH, Mayo W, Koob GF, Le Moal M. 2003. Neuroactive steroids: new biomarkers of cognitive aging. J Steroid Biochem Mol Biol 85: 329-335.

Vallée M, Rivera JD, Koob GF, Purdy RH, Fitzgerald RL. 2000. Quantification of neurosteroids in rat plasma and brain following swim stress and allopregnanolone administration using negative chemical ionization gas chromatography/mass spectrometry. Anal Biochem 287: 153-166.

Watson D, Taylor GW, Murry S. 1985. Thermospray liquid chromatography negative ion mass spectrometry of steroid sulphates conjugates. Biomed Mass Spectrom 12: 610-615.

Wei JQ, Wei JL, Zhou XT. 1990. Optimization of an isocratic reversed phase liquid chromatographic system for the separation of fourteen steroids using factorial design and computer simulation. Biomed Chromatogr 4: 34-38.

Weill-EngererS, DavidJ-P, Sazdovitch V, Lière P, Schumacher M, et al. 2003. In vitro metabolism of dehydropiandrosterone (DHEA) to 7α-hydroxy-DHEA and Δ5-androstene-3β,17β-diol in specific regions of the aging brain from Alzheimer's and non-demented patients. Brain Res 969: 117-125.

Whitehouse CM, Dreyer RN, Yamashita M, Fenn JB. 1985. Electrospray interface for liquid chromatographs and mass spectrometers. Anal Chem 57: 675-679.

Wood TD, Moy MA, Dolan AR, Bigwarfe PM Jr, White TP, et al. 2003. Miniaturization of electrospray ionization mass spectrometry. Applied Spectroscopy Reviews 38: 187-244.

8 Methods in Immunochemistry

L. L. Jantzie, V. A-M. I. Tanay · K. G. Todd

Abstract: Laboratory techniques using the foundations of immunology have been available for many years and the introduction to and use of immunochemistry in neuroscience, as a tool to localize and identify proteins, has revolutionized clinical diagnostic and basic science practices. Immunohistochemistry uses antibodies to map the nervous system and to pinpoint antigens to specific locations and subpopulations of cells in the brain. Western blotting is a powerful tool based on protein separation and is routinely used for the detection and characterization of specific proteins. In this case, the detection process that selectively identifies the proteins of interest also relies on the specificity and the affinity of antibodies for an antigen. Enzyme-linked immunosorbent assays (ELISA) also use antibodies or antigens, coupled to an easily assayed enzyme, to provide a quantitative concentration measurement of the antigen or antibody of interest. The commonality of all three techniques is the crucial antibody-antigen reaction. This chapter discusses the intricacies of this immunoreaction and focuses on the use of immunohistochemistry, Western blotting and enzyme-linked immunosorbent assays in basic science. All three of these common laboratory practices are discussed in great detail with specific focus on general principles, set-up, reagents, procedure, detection, visualization, quantification, quality control, and troubleshooting.

List of Abbreviation: 4CN, 4-chloro-1-naphthol; ABC, Avidin-Biotin Complex; ABTS, 2,2′-azinodiethyl-benzthiazoline sulfonate; AEC, 3-amino-9-ethylcarbazole; AMCA, Aminomethylcoumarin; AP, Alkaline Phosphatase; BCIP/NBT, 5-bromo-4-chloro-3-indolyl phosphate/nitroblue tetrazolium; BSA, Bovine Serum Albumin; CAPS, N-cyclohexyl-3-aminopropanesulfonic acid; Cy2, Cyanine; Cy3, Indocarbocyanine; DAB, 3,3'-diaminobenzidine; DMF, Dimethyl Formamide; DNA, Deoxyribonucleic Acid; DTT, Dithiothreitol; ECL, Enhanced Chemiluminescence; EDTA, Ethylenediaminetetraacetic acid; EIER, Enzyme-Induced Epitope Retrieval; ELISA, Enzyme-Linked Immunosorbent Assay; FITC, Fluorescein Isothiocyanate; GFAP, Glial Fibrillary Acidic Protein; HIER, Heat-Induced Epitope Retrieval; HRP, Horseradish Peroxidase; I, Current; IHC, Immunohistochemistry; LAB, Labeled avidin-biotin; NeuN, Neuronal Nuclei; NP-40, Nonidet P-40; OCT, Optimal Cutting Temperature Medium; OD, Optical Density; OPD, o-Phenylenediamine; P, Power; PAGE, Polyacrylamide denaturing gel electrophoresis; PAP, Peroxidase-Antiperoxidase; PBS, Phosphate Buffered Saline; PBST, Phosphate Buffered Saline with 0.05–0.1% Tween-20; PMSF, Phenylmethylsulfonylfluoride; pNPP, *p*-Nitrophenyl Phosphate; PVDF, Polyvinylidene Difluoride; R, Resistance; RIPA, Radioimmune Precipitation Buffer; RNA, Ribonucleic Acid; SDS, Sodium dodecylsulfate; TBS, Tris Buffered Saline; TBST, Tris Buffered Saline with 0.05–0.1% Tween-20; TMB, 3,3′, 5,5′-tetramethylbenzidene; TNFα, Tumor Necrosis Factor Alpha; TRITC, Tetramethyl rhodamine; TWEEN-20, Polyoxyethylene-sorbitan monolaurate; V, Voltage; WB, Western Blot

1 Introduction

Throughout the evolution of diagnostic and research science, precise quantitative and cellular localization techniques have become increasingly desirable. Techniques using the foundation of immunology, including immunofluorescence and enzyme immunoassays, have been available for about 40 years, but the development of immunodiagnostic methods has been much slower due to the unpredictable quality of polyclonal antibodies and the success of powerful techniques including silver impregnation, degenerative fiber staining, and ortho- or retrograde transport of labeled substances (Pool et al., 1983; Carson, 1997). The introduction and use of immunochemistry to neuroscience as a tool to study the localization and identification of proteins has revolutionized clinical diagnostics and basic science techniques. In this chapter, we focus on the use of immunohistochemistry, Western blotting, and enzyme-linked immunosorbent assay (ELISA).

2 Immunohistochemistry

In neurohistochemistry, antibodies are used to provide extensive maps of the nervous system and to pinpoint antigens to specific locations and subpopulations of cells in the brain (Bloom et al., 1986; MacMillan and Cuello, 1986). In most laboratories, the use of immunohistochemical stains has become as popular as the use of any special stain (e.g., hematoxylin, cresyl violet), and in some instances the use of immunohistochemistry (IHC) has entirely replaced the older histochemical or empirical methods (Carson, 1997). With dramatic advances in immunology, IHC has become a molecularly oriented scientific art form and, most importantly, has established itself as one of the most powerful and widely used methods to identify cell types, transmitter-specific neurons, neuropeptides, and biosynthetic enzymes (Cuello et al., 1983; Bloom et al., 1986).

Many factors influence the final outcome in IHC. Before beginning any protocol, it is very important to choose the correct immunochemical staining reagents, procedure, and method for visualization. It is also vital to section the tissue and employ appropriate fixation, and as part of the technique, preserve antigenicity and cellular morphology, maintain low background staining, and use proper controls within the procedure. The antigen–antibody reaction is the backbone of IHC, and is governed by the state of fixation of the antigen in the tissue section, the relative freedom of the antibody to move in the dilution buffer covering the sections, and the pH of the buffers used, in addition to other biological and chemical factors (Montero, 2003). Each of these factors are discussed in this chapter in great detail, to illustrate important aspects for optimal IHC procedures, and to highlight areas important for successful trouble-shooting.

2.1 General Principles

IHC was conceptualized on the principle that the immune system produces antibodies directed against foreign antigenic substances. Antibodies produced against a specific antigen are used to visually localize the antigen in fixed tissue preparations (MacMillan and Cuello, 1986). In IHC, the antigen–antibody reaction takes place between two protein macromolecules (Montero, 2003). The antigen is any substance capable of inducing a detectable immune response, and is usually a protein but can also be a polysaccharide, nucleic acid, or other polymer (Carson, 1997). Antibodies or immunoglobulins are glycoproteins. They are produced by B-lymphocytes in response to antigenic stimulation and can be divided into five major classes (IgG, IgM, IgA, IgD, and IgE) (Carson, 1997; Montero, 2003). The five classes differ in structure and function and are antigenically distinct, but do maintain a conserved "Y" shaped morphology. The "Y" shaped morphology is important, as it is the upper arm region of the antibody that binds to its specific antigen. It should be emphasized that in IHC, one of the macromolecules, the antigen, is localized and immobilized in a fixed section of tissue, whereas the antibody is diluted in buffer, is covering the tissue section, and is free during the incubation period (Montero, 2003). This is significant because the final result of the IHC is dependent on optimizing the reactions and interactions between these macromolecules and amplifying the signal to make it visible to the eye using microscopy (see *Figure 8-1* for a schematic diagram of IHC).

Traditionally, immunohistochemical studies have been performed with antibodies of polyclonal origin. Antibodies are said to be polyclonal when they are collected from different clones of cells, after a significant immune response. Specifically, the presence of an antigen causes lymphocytes in an exposed area to

Figure 8-1
Schematic representation of antibody labeling

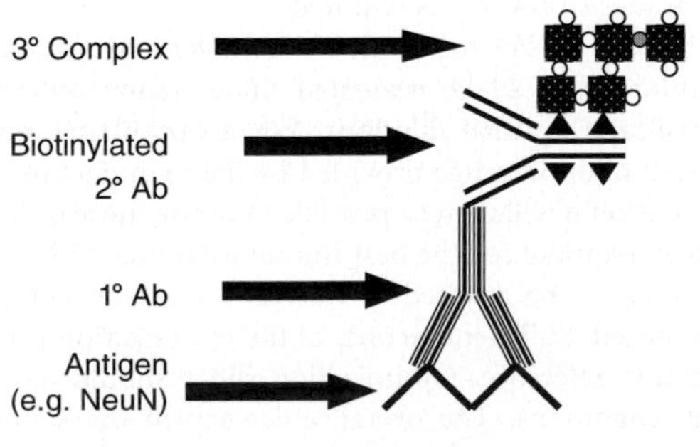

3° Complex
Biotinylated 2° Ab
1° Ab
Antigen (e.g. NeuN)

proliferate, and each of these lymphocytes then forms clones of cells (Carson, 1997). All cells within a single clone produce an identical antibody, but different clones produce different classes of antibodies with specificities to different molecular sites (epitopes) on the antigen (Carson, 1997). The result of antigenic stimulation is the production of a mixture of antibodies from many clones of lymphocytes, and these pools of antibodies are known as polyclonal antibodies. In terms of IHC, polyclonal antibodies often require more concentrated dilutions because they recognize different determinants with different affinities (Cuello et al., 1983) and can be difficult to standardize. Polyclonals can also be difficult to work with because they lack chemical homogeneity and a continuous supply for use.

The development of monoclonal antibodies has transformed immunology and immunochemistry, as they have allowed another element of specificity in the identification and localization of proteins. Monoclonal antibodies have the advantage of being able to immortalize difficult or rare antibodies, and also offer more specific cross reactivity to diverse antigens (Cuello et al., 1983). Monoclonals are prepared by injecting mice with antigens and by harvesting B-lymphocytes making the desired antibody. These harvested lymphocytes are then fused with nonsecreting myeloma cells, and this in vitro fusion yields hybrid cells that retain the antibody-secreting capability of the B-cell and the immortality of the tumor cells (Carson, 1997). These hybrid cells (hybridomas) can be cloned, and a single clone can be used to produce an antibody that is identical in molecular structure to the original. Monoclonal antibodies can be characterized, standardized, and produced in unlimited quantities and they offer many other advantages including high homogeneity, absence of nonspecific antibodies, low batch-to-batch variability, and high purity, affinity, and selectivity. To view the characteristics of mono- and polyclonal antibodies, please refer to *Table 8-1.*

Table 8-1
Specificity and characteristics of monoclonal and polyclonal antibodies

Monoclonal	Polyclonal
Specific for epitope	Possible cross-reactivity
Possible low affinity	Large population of affinities and epitope specificities
May be difficult to produce desired antibody	
Time-consuming to produce	Relatively fast to produce
Quantities theoretically unlimited	Limited quantities
Expensive	Inexpensive
Possible problems with detection on Westerns	May be difficult to reproduce antibodies for some antigens
Can use impure antigens to produce	

Storage and preparation of polyclonal and monoclonal antibodies are very important for successful immunostaining. When a new antibody arrives in a laboratory, special attention should be paid to its storage instructions and optimal temperature. Detailed records of arrival and expiry dates, and whether it needs to be aliquoted to maintain efficacy by avoiding freeze-thaw effects, should be kept. Attention to these details will preserve the efficacy of the antibody and its shelf life, prevent damaging freeze-thaw cycles, and will make optimization of the immunohistochemical procedures easier. It is often necessary to aliquot antibodies, especially when storage below 4°C is required.

When antibodies are to be used in IHC for the first time, extensive optimization of the procedure may be required. Every new antibody should be evaluated under many different conditions, taking into consideration antigen retrieval and optimal dilutions. When optimizing antibody concentrations, it is best to start with the suggested dilution range provided by the manufacturer (dilutions are expressed as ratios). From the suggested dilution it will then be possible to titrate more or less concentrated and find the final, optimal working dilution required for the best immunostaining. Often, degrees of antigen retrieval and tissue penetration will need to be assessed at this time and concentrations of blocking sera and detergents may need to be adjusted. Sufficient records of the optimization procedures should be kept and all trials documented for future reference. Optimization slides should also be kept to document the established optimal working conditions. The overall efficiency of the staining procedure will depend upon antigen preservation and retention, antigen size and spacing (with respect to steric hindrance), tissue permeability, the avidity of the antibodies, and the detection system used (Larsson, 1993).

2.2 Tissue Sectioning and Processing

When performing IHC on brain tissue it is very important to process the tissue and individual brain slices properly. The focus here should be on the preservation of immunoreactivity, tissue structure, and cellular morphology (Pool et al., 1983). In reality, all forms of tissue processing, whether it be embedding or fixation of the brain sections, will change the immunoreactivity of an antigen (Montero, 2003). It is also important to consider that embedding tissue in any media (paraffin, optimal cutting temperature medium [OCT], etc.) and the long incubations during IHC in general, inevitably change cellular morphology within brain sections (Larsson, 1993). Fresh frozen brain sections are commonly used in IHC, and with this method, brains are rapidly removed from the skull after decapitation and flash frozen (e.g., in cold isopentane). Frozen brains are then sectioned on a cryostat (frozen sliding microtome), and sections are mounted on charged microscope, poly-L-lysine or gelatin coated slides. Fresh, frozen sections prepared in this manner preserve antigenicity very well, but before continuing with IHC, the tissue will need to be fixed. Fresh, frozen brain tissue mounted on slides can be successfully stored at −20°C until fixation and subsequent use, whereas fixed tissue on microscope slides should be stored at 4°C. Using tissue from cardiac-perfused animals is an alternative way to prepare for IHC. With cardiac perfusions, brains are fixed with formalin as part of the procedure before removal from the skull and are then postfixed in formalin and then incubated in a sucrose solution. Preparation and fixation in this way maintains cellular morphology better than fresh frozen preparations and may have a lower incidence of background staining, as all blood is removed during the perfusion process. Fresh frozen tissue is useful because it maintains brain chemistry, and allows tissue to be used not only for IHC, but also for other techniques such as high-performance liquid chromatography, Western blotting, and analyses of RNA or DNA.

Most antigens must be fixed in situ as part of the immunohistochemical procedure (Carson, 1997) because they are soluble in aqueous solutions. Fixation retains the antigen's ability to bind antibody, preserves cellular morphology, and preserves the cytoarchitecture of cells, while maintaining the antigens in their original locations (MacMillan and Cuello, 1986; Larsson, 1993; Kumar, 1994). The most common fixatives in IHC are alcohol, acetone, buffered formalin, gluteraldehyde, and formaldehyde solutions. Acetone and alcohol are protein precipitation fixatives, whereas formalin and gluteraldehyde are cross-linking fixatives (Larsson, 1993). Fixation can affect the tertiary structure of proteins and masks, but not necessarily destroys the antigen (Cattoretti et al., 1993). Fixation may also cause problems with antigen extraction or dislocation, and may be denaturing (Larsson, 1993). Thus, with all types of fixation, there is a compromise between structural preservation and retention of antigenicity (Cuello et al., 1983; MacMillan and Cuello, 1986). In addition to compromising antigenicity, fixing may render the cell impermeable to reactions with antibodies, or enhance nonspecific background reactions. It has long been known that the duration of fixation is a very important variable in successful IHC (Swaab et al., 1975; Buijs et al., 1978). Overfixation can hamper antibody penetration into the cells and limit access to desired antigens. When tissue has been overfixed, proteins become excessively cross-linked, and antibodies are barred access to their given epitopes (Carson, 1997). It is important to optimize the fixation time on a few slides and then adhere to the fixation protocol diligently for optimal antigen retrieval. Antigen retrieval in fresh frozen sections can become a problem early in IHC, as cell membrane integrity is an important consideration and can predict the success of antibody penetration into the tissue (Pool et al., 1983). A good way to combat minor overfixation or to increase penetrance of the antibodies across intact cell membranes is to use a detergent such as Triton X-100 or Tween-20 when blocking with serum or with the primary antibody solution. The application of too much Triton on sections, however, can degrade brain tissue and compromise the overall structure of the section. Using thinner brain sections can also help with this type of antigen retrieval but this may not always be a practical solution, as tissue integrity and cellular morphology are not maintained if the section is too thin.

The sensitivity of IHC has been improved by developing formal methods of antigen retrieval and signal amplification (Kim et al., 2003). Knowledge of the exact localization of an antigen in tissue is critical to interpret not only the accuracy of IHC staining results, but also the reliability of an antigen retrieval method (Shi et al., 2001). Poor antigen retrieval is often a major roadblock to successful IHC and can also occur when antigens are nuclear proteins, or as mentioned earlier, when tissue is overfixed or when proteins have become denatured. Several methods of antigen retrieval that include heating, enzyme digestion, and treatment with detergents have been reported (Shi et al., 2001; Ino, 2003). When antigen retrieval becomes a serious problem and a series of optimizations with different concentrations of a detergent like Triton have proved unsuccessful, the use of an enzyme-based protein digesting serum like Proteinase-K may be used. Proteinase-K or trypsin gently reduce the amount of cross-linked proteins and help recover antigenicity. Formal methods of antigen retrieval have also been developed, and they include heat-induced epitope retrieval (HIER) and enzyme induced epitope retrieval (EIER). Newer methods for antigen retrieval on frozen sections include using tyramine, heat, and special retrieval solutions, in combination with dry ice (Kim et al., 2002; Ino, 2003).

2.3 Blocking Reactions

Blocking reactions are important in most IHC procedures and are used as a means to reduce nonspecific binding of antibodies and to limit background staining. All peroxide-based IHC techniques suffer from endogenous peroxidase activities as well as from the pseudo-peroxidase activity of hemoglobin (Larsson, 1993). Therefore, the first blocking solution used in IHC is usually a solution of hydrogen peroxide diluted in methanol (Streefkerk and van der Ploeg, 1976; Kim et al., 2003). The second blocking step in IHC is used to reduce the nonspecific background staining caused by primary and secondary antibodies binding to proteins other than the antigen of interest. The best way to prevent this type of nonspecific staining is to incubate the sections with a mild protein solution (e.g., normal horse serum, bovine serum albumin [BSA], rabbit serum, donkey serum, commercial protein serum block, etc.), as the first protein solution applied to tissue sections will absorb the nonspecificity (Sternberger, 1979). Proteins in the blocking serum will bind to charged sites in the tissue (e.g., collagen and connective tissue) and limit nonspecific primary binding. Sections are usually incubated with the blocking serum for 10 to 60 min, and the excess is tapped off (as opposed to rinsed off). Following the blocking serum step, the primary antibody can be applied.

2.4 Visualization: Enzyme and Fluorescent Immunohistochemistry

Primary antibodies can be visualized in many ways. The simplest method is direct conjugation of an enzymatic or fluorochrome label to the primary antibody itself. This procedure, although once quite common, has fallen out of favor due to lack of sensitivity and insufficient signal amplification, but with the advent of monoclonal antibodies this once again has become a quick and precise tool for antigen tracing and is very useful in double and triple staining (Larsson, 1993). Indirect methods to visualize primary antibodies are more commonly used in laboratories utilizing IHC today. The indirect method, as developed by Weller and Coons (Weller and Coons, 1954) is more sensitive and offers intense signal amplification. Sometimes called the sandwich technique, the indirect method of immunohistochemical staining entails an application of a second layer of antibody that is specific to the IgG of the species in which the primary antisera were raised. The secondary antibody applied is conjugated to an enzymatic or fluorescent label and has the effect of amplifying the signal from the primary antibody because each primary antibody molecule will bind more than one secondary antibody molecule (MacMillan and Cuello, 1986).

2.5 Enzyme Immunohistochemistry

If antibodies conjugated to fluorochromes are not desirable, enzyme-labeled antibodies can also be used. In this case, in the presence of a substrate and a chromagen, an enzyme provides the indicator system necessary to visualize the location of the antibody (Carson, 1997). Horseradish peroxidase (HRP) and alkaline phosphatase (AP) are the enzymes most commonly conjugated to antibodies not labeled by fluorochromes. When HRP is in the presence of hydrogen peroxide (substrate) and a chromagen such as DAB (3,3′-diaminobenzidine), it will identify the sites of antibody binding by forming an insoluble colored compound (Carson, 1997). Many methods are available to intensify the DAB reaction product, if decreased background or increased antigen signal-to-noise ratio is required. The most common method of intensification is to use heavy metals (nickel chloride, copper chloride, or cobalt chloride) or glucose oxidase to supplement the DAB reaction (Mullink et al., 1992). Phenol-tetrazolium methods also exist, and represent useful alternatives to DAB and may offer significant advantages in specific applications (Murray et al., 1991).

A common and relatively new technique for the localization of antigens uses both avidin and biotin as part of a tertiary complex (avidin–biotin complex or ABC). A tertiary complex is used in IHC to amplify the signals of both the primary and secondary antibodies. The ABC method may produce the most intense staining and least background staining of any previous enzyme method available (i.e., the peroxidase–antiperoxidase [PAP] method) (Hsu and Raine, 1981; Hsu et al., 1981). Avidin is a large glycoprotein from egg white that has very high affinity (four binding sites per molecule) for biotin, a vitamin of low molecular weight found in egg yolk (MacMillan and Cuello, 1986). The binding between avidin and biotin is strong and essentially irreversible. Biotinylated secondary antibodies (biotin conjugated to IgG) are widely available, and these antibodies are used in conjunction with the avidin–biotin method. Specifically in the ABC technique, a biotinylated secondary antibody is applied, followed by a complex of HRP–biotin–avidin that has one biotin-binding site on avidin free (Hsu and Raine, 1981; Hsu et al., 1981). The application of the ABC tertiary complex further amplifies the building signal and allows for subsequent interaction with a chromagen like DAB. The ABC method has low background and sensitivity up to 40 times that of other immunoperoxidase methods (Hsu et al., 1981), and the antibodies used may be at higher dilutions than in other techniques. Another ABC method, known as the labeled avidin–biotin (LAB) method, is also widely used. It has been reported to be up to eight times more sensitive than the ABC method (Giorno, 1984) and entails the application of enzyme-labeled avidin as the third step of the IHC procedure as opposed to the application of the preformed avidin–biotin–enzyme complex. ➋ *Figure 8-2* displays photomicrographs of IHC analyses of various cell markers obtained from our laboratory.

2.6 Fluorescent Immunohistochemistry

Coons and colleagues (Coons et al., 1941) were the first group to use an antibody conjugated to a fluorescent dye for the localization of an antigen in a tissue section. Since then, IHC performed with fluorochromes has become a valuable and extensively used technique in clinical diagnostics and research science. Immunofluorescence is pathology's oldest immunohistochemical tool, making it possible to visualize very small amounts of antigens in tissue sections or cell suspensions (Carson, 1997). Fluorescent immunostaining is similar to enzyme immunohistochemistry, except that a fluorochrome, instead of an enzyme, is used to label an antibody (Kumar, 1994). With fluorescent microscopy, the specimen to which the fluorochrome is bound is illuminated with light of a short wavelength. Part of this light is absorbed by the specimen and emitted as fluorescence at a wavelength that is longer than that of the incident light (Carson, 1997). As the fluorochrome has the property of emitting fluorescence on excitation, it possesses advantages over enzyme immunohistochemistry in that it can be directly examined without having to develop a chromogenic reaction by incubation with the substrate (Kumar, 1994). Fluorochromes commonly used in IHC include, FITC (fluoroscein isothiocyanate), TRITC (tetramethyl rhodamine), Rhodamine, Texas Red, Cy2 (cyanine), Cy3 (indocarbocyanine), and AMCA (aminomethylcoumarin). All of these dyes absorb light invisible to the human eye and emit light that is visible, even when the antigen is present in minute amounts. Limitations of fluorochrome-conjugated antibodies and fluorescent microscopy include fading or bleaching of the fluorescence under irradiation and the need for a fluorescence microscope (Kumar, 1994). When performing fluorescent immunochemistry, it is important to use a mounting medium that is compatible with the immunochemical marker employed and does not autofluoresce (Cote et al., 1994). Often the organic based Permount is used to mount slides in enzyme labeled IHC and is very useful if the enzyme-reaction end product is not soluble in alcohol or xylene (Cote

Figure 8-2
Representative photomicrographs of: neuronal immunostaining with antibodies recognizing neuron specific enolase (a) and tumor necrosis factor alpha (b); in vitro (c) and ex vivo (d) immunostaining of astrocytes with antibodies recognizing glial fibrillary acidic protein; microglial cells with ED-1 (e) and OX-42 (f) antibodies

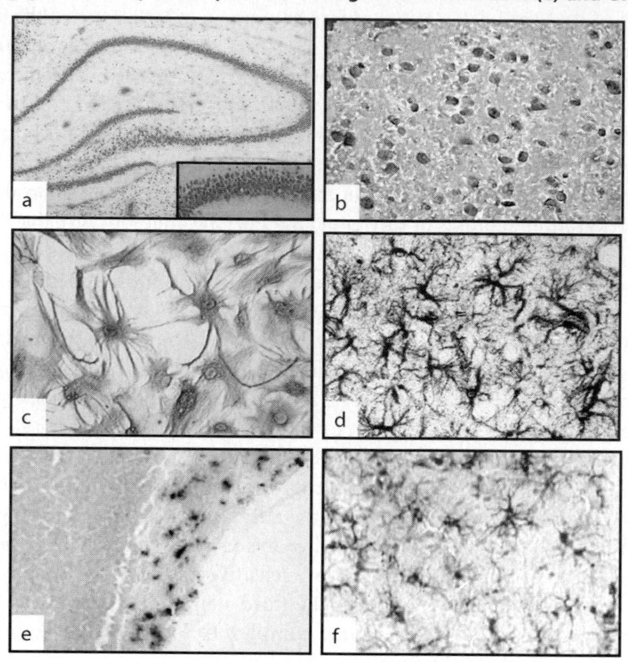

et al., 1994). Permount or similar organic mounting media are known to autofluoresce and thus, are a poor choice in fluorescent microscopy. Autofluorescence can be minimized by using aqueous mounting media such as Aquamount or Geltol. Autofluorescence can also happen when natural substances in a brain section fluoresce on excitation with UV light (Kumar, 1994). Collagen and elastic fibers often autofluoresce in sections and this fluorescence must be separated from antibody–antigen specific staining (*Table 8-2*).

Table 8-2
Peak wavelengths of absorption and emission for different fluoro-conjugated antibodies

Fluorochome	Absorption (nm)	Emission (nm)	Color
Fluoroscein (FITC)	492	520	Green/yellow
Cyanine (Cy2)	492	510	Green
Texas Red (TR)	596	620	Red
Rhodamine Red-X (RRX)	570	590	Red
Tetramethyl Rhodamine (TRITC)	550	570	Red
Aminomethylcoumarin (AMCA)	350	450	Blue
Indocarbocyanine (Cy3)	550	570	Red

2.7 Double Staining

When using IHC procedures, it is often desirable to demonstrate that more than one antigen are present in the same cell. This is especially true in neurohistochemistry if it is desirable to colocalize specific proteins to certain cell types or to examine release of factors and changes in proteins as neurological injury evolves. Double staining allows the simultaneous visualization of more than one antigen on single tissue sections and is a very convenient method to detect the colocalization of two antigens (Vandesande, 1983). Double staining can be performed by sequential application of primary antibodies and fluorescent conjugated secondaries. Also known as indirect immunofluorescence, this procedure involves the use of fluorochrome-labeled, rather than enzyme-labeled, secondary antibodies. These secondaries are free to bind to the primary antibody that has already bound the antigen against which they were raised. The success of a good double-staining protocol lies in the selection of reagents devoid of any cross-reactivity and the formulation of two highly contrasting colors of fluorescence. Another method of double staining is called double direct immunofluorescence and may be the least sensitive method for double labeling (Vandesande, 1983). The direct double-immunofluorescence method entails the use of two or more distinct antibodies to different antigens, conjugated to different fluorochromes (Kumar, 1994).

2.8 Quality Control

All immunohistochemical procedures should be standardized, optimized, and performed exactly the same way each time to maintain consistency and eliminate day-to-day error. Once the method of epitope enhancement and signal amplification is chosen, controls must be run each time immunohistochemical staining is done (Carson, 1997). A negative control is the most important control to run during an IHC procedure. Its purpose is to distinguish positive staining from background staining, to check reliability of reagents, and to detect unintended antibody cross-reactivity. Negative controls are commonly run by substituting the primary antibody for either nonimmune serum from the same species as the primary antibody, or by incubating the sections with the diluent used for the primary antibody (Carson, 1997). The most important aspect of preparing a negative control is to keep the entire procedure exactly the same as for all other slides, without the application of a primary antibody.

2.9 Summary of ABC–Immunoperoxidase Procedure

The procedure given here summarizes the localization of tissue antigens using a primary antibody, biotinylated secondary antibody, avidin–biotin complex, and DAB chromagen on fresh frozen brain sections.

1. Tissue sectioning: Set cryostat at –16°C and slice brain or desired tissue into sections 16–20 μm thin. Mount sections on poly-L-lysine coated or positively charged slides. Allow sections to dry overnight at room temperature. Tissue at this stage (fresh frozen, unfixed) should be stored at –20°C.
2. Fixation: Fix tissue in 10% neutral buffered formalin (or equivalent) by placing slides in a slide rack and incubating them in a staining dish containing the fixative. Upon removal from formalin solution, rinse with tap and double distilled water (ddH_2O) to remove salts.
3. Defat and dehydrate/rehydrate: Take slides through a series of graded ethanol washes and xylene, and back through the ethanols to defat, dehydrate, and rehydrate the tissue sections.
4. Quenching of endogenous peroxidase activity: Move slides to a humidifying chamber (make sure there is about 1 cm of warm water in the bottom of the chamber). Incubate sections in hydrogen peroxide made in methanol to quench endogenous peroxidase. Incubation time to be determined individually, but often 10–30 min is sufficient. Less time is required when a higher concentration of hydrogen peroxide is being used. Rinse with ddH_20. Remember it is very important to never let the sections dry.
5. Protein block: Incubate slides in nonimmune serum (e.g., 10% normal horse serum in phosphate buffered saline [PBS]), for 20–30 min to prevent nonspecific binding. An appropriate concentration (e.g., 0.1%–1.0%) of detergent like triton X-100 may be used during this step to increase tissue penetration. Shake off rather than rinse off blocking serum.
6. Primary antibody: Apply the primary antibody to the sections. Incubation times may range from 1 to 24 h and this incubation time should be optimized along with the optimal end dilution for each

primary antibody. Primary antibodies should be diluted in neutral buffers like PBS or Tris buffered saline (TBS). Low concentrations of blocking sera and/or triton may be included in this solution. Leave the blocking serum on the negative control slide or apply buffer to the negative control at this point. Rinse primary off with the chosen neutral buffer (e.g., PBS).

7. Secondary antibody: Incubate sections with a biotinylated, species-specific secondary antibody for 30 min. The species of secondary IgG antibody used should correspond to the host used for growing the primary antibody. For example, if the primary antibody used is mouse anti-NeuN, the secondary antibody to be used is anti-mouse IgG. Rinse with buffer after incubation is complete.
8. Avidin–biotin complex: Prepare the ABC according to the instructions provided by the manufacturer and allow the solution to conjugate for at least 30 min at room temperature before applying to slides. Sections should incubate with the ABC for at least 30 min. Rinse sections with buffer.
9. Develop sections with DAB: Often one tablet of DAB makes 5 mL of solution and hydrogen peroxide is needed to activate the solution. Follow the instructions provided by the manufacturer.
10. Rinse sections well: Use a series of buffer and water washes.
11. Dehydrate sections: Sections should again be dehydrated in a series of graded ethanol washes before permanent mounting.
12. Mount and coverslip: Mount slides using clean cover slips and dry overnight. Slides can be cleaned the following day using xylene or chloroform. Remember, it is important to remove all bubbles from under the cover slip. Be conservative with the amount of Permount used in mounting, as using too much can obstruct the view of the sections under a microscope.

2.10 Common Problems and Troubleshooting

The following list comprises several common problems encountered in IHC and some avenues for investigation.

1. Lack of staining
   - No antigen present
   - Inactive primary or secondary antibodies (expired or stored improperly)
   - Inadequate or overfixation of tissue
   - Incompatible primary and secondary antibody
   - Antigen destroyed before primary antibody applied
   - Antibody solutions are too dilute or too concentrated
   - Inadequate antibody penetration into tissue

Storing and titrating antibodies correctly are often easy solutions when no staining is observed. Some antibodies are extremely sensitive to repeat freeze-thaw cycles and others have a limited shelf life, as evidenced by a short expiry dates. It is important to check the compatibility of the primary and secondary antibodies before starting IHC, as using the wrong secondary IgG is a common but easily corrected problem. As mentioned, antigen retrieval can be a problem when tissue has been overfixed, so special attention should be paid to fixation time and, once optimized, should be held constant for all subsequent runs.

2. Tissue is nonuniform in appearance, has floated away, is bubbling, or appears damaged
   - Tissue has been overdigested by proteinase-K or Triton
   - Concentration of hydrogen peroxide is too strong
   - Hydrogen peroxide has been left on the tissue section too long
   - Force/buffer stream used in washing steps is too strong

Overdigestion of tissue is common when proteinase-K or triton is used to improve antigen retrieval penetrance of the primary antibody. The easiest correction is to dilute triton solutions and decrease the time of the proteinase-K incubation. Tissue can also be digested when the hydrogen peroxide solution, used to quench endogenous peroxidase activity, is left on too long or is too concentrated. To correct this, check calculations and dilutions and watch the sections to ensure bubbling is not occurring during the incubation. Antigens may be destroyed in the hydrogen peroxide incubation step, so it is important to monitor this closely.

3. Overstaining and/or high background staining
   - Concentration of primary antibody is too high
   - Antibody incubation times are too lengthy
   - Primary and/or secondary antibodies are binding nonspecifically
   - Insufficient blocking of endogenous binding sites
   - Insufficient quenching of endogenous peroxidase
   - Insufficient washing between steps in the procedure

Again, check the dilutions of the primary and secondary antibodies and reevaluate the incubation times of the primary, secondary, tertiary complex, and chromagen steps. Change blocking sera used to reduce nonspecific background and/or include blocking serum in primary and secondary antibody solutions. Ensure that the endogenous peroxidase is sufficiently quenched.

3 Western Blotting

The process of protein separation and identification has evolved since Laemmli (1970) first described the principle of protein separation on sodium dodecylsulfate (SDS) polyacrylamide denaturing gel electrophoresis (SDS-PAGE). The electrophoretic transfer of proteins from gel to membrane and the use of the membrane as a solid support for immunoassay was first introduced by Towbin and colleagues (1979), and further modified by Burnette (1981), who introduced a sample lysis step (for protein isolation) and coined the term Western blotting (WB). Today, WB is a powerful tool routinely used for the detection and characterization of proteins. The detection process in WB that selectively identifies proteins relies on the specificity and the affinity of antibodies for an antigen. The protein/antigen identification involves the sequential recognition of the target protein by a primary antibody, which, in turn, is recognized by a secondary antibody tagged by a signal-generating molecule. The secondary antibody can be labeled with a radioisotope, a fluorophore, or an enzyme such as HRP or AP. Thus, it is the presence of the secondary antibody that is detected by the emission of a signal.

The affinity and specificity of the antibodies used will have a great impact on the sensitivity of a blotting system, and the choice between monoclonal or polyclonal antibodies must be made (refer to ➋ *Table 8-1*). An overview of the variety of approaches that can be taken to isolate proteins for antibody production can be found in Diano et al. (1987), Leppard et al. (1983), and Knudsen (1985).

3.1 Protein Sample Preparation

3.1.1 Buffers for Protein Isolation

In his original report, Laemmli (1970) simply denatured pelleted cells in a gel loading buffer of the following final concentration: 0.0625M Tris-HCl pH6.8, 2% SDS (w:v), 10% glycerol (v:v), 5% 2-mercaptoethanol (v:v), and 0.001% bromophenol blue (w:v) and boiled the sample to ensure complete protein dissociation prior to loading onto the SDS-PAGE. Burnette (1981) introduced the concept of protein solubilization (isolation) by using the following radioimmune precipitation (RIPA) buffer: 150 mM NaCl, 1% Igepal CA-630, 0.5% sodium deoxycholate, 1% sodium dodecylsulfate, and 50 mM Tris pH 8.0 as modified from Gilead et al. (1976). For the proteins present in large amounts or ubiquitously distributed throughout cells, whole cell protein extracts using buffers such as RIPA, NP-40 (150mM NaCl, 1% NP-40, 50 mM Tris pH 8.0; modified from Sefton (1980)), high salt lysis (500 mM NaCl, 1% NP-40, 50 mM Tris pH 8.0; Harlow (1988)), or low salt lysis (1% NP-40, 50 mM Tris pH 8.0; Harlow (1988)) may provide adequate protein solubilization. The choice of a lysis buffer depends upon an efficient solubilization of the antigen of interest and the fact that after solubilization the antigen should remain recognizable by the primary antibody. It is advisable to start with a strong buffer, then to decrease or alter the composition of the buffer until optimal solubilization conditions are found. The optimization can be monitored by checking for the presence of the antigen in the lysate and the remaining cell debris.

In contrast, the study of proteins present in minute amounts and/or of restricted localization may require selective protein extraction. Several procedures have been described including use of nuclei (Sonnenberg et al., 1989); modified by Hope (Hope et al., 1994), sarcolemmal-enriched membrane fractions (Tuana et al., 1987), apolipoprotein (Peitsch et al., 1989), or microsome fractions (Kumar et al., 1985) for protein extractions. Numerous others exist and as a general rule, they rely first on the isolation of the cellular compartment of interest followed by the solubilization of the proteins contained in that cellular fraction.

3.1.2 Protease Inhibitors

During the protein isolation step, the presence of protease inhibitors is crucial to preserve the integrity of the antigen under study. Some protease inhibitors, such as phenylmethylsulfonylfluoride (PMSF), have a wide spectrum of activity and are therefore commonly used, whereas other inhibitors, such as okadaic acid are used primarily when the phosphorylated form of protein is of interest. ➋ *Table 8-3* contains a list of some commonly used protease inhibitors and their activity spectra.

Table 8-3
Protease inhibitors and their selectivity

Protease Inhibitor	Spectrum of activity
Aprotinin	Serine protease inhibitor that inhibits trypsin, chymotrypsin, kallikrein, and plasmin Binding is reversible, with most aprotinin-protease complexes dissociating at pH>10 or <3
Benzamidine	Peptidase inhibitor
Leupeptin	Inhibitor of serine and cysteine proteases: plasmin, trypsin, papain and cathepsin B
Okadaic acid	Inhibitor of type 1 and type 2A protein phosphatases
Pepstatin A	Potent inhibitor of acid proteases, including pepsin, renin and cathepsin D and many microbial aspartic proteases
Phenylmethanesulfonyl fluoride (PMSF)	Inhibits serine proteases such as trypsin and chymotrypsin. Also inhibits cysteine proteases (reversible by reduced thiols) and mammalian acetylcholinesterase
Sodium orthovanadate	Inhibits ATPase, alkaline phosphatase and tyrosine phosphatase
Dithiothreitol (DTT)	Reagent for maintaining –SH groups in the reduced state. Effective for reducing protein disulfide bonds prior to SDS-PAGE

3.1.3 Markers

To verify that the protein being detected by the antibody set is due to the desired specific immune reaction, or to estimate the size of a protein, the mobility of the protein of interest in gel is compared to that of protein of known molecular weight (referred to as protein molecular markers). The protein molecular weight markers are loaded on the gel alongside the samples to be analyzed. Many commercially available markers are labeled with dyes so that the bands are readily visible during electrophoresis (to check the running of the gel) and after transfer (if the marker lane is transferred, then the samples should have been transferred too), and critically, at detection (to ensure that the band labeled by the antibody is of the expected size). More recently, companies have developed molecular weight markers labeled with biotin. Incubation of the blot with streptavidin-HRP followed by detection by chemiluminescence or fluorescence will result in a ladder of bands at the position of each individual marker band, allowing direct positioning of the weight markers on the autoradiographic film.

3.2 Electrophoresis

The most widely used technique for the fractionation of proteins is the SDS-polyacrylamide gel because it is relatively quick and convenient to use. The aim of SDS-PAGE is to separate proteins from a heterogeneous mixture based solely on their molecular weight. Amino acids are ionized in aqueous solutions; hence, the protein that they compose will carry a net charge, which is the sum of all the individual amino acid charges. To separate proteins by weight rather than by a combination of their charge and weight, the proteins are denatured by heat in the presence of SDS and a reducing agent such as β-mercaptoethanol. SDS, which carries a negative charge, coats the protein, masking the positively charged amino acids, thus causing a net negative charge. The number of SDS molecules required to coat a protein is proportional to peptide length, maintaining a constant charge: mass ratio. β-Mercaptoethanol reduces the sulfur bonds that link polypeptide chains making up multi-subunit protein complexes, thus allowing the separation of each individual subunit. After denaturing, the protein sample is run on an SDS polyacrylamide gel. To obtain a sharp banding of the protein sample by SDS-PAGE, a discontinuous gel system is used. The protein sample runs consecutively through a stacking layer, which allows the proteins to "concentrate" in a unique sharp band followed by a separating layer where proteins separate due to the sieving effect of the gel matrix. The two gel layers can differ by either their salt concentration or acrylamide concentration, or a combination of these factors. In general, the composition of the stacking gel is set at 5% acylamide (w:v), whereas that of the separating gel depends mainly on the weight of the protein of interest and can vary from 6% to 15% acrylamide (w:v). The most widely used buffer in Western blotting is the Tris-glycine electrophoresis buffer: 25 mM Tris, 250 mM glycine (electrophoresis grade, pH 8.3), 0.1% SDS.

3.3 Transfer

After electrophoresis, protein bands are transferred onto a solid support. Many aspects of a transfer can affect antigen detection. Some of these factors are specific to the transfer method and choice of membrane, whereas others apply to the entire blotting procedure. For example, the transfer efficiency may be affected by the presence of SDS in the gel and whether the gel was stained prior to transfer.

3.3.1 Choice of Membrane

Antigens can be immobilized on a number of different membranes using one of several methods, and antigens bind differently to different types of membranes. The choice of a membrane is dependent on many factors including the particular antigen to be detected and the requirements of the experiment. To maximize detection sensitivity, it is advisable to use "mock" samples to test different types of membranes with the particular antigen–antibody combination to be used.

Pure nitrocellulose cast on both sides of an inert polyester support material is the most frequently used membrane support for the transfer of proteins and nucleic acids. Supported nitrocellulose membranes combine sensitivity and strength, are compatible with all standard detection methods, and exhibit good binding capacity and reduced background interference. The supported membrane is ideal for repeated screening and reprobing, can be autoclaved, and is available in two pore sizes. The selection of a particular pore size depends on the molecular weight of the protein of interest. For most analytical protein transfers, 0.45 μm pore size is adequate, but for the transfer of low molecular weight proteins (≤15 kDa) and nucleic acids, 0.2 μm is recommended. Methanol must be used in the transfer buffer when using nitrocellullose.

Polyvinylidene difluoride (PVDF) membranes were originally introduced as an ideal medium that could sustain the harsh chemical environment of N-terminal or Edman degradation protein sequencing. It is naturally a very hydrophobic support that requires a prewetting step in 100% methanol. PVDF membranes have a high protein binding capacity, target retention, and resistance to cracking that made them appealing for general laboratory techniques. PVDF membranes have a higher protein retention capacity than nitrocellulose membranes, which helps prevent the transfer of proteins through the membrane, especially if SDS is present in the transfer buffer. Although this type of membrane has to be prewetted in alcohol prior to use, it may be used with a transfer buffer that does not contain methanol. Additionally, the mechanical resistance of PVDF membranes allows multiple reprobing of blots.

Nylon can also be a support to protein transfer. The protein-binding capacity of nylon membranes depends upon whether the membrane contains nylon, charge modified nylon, or activated nylon. In general, immunoblotting with nylon membranes requires more blocking to overcome higher background signals, and they do not bind proteins as well as other membranes. However, nylon membranes are less brittle than those composed of nitrocellulose.

3.3.2 Types of Transfer

There are two types of transfer conditions, wet and semidry. Wet transfer can be done under high intensity or standard field conditions. High intensity wet transfer is performed when the electrodes are 4 cm apart, requires efficient cooling, and can be accomplished in as little as 30 min or run overnight at low voltages. Standard field wet transfer is performed when the electrodes are 8 cm apart and may be run overnight. Semidry transfer provides the highest field strength as the electrode distance is limited only by the gel and filter paper. This method is used for fast, high-intensity transfers and is most suited for medium-range (10–100 kDa) proteins as very small proteins can be forced through the membrane and very large proteins may not transfer completely.

3.3.3 Transfer Buffers

During wet transfer, the transfer buffer heats, and as it overheats it breaks down, causing even greater overheating. This phenomenon can quickly become a serious problem. To avoid buffer breakdown, it is important to (1) have adequate cooling; (2) use the buffer that is recommended for the transfer unit; (3) avoid excessive power conditions. For most transfer apparatuses, the buffer should be cooled by recirculating either chilled water or buffer through the transfer tank. In general, it is not recommended to transfer in a cold room, or to place the transfer chamber in an ice bath, since these procedures do not provide sufficient cooling. Indeed, most transfer cells are made of plastic, which is a poor heat transducer, and the buffer in contact with the gel is held static by the gel-membrane sandwich. Hence, the buffer should be remixed during the transfer to maintain buffer homogeneity and avoid localized overheating.

The ionic concentration of the transfer buffer used must be known to avoid overheating (if the transfer is held at constant voltage) or to avoid artificially low voltage (if the ionic concentration of the buffer is too high and the transfer held at constant current). In general, the presence of SDS (0.01%) in the transfer buffer improves the transfer efficiency of large proteins (>100 kDa), whereas methanol improves the binding of smaller proteins to the membrane. Low pH (<8) of the polyacrylamide gel or transfer buffer or high concentrations of SDS (>0.1%) can lead to inefficient binding of the antigen. The composition of several frequently used transfer buffers is provided in ➲ *Table 8-4*.

Table 8-4
Composition of several common transfer buffers

Buffer composition	Application	Reference
25 mM Tris, 192 mM glycine, pH 8.3, 20% methanol	SDS-PAGE	Towbin et al. (1979)
48 mM Tris, 39 mM glycine, pH 9.2, 0.0375% SDS, 20% methanol	SDS-PAGE	Bjerrum et al. (1988)
10 mM $NaHCO_3$, 3 mM Na_2CO_3, pH 9.9, 20% methanol	Basic protein from SDS-PAGE	Dunn (1986)

3.4 Power Conditions

According to the general principles of electricity, heat generated in a transfer buffer is directly proportional to the power applied to the buffer. The power in watts is equal to the voltage in volts multiplied by the current in amperes—$P = IV$, where P represents power, I is the current, and V stands for voltage. Voltage, current, and resistance are related by the equation, $V = RI$. Therefore, $P = RI^2$.

The voltage and current of a transfer are determined by the power conditions set on the power supply for the transfer and the resistance of the circuit (basically the buffer). As the buffer breaks down, its resistance drops. If the power or voltage is set to be constant, the current will increase as the resistance drops, resulting in heating. However, if the current is held constant, and the resistance drops, so will the power and the voltage, causing the proteins to transfer more slowly.

After transfer, the efficiency of the procedure can be monitored by staining the membrane with general protein stains. Although this additional step bears the advantage of enabling early troubleshooting, it should be kept in mind that this procedure may affect the later antigen–antibody reaction. Alternatively, efficient transfer of proteins can be monitored by staining a test lane excised from the membrane. For PVDF membranes, India ink (Thompson and Larson, 1992), 0.2% Coomassie in 45% methanol, 5% acetic acid, or Ponceau S are popular stains. India ink, or 0.1% amido black in 20% methanol, 10% acetic acid, or Ponceau S can also be used to stain nitrocellulose membranes. A benefit of Ponceau S is that it can be used directly on either type of membrane without any adverse effects on subsequent antigen detection. The membrane will have to be destained (incubating with gentle agitation with a wide variety of detaining solutions [e.g., 250 mM $NaHCO_3$, pH 9.0]) before proceeding to the next step.

3.5 Immunoreaction

At this stage, the membranes can be wrapped in plastic, kept moist, and stored at 4°C. Once the following detection procedures have commenced, the membranes should not be allowed to dry. Blocking, antibody incubation, and washing steps are commonly performed in buffered saline solutions. The blocking step corresponds to the incubation of the blot in a solution that contains an agent that binds to the nonspecific binding sites present on the membrane to prevent background staining. Washing removes unbound antibody and blocking agents prior to and should be undertaken between each step in the protocol. As a general rule, blotting is performed at room temperature, with gentle shaking, in a shallow container that is slightly larger than the membrane. The solutions commonly used are TBST (20 mM Tris-HCl, pH 7.5, 150 mM NaCl, 0.05-0.1% Tween 20 v/v/v) or PBST (50 mM sodium phosphate, pH7.4, 150 mM NaCl, 0.05-1% Tween 20 –v:v-) as a buffer.

To avoid nonspecific binding of the antibody, the membrane is incubated with blocking proteins that saturate all free sites. For nitrocellulose, it is recommended to use 3% gelatin (w/v) in TBS (TBST without Tween-20). Gelatin cannot be used at 4°C; therefore, if blocking is to be performed at this temperature, bovine serum albumin can substitute for gelatin. For PVDF membranes, the blocking agent most commonly used is 5% (w/v) nonfat dry milk in TBST or PBST buffer; other blocking agents include dextran, gelatin, or casein. Alternatively, when the phosphorylated form of an antigen is being detected, primary antibody manufacturers often recommend the use of bovine serum albumin at a concentration of 1–5% (w/v).

Regardless of the detection method used, to achieve high sensitivity with low background, it is essential to optimize the concentration of both primary and secondary antibodies. Some primary antibodies (particularly monoclonals) recognize epitopes that are part of secondary or tertiary structures and become altered when the antigen is denatured or transferred; as such these antibodies often give weak signals. High background staining can also result from primary antibody recognition of epitopes shared by other protein species in the sample, or primary antisera containing a mixture of antibodies with multiple specificities. The latter problem can be overcome by preadsorbing the antisera to remove the cross-reacting antibodies. Generally, primary antibody optimization is performed by blotting a series of known amounts of antigen to test several dilutions of primary antibody. If the primary antibody is too concentrated, a high nonspecific signal results. At too low concentrations, the specific signal will be weak. When using a low titer or low affinity antibody, the signal can be increased by a variety of means including eliminating Tween 20 from the buffers, increasing the incubation time, or increasing the concentration of the primary antibody solution. Once an optimal primary antibody dilution is determined, the same process is repeated to test appropriate concentrations of secondary antibodies. In general, the background will increase when the conjugated secondary antibody is used at dilutions lower than 1:2,500. Finally, it is important to note that a few IgGs and other immunoglobulins (e.g., IgM) have a tendency to stick nonspecifically to membranes, thus generating higher backgrounds. This can be overcome by diluting the secondary antibody in blocking solution prior to incubation or by increasing the number and duration of washes after incubation with the secondary antibody.

The number and duration of washing steps can help with the reduction of background. As a rule of thumb, the volume of the wash should be twice the antibody dilution volume, and standard wash time should be 5x 4 min or 3x 10 min. Washing solutions used are PBST or TBST as buffering base, but Tween-20 is sometimes omitted when there is the chance that a low affinity antibody or a weakly bound antigen could be washed away.

3.6 Detection

For many years, due to the availability and low cost of radioisotope-labeled secondary antibodies, radioactive detection was the method of choice in Western blotting. Newer methods that are less hazardous and easier to use, while maintaining comparable sensitivity, have been developed. Today, Western blotting detection methods can be light-based, (chemiluminescence, bioluminescence, chemifluorescence, and fluorescence), radioactivity-based, or color-based. It is important to note that the detection sensitivity depends on the affinity of the primary antibody for the antigen and on the affinity of the secondary antibody for the primary antibody and can therefore vary considerably from one protein sample to another and from one antibody batch to another.

3.6.1 Radioactive Detection

Radioisotopes such as ^{35}S, ^{32}P, ^{33}P, ^{14}C, and ^{125}I can be used to label secondary antibodies that are detected by autoradiography on X-ray film. This method of detection is sensitive and provides good resolution at a reasonable price. For direct autoradiography without intensifying screens or scintillators, the response of the film is linear only for 1–2 orders of signal magnitude. When the sensitivity is increased with intensifying screens or fluorographic scintillators, the linear response of the film can be extended by preexposing the film to a flash of light. Phosphor imagers offer an alternative radioactive detection method. This is a relatively costly approach, with instrumentation for scanning the imaging plate and the high cost of the phosphor imager plate itself, but the financial drawback is offset by increased sensitivity and linear range, (4.8 orders of magnitude), reusable plates, and time savings (exposure time 10 to 20 times shorter than for a film) when compared to X-ray films. The extended linear dynamic range of the phosphor imager contributes to its reliability in accurate quantitative data. The main disadvantages to the isotopic detection are linked to the use of radioactivity—exposure of personnel to radiation, disposal costs, and environmental concerns.

3.6.2 Colorimetric Detection

Colorimetric methods were developed as a nonradioactive alternate method of detection. Enzymes such as HRP or AP, which are usually conjugated to the secondary antibody, can convert substrates to colored precipitates that accumulate on the blot and generate a colored signal. Such substrates include 5-bromo-4-chloro-3-indolyl phosphate/nitroblue tetrazolium (BCIP/NBT) for AP and 3,3′-diaminobenzidine

(DAB), 3,3′,5,5′-tetramethylbenzidene (TMB) or 4-chloro-1-naphthol (4CN) for HRP. Visual monitoring of the colored signal allows the reaction to be stopped when a desirable signal-to-noise ratio has been reached. Because constant monitoring of the reaction is possible, this detection method is more easily controlled than any film-based method. In addition, colorimetric methods do not require costly materials such as X-ray films and darkroom chemicals. The colored blots can be scanned by a densitometer to generate a digitized copy of the blot, which can then be quantified by densitometric analysis. Traditionally, colorimetric methods are considered to be of medium sensitivity compared to radioactive or chemiluminescent detection methods and, due to the nature of the detection process, stripping and reprobing are not possible.

3.6.3 Chemiluminescence

In general, luminescence is characterized by the emission of light generated from energy released after the substance, in an excited state, returns to a more stable (lower energy) state. In the case of chemiluminescence, the excitation energy is generated by a chemical reaction. One of the most widely used systems is the HRP/hydrogen peroxide oxidation of cyclic diacylhydrazides, such as luminal, in alkaline conditions. The oxidation of luminol elevates its energy to an excited state, and the return to ground state is accompanied by light emission. Enhanced chemiluminescence (also known as ECL) refers to the catalyzation of the reaction in the presence of chemical enhancers such as phenols. Chemical enhancers can increase the amount of light emitted by about 1,000 times and also extend the time period over which the light is emitted. Generally, maximum light emission is reached after 5–20 min of reaction and subsequently decays slowly with a decay half-life of approximately 60 min. The light emitted peaks at a wavelength of 428 nm and can therefore be detected by exposure to blue light-sensitive autoradiography film.

Chemiluminescence allows the nonradioactive detection of antigens with HRP-labeled antibodies, which is as sensitive as radioactive detection systems (low picogram amounts of proteins) and ten times more sensitive than colorimetric detection. The generated autoradiographic images are of high resolution due to the high signal-to-noise ratio of the reaction, and data collection is fast; generally, less than 1-min exposure to film will give rise to an autoradiogram that can be quantified with a densitometer or image analysis program. In addition to the stability of the data (autoradiograms can be stored forever), an advantage of chemiluminescence is the economically small amount of sample and antibodies required to achieve good detection. In addition, several antigens can be detected from the same blot after stripping and reprobing the membrane with different antibodies.

3.6.4 Bioluminescence

In bioluminescence, the light emission phenomenon is generated by a bioluminogenic substrate (luciferin) extracted from an organism (firefly). This detection system differs from chemiluminescence by the type of enzymes and cofactors involved as well as by the mechanism through which light is emitted. In this method, membranes blotted with the antibody–enzyme complex are incubated with the bioluminogenic substrate (luciferin-based derivative) and the light emitted is immediately measured by a photon-counting camera that visualizes the blotted proteins as bright spots. Although very similar to chemiluminescence in sensitivity and speed of detection, this method is not as widely used due to the paucity of commercially available bioluminogenic substrates. PVDF is the membrane of choice for bioluminescence detection since nitrocellulose membranes may contain inhibitors of luciferase activity.

3.6.5 Chemifluorescence

Chemifluorescence detection relies on the conversion of a fluorogenic compound to a fluorescent product by an enzymatic reaction. The conversion of the nonfluorescent or weakly fluorescent substrates to fluorescent compounds can be catalyzed by several enzymes, including AP and HRP. In general, the enzymatic reaction involves the cleavage of a phosphate group from the fluorogenic precursor. The fluorescence can be detected by a fluorescence imager and quantified. The chemifluorescence method generates fluorescent reaction products that are stable enough to allow the scanning of blots at a later, more convenient time. The method is compatible with standard stripping and reprobing procedures.

3.6.6 Fluorescence

Fluorescence detection relies on the visualization of a secondary antibody that has been labeled with a fluorophore such as fluorescein (FITC), Texas Red, Tetramethyl rhodamine (TRITC), or R-phycoerythrin. Although this method of detection has a reduced sensitivity of twofold to fourfold compared to chemiluminescence detection, it presents a tenfold greater linear dynamic range, thus providing better linearity and better quantification within the detection limits. Since secondary antibodies can be labeled with fluorophores of distinct colors, multiplexing (simultaneous detection of several antigens) of the same blot is feasible.

3.7 Troubleshooting

One question frequently asked in Western blotting is, at which step is it possible to pause in the protocol and for how long? The best step to make an overnight stop is after protein transfer. At this stage, the blot can be either stored in TBS or blocked overnight at 4°C. Alternatively, some primary antibody manufacturers recommend (due to the specificity and the affinity of their antibody) probing overnight with the primary antibody, making this step ideal for an end-of-day break. If the procedure has to be postponed for several days, it is recommended to wash the membranes in TBS after transfer, dry, and store them between pieces of filter paper until ready to continue. Other common problems and solutions are summarized in ➋ *Table 8-5*.

3.8 Extensions of the Technique

3.8.1 Quantification

The existence of a linear dynamic range for a given detection method or process indicates that there is a direct relationship between the signal intensity and the actual quantity of the material being detected. This direct relationship enables accurate quantitation and the elimination of overexposure and saturated signals as well as the signals, which, although detectable, are below the minimum level of quantification.

Autoradiographic films exhibit a linear response to the light produced from enhanced chemiluminescence (Johnstone and Thorpe, 1982) or to the amount of radioactivity to which they are apposed. This relationship between the amount of light or radioactivity emitted and the density of the antigen can be used for the accurate quantification of proteins by densitometry. The range of densities over which the film is linear can be extended by preflashing the film prior to exposure (Laskey and Mills, 1975, 1977), making quantification of lower levels of protein, in particular, more accurate. Preflashing is performed with a modified flash unit that has been calibrated (by adjusting its distance from the film) to raise the film optical density by 0.1 to 0.2 OD unit above that of the unexposed film. The flash duration should be about 1 msec.

3.8.2 Absolute versus Relative Quantification

The availability of synthetic peptides presenting an epitope that can be recognized by the primary antibodies used in Western blots allows their use as absolute standards when loaded in range of known amounts. Standard Western blotting procedures are performed on the sample containing the protein to be quantified along with a set of standards (known amounts of the epitope-bearing synthetic peptide). It is suggested to use at least five different dilutions of standard and that the dilution range should not be greater than one order of magnitude. For accurate quantification, it is important that the amount of protein to be quantified lies within the range of standards and that the amount of light emitted by the detection method lies within the linear range of the film. This is achieved by running more than one dilution of the protein sample to be quantified and by making several exposures to film for varying lengths of time. If the lowest concentration of standard is barely visible on the film, it can be expected that all the other standards should be in the linear range of the film. The autoradiographic images generated on the film can be digitized by scanning, and several software packages now allow the densitometric analysis of the image.

For the majority of the proteins, however, no synthetic peptides are available; therefore, relative quantification must be performed. To that end, the amount of the protein of interest is assessed relative to that of an internal standard (known protein present or loaded in known amount in the sample to be analyzed). The detection of these proteins can be simultaneous if they are well separated on the SDS-PAGE and if the reaction patterns of the primary antibodies used do not overlap. This can be determined in preliminary Western blots where single and dual probes are performed to investigate whether the band pattern generated by either primary antibody is altered by the presence of the second one. As a rule of thumb, the internal standard should be at least 20 kDa different from the protein of interest, and the two primary antibodies used should be generated in different host species so that the dilutions of the secondary antibodies used can be tailored to lie within the linearity range of the film. Alternatively, the presence of the two proteins can be determined sequentially by stripping the blot after the first detection and reprobing with the second set of antibodies. Subsequently, the two autoradiographic images scanned will have to be realigned for analysis by imaging software.

4 Enzyme-Linked Immunosorbent Assays

The term "enzyme-linked immunosorbent assay" was first coined by Engvall and Perlmann (Engvall et al., 1971) to describe an enzyme-based immunological method to determine concentrations of a variety of antigens or antibodies. This quantitative immunoassay uses antibodies or antigens, coupled to an easily assayed enzyme, to provide a measurement of the concentration of the antigen or antibody of interest. A variety of protocols is available and there are two main variations on the theme— ELISA can be used to detect and quantitate antigens by using specific antibodies or it can be used to investigate antibodies that recognize an antigen. Examples of ELISA protocols include indirect, direct competitive, antibody sandwich, double antibody sandwich, direct cellular, and indirect cellular; the latter two use cell-associated molecules bound to a solid surface rather than the antigen or antibody used in the other four methods. In immunoassays for research purposes, the analyte of interest is generally the antigen (or hapten), not an antibody. Qualitative identification of specific antibodies is often of clinical interest but it is relatively rare that their precise quantification is required.

Currently, enzymes are the most widely used labels for immunoassays due to catalytic amplification whereby a single enzyme label can provide multiple copies of detectable species. This amplification results in very low limits of detection that rival those of radioimmunoassay (from which enzyme immunoassays were adapted) without the problems associated with the use of radioligands (Wolters et al., 1976). ELISA methods are a type of heterogeneous immunoassay in that they require a separation step prior to quantitation as a means to separate the bound and free fractions of the enzyme-labeled species. In ELISA, separation is accomplished by conducting the assay with antibodies immobilized onto solid supports. Antibodies bind strongly and spontaneously to glass and some plastics; thus, they can be attached to a number of substrates including beads, tubes, paper discs, or the wells of microtiter plates. This immobilization of the antibody onto the solid support of, for example, a microtiter plate, results in all bound species also being immobilized. All unbound material is then easily removed by decanting and rinsing the plate.

ELISA assays may be competitive or noncompetitive. As the name implies, in a competitive ELISA, enzyme-labeled antigen competes with free antigen (the analyte of interest) for a fixed and limited quantity of immobilized antibody binding sites. After incubation, the microtiter plate (solid support) is rinsed to remove all unbound species and the enzyme substrate is added in saturating concentration. The conversion of substrate to produce can be measured continuously (kinetic assay) or, more commonly,

Table 8-5
Common problems encountered with Western blotting and their solutions

Problem	Origin	Solution
No bands or faint bands on the blot	Gel and membrane reversed during transfer	In standard basic transfers from SDS-containing gels, the gel should be on the cathode (negative) side of the sandwich, and the membrane on the anode (positive) side of the sandwich. For acidic transfers, the gel and the membrane positions should be reversed
	Poor transfer efficiency	Transfer efficiency is affected by the size of the protein, the percentage of acrylamide in the gel, the strength of the electric field, the duration of transfer and the pH of the buffer. The larger the protein, the more slowly it will transfer. Transfer of large proteins requires a high field strength. Smaller proteins may be forced through the membrane if blotted for a long time in a high field strength: avoid this problem by blotting onto 0.2μm pore size nitrocellulose or PVDF. If the isoelectric point of the protein is close to the pH of the buffer, the protein will not carry much charge, and therefore will not move in the electrophoretic field: use either a carbonate buffer (pH 9.9), a CAPS buffer (pH 11) or an acidic buffer Check transfer efficiency by staining the gel after transfer, or by staining a second blot with a total protein stain, such as coomassie blue or ponceau red. Alternatively, use commercially available prestained protein standards that are run along the samples of interest and that are visible during both the separation electrophoresis and on the membrane after transfer
	Low amount of antigen bound per unit area of membrane	Epitopes recognized by the antibody may become buried or denatured when the protein is transferred to surfaces such as nitrocellulose. This effect may be enhanced when blotting protein from SDS-containing gels, Insufficient contact between the gel and the membrane during transfer (e.g. air bubbles) will also reduce the amount of antigen available for recognition
	Old or improperly stored reagents	Antibodies degrade with time, and will break down very rapidly with repeated freeze-thaw cycles. Aliquot antibody stocks into volumes that would not need to be thawed more than 3 times, and keep the antibody stock solution on ice when using it. For other reagents, follow closely storage instructions from manufacturer

Table 8-5 (continued)

Problem	Origin	Solution
	Impure or low titer antibodies	Antibody concentrations and affinities vary considerably. The optimal dilution for a given primary antibody must be determined empirically (although most companies will give an indicative range of dilutions). In general, early bleed serum or tissue culture supernatant are used at 1:100–1:1,000 dilution, and ascites fluid or serum from hyperimmunized animals at 1:1,000–1:100,000 dilution. Secondary antibodies are used at dilutions ranging from 1:2,000 to 1:10,000
	Detection enzyme is inactivated	Sodium azide is a powerful inhibitor of HRP, but sodium azide can be used with alkaline phosphatase conjugated antibodies without harmful effects. In addition, tap water or water deionized with polystyrene resins may inactivate the enzyme conjugate. Only use distilled, deionized water
	Tween-20 in washes	Tween-20 may interfere with some antibody-antibody reactions or may wash the protein of interest off the membrane. Tween-20 may be left out of the washes, but this may result in increased background
	Detection system lacks sensitivity necessary to detect the amount of protein loaded	Ensure that the amount of protein loaded lies within the range of sensitivity of the detection system used Horseradish peroxidase 500pg/band Alkaline phosphatase 100pg/band Amplified alkaline phosphatase 5pg/band Colloidal gold 100pg/band Enhanced colloidal gold 10pg/band Chemiluminescence 10pg/band
Bands detected of lower size than expected	Protein degradation or metabolism	Protease inhibitors must be present throughout the protein isolation procedure, it may be also of interest to add specific phosphatase inhibitors to the cocktail of protease inhibitors, especially when aiming at the detection of the phosphorylated form of the protein of interest. In addition, great care should be taken with the storage and the use of the antibodies and conjugating reagents. Freeze-thaw cycles should be kept to a minimum by either aliquoting antibody stock solutions into working aliquots or into aliquots which can be kept at 4°C. It is also always a good idea to keep the antibody aliquots on ice while in use
High background staining hiding the bands	Insufficient blocking	Any area on which either the primary or secondary antibody can bind to the membrane will show positive signal upon subsequent signal development

Table 8-5 (continued)

Problem	Origin	Solution
	Excessive signal development	The length of time for signal development varies with the blot, the number or protein bands, and the desired sensitivity. If the membrane is left in the developing solution (colorimetric) or apposed to film for too long, excess precipitate or light formed by the enzyme can cause high background
	Insufficient washing	The membrane should be washed at least twice after each antibody step in a large volume of TBST (0.05–0.1% Tween-20). Other detergents can be used at very low concentrations, but these stronger detergents can decrease the binding of antibodies, and cause the antigen to be washed off the membrane. This approach is generally not recommended
	Contaminated transfer buffer or apparatus	The transfer apparatus and the fiber pad used to hold the gel in the gel cassette should be thoroughly cleaned with detergents between use. Dirty fiber pads are a major cause of blot contamination. Also do not re-use transfer buffer, prepare fresh solution
	Improper antibody dilution	The dilution of the primary antibody that will give maximum sensitivity with low background depends upon the source of the antibody and must be empirically determined. Insufficient dilution of either the primary antibody or the secondary antibody can cause non-specific binding and high background
	Low grade secondary antibody	The secondary antibody used should be affinity purified and of blotting grade. Such antibodies are affinity purified and cross absorbed against other species of IgG to eliminate non-specific binding

using a fixed-time approach. In the fixed-time approach, after a given incubation time, the enzymatic reaction is stopped by the addition of a denaturing acid or base. Quantitation of the product yields a calibration curve in which product concentration decreases with increasing free antigen concentration.

Noncompetitive ELISA methods are based on "sandwich assays" in which an excess supply of immobilized primary antibody, the "capture antibody," quantitatively binds the antigen of interest and an enzyme-labeled secondary antibody is then allowed to react with the bound antigen forming a sandwich. A color reaction product produced by the enzyme is then used to measure the enzyme activity that is bound to the surface of the microtiter plate. Sandwich ELISA (noncompetitive) methods yield calibration curves in which enzyme activity increases with increasing free antigen concentration.

4.1 Antibody Sandwich ELISA Assays

Given their high degree of sensitivity, which is 2–5 times more than assays in which the antigen is directly bound to the solid phase, antibody sandwich ELISA is arguably the most useful of the immunosorbent assays. This fast and accurate assay is used to determine concentrations of antigens in unknown samples and, if a purified antigen standard is acquired, this assay allows the investigator to determine the absolute amount of antigen in a sample. Antibody sandwich ELISA is the most commonly used procedure to screen and detect soluble antigen. The general procedures involved for detection of the antigen of interest are to coat the wells of microtiter plates with the capture antibody and incubate with the test solution (e.g., tissue homogenate), which contains the antigen analyte. All unbound antigen is then washed away and second antigen-specific antibody conjugated to an enzyme is added and incubated. The unbound conjugate is washed away and a color substrate is added. After another incubation period, the degree of substrate hydrolysis is measured spectrophotometrically on a microtiter plate reader. In this assay, the amount of substrate hydrolyzed is proportional to the quantity of antigen in the test solution. Many kits are commercially available that include microtiter plates precoated with the specific capture antibody/antibodies, enzyme conjugates, and color substrate. However, often the kits are very expensive. When coating your own plates, the optimal concentrations of the capture antibody, conjugate, and substrate will need to be determined. This can be accomplished by using crisscross serial dilution analyses. Standard curves are run with each assay to determine the concentrations of antigen in the test solution. The antibody sandwich assay requires relatively large amounts of pure or semipure specific capture antibodies.

The double or two antibody sandwich ELISA uses two antibodies that bind to epitopes that do not overlap on the antigen. This can be achieved using two monoclonal antibodies that recognize discrete sites or affinity-purified polyclonal antibodies. The two antibody sandwich ELISA is most useful for antigen screening and epitope mapping. It requires a capture antibody that is specific for immunoglobulin from the immunized species and takes longer to perform than the antibody sandwich method. However, the two antibody method does not require a purified antigen.

The major advantages of sandwich ELISA are that the antigen does not need to be purified prior to use, and that these assays are very specific. The major disadvantage of these assays is that not all antibodies can be used. It is important that if monoclonal antibody combinations are considered for use, they must first be qualified as "matched pairs," meaning that they can recognize separate epitopes on the antigen such that they do not hamper each other's binding. In contrast to Western blots, which use precipitating substrates, ELISA protocols use substrates that produce soluble products. Preferably the enzyme substrates should be stable, safe, and inexpensive. The most popular enzymes used are those that convert a colorless substrate to a colored product such as p-nitrophenylphosphate (pNPP), which is converted to the yellow colored p-nitrophenol by alkaline phosphatase. The most common substrates used with peroxidase include 2,2′-azinodiethyl-benzthiazoline sulfonate (ABTS), o-phenylenediamine (OPD) and 3,3′5,5′- tetramethylbenzidine base (TMB), which yield green, orange, and blue colors, respectively. Please see ➋ *Table 8-6* for commonly used enzyme–substrate combinations.

In general, four factors help to determine the sensitivity of the sandwich ELISA. These factors are (1) the number of molecules of the first antibody that are bound to the solid phase; (2) the avidity of the first antibody for the antigen; (3) the avidity of the second antibody for the antigen; (4) the specific activity of the second antibody. By diluting or concentrating the antibody solution, the amount of capture antibody that is bound to the solid phase can be adjusted. In contrast, the avidity of the antibodies for the antigen can be altered only by substituting other antibodies. The specific activity of the second antibody is determined by the number and type of labeled moieties it contains.

4.2 Troubleshooting ELISA Assays

- Negative controls are positive. Check for contamination of the substrate solution, enzyme-labeled antibody, or the controls themselves.
- No color in samples and positive controls. Check that reagents have not expired or been inappropriately stored. Ensure appropriate concentrations of reagents. Assess antibody integrity.
- Low level of color in samples and positive controls. Check dilution of enzyme-labeled antibody and concentration of the color substrate.

Table 8-6
ELISA enzyme substrates

Substrate	Reagent to stop reaction	Color of product/wavelength for quantification
Alkaline Phosphatase		
p-Nitrophenyl Phosphate (pNPP)	2M NaOH	Yellow/405 nm
Bromochloroindolyl Phosphate-Nitro blue Tetrazolium (BCIP/NBT)	EDTA Purple	Black/NA
Horseradish Peroxidase		
3,3′,5,5′-Tetramethyl-benzidine (TMB)	1 M Sulfuric Acid (H_2SO_4)	Yellow/450 nm
o-Phenylene Diamine (OPD)	H_2SO_4	Orange-brown/492 nm
2,2′-azinodiethyl-benzthiazoline sulfonate (ABTS)	20% SDS / 50% DMF	Green/410, 650 nm
Chlornaphthol	PBS	Blue-black/NA
3-Amino-9-ethylcarbazole (AEC)	PBS	Red/NA
Diaminobenzidine (DAB)	PBS	Brown/NA

- Color developed in test samples but not positive or negative controls. Likely that the positive control is the source of the problem. Check that it has been stored properly and not expired.
- Color observed by absorbance is unusually low. Check setting of wavelength.
- When rerunning an assay while troubleshooting, change only one factor at a time.

As the variety of commercially available ELISA kits grows seemingly every day, so does the availability of Web sites and technical data sheets available to the researcher. There are many excellent sites available and the ones we often use include, but are not limited to, those of CHEMICON International, BD Biosciences, Protocol Online, and Invitrogen. Additionally, very helpful resources for any laboratory are the volumes of Current Protocols such as Current Protocols in Molecular Biology and Current Protocols in Cell Biology.

Acknowledgments

The authors are grateful to the Canadian Institutes for Health Research, Natural Sciences and Engineering Research Council and Davey Fund for Brain Research for research funding. We are also very grateful to Dr. G.B. Baker and his eternal patience and outstanding grammatical skills.

References

Bjerrum OJ, Heegaard NHH. 1988. CRC handbook of immunoblotting of proteins. Boca Raton, FL: CRC Press; pp 2 v.

Bloom FE, Battenberg E, Milner R, Sutcliffe G. 1986. Molecular biology and histochemical mapping of the nervous system. Neurohistochemistry: modern methods and applications, Vol. 16. Panula R, Paivarinta H, Soinila S, editors. New York: Alan R. Liss Inc.; pp. 3-20.

Buijs RM, Swaab DF, Dogterom J, van Leeuwen FW. 1978. Intra- and extrahypothalamic vasopressin and oxytocin pathways in the rat. Cell Tissue Res 186: 423-433.

Burnette WN. 1981. "Western blotting": electrophoretic transfer of proteins from sodium dodecyl sulfate–polyacrylamide gels to unmodified nitrocellulose and radiographic detection with antibody and radioiodinated protein A. Anal Biochem 112: 195-203.

Carson FL. 1997. Histotechnology: a self-instructional text. Second Edition. Chicago: ASCP Press.

Cattoretti G, Pileri S, Parravicini C, Becker MH, Poggi S, et al. 1993. Antigen unmasking on formalin-fixed, paraffin-embedded tissue sections. J Pathol 171: 83-98.

Coons AH, Creech HJ, Jones RN. 1941. Immunological properties of an antibody containing a fluorescent group. Proceedings of the Society for Experimental Biology and Medicine 47: 200-202.

Cote SL, Ribero-da-silva A, Cuello AC. 1994. Current protocols for light microscopy immunocytochemistry. Immunohistochemistry II, Cuello AC, editor. Chichester : John Wiley & Sons.

Cuello AC, Milstein C, Galfre G. 1983. Preparation and application of monoclonal antibodies for immunohistochemistry and immunocytochemistry. Immunohistochemistry, Cuello AC, editor. Chichester : John Wiley & Sons.

Diano M, Le Bivic A, Hirn M. 1987. A method for the production of highly specific polyclonal antibodies. Anal Biochem 166: 224-229.

Dunn SD. 1986. Effects of the modification of transfer buffer composition and the renaturation of proteins in gels on the recognition of proteins on Western blots by monoclonal antibodies. Anal Biochem 157: 144-153.

Engvall E, Jonsson K, Perlmann P. 1971. Enzyme-linked immunosorbent assay. II. Quantitative assay of protein antigen, immunoglobulin G, by means of enzyme-labelled antigen and antibody-coated tubes. Biochim Biophys Acta 251: 427-434.

Gilead Z, Jeng YH, Wold WS, Sugawara K, Rho HM, Harter ML, Green M. 1976. Immunological identification of two adenovirus 2-induced early proteins possibly involved in cell transformation. Nature 264: 263-266.

Giorno R. 1984. A comparison of two immunoperoxidase staining methods based on the avidin–biotin interaction. Diagn Immunol 2: 161-166.

Harlow E, Lane D. 1988. Antibodies: a laboratory manual. Cold Spring Harbor, NY: Cold Spring Harbor Laboratory; pp. xiii, 726.

Hope BT, Nye HE, Kelz MB, Self DW, Iadarola MJ, et al. 1994. Induction of a long-lasting AP-1 complex composed of altered Fos-like proteins in brain by chronic cocaine and other chronic treatments. Neuron 13: 1235-1244.

Hsu SM, Raine L. 1981. Protein A, avidin, and biotin in immunohistochemistry. J Histochem Cytochem 29: 1349-1353.

Hsu SM, Raine L, Fanger H. 1981. Use of avidin–biotin–peroxidase complex (ABC) in immunoperoxidase techniques: a comparison between ABC and unlabeled antibody (PAP) procedures. J Histochem Cytochem 29: 577-580.

Ino H. 2003. Antigen retrieval by heating en bloc for pre-fixed frozen material. J Histochem Cytochem 51: 995-1003.

Kim SH, Shin YK, Lee KM, Lee JS, Yun JH, et al. 2003. An improved protocol of biotinylated tyramine-based immunohistochemistry minimizing nonspecific background staining. J Histochem Cytochem 51: 129-132.

Kim SH, Jung KC, Shin YK, Lee KM, Park YS, et al. 2002. The enhanced reactivity of endogenous biotin-like molecules by antigen retrieval procedures and signal amplification with tyramine. Histochem J 34: 97-103.

Knudsen KA. 1985. Proteins transferred to nitrocellulose for use as immunogens. Anal Biochem 147: 285-288.

Kumar BV, Lakshmi MV, Atkinson JP. 1985. Fast and efficient method for detection and estimation of proteins. Biochem Biophys Res Commun 131: 883-891.

Kumar V. 1994. Immunofluorescence and enzyme immunomicroscopy methods. Immunochemistry. van Oss CJ, van Regenmortel MHV, editors. New York: Marcel Dekker Inc.

Laemmli UK. 1970. Cleavage of structural proteins during the assembly of the head of bacteriophage T4. Nature 227: 680-685.

Larsson L-I. 1993. Antibody specificity in immunohistochemistry. Immunohistochemistry II. Cuello AC, editors. Chichester: John Wiley & Sons.

Laskey RA, Mills AD. 1975. Quantitative film detection of ^{3}H and ^{14}C in polyacrylamide gels by fluorography. Eur J Biochem 56: 335-341.

Laskey RA, Mills AD. 1977. Enhanced autoradiographic detection of ^{32}P and ^{125}I using intensifying screens and hypersensitized film. FEBS Lett 82: 314-316.

Leppard K, Totty N, Waterfield M, Harlow E, Jenkins J, Crawford L. 1983. Purification and partial amino acid sequence analysis of the cellular tumour antigen, p53, from mouse SV40-transformed cells. Embo J 2: 1993-1999.

MacMillan FM, Cuello AC. 1986. Monoclonal antibodies in neurohistochemistry: the state of the art. Neurohistochemistry: modern methods and applications. Chan-Palay V, Palya SL, editors. New York: Alan R Liss Inc.

Montero C. 2003. The antigen–antibody reaction in immunohistochemistry. J Histochem Cytochem 51: 1-4.

Mullink H, Vos W, Jiwa M, Horstman A, der Valk van P, Walboomers JM, Meijer CJ. 1992. Application and comparison of silver intensification methods for the diaminobenzidine and diaminobenzidine-nickel endproduct of the peroxidation reaction in immunohistochemistry and in situ hybridization. J Histochem Cytochem 40: 495-504.

Murray GI, Foster CO, Ewen SW. 1991. A novel tetrazolium method for peroxidase histochemistry and immunohistochemistry. J Histochem Cytochem 39: 541-544.

Peitsch MC, Kress A, Lerch PG, Morgenthaler JJ, Isliker H, Heiniger HJ. 1989. A purification method for apolipoprotein A-I and A-II. Anal Biochem 178: 301-305.

Pool CW, Buijs RM, Swaab DF, Boer GJ, van Leeuwen FW. 1983. On the way to a specific immunocytochemical localization. Immunohistochemistry. Cuello AC, editor. Chichester : John Wiley & Sons.

Sefton BM, Hunter T, Beemon K. 1980. Temperature-sensitive transformation by Rous sarcoma virus and temperature-sensitive protein kinase activity. J Virol 33: 220-229.

Shi SR, Cote RJ, Taylor CR. 2001. Antigen retrieval techniques: current perspectives. J Histochem Cytochem 49: 931-937.

Sonnenberg H, Milojevic S, Yip C, Veress AT. 1989. Basal and stimulated ANF secretion: role of tissue preparation. Can J Physiol Pharmacol 67: 1365-1368.

Sternberger LA. 1979. *Immunocytochemistry,* Wiley. New York: Second edition.

Streefkerk JG, van der Ploeg M. 1976. Model studies on quantitative aspects of histochemical reactions on peroxidase, with special reference to the effect of methanol on peroxidase activity. Annal Histochem 21: 67-75.

Swaab DF, Pool CW, Nijveldt F. 1975. Immunofluorescence of vasopressin and oxytocin in the rat hypothalamo-neurohypophypopseal system. J Neural Trans 36: 195-215.

Thompson D, Larson G. 1992. Western blots using stained protein gels. Biotechniques 12: 656-658.

Towbin H, Staehelin T, Gordon J. 1979. Electrophoretic transfer of proteins from polyacrylamide gels to nitrocellulose sheets: procedure and some applications. Proc Natl Acad Sci USA 76: 4350-4354.

Tuana BS, Murphy BJ, Yi Q. 1987. Subcellular distribution and isolation of the Ca^{2+} antagonist receptor associated with the voltage regulated Ca^{2+} channel from rabbit heart muscle. Mol Cell Biochem 76: 173-184.

Vandesande F. 1983. Immunohistochemical double staining techniques. Immunohistochemistry. Cuello AC, editors. Chichester: John Wiley & Sons.

Weller TH, Coons AH. 1954. Fluorescent antibody studies with agents of varicella and herpes zoster propagated in vitro. Proc Soc Exp Biol Med 86: 789-794.

Wolters G, Kuijpers L, Kacaki J, Schuurs A. 1976. Solid-phase enzyme-immunoassay for detection of hepatitis B surface antigen. J Clin Pathol 29: 873-879.

9 In Vivo Microdialysis: A Method for Sampling Extracellular Fluid in Discrete Brain Regions

D. L. Krebs-Kraft · K. J. Frantz · M. B. Parent

Abstract: In vivo microdialysis is a prominent method for sampling extracellular fluid (ECF) from the brain. A major strength of this method is that it allows for neurochemical sampling in living, free-moving organisms, rather than in post-mortem tissue. Furthermore, in vivo microdialysis affords researchers the ability to monitor extracellular concentrations of endogenous compounds, administer drugs into the extracellular space of discrete brain regions, sample drug- induced changes in neurotransmitters and/or their metabolites, and correlate behavioural changes with variations in neurochemistry. This chapter reviews the principles underlying the in vivo microdialysis method and will discuss the substances that can and cannot be sampled by this technique. Then, several technical considerations, including probe construction and specifications, perfusate composition, operating parameters, and the use of surgically-implanted versus guided probes will be discussed. Subsequently, quantitative methods for estimating extracellular concentrations of substances and the ways in which this technique can be adapted or combined with other methods, such as drug administration, electrophysiology, and behavioral observation will be reviewed. To illustrate how the in vivo microdialysis method can be applied, the use of this technique in memory and drug addiction research, in studies of ontological development, and in clinical settings will be considered.

List of Abbreviations: ECF, extracellular fluid; HPLC, high performance liquid chromatography; CE, capillary electrophoresis; UV, ultraviolet; EC, electrochemical; LIF, laser-induced fluorescence; TTX, tetrodotoxin; ACh, acetylcholine; Ca^{2+}, calcium; GABA, gamma amino-butyric acid; K^+, potassium; PLZ, phenelzine; aCSF, artificial cerebrospinal fluid; $5HT_6$, serotonin, type 6 receptor; NK(1), non-peptidergic neurokinin; CRF, corticotrophin-releasing factor; PN10, postnatal day 10; c.c., cubic centimeter or milliliter; ° C, degrees Celsius; i.m., intramuscular; i.p., intraperitioneal; s.c., subcutaneous; B.A.S., Bioanalytical Systems, Inc.; NaCl, sodium chloride; KCl, potassium chloride; $CaCl_2$, calcium chloride; $MgCl_2$, magnesium chloride; NaH_2PO_4, sodium phosphate monobasic; Na_2HPO_4, sodium phosphate dibasic anhydrous; S.E.M., standard error of the mean; DA, dopamine; hr, hour; min, minute; nmol, nanomole; μM, micromolar; mM, millimolar; g, gram; mg/kg, milligram per killigram

1 Definition and Description

In the past two decades, in vivo microdialysis has been the foremost method for sampling extracellular fluid (ECF) from the brain. A major strength of this method is that it allows for neurochemical sampling in living organisms, rather than in postmortem tissue. Moreover, recent advances have allowed this procedure to be performed in conscious, freely moving animals, rather than in anesthetized subjects. In addition to affording researchers the ability to monitor extracellular concentrations of endogenous compounds, in vivo microdialysis can also be used to administer drugs into the extracellular space of discrete brain regions, sample drug-induced changes in neurotransmitters and/or their metabolites, and correlate behavioral changes with variations in neurochemistry (Lonnroth et al., 1987; Benveniste, 1989; Benveniste and Huttemeier, 1990; Herregodts, 1991; Gardner et al., 1993; Dawson et al., 1994; Hernandez et al., 1994; Adell and Artigas, 1998b; Parent et al., 2001).

This chapter reviews the principles underlying the in vivo microdialysis method and discusses the substances that can and cannot be sampled by this technique. Then, several technical considerations, including probe construction and specifications, perfusate composition, operating parameters, and the use of surgically implanted versus guided probes will be discussed. Subsequently, quantitative methods for estimating extracellular concentrations of substances and the ways in which this technique can be adapted or combined with other methods, such as drug administration, electrophysiology, and behavioral observation will be reviewed. To illustrate how the in vivo microdialysis method can be applied, the use of this technique in memory and drug addiction research, in studies of ontological development, and in clinical settings is considered. Finally, a sample protocol is provided. It should be noted, that in most cases, the discussion centers around the use of the in vivo microdialysis procedure in the rat.

In vivo microdialysis is based on the principle of dialysis, the process whereby concentration gradients drive the movement of small molecules and water through a semipermeable membrane. In vivo microdialysis involves the insertion of a small semipermeable membrane into a specific region of a living animal, such as the brain. The assembly that contains this semipermeable membrane is called a probe, which is composed of an inlet and an outlet compartment surrounded by a semipermeable membrane (see *Figure 9-1*). Using a microinfusion pump set at a low flow rate (0.2–3 μL/min), an aqueous solution known as the perfusate is pumped into the inlet compartment of the microdialysis probe. Ideally, the

Figure 9-1
Schematic illustration of an in vivo microdialysis set-up

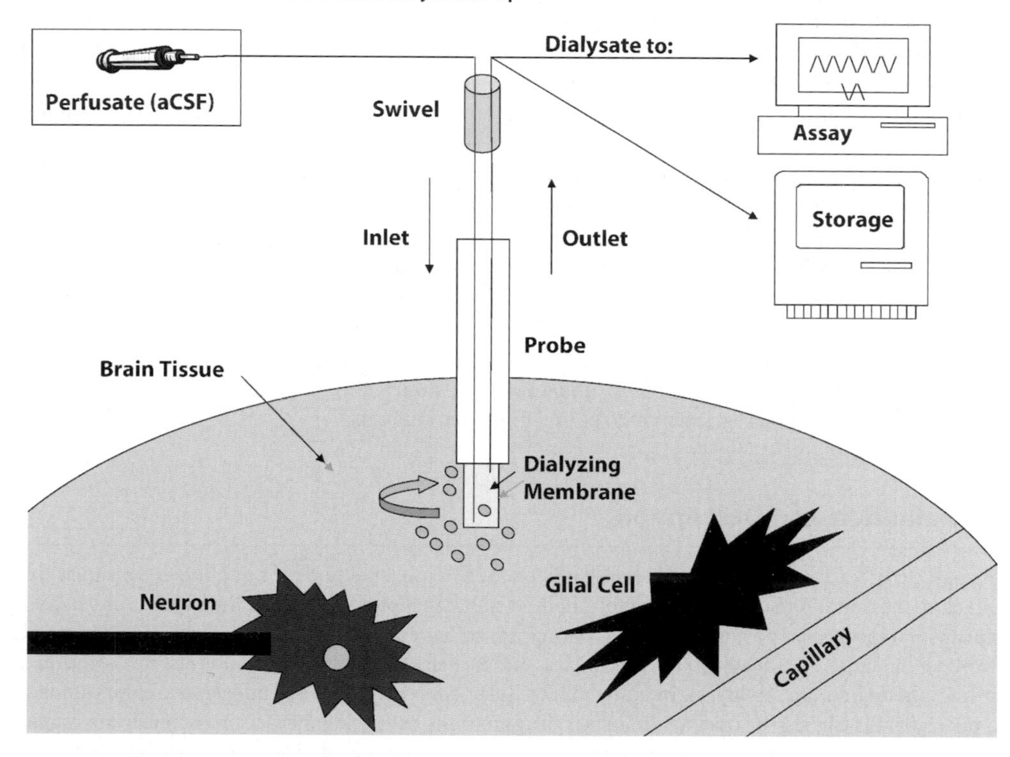

perfusate mimics the ionic composition of the brain ECF to prevent unsolicited changes in the composition and activity of neurochemicals in the ECF surrounding the probe. Endogenous substances that are able to pass across the semipermeable membrane, such as neurotransmitters, diffuse down their concentration gradients through the semipermeable membrane, into the outlet compartment, and out of the probe. The resulting solution that is collected is referred to as the dialysate, which consists of a mixture of components from both the perfusate and the ECF. For a number of reasons, the concentration of an endogenous substance in the dialysate will be lower than that in the ECF surrounding the dialysis probe. The term "recovery" is used to describe this relationship between in vivo and dialysate concentrations.

Microdialysis is a sampling technique that must be coupled with an analytical method to identify and quantify chemical components of the dialysate. The samples can be analyzed immediately upon collection (i.e., online), or they can be stored (−80°C) for future analysis. Only analytical techniques sensitive enough to measure both small sample volumes and low concentrations of substances can be used to measure compounds in dialysate samples. High-performance liquid chromatography (HPLC) or capillary electrophoresis (CE) combined with ultraviolet (UV), electrochemical (EC), or laser-induced fluorescence (LIF) detection procedures have been employed successfully for the analysis of neurochemicals (Gardner et al., 1993; Dawson et al., 1994, 1997; Hernandez et al., 1994; Kostel and Lunte, 1997; Chen et al., 2001; Kennedy et al., 2002b).

2 What is Sampled?

Neurotransmitters, peptides, and other endogenous substances in the extracellular space can be sampled using in vivo microdialysis. The dialysis membrane typically excludes the transport of larger molecules and enzymes that could otherwise interfere with analysis of the substances of interest. In the case of neurotransmitters, the levels in dialysates are the net result of the interaction between processes affecting release into and removal from the extracellular space. Consequently, in vivo microdialysis can only sample a neurotransmitter that has not yet been removed by clearing mechanisms and this method does not provide information regarding intracellular levels of substances.

Before the origin of neurotransmitters in dialysate can be considered neuronal, specific experimental criteria must be fulfilled (Westerink et al., 1987; Westerink, 1995; Timmerman and Westerink, 1997). Two methods that can be used effectively to verify that an increase in a neurotransmitter is derived from neuronal sources examine the dependency of this elevation on action potentials and calcium-dependent exocytosis. If an increase in ECF neurotransmitter levels is stimulated by action potentials, then adding a sodium channel antagonist, such as tetrodotoxin (TTX), to the perfusate should significantly decrease dialysate levels of that neurotransmitter. Similarly, basal and stimulated levels of neurochemicals that are neuronal in origin should be decreased when calcium (Ca^{2+}) is omitted from the perfusate and/or when Ca^{2+} chelators are added. These two criteria have been fulfilled most clearly for dopamine and acetylcholine (Westerink and De Vries, 1988; Imperato et al., 1989; Drew et al., 1990; Osborne et al., 1990; Anderson et al., 1993; Consolo et al., 1994; Frantz et al., 2002). On the other hand, dialysate levels of serotonin, norepinephrine, glutamate, gamma amino-butyric acid (GABA), and glycine appear to be partially or entirely Ca^{2+}-independent and TTX-insensitive, suggesting that their efflux is at least partially nonneuronal and nonexocytotic in origin (Westerink et al., 1987; Westerink, 1995; Timmerman and Westerink, 1997; Del Arco et al., 2003). In fact, the majority of basal and stimulated GABA and glutamate dialysate levels are TTX- and Ca^{2+}-insensitive, although recent evidence raises the possibility that chemical conditions of the analytical techniques used to measure GABA levels may mask the TTX- and Ca^{2+}-dependence of GABA levels in dialysates (Rea et al., 2003). Despite these latter findings, the bulk of the current evidence suggests that GABA and glutamate in dialysates are derived from nonneuronal and nonexocytotic sources, such as glia, protein metabolism, or reversal of reuptake mechanisms (Westerink et al., 1987; Herrera-Marschitz et al., 1996; Jabaudon et al., 1999; Parent et al., 2001; Frantz et al., 2002). Consistent with this possibility, functional and structural evidence suggests that neuronal glutamate and GABA are involved in only short distance signaling and therefore may not even reach the microdialysis probe (Del Arco et al., 2003).

Another method that has been used to investigate the neuronal origin of neurotransmitters is the addition of high, depolarizing concentrations of potassium (K^+) to the perfusate to induce exocytotic release of neurotransmitters (Blandina et al., 1996; Herzog et al., 2003; Tanaka et al., 2004). The results of such studies have shown K^+-induced increases in extracellular levels of acetylcholine (Nilsson et al., 1990; Blandina et al., 1996; Herzog et al., 2003; Tanaka et al., 2004), dopamine (Galindo et al., 1999; Stanford et al., 2001; Alfonso et al., 2003; Carboni et al., 2003), GABA (Tossman and Ungerstedt, 1986; Timmerman et al., 1992; Hada et al., 2003), glutamate (Tossman and Ungerstedt, 1986; Szerb, 1991; Fujikawa et al., 1996; Galvan et al., 2003; Takeda et al., 2003), and serotonin (Kalen et al., 1988; Mateo et al., 1990; Inoue et al., 2002; de Groote et al., 2003). However, it is important to keep in mind that such findings do not unequivocally indicate neuronal sources for the neurotransmitter increase, particularly in the case of GABA. Elevated K^+ may cause glial cells to release GABA (Paulsen and Fonnum, 1989; Westerink and de Vries, 1989; Silverstein and Naik, 1991; Hondo et al., 1995) or may disturb the osmotic balance of cells (Szerb, 1991; Liu and MCAdoo, 1993), which could lead to elevations of neurotransmitters in the ECF (Westerink et al., 1987; Timmerman and Westerink, 1995).

Thus, when interpreting data from in vivo microdialysis studies, it is important to be cognizant that one is sampling extracellular levels of a substance, that changes can reflect clearance and/or release, and that the source of the change is not necessarily neuronal. It is equally important, however, to keep in mind that experimentally induced alterations in dialysate levels of a neurochemical are meaningful, regardless of the source of that change. For instance, the antidepressant phenelzine causes dramatic increases in brain extracellular GABA levels (Parent et al., 2002) (see *Figure 9-2*). This finding is important for understanding the mechanisms underlying the therapeutic effects of phenelzine, and the possibility that the increase in GABA could be derived primarily from glial sources does not detract from the significance of this result.

Figure 9-2

(a). Schematic illustration of coronal sections of the rat brain showing the approximate location of microdialysis probes in the caudate-putamen. Atlas plates were adapted from Paxinos & Watson (1986). (b). Effects of systemic administration of vehicle or phenelzine (PLZ; 15 or 30 mg/kg) on mean (+/−S.E.M.) % of baseline extracellular GABA levels in the caudate-putamen. Systemic infusions of PLZ increased extracellular GABA levels in the caudate-putamen (*$p < 0.05$ vs. vehicle-same sample period). Figure from Parent, M.B., Master, S., Kashlub, S., & Baker, G.B. (2002). Effects of the antidepressant/antipanic drug phenelzine and its putative metabolite phenylethylidenehydrazine on extracellular γ-aminobutryic acid levels in the striatum. *Biochemical Pharmacology, 63*, 57-64

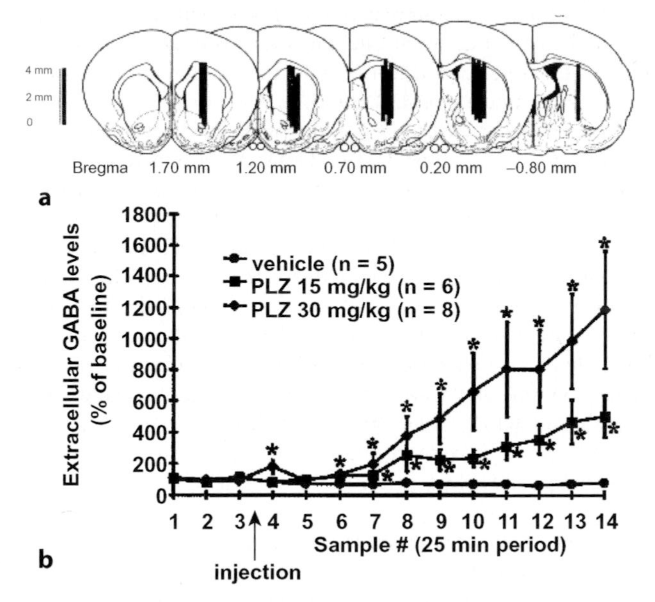

3 Advantages of Conscious Preparations

When in vivo microdialysis procedures were first introduced, the experiments were typically performed in anesthetized preparations. One benefit was that anesthesia reduced the influence of exogenous factors that could alter neural activity and allowed drugs to be administered with little difficulty (Adell and Artigas, 1998b). However, the use of an anesthetic is ultimately undesirable because the anesthetic agent may interact with the compounds of interest. In addition, experiments with anesthetized animals typically have not included a sufficient stabilization period (i.e., the interval between the time a probe is inserted and samples are collected; see Section 4.4). The introduction of mechanical fluid swivel mounts to minimize movement-induced tangling in the dialysis tubing has facilitated microdialysis sampling procedures in conscious, freely moving animals. Early comparative analyses indicated that different patterns of results were obtained in anesthetized versus conscious preparations (Albrecht et al., 2000; Bongiovanni et al., 2003; Ramirez-Munguia et al., 2003). In addition to avoiding artifacts produced by anesthesia, the use of conscious, freely moving animals allows researchers to correlate behavioral changes with neurochemical variations (see Sections 6 and 7).

4 Technical Considerations

4.1 Probe Construction and Membrane Specifications

Microdialysis probes can be assembled either in individual laboratories or made commercially to suit experimental requirements. The two principal types of dialysis probes are transcerebral and vertical. The transcerebral probe provides a large area for dialysis and thus can sample from sizeable brain regions, such as cortex and striatum. The recovery of endogenous substances with such probes is high because of the large surface area. However, problems with accessing deep brain structures as well as the particularly invasive nature of the associated surgical procedures discourage the use of this type of probe. Vertical probes, which can be subdivided further into concentric or loop subtypes, do not have these limitations of the transcerebral probe and are thus more commonly used (Adell and Artigas, 1998a). The loop (also known as U-shaped) probes have a larger surface area and provide greater recovery of neurochemicals than do concentric probes (Horn and Engelmann, 2001). However, the loop-type probes cause more tissue damage than the concentric types (Horn and Engelmann, 2001).

The length and diameter of the semipermeable probe membrane determine the spatial resolution of the in vivo microdialysis procedure. Typical membranes range from 1 to 4 mm in length and from 150 to 500 mm in diameter. The choice of membrane length and diameter is a balance between analyte recovery and anatomical specificity. A longer membrane with a wider diameter is preferred because an increase in membrane surface area augments the recovery of extracellular compounds (Benveniste et al., 1984; Tossman and Ungerstedt, 1986; Ungerstedt, 1991b; Kehr, 1993). However, the size of the brain region under investigation also influences the appropriate membrane length and diameter (Tossman and Ungerstedt, 1986; Parsons and Justice, 1994; Khan and Shuaib, 2001). For instance, to restrict sampling to a small brain region, such as the substantia nigra, ventral tegmental area, or median eminence in the rat brain, a small 1-mm membrane length should be used. It is more difficult to preserve specificity with even smaller brain regions or subregions of nuclei, and the use of smaller membranes yields poor recovery (Westerink, 1995). Of course, it would be possible to sample more accurately and specifically from such brain regions in a larger species, such as nonhuman primates or humans in clinical settings.

The active dialyzing membranes of the probes are made of various materials, such as cellulose, copolymer, acrylic copolymer, and polysulfone (Levine and Powell, 1989; de Lange et al., 2000). The membrane should be inert to the neurochemical being recovered, so that the diffusion properties of the neurochemical are not affected (Ungerstedt, 1991a; Khan and Shuaib, 2001). To date, evidence regarding the effects of different semipermeable membranes on recovery is equivocal (Hsiao et al., 1990; McNay and Sherwin, 2004). For instance, the manufacturer of one commercially available microdialysis probe has shown that cellulose membranes provide better in vitro recovery than other membranes (Bioanalytical Systems, 1988; Hsiao et al., 1990). In vitro recovery refers to the efficiency of the probe in extracting a substance when the probe is immersed in a solution containing a known concentration of that substance. However, it is not clear whether in vitro recovery reflects in vivo recovery from the brain; there are marked differences in the diffusion characteristics of in vitro solutions versus those of brain ECF (Nicholson and Rice, 1986; Hsiao et al., 1990). For instance, polycarbonate-ether, regenerated cellulose, and polyacrylonitrile membranes do not exhibit differential recovery of neurochemicals in the ECF, whereas the polyacrylonitrile membrane has the best recovery in vitro (Hsiao et al., 1990). Recent evidence suggests that membrane types may also affect recovery in vivo; a cuprophan membrane provides significantly more efficient recovery of glucose from the ECF of the hippocampus than does a polycarbonate membrane (McNay and Sherwin, 2004). Whether the difference in glucose recovery is due entirely to membrane material is unclear, however, because the probes also varied in diameter in that study (McNay and Sherwin, 2004).

4.2 Perfusate Composition

Solutions that have been used as perfusates for brain in vivo microdialysis experiments include 0.9% saline, Ringer's solution, modified Ringer's solution, and artificial cerebrospinal fluid (aCSF) (Benveniste and Huttemeier, 1990; Khan and Shuaib, 2001). For several reasons, it is critical that the composition, pH, and temperature of the perfusate mimic brain ECF (Hansen, 1985; de Lange et al., 2000; Khan and Shuaib, 2001). For example, if a substance present in the ECF is omitted from the perfusate, then this substance will be drawn out of the ECF and into the probe because of the concentration gradient. One critical ECF constituent that is commonly omitted is glucose, which likely results in the extraction of glucose from the ECF and changes in neural functioning. Similarly, the omission of Ca^{2+} from the perfusate will deplete Ca^{2+} in the ECF and consequently impair synaptic transmission (Imperato and Di Chiara, 1984; Westerink and De Vries, 1988). In some instances, the appropriate substances are included, but their concentration in the perfusate does not mimic that found in the ECF. For example, extracellular glucose levels are estimated to range from 1 to 2 mM (Fellows et al., 1992; McNay and Gold, 1999; McNay et al., 2000a, 2001; McNay and Sherwin, 2004). Yet, some researchers have included as high as 7–10 mM glucose in the perfusate. The use of such high glucose concentrations is a concern, because similar amounts (6.6. mM) increase extracellular levels of various neurotransmitters (Ragozzino et al., 1998; Parent et al., 2003). In addition, using perfusates at room temperature rather than at brain temperature creates a gradient that changes diffusion characteristics and influences recovery from the probe. Increasing the temperature of the perfusate augments the movement of molecules and thus promotes the diffusion of substances across the membrane and elevates recovery. Following acetaminophen administration, elevating the temperature from 24 to 38°C enhances or impairs changes in the recovery of acetaminophen from the brain ECF by 96% (de Lange et al., 1994). Similarly, the pH of the perfusate also influences the recovery by affecting Na gradients (Imperato and Di Chiara, 1984; Hansen, 1985; Solis et al., 1986; Vezzani et al., 1988; Westerink and De Vries, 1988; Moghaddam and Bunney, 1989; Herregodts, 1991; Khan and Shuaib, 2001).

4.3 Operating Parameters

The interaction between perfusion flow rate and the interval at which samples are collected determines the volume of each dialysate sample. The volume needed, in turn, is dictated by the demands of the analytical technique that will be used to quantitate the samples. The perfusion flow rate will influence the ability of the probe to extract substances (i.e., recovery) and the sampling interval will determine the temporal resolution of the in vivo microdialysis procedure. Consequently, perfusate flow rates are chosen on the basis of a cost-benefit analysis and typically range between 0.1 and 5 ml/min. Lower flow rates are preferred because they minimize tissue disturbance and disequilibrium and increase recovery from the probe (Wages et al., 1986; Benveniste, 1989; Gonzalez-Mora et al., 1991; Kehr, 1993; Khan and Shuaib, 2001). Nevertheless, higher flow rates are often employed because they correlate with higher absolute recovery (the amount of the analyte in the dialysate per unit time). More importantly, higher flow rates provide the sample volume necessary to conduct reliable chemical separation and detection (Wages et al., 1986). HPLC separation techniques coupled with EC or UV detection of analytes generally require 4–20 μL sample volumes, resulting in limited temporal resolution of usually at least 10 min per sample, depending on the flow rate (Westerink et al., 1987; Kennedy et al., 2002b; Parrot et al., 2003).

As a result of these long sampling periods, a major criticism of the in vivo microdialysis procedure has been a lack of temporal sensitivity (Westerink et al., 1987). This criticism is particularly relevant for amino acid sampling. Relative to other neurotransmitters and to other sources of GABA and glutamate, neuronally derived GABA and glutamate appear to be present in the synapse in very small amounts for only brief periods. As a result, the sensitivity of dialysis probes and the typical length of sampling intervals do not allow the detection of extracellular GABA and glutamate involved in synaptic transmission (Westerink et al., 1987). The lack of effective temporal resolution may be overcome by analyzing samples via separation by CE and detection by LIF (Bergquist et al., 1994, 1996; Zhou et al., 1995; Lada et al., 1997; Bowser and Kennedy, 2001; Kennedy et al., 2002a, b; Parrot et al., 2003). CE–LIF can be performed on small sample volumes (several nanoliters), at short sampling intervals (several seconds), with low thresholds of detection (attomolar), and high reproducibility. To date, CE–LIF is used mainly for amino acid detection, although methodologies for catecholamines are reported (Parrot et al., 2003). An additional modification of the microdialysis methodology involves direct sampling of ECF through a capillary rather than through a membrane. This confers even greater spatial and temporal resolution, less tissue damage, lower flow rates, and no loss in sensitivity to in vivo neurotransmitter dynamics, in comparison with more traditional dialysis probes and analytical methods (Kennedy et al., 2002a).

4.4 Surgically Implanted Versus Guided Probes

In vivo microdialysis probes can be introduced into the brain via one of two common methods. One way is to surgically implant a probe, which entails lowering the probe into the target brain region and permanently affixing it to the skull at the time of surgery. In vivo microdialysis procedures can be performed immediately after recovery from surgery or while the subject is maintained under anesthesia. The second method involves implanting a guide cannula above the brain region of interest while the subject is under anesthesia (Benveniste et al., 1987; Benveniste, 1989; Benveniste and Huttemeier, 1990). Following a period of recovery, the probe is inserted through the guide into the brain of the conscious animal. For a number of reasons, guided probes are preferable (although there are exceptions, such as dialysis in very young animals—see ➋ Section 3). Guided probes minimize the likelihood that the anesthetic agent interferes with the dialysate composition or the pharmacological action of administered compounds. The use of a guide cannula also affords the ability to insert a probe on more than one occasion (Wellman, 1990; Devine et al., 1993; Kolachana et al., 1994).

Although the diameter of in vivo microdialysis probes is small (100–300 mm), inserting a probe into the brain disrupts neural processes nonetheless. Tissue disruptions immediately after insertion of a surgically implanted probe include initial formation of eicosanoids, local disruption in cerebral blood flow and glucose metabolism, disruption of the blood–brain barrier produced by the probe implantation, and the release of excitatory amino acids resulting from the surgical procedure (Benveniste and Diemer, 1987; Yergey and Heyes, 1990; Liu et al., 1991; Panter and Faden, 1992; Adell and Artigas, 1998b; Khan and Shuaib, 2001). Consequently, it is necessary to include a stabilization period following probe insertion before baseline samples are collected. Performing the chemical assays online as the samples are being collected is one way to identify the necessary stabilization period for the analyte of interest. Otherwise, if the samples are to be assayed at a later time, then a set stabilization period is used. The amount of time needed to attain equilibrium depends in large part on whether a probe is surgically implanted or inserted via a guide cannula. When using surgically implanted probes, the stabilization period must account for recovery from anesthesia and tissue disruptions. Most of these changes normalize within 24 h, indicating that the stabilization period should be preferably at least 24 h after the implantation procedure. Longer periods are not ideal, however, because the presence of gliosis, 2–3 days after surgery, in a small region around the surgically implanted probe can affect the diffusion characteristics of the probe and tissue (Imperato and Di Chiara, 1984; L'Heureux et al., 1986; Benveniste and Diemer, 1987; Ruggeri et al., 1990; Shuaib et al., 1990; de Lange et al., 1995).

Guided probes can dramatically reduce the length of the stabilization period. The use of a guide cannula allows researchers to wait days or weeks after surgery before inserting the probe. By then, the effects of the anesthesia have dissipated, the changes produced by the surgery and implantation have stabilized, and gliosis is not an issue because usually the tip of the guide ends dorsal to the target brain region. Nonetheless, introduction of the probe into the area beyond the guide will disrupt neurotransmitter levels in the vicinity of the probe (Camp and Robinson, 1991; Robinson and Camp, 1991; Georgieva et al., 1993; Kolachana et al., 1994). For instance, basal extracellular dopamine levels increase dramatically after guided probe insertion and stabilize within 20–40 min and metabolites stabilize 2–3 h later (Kalivas and Duffy, 1990; Pettit and Justice, 1991). Most of these neurochemical changes stabilize within 2 h after guided probe insertion, suggesting the ideal length for the stabilization period is 2–3 h after probe introduction. However, the stabilization period may be different for each analyte and should ideally be empirically determined.

The method of probe insertion also influences the total length of the experimental procedure and the number of times that samples can be obtained from the same animal. For instance, the use of a guide cannula allows for multiple insertions of the probe at different time points, whereas surgically implanted probes are typically used for continuous sampling. Overall, the evidence suggests that repeated insertion is less disruptive than continuous sampling over days; gliosis is minimized and recovery remains more stable (Imperato and Di Chiara, 1984; Korf and Venema, 1985; Westerink and Tuinte, 1986; Benveniste and Diemer, 1987; Benveniste et al., 1987; Reiriz et al., 1989; Wellman, 1990; Devine et al., 1993; Georgieva et al., 1993; Kolachana et al., 1994). For example, basal levels of dopamine measured in the mesostriatal system do not differ when the same probe is inserted twice, 1-week apart (Kalivas and Duffy, 1990; Pettit and Justice, 1991; Georgieva et al., 1993). In contrast, the amount of dopamine recovered declines progressively under conditions of continuous sampling from the brain over 7–10 days (Imperato and Di Chiara, 1984; Korf and Venema, 1985; Westerink and Tuinte, 1986; Reiriz et al., 1989; Camp and Robinson, 1992).

5 Quantitative Dialysis: Estimating Extracellular Levels

5.1 Introduction

For the majority of researchers, knowledge of the actual concentration of a neurotransmitter or other analyte in vivo is not necessary; the change in concentration from baseline levels is of primary interest (Glick et al., 1994). Under conditions in which extracellular concentrations do need to be reported, multiple factors that influence in vivo recovery from probes must be taken into account (Justice, 1993; Kehr, 1993; Parsons and Justice, 1994). Influential factors include probe characteristics such as membrane composition, dialysis operating parameters such as flow rate, and biological processes such as analyte reuptake from ECF. To date, the most accurate approaches in quantitative microdialysis use the variation in flow rate to plot functions that reveal extracellular concentrations of analytes. These methods include interpolation to the point of no net flux of analyte across the membrane, extrapolation to zero flow across the membrane, and the low flow method. These methods are applied to the quantification of exogenous as well as endogenous compounds (Hurd et al., 1988; Menacherry et al., 1992). Several comprehensive papers are available for researchers interested in mathematical models for most accurately determining extracellular concentrations of neurotransmitters (Benveniste, 1989; Benveniste and Huttemeier, 1990; Parsons and Justice, 1994; Adell and Artigas, 1998b; Peters and Michael, 1998; Stahle, 2000).

5.2 Adjustments for In Vitro Recovery

Previously, in vitro recovery was the most commonly used method for estimating ECF concentrations of a substance (Benveniste, 1989; Stahle et al., 1991). To determine in vitro recovery, the probe is immersed in a known concentration of the analyte, preferably at brain temperature, and perfused with a medium free of the analyte. Percent recovery (or relative recovery) is defined as the ratio between two measures: (a) the concentration of the analyte that is recovered from the probe and (b) the known concentration. In vitro calibration is limited and no longer considered appropriate, because it fails to factor in physiological factors, such as extracellular tortuosity and neurochemical reuptake, which influence in vivo but not in vitro recovery (Benveniste, 1989; Benveniste and Huttemeier, 1990; Bungay et al., 1990; Hsiao et al., 1990; Morrison et al., 1991; Parsons et al., 1991b; Parsons and Justice, 1992; Stahle, 2000).

5.3 Interpolation to No Net Flux

Sampling from ECF is based on laws of diffusion, such that molecules in greater concentration outside the probe flow across the membrane into the dialysate, whereas molecules in greater concentration inside the probe (i.e., in the perfusate medium) flow across the membrane into the brain tissue (Ungerstedt, 1984; Sharp et al., 1986). Taking advantage of this process, the no net flux method (also known as the difference method) involves perfusing concentrations of an analyte estimated to be below or above the extracellular concentration in tissue through the probe in random order, thereby creating conditions in which analyte molecules are either gained from or lost to the tissue (Lonnroth et al., 1987, 1989; Peters and Michael, 1998). Regression analysis of the relationship between the analyte concentration in the perfusate and the mathematical difference between [analyte in perfusate]−[analyte in dialysate] reveals a curve that crosses zero when analyte concentration in the perfusate equals that in the dialysate (i.e., the point of no net flux; see *Figure 9-3*). In addition, the slope of the no net flux regression line may represent the rate of recovery from the probe (Lonnroth et al., 1987, 1989; Parsons and Justice, 1994; Peters and Michael, 1998). The

Figure 9-3
Graph used to calculate the point of no net flux for dopamine (DA). Using regression analysis, the extracellular concentration of DA is estimated via the difference method: [the DA concentration in the perfusate minus the concentration of DA in the dialysate] plotted against the DA concentration in the perfusate. Values above the zero on the y-axis indicate diffusion to the brain, whereas values below the zero indicate diffusion from the brain. The zero point on the y-axis represents a steady state, at which no net flux of DA occurs across the dialysis membrane and represents the extracellular concentration of DA on the x-axis. Figure from Parsons, L.H., Justice, J.B., Jr. (1994). Quantitative approaches to in vivo brain microdialysis. *Crit Rev Neurobiol.* 8(3): 189-220

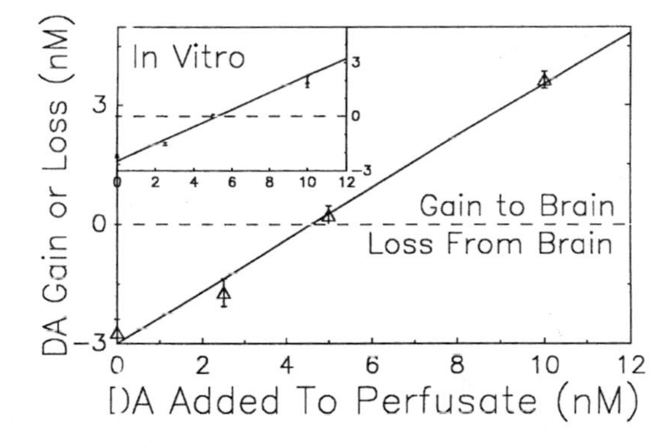

accuracy of the method has been confirmed by comparing results obtained with the more challenging methodologies described later. The technique is relatively straightforward, but requires lengthy experimental protocols due to multiple changes in perfusate solution and equilibration times for each new solution. Sequential presentation of multiple perfusate solutions to a single subject prohibits the use of this method for capturing dynamic changes in concentration or recovery; no net flux is a steady-state model. On the other hand, a variant, using no net flux between subjects, responds effectively to this limitation by presenting individual subjects in a group with different analyte concentrations in the perfusate and plotting their dialysate concentrations together in a no net flux regression at varying time points before and after the experimental manipulation (Olson and Justice, 1993; Yim and Gonzales, 2000).

5.4 Extrapolation to Zero Flow

An alternative approach to estimating the point at which dialysis sampling does not influence extracellular analyte concentrations is extrapolation to zero flow (Jacobson et al., 1985). The flow rate within or between subjects is gradually decreased, and analyte concentrations measured in dialysates are plotted against decreasing flow rates such that a regression analysis approximates the analyte concentration at the point of zero flow. This technique is also quite straightforward, but increasingly low flow rates provide decreasingly utilizable sample volumes or require increased sampling intervals. In the former case, sample analysis is compromised; in the latter case, temporal resolution is sacrificed. Moreover, the mathematical extrapolation is less reliable than is the interpolation of the no net flux approach.

5.5 Low Flow Method

As flow rates decrease, the perfusion medium in the probe approaches equilibrium with the ECF (Wages et al., 1986). Therefore, the dialysate concentration of an analyte sampled at very low flow rates more closely approximates the concentration in the extracellular environment (Menacherry et al., 1992). Like no net flux and the zero flow models, this is another steady-state analysis with limited application to transient changes based on behavior or pharmacological manipulations. However, the advent of new techniques in analytical chemistry requiring only small sample volumes from short sampling intervals may signal a potential return to the low flow method.

6 Combined with Other Methods

6.1 Drug Administration

Pharmacological manipulations are often incorporated into in vivo microdialysis experiments. Coupling in vivo microdialysis with systemic drug administration permits the examination of the effect(s) of the drug on extracellular levels of neurochemicals in discrete neural regions. In vivo microdialysis sampling can also be used to characterize the diffusion of systemically administered drugs across the blood–brain barrier. For example, several studies have used in vivo microdialysis to quantify the entry of cocaine into the brain following systemic administration (Hurd et al., 1988; Nicolaysen et al., 1988; Pettit et al., 1990). In experiments involving systemic drug administration, it is important to include vehicle and handled controls, because the drug administration procedure may influence neurotransmitter function through stress or the activation of neuronal groups sensitive to sensory stimuli (Abercrombie and Zigmond, 1995; Westerink, 1995; Fornal et al., 1996; Adell et al., 1997; Timmerman et al., 1999).

One limitation of systemic drug administration is that the effect of the drug on brain function is not restricted to specific neural regions. To identify neural substrates involved in drug effects, in vivo microdialysis probes can be used to apply chemicals to restricted brain areas through "reverse dialysis" or "retrodialysis." In this case, the direction of movement is reversed because the concentration of the exogenous substance is intentionally higher in the perfusate than in the ECF. As a result, the exogenous substance is transferred into the extracellular space from the perfusate. In such experiments, baseline samples are collected and then a control perfusion medium is replaced with a medium containing the compound(s) of interest or the concentration of the compound is increased. For example, elevating the concentration of glucose in the perfusate (from 2 to 6.6 mM) increases hippocampal extracellular taurine levels (Parent et al., 2003; see ➲ *Figure 9-4*). In addition to being valuable for assessing regional differences in drug effects, reverse dialysis is also useful for examining the brain-related effects of substances that do not readily cross the blood–brain barrier (Adell and Artigas, 1998a).

One drawback of delivering drugs to the brain via reverse dialysis is that the amount of substance that is actually delivered is difficult to determine. Instead, known quantities of drugs can be injected using a microcannula that is either attached to the probe or implanted into a different brain region (Frothingham and Basbaum, 1992; Yadid et al., 1993; Ragozzino and Gold, 1995; de Groot et al., 2003). For instance, we have conducted experiments in which drugs are delivered into the medial septum via a cannula, and dialysate samples are collected from the hippocampus, a brain region that receives many efferent projections from the septum (de Groot et al., 2003; Parent et al., 2003). The findings of these experiments reveal

Figure 9-4
Effects of hippocampal perfusions of glucose and septal infusions of muscimol on mean (+/−S.E.M.) % of baseline extracellular taurine levels in the hippocampus. Elevating hippocampal glucose (6.6 mM) increases hippocampal taurine extracellular levels. Septal infusions of muscimol (5 but not 1 nmol) prevent these glucose-induced increases in extracellular taurine

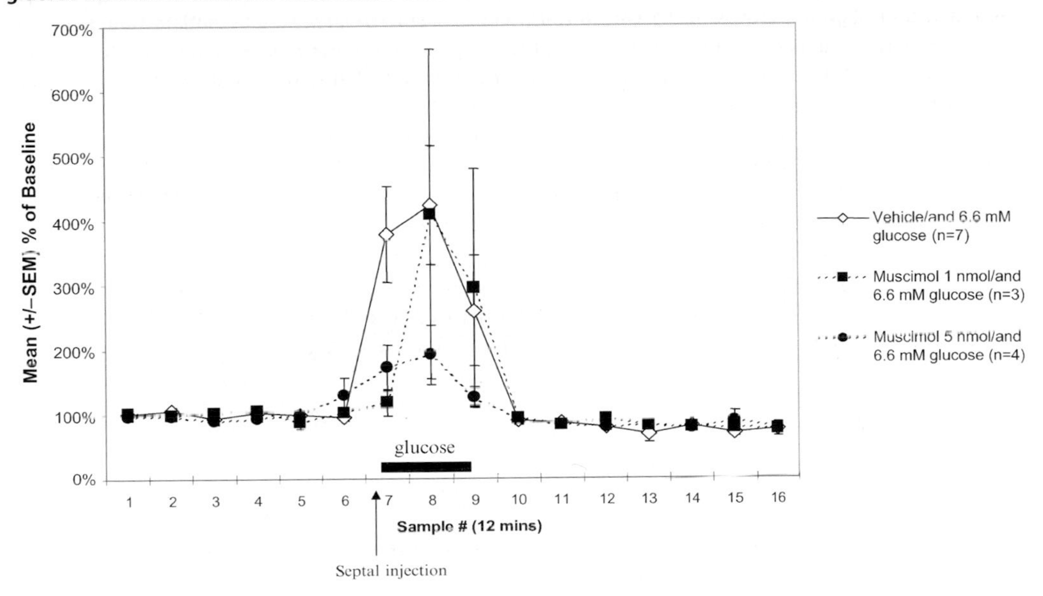

neurochemical interactions between the two brain regions. For example, septal infusions of the GABA agonist muscimol prevent glucose induced increases in hippocampal extracellular taurine levels (Parent et al., 2003; see *Figure 9-4*). The microcannula approach is also useful for the delivery of high-molecular-weight compounds that cannot cross the dialysis membrane or for the delivery of certain substances that would adhere to the probe (Frothingham and Basbaum, 1992; Yadid et al., 1993; Adell and Artigas, 1998a).

Another approach that can be used to study neurochemical circuitry is a dual probe design (Moor et al., 1994; Bianchi et al., 1998; Enrico et al., 1998; Mateo et al., 1998; Moor et al., 1998; Westerink et al., 1998; Matsumoto et al., 2003; Geranton et al., 2004). In this case, two microdialysis probes are inserted into two different brain regions. Reverse dialysis can be used to deliver drugs in one or both regions, and both areas can be sampled simultaneously. Another application of the dual probe approach is to analyze region-dependent effects of systemically administered drugs. For instance, systemic administration of the serotonin [5-HT(6)] receptor antagonist drug SB 258510A enhances amphetamine-stimulated increases in extracellular dopamine levels to a greater degree in the frontal cortex than in the nucleus accumbens, two terminal regions of the mesocorticolimbic dopamine projection pathway (see *Figure 9-5*).

6.2 Electrophysiology

Functional interactions between brain regions can also be characterized by combining electrical stimulation with in vivo microdialysis. A prototypical experiment in this area involves using an electrode to stimulate the cell body region of neurons and a dialysis probe to monitor analytes in the terminal region (Taber and Fibiger, 1993, 1994, 1995; Adell and Artigas, 1998b; You et al., 1998). Moreover, a modified probe incorporating a chlorided silver electrode permits simultaneous recording of extracellular currents and measurement of analytes in the same neural region (Ludvig et al., 1994; Obrenovitch et al., 1994). Similarly, combining reverse dialysis with electrophysiology to record extracellular currents extends the application of in vivo microdialysis to include electrophysiological effects of pharmacological manipulations (Ludvig and

Figure 9-5

Effects of systemic pre-treatment with 3 mg/kg SB 258510A on amphetamine-induced increases in mean (+/−S.E.M.) % of baseline extracellular dopamine in the frontal cortex and nucleus accumbens (using dual-probe in vivo microdialysis procedures in the same animal). Pretreatment with SB 258510A significantly potentiated the amphetamine-induced increase in extracellular dopamine in the frontal cortex (*p < .05 vs. pretreated control group). Pretreatment with SB 258510A produced a trend toward potentiated amphetamine-induced increases in dopamine in the nucleus accumbens. Figure from Frantz, K. J., Hansson, K.J., Stouffer, D.G., and Parsons, L.H. (2002). 5-HT (6) receptor antagonism potentiates the behavioural and neurochemcial effects of amphetamine but not cocaine. *Neuropharmacology*, 42(2): 170-180.

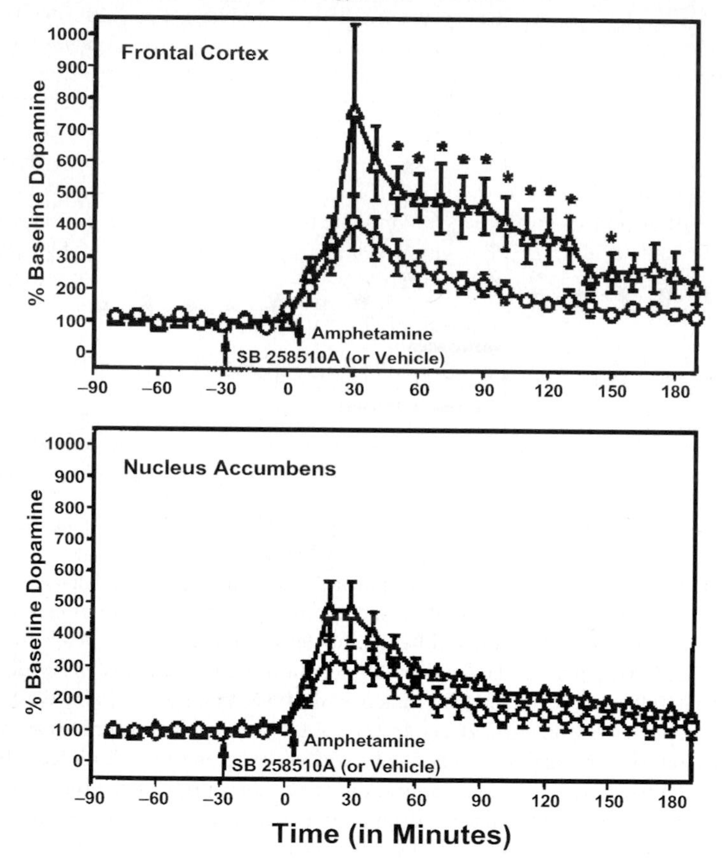

Tang, 2000; Sakai and Crochet, 2000). Another modification of in vivo microdialysis probes involves the attachment of a stimulating electrode to the microdialysis probe, which permits the examination of an electrically evoked neurotransmitter release under in vivo conditions (Adell and Artigas, 1998b; West et al., 2002; Thakkar et al., 2003). Other investigators use the two methodologies in parallel to maximize data collection. For example, recent findings demonstrate that systemic infusions of a drug affect electrophysiological properties in one brain region and neurochemical levels in another (Lejeune et al., 2002). Peripheral administration of the selective, nonpeptidergic neurokinin [nk(1)] receptor antagonist GR205, 171 dose-dependently enhances the firing rate of ventral tegmental area neurons and increases extracellular dopamine levels in the frontal cortex, but not in the striatum and nucleus accumbens. Furthermore, the antagonist is selective for the dopaminergic pathways so that the firing rate of serotonergic neurons and serotonin extracellular levels are unaffected by the antagonist (Lejeune et al., 2002).

6.3 Behavioral Analysis

As mentioned previously, the use of conscious, freely moving animals permits in vivo microdialysis procedures to be coupled with behavioral analyses, thus providing neurochemical correlates of various behaviors. This powerful combination has led to significant findings in many areas, including circadian rhythms, food intake, sexual behavior and reproduction, stress, addiction, and learning and memory (Robinson, 1995; Westerink, 1995; Ebihara et al., 1997; Castaneda et al., 2004). The results of these studies have revealed that the functioning of several neurochemical systems varies considerably with the behavioral state (Day et al., 1991; Sakai, 1991; Jacobs and Azmitia, 1992; Fornal et al., 1996; Adell and Artigas, 1998b). However, simple handling procedures and motor activity have neurochemical correlates as well (Inglis and Fibiger, 1995; Acquas et al., 1998; Thiel et al., 1998b; Himmelheber et al., 2000; Giovannini et al., 2001). For instance, arousing stimuli such as handling, injecting drugs, or placing a rat in a novel testing situation increase extracellular acetylcholine in the cortex, hippocampus, nucleus accumbens, and the amygdala (Nilsson et al., 1990; Inglis et al., 1994; Pfister et al., 1994; Acquas et al., 1996; Aloisi et al., 1997; Hajnal et al., 1998; Pallares et al., 1998; Thiel et al., 1998a, b; Ceccarelli et al., 1999). In some cases, the effects of manipulations on neurochemical function differ in resting versus active animals, such as animals learning in a maze, walking on a treadmill, feeding, or mating (Westerink, 1995; Ragozzino et al., 1996, 1998; Etgen and Morales, 2002; Rada et al., 2003). For instance, although systemic infusions of glucose do not affect hippocampal extracellular acetylcholine levels in a rat in its home cage, the same injections produce a large increase when the rat is exploring a maze (Ragozzino et al., 1996; see ➋ *Figure 9-6*).

7 Applications

The following section illustrates advances made using in vivo microdialysis by describing the contributions of this technique to the study of memory and drug addiction, which are our areas of research interest. In addition, we have chosen two areas that present unique challenges and new applications of this procedure, specifically the study of ontological development and the use of in vivo microdialysis in human research and clinical analyses.

7.1 The Use Of In Vivo Microdialysis in Memory Research

7.1.1 Introduction

The in vivo microdialysis procedure has corroborated many theories of memory function that were derived originally from studies examining the effects of drugs or lesions on behavioral measures of memory. In vivo microdialysis has provided neurochemical correlates of memory and has also proven to be a useful tool for studying pharmacological interactions within and between brain areas during memory-related processes.

7.1.2 Neurochemical Correlates

The results of lesion and pharmacological studies indicate that various brain areas participate in different types of memory. For example, evidence indicates that the prefrontal cortex is preferentially involved in working memory (Goldman-Rakic, 1990; Funashashi and Kubota, 1994; Kane and Engle, 2002). Furthermore, the hippocampus is involved in the formation of associations that underlie spatial memory and the striatum is involved in procedural or habit memory (Jarrard, 1993; O'Keefe, 1993; Hasselmo et al., 2002; Packard and Knowlton, 2002; White and McDonald, 2002; Schultz et al., 2003). In contrast, the amygdala is thought to be involved in emotional memory and to play a more general role in regulating memory mediated by other brain regions (Dalmaz et al., 1993; Gallagher and Chiba, 1996; Packard and Cahill, 2001; McGaugh et al., 2002). In vivo microdialysis has proven to be useful in dissociating the contribution of different brain regions to various types of memory, corroborating this theory of multiple memory systems.

Figure 9-6

(a) Effects of systemic glucose (100, 250, & 1000 mg/kg) infusions on mean (+/−S.E.M.) % of baseline extracellular acetylcholine (ACh) levels in the hippocampus of a "resting rat". Systemic saline and glucose treatment did not significantly affect extracellular ACh levels in the hippocampus during the resting condition. (b). Effects of systemic glucose (100, 250, & 1000 mg/kg) administration on mean (+/−S.E.M.) % of baseline extracellular hippocampal ACh levels in rats exploring a maze (i.e., spontaneous alternation). Although glucose did not affect ACh in rats in the resting condition (see Figure 6-a.) it did increase ACh levels in a dose-dependent manner in rats behaving in the maze. Adapted from Ragozzino, M.E., Unick, K.E., and Gold, P.E. (1996). Hippocampal acetylcholine release during memory testing in rats: Augmentation by glucose. *Proc Natl Acad Sci USA*, 93(10): 4693-4698

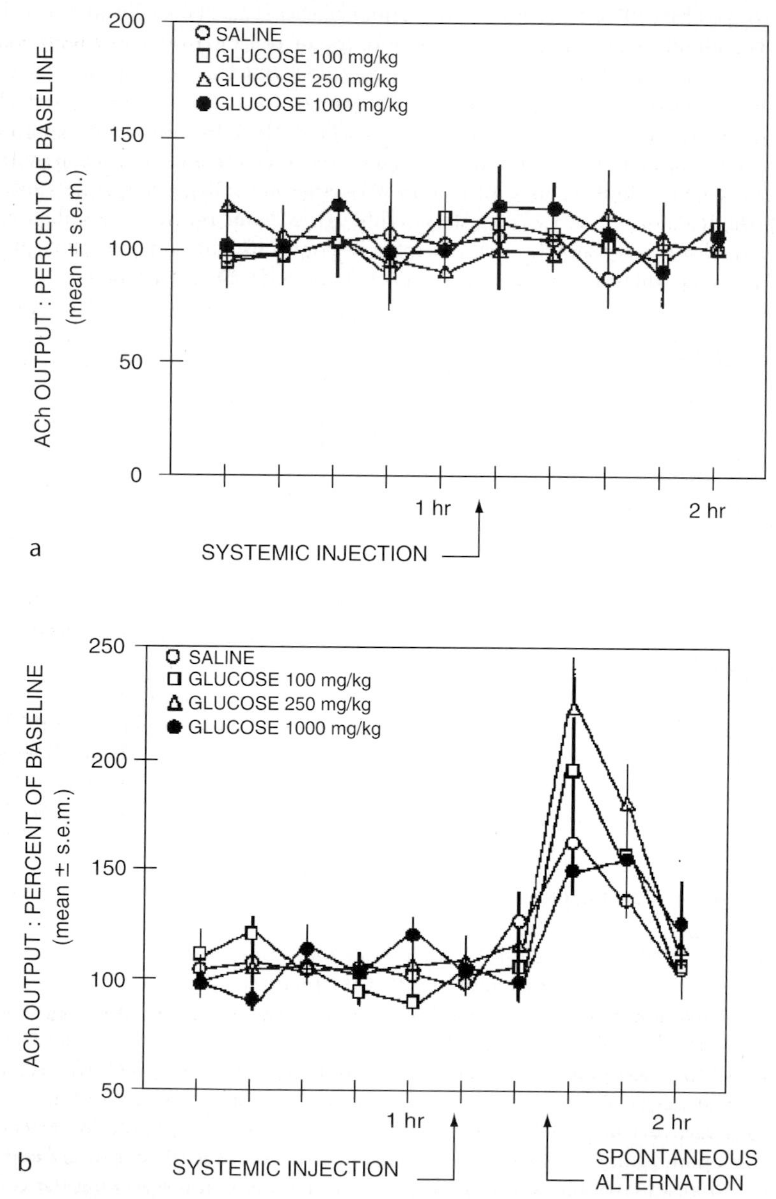

For instance, performance on a working memory task increases dopamine levels in prefrontal cortex (Watanabe et al., 1997; Stark et al., 2000; Kodama et al., 2002) and the elevation in dopamine correlates with the degree of memory (Phillips et al., 2004). Similarly, performance in a spatial task increases acetylcholine in the hippocampus, but not the striatum; whereas performance in a procedural or habit task elevates acetylcholine in the striatum (Ragozzino et al., 1996, 1998; Fadda et al., 2000; Stefani and Gold, 2001; Chang and Gold, 2003; McIntyre et al., 2003b). Acetylcholine levels in the amygdala increase during the performance of emotional, spatial, or procedural memory (Cangioli et al., 2002; Gold, 2003; McIntyre et al., 2003a). These latter findings support the hypothesis that the role of the amygdala in memory is not restricted to a specific type of memory, but rather that it modulates memory processes mediated by other brain systems (Dalmaz et al., 1993; Gallagher and Chiba, 1996; Packard and Cahill, 2001; McGaugh et al., 2002; Gold, 2003).

Additional findings from in vivo microdialysis suggest that some brain regions both cooperate and compete with each other for control of memory processes (McIntyre et al., 2002a, 2003a; Chang and Gold, 2003). For instance, hippocampal levels of acetylcholine increase when rats are tested in either a hippocampal- or an amygdala-dependent task (McIntyre et al., 2002a). Moreover, the amount of acetylcholine in the hippocampus is negatively correlated with good performance in the amygdala-dependent task, suggesting competition between the two brain regions (McIntyre et al., 2002a). In contrast, the fact that increased levels of extracellular acetylcholine in the amygdala correlate positively with performance on a hippocampal-dependent task suggests that these brain regions cooperate on some level (McIntyre et al., 2003a). In addition, acetylcholine in the hippocampus and striatum increases at different points when rats are trained on a task that can be solved by either brain area, corroborating the possibility that each brain region contributes separately to different types of memory (Chang and Gold, 2003). Interestingly, patterns of acetylcholine increases in the hippocampus and striatum predict individual differences in the strategies that an animal uses to solve the task (McIntyre et al., 2003b). In sum, the results of these different studies using in vivo microdialysis reveal the complexity and interdependence of the neural circuits underlying learning and memory.

The results of lesion and pharmacological studies have led to the theory that emotional arousal influences memories through a process that involves the amygdala noradrenergic system (McGaugh et al., 2002; McIntyre et al., 2003c) and recent studies using the in vivo microdialysis procedure have substantiated this theory (McIntyre et al., 2002b). Emotional arousal increases a variety of peripheral hormones, such as epinephrine and glucocorticoids (Fleshner et al., 1993; Tsigos and Chrousos, 2002), which in turn enhance or impair memory (Sternberg et al., 1985; Izquierdo et al., 1988; Roozendaal, 2002). Moreover, amygdala lesions and pharmacological blockade of amygdala noradrenergic receptors prevent the mnemonic effects of epinephrine and glucocorticoids (Liang et al., 1986; Roozendaal and McGaugh, 1996; Quirarte et al., 1997). Findings from experiments using in vivo microdialysis have corroborated the theory that these memory-modulating effects of peripheral hormones are mediated via an influence on the amygdala noradrenergic system. Specifically, drugs such as epinephrine, naloxone, and picrotoxin, which are known to enhance memory (Breen and McGaugh, 1961; Gallagher and Kapp, 1981; Introini et al., 1985), also increase extracellular levels of norepinephrine in the amygdala (Quirarte et al., 1998; Williams et al., 1998, 2000). More directly, rats that are trained on a shock avoidance task show increased levels of norepinephrine in the amygdala and the amount of the increase is correlated positively with memory scores (McIntyre et al., 2002b).

7.1.3 Pharmacological Manipulations and Interactions Between Brain Regions

Neurochemical correlates of memory have also been investigated by combining lesions or pharmacological manipulations with in vivo microdialysis procedures. In many experiments, the effects of a pharmacological manipulation on neurochemistry and behavior are examined in separate groups of animals (Darnaudery et al., 2002; de Groot et al., 2003; Gibbs et al., 2004). Often the findings of these studies show that that the effects of a drug or lesion on memory and neurochemistry are correlated. For instance, infusions of pregnenolone into the medial septum enhance acetylcholine release in the hippocampus in one set of animals and the same infusions improve memory in another group of animals (Darnaudery et al., 2002). However, simultaneously sampling brain ECF while an animal is performing in a memory task is a major strength of the dialysis procedure (Orsetti et al., 1996; Ragozzino et al., 1998; McNay et al., 2000b; Giovannini et al., 2001; McNay and Gold, 2001; Stefani and Gold, 2001; McIntyre et al., 2002a, 2003a, b; Chang and Gold, 2003; Pepeu and Giovannini, 2004). The results of such studies often show that the effects of a pharmacological manipulation on neurochemistry are dependent on whether or not the animal is performing in a behavioral task (Ragozzino et al., 1996, 1998; Sarter et al., 1996; see ➊ *Figure 9-6*).

In vivo microdialysis has been employed to investigate neurochemical interactions between brain regions during memory. For example, this technique has been useful for studying communication between the septum and hippocampus (Moor et al., 1995; Yamamuro et al., 1995; Orsetti et al., 1996). The septum is a major source of cholinergic input to the hippocampus (Frotscher and Leranth, 1985) and the septum may influence memory through an effect on hippocampal acetylcholine function (Gorman et al., 1994; Ragozzino and Gold, 1995; Herzog et al., 2000; Darnaudery et al., 2002). The results of in vivo microdialysis experiments show that neurochemical manipulations of the septum produce parallel changes in memory and hippocampal acetylcholine. For example, septal infusions of glucose prevent both opioid agonist-induced memory deficits and opioid-induced decreases in extracellular hippocampal acetylcholine levels (Ragozzino and Gold, 1995). Similarly, septal infusions of a GABA agonist impair various measures of memory (Brioni et al., 1990; Durkin, 1992; Parent and Gold, 1997; Degroot and Parent, 2001) and decrease hippocampal extracellular acetylcholine levels (Gorman et al., 1994; Moor et al., 1998; de Groot et al., 2003). Moreover, infusions of glucose into the hippocampus reverse the memory deficits produced by septal infusions of the GABA agonist and increase hippocampal extracellular acetylcholine levels when septal GABA receptors are activated (Parent et al., 1997; de Groot et al., 2003).

7.1.4 Limitations

Due to the relatively poor temporal resolution of current in vivo microdialysis sampling techniques, any short-lived changes in neurochemical release that might accompany distinct cognitive or behavioral events are likely masked. Similarly, the poor temporal resolution makes it difficult to dissociate the events associated with encoding, storage, or retrieval of memory. Finally, it is important to take into account that handling, movement, or sensory stimuli can promote a change in the extracellular concentrations of neurochemicals that may be mistakenly construed as reflecting memory (Thiel et al., 1998a; Giovannini et al., 2001). Nevertheless, the in vivo microdialysis procedure has made a significant contribution to our understanding of the neurochemistry of memory.

7.2 The Use of In Vivo Microdialysis in Addiction Research

7.2.1 Introduction

In vivo microdialysis provides an excellent mechanism for defining drug effects under diverse behavioral and pharmacological conditions. It also aids in delineating neural substrates of drug-related behavior, identifying new molecules associated with drug intake, and comparing unique drug responses among genetically diverse subjects. Advances in these areas have informed further basic research as well as clinical experimentation.

7.2.2 Drug Dosing and Behavioral Conditions

In vivo microdialysis sampling coupled with animal models of drug use provides data regarding the neurochemical underpinnings of drug intake, as well as information regarding neurochemical adaptations associated with the trajectory from "recreational use" to compulsive drug seeking and drug dependence. Arguably, the most important in vivo microdialysis-facilitated advance in this area has been the combination of in vivo microdialysis with drug self-administration to create a profile of neurochemical correlates of drug-seeking behavior (Pettit and Justice, 1989, 1991; Fibiger et al., 1992; Hemby et al., 1995; Meil et al., 1995; Wise et al., 1995a, b). Generally, mesolimbic dopamine transmission increases during drug intake, particularly psychomotor stimulant drugs (Pettit and Justice, 1989, 1991). Specifically, phasic fluctuations in elevated dopamine levels correlate with patterns of cocaine-seeking behavior; a slight decline in dopamine triggers a lever-press resulting in cocaine infusion (Wise et al., 1995b). Moreover, dialysis sampling has helped define the drug-specificity of such correlations. In contrast with cocaine self-administration, dopamine fluctuations are not consistently correlated with heroin self-administration (Hemby et al., 1995; Wise et al., 1995a). Moreover, in vivo microdialysis sampling from the brains of animals self-administering cocaine compared with yoked controls reveals significantly greater nucleus accumbens dopamine and acetylcholine efflux when drug intake is contingent on animal behavior rather than when it is noncontingent (Hemby et al., 1997; Mark et al., 1999; Sizemore et al., 2000). These results corroborate and extend findings from other methodologies (Smith et al., 1980, 1982, 1984; Dworkin et al., 1984; Gratton and Wise, 1994) and have contributed to the ongoing discussion regarding the role of dopamine in behavioral reinforcement (see the next section). Beyond dopamine, prefrontal glutamate efflux is triggered by a priming injection of cocaine in rats withdrawn from self-administration of the drug but not rats withdrawn from yoked administration of cocaine on the same schedule (McFarland et al., 2003). Further parsing of the neurochemical bases of drug self-administration such as this can be achieved by presenting various schedules of drug access, such as "binge" sessions of drug self-administration (Parsons et al., 1996; Zhang et al., 2001, 2003; Zorrilla et al., 2001), escalation of drug intake under conditions of extended daily access to drugs (Ahmed et al., 2003), drug withdrawal (Parsons et al., 1991a, 1995, 1996; Hildebrand et al., 1998; Zorrilla et al., 2001; Gerrits et al., 2002), or reinstatement of drug-seeking behavior after extended withdrawal (Neisewander et al., 1996; Tran-Nguyen et al., 1998; Katner and Weiss, 1999; Ranaldi et al., 1999)

Other analyses of repeated drug exposure reveal sensitization, tolerance, and additional neuroadaptations in neurochemical drug effects. For example, stimulant-induced dopamine efflux in the nucleus accumbens increases with repeated exposure to the same drug, another drug, or stress (Kalivas and Duffy, 1990; Pettit et al., 1990; Sorg and Kalivas, 1991; Hamamura and Fibiger, 1993; Morgan et al., 1997; Tidey and Miczek, 1997; Vanderschuren and Kalivas, 2000; Zapata et al., 2003; Vezina, 2004). As noted above, neurochemical responsivity can depend on whether drug administration is contingent on behavioral output; this relationship also extends to sensitized nucleus accumbens dopamine responses to a challenge cocaine injection after cocaine self-administration or yoked cocaine injection (Hooks et al., 1994). Alternative neuroadaptations to chronic drug exposure may take the form of "cellular switches," such as opposite responses to specific receptor ligands in the ventral tegmental area of drug-naïve versus drug-dependent subjects (Laviolette et al., 2004).

7.2.3 Defining Roles of Relevant Neural Substrates

Microdialysis probes aimed at specific target nuclei or subregions have contributed to our understanding of the roles of these areas in particular aspects of drug intake, as well as clarifying specific mechanisms of drug action. An extraordinary challenge has been defining reward versus reinforcement (White, 1989; Joseph et al., 2003), as well as resolving the debate on the role of nucleus accumbens dopamine (Koob and Le Moal, 2001; Gerrits et al., 2002; Wightman and Robinson, 2002; Joseph et al., 2003). Recent theories define the nucleus accumbens as being more involved in behavioral reinforcement than in drug reward per se, a conclusion that fits the aforementioned greater elevation in dopamine during contingent drug self-administration compared with noncontingent drug exposure. In vivo microdialysis has contributed to conclusions such as these in part by helping to delineate specific characteristics of the nucleus accumbens shell and core subregions. The shell is now recognized as being involved mainly in behavioral reinforcement, while the core regulates motor output from mesocorticolimbic circuits (Hemmati et al., 2001; Baldo et al., 2002; Cadoni et al., 2003; Zocchi et al., 2003; Lecca et al., 2004). A similar approach has indicated that the central nucleus of the amygdala has a greater role in drug-seeking behavior than the basolateral nucleus of the amygdala (Panagis et al., 2000; McFarland et al., 2004). At the mechanistic level, dual-probe microdialysis is particularly useful for tracing region-specific drug effects and drug-related neural circuitry (Yoshida et al., 1993; Hedou et al., 1999; Kawahara et al., 1999). For example, as was mentioned previously, a specific serotonin receptor antagonist enhances amphetamine-stimulated dopamine release selectively in the frontal cortex, but not in the nucleus accumbens (Frantz et al., 2002). Moreover, in vivo microdialysis combined with electrophysiological recording or stimulation of cell firing patterns aids our understanding of whether drug-related neurotransmitter efflux causes or results from drug intake (Wightman and Robinson, 2002).

7.2.4 Identification of New Molecules with Roles in Addiction

Oftentimes, techniques in analytical chemistry provide data on not only the analyte of interest but also related chemicals detected by the analytical method. Thus, new neurochemical correlates of drug-related behavior are identified with some regularity. A recent example is the ubiquitous amino acid, taurine. Originally identified in assays aimed at glutamate and aspartate quantification (Dahchour et al., 1994; De Witte et al., 1994), changes in taurine correlate with ethanol intake and are now thought to provide neuroprotection against long-term cellular damage induced by ethanol (Olive, 2002). On the other hand, some molecules hypothesized to be involved in drug-seeking behavior have been challenging for microdialysis until recently. For instance, since the early 1990s, the corticotropin-releasing factor (CRF) has been known to be involved in drug-seeking behavior (Rassnick et al., 1993), but was only recently identified in ECF samples from awake and freely moving animals. The inventive combination of dialysis sampling with a highly sensitive immunoassay for CRF-like immunoreactivity has enabled tracking of robust increases in CRF in the amygdala or bed nucleus of the stria terminalis during alcohol or cocaine withdrawal (Merlo Pich et al., 1995; Richter and Weiss, 1999; Olive et al., 2002). Other relatively new molecules to the addiction field, such as neuropeptides and communicative fatty acids (endocannabinoids), are readily investigated with reverse dialysis of receptor agonists, antagonists, or uptake inhibitors. However, sampling them as dialysate analytes remains challenging (Giuffrida et al., 1999; Tzavara et al., 2001; Wade et al., 2004).

7.2.5 Phenotypic Comparisons of Genetically Diverse Subjects

The genomic revolution provides further applications of in vivo microdialysis in addiction research. Interspecies comparisons underscore the generalizability of drug activity and reveal species-specific neuroadaptations to chronic drug intake. Identical dialysis protocols carried out in different rodent strains, such as alcohol-preferring versus nonpreferring lines or high versus low novelty responders, reveal neurochemical predispositions to drug reinforcement (Hooks et al., 1992; Katner and Weiss, 2001; Zapata et al., 2003; Lecca et al., 2004). Similarly, in vivo microdialysis in genetically engineered mice facilitates functional analysis of controlled genetic variations (Carboni et al., 2001; Rocha et al., 2002; Bohn et al., 2003). For example, knock-out of the serotonin 2C receptor subtype in mice results in enhanced cocaine-induced dopamine release in the nucleus accumbens, along with increased locomotor and reinforcing effects of cocaine (Rocha et al., 2002). Finally, sex- and age-dependent differences in acute and chronic drug effects are readily defined by the use of in vivo microdialysis in specific populations within a single animal strain (Blanchard and Glick, 1995; Becker, 1999; Laviola et al., 2001; Robinson et al., 2002).

7.3 In Vivo Microdialysis Across Ontogeny

7.3.1 Introduction

To date, the majority of basic research involving in vivo microdialysis has been conducted using adult subjects, leaving a significant gap in current knowledge of neurobiological development. Establishment of full neurochemical profiles of animals across developmental phases would contribute significantly to research and treatment for pediatric conditions and developmental disorders. Although a full survey of existing and potential applications of developmental dialysis is beyond the scope of this review, the highlights below introduce innovative solutions to methodological challenges, along with some important advances resulting from the experimentation.

7.3.2 Methodological Challenges

The use of in vivo microdialysis across ontogeny presents unique methodological challenges in anatomical, physiological, and temporal domains. Rapid developmental change in neonatal rodent brain anatomy discourages the implantation of guide cannulae weeks or days in advance of probe insertion, so investigators prefer surgical implantation of the probe itself without a guide (Andersen and Gazzara, 1994; Gazzara and Andersen, 1994a, b) or surgical implantation of guides within several hours of probe insertion (Ishida et al., 1997; Ogasawara et al., 1999; Kosten et al., 2003). Accurate probe placement in the small and rapidly changing brain can be facilitated by site-specific guidelines such as those recently developed for the nucleus accumbens (Philpot et al., 2001). With regard to anesthesia during probe or cannula implantation surgery, cold anesthesia may be the best approach to safe surgery (Frantz and Van Hartesveldt, 1999; Kosten et al., 2003). Concerns regarding the use of anesthetic agents in temporal proximity to in vivo microdialysis sampling, as well as developmental differences in responsiveness to anesthetic agents, caution against the use of common anesthetics. However, cold anesthesia is reported only up to postnatal day (PN) 10 subjects, creating a confounding variable between age groups in ontological comparisons. Other investigators have used ether, a xylazine/ketamine cocktail, or urethane (Andersen and Gazzara, 1994; Gazzara and Andersen 1994a, b; Ishida et al., 1997; Frantz and Van Hartesveldt, 1999; Ogasawara et al., 1999; Philpot et al., 2001; Kosten et al., 2003) in neonatal or older developing rats. Additional adaptations in surgical technique for developing rodents are required, including a mouse adapter for stereotaxic surgery in young rats, ear bar variants and/or clay or plastic molds to secure extremely small subjects during surgery, and/or using ethyl-2-cyanoacrylate (e.g., Superglue) with dental acrylic rather than skull screws to anchor cannulae and/or probes (Ishida et al., 1997; Frantz and Van Hartesveldt, 1999; McCormick et al., 2002; Kosten et al., 2003). The use of a heating pad during surgery and an incubator during dialysate sampling is especially important for young animals with immature temperature regulatory mechanisms. Confounds related to maternal separation, neonatal isolation, and handling present significant concerns for in vivo microdialysis procedures in individual pups. For pups tested before weaning, dialysis sampling usually entails complete food-deprivation, a procedure known to decrease striatal dopamine by 50% (significantly different from 20% reduction in fed controls; Ishida et al., 1997). The following attempts to resolve this confound have met with limited success: allowing pups to suckle right up to the time of sampling, allowing suckling of restrained dams during sampling, or feeding pups manually through intragastric infusions Ishida et al., 1997). We suggest instead using intraoral cannulae through which milk solutions can be perfused (Capuano et al., 1992); cannula lines could easily be attached to probe lines and hooked up to separate syringes and pumps for nutrient infusion during in vivo microdialysis sampling. Other challenges remaining to be addressed include differential sampling regions in rats of different ages (anatomical specificity/spatial resolution) and the influence of the higher ratio of probe diameter and length to brain region size in younger versus older subjects (Stahle, 2000; Westerink and De Vries, 2001).

7.3.3 Applications

Innovative methodologies for in vivo microdialysis in immature subjects have facilitated research in multiple areas. Clinically driven experimentation on neonatal anoxia, hypoxia, or ischemia indicates that perinatal manipulations of oxygen and blood flow result in acute and chronic disruptions of neurotransmission and transmitter turnover (Chen et al., 1997; Nakajima et al., 1999; Ogasawara et al., 1999). Recently, a role for toxic free radicals in brain damage induced by prenatal infection was also delineated by in vivo microdialysis in rat pups (Cambonie et al., 2000, 2004). More subtle neonatal manipulations, such as maternal separation or periodic neonatal isolation, coupled with subsequent in vivo microdialysis reveal long-term changes in stress-responsivity or drug-stimulated neurotransmitter efflux in mesocorticolimbic circuits (Kehoe et al., 1998; McCormick et al., 2002; Kosten et al., 2003). On a mechanistic level, in vivo microdialysis across ontogeny defines developmental parameters of neurotransmitter release and reuptake via measures such as K^+-evoked release, Ca^{2+} dependency, and TTX sensitivity (Andersen and Gazzara, 1994; Gazzara and Andersen, 1994a; Nakajima et al., 1998). For example, a controversy existed as to whether or not dopamine autoreceptors were functional before puberty in young rats (Spear and Brake, 1983; Andersen and Gazzara, 1994; Frantz and Van Hartesveldt, 1995). It was resolved in part by data indicating that low doses of the dopamine D2/3 receptor agonist quinpirole delivered through a microdialysis probe in the developing rat striatum significantly attenuated K^+-evoked dopamine release in 5–22-day-old rat pups, indicating that autoreceptors were indeed functional already at these prepubertal ages (Spear and Brake, 1983; Andersen and Gazzara, 1994; Frantz and Van Hartesveldt, 1995).

7.4 Human In Vivo microdialysis/Clinical Applications

7.4.1 Introduction

The use of the in vivo microdialysis procedure in humans has transitioned from experimental sampling to preclinical evaluation, and has recently been validated for clinical application (Benveniste and Huttemeier, 1990; Lonnroth and Smith, 1990; Lonnroth, 1991; Ungerstedt, 1991b; Justice, 1993; Mendelowitsch, 2001). In vivo microdialysis was first applied to humans in a study aimed at determining extracellular glucose levels in adipose tissue (Lonnroth et al., 1987; Bolinder et al., 1993). Since then, numerous studies have been conducted using in vivo microdialysis in other tissues in humans, such as the brain (Meyerson et al., 1990; Persson and Hillered, 1992; During and Spencer, 1993; Johnston et al., 2003; Melani et al., 2003).

7.4.2 Types of In Vivo Microdialysis Procedures

Two in vivo microdialysis techniques are employed in the clinical setting: open and closed, which refer to the degree of exposure of the brain (Kanthan and Shuaib, 1995). In the open technique, a craniotomy is performed to expose the brain and the microdialysis probe is then implanted into the brain region of interest. In the closed technique, a burr hole is made in the skull and a modified probe is then inserted into the brain via the burr hole. The closed technique permits multiple insertions of the dialysis probe into the brain tissue and consequently minimizes problems associated with prolonged probe placement, such as edema and gliosis (Kochs, 1997). Therefore, the closed technique allows for continuous but intermittent sampling and has lower risks of tissue damage and infection than the open method (Kanthan and Shuaib, 1995).

7.4.3 Clinical Applications and Research

In the clinical setting, in vivo microdialysis is used to study and monitor the chemical changes associated with numerous brain disorders, such as Parkinson's disease, epilepsy, malignant neoplasia, and brain injury in the neurointensive care setting (Meyerson et al., 1990; Ronquist et al., 1992; During and Spencer, 1993). Since the late 1980s, the findings of in vivo microdialysis procedures in the neurointensive care unit show that chemical substances such as glucose, amino acids, and energy metabolites are measurable in the human brain (Landolt and Langemann, 1996). In vivo microdialysis is often used as a diagnostic tool and to monitor patients for ischemia after brain trauma (Persson and Hillered, 1992; Landolt et al., 1994, 1996; Kanthan and Shuaib, 1995; Hillered and Persson, 1999; Gopinath et al., 2000; Naredi et al., 2001; Marion et al., 2002; Nordstrom et al., 2003; Sarrafzadeh et al., 2003). After brain injury, the main objectives in the clinical setting are to measure brain energy metabolites, such as lactate and pyruvate as a signal of secondary ischemia and to monitor secondary injury processes such as increased glutamate activity and free radical production (Bullock et al., 1995; Hillered and Persson, 1999; Gopinath et al., 2000; Hlatky et al., 2002).

In the individual patient, in vivo microdialysis procedures permit the utilization of "Lund therapy" and maintenance of normal concentrations of energy metabolites, fluid balance, cerebral blood volume, and osmotic pressure (Stahl et al., 2001b; Nordstrom, 2003; Nordstrom et al., 2003). Changes in cerebral energy metabolism that accompany cerebral ischemia follow a certain pattern and may be detected by in vivo brain microdialysis before the secondary damage causes an increase in intracranial pressure (Stahl et al., 2001a). Experimental studies have shown severe hyperglycemia increases concentrations of the energy related metabolites that are negatively correlated with recovery from brain trauma suggesting in vivo microdialysis should be used to monitor and control blood-glucose levels in the intensive care unit (Diaz-Parejo et al., 2003). Most importantly, in vivo microdialysis procedures detect changes associated with intracranial pressure 24 h before pressure increases and subsequent tissue loss. This finding suggests that in vivo microdialysis procedures may be invaluable for improved therapies during this 24-h time frame before brain death (Berger et al., 1999). In vivo microdialysis methods can also be used to evaluate treatment strategies after brain trauma online (Stahl et al., 2001b). For example, intravenous infusions of a beta1-antagonist and an alpha2-agonist reduce cerebral pressure (Stahl et al., 2001b).

Online measurement of dialysates can also be used to monitor patients during surgery along with techniques such as somatosensory-evoked potentials and electroencephalography (Landolt et al., 1993; Mendelowitsch et al., 1998; Fried et al., 1999; Mendelowitsch, 2001). If samples are not quantified immediately as they are collected (i.e., online), then the time between sampling and analysis may be too long to have any clinical benefit to patients. The applications of in vivo microdialysis procedures to the clinical setting are limited because assays reflect only the metabolic disturbances and neurochemical changes within the brain area surrounding the probe, and the invasiveness of the procedures poses a potential risk of infection and bleeding in the patient (Peerdemann et al., 2000, 2003). In addition to informing clinical studies, in vivo microdialysis has been used experimentally to measure several drug concentrations in human brain tissue, to quantify drug distribution in tissues, to determine the chemical basis for seizures, and to identify the neurochemical correlates of cognitive performance (Ronquist et al., 1992; During and Spencer, 1993; Scheyer et al., 1994; Muller et al., 1997; Muller, 2000; Joukhader et al., 2001).

8 Sample Protocol

8.1 Subjects

Male Sprague–Dawley rats (200–250 g at the start of experimentation) are used. Each rat is individually housed in a polycarbonate cage and maintained on a 12-h light–dark cycle (lights on at 0700 h) with water and food available ad libitum.

8.2 Surgery

At least 1 week after their arrival, the rats are given atropine sulfate (0.4 mg/kg, i.p.) and anesthetized with sodium pentobarbital (50 mg/kg; i.p.) or isoflurane gas (5%). When necessary, supplemental doses of sodium pentobarbital are given or isoflurane gas is adjusted to maintain anesthesia. The rats are hydrated with 0.9% saline (3.0 cc, s.c.) and given penicillin (0.05 cc; 1500 units, i.m.).

Stereotaxic surgical procedures are used to implant a permanent intracerebral guide cannula (Bioanalytical Systems [BAS]) into the region immediately dorsal to the brain region of interest using the atlas of Paxinos and Watson (1986). The cannula is secured to the skull with four jeweller's screws and cranioplastic cement. In addition, a wire loop is embedded in the cement to be used later to attach the rat to the tether during the microdialysis sampling procedures. A dummy cannula is inserted into the guide cannula to keep it free from debris. The skin is sutured rostral and caudal to the cemented area (5–0 braided silk, Ethicon Sutures) and the incision is sealed with an adhesive (Vetbond; 3M Animal Care Products). Following surgery, the rats are kept in a warm temperature-controlled environment until recovery from anesthesia.

If the rat experiences respiratory problems during recovery, doxapram hydrochloride is administered (Dopram-V; 0.10 cc, i.p.; Ayerst Laboratories). Two days following surgery, the patency of each cannula is checked and betadine is applied to the surgical wound (10% Povidone-iodine; Purdue Frederick).

8.3 In Vivo Microdialysis

Each rat is allowed to recover from surgery for at least 1 week and is handled for 3 min on two separate occasions during that period. Experiments are conducted during the midlight phase. On the day of in vivo microdialysis, the rat is placed in a round Plexiglas bowl (BAS) containing a mixture of clean bedding and bedding from the rat's cage. After 5 min, the rat is attached to the tether (BAS). Following a 1-h habituation period, a microdialysis probe (2 mm membrane for medial septum and nucleus accumbens, 3 mm membrane for caudate nucleus and hippocampus; BAS) is inserted into the guide cannula. The probe is perfused at the rate of 1 mL/min with artificial cerebrospinal fluid [m*M*: Sodium chloride (NaCl) 145.0, potassium chloride (KCl) 3.0, calcium chloride ($CaCl_2$) 1.5, sodium phosphate monobasic (NaH_2PO_4) 2.0, sodium phosphate dibasic anhydrous (Na_2HPO_4) 2.0, dextrose 2.0; pH 7.3; filtered and degassed]. After a 2-h stabilization period, three 25-min baseline samples are collected. Then, the vehicle or the agent of interest is administered intraperitoneally or directly into the probe or into a cannula implanted into another neural region. Following the injection, a series of 25-min samples are collected. The samples are either assayed immediately using HPLC, CE combined with UV, EC, or LIF detection procedures, or kept on dry ice during the experiment and then transferred to a −80°C freezer for long-term storage. The overall motor activity of each rat is assessed at the midway point of each sample using a 5 point scale [0 = no obvious movement, 1 = head movement, 2 = head and forelimb movement, 3 = infrequent movement of all four limbs with minimal movement (i.e., burrowing into bedding), and 4 = movement of all four paws with locomotion and/or rearing (Moore et al., 1992).

8.4 Histology

After the completion of the in vivo microdialysis experiment, rats are euthanized with an overdose of sodium pentobarbital (100 mg/kg) and perfused intracardially with 0.9% saline followed by 10% formalin. The brains are extracted and stored in a 10% formalin solution until they are sectioned into slices (40–60 mm) consecutively through the guide cannula tract. The sections are mounted onto gelatin-coated glass slides and stained with thionin. An observer unaware of the rat's treatment or results verifies the cannula location for each rat.

Acknowledgments

This material is based on work supported in part by grants funded by NINDS-NIDDK-JDF (RO1NS41173–02) and the STC Program of the National Science Foundation under Agreement No. IBN-9876754.

References

Abercrombie E, Zigmond M. 1995. Pharmacology: the fourth generation of progress. Raven; New York: 355-361.

Acquas E, Wilson C, Fibiger HC. 1996. Conditioned and unconditioned stimuli increase frontal cortical and hippocampal acetylcholine release: effects of novelty, habituation, and fear. J Neurosci 16(9): 3089-3096.

Acquas E, Wilson C, Fibiger HC. 1998. Pharmacology of sensory stimulation-evoked increases in frontal cortical acetylcholine release. Neuroscience 85(1): 73-83.

Adell A, Artigas F. 1998a. In vivo neuromethods. Neuromethods. Bateson A, editor. Totowa, NJ: Human Press; 32:1-33.

Adell A, Artigas F. 1998b. A microdialysis study of the in vivo release of 5-HT in the median raphe nucleus of the rat. Br J Pharmacol 125(6): 1361-1367.

Adell A, Casanovas JM, Artigas F. 1997. Comparative study in the rat of the actions of different types of stress on the release of 5-ht in raphe nuclei and forebrain areas. Neuropharmacology 36(4–5): 735-741.

Ahmed SH, Lin D, Koob GF, Parsons LH. 2003. Escalation of cocaine self-administration does not depend on altered cocaine-induced nucleus accumbens dopamine levels. J Neurochem 86(1): 102-113.

Albrecht J, Hilgier W, Zielinska M, Januszewski S, Hesselink M, et al. 2000. Extracellular concentrations of taurine, glutamate, and aspartate in the cerebral cortex of rats at the asymptomatic stage of thioacetamide-induced hepatic failure: modulation by ketamine anesthesia. Neurochem Res 25(11): 1497-1502.

Alfonso M, Duran R, Campos F, Perez-Vences D, Faro LR, et al. 2003. Mechanisms underlying domoic acid-induced dopamine release from striatum: an in vivo microdialysis study. Neurochem Res 28(10): 1487-1493.

Aloisi AM, Casamenti F, Scali C, Pepeu G, Carli G. 1997. Effects of novelty, pain and stress on hippocampal extracellular acetylcholine levels in male rats. Brain Res 748(1–2): 219-226.

Andersen SL, Gazzara RA. 1994. The development of D2 autoreceptor-mediated modulation of K(+)-evoked dopamine release in the neostriatum. Brain Res Dev Brain Res 78(1): 123-130.

Anderson JJ, Chase TN, Engber TM. 1993. Substance P increases release of acetylcholine in the dorsal striatum of freely moving rats. Brain Res 623(2): 189-194.

Baldo BA, Sadeghian K, Basso AM, Kelley AE. 2002. Effects of selective dopamine D1 or D2 receptor blockade within nucleus accumbens subregions on ingestive behavior and associated motor activity. Behav Brain Res 137(1–2): 165-177.

Becker JB. 1999. Gender differences in dopaminergic function in striatum and nucleus accumbens. Pharmacol Biochem Behav 64(4): 803-812.

Benveniste H. 1989. Brain microdialysis. J Neurochem 52(6): 1667-1679.

Benveniste H, Diemer NH. 1987. Cellular reactions to implantation of a microdialysis tube in the rat hippocampus. Acta Neuropathol (Berl) 74(3): 234-238.

Benveniste H, Huttemeier PC. 1990. Microdialysis—theory and application. Prog Neurobiol 35(3): 195-215.

Benveniste H, Drejer J, Schousboe A, Diemer NH. 1984. Elevation of the extracellular concentrations of glutamate and aspartate in rat hippocampus during transient cerebral ischemia monitored by intracerebral microdialysis. J Neurochem 43(5): 1369-1374.

Benveniste H, Drejer J, Schousboe A, Diemer NH. 1987. Regional cerebral glucose phosphorylation and blood flow after insertion of a microdialysis fiber through the dorsal hippocampus in the rat. J Neurochem 49(3): 729-734.

Berger C, Annecke A, Aschoff A, Spranger M, Schwab S. 1999. Neurochemical monitoring of fatal middle cerebral artery infarction. Stroke 30(2): 460-463.

Bergquist J, Gilman SD, Ewing AG, Ekman R. 1994. Analysis of human cerebrospinal fluid by capillary electrophoresis with laser-induced fluorescence detection. Anal Chem 66 (20): 3512-3518.

Bergquist J, Vona MJ, Stiller CO, O'Connor WT, Falkenberg T, et al. 1996. Capillary electrophoresis with laser-induced fluorescence detection: a sensitive method for monitoring extracellular concentrations of amino acids in the periaqueductal grey matter. J Neurosci Methods 65(1): 33-42.

Bianchi L, Colivicchi MA, Bolam JP, Della Corte L. 1998. The release of amino acids from rat neostriatum and substantia nigra in vivo: a dual microdialysis probe analysis. Neuroscience 87(1): 171-180.

Blanchard BA, Glick SD. 1995. Sex differences in mesolimbic dopamine responses to ethanol and relationship to ethanol intake in rats. Recent Dev Alcohol 12: 231-241.

Blandina P, Giorgetti M, Bartolini L, Cecchi M, Timmerman H, et al. 1996. Inhibition of cortical acetylcholine release and cognitive performance by histamine H3 receptor activation in rats. Br J Pharmacol 119(8): 1656-1664.

Bohn LM, Gainetdinov RR, Sotnikova TD, Medvedev IO, Lefkowitz RJ, et al. 2003. Enhanced rewarding properties of morphine, but not cocaine, in beta(arrestin)-2 knock-out mice. J Neurosci 23(32): 10265-10273.

Bolinder J, Ungerstedt U, Arner P. 1993. Long-term continuous glucose monitoring with microdialysis in ambulatory insulin-dependent diabetic patients. Lancet 342(8879): 1080-1085.

Bongiovanni R, Yamamoto BK, Simpson C, Jaskiw GE. 2003. Pharmacokinetics of systemically administered tyrosine: a comparison of serum, brain tissue and in vivo microdialysate levels in the rat. J Neurochem 87(2): 310-317.

Bowser MT, Kennedy RT. 2001. In vivo monitoring of amine neurotransmitters using microdialysis with on-line capillary electrophoresis. Electrophoresis 22(17): 3668-3676.

Breen RA, McGaugh J. 1961. Facilitation of maze learning with posttrial injections of picrotoxin. J Fr Med Chir Thorac 54: 498-501.

Brioni JD, Decker MW, Gamboa LP, Izquierdo I, McGaugh JL. 1990. Muscimol injections in the medial septum impair spatial learning. Brain Res 522(2): 227-234.

Bullock R, Zauner A, Woodward J, Young HF. 1995. Massive persistent release of excitatory amino acids following human occlusive stroke. Stroke 26(11): 2187-2189.

Bungay PM, Morrison PF, Dedrick RL. 1990. Steady-state theory for quantitative microdialysis of solutes and water in vivo and in vitro. Life Sci 46(2): 105-119.

Cadoni C, Solinas M, Valentini V, Di Chiara G. 2003. Selective psychostimulant sensitization by food restriction: differential changes in accumbens shell and core dopamine. Eur J Neurosci 18(8): 2326-2334.

Cambonie G, Laplanche L, Kamenka JM, Barbanel G. 2000. N-Methyl-D-aspartate but not glutamate induces the release of hydroxyl radicals in the neonatal rat: modulation by group I metabotropic glutamate receptors. J Neurosci Res 62(1): 84-90.

Cambonie G, Hirbec H, Michaud M, Kamenka JM, Barbanel G. 2004. Prenatal infection obliterates glutamate-related protection against free hydroxyl radicals in neonatal rat brain. J Neurosci Res 75(1): 125-132.

Camp DM, Robinson TE. 1991. The feasibility of repeated intracerebral microdialysis for within-subjects design experiments on the mesostriatal dopamine system. Curr Sep 10: 78.

Camp DM, Robinson TE. 1992. On the use of multiple probe insertions at the same site for repeated intracerebral microdialysis experiments in the nigrostriatal dopamine system of rats. J Neurochem 58(5): 1706-1715.

Cangioli I, Baldi E, Mannaioni P, Bucherelli C, Blandina P, et al. 2002. Activation of histaminergic H3 receptors in the rat basolateral amygdala improves expression of fear memory and enhances acetylcholine release. Eur J Neurosci 16 (3): 521-528.

Capuano CA, Leibowitz SF, Barr GA. 1992. The pharmaco-ontogeny of the paraventricular alpha 2-noradrenergic receptor system mediating norepinephrine-induced feeding in the rat. Brain Res Dev Brain Res 68(1): 67-74.

Carboni E, Silvagni A, Valentini V, Di Chiara G. 2003. Effect of amphetamine, cocaine and depolarization by high potassium on extracellular dopamine in the nucleus accumbens shell of SHR rats. An in vivo microdyalisis study. Neurosci Biobehav Rev 27(7): 653-659.

Carboni E, Spielewoy C, Vacca C, Nosten-Bertrand M, Giros B, et al. 2001. Cocaine and amphetamine increase extracellular dopamine in the nucleus accumbens of mice lacking the dopamine transporter gene. J Neurosci 21(9): RC141-RC144.

Castaneda TR, de Prado BM, Prieto D, Mora F. 2004. Circadian rhythms of dopamine, glutamate and GABA in the striatum and nucleus accumbens of the awake rat: modulation by light. J Pineal Res 36(3): 177-185.

Ceccarelli I, Casamenti F, Massafra C, Pepeu G, Scali C, et al. 1999. Effects of novelty and pain on behavior and hippocampal extracellular ach levels in male and female rats. Brain Res 815(2): 169-176.

Chang Q, Gold P. 2003. Switching memory systems during learning: Changes in patterns of brains acetylcholine release in the hippocampus and striatum in rats. The Journal of Neuroscience 23(7): 3001-3005.

Chen Y, Engidawork E, Loidl F, Dell'Anna E, Goiny M, et al. 1997. Short- and long-term effects of perinatal asphyxia on monoamine, amino acid and glycolysis product levels measured in the basal ganglia of the rat. Brain Res Dev Brain Res 104(1–2): 19-30.

Chen Z, Wu J, Baker GB, Parent M, Dovichi NJ. 2001. Application of capillary electrophoresis with laser-induced fluorescence detection to the determination of biogenic amines and amino acids in brain microdialysate and homogenate samples. J Chromatogr A 914(1–2): 293-298.

Consolo S, Baldi G, Russi G, Civenni G, Bartfai T, et al. 1994. Impulse flow dependency of galanin release in vivo in the rat ventral hippocampus. Proc Natl Acad Sci USA 91(17): 8047-8051.

Dahchour A, Quertemont E, De Witte P. 1994. Acute ethanol increases taurine but neither glutamate nor GABA in the nucleus accumbens of male rats: a microdialysis study. Alcohol Alcohol 29(5): 485-487.

Dalmaz C, Introini-Collison IB, McGaugh JL. 1993. Noradrenergic and cholinergic interactions in the amygdala and the modulation of memory storage. Behav Brain Res 58(1–2): 167-174.

Darnaudery M, Pallares M, Piazza P, Le Moal M, Mayo W. 2002. The neurosteroid pregnenolone sulfate infused into the medial septum nucleus increases hippocampal acetylcholine and spatial memory in rats. Brain Research 951(2): 237-242.

Dawson LA, Stow JM, Dourish C, Routledge C 1994. Monitoring Molecules in Science. Cador M, UNiverstiy of Bourdeaux: 27-28.

Dawson LA, Stow JM, Palmer AM. 1997. Improved method for the measurement of glutamate and aspartate using capillary electrophoresis with laser induced fluorescence detection and its application to brain microdialysis. J Chromatogr B Biomed Sci Appl 694(2): 455-460.

Day J, Damsma G, Fibiger HC. 1991. Cholinergic activity in the rat hippocampus, cortex and striatum correlates with locomotor activity: an in vivo microdialysis study. Pharmacol Biochem Behav 38(4): 723-729.

de Groote L, Klompmakers AA, Olivier B, Westenberg HG. 2003. Role of extracellular serotonin levels in the effect of 5-HT1B receptor blockade. Psychopharmacology (Berl) 167(2): 153-158.

de Lange EC, de Boer AG, Breimer DD. 2000. Methodological issues in microdialysis sampling for pharmacokinetic studies. Adv Drug Deliv Rev 45(2–3): 125-148.

de Lange EC, Danhof M, de Boer AG, Breimer DD. 1994. Critical factors of intracerebral microdialysis as a technique to determine the pharmacokinetics of drugs in rat brain. Brain Res 666(1): 1-8.

de Lange EC, de Vries JD, Zurcher C, Danhof M, de Boer AG, et al. 1995. The use of intracerebral microdialysis for the determination of pharmacokinetic profiles of anticancer drugs in tumor-bearing rat brain. Pharm Res 12(12): 1924-1931.

De Witte P, Dahchour A, Quertemont E. 1994. Acute and chronic alcohol injections increase taurine in the nucleus accumbens. Alcohol Alcohol Suppl 2: 229-233.

Degroot A, Parent M. 2001. Infusions of physostigmine into the hippocampus or the entorhinal cortex attenuate avoidance retention deficits produced by intra-septal infusions of the GABA agonist muscimol. Brain Research 920(1-2): 10-18.

Del Arco A, Segovia G, Fuxe K, Mora F. 2003. Changes in dialysate concentrations of glutamate and GABA in the brain: An index of volume transmission mediated actions? J Neurochem 85(1): 23-33.

Devine DP, Leone P, Wise RA. 1993. Striatal tissue preparation facilitates early sampling in microdialysis and reveals an index of neuronal damage. J Neurochem 61(4): 1246-1254.

Diaz-Parejo P, Stahl N, Xu W, Reinstrup P, Ungerstedt U, et al. 2003. Cerebral energy metabolism during transient hyperglycemia in patients with severe brain trauma. Intensive Care Med 29(4): 544-550.

Drew KL, O'Connor WT, Kehr J, Ungerstedt U. 1990. Regional specific effects of clozapine and haloperidol on GABA and dopamine release in rat basal ganglia. Eur J Pharmacol 187(3): 385-397.

During MJ, Spencer DD. 1993. Extracellular hippocampal glutamate and spontaneous seizure in the conscious human brain. Lancet 341(8861): 1607-1610.

Durkin T. 1992. GABAergic mediation of indirect transsynaptic control over basal and spatial memory testing-induced activation of septo-hippocampal cholinergic activity in mice. Behavioral Brain Research 50(1-2): 155-165.

Dworkin SI, Lane JD, Smith JE. 1984. An investigation of brain reinforcement systems involved in the concurrent self-administration of food, water, and morphine. NIDA Res Monogr 49: 165-171.

Ebihara S, Adachi A, Hasegawa M, Nogi T, Yoshimura T, et al. 1997. In vivo microdialysis studies of pineal and ocular melatonin rhythms in birds. Biol Signals 6(4–6): 233-240.

Enrico P, Bouma M, de Vries JB, Westerink BH. 1998. The role of afferents to the ventral tegmental area in the handling stress-induced increase in the release of dopamine in the medial prefrontal cortex: a dual-probe microdialysis study in the rat brain. Brain Res 779(1–2): 205-213.

Etgen AM, Morales JC. 2002. Somatosensory stimuli evoke norepinephrine release in the anterior ventromedial hypothalamus of sexually receptive female rats. J Neuroendocrinol 14(3): 213-218.

Fadda F, Cocco S, Stancampiano R. 2000. Hippocampal acetylcholine release correlates with spatial learning performance in freely moving rats. NeuroReport 11(10): 2265-2269.

Fellows LK, Boutelle MG, Fillenz M. 1992. Extracellular brain glucose levels reflect local neuronal activity: a microdialysis study in awake, freely moving rats. J Neurochem 59(6): 2141-2147.

Fibiger HC, Phillips AG, Brown EE. 1992. The neurobiology of cocaine-induced reinforcement. Ciba Found Symp 166: 96-111; discussion 111–124.

Fleshner M, Watkins L, Lockwood L, Grahn R, Gerhardt GA, et al. 1993. Blockade of the hypothalamic–pituitary–adrenal response to stress by intracentricular injection of dexamethasone: a method for studying the stress-induced peripheral effects of glucocorticoids. Psychoneuroendocrinology 18(4): 251-263.

Fornal CA, Metzler CW, Marrosu F, Ribiero-do-Valle LE, Jacobs BL. 1996. A subgroup of dorsal raphe serotonergic neurons in the cat is strongly activated during oral-buccal movements. Brain Res 716(1–2): 123-133.

Frantz K, Van Hartesveldt C. 1999. The locomotor effects of MK801 in the nucleus accumbens of developing and adult rats. Eur J Pharmacol 368(2–3): 125-135.

Frantz KJ, Van Hartesveldt C. 1995. Sulpiride antagonizes the biphasic locomotor effects of quinpirole in weanling rats. Psychopharmacology (Berl) 119(3): 299-304.

Frantz KJ, Hansson KJ, Stouffer DG, Parsons LH. 2002. 5-HT(6) receptor antagonism potentiates the behavioral and neurochemical effects of amphetamine but not cocaine. Neuropharmacology 42(2): 170-180.

Fried I, Wilson CL, Maidment NT, Engel J Jr, Behnke E, et al. 1999. Cerebral microdialysis combined with single-neuron and electroencephalographic recording in neurosurgical patients. Technical note. J Neurosurg 91(4): 697-705.

Frothingham EP, Basbaum AI. 1992. Construction of a microdialysis probe with attached microinjection catheter. J Neurosci Methods 43(2–3): 181-188.

Frotscher M, Leranth C. 1985. Cholinergic innervation of the rat hippocampus as revealed by choline acetyltransferase immunocytochemistry: a combined light and electron microscopic study. J Comp Neurol 239(2): 237-246.

Fujikawa DG, Kim JS, Daniels AH, Alcaraz AF, Sohn TB. 1996. In vivo elevation of extracellular potassium in the rat amygdala increases extracellular glutamate and aspartate and damages neurons. Neuroscience 74(3): 695-706.

Funashashi S, Kubota K. 1994. Working memory and prefrontal cortex. Neurosci Res 21(1): 1-11.

Galindo A, Del Arco A, Mora F. 1999. Endogenous GABA potentiates the potassium-induced release of dopamine in striatum of the freely moving rat: a microdialysis study. Brain Res Bull 50(3): 209-214.

Gallagher M, Chiba AA. 1996. The amygdala and emotion. Curr Opin Neurobiol 6(2): 221-227.

Gallagher M, Kapp B. 1981. Effects of phentolamine administration into the amygdala complex of rats on time-dependent memory processes. Behav Neural Biol 31(1): 90-95.

Galvan A, Smith Y, Wichmann T. 2003. Continuous monitoring of intracerebral glutamate levels in awake monkeys using microdialysis and enzyme fluorometric detection. J Neurosci Methods 126(2): 175-185.

Gardner EL, Chen J, Paredes W. 1993. Overview of chemical sampling techniques. J Neurosci Methods 48(3): 173-197.

Gazzara RA, Andersen SL. 1994a. Calcium dependency and tetrodotoxin sensitivity of neostriatal dopamine release in 5-day-old and adult rats as measured by in vivo microdialysis. J Neurochem 62(5): 1741-1749.

Gazzara RA, Andersen SL. 1994b. The ontogeny of apomorphine-induced alterations of neostriatal dopamine release: effects on potassium-evoked release. Neurochem Res 19(3): 339-345.

Georgieva J, Luthman J, Mohringe B, Magnusson O. 1993. Tissue and microdialysate changes after repeated and permanent probe implantation in the striatum of freely moving rats. Brain Res Bull 31(5): 463-470.

Geranton SM, Heal DJ, Stanford SC. 2004. 5-HT has contrasting effects in the frontal cortex, but not the hypothalamus, on changes in noradrenaline efflux induced by the monoamine releasing-agent, D-amphetamine, and the reuptake inhibitor, bts 54 354. Neuropharmacology 46(4): 511-518.

Gerrits MA, Petromilli P, Westenberg HG, Di Chiara G, van Ree JM. 2002. Decrease in basal dopamine levels in the nucleus accumbens shell during daily drug-seeking behaviour in rats. Brain Res 924(2): 141-150.

Gibbs R, Gabor R, Cox T, Johnson D. 2004. Effects of raloxifen and estradiol on hippocampal acetycholine release and spatial learning in the rat. Psychoneuroendocrinology 29(6): 741-748.

Giovannini M, Rakovska A, Benton R, Pazzagli M, Bianchi L, et al. 2001. Effects of novelty and habituation on acetylcholine, GABA, and glutamate release from the frontal cortex and hippocampus of freely moving rats. Neuroscience 106(1): 43-53.

Giuffrida A, Parsons LH, Kerr TM, Rodriguez de Fonseca F, Navarro M, et al. 1999. Dopamine activation of endogenous cannabinoid signaling in dorsal striatum. Nat Neurosci 2(4): 358-363.

Glick SD, Dong N, Keller RW Jr, Carlson JN. 1994. Estimating extracellular concentrations of dopamine and 3,4-dihydroxyphenylacetic acid in nucleus accumbens and striatum using microdialysis: relationships between in vitro and in vivo recoveries. J Neurochem 62(5): 2017-2021.

Gold PE. 2003. Acetylcholine modulation of neural systems involved in learning and memory. Neurobiol Learn Mem 80(3): 194-210.

Goldman-Rakic P. 1990. Cellular and circuit basis of working memory in prefrontal cortex of nonhuman primates. Prog Brain Res 85: 325-335.

Gonzalez-Mora JT, Fumero B, Mas M. 1991. Mathematical resolution of mixed in vivo voltammetry signals: models, equipment, assesment by simultaneous microdialysis sampling. J neurosci methods 231–244.

Gopinath SP, Valadka AB, Goodman JC, Robertson CS. 2000. Extracellular glutamate and aspartate in head injured patients. Acta Neurochir Suppl 76: 437-438.

Gorman L, Pang K, Frink K, Givens B, Olton D. 1994. Acetylcholine release in the hippocampus: Effects of cholinergic and GABAergic compounds in the medial septal area. Neuroscience Letters 166(2): 199-202.

Gratton A, Wise RA. 1994. Drug- and behavior-associated changes in dopamine-related electrochemical signals during intravenous cocaine self-administration in rats. J Neurosci 14(7): 4130-4146.

Hada J, Kaku T, Jiang MH, Morimoto K, Hayashi Y. 2003. Inhibition of high K^+-evoked gamma-aminobutyric acid release by sodium nitroprusside in rat hippocampus. Eur J Pharmacol 467(1–3): 119-123.

Hajnal A, Pothos EN, Lenard L, Hoebel BG. 1998. Effects of feeding and insulin on extracellular acetylcholine in the amygdala of freely moving rats. Brain Res 785(1): 41-48.

Hamamura T, Fibiger HC. 1993. Enhanced stress-induced dopamine release in the prefrontal cortex of amphetamine-sensitized rats. Eur J Pharmacol 237(1): 65-71.

Hansen AJ. 1985. Effect of anoxia on ion distribution in the brain. Physiol Rev 65(1): 101-148.

Hasselmo M, Hay J, Ilyn M, Gorchetchniko A. 2002. Neuromodulation, theta rhythm and rat spatial naviation. Neural Networks 15: 689-707.

Hedou G, Feldon J, Heidbreder CA. 1999. Effects of cocaine on dopamine in subregions of the rat prefrontal cortex and their efferents to subterritories of the nucleus accumbens. Eur J Pharmacol 372(2): 143-155.

Hemby SE, Co C, Koves TR, Smith JE, Dworkin SI. 1997. Differences in extracellular dopamine concentrations in the nucleus accumbens during response-dependent and response-independent cocaine administration in the rat. Psychopharmacology (Berl) 133(1): 7-16.

Hemby SE, Martin TJ, Co C, Dworkin SI, Smith JE. 1995. The effects of intravenous heroin administration on extracellular nucleus accumbens dopamine concentrations as determined by in vivo microdialysis. J Pharmacol Exp Ther 273(2): 591-598.

Hemmati P, Shilliam CS, Hughes ZA, Shah AJ, Roberts JC, et al. 2001. In vivo characterization of basal amino acid levels in subregions of the rat nucleus accumbens: effect of a dopamine D(3)/D(2) agonist. Neurochem Int 39(3): 199-208.

Hernandez L, Paez X, Tucci S, Murzi E, Rodriguez N, et al. 1994. Monitoring Molecules in Neuroscience. Bordeaux: University of Bordeaux.

Herregodts P. 1991. Neurochemical studies of monoaminergic neurotransmitters in the central nervous system. VUB Press; Brussels: pp. 219-231.

Herrera-Marschitz M, You ZB, Goiny M, Meana JJ, Silveira R, et al. 1996. On the origin of extracellular glutamate levels monitored in the basal ganglia of the rat by in vivo microdialysis. J Neurochem 66(4): 1726-1735.

Herzog C, Gahndi C, Bhattacharya P, Walsh TJ. 2000. Effects of intraseptal zolpidem and chlordiazepoxide on spatial working memory and high-affinity choline uptake in the hippocampus. Neurobiology of Learning and Memory 73: 168-179.

Herzog CD, Nowak KA, Sarter M, Bruno JP. 2003. Microdialysis without acetylcholinesterase inhibition reveals an age-related attenuation in stimulated cortical acetylcholine release. Neurobiol Aging 24(6): 861-863.

Hildebrand BE, Nomikos GG, Hertel P, Schilstrom B, Svensson TH. 1998. Reduced dopamine output in the nucleus accumbens but not in the medial prefrontal cortex in rats displaying a mecamylamine-precipitated nicotine withdrawal syndrome. Brain Res 779(1–2): 214-225.

Hillered L, Persson L. 1999. Neurochemical monitoring of the acutely injured human brain. Scand J Clin Lab Invest Suppl 229: 9-18.

Himmelheber AM, Sarter M, Bruno JP. 2000. Increases in cortical acetylcholine release during sustained attention performance in rats. Brain Res Cogn Brain Res 9(3): 313-325.

Hlatky R, Furuya Y, Valadka AB, Goodman JC, Robertson CS. 2002. Comparison of microdialysate arginine and glutamate levels in severely head-injured patient. Acta Neurochir Suppl 81: 347-349.

Hondo H, Nakahara T, Nakamura K, Hirano M, Uchimura H, et al. 1995. The effect of phencyclidine on the basal and high potassium evoked extracellular GABA levels in the striatum of freely-moving rats: an in vivo microdialysis study. Brain Res 671(1): 54-62.

Hooks MS, Colvin AC, Juncos JL, Justice JB Jr. 1992. Individual differences in basal and cocaine-stimulated extracellular dopamine in the nucleus accumbens using quantitative microdialysis. Brain Res 587(2): 306-312.

Hooks MS, Duffy P, Striplin C, Kalivas PW. 1994. Behavioral and neurochemical sensitization following cocaine self-administration. Psychopharmacology (Berl) 115(1–2): 265-272.

Horn T, Engelmann M. 2001. In vivo microdialysis for nonpeptides in rat brain, a practical guide. Methods 23: 41-53.

Hsiao JK, Ball BA, Morrison PF, Mefford IN, Bungay PM. 1990. Effects of different semipermeable membranes on in vitro and in vivo performance of microdialysis probes. J Neurochem 54(4): 1449-1452.

Hurd YL, Kehr J, Ungerstedt U. 1988. In vivo microdialysis as a technique to monitor drug transport: correlation of extracellular cocaine levels and dopamine overflow in the rat brain. J Neurochem 51(4): 1314-1316.

Imperato A, Di Chiara G. 1984. Trans-striatal dialysis coupled to reverse phase high performance liquid chromatography with electrochemical detection: a new method for the study of the in vivo release of endogenous dopamine and metabolites. J Neurosci 4(4): 966-977.

Imperato A, Ramacci MT, Angelucci L. 1989. Acetyl-L-carnitine enhances acetylcholine release in the striatum and hippocampus of awake freely moving rats. Neurosci Lett 107(1–3): 251-255.

Inglis FM, Fibiger HC. 1995. Increases in hippocampal and frontal cortical acetylcholine release associated with presentation of sensory stimuli. Neuroscience 66(1): 81-86.

Inglis FM, Day JC, Fibiger HC. 1994. Enhanced acetylcholine release in hippocampus and cortex during the anticipation and consumption of a palatable meal. Neuroscience 62(4). 1049-1056.

Inoue S, Kita T, Yamanaka T, Ogawa Y, Nakashima T, et al. 2002. Measurement of 5-hydroxytryptamine release in the rat medial vestibular nucleus using in vivo microdialysis. Neurosci Lett 323(3): 234-238.

Introini I, McGaugh JL, Baratti C. 1985. Pharmacological evidence of a central effect of naltexone, morphine, and beta-endorphin and a peripheral effect of met- and leu-enkephalin on retention of an inhibitory response in mice. Behav Neural Biol 44(3): 434-446.

Ishida A, Nakajima W, Takada G. 1997. Short-term fasting alters neonatal rat striatal dopamine levels and serotonin metabolism: an in vivo microdialysis study. Brain Res Dev Brain Res 104(1–2): 131-136.

Izquierdo I, Dalmaz C, Dias R, Godoy M. 1988. Memory facilitation by posttraining and pretest acth, epinephrine, and vasopressin administration: two separate effects. Behav Neurosci 102(5): 803-806.

Jabaudon D, Shimamoto K, Yasuda-Kamatani Y, Scanziani M, Gahwiler BH, et al. 1999. Inhibition of uptake unmasks rapid extracellular turnover of glutamate of nonvesicular origin. Proc Natl Acad Sci U S A 96(15): 8733-8738.

Jacobs BL, Azmitia EC. 1992. Structure and function of the brain serotonin system. Physiol Rev 72(1): 165-229.

Jacobson I, Sandberg M, Hamberger A. 1985. Mass transfer in brain dialysis devices–a new method for the estimation of extracellular amino acids concentration. J Neurosci Methods 15(3): 263-268.

Jarrard L. 1993. On the role of hippocampus in learning and memory in the rat. Behav Neural Biol 60(1): 9-26.

Johnston AJ, Steiner LA, Chatfield DA, Coleman MR, Coles JP, et al. 2003. Effects of propofol on cerebral oxygenation and metabolism after head injury. Br J Anaesth 91(6): 781-786.

Joseph MH, Datla K, Young AM. 2003. The interpretation of the measurement of nucleus accumbens dopamine by in vivo dialysis: the kick, the craving or the cognition? Neurosci Biobehav Rev 27(6): 527-541.

Joukhader C, Derendorf H, Muller M. 2001. Microdialysis. A novel tool for clinical studies of anti-infective agents. Eur J Clin Pharmacol 57(3): 211-219.

Justice JBJ. 1993. Quantitative microdialysis of transmitters. J Neurosci Methods 48(3): 263-276.

Kalen P, Strecker RE, Rosengren E, Bjorklund A. 1988. Endogenous release of neuronal serotonin and 5-hydroxyindoleacetic acid in the caudate-putamen of the rat as revealed by intracerebral dialysis coupled to high-performance liquid chromatography with fluorimetric detection. J Neurochem 51(5): 1422-1435.

Kalivas PW, Duffy P. 1990. Effect of acute and daily cocaine treatment on extracellular dopamine in the nucleus accumbens. Synapse 5(1): 48-58.

Kane M, Engle R. 2002. The role of prefrontal cortex in working-memory capacity, executive attention, and general fluid intelligence: an individual-differences perspective. Psychon Bull Rev 9(4): 637-671.

Kanthan R, Shuaib A. 1995. Clinical evaluation of extracellular amino acids in severe head trauma by intracerebral in vivo microdialysis. J Neurol Neurosurg Psychiatry 59(3): 326-327.

Katner SN, Weiss F. 1999. Ethanol-associated olfactory stimuli reinstate ethanol-seeking behavior after extinction and modify extracellular dopamine levels in the nucleus accumbens. Alcohol Clin Exp Res 23(11): 1751-1760.

Katner SN, Weiss F. 2001. Neurochemical characteristics associated with ethanol preference in selected alcohol-preferring and -nonpreferring rats: a quantitative microdialysis study. Alcohol Clin Exp Res 25(2): 198-205.

Kawahara Y, Kawahara H, Westerink BH. 1999. Tonic regulation of the activity of noradrenergic neurons in the locus coeruleus of the conscious rat studied by dual-probe microdialysis. Brain Res 823(1–2): 42-48.

Kehoe P, Shoemaker WJ, Arons C, Triano L, Suresh G. 1998. Repeated isolation stress in the neonatal rat: relation to brain dopamine systems in the 10-day-old rat. Behav Neurosci 112(6): 1466-1474.

Kehr J. 1993. A survey on quantitative microdialysis: theoretical models and practical implications. J Neurosci Methods 48(3): 251-261.

Kennedy RT, Thompson JE, Vickroy TW. 2002a. In vivo monitoring of amino acids by direct sampling of brain extracellular fluid at ultralow flow rates and capillary electrophoresis. J Neurosci Methods 114(1): 39-49.

Kennedy RT, Watson CJ, Haskins WE, Powell DH, Strecker RE. 2002b. In vivo neurochemical monitoring by microdialysis and capillary separations. Curr Opin Chem Biol 6 (5): 659-665.

Khan SH, Shuaib A. 2001. The technique of intracerebral microdialysis. Methods 23(1): 3-9.

Kochs E. 1997. Monitoring of cerebral ischaemia. Acta Anaesthesiol Scand Suppl 111: 92-95.

Kodama T, Hikosaka K, Watanabe M. 2002. Differential changes in glutamate concentration in the primate prefrontal cortex during spatial delayed alternation and sensory-guided tasks. Exp Brain Res 145(2): 133-141.

Kolachana BS, Saunders RC, Weinberger DR. 1994. An improved methodology for routine in vivo microdialysis in non-human primates. J Neurosci Methods 55(1): 1-6.

Koob GF, Le Moal M. 2001. Drug addiction, dysregulation of reward, and allostasis. Neuropsychopharmacology 24(2): 97-129.

Korf J, Venema K. 1985. Amino acids in rat striatal dialysates: methodological aspects and changes after electroconvulsive shock. J Neurochem 45(5): 1341-1348.

Kostel KL, Lunte SM. 1997. Evaluation of capillary electrophoresis with post-column derivatization and laser-induced fluorescence detection for the determination of substance P and its metabolites. J Chromatogr B Biomed Sci Appl 695(1): 27-38.

Kosten TA, Zhang XY, Kehoe P. 2003. Chronic neonatal isolation stress enhances cocaine-induced increases in ventral striatal dopamine levels in rat pups. Brain Res Dev Brain Res 141(1–2): 109-116.

Lada MW, Vickroy TW, Kennedy RT. 1997. High temporal resolution monitoring of glutamate and aspartate in vivo using microdialysis on-line with capillary electrophoresis with laser-induced fluorescence detection. Anal Chem 69 (22): 4560-4565.

Landolt H, Langemann H. 1996. Cerebral microdialysis as a diagnostic tool in acute brain injury. Eur Anaesthesiol 13 (3): 269-278.

Landolt H, Langemann H, Alessandri B. 1996. A concept for the introduction of cerebral microdialysis in neurointensive care. Acta Neurochir Suppl (Wien) 67: 31-36.

Landolt H, Langemann H, Gratzl O. 1993. On-line monitoring of cerebral pH by microdialysis. Neurosurgery 32(6): 1000-1004; discussion 1004.

Landolt H, Langemann H, Mendelowitsch A, Gratzl O. 1994. Neurochemical monitoring and on-line pH measurements using brain microdialysis in patients in intensive care. Acta Neurochir Suppl (Wien) 60: 475-478.

Laviola G, Pascucci T, Pieretti S. 2001. Striatal dopamine sensitization to D-amphetamine in periadolescent but not in adult rats. Pharmacol Biochem Behav 68(1): 115-124.

Laviolette SR, Gallegos RA, Henriksen SJ, Kooy van der D. 2004. Opiate state controls bi-directional reward signaling via $GABA_A$ receptors in the ventral tegmental area. Nat Neurosci 7(2): 160-169.

Lecca D, Piras G, Driscoll P, Giorgi O, Corda MG. 2004. A differential activation of dopamine output in the shell and core of the nucleus accumbens is associated with the motor responses to addictive drugs: a brain dialysis study in roman high- and low-avoidance rats. Neuropharmacology 46(5): 688-699.

Lejeune F, Gobert A, Millan MJ. 2002. The selective neurokinin (nk)(1) antagonist, GR205,171, stereospecifically enhances mesocortical dopaminergic transmission in the rat: a combined dialysis and electrophysiological study. Brain Res 935(1–2): 134-139.

Levine JE, Powell KD. 1989. Microdialysis for measurement of neuroendocrine peptides. Methods Enzymol 168: 166-181.

L'Heureux R, Dennis T, Curet O, Scatton B. 1986. Measurement of endogenous noradrenaline release in the rat cerebral cortex in vivo by transcortical dialysis: effects of drugs affecting noradrenergic transmission. J Neurochem 46(6): 1794-1801.

Liang KC, Juler RG, McGaugh JL. 1986. Modulating effects of posttraining epinephrine on memory: involvement of the amygdala noradrenergic system. Brain Res 368(1): 125-133.

Liu D, McAdoo D. 1993. An experimental model combining microdialysis with electrophysiology, histology, and neurochemistry for exploring mecahnisms of secondary damage in spainal cord injury: effects of potassium. J Neurotrauma 10(3): 349-362.

Liu D, Thangnipon W, McAdoo DJ. 1991. Excitatory amino acids rise to toxic levels upon impact injury to the rat spinal cord. Brain Res 547(2): 344-348.

Lonnroth P. 1991. Microdialysis—a new and promising method in clinical medicine. J Int Med 230(4): 63-64.

Lonnroth P, Smith U. 1990. Microdialysis—a novel technique for clinical investigations. J Int Med 227(5): 295-300.

Lonnroth P, Jansson PA, Smith U. 1987. A microdialysis method allowing characterization of intercellular water space in humans. Am J Physiol 253(2 Pt 1): E228-E231.

Lonnroth P, Jansson PA, Fredholm BB, Smith U. 1989. Microdialysis of intercellular adenosine concentration in subcutaneous tissue in humans. Am J Physiol 256(2 Pt 1): E250-E255.

Ludvig N, Tang HM. 2000. Cellular electrophysiological changes in the hippocampus of freely behaving rats during local microdialysis with epileptogenic concentration of N-methyl-D-aspartate. Brain Res Bull 51(3): 233-240.

Ludvig N, Potter PE, Fox SE. 1994. Simultaneous single-cell recording and microdialysis within the same brain site in freely behaving rats: a novel neurobiological method. J Neurosci Methods 55(1): 31-40.

Marion D, Puccio A, Wisniewski S, Kochanek P, Dixon C, et al. 2002. Effect of hyperventilation on extracellular concentrations of glutamate, lactate, pyruvate, and local cerebral blood flow in patients with severe traumatic brain injury. Crit Care Med 30(12): 2619-2625.

Mark GP, Hajnal A, Kinney AE, Keys AS. 1999. Self-administration of cocaine increases the release of acetylcholine to a greater extent than response-independent cocaine in the nucleus accumbens of rats. Psychopharmacology (Berl) 143(1): 47-53.

Mateo F, Rollema H, Basbaum A. 1990. Characterization of monoamine release in the lateral hypothalmus of awake, freely moving rats using in vovo microdialyisis. Brain Res 528: 39-47.

Mateo Y, Pineda J, Meana JJ. 1998. Somatodendritic alpha2 adrenoceptors in the locus coeruleus are involved in the in vivo modulation of cortical noradrenaline release by the antidepressant desipramine. J Neurochem 71(2): 790-798.

Matsumoto M, Kanno M, Togashi H, Ueno K, Otani H, et al. 2003. Involvement of gabaa receptors in the regulation of the prefrontal cortex on dopamine release in the rat dorsolateral striatum. Eur J Pharmacol 482(1–3): 177-184.

McCormick CM, Kehoe P, Mallinson K, Cecchi L, Frye CA. 2002. Neonatal isolation alters stress hormone and mesolimbic dopamine release in juvenile rats. Pharmacol Biochem Behav 73(1): 77-85.

McFarland K, Lapish CC, Kalivas PW. 2003. Prefrontal glutamate release into the core of the nucleus accumbens mediates cocaine-induced reinstatement of drug-seeking behavior. J Neurosci 23(8): 3531-3537.

McFarland K, Davidge SB, Lapish CC, Kalivas PW. 2004. Limbic and motor circuitry underlying footshock-induced reinstatement of cocaine-seeking behavior. J Neurosci 24 (7): 1551-1560.

McGaugh JL, McIntyre CK, Power AE. 2002. Amygdala modulation of memory consolidation: interaction with other brain systems. Neurobiology of Learning Memory 78(3): 539-552.

McIntyre C, Marriott L, Gold P. 2003a. Cooperation between memory systems: Acetylcholine release in the amygdala correlates positively with performance on a hippocampus-dependent task. Behavioral Neuroscience 117(2): 320-326.

McIntyre C, Marriott L, Gold P. 2003b. Patterns of brain acetylcholine release predict individual differences in preferred learning strategies in rats. Neurobiology of Learning and Memory 79(2): 177-183.

McIntyre C, Pal S, Marriott L, Gold P. 2002a. Competition between memory systems: Acetylcholine release in the hippocampus correlates negatively with good performance on an amygdala-dependent task. The Journal of Neuroscience 22(3): 1171-1176.

McIntyre CK, Hatfield T, McGaugh JL. 2002b. Amygdala norepinephrine levels after training predict inhibitory avoidance retention performance in rats. Eur J Neurosci 16(7): 1223-6.

McIntyre CK, Power AE, Roozendaal B, McGaugh JL. 2003c. Role of the basolateral amygdala in memory consolidation. Ann NY Acad Sci 985: 273-293.

McNay EC, Fries TM, Gold PE. 2000. Decreases in rat extracellular hippocampal glucose concentration associated with cognitive demand during a spatial task. Proc Natl Acad Sci USA 97(6): 2881-2885.

McNay EC, Gold PE. 1999. Extracellular glucose concentrations in the rat hippocampus measured by zero-net-flux: effects of microdialysis flow rate, strain, and age. J Neurochem 72(2): 785-790.

McNay EC, Gold PE. 2001. Age-related differences in hippocampal extracellular fluid glucose concentration during behavioral testing and following systemic glucose administration. J Gerontol A Biol Sci Med Sci 56(2): B66-B71.

McNay EC, Sherwin RS. 2004. From artificial cerebro-spinal fluid (acsf) to artificial extracellular fluid (aecf): microdialysis perfusate composition effects on in vivo brain ecf glucose measurements. J Neurosci Methods 132(1): 35-43.

McNay EC, Fries TM, Gold PE. 2000a. Decreases in rat extracellular hippocampal glucose concentration associated with cognitive demand during a spatial task. Proc Natl Acad Sci U S A 97(6): 2881-2885.

McNay EC, McCarty RC, Gold PE. 2001. Fluctuations in brain glucose concentration during behavioral testing: dissociations between brain areas and between brain and blood. Neurobiol Learn Mem 75(3): 325-337.

Meil WM, Roll JM, Grimm JW, Lynch AM, See RE. 1995. Tolerance-like attenuation to contingent and noncontingent cocaine-induced elevation of extracellular dopamine in the ventral striatum following 7 days of withdrawal from chronic treatment. Psychopharmacology (Berl) 118(3): 338-346.

Melani A, De Micheli E, Pinna G, Alfieri A, Corte LD, et al. 2003. Adenosine extracellular levels in human brain gliomas: an intraoperative microdialysis study. Neurosci Lett 346(1–2): 93-96.

MenacherryS, Hubert W, Justice JB Jr. 1992. In vivo calibration of microdialysis probes for exogenous compounds. Anal Chem 64(6): 577-583.

Mendelowitsch A. 2001. Microdialysis: intraoperative and posttraumatic applications in neurosurgery. Methods 23 (1): 73-81.

Mendelowitsch A, Sekhar LN, Caputy AJ, Shuaib A. 1998. Intraoperative on-line monitoring of cerebral pH by microdialysis in neurosurgical procedures. Neurol Res 20(2): 142-148.

Merlo Pich E, Lorang M, Yeganeh M, Rodriguez de Fonseca F, Raber J, et al. 1995. Increase of extracellular corticotropin-releasing factor-like immunoreactivity levels in the amygdala of awake rats during restraint stress and ethanol withdrawal as measured by microdialysis. J Neurosci 15(8): 5439-5447.

Meyerson BA, Linderoth B, Karlsson H, Ungerstedt U. 1990. Microdialysis in the human brain: extracellular measurements in the thalamus of parkinsonian patients. Life Sci 46(4): 301-308.

Moghaddam B, Bunney BS. 1989. Ionic composition of microdialysis perfusing solution alters the pharmacological responsiveness and basal outflow of striatal dopamine. J Neurochem 53(2): 652-654.

Moor E, de Boer P, Auth F, Westerink B. 1995. Characterisation of muscarinic autoreceptors in the septo-hippocampal system of the rat: a microdialysis study. Eur J Pharmacol 294(1): 155-161.

Moor E, de Boer P, Beldhuis HJ, Westerink BH. 1994. A novel approach for studying septo-hippocampal cholinergic neurons in freely moving rats: a microdialysis study with dual-probe design. Brain Res 648(1): 32-38.

Moor E, Schirm E, Jacso J, Westerink BH. 1998. Involvement of medial septal glutamate and GABAA receptors in behaviour-induced acetylcholine release in the hippocampus: a dual probe microdialysis study. Brain Res 789(1): 1-8.

Morgan AE, Horan B, Dewey SL, Ashby CR Jr. 1997. Repeated administration of 3,4-methylenedioxymethamphetamine augments cocaine's action on dopamine in the nucleus accumbens: a microdialysis study. Eur J Pharmacol 331(1): R1-R3.

Morrison PF, Bungay PM, Hsiao JK, Ball BA, Mefford IN, et al. 1991. Quantitative microdialysis: analysis of trasients and application to pharmacokinetics in brain. Microdialysis in the neurosciences. Robinson TE, Justice JB, editors. New York: Elsevier; pp. 47.

Muller M. 2000. Microdialysis in clinical drug delivery studies. Adv Drug Deliv Res 45(2–3): 255-269.

Muller M, Mader R, Steiner B, Steger G, Jansen B, et al. 1997. 5-Flourouracil kinetics in the interstitial tumor space: clinical response in breast cancer patients. Cancer Res 57(13): 2598-2601.

Nakajima W, Ishida A, Takada G. 1999. Anoxic and hypoxic immature rat model for measurement of monoamine using in vivo microdialysis. Brain Res Brain Res Protoc 3 (3): 252-256.

Nakajima W, Ishida A, Ogasawara M, Takada G. 1998. Effect of N-methyl-D-aspartate and potassium on striatal monoamine metabolism in immature rat: an in vivo microdialysis study. Neurochem Res 23(9): 1159-1165.

Naredi S, Olivecrona M, Lindgren C, Ostlund AL, Grande PO, et al. 2001. An outcome study of severe traumatic head injury using the "lund therapy" with low-dose prostacyclin. Acta Anaesthesiol Scand 45(4): 402-406.

Neisewander JL, O'Dell LE, Tran-Nguyen LT, Castaneda E, Fuchs RA. 1996. Dopamine overflow in the nucleus accumbens during extinction and reinstatement of cocaine self-administration behavior. Neuropsychopharmacology 15(5): 506-514.

Nicholson C, Rice ME. 1986. The migration of substances in the neuronal microenvironment. Ann NY Acad Sci 481: 55-71.

Nicolaysen LC, Pan HT, Justice JB Jr. 1988. Extracellular cocaine and dopamine concentrations are linearly related in rat striatum. Brain Res 456(2): 317-323.

Nilsson OG, Kalen P, Rosengren E, Bjorklund A. 1990. Acetylcholine release in the rat hippocampus as studied by microdialysis is dependent on axonal impulse flow and increases during behavioural activation. Neuroscience 36(2): 325-338.

Nordstrom CH. 2003. Assessment of critical thresholds for cerebral perfusion pressure performing bedside monitoring of cerebral energy metabolism. Neurosurg Focus 15(6): E5

Nordstrom CH, Reinstrup P, Xu W, Gardenfors A, Ungerstedt U. 2003. Assessment of the lower limit for cerebral perfusion pressure in severe head injuries by bedside monitoring of regional energy metabolism. Anesthesiology 98(4): 809-814.

Obrenovitch TP, Urenjak J, Zilkha E. 1994. Intracerebral microdialysis combined with recording of extracellular field potential: a novel method for investigation of depolarizing drugs in vivo. Br J Pharmacol 113(4): 1295-1302.

Ogasawara M, Nakajima W, Ishida A, Takada G. 1999. Striatal perfusion of indomethacin attenuates dopamine increase in immature rat brain exposed to anoxia: an in vivo microdialysis study. Brain Res 842(2): 487-490.

O'Keefe J. 1993. Hippocampus, theta, and spatial memory. Curr Opin Neurobiol 3(6): 917-924.

Olive MF. 2002. Interactions between taurine and ethanol in the central nervous system. Amino Acids 23(4): 345-357.

Olive MF, Koenig HN, Nannini MA, Hodge CW. 2002. Elevated extracellular crf levels in the bed nucleus of the stria terminalis during ethanol withdrawal and reduction by subsequent ethanol intake. Pharmacol Biochem Behav 72 (1–2): 213-220.

Olson RJ, Justice JB Jr. 1993. Quantitative microdialysis under transient conditions. Anal Chem 65(8): 1017-1022.

Orsetti M, Casamenti F, Pepeu G. 1996. Enhanced acetylcholine release in hippocampus and cortex during acquistion of an operant behavior. Brain Research 724: 89-96.

Osborne PG, O'Connor WT, Drew KL, Ungerstedt U. 1990. An in vivo microdialysis characterization of extracellular dopamine and GABA in dorsolateral striatum of awake freely moving and halothane anaesthetised rats. J Neurosci Methods 34(1–3): 99-105.

Packard M, Cahill L. 2001. Affective modulation of multiple memory systems. Curr Opin Neurobiol 11(6): 752-756.

Packard M, Knowlton B. 2002. Learning and memory functions of the basal ganglia. Annu Rev Neurosci 25: 563-593.

Pallares M, Darnaudery M, Day J, Le Moal M, Mayo W. 1998. The neurosteroid pregnenolone sulfate infused into the nucleus basalis increases both acetylcholine release in the frontal cortex or amygdala and spatial memory. Neuroscience 87(3): 551-558.

Panagis G, Hildebrand BE, Svensson TH, Nomikos GG. 2000. Selective c-fos induction and decreased dopamine release in the central nucleus of amygdala in rats displaying a mecamylamine-precipitated nicotine withdrawal syndrome. Synapse 35(1): 15-25.

Panter S, Faden A. 1992. Pretreatment with NMDA antagonists limits release of excitatory amino acids following traumatic brain injury. Neurosci Lett 136(2): 165-168.

Parent M, Anderson A, Baker G. 2003. Septal GABA receptor activation prevents glucose-induced increases in hippocampal extracellular glutamate, GABA, and taurine levels. Society for Neuroscience Abstracts, New Orleans, LA.

Parent M, Bush D, Rauw G, Master S, Vaccarino F, et al. 2001. Analysis of amino acids and catecholamines, 5-hydroxytryptamine and their metabolites in brain areas in the rat using in vivo microdialysis. Methods 23(1): 11-20.

Parent M, Laurey P, Wilkniss S, Gold P. 1997. Intraseptal infusions of muscimol impair spontaneous alternation performance: Infusions of glucose into the hippocampus, but not the medial septum, reverse the deficit. Neurobiology of Learning and Memory 68(1): 75-85.

Parent MB, Gold PE. 1997. Intra-septal infusions of glucose potentiate inhibitory avoidance deficits when co-infused with the GABA agonist muscimol. Brain Res 745(1–2): 317-320.

Parent MB, Master S, Kashlub S, Baker GB. 2002. Effects of the antidepressant/antipanic drug phenelzine and its putative metabolite phenylethylidenehydrazine on extracellular gamma-aminobutyric acid levels in the striatum. Biochem Pharmacol 63(1): 57-64.

Parrot S, Bert L, Mouly-Badina L, Sauvinet V, Colussi-Mas J, et al. 2003. Microdialysis monitoring of catecholamines and excitatory amino acids in the rat and mouse brain: recent developments based on capillary electrophoresis with laser-induced fluorescence detection—a mini-review. Cell Mol Neurobiol 23(4–5): 793-804.

Parsons LH, Justice JB Jr. 1992. Extracellular concentration and in vivo recovery of dopamine in the nucleus accumbens using microdialysis. J Neurochem 58(1): 212-218.

Parsons LH, Justice JB Jr. 1994. Quantitative approaches to in vivo brain microdialysis. Crit Rev Neurobiol 8(3): 189-220.

Parsons LH, Koob GF, Weiss F. 1995. Serotonin dysfunction in the nucleus accumbens of rats during withdrawal after unlimited access to intravenous cocaine. J Pharmacol Exp Ther 274(3): 1182-1191.

Parsons LH, Koob GF, Weiss F. 1996. Extracellular serotonin is decreased in the nucleus accumbens during withdrawal from cocaine self-administration. Behav Brain Res 73(1–2): 225-228.

Parsons LH, Smith AD, Justice JB Jr. 1991a. Basal extracellular dopamine is decreased in the rat nucleus accumbens during abstinence from chronic cocaine. Synapse 9(1): 60-65.

Parsons LH, Smith AD, Justice JB Jr. 1991b. The in vivo microdialysis recovery of dopamine is altered independently of basal level by 6-hydroxydopamine lesions to the nucleus accumbens. J Neurosci Methods 40(2–3): 139-147.

Paulsen RE, Fonnum F. 1989. Role of glial cells for the basal and Ca^{2+}-dependent K^{+}-evoked release of transmitter amino acids investigated by microdialysis. J Neurochem 52(6): 1823-1829.

Paxinos G, Watson C. 1986. The rat brain in stereotaxic coordinates. New York: Academic Press.

Peerdemann S, Girbes A, Vandertop W. 2000. Cerebral microdialysis as a new tool for neurometabolic monitoring. Intensive Care Med 26(6): 662-669.

Peerdemann S, van Tulder M, Vandertop W. 2003. Cerebral microdialysis as a monitoring method in subarachnoid hemorrhage patients, and correlation with clinical events—a systematic review. J Neurol 250(7): 797-805.

Pepeu G, Giovannini MG. 2004. Changes in acetylcholine extracellular levels during cognitive processes. Learn Mem 11(1): 21-27.

Persson L, Hillered L. 1992. Chemical monitoring of neurosurgical intensive care patients using intracerebral microdialysis. J Neurosurg 76(1): 72-80.

Peters JL, Michael AC. 1998. Modeling voltammetry and microdialysis of striatal extracellular dopamine: the impact of dopamine uptake on extraction and recovery ratios. J Neurochem 70(2): 594-603.

Pettit HO, Justice JB Jr. 1989. Dopamine in the nucleus accumbens during cocaine self-administration as studied by in vivo microdialysis. Pharmacol Biochem Behav 34(4): 899-904.

Pettit HO, Justice JB Jr. 1991. Effect of dose on cocaine self-administration behavior and dopamine levels in the nucleus accumbens. Brain Res 539(1): 94-102.

Pettit HO, Pan HT, Parsons LH, Justice JB Jr. 1990. Extracellular concentrations of cocaine and dopamine are enhanced during chronic cocaine administration. J Neurochem 55(3): 798-804.

Pfister M, Boix F, Huston JP, Schwarting RK. 1994. Different effects of scopolamine on extracellular acetylcholine levels in neostriatum and nucleus accumbens measured in vivo: possible interaction with aversive stimulation. J Neural Transm Gen Sect 97(1): 13-25.

Phillips AG, Ahn S, Floresco SB. 2004. Magnitude of dopamine release in medial prefrontal cortex predicts accuracy of memory on a delayed response task. J Neurosci 24(2): 547-553.

Philpot RM, McQuown S, Kirstein CL. 2001. Stereotaxic localization of the developing nucleus accumbens septi. Brain Res Dev Brain Res 130(1): 149-153.

Quirarte G, Roozendaal B, McGaugh J. 1997. Glucocorticoid enhancement of memory storage involves noradrenergic activation in the basolateral amygdala. Proc Natl Acad Sci USA 94(25): 14048-14053.

Quirarte G, Galvez R, Roozendaal B, McGaugh JL. 1998. Norepinephrine release in the amygdala in response to footshock and opiod peptidergic drugs. Brain Res 808(2): 134-140.

Rada P, Mendialdua A, Hernandez L, Hoebel BG. 2003. Extracellular glutamate increases in the lateral hypothalamus during meal initiation, and GABA peaks during satiation: microdialysis measurements every 30 s. Behav Neurosci 117(2): 222-227.

Ragozzino ME, Unick KE, Gold PE. 1996. Hippocampal acetylcholine release during memory testing in rats: augmentation by glucose. Proc Natl Acad Sci USA 93(10): 4693-4698.

Ragozzino ME, Gold PE. 1995. Glucose injections into the medial septum reverse the effects of intraseptal morphine infusions on hippocampal acetylcholine output and memory. Neuroscience 68(4): 981-988.

Ragozzino ME, Pal SN, Unick K, Stefani MR, Gold PE. 1998. Modulation of hippocampal acetylcholine release and spontaneous alternation scores by intrahippocampal glucose injections. The Journal of Neuroscience 18(4): 1595-1601.

Ramirez-Munguia N, Vera G, Tapia R. 2003. Epilepsy, neurodegeneration, and extracellular glutamate in the hippocampus of awake and anesthetized rats treated with okadaic acid. Neurochem Res 28(10): 1517-1524.

Ranaldi R, Pocock D, Zereik R, Wise RA. 1999. Dopamine fluctuations in the nucleus accumbens during maintenance, extinction, and reinstatement of intravenous D-amphetamine self-administration. J Neurosci 19(10): 4102-4109.

Rassnick S, Heinrichs SC, Britton KT, Koob GF. 1993. Microinjection of a corticotropin-releasing factor antagonist into the central nucleus of the amygdala reverses anxiogenic-like effects of ethanol withdrawal. Brain Res 605(1): 25-32.

Rea K, Cremers T, Aarnouste E, Westerink B. 2003. The detection of TTX- and calcium dependent GABA by microdialysis. Society for Neuroscience Abstracts, New Orleans, LA.

Reiriz J, Mena M, Bazan E, Muradas V, Lerma J, et al. 1989. Temporal profile of levels of monoamines and their metabolites in striata of rats implanted with dialysis tubes. J Neurochem 53(3): 789-792.

Richter RM, Weiss F. 1999. In vivo crf release in rat amygdala is increased during cocaine withdrawal in self-administering rats. Synapse 32(4): 254-261.

Robinson DL, Brunner LJ, Gonzales RA. 2002. Effect of gender and estrous cycle on the pharmacokinetics of ethanol in the rat brain. Alcohol Clin Exp Res 26(2): 165-172.

Robinson JE. 1995. Microdialysis: a novel tool for research in the reproductive system. Biol Reprod 52(2): 237-245.

Robinson TE, Camp DM. 1991. The effects of four days of continuous striatal microdialysis on indices of dopamine and serotonin neurotransmission in rats. J Neurosci Methods 40(2–3): 211-222.

Rocha BA, Goulding EH, O'Dell LE, Mead AN, Coufal NG, et al. 2002. Enhanced locomotor, reinforcing, and neurochemical effects of cocaine in serotonin 5-hydroxytryptamine 2C receptor mutant mice. J Neurosci 22(22): 10039-10045.

Ronquist G, Hugosson R, Sjolander U, Ungerstedt U. 1992. Treatment of malignant glioma by a new therapeutic principle. Acta Neurochir (Wien) 114(1–2): 8-11.

Roozendaal B. 2002. Stress and memory: opposing effects of glucocorticoids on memory consolidation and memory retreival. Neurobiol Learn Mem 78(3): 578-595.

Roozendaal B, McGaugh JL. 1996. Amygdaloid nuclei lesions differentially affect glucocorticoid-induced memory enhancement in an inhibitory avoidance task. Neurobiol Learn Mem 65(1): 1-8.

Ruggeri M, Merlo Pich E, Zini I, Fuxe K, Ungerstedt U, et al. 1990. Indole-pyruvic acid increases 5-hydroxyindoleacetic acid levels in the cerebrospinal fluid and frontoparietal cortex of the rat: a microdialysis study. Acta Physiol Scand 138(1): 97-98.

Sakai K. 1991. Physiological properties and afferent connections of the locus coeruleus and adjacent tegmental neurons involved in the generation of paradoxical sleep in the cat. Prog Brain Res 88: 31-45.

Sakai K, Crochet S. 2000. Serotonergic dorsal raphe neurons cease firing by disfacilitation during paradoxical sleep. Neuroreport 11(14): 3237-3241.

Sarrafzadeh AS, Kiening KL, Callsen TA, Unterberg AW. 2003. Metabolic changes during impending and manifest cerebral hypoxia in traumatic brain injury. Br J Neurosurg 17(4): 340-346.

Sarter M, Bruno J, Givens B, Moore H, McGaughy J, et al. 1996. Neuronal mechanisms mediating drug-induced cognition enhancement: Cognitive activity as a necessary intervening variable. Cognitive Brain Research 3: 329-343.

Scheyer RD, During MJ, Spencer DD, Cramer JA, Mattson RH. 1994. Measurement of carbamazepine and carbamazepine epoxide in the human brain using in vivo microdialysis. Neurology 44(8): 1469-1472.

Schultz W, Tremblay L, Hollerman J. 2003. Changes in behavior-related neuronal activity in the striatum during learning. Trends Neurosci 26(6): 321-328.

Sharp T, Ljungberg T, Zetterstrom T, Ungerstedt U. 1986. Intracerebral dialysis coupled to a novel activity box—a method to monitor dopamine release during behaviour. Pharmacol Biochem Behav 24(6): 1755-1759.

Shuaib A, Xu K, Crain B, Siren AL, Feuerstein G, et al. 1990. Assessment of damage from implantation of microdialysis probes in the rat hippocampus with silver degeneration staining. Neurosci Lett 112(2–3): 149-154.

Silverstein FS, Naik B. 1991. Effect of depolarization on striatal amino acid efflux in perinatal rats: an in vivo microdialysis study. Neurosci Lett 128(1): 133-136.

Sizemore GM, Co C, Smith JE. 2000. Ventral pallidal extracellular fluid levels of dopamine, serotonin, gamma amino butyric acid, and glutamate during cocaine self-administration in rats. Psychopharmacology (Berl) 150(4): 391-398.

Smith JE, Co C, Lane JD. 1984. Limbic acetylcholine turnover rates correlated with rat morphine-seeking behaviors. Pharmacol Biochem Behav 20(3): 429-442.

Smith JE, Co C, Freeman ME, Lane JD. 1982. Brain neurotransmitter turnover correlated with morphine-seeking behavior of rats. Pharmacol Biochem Behav 16(3): 509-519.

Smith JE, Co C, Freeman ME, Sands MP, Lane JD. 1980. Neurotransmitter turnover in rat striatum is correlated with morphine self-administration. Nature 287(5778): 152-154.

Solis JM, Herranz AS, Herreras O, Munoz MD, Martin del Rio R, et al. 1986. Variation of potassium ion concentrations in the rat hippocampus specifically affects extracellular taurine levels. Neurosci Lett 66(3): 263-268.

Sorg BA, Kalivas PW. 1991. Effects of cocaine and footshock stress on extracellular dopamine levels in the ventral striatum. Brain Res 559(1): 29-36.

Spear LP, Brake SC. 1983. Periadolescence: age-dependent behavior and psychopharmacological responsivity in rats. Dev Psychobiol 16(2): 83-109.

Stahl N, Mellergard P, Hallstrom A, Ungerstedt U, Nordstrom CH. 2001a. Intracerebral microdialysis and bedside biochemical analysis in patients with fatal traumatic brain lesions. Acta Anaesthesiol Scand 45(8): 977-985.

Stahl N, Ungerstedt U, Nordstrom CH. 2001b. Brain energy metabolism during controlled reduction of cerebral perfusion pressure in severe head injuries. Intensive Care Med 27(7): 1215-1223.

Stahle L. 2000. On mathematical models of microdialysis: geometry, steady-state models, recovery and probe radius. Adv Drug Deliv Res 45(2–3): 149-167.

Stahle L, Segersvard S, Ungerstedt U. 1991. A comparison between three methods for estimation of extracellular concentrations of exogenous and endogenous compounds by microdialysis. J Pharmacol Methods 25(1): 41-52.

Stanford JA, Currier TD, Purdom MS, Gerhardt GA. 2001. Nomifensine reveals age-related changes in K(+)-evoked striatal DA overflow in F344 rats. Neurobiol Aging 22(3): 495-502.

Stark H, Bischof A, Wagner T, Scheich H. 2000. Stages of avoidance strategy formation in gerbils are correlated with dopaminergic transmission activity. Eur J Pharmacol 405 (1–3): 263-275.

Stefani MR, Gold PE. 2001. Intrahippocampal infusions of K-ATP channel modulators influence spontaneous alternation performance: Relationship to acetylcholine release in the hippocapus. The Journal of Neuroscience 21(2): 609-614.

Sternberg D, Martinez J, Gold PE, McGaugh JL. 1985. Age-related memory deficits in rats and mice: enhancement with peripheral injections of epinephrine. Behav Neural Biol 44(2): 213-220.

Szerb JC. 1991. Glutamate release and spreading depression in the fascia dentata in response to microdialysis with high K^+: role of glia. Brain Res 542(2): 259-265.

Taber M, Fibiger H. 1994. Cortical regulation of acetylcholine release in rat striatum. Brain Research 639(2): 354-356.

Taber MT, Fibiger HC. 1993. Electrical stimulation of the medial prefrontal cortex increases dopamine release in the striatum. Neuropsychopharmacology 9(4): 271-275.

Taber MT, Fibiger HC. 1995. Electrical stimulation of the prefrontal cortex increases dopamine release in the nucleus accumbens of the rat: modulation by metabotropic glutamate receptors. J Neurosci 15(5 Pt 2): 3896-3904.

Takeda A, Hirate M, Tamano H, Oku N. 2003. Release of glutamate and GABA in the hippocampus under zinc deficiency. J Neurosci Res 72(4): 537-542.

Tanaka Y, Han H, Hagishita T, Fukui F, Liu G, et al. 2004. Alpha-sialylcholesterol enhances the depolarization-induced release of acetylcholine and glutamate in rat hippocampus: in vivo microdialysis study. Neurosci Lett 357(1): 9-12.

Thakkar MM, Delgiacco RA, Strecker RE, McCarley RW. 2003. Adenosinergic inhibition of basal forebrain wakefulness-active neurons: a simultaneous unit recording and microdialysis study in freely behaving cats. Neuroscience 122(4): 1107-1113.

Thiel C, Huston J, Schwarting R. 1998a. Hippocampal acetylcholine and habituation learning. Neuroscience 85(4): 1253-1262.

Thiel CM, Huston JP, Schwarting RK. 1998b. Cholinergic activation in frontal cortex and nucleus accumbens related to basic behavioral manipulations: handling, and the role of post-handling experience. Brain Res 812(1–2): 121-132.

Tidey JW, Miczek KA. 1997. Acquisition of cocaine self-administration after social stress: role of accumbens dopamine. Psychopharmacology (Berl) 130(3): 203-212.

Timmerman W, Westerink BH. 1995. Extracellular gamma-aminobutyric acid in the substantia nigra reticulata measured by microdialysis in awake rats: effects of various stimulants. Neurosci Lett 197(1): 21-24.

Timmerman W, Westerink BH. 1997. Brain microdialysis of GABA and glutamate: what does it signify? Synapse 27(3): 242-261.

Timmerman W, Zwaveling J, Westerink BH. 1992. Characterization of extracellular GABA in the substantia nigra reticulata by means of brain microdialysis. Naunyn Schmiedeberg's Arch Pharmacol 345(6): 661-665.

Timmerman W, Cisci G, Nap A, de Vries JB, Westerink BH. 1999. Effects of handling on extracellular levels of glutamate and other amino acids in various areas of the brain measured by microdialysis. Brain Res 833(2): 150-160.

Tossman U, Ungerstedt U. 1986. Microdialysis in the study of extracellular levels of amino acids in the rat brain. Acta Physiol Scand 128(1): 9-14.

Tran-Nguyen LT, Fuchs RA, Coffey GP, Baker DA, O'Dell LE, et al. 1998. Time-dependent changes in cocaine-seeking behavior and extracellular dopamine levels in the amygdala during cocaine withdrawal. Neuropsychopharmacology 19 (1): 48-59.

Tsigos C, Chrousos G. 2002. Hypothalamic–pituitary–adrenal axis, neuroendocrine factors and stress. J Psychosom Res 53 (4): 865-871.

Tzavara ET, Perry KW, Rodriguez DE, Bymaster FP, Nomikos GG. 2001. The cannabinoid CB(1) receptor antagonist SR141716a increases norepinephrine outflow in the rat anterior hypothalamus. Eur J Pharmacol 426(3): R3-R4.

Ungerstedt. 1984. Measurement of neurotransmitter release by intracranial dialysis. Measurement of neurotransmitter release in vivo. Marsden CA, editor. New York: John Wiley & Sons Ltd; pp. 81-105.

Ungerstedt U. 1991a. Introduction to intracerebral microdialysis. Microdialysis in the neurosciences. Robinson TE, Justice JB Jr, editors. New York: Elsevier Science Publishers; pp. 3-22.

Ungerstedt U. 1991b. Microdialysis in the neurosciences. New York: Elsevier Science Publishers B.V.

Vanderschuren LJ, Kalivas PW. 2000. Alterations in dopaminergic and glutamatergic transmission in the induction and expression of behavioral sensitization: a critical review of preclinical studies. Psychopharmacology (Berl) 151(2–3): 99-120.

Vezina P. 2004. Sensitization of midbrain dopamine neuron reactivity and the self-administration of psychomotor stimulant drugs. Neurosci Biobehav Rev 27(8): 827-839.

Vezzani A, Wu HQ, Angelico P, Stasi MA, Samanin R. 1988. Quinolinic acid-induced seizures, but not nerve cell death, are associated with extracellular Ca^{2+} decrease assessed in the hippocampus by brain dialysis. Brain Res 454(1–2): 289-297.

Wade MR, Tzavara ET, Nomikos GG. 2004. Cannabinoids reduce cAMP levels in the striatum of freely moving rats: an in vivo microdialysis study. Brain Res 1005(1–2): 117-123.

Wages SA, Church WH, Justice JB Jr. 1986. Sampling considerations for lin-line microbore liquid chromatography of brain dialysate. Anal Chem 58: 1649-1654.

Watanabe M, Kodama T, Hikosaka K. 1997. Increase of extracellular dopamine in primate prefrontal cortex during a working memory task. J Neurophysiol 78(5): 2795-2798.

Wellman PJ. 1990. An inexpensive guide cannula and collar for microdialysis experiments. Brain Res Bull 25(2): 345-346.

West AR, Moore H, Grace AA. 2002. Direct examination of local regulation of membrane activity in striatal and prefrontal cortical neurons in vivo using simultaneous intracellular recording and microdialysis. J Pharmacol Exp Ther 301(3): 867-877.

Westerink BH. 1995. Brain microdialysis and its application for the study of animal behaviour. Behav Brain Res 70(2): 103-124.

Westerink BH, De Vries JB. 1988. Characterization of in vivo dopamine release as determined by brain microdialysis after acute and subchronic implantations: methodological aspects. J Neurochem 51(3): 683-687.

Westerink BH, De Vries JB. 1989. On the origin of extracellular GABA collected by brain microdialysis and assayed by a simplified on-line method. Naunyn Schmiedeberg's Arch Pharmacol 339(6): 603-607.

Westerink BH, De Vries JB. 2001. A method to evaluate the diffusion rate of drugs from a microdialysis probe through brain tissue. J Neurosci Methods 109(1): 53-58.

Westerink BH, Tuinte MH. 1986. Chronic use of intracerebral dialysis for the in vivo measurement of 3,4-dihydroxyphenylethylamine and its metabolite 3,4-dihydroxyphenylacetic acid. J Neurochem 46(1): 181-185.

Westerink BH, Enrico P, Feimann J, De Vries JB. 1998. The pharmacology of mesocortical dopamine neurons: a dual-probe microdialysis study in the ventral tegmental area and prefrontal cortex of the rat brain. J Pharmacol Exp Ther 285 (1): 143-154.

Westerink BH, Damsma G, Rollema H, De Vries JB, Horn AS. 1987. Scope and limitations of in vivo brain dialysis: a comparison of its application to various neurotransmitter systems. Life Sci 41(15): 1763-1776.

White N. 1989. Reward or reinforcement: what's the difference? Neurosci Biobehav Rev 13: 181-186.

White N, McDonald R. 2002. Multiple parallel memory systems in the brain of the rat. Neurobiol Learn Mem 77(2): 125-184.

Wightman RM, Robinson DL. 2002. Transient changes in mesolimbic dopamine and their association with "reward." J Neurochem 82(4): 721-735.

Williams CL, Men D, Clayton EC. 2000. The effects of noradrenergic activation of the nucleus tractus solitarius on memory and in potentiating norepinephrine release in the amygdala. Behav Neurosci 114(6): 1131-1144.

Williams CL, Men D, Clayton EC, Gold PE. 1998. Norepinephrine release in the amygdala after systemic injection of epinephrine or escapable footshock: contribution of the nucleus of the solitary tract. Behav Neurosci 112(6): 1414-1422.

Wise RA, Leone P, Rivest R, Leeb K. 1995a. Elevations of nucleus accumbens dopamine and dopac levels during intravenous heroin self-administration. Synapse 21(2): 140-148.

Wise RA, Newton P, Leeb K, Burnette B, Pocock D, et al. 1995b. Fluctuations in nucleus accumbens dopamine concentration during intravenous cocaine self-administration in rats. Psychopharmacology (Berl) 120(1): 10-20.

Yadid G, Pacak K, Kopin IJ, Goldstein DS. 1993. Modified microdialysis probe for sampling extracellular fluid and administering drugs in vivo. Am J Physiol 265(5 Pt 2): R1211-R1211.

Yamamuro Y, Hori K, Tanaka J, Iwano H, Nomura M. 1995. Septo-hippocampal cholinergic system under the discrimination learning task in the rat: a microdialysis study with the dual-probe approach. Brain Res 684(1): 1-7.

Yergey JA, Heyes MP. 1990. Brain eicosanoid formation following acute penetration injury as studied by in vivo microdialysis. J Cereb Blood Flow Metab 10(1): 143-146.

Yim HJ, Gonzales RA. 2000. Ethanol-induced increases in dopamine extracellular concentration in rat nucleus accumbens are accounted for by increased release and not uptake inhibition. Alcohol 22(2): 107-115.

Yoshida M, Yokoo H, Tanaka T, Mizoguchi K, Emoto H, et al. 1993. Facilitatory modulation of mesolimbic dopamine neuronal activity by a mu-opioid agonist and nicotine as examined with in vivo microdialysis. Brain Res 624(1–2): 277-280.

You ZB, Tzschentke TM, Brodin E, Wise RA. 1998. Electrical stimulation of the prefrontal cortex increases cholecystokinin, glutamate, and dopamine release in the nucleus accumbens: an in vivo microdialysis study in freely moving rats. J Neurosci 18(16): 6492-6500.

Zapata A, Chefer VI, Ator R, Shippenberg TS, Rocha BA. 2003. Behavioural sensitization and enhanced dopamine response in the nucleus accumbens after intravenous cocaine self-administration in mice. Eur J Neurosci 17(3): 590-596.

Zhang Y, Schlussman SD, Ho A, Kreek MJ. 2001. Effect of acute binge cocaine on levels of extracellular dopamine in the caudate putamen and nucleus accumbens in male c57bl/6j and 129/j mice. Brain Res 923(1–2): 172-177.

Zhang Y, Schlussman SD, Ho A, Kreek MJ. 2003. Effect of chronic "binge cocaine" on basal levels and cocaine-induced increases of dopamine in the caudate putamen and nucleus accumbens of c57bl/6j and 129/j mice. Synapse 50(3): 191-199.

Zhou SY, Zuo H, Stobaugh JF, Lunte CE, Lunte SM. 1995. Continuous in vivo monitoring of amino acid neurotransmitters by microdialysis sampling with on-line derivatization and capillary electrophoresis separation. Anal Chem 67(3): 594-599.

Zocchi A, Girlanda E, Varnier G, Sartori I, Zanetti L, et al. 2003. Dopamine responsiveness to drugs of abuse: a shell-core investigation in the nucleus accumbens of the mouse. Synapse 50(4): 293-302.

Zorrilla EP, Valdez GR, Weiss F. 2001. Changes in levels of regional CRF-like-immunoreactivity and plasma corticosterone during protracted drug withdrawal in dependent rats. Psychopharmacology (Berl) 158(4): 374-381.

10 Characterization of Receptors by Radiolabelled Ligand Binding Techniques

I. L. Martin · M. Davies · S. M. J. Dunn

Abstract: Radiolabelled ligand binding studies remain one of the most widely used techniques in characterizing the biochemical and pharmacological properties of receptors. In this chapter, we provide an overview of radiolabelled ligand binding, discussing both theoretical and practical aspects of these experiments. Specific protocols that have been used in our laboratories to study the binding of radiolabelled ligands to two members of a ligand-gated ion channel family i.e., the nicotinic and $GABA_A$ receptors are included.

List of Abbreviations: ACh, acetylcholine; DNPP, diethyl-p-nitrophenyl phosphate; GABA, γ–aminobutyric acid; nAChR, nicotinic acetylcholine receptor

1 Introduction

The introduction of radioligand binding techniques in the 1970s proved to be a watershed in the study of the interaction between ligands and the receptors to which they must bind in order to exert their biological effects. In the words of Paul Ehrlich (1913) *corpora non agunt nisi fixata* ("agents cannot act unless they are bound"). For more than three decades, radiolabelled ligand binding studies have been one of the most widely used approaches for the characterization of not only receptor–ligand interactions *per se* but also the properties of the interactants, i.e., the ligands and their receptors.

The underlying principle of the technique is simple: an isotopically labelled ligand is allowed to bind to a receptor preparation in order to provide a tag for the receptor of interest. The properties of the receptor–ligand complex can then be explored directly using the radioactivity of the bound ligand as a marker. As described below, the use of appropriate radioligand binding methodology can provide many insights into the recognition characteristics of the receptor.

In this chapter, we provide an overview of radioligand binding, discussing both the theoretical and practical aspects of the technique. The first section deals with the equations that are commonly used to describe receptor–ligand interactions, while the second section discusses methodological aspects of radioligand binding studies. Throughout, we also provide brief discussions of the limitations of the approach. Our laboratories are primarily interested in the study of ligand-gated ion channels, particularly the properties of the nicotinic acetylcholine receptor (nAChR) and the structurally related γ-aminobutyric acid type A ($GABA_A$) receptor. The nAChR is the prototype of this receptor ion channel family, mainly because its abundance in the electric organs of *Torpedo* species has facilitated detailed biochemical studies. The $GABA_A$ receptor is a major inhibitory neurotransmitter receptor in the mammalian brain and is also the target for a number of therapeutic agents, including the widely prescribed anxiolytics, the benzodiazepines. To illustrate the principles of radiolabelled ligand binding techniques, we have included several protocols that are routinely used in our laboratories.

2 Ligand Binding: Simple Theory

2.1 Equilibrium Binding: Saturation Analysis

In the simplest case of a reversible bimolecular association of a ligand with its receptor, the interaction may be described by:

$$R + L \underset{k_{-1}}{\overset{k_1}{\rightleftharpoons}} RL$$

where R, L and RL represent the receptor, ligand and receptor–ligand complex, respectively, with k_1 and k_{-1} being the association and dissociation rate constants of the interaction. At equilibrium, there is no net change in the concentration of RL. Thus the rate of the forward reaction is equal and opposite to the rate of the reverse reaction, i.e.,

$$k_1[R][L] = k_{-1}[RL] \tag{1}$$

This can be rearranged to:

$$\frac{[R][L]}{[RL]} = \frac{k_{-1}}{k_1} = K_d \tag{2}$$

where K_d is the equilibrium dissociation constant.

The Law of Mass Action dictates that the total concentration of receptor binding sites (R_0) in the assay is invariant and it is the sum of the concentrations of receptor that is bound to the ligand (RL), and the receptor that remains free (R), i.e.,

$$[R_0] = [R] + [RL] \tag{3}$$

Substitution for [R], from Eq. (3) into Eq. (2) yields:

$$\frac{([R_0] - [RL])[L]}{[RL]} = K_d \tag{4}$$

which, by simple rearrangement, gives the classical equation describing the formation of the receptor–ligand complex:

$$[RL] = \frac{[R_0][L]}{K_d + [L]} \tag{5}$$

There are two important conclusions that can be drawn from this equation. Firstly, when the ligand concentration, [L], is very much greater that the equilibrium dissociation constant, K_d ($[L] \gg K_d$), Eq. (5) simplifies to:

$$[RL] \approx [R_0] \tag{6}$$

illustrating that, at high ligand concentrations, the receptor sites will become saturated. Secondly when the ligand concentration, [L], equals the K_d, 50% of the available sites will be occupied by the ligand:

$$[RL] = [R_0]/2 \tag{7}$$

Both of these conditions can readily be appreciated when [RL] is plotted against [L], as shown in *Figure 10-1a.*

Figure 10-1

The effects of ligand concentration on the formation of the receptor–ligand complex according to Eq. 5 in the text. (a) Simulated data using values of 10 nM for both R_0 and K_d. (b) Simulated data for total, specific and nonspecific binding using the same parameters as in (a) but including nonspecific binding (ns) of 0.03 nM/nM free ligand (see Section 10.2.2)

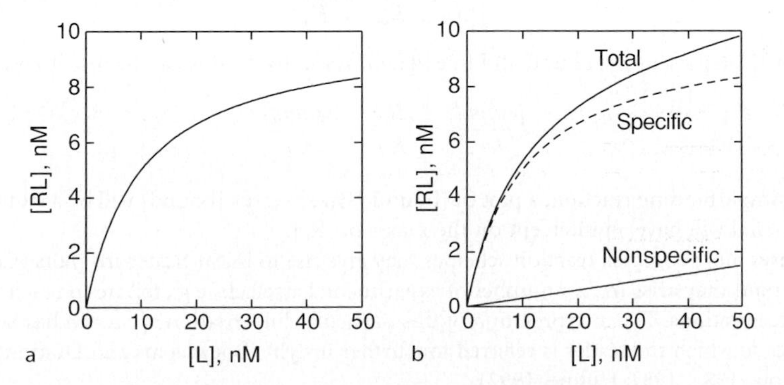

2.2 Nonspecific Binding

Experimentally, the ligand frequently binds not only to the receptor but also to other components present in the reaction mix, such as the membrane in which the receptor of interest is embedded. This "nonspecific binding" (nsb) exhibits quite distinct characteristics; it is generally of low affinity and nonsaturable and can be modelled by the equation:

$$nsb = ns[L] \tag{8}$$

where ns is defined as the slope of the line representing nonspecific binding in ➋ *Figure 10-1b.*

In practice nonspecific binding can be determined experimentally by making use of the saturability of the specific binding to the receptor. This normally involves carrying out a parallel experiment in which a high concentration of a nonradiolabelled ligand for the receptor of interest (sufficient to saturate the specific receptor sites) is included. Under these conditions, the binding of the radioligand to the receptor population of interest will be reduced to zero while the nonspecific binding will remain unchanged (since there is essentially an infinite number of nonspecific sites). Thus we can modify ➋ Eq. (5) to ➋ Eq. (9), where now the total binding of the radioligand $[RL_T]$ is defined as:

$$[RL_T] = \frac{[R_0][L]}{K_d + [L]} + ns[L] \tag{9}$$

since:

$$[RL_T] = [RL] + ns[L] \tag{10}$$

Thus the concentration of the specifically bound complex, [RL] is the difference between the total and nonspecific binding curves, as shown in ➋ *Figure 10-1b.*

The appropriate definition of nonspecific binding is essential prior to characterization of the kinetic and equilibrium properties of the binding interaction. As a rule, nonspecific binding can be defined using a concentration of the unlabelled ligand that is 100 times its K_d value for the sites of interest. Failure to appropriately define nonspecific binding will invalidate the determination of the binding parameters.

2.3 Analysis of Saturation Binding Isotherms: The Scatchard Transformation

A historically important graphical technique to describe binding data (see ➋ *Figure 10-2*) is the Scatchard plot (1949). This is simply a linearization of the function:

$$K_d = \frac{[R][L]}{[RL]} = \frac{([R_0] - [RL])[L]}{[RL]} \tag{11}$$

for which rearrangement gives:

$$\frac{[RL]}{[L]} = \frac{[R_0]}{K_d} - \frac{[RL]}{K_d} \tag{12}$$

Since [RL] and [L] represent the bound and free ligand, respectively, this can be rewritten as:

$$\frac{[bound]}{[free]} = \frac{[R_0]}{K_d} - \frac{[bound]}{K_d} \tag{13}$$

Thus, for this simple binding reaction, a plot of [bound]/[free] versus [bound] will be a straight line with a slope of $-1/K_d$ and will have an intercept on the x-axis of $[R_0]$.

In some cases more complex reaction schemes may give rise to linear Scatchard plots (Conners, 1987), and nonlinear plots may arise from a number of experimental artefacts, e.g., failure to reach equilibrium at low ligand concentrations. The interpretation of this particular linearisation approach has been the subject of many articles to which the reader is referred for further insight (Boeynaems and Dumont, 1975; Norby et al., 1980; Klotz, 1982, 1983; Hulme, 1992).

Figure 10-2
Scatchard plot of the binding data shown in Figure 10-1a, illustrating how the K_d and R_0 can be obtained from such a linearization procedure

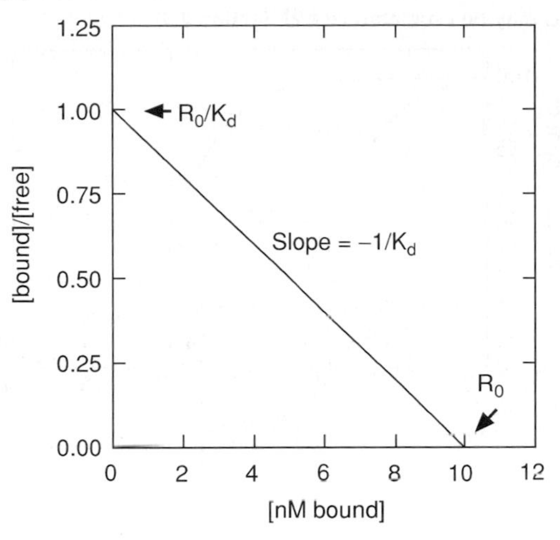

2.4 Equilibrium Binding: Competition Studies

The affinity of nonradiolabelled ligands may be estimated by their ability to compete with a radioligand for binding to the receptor of interest. Here the equations outlined above hold for both the radioactive and nonradioactive ligands. Receptor occupation can be determined only for the proportion that is tagged with the radioactive ligand. In these so-called "displacement studies", the unlabelled ligand displaces the radioactive ligand from its binding sites and the concentration dependence of this effect can lead to an indirect measurement of its binding affinity. Experimentally, it is common to use a concentration of the radiolabelled compound which is close to its K_d value. In these experiments, the IC_{50} value for the competing ligand is defined as the concentration at which it reduces the specific binding of the radioligand to 50% of that obtained in its absence (Figure 10-3a). Under equilibrium conditions, with a single homogeneous population of binding sites, the equilibrium dissociation constant for the competitor (K_I) is defined by the Cheng–Prusoff equation (Cheng and Prusoff, 1973):

$$K_I = \frac{IC_{50}}{1 + [L]/K_d} \tag{14}$$

where [L] and K_d refer to the concentration of the radiolabelled ligand and its equilibrium dissociation constant for binding. This equation is valid only if the bound ligand is <10% of the free ligand concentration and when $K_d \gg$ the concentration of the receptor sites. If this is not the case, a more complex correction factor must be applied (see Wieland and Molinoff, 1981).

2.5 Analysis of Competition Curves: The Logit–Log Plot

Linear transformation of competition data is usefully carried out using the logit–log plot, also referred to as the indirect Hill plot:

$$\log\left[\frac{[RL]_I}{[RL] - [RL]_I}\right] = n\log[I] + n\log IC_{50} \tag{15}$$

Figure 10-3
Competition curves for displacement of a radiolabelled ligand by a competitive inhibitor. (a) Typical displacement curves for two inhibitors having IC_{50} values of 10 nM and 1 μM. (b) Indirect logit–log plots of the data in (c) showing how IC_{50} values may be estimated (see Section 2.5)

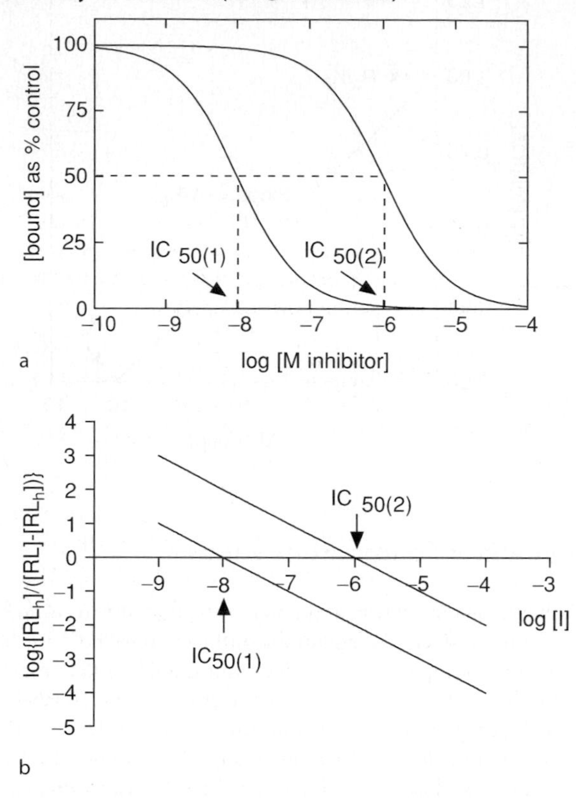

where $[RL]_I$ is the specific binding in the presence of the inhibitor at concentration [I] and [RL] is the specific binding in its absence. When n = 1, the data are compatible with displacement from a homogeneous population of sites and under these conditions the IC_{50} is the intercept on the x-axis (*Figure 10-3b*). This can then be used with Eq. 14 to calculate the K_I of the displacer as long as the K_d for the radiolabelled ligand is known.

2.6 Complex Receptor–Ligand Interactions

2.6.1 Induced Change in Conformation

In the above discussion, we have considered the simplest case in which a ligand binds to a single homogeneous population of binding sites and that simple formation of the receptor–ligand complex underlies the functional effect. However, we know that the binding of an agonist to its receptor is only the first step in the efficacy chain that leads to a response. This can be modelled by supplementing the scheme in 2.1 with an additional step to give:

$$R + L \underset{k_{-1}}{\overset{k_1}{\rightleftharpoons}} RL \underset{k_{-2}}{\overset{k_2}{\rightleftharpoons}} R^*L$$

where R* represents the receptor R in a distinct (activated) conformational state. The overall K_d for the interaction now becomes:

$$K_d = \frac{K_1 K_2}{1 + K_2} \tag{16}$$

where the equilibrium constants, K_1 and K_2 are defined as the ratio of the forward and backward rate constants (k_{-1}/k_1 and k_{-2}/k_2, respectively).

2.6.2 Multiple Binding Sites

A further complication that is often encountered in equilibrium binding experiments is that the ligand binds to two populations of receptors and that each interaction has a distinct binding affinity. Assuming that these populations can be described by the simple interactions rehearsed above, then binding can be described simply by the sum of binding to each population of sites:

$$[RL] = \frac{[R_1][L]}{K_{d1} + [L]} + \frac{[R_2][L]}{K_{d2} + [L]} \tag{17}$$

where R_1 and R_2 are the binding capacities of the two sites for which the ligand has affinities of K_{d1} and K_{d2}, respectively (*Figure 10-4*).

Figure 10-4
An example of ligand binding heterogeneity. The curve in (a) was generated assuming two populations of sites differing in affinity by 10-fold (see Eq. 17). The parameters used were R_1=0.5 nM, K_{d1}=10 nM, R_2=0.5 nM, K_{d2}=100 nM, The Scatchard plot in (b) was generated using the data in (a). The dashed lines illustrate how curved Scatchard plots may be analysed to yield binding parameters for the two classes of sites

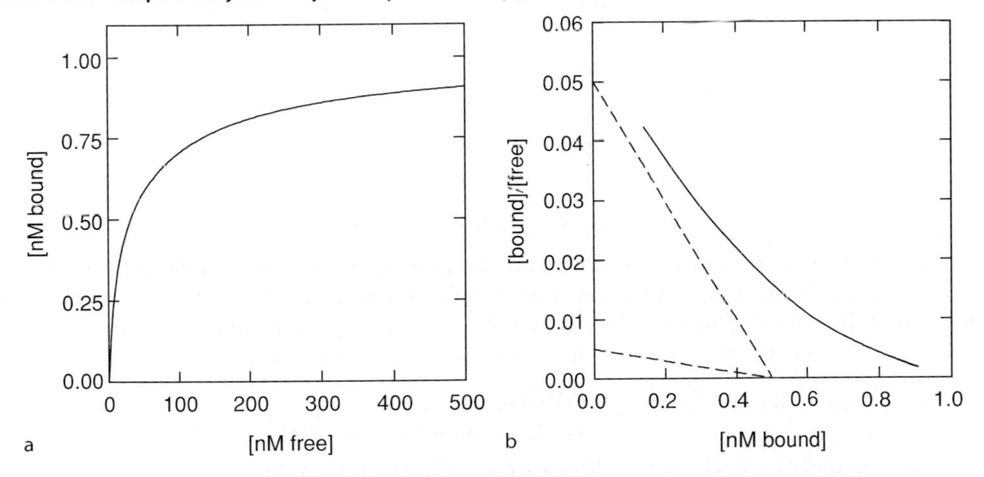

2.7 Limits on the Interpretation of Equilibrium Radioligand Binding Studies

In the above discussion, we have assumed that the interaction between the receptor and the ligand meet a number of criteria, i.e., that

- the receptor–ligand interaction is reversible
- the association process is bimolecular
- the dissociation process is monomolecular

- receptors in the population are equivalent and independent of each other
- measurements are made after the interaction has reached equilibrium
- the ligand can exist only free or bound to the receptor

Using these assumptions, radiolabelled ligand binding techniques can provide specific information on binding site densities and ligand binding affinities. These approaches have proved invaluable in studies of ligand–receptor interactions and, prior to molecular cloning, provided the first evidence for the existence of receptor subtypes. In more complex analyses, ligand binding studies can be useful for detecting cooperativity between binding sites, for studying inhibitory mechanisms (e.g., competitive versus noncompetitive inhibitors) and for elucidating the effects of allosteric modulators that may be of both physiological or pharmacological importance.

Equilibrium binding assays do, however, suffer from a number of limitations. Many important ligand–receptor interactions, for example, may occur under preequilibrium conditions. In the case of ligand-gated ion channels, the binding of an agonist to the receptor rapidly induces a conformational change to open the ion channel. If the agonist remains bound, the receptor undergoes further slow conformational transitions that lead to a refractory, desensitized state in which the ion channel is closed. Under equilibrium conditions only the desensitized receptor–agonist complex is accessible to measurement. Often one would like information on the agonist binding events that cause the channel to open. This requires details of the kinetic processes that lead to the final equilibrium state. In the following section we, therefore, review briefly some elementary kinetics of receptor–ligand interactions.

2.8 Receptor–Ligand Interactions: Simple Kinetic Theory

Kinetic studies can be invaluable in providing detailed information concerning the mechanisms that lead to the final equilibrium state. The design of kinetic experiments, and the analysis of the data obtained from them are frequently complicated. Below we describe only the most elementary kinetic studies and the interested reader should consult specialist texts for more comprehensive descriptions.

2.8.1 Association Kinetics

The rate of complex formation in a bimolecular interaction described in ➋ Section 2.1 is simply described by the equation:

$$d[RL]/dt = k_1[R][L] - k_{-1}[RL] \tag{19}$$

Clearly throughout the kinetic experiments the concentration of each of the components RL, R and L are changing but, if the initial ligand concentration $[L_0]$ is very much greater than the total receptor concentration $[R_0]$, the free ligand concentration will suffer little depletion as the association continues to equilibrium. This is referred to as 'pseudo first-order kinetics' which are easier to analyse:

$$\begin{aligned} d[RL]/dt &= k_1[R][L_0] - k_{-1}[RL] \\ &= k_1([R_0] - [RL])[L_0] - k_{-1}[RL] \\ &= k_1[R_0][L_0] - [RL](k_1[L_0] + k_{-1}) \end{aligned} \tag{20}$$

Since $[R_0]$, $[L_0]$, k_1 and k_{-1} are constants, the reaction is approximately first order with respect to [RL]. The reaction thus proceeds exponentially:

$$[RL]_t = [RL]_{eq}(1 - e^{-k_{app}t}) \tag{21}$$

where $[RL]_t$ and $[RL]_{eq}$ are the concentrations of the complex at time t and equilibrium, respectively, and the apparent rate constant, k_{app}, is given by:

$$k_{app} = k_1[L_0] + k_{-1} \tag{22}$$

A plot of the apparent rate constant against the ligand concentration will be linear with a slope of k_1 and it will intercept the y-axis at k_{-1} as shown in *Figure 10-5a.*

Figure 10-5

Effect of ligand concentration on the apparent rate constant (k_{app}) for (a) a pseudo first order reaction (Eq. 22) or a two-step reaction (Eq. 23). The data in (a) were generated using k_1 and k_{-1} values of $2 \times 10^6\ M^{-1}\text{-}s^{-1}$ and $1.0\ s^{-1}$, respectively. The data in (b) were generated using $K_1=40$ nM, $k_2=0.2\ s^{-1}$ and $k_{-2}=0.05\ s^{-1}$

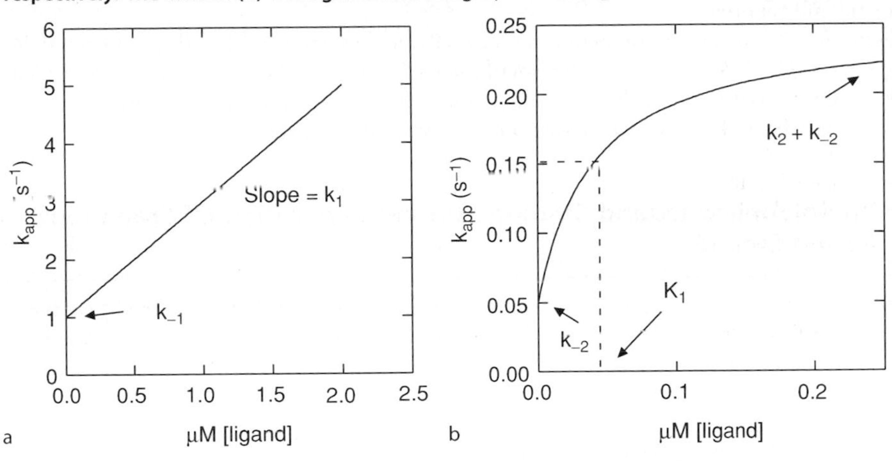

Experiments of this type can provide some valuable pieces of information:

1. According to Eq. 21, the association reaction is a single exponential process. Deviations from this indicate that the reaction is more complex.
2. If there is a nonlinear dependence of k_{app} on $[L_0]$ then the reaction cannot be a simple bimolecular process. However, the extended equation described in Section 2.6.1 will give rise to a hyperbolic curve as shown in *Figure 10-5b*, where K_1 is given by k_{-1}/k_1, and k_{app} is described by:

$$k_{app} = \frac{k_2[L]}{K_1 + [L]} + k_{-2} \tag{23}$$

3. The maximum value of k_1 is limited by the diffusion-controlled rate for collision between two molecules. For a small molecule binding to a macromolecular receptor, this rate is of the order of 10^7 to 10^8 M^{-1}-s^{-1}. If the rates deviate from this markedly it suggests that the interaction of ligand with receptor is more complex.
4. The estimated value of k_{-1} obtained in such an experiment should agree with that determined directly in dissociation experiments.
5. A lack of agreement between the K_d estimated in equilibrium experiments and that determined in kinetic experiments ($k_{-1}/k_1=K_d$) suggests a more complex ligand–receptor interaction.

2.8.2 Dissociation Kinetics

After the interaction between the ligand and receptor has reached equilibrium, dissociation kinetics are initiated either by dilution or by the introduction of unlabelled ligand which competes for the receptor binding sites. The purpose, in both cases, is to ensure that once the radiolabelled ligand has dissociated from the receptor, it is not able to reassociate with the receptor. This allows the following reaction to occur in isolation:

$$RL \xrightarrow{k_{-1}} R + L$$

The rate of dissociation is given by:

$$d[RL]/dt = k_{-1}[RL] \quad (24)$$

and the reaction exhibits a single exponential decay:

$$[RL]_t = [RL]_{eq} \exp^{-k_{-1}t} \quad (25)$$

In a similar manner to the association kinetic experiments, dissociation studies can provide important mechanistic information:

1. If dissociation does not occur mono-exponentially, it may indicate that there are multiple ligand–receptor complexes or that the mechanism of interaction of ligand with receptor is more complex.
2. If the method used to initiate the dissociation affects the measured dissociation rate constant, allosteric interactions within the ligand–receptor complex may be implicated.

3 Radiolabelled Ligand Binding Studies: Separation Of Free From Bound Ligand

In radioligand binging studies, it is essential to experimentally distinguish between the ligand bound to the receptor and that which remains free in solution. There are three methods that are commonly used to accomplish this physical separation. Below we discuss briefly the advantages and disadvantages of each.

3.1 Equilibrium Dialysis

Theoretically, equilibrium dialysis is the most accurate way to differentiate bound from free ligand. It is most commonly carried out using two chambers that are separated by a dialysis membrane. This membrane is normally permeable to the ligand, L, but not to the receptor, R. Over a period of time, the ligand will diffuse into the chamber containing the receptor, and this chamber will eventually contain three components (R, RL and L) while the other chamber will contain only the free ligand, L. By sampling the two chambers after equilibrium has been reached, it is thus possible to calculate the concentration of the ligand–receptor complex and the free ligand concentration as shown in *Figure 10-6*.

The most common artefact in equilibrium dialysis techniques is the absorption of ligand to the membrane and/or the chamber material. This results in an overestimation of the concentration of the receptor–ligand complex. Another consideration is the Donnan effect that arises from charges on the ligand and/or the receptor as the system approaches equilibrium. The easiest way to counter this is to manipulate the salt concentration of the dialysis buffer such that these charges are masked. Although equilibrium dialysis is a simple technique that can be used for both membrane bound and soluble receptors, a major drawback is the time required for equilibrium to be reached. This is particularly true when the ligand concentration is low and can significantly compromise such studies with labile receptor preparations.

3.2 Centrifugation Assays

This method relies on centrifugal force to separate the receptor–ligand complex from the free radioligand. The complex must therefore have a sufficiently high sedimentation coefficient in order that it can be pelleted while leaving the free ligand in the supernatant. Centrifugation is thus useful for particulate samples such as membrane-bound receptors but is less useful for soluble proteins, which can be sedimented only by very high centrifugal forces. In a typical assay, the receptor and radioligand are incubated for a sufficient time to allow the reaction to reach equilibrium, after which the samples are centrifuged. This approach requires that there is significant depletion in the ligand concentration due to binding and to allow a reliable measure of the bound ligand. Centrifugation is thus best suited for studies of high affinity binding sites where very low ligand concentrations can be used or, in a few cases, when an abundant receptor

Figure 10-6
At equilibrium, the concentration of free ligand, [L], is the same on both sides of the dialysis membrane. Thus, the difference in the amount of radioactivity measured in aliquots taken from both sides of the membrane can be used to estimate [RL] (see Protocol 4.1)

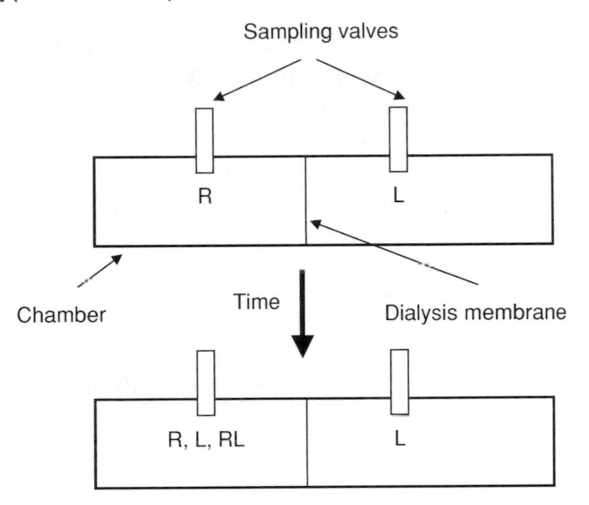

population is available. We routinely use centrifugation techniques to study the binding of nicotinic acetylcholine receptor radioligands to *Torpedo* membrane preparations (see Protocol 4.2) where the receptor density is several orders of magnitude higher than that of most other neurotransmitter receptors in their native membrane environment.

3.3 Filtration Assays

The most commonly used method for the separation of bound from free ligand is the filtration assay (see Protocol 4.3). This involves passing the reaction mixture through a filter made of, for example, glass fiber or cellulose. Particulate fractions, e.g., membranes and membrane-bound radioligand are selectively retained on the filter while the rest of the incubation mixture, containing free ligand, is allowed to pass through. Filtration is usually performed under reduced pressure in order to increase the speed of filtration, thus reducing nonspecific binding to the filter. To reduce further any nonspecific retention of free ligand by the filter, it is usually necessary to include one or more filter washing steps. A significant problem in such assays is that dissociation of bound ligand may occur during this washing. If, for example, washing of the filters takes 10 seconds, then a receptor–ligand complex whose dissociation half-time ($t_{½}$) is 10 seconds will have dissociated by 50% by the end of the wash. Manual filtration assays are thus limited to high affinity ligands with slow rates of dissociation. However, it is possible to measure lower affinity binding using an automated modification of the filtration technique, which allows the filter washing times to be dramatically reduced. In such studies, we use a "Biologic" Rapid Filtration System (Dupont, 1984) sold by Molecular Kinetics Inc. This instrument, which relies on a microcomputer-controlled syringe to deliver a specific volume of wash buffer within a specified time, is capable of performing the washing step in tens of milliseconds. We routinely use this device for reliable measurements of low affinity GABA sites (K_ds in the micromolar range) on the $GABA_A$ receptor (see Protocol 4.4). One drawback of the Rapid Filtration System is that, since only one filter can be processed at a time, the experimental procedures can become very lengthy. However, the system has many advantages and we find that it is useful not only for studying low affinity binding sites, but also for studies of rapid dissociation kinetics and ion efflux experiments.

4 Radiolabelled Ligand-Binding Protocols

4.1 Equilibrium Dialysis

To illustrate the principles of an equilibrium dialysis experiment, we will describe the binding of [^{3}H] acetylcholine to the nicotinic acetylcholine receptor (nAChR) in native membranes from *Torpedo* electroplax. The data obtained from such an experiment are shown in *Figure 10-7* where they are compared with data obtained from a centrifugation assay described below (Protocol 4.2).

Figure 10-7

Representative data for [^{3}H]acetylcholine binding to the membrane-bound *Torpedo* nAChR. Bindng was measured either by equilibrium dialysis (closed circles) as described in Protocol 4.1 or by centrifugation (open squares, see Protocol 4.2). Estimated K_d values from nonlinear regression curve fitting were 12 nM and 10 nM, respectively with corresponding R_0 values of 0.14 μM and 0.135 μM

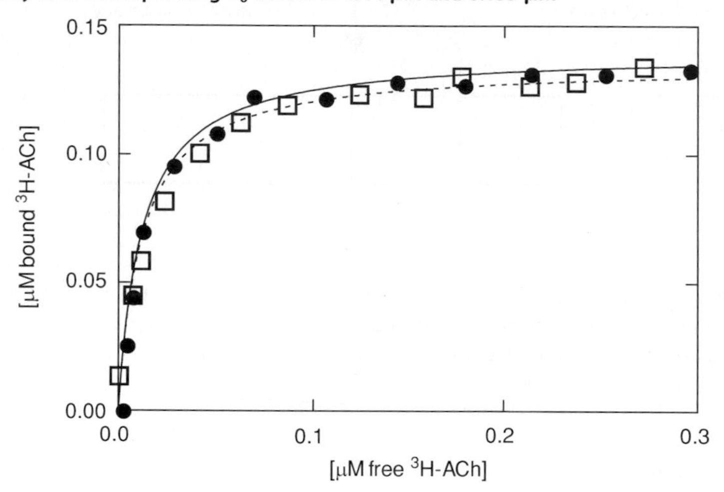

1. *Torpedo* membranes that are highly enriched in the nAChR are first prepared according to published procedures (Elliott et al., 1980).
2. An essential step prior to binding experiments is to inhibit acetylcholinesterase in the receptor preparation. *Torpedo* membranes are, therefore, first treated with an acetylcholinesterase inhibitor such as diethyl-*p*-nitrophenyl phosphate (DNPP). It is important to recognize that that DNPP is highly toxic and must be handled with care. DNPP is prepared as a 300 mM stock in 2-propanol and is added to concentrated *Torpedo* membranes (approximately 5 mg protein/ml) to give a final concentration of 1.5 mM. After incubation for 3 minutes at room temperature, the membranes are diluted with about 10 volumes of buffer to reduce both the DNPP and solvent concentration. The membranes are then kept on ice until use.
3. The membranes are diluted to the desired final concentration (usually 0.1–0.5 μM in receptor binding sites) in *Torpedo* Ringers (20 mM Hepes, 250 mM NaCl, 5 mM KCl, 4 mM $CaCl_2$, 2 mM $MgCl_2$, pH 7.4). For measurement of nonspecific binding, an aliquot of these membranes is incubated with an excess of unlabelled acetylcholine (100 μM–1 mM).
4. A series of dilutions of [^{3}H]acetylcholine is prepared. Note that the concentrations required to achieve the desired final concentrations should take into account the volume of the two chambers of the dialysis apparatus assuming that the free ligand will equilibrate between the two compartments.
5. The dialysis chambers are assembled. We use microdialysis chambers in which the volume of the two compartments is 0.35–0.4 ml. The chambers are separated by 50,000 molecular weight cutoff SpectraPor dialysis tubing.

6. Aliquots of membranes (0.35 ml) are introduced into one compartment of the microdialysis cell and aliquots of [^{3}H]acetylcholine are introduced into the other compartment.
7. The chambers are sealed and allowed to shake or rock for 6–12 hours or whatever time is necessary to ensure ligand equilibration. Aliquots from each chamber are removed and measured for total ([RL] + [L]) and free ([L]) radioligand.
8. The concentration of bound ligand is calculated from the difference between total and free ligand and the data are analysed as above.

4.2 Centrifugation

This protocol describes the binding of [^{3}H]acetylcholine to *Torpedo* membranes using a centrifugation assay. Some typical data, obtained in such an assay, are shown in *Figure 10-7* for comparison to those obtained by equilibrium dialysis.

1. *Torpedo* membranes are first treated with an acetylcholinesterase inhibitor as described in Step 2 of Protocol 4.1. The membranes are then kept on ice until use.
2. A series of microcentrifuge tubes is set up, containing the membranes (usually 0.1–0.5 μM in receptor binding sites; approximately 0.05–0.25 mg protein/ml) in *Torpedo* membrane protein and [^{3}H]acetylcholine concentrations ranging from 0 to 2.0 μM. Nonspecific binding is determined in the presence of excess unlabelled acetylcholine (100 μM–1 mM).
3. The tubes are vortexed, and the binding reaction is allowed to come to equilibrium (45 minutes at room temperature).
4. The samples are vortexed again, and an aliquot is taken and used to determine the total [^{3}H]acetylcholine present ([L_0]) by counting for [^{3}H] after addition of scintillation fluid.
5. The samples are centrifuged in a bench-top microfuge at 13,000 × g for 15 minutes at 4°C.
6. Aliquots of the supernatant are taken and used to estimate the concentration of free radiolabelled ligand.
7. The amount of bound radiolabelled ligand is estimated from the difference between total and free ligand ([RL]=[L_0] – [L]). The data are corrected for nonspecific binding and are analysed as appropriate.

4.3 Manual Filtration

The majority of our radioligand binding assays involve filtration assays. For basic binding experiments, we use a vacuum manifold (Hoefer) with multiple filtration ports (or a Brandel cell harvester for larger numbers of samples). The following is a general procedure that we use for measuring the binding of radiolabelled benzodiazepines to the $GABA_A$ receptor in bovine brain membranes (see *Figure 10-8a*).

1. A series of tubes containing assay buffer (50 mM Tris/HCl, 150 mM KCl, pH 7.4), radiolabelled benzodiazepine (typically 0–50 nM) and bovine brain membranes (final concentration 0.5 mg/ml or approximately 1 nM in benzodiazepine binding sites) is set up. Nonspecific binding is determined in the presence of an excess of unlabelled benzodiazepine. As some benzodiazepines are light sensitive, we usually use opaque amber tubes for these assays.
2. The contents are mixed thoroughly, and are allowed to come to equilibrium (1 hour at 4°C). During this incubation period, the appropriate number of filters are soaked in ice-cold buffer. We routinely use Whatman GF/C filters since we have found that these yield the best specific/nonspecific binding ratio in our experiments. For other applications, the suitability of other types of filters may have to be experimentally determined.
3. The filters are mounted in the filtration manifold, vacuum is applied and an aliquot of the first sample is applied to the first filter. Once the sample compartment drains, the filter is washed immediately by the

Figure 10-8

Representative data for radiolabelled ligand binding to $GABA_A$ receptors in bovine brain membranes using filtration techniques. (a) Results from a manual filtration assay (Protocol 4.3) in which the binding of [^{3}H]$R_0$15-1788 was measured. Binding constants were estimated from the direct binding curve (main graph) or from a Scatchard plot (inset). Best fit parameters were 2.07 nM and 0.06 nM for the K_d and R_0, respectively. (b) The binding of [^{3}H] muscimol to brain membranes was measured using a BioLogic rapid filtration device (Protocol 4.4) to reveal heterogeneity in binding as shown by the curved Scatchard plot presented here. Data fitting gave two classes of sites with $R_1 = 1.09$ nM, $K_{d1} = 19$ nM, $R_2 = 2.34$ nM and $K_{d2} = 200$ nM

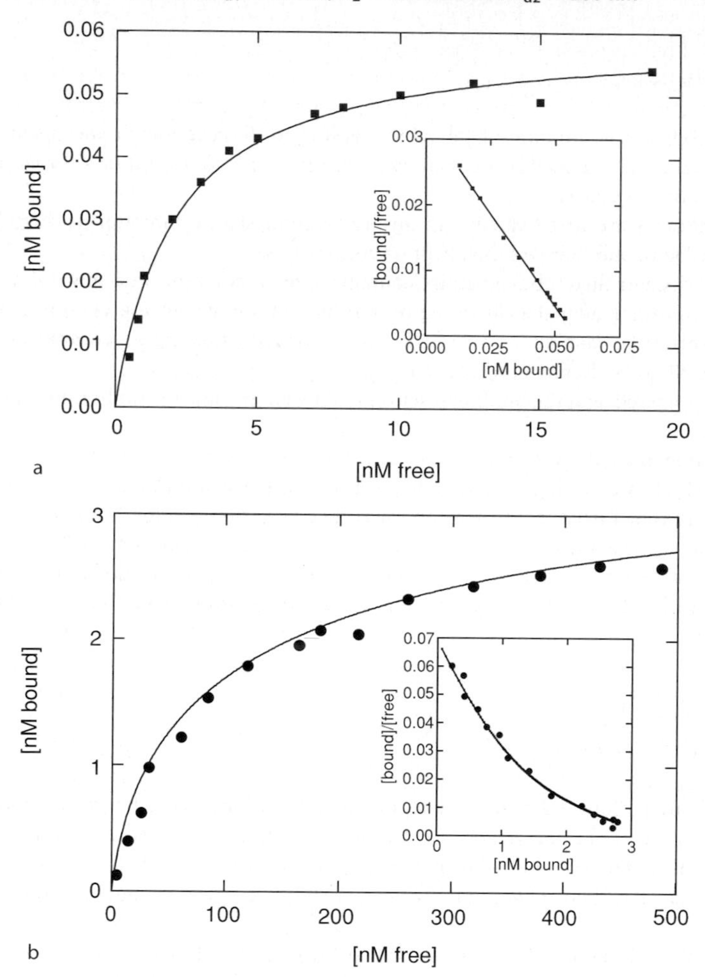

rapid addition of two 5 ml volumes of ice cold sample buffer. This is repeated for each of the remaining samples.

4. Aliquots of the residual samples are taken and used to determine the total amount of radioligand present.
5. The filters are dried under a heat lamp, transferred to scintillation vials and counted for radioactivity after addition of 5 ml of scintillation fluid. This is used to estimate the amount of bound radioligand.
6. The concentration of free radiolabelled ligand is estimated from the difference between total and bound ligand and the data are then analyzed as above.

4.4 Automated Rapid Filtration

As described above, automated rapid filtration techniques may be useful to examine the binding characteristics of a relatively rapidly dissociating ligand. The principles of the technique are similar to those of the manual filtration assays described above but the difference is that after application of the sample, the time and flow rate of filter washing is precisely controlled. The filter is first mounted in the filter holder, the vacuum is turned on and the sample is applied as above. The filter and filter holder is then mechanically brought in contact with a syringe delivering wash buffer. The filters are then washed during forced filtration for a preset time (0.01–9.0 s) and at a preset flow rate (0.1–10 ml/s). Below, we illustrate the general approach by describing the binding of [^{3}H]muscimol binding to high and low affinity $GABA_A$ receptor sites in bovine brain membranes. Data from a typical experiment are shown in *Figure 10-8b*. It should be noted that this is an equilibrium experiment and the purpose of using the rapid filtration system is to reduce the time of filter washing for better definition of the lower affinity, rapidly dissociating sites. The system is, however, designed for kinetic analysis in which changes in bound ligand are measured as a function of time.

1. The experiment is initiated as described above (Protocol 4.3) by mixing the radioligand with the receptor preparation and allowing the reaction to come to equilibrium. The samples typically contain bovine brain membranes at a concentration of 0.5 mg/ml in a total volume of 1 ml. For reliable measurements of [^{3}H]muscimol binding, it is essential to remove all endogenous GABA from the membrane preparations and this requires thorough washing of the membranes and freeze-thaw cycles prior to assay. The concentration range of [^{3}H]muscimol investigated ranges between 1 and 500 nM to allow reliable measurements of both the high and low affinity binding sites of the $GABA_A$ receptor. We generally include 14–20 concentrations of [^{3}H]muscimol so that heterogeneous binding curves are reasonably well defined and thus more easily analyzed.
2. The filtration conditions (i.e., the rate and time of filtration) should be established in preliminary experiments. The two major considerations are: (1) filter washing must be sufficient to minimise the nonspecific retention of radiologand by the filter and (2) the time course of filtration should be sufficiently short that dissociation of the radioligand from its receptor is avoided. In these studies, it is important to have a preliminary estimate of the dissociation rate, i.e., k_{-1} in equation 23. For a simple exponential reaction, the half-time of dissociation ($t_{½}$) is related to the dissociation rate constant by: $t_{½} = 0.693/k_{-1}$
3. Since adequate washing cannot realistically be obtained in less than about 50 msec, the complex half-life must be at least 250 msec ($k_{-1} < 2.8\ s^{-1}$) for reasonable binding measurements to be obtained. Unfortunately, the dissociation rates of many important receptor–ligand complexes are faster than this and these are not amenable to measurement by these techniques. To study the equilibrium binding of [^{3}H]muscimol, we routinely use a filtration (wash) time of 0.5 sec and a wash volume of 2 ml.
4. After equilibration, the sample is applied to a GF/C filter under reduced pressure. The filter is dried under vacuum such that the sample equilibrates with the filter hydration volume. Reproducible sample application to the central part of the filter is facilitated by the use of a special adapter.
5. The wash step is initiated under microprocessor control. The syringe reservoir assembly is forced against the filter holder, forming a tight seal. The syringe dispenses the preset volume of buffer at the set flow rate. The buffer passes through the filter under vacuum and into a waste container. In control experiments it is important to establish the best filters to use since both the porosity and the mechanical stability of the filter will affect the washing efficiency and the reproducibility of the measurements. At the end of filtration, the filter holder is physically separated from the wash assembly and the filter is removed and processed as above.

4.5 Association Kinetics

The rate of formation of a receptor–ligand complex depends on the concentration of the two species, the affinity of the complex and the nature of the underlying mechanism. As a first approximation, if one assumes that the association rate constant approaches diffusion control (1×10^7 M^{-1}-s^{-1}) and one has an estimate of the equilibrium dissociation constant then, the association rate can be estimated from the equation:

$$k_{app} = k_1[L] + k_{-1} \quad (29)$$

If, for example, the K_d is 1 nM, then an initial guess for k_{-1} (= $K_d \times k_1$) is 0.01 s^{-1}. From the above equation, one may predict that k_{app} will be 0.02 s^{-1} at 1 nM ligand, 0.11 s^{-1} at 10 nM and 1.01 s^{-1} at 100 nM ligand. The corresponding half-times of the reaction (0.693/k_{app}) are 35 s, 6.3 s and 0.69 s, respectively. This simple example illustrates that the nature of the interaction places strict limitations on the ligand concentration ranges that are suitable for study using manual techniques and also shows that normally only high affinity binding sites can be studied by these approaches. Usually preliminary studies are necessary to establish the feasibility of the experiments and the appropriate time points to use. In the example below, we describe the binding of a radiolabelled benzodiazepine to the $GABA_A$ receptor in bovine brain membranes where conditions can be chosen such that the reaction occurs in the experimentally accessible second to minute range (see *Figure 10-9a*). In these studies, it is important to follow the reaction for a sufficient time period that it reaches saturation.

Figure 10-9

Kinetic analysis of the interaction of [^{3}H]flunitrazepam with the bovine brain $GABA_A$ receptor. (a) The association reaction can be desribed by a single exponential function as described by Eq.22. At the ligand concentration used (2 nM), the measured half-time ($t_{1/2}$) obtained from nonlinear regression curve fitting was 16.8 sec. (b) The dissociation of [^{3}H]flunitrazepam from the bovine brain GABAA receptor also occurs in a mono-exponential fashion with a half-time of 50.6 sec

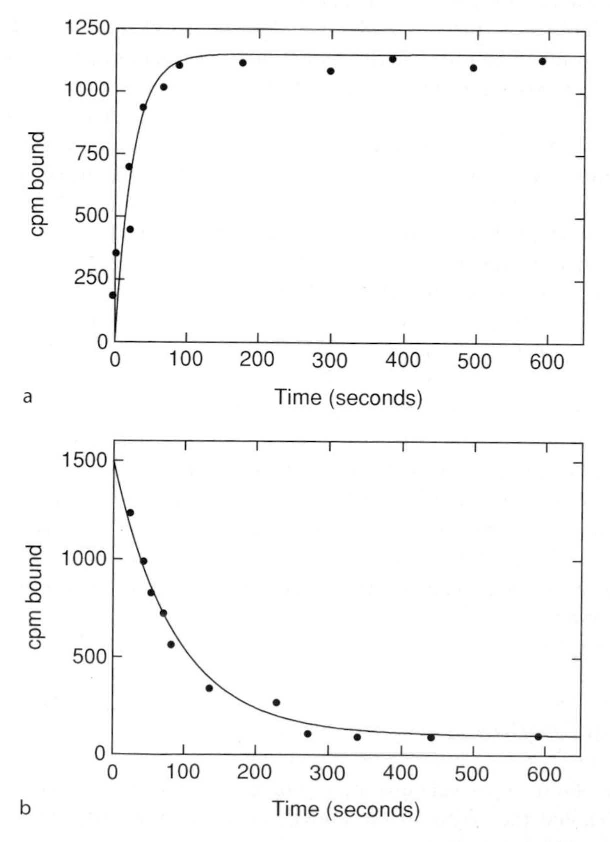

1. The membrane preparation is diluted to an appropriate final concentration in the incubation buffer and the mixture is stirred with a magnetic stirrer. Because multiple aliquots will be taken from the incubation mixture, it is important to start with an adequate total volume. To initiate association, the radioligand is added to the mixture under constant stirring.
2. At the appropriate times, aliquots are filtered through GF/C filters and the filters are washed twice with 5 ml of ice-cold buffer.
3. After counting for bound radioligand the data are analyzed as described in Section 2.8.1.

4.6 Dissociation Kinetics

To illustrate the principles of dissociation experiments, we describe below the dissociation of a radiolabelled benzodiazepine from $GABA_A$ receptors in bovine brain membranes. This reaction apparently proceeds in a simple exponential decay as described by equation 23 above. Sample data are illustrated in *Figure 10-9b*.

1. As for studies of association kinetics, preliminary experiments should be performed to establish the appropriate time points to use in the assay.
2. The radioligand is incubated with the receptor and the reaction is allowed to reach equilibrium. In the example described, the incubation time is usually 1 hour at 4°C. Again, since samples are to be taken from a single incubation mixture, the initial volume must be sufficient for sampling at all desired time points.
3. The dissociation reaction is initiated either by dilution of the sample with a large volume of assay buffer or by the addition of excess unlabelled ligand that competes for the radioligand binding sites. As described in Section 2.8.2, the conditions chosen should be adequate to prevent significant rebinding of the radiolabelled ligand. In dilution experiments, ideally one attempts to approach "infinite dilution" but due to the technical restraints imposed by handling of very large volumes, it is usually necessary to compromise by using dilution factors of approximately 100–200.
4. At the appropriate time points, aliquots are removed from the mixture and filtered under vacuum through GF/C filters. Filters are then washed twice with 5 ml of ice-cold buffer and processed as described above.
5. The data are then analyzed with the appropriate method (see Section 2.8.2).

5 Data Analysis and Interpretation

Most laboratories now have access to powerful computers and an extensive array of commercially available data analysis software (e.g., Prism (GraphPad, San Diego, CA), Sigma Plot (San Rafael, CA)). This provides ready access to the use of nonlinear regression techniques for the direct analysis of binding data, together with appropriate statistical analyses. However, there remains a valuable place for the manual methods, which involve linearisation, particularly in the undergraduate arena and these have been rehearsed in the above text.

There are a plethora of criteria that should be applied to ensure that the experimentally determined parameters provide a true reflection of the physical interactions that they represent. However, if the data are to be credible they must demonstrate an internal consistency. The equilibrium dissociation constant should, for example, be the same if it has been determined from equilibrium saturation assays or by calculation from the appropriate kinetic constants; if it is not, this implies that the physical characteristics of the interaction are outside the criteria for which the equations have been developed, i.e., those rehearsed in Section 2.7. Statistical comparison of data sets must also be carefully assessed; here the availability of the powerful computation facilities available on most laboratory desks has taken much of the drudgery out of such analysis.

6 Conclusions

Radioligand binding techniques have proved to be enormously valuable not only for the characterisation of receptors that are responsible for the mediation of physiological and pharmacological responses, but also for the characterisation of new chemical entities during development. In the latter regard, the approach has been modified significantly: the radioligands have been replaced by alternative tagging procedures that allow for the automated detection of ligand–receptor complexes. These include the use of techniques such as fluorescence to facilitate the high-throughput screening procedures that are the life-blood of industrial science. In terms of receptor biochemistry, the advent of site-directed mutagenesis has facilitated the dissection of the unique recognition properties of individual proteins by allowing a comparison between the interactions of wild type and mutant receptors.

While radioligand binding methods have proved to be of huge importance over many years, they do have several important limitations, some of which have been referred to above. In only a few cases can the functionality of the ligands be implied from radioligand binding experiments. They provide information on ligand–protein recognition, the kinetic parameters that underlie their interaction and the density of the binding site population. However, the consequences of binding site occupancy on the function of the protein can be ascertained in only a very few cases.

In this article we have provided a brief overview of the methods as they are commonly used today. Above this are layers of sophistication associated with more complex interactions; the principles remain the same, only the analysis changes.

References

Agey MW, Dunn SMJ. 1989. Kinetics of [^{3}H]muscimol binding to the $GABA_A$ receptor in bovine brain membranes. Biochemistry 28: 4200.

Boeynaems JM, Dumont JE. 1975. Quantitative analysis of the binding of ligands to their receptors. J Cyclic Nucl Res 1: 123.

Cheng Y, Prusoff W. 1973. Relationship between the inhibition constant (K_I) and the concentration of inhibitor which causes 50 per cent inhibition (I_{50}) of an enzymatic reaction. Biochem Pharmacol 22: 3099.

Connors KA. 1987. Binding constants: the measurement of molecular complex stability. New York: John Wiley and Sons.

Cornish-Bowden A, Koshland DE. 1975. Diagnostic uses of the Hill (logit and Nernst) plots. J Mol Biol 95: 201.

Dupont Y. 1984. A rapid filtration technique for membrane fragments or immobilized enzymes: measurements of substrate binding or ion fluxes with a few millisecond time resolution. Anal Biochem 142: 504.

Ehrlich P. 1913. Address to the International Medical Congress, London. Lancet 2: 445.

Elliott J, Blanchard SG, Wu W, Miller J, Strader CD, et al. 1980. Purification of *Torpedo californica* post-synaptic membranes and fractionation of their constituent proteins. Biochem J 185: 667.

Hulme EC. 1992. Receptor–ligand interactions: a practical approach. Oxford: IRL Press.

Klotz IM. 1982. Numbers of receptor sites from Scatchard graphs: facts and fantasies. Science 217: 1247.

Klotz IM. 1983. Ligand–receptor interactions: what we can and cannot learn from binding measurements. TIPS 4: 253.

Norby JG, Ottolenghi P, Jensen J. 1980. Scatchard plot: common misinterpretation of binding experiments. Anal Biochem 102: 318.

Scatchard G. 1949. The attractions of proteins for small molecules and ions. Ann NY Acad Sci 51: 660.

Wieland GA, Molinoff PB. 1981. Quantitative analysis of drug–receptor interactions: I. Determination of kinetic and equilibrium properties. Life Sci 29: 313.

11 Analysis of Receptor Localization in the Central Nervous System Using In Vitro and In Vivo Receptor Autoradiography

S. Kar · C. Hawkes

Abstract: Quantitative receptor autoradiography methods have been widely used over the last three decades to study the distribution and physiological role of a receptor in various tissues. This review provides an overview of in vivo and in vitro receptor autoradiography methods and their advantages as well as disadvantages in the study of receptors in the central nervous system. Comparison with immunohistochemical and in situ hybridization methods is also highlighted in relation to the study of a given receptor in the nervous sytem.

List of Abbreviations: BSA, bovine serum albumin; CNS, central nervous sytem; EDTA, ethylenediaminetetraacetic acid; EGTA, ethyleneglycol-bis-(β-aminoethyl ether) N,N,N′,N′-tetraacetic acid; GDP, guanosine diphosphate; GTP, guanosine triphosphate; IGF-I, insulin-like growth factor-I; IGF-II, insulin-like growth factor-II; PEI, polyethyleneimine; PET, positron emission tomography; SPECT, single photon emission compound tomography

1 Introduction

An increasing number of peptide and growth factor families have been characterized over the past three decades and this has markedly enhanced the apparent complexity of the various regulatory systems, which act as neuromodulators, neurotransmitters, or trophic substances. The recognition of the extensive colocalization of peptides and growth factors with classical neurotransmitters has also further demonstrated the heterogeneous nature of phenotypic expression of the central nervous system (CNS) (Kuhar et al., 1986; Hokfelt et al., 1987, 2000; Thoenen 1991; Dechant, 1994; Dreyfus, 1998). At present, the information available with regard to the precise role of various growth factors or neuropeptides is somewhat limited partly due to the incomplete knowledge about the organization of their receptors (or receptor subtypes) in the CNS and/or lack of selective agonists and antagonists to directly assess their functional relevance. The situation will, however, change not only with the development of highly potent selective antagonists or agonists but also with a better understanding of the receptor and the associated intracellular signaling mechanisms involved in mediating the physiological or pathophysiological roles in the nervous system.

2 Receptor Autoradiography

Over the years, the use of radioligand membrane binding techniques has permitted the detailed pharmacological characterization of receptor sites in both peripheral tissue and in the brain. This method is quite rapid, highly quantitative, and is extremely useful in the development of selective agonists and/or antagonists to distinguish different receptor types or subtypes for a given receptor family (Yamamura et al., 1978; Kuhar et al., 1986; Quirion et al., 1989; Keen and MacDermot, 1993; Qume, 1999). However, one of the major limitations of this technique is the lack of resolution at the anatomical level. The differential distribution of binding sites in a heterogeneous tissue like the CNS is practically impossible to obtain from membrane binding assays. This limitation is usually overcome by using receptor autoradiography, which enables one to localize and quantify receptor density in small but discrete regions of the CNS or in tissues where overall quantities of receptors are relatively low. The usual sequence for localizing receptors by autoradiography is to label or tag the receptor of interest with a radiolabeled ligand, in an intact animal or tissue section, and then to generate an autoradiogram by an appropriate means, which reveals the receptor-bound ligand (Kuhar et al., 1986; Palacios and Dietl, 1987; Kuhar and Unnerstall, 1990; Quirion et al., 1993; Sharif, 1996; Chabot et al., 1999). In this method, the radioactive decay occurring within the sample emits radiation particles, causing changes within the detection system that are then amplified to produce a detectable signal. The steps involving selective labeling and appropriate visualization of the receptor are the two main factors that are critical for the success of the experiment.

The accuracy of the results depends on many factors, including the type of radioligand used (irreversible or reversible ligands), the autoradiographic method (in vivo or in vitro labeling) and the photographic medium (wet photographic emulsion or dry emulsion-coated coverslips or radiosensitive films) employed to record the radioactive emissions. One of the most important issues to be considered is the avoidance of any diffusion of the radiolabeled molecules bound to the receptor of interest until the autoradiogram is generated (Kuhar and Unnerstall, 1990; Quirion et al., 1993; Chabot et al., 1999). One way to overcome this problem is to use irreversible radioligands, which can be covalently bound to the receptor, thereby allowing for various photographic media to be selected for signal detection. However, given the relatively high level of nonspecific binding produced by irreversible ligands, the majority of receptor autoradiographic studies use reversible ligands, which can readily dissociate from the binding sites in an aqueous environment. To avoid radioligand dissociation, two alternative approaches are generally used. A radioligand possessing a free amino group not involved in ligand–receptor interactions may be covalently linked to the receptor by the use of a chemical fixation method (e.g., aldehyde fixing agents such as paraformaldehyde or glutaraldehyde) prior to the application of the liquid photographic emulsion. Alternatively, Young and Kuhar (1979) have modified the use of the dry apposition method of Stumpf and Roth (1966) to visualize receptors using autoradiography. This involves the application of a liquid photographic emulsion onto a glass coverslip, followed by the apposition of the dried emulsion-coated coverslip to the labeled tissue section. However, the development of commercial emulsion-coated plastic films has allowed the widespread use of the dry apposition method for the visualization of diffusible radioactive ligands with a reasonable degree of resolution, permitting the macroscopic localization of binding sites and the quantification of the receptor density with the help of a computer-assisted image analysis system. The choice of detection methods to visualize radiolabeled ligand/binding site complexes is determined by the degree of resolution required. For gross anatomical resolution, autoradiograms are usually generated using emulsion-coated films or digital images derived from phosphor plate imagers. The dry emulsion-coated coverslip technique and the direct dipping of prefixed tissues followed by staining are methods of choice for greater cellular resolution. Of critical importance for all these strategies is the verification that the observed radioligand binding or regional accumulation in the brain or in any other tissues is specific and not due to an interaction with some nonspecific binding sites. Details of the procedure used to localize various neurotransmitter/modulator and growth factor receptors have been described extensively in several earlier publications (Kuhar et al., 1986; Palacios and Dietl, 1987; Quirion et al., 1989, 1993; Kuhar and Unnerstall, 1990; Sharif, 1996; Chabot et al., 1999; Qume, 1999). This review presents an overview of in vivo and in vitro receptor autoradiography methods and their advantages as well as disadvantages in the study of receptors in the CNS. Some examples of the application of in vitro receptor autoradiography using studies on insulin-like growth factor receptors are given.

3 In Vivo Receptor Autoradiography

The term "in vivo receptor labeling" is applied to the procedure in which receptors are labeled in intact living tissues in vivo after systemic administration of the radiolabeled ligand or drug. In this procedure, high-affinity radioligands are usually injected intravenously and after a relatively short time interval, the ligand is carried to the brain by the blood, diffuses into the brain, and binds to the target receptors. The nonreceptor-bound ligand is then removed from the brain and other tissues by various excretory processes. Specifically bound ligand, on the other hand, is identified by generating autoradiograms to provide information on the distribution of the receptor at the regional and cellular level (Kuhar et al., 1986; Quirion et al., 1993; Sharif, 1996; Chabot et al., 1999). This method has been used to demonstrate several receptor types in the brain, including the unique *sigma* (σ) receptor that may have a role in psychosis. Using (+)-[^{3}H]SKF 10047 as a radioligand, the highest levels of σ-labeling were reported to be localized in various cranial nerve nuclei, whereas lower but significant amounts of labeling were evident in the cortex, hippocampus, various hypothalamic nuclei, red nucleus, substantia nigra, central grey, and cerebellum (Bouchard et al., 1996). This in vivo distribution of σ-receptor sites corresponds rather well to the results obtained from the membrane binding assay (Compton et al., 1987).

In vivo receptor labeling offers binding conditions similar to those prevailing in the physiological state and it provides anatomical resolutions compatible with perfusion fixation of target tissue. This technique is useful for the evaluation of drug distribution, uptake, metabolism, receptor occupancy, and binding modulation; however, this procedure has some serious limitations. This method requires high-affinity radioligands that are metabolically stable and in the case of CNS studies, these molecules should be able to cross the blood–brain barrier. Also, the binding conditions that are so critically important to control the specificity of labeling cannot be modified or controlled very easily in vivo. Moreover, this procedure is somewhat expensive, as a large quantity of radioligand is required not only to overcome potential in vivo degradation of the ligand, but also to label an entire animal even though only small portions of a tissue, such as the brain, are actually needed in a given experiment (Kuhar et al., 1986; Quirion et al., 1993; Qume, 1999). Notwithstanding these issues, the in vivo labeling method provides a basis for the development of noninvasive imaging techniques such as positron emission tomography (PET) and single photon emission compound tomography scanning (SPECT). These procedures use in vivo labeling to visualize receptor binding sites as well as to measure other parameters in living human CNS (Kuhar et al., 1986; Chabot et al., 1999; Ichise et al., 2001).

4 In Vitro Receptor Autoradiography

Among the various autoradiographic methods currently available, in vitro receptor autoradiography has become the technique of choice for studying the discrete localization of receptor sites in the nervous system. The procedure is rather simple and the main steps involved in the method include: (1) incubation of slide-mounted tissue sections in the presence of the radioligand under optimized binding conditions; (2) washing steps determined empirically to remove excess unbound ligand; (3) a very rapid wash in deionized water to remove salts; and (4) the generation of autoradiograms using photographic emulsion or commercially available highly sensitive films. This method is not only sensitive and cost-effective, but also allows the use of peptides as well as nonpeptide ligands that cannot readily cross the blood–brain barrier (Kuhar et al., 1986; Quirion et al., 1993; Sharif, 1996; Chabot et al., 1999). Various receptor distributions can be easily compared using this procedure, as it is possible to use different ligands with consecutive tissue sections. Also, in vitro labeling allows for precise control of the binding parameters, which enables one to obtain high ratios of specific to nonspecific binding in very small pieces of tissue or in a complex structure like the brain where anatomical information is often desirable. Additionally, radioligand binding kinetics and pharmacological specificity can also be determined when combined with a computer image analysis system (Palacios and Dietl, 1987; Kuhar and Unnerstall, 1990; Qume, 1999). However, when labeling peptide or growth factor receptors in mounted sections, the following parameters have to be taken into account: (1) the chemical stability of the ligand; (2) the presence of endogenous ligands; (3) the presence of peptidases in the tissue; and (4) the possibility of ligand binding to sites unrelated to the receptor itself.

4.1 Technical Issues Associated with In Vitro Receptor Autoradiography

The majority of radiolabeled probes available to study peptide and growth factor receptors are peptidergic in nature and therefore highly sensitive to enzymatic degradation and radiolysis. Additionally, they often adhere to glass and plastic surfaces because of charged residues. To limit radiolysis and denaturation, peptide radioligands should be stored in small quantities at −20°C to −80°C for only a short period of time (Quirion et al., 1993). The loss of a radioligand due to adhesion to glass or plastic surfaces can be minimized by coating the glassware and slides with bovine serum albumin (BSA, 0.01–0.1%), polylysine (0.01–0.05%), gelatin (0.5%), or polyethyleneimine (PEI, 0.01–0.1%) and also by the addition of BSA to the incubation buffer. As most ligands bind reversibly, diffusion and loss of the ligand from the receptor sites at the end of the incubation period can be attenuated by limiting the exposure of the radiolabeled sections to the aqueous medium (Kuhar et al., 1986; Quirion et al., 1989; Sharif 1996; Chabot et al., 1999). Endogenous ligands that may interfere with binding are usually eliminated by preincubating the tissue sections in the buffer. Dissociation of the endogenously bound ligands can also be favored, in the case of G-protein coupled receptors, by adding guanine nucleotide [e.g., GTP or its stable analogue GPP(NH)p] into the preincubation medium (Kuhar and Unnerstall, 1990; Quirion et al., 1993; Qume, 1999; Sovago et al., 2001).

A mixture of peptidase inhibitors is usually used to reduce or block the degradation of radiolabeled peptide ligands or unlabeled peptides during the incubation period. The nature of the cocktail of enzyme inhibition varies according to each peptide family and may often be empirical, as metabolic pathways are mostly unknown for the majority of peptides and trophic factors. The enzyme inhibitors routinely used in receptor autoradiography include bacitracin (a nonselective peptidase inhibitor), chymostatin (inhibitor of serine proteases), aprotinin (a trypsin inhibitor), captopril (inhibitor of angiotensin converting enzyme), and leupeptin (inhibitor of leucine aminopeptidase) (Kuhar et al., 1986; Quirion et al., 1993; Sharif, 1996). Chelating agents such as EDTA and EGTA are also used to inhibit the action of metalloproteases and calcium-activated proteases, although caution should be exercised, as these may alter ligand interactions by removing regulatory metal cations. Divalent cations (Mn^{+2} and Mg^{+2}) are often used to ensure that receptor binding sites are in a high-affinity state, as most neurotransmitter and peptide receptors belong to the seven-transmembrane, G protein-coupled receptor family (Palacios and Dietl, 1987; Quirion et al., 1993; Qume, 1999). This usually improves the specific binding and decreases the dissociation rate of the ligand from its receptor sites. This constitutes a major advantage for in vitro receptor autoradiography, as it allows for longer washes and rinsing periods at the end of an assay. The influence of cations and peptidase inhibitors on the characteristics and density of the receptor being studied has to be carefully monitored in preliminary experiments to ensure a high signal-to-noise ratio and a high quality of autoradiograms generated. As no general rule truly applies, it is advisable to check each individual ligand's binding parameters for all the conditions mentioned above (Kuhar et al., 1986; Quirion et al., 1989; 1993; Chabot et al., 1999; Quam, 1999).

4.2 Dry and Wet Emulsion Autoradiography

Autoradiograms are generated by placing the labeled tissue sections in close contact with various photographic emulsions. This can be done either by dipping the tissue sections in liquid emulsion or by using a dry emulsion in the form of an emulsion-coated coverslip or films such as tritium-sensitive or dry X-ray films (Palacios and Dietl, 1987; Quirion et al., 1993; Sharif, 1996). The main advantage of using dipping procedures or emulsion-coated coverslips is the possibility of visualizing the autoradiographic grains together with the tissue. On the other hand, the use of dry film allows for easy quantification, although there is a loss of anatomical register between the localization of autoradiograms and the actual tissue structure (Kuhar et al., 1986; Sharif, 1996; Chabot et al., 1999).

4.3 Quantitative Analysis of Autoradiograms

Once generated, autoradiograms can be analyzed using a variety of computer-assisted image analysis systems (e.g., Microcomputer Imging Device, MCID, Imaging Research, Ontario, Canada). Basically, autoradiographic films are digitalized using a camera and then analyzed by microcomputers. Receptor densities (expressed in fmol/mg of tissue wet weight) are calculated from captured optical densities, which are calibrated according to appropriate, commercially available standards. One of the most characterized problems using receptor autoradiography with tritiated ligands has been the "quenching" caused by the differential absorption of β-radiation by different tissue elements (Kuhar and Unnerstall, 1990; Quirion et al., 1993; Chabot et al., 1999). A number of procedures have been designed to overcome this problem. This difficulty is not encountered with [^{125}I]labeled ligands, which are the most popular in studying growth factor receptors. In contrast to films, the quantification of autoradiograms generated from emulsion procedures is rather difficult and subject to a number of technical problems. Most commonly, the quantification of emulsion autoradiographic silver grains is carried out by visual or computer-assisted counting. The use of appropriate standards subsequently allows for the full transformation of grain density into receptor-related units (Kuhar et al., 1986; Quirion et al., 1989, 1993; Chabot et al., 1999). Both grain counting and densitometric analysis have inherent difficulties including image resolution, tissue quenching, and efficiency. New systems are currently evolving that allow the imaging of radioisotopes without the need for photographic emulsions (i.e., Micro and Beta Imagers, Biospace Mesures, France). Such systems promise shorter exposure times (hours rather than days), increased sensitivity for ^{3}H compared to film, increased spatial resolution, real-time accumulation of data, and double-labeling detection. The major drawback with this seems to be the low number of samples that can be measured at one time.

5 In Vitro Autoradiography and Growth Factor Receptors in the CNS

In vitro receptor autoradiography has been used extensively to study the regional and anatomical distribution of neurotransmitter/neuromodulator receptor families in different tissues, including the CNS (Kuhar et al., 1986; Palacios and Dietl, 1987; Quirion et al., 1993; Sharif, 1996; Chabot et al., 1999). The physiological and pharmacological effects of growth factors, like those of neurotransmitters and modulators, are also mediated by binding to specific receptors. With a steady increase in the number of trophic factors, multiple receptors and/or their subtypes have been characterized both in peripheral and nervous tissues (Quirion et al., 1989, 1993; Thoenen, 1991; Reynolds and Weiss, 1993; Dechant et al., 1994; Lindsay, 1994; Henderson, 1996; Dupont and LeRoith, 2001; Teng and Hempstead, 2004). Additionally, there is evidence that a given family of growth factors usually interacts with multiple receptor subtypes to induce their biological effects, thus adding another layer of complexity. Over the last decade, we have been studying to characterize the distributional and pharmacological profile of insulin-like growth factor-I and insulin-like growth factor-II (IGF-I and IGF-II) and insulin receptors in the CNS using in vitro receptor autoradiography (and membrane binding assays) to better understand their roles in brain functions (Kar et al. 1993a, b, 1997a, b; Dore et al., 1997; Jafferali et al., 2000; Hawkes and Kar, 2004). Using radiolabeled molecules with a high specific activity such as, for example, the same growth factor molecule or its analogue, it is possible to study the binding sites of these receptors in discrete regions of the brain.

5.1 Insulin-Like Growth Factor Receptor Family

The IGFs constitute a family of three structurally related polypeptide growth factors, i.e., IGF-I, IGF-II, and insulin, which are widely distributed in the CNS. Functionally, IGFs and insulin are considered to play an important role in neural growth and differentiation, possibly involving dendritic maturation, synaptogenesis, and myelinization. This is primarily supported by a number of in vitro and some in vivo studies that have shown that IGFs and/or insulin stimulate DNA synthesis and cell proliferation in fetal brain cells, differentiation of astroglial, oligodendroglial, and neuronal precursor cells, as well as neurite formation and outgrowth (Adamo et al., 1989; Jones and Clemmons, 1995; Pablo and de la Rosa, 1995; Dore et al., 1997; Dupont and LeRoith, 2001; LeRoith, 2003). Evidence further supports a role for these growth factors in the maintenance of normal and activity-dependent functioning of the adult brain (Kar et al. 1993b, 1997b; Guthrie et al., 1995; Guan et al., 1996; Anderson et al., 2002). The physiological responses of these growth factors are mediated by three distinct transmembrane receptors designated as the IGF-I, IGF-II, and insulin receptors (Adamo et al., 1989; Kornfeld 1992; Jones and Clemmons 1995; Dore et al., 1997; Hawkes and Kar, 2004). The IGF-I and insulin receptors are structurally related members of the tyrosine kinase receptor family and each of this receptor exists at the cell surface as a heterotetramer consisting of two α and two β subunits joined by disulfide bonds. Activation of the IGF-I receptor leads to intracellular signaling that promotes growth, proliferation, and survival of cells in a variety of tissues including the CNS, whereas insulin receptor stimulation is believed to be associated with a variety of metabolic events including glucose uptake, activation of pyruvate dehydrogenase, and synthesis of protein and lipid (Adamo et al., 1989; Jones and Clemmons, 1995; Dore et al., 1997; Dupont and LeRoith, 2001). The IGF-II receptor is structurally distinct from both the IGF-I and insulin receptors and has no intrinsic tyrosine kinase activity. It comprises a single polypeptide chain with a large extracellular domain and a short cytoplasmic tail. Several lines of evidence have clearly indicated a role for this receptor in lysosomal enzyme trafficking, clearance, and/or activation of a variety of growth factors and endocytosis-mediated degradation of IGF-II. There is also a growing body of evidence supporting a possible role for this receptor in transmembrane signal transduction in response to IGF-II binding, but its relevance remains unclear (Kornfeld, 1992; Hille-Rehfeld, 1995; Dahms and Hancock, 2002; Hawkes and Kar, 2004).

5.2 Characterization of Insulin-Like Growth Factor Receptors Using Film Autoradiography

Using in vitro receptor autoradiography, we and others have shown that IGF-I, IGF-II, and insulin receptor binding sites are widely but selectively distributed throughout the adult rat brain (Hill et al., 1988; Lesniak et al., 1988; Smith et al., 1988; Werther et al., 1990; Kar et al., 1993a). Under our conditions, specific binding for each ligand represents 80–95% of total binding (Kar et al., 1993a). A high density of [^{125}I]IGF-I receptor binding, as reported earlier, is evident in the olfactory bulb, choroid plexus, CA2–CA3 regions of the hippocampus, piriform cortex, median eminence, and in the molecular layer of the cerebellum, whereas moderate labeling is noted in the various thalamic nuclei, cortex, selected layers of the hippocampal formation, and in the brainstem. The distributional profile of [^{125}I]IGF-II receptor binding showed that olfactory bulb, pyramidal, and granule cells of the hippocampus and the granule cell layer of the cerebellum exhibit a high density of specific labeling. A moderate concentration of receptor was associated with the cortex, hypothalamus, thalamus, striatum, and amygdaloidal body and in certain nuclei of the brainstem. The brain regions enriched with [^{125}I]insulin receptor binding sites include the olfactory bulb, choroid plexus, CA1 regions of the hippocampus, entorhinal cortex, and the molecular layer of the cerebellum. Moderate levels of binding are evident in the neocortex, substantia nigra, selected layers of the hippocampus, and certain nuclei of the hypothalamus and thalamus.

To characterize the specificity of [^{125}I]IGF-I, [^{125}I]IGF-II, and [^{125}I]insulin binding sites, we subsequently performed competition binding experiments with unlabeled IGF-I, IGF-II, and insulin, using in vitro receptor autoradiography in the hippocampus. All three receptor binding sites, as previously mentioned, exhibited different distributional and pharmacological profiles within the hippocampus (➋ *Figure 11-1a–c*) (Kar et al., 1997a; Hawkes et al., 2004). Specific [^{125}I]IGF-I binding sites, as evident from the autoradiograms, were competed potently by IGF-I>IGF-II>insulin (➋ *Figure 11-1a, d, g, j*). In contrast to [^{125}I]IGF-I, specific [^{125}I]IGF-II binding was inhibited in a concentration-dependent manner by IGF-II>IGF-I (➋ *Figure 11-1b, e, h*). Unlabeled insulin was found to be ineffective in displacing [^{125}I]IGF-II binding (➋ *Figure 11-1k*). Competition studies with [^{125}I]insulin receptor binding sites revealed that insulin competed more potently than IGF-II or IGF-I (➋ *Figure 11-1c, f, i, l*). Thus in summary, the IGF-I and insulin receptors, as evident from the competition-inhibition experiments, preferentially bind their own ligands and interact with other ligands at lower affinity. Similarly, the IGF-II receptor recognizes IGF-II with higher affinity than IGF-I and does not interact with insulin.

5.3 Emulsion Autoradiography of Insulin-Like Growth Factor Receptors

To determine the cellular localization of IGF-I and IGF-II receptors, we also performed emulsion autoradiography using [^{125}I]IGF-I and [^{125}I]IGF-II in the adult rat hippocampal formation (Kar et al., 1997a). High levels of [^{125}I]IGF-I binding sites were localized primarily in the CA2-CA3 subfields and the dentate gyrus. Within the Ammon's horn, a high density of silver grains was evident particularly in the stratum oriens and polymorphic neurons of the hilar region, whereas the stratum radiatum and lacunosum-moleculare showed only moderate amounts of grains (➋ *Figure 11-2a–h*). The pyramidal neurons of the Ammon's horn exhibited a low density of silver grains (➋ *Figure 11-2d, e*). In the dentate gyrus, silver grains were predominantly localized in the molecular layer rather than the granular cell layer (➋ *Figure 11-2f, g*). High amounts of [^{125}I]IGF-II binding sites were evident primarily in the pyramidal cell layer of the CA1-CA3 subfields, whereas other layers of the Ammon's horn such as the strata oriens, radiatum, and lacunosum-moleculare showed apparently low levels of specific labeling (➋ *Figure 11-2i–p*). The amounts of silver grains in the CA2 or CA3 pyramidal cell layer were higher than in the CA1 subfield (➋ *Figure 11-2j–m*). In the dentate gyrus, the granular cell layer, as opposed to the molecular layer, showed relatively

Figure 11-1
Photomicrographs of the autoradiographic distribution of [^{125}I]IGF-I (a, d, g, j), [^{125}I]IGF-II (b, e, h, k), and [^{125}I] insulin (c, f, i, l) receptor binding sites in the absence or presence of 100 nM IGF-I, IGF-II, and insulin in the adult rat hippocampal formation. [^{125}I]IGF-I binding was competed for potently by IGF-I>IGF-II>insulin (d, g, j). [^{125}I] IGF-II binding in various layers of the hippocampus was competed for potently by IGF-II>IGF-I (e, h), whereas unlabeled insulin did not displace [^{125}I]IGF-II binding (k). [^{125}I]Insulin binding, on the other hand, was competed for by insulin>IGF-II>IGF-I (f, i, l). NS, nonspecific binding; Or, stratum oriens; Py, pyramidal cell layer; Rad, stratum radiatum; Lmol, stratum lacunosum moleculare; DG, dentate gyrus; Mol, molecular layer of the DG; GrDG, granular cell layer of the DG

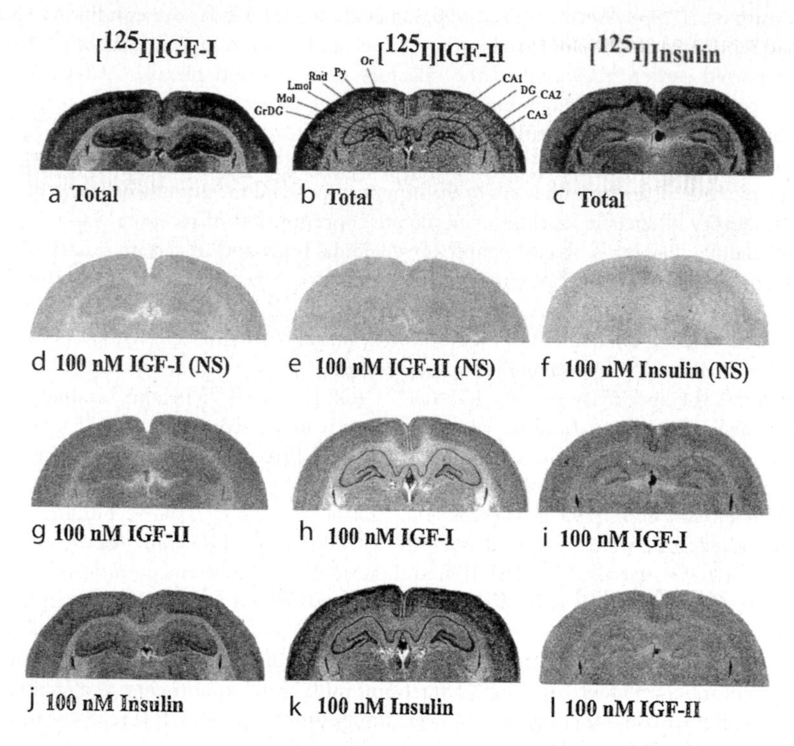

high levels of [^{125}I]IGF-II binding sites (*Figure 11-2n, o*). High amounts of silver grains were also evident in the polymorphic neurons of the hilar region.

5.4 Regulation of Insulin-Like Growth Factor Receptors

In the nervous system, in vitro receptor autoradiography methods have proven useful to evaluate the plasticity of neurotransmitter/growth factor receptors following various surgical/pharmacological manipulations, as well as under different neuropathological conditions. Multiple examples of receptor plasticity have been reported in the literature (see Kuhar et al., 1986; Palacios and Dietl, 1987; Quirion et al., 1993; Chabot et al., 1999). We have shown that IGFs and insulin receptor binding sites are selectively altered in discrete brain regions following surgical or pharmacological manipulations. Electrolytic lesioning of the entorhinal cortex has been reported to increase hippocampal [^{125}I]IGF-I, [^{125}I]IGF-II, and [^{125}I]insulin receptor binding sites at different times over a 28-day postlesion period (Kar et al., 1993b). A single systemic injection of kainic acid, on the other hand, evoked a significant decrease in hippocampal IGFs and insulin receptor binding sites over a 30-day experimental paradigm. [^{125}I]IGF-I receptor binding sites were transiently decreased at 12 h following treatment with kainic acid in almost all regions of the hippocampal

Figure 11-2

Photomicrographs showing the distributions of [^{125}I]IGF-I (a-g) and [^{125}I]IGF-II (i–o) receptor sites in the hippocampus (a, i) and in the CA1 (b, c, j, k), CA2 (d, e, l, m) and the dentate gyrus (DG) (f, g, n, o) regions of the adult rat. H and P represent nonspecific labeling in the presence of excess unlabeled IGF-I and IGF-II, respectively. Note the high density of [^{125}I]IGF-I labeled silver grains in the stratum oriens (Or) and in the molecular layers (Mol) of the dentate gyrus, whereas pyramidal neurons (Py) and the granule cells (GrDG) are relatively spared (b–g). c, e, and g are the brightfields representative of b, d, and f respectively. [^{125}I]IGF-II labeled silver grains, in contrast to [^{125}I]IGF-I, are predominantly concentrated in the pyramidal cells (Py) and in the granule cells of the dentate gyrus (GrDG), whereas the adjacent layers are rather weakly labeled (j–o). k, m, and o are the brightfield representative of j, l, and n, respectively. (Photomicrographs are from Kar et al. 1997. Proc Natl Acad Sci USA 94: 14054–14059)

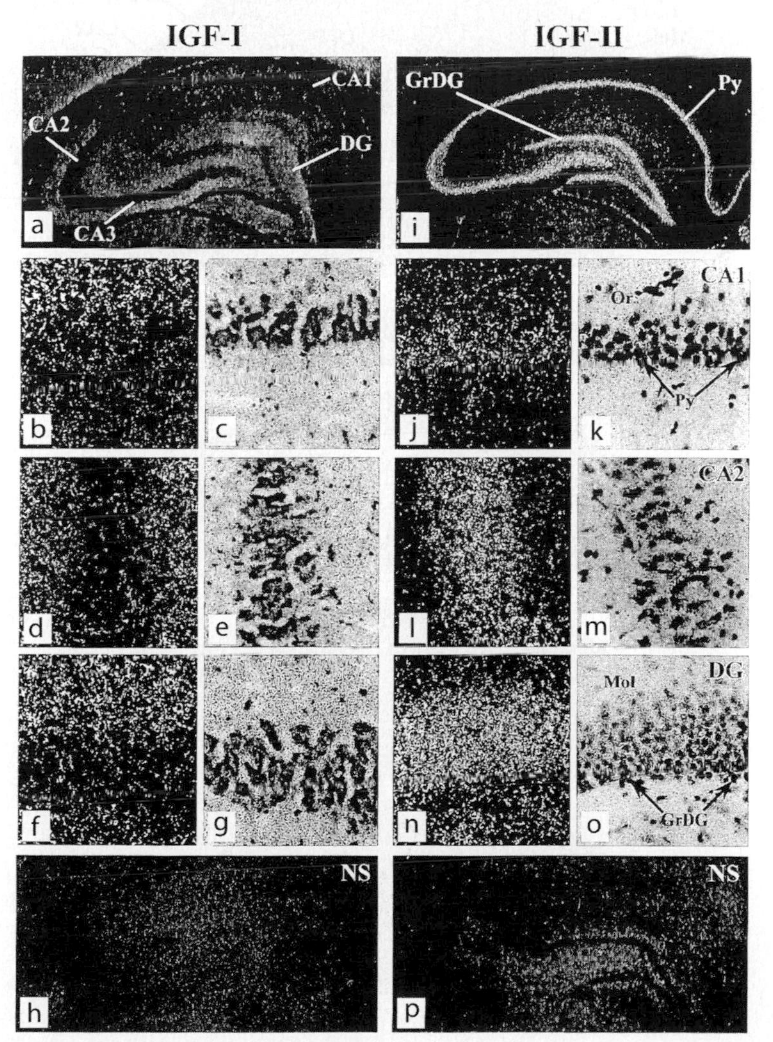

formation. [^{125}I]IGF-II receptor binding sites, on the other hand, showed a significant decrease in the CA1 subfield and pyramidal cell layer of Ammon's horn at all time points studied, whereas the hilar region and the stratum radiatum did not display any alterations at anytime (*Figure 11-3a–f*). Interestingly, [^{125}I] insulin receptor binding was found to be decreased at all time points in the molecular layer of the dentate

Figure 11-3

Photomicrographs of [^{125}I]IGF-II binding sites in rat hippocampal formation in normal (a), 12 h (b), 2 days (c), 12 days (d), and 30 days (e) following a single dose of 12 mg/kg systemic kainic acid administration. f represents [^{125}I]IGF-II binding sites in the presence of 100 nM unlabeled human IGF-II. Note the decrease in [^{125}I]IGF-II receptor sites in the pyramidal cell layer of the hippocampal formation observed at all time points (b, c, d, e) studied. However, other laminae were found to be differentially altered following administration of kainic acid. Abbreviations are the same as in *Figure 11-1*. (Photomicrographs are from Kar et al. 1997. Neuroscience 80: 1041–1055.)

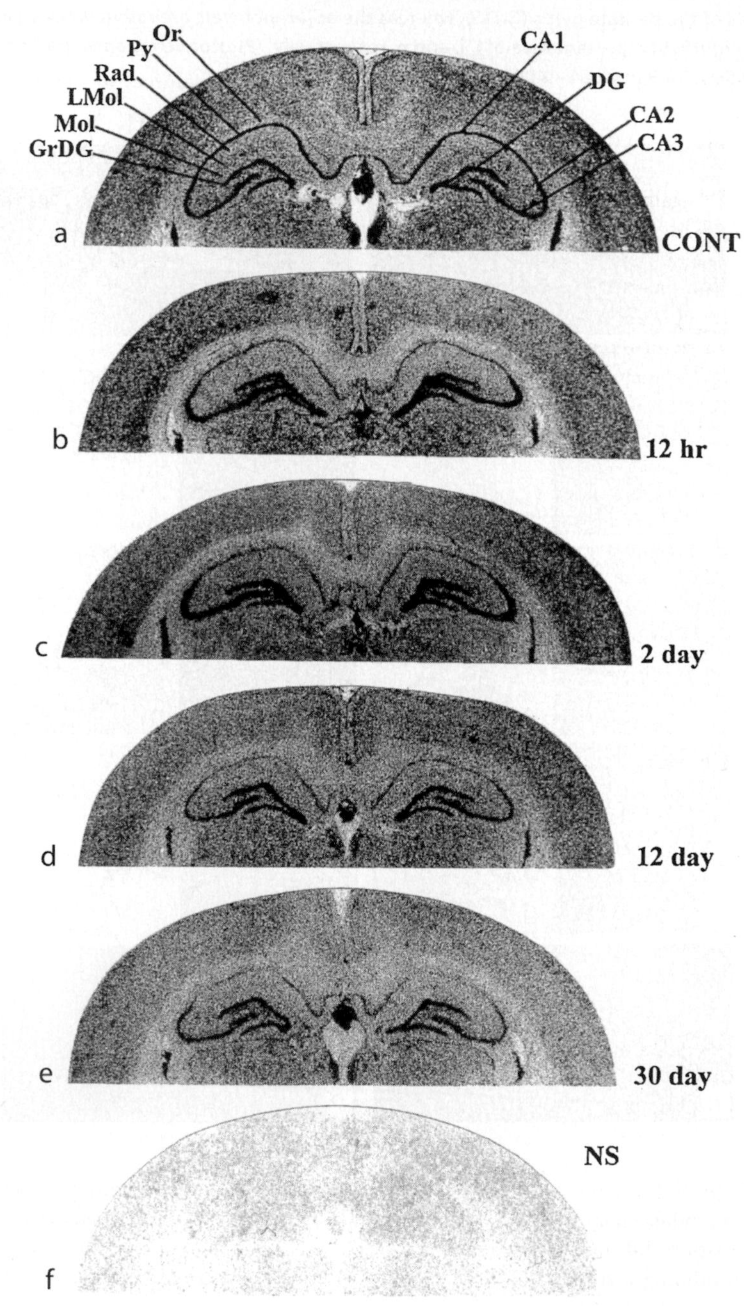

gyrus, while binding in CA1–CA3 subfields and discrete layers of the Ammon's horn was found to be affected only at 12 h following the treatment (Kar et al., 1997b). The temporal profile of these alterations may be attributed to the various operative mechanisms aiming at maintaining neuronal viability and circuitry within the hippocampal formation that follows kainate-induced seizures.

6 Receptor Immunohistochemistry

While receptor autoradiography is an excellent method to visualize and quantify receptor binding sites in the nervous and other tissues, it is unsuitable for tract-tracing and colocalization studies. In this procedure, it is often difficult to delineate whether receptor binding sites are localized on the neuronal cell body, dendrites, and/or axon-terminals. Another limitation relates to the lack of full selectivity of the vast majority of ligands for a given receptor type or subtypes (Kuhar et al., 1986; Quirion et al., 1993; Chabot et al., 1999). With the molecular cloning of various receptor genes as well as the increased availability of antibodies directed against the receptors, it is possible to localize receptor proteins at the cellular level using different immunohistochemical methods, which are based on the principle of detecting specific antigenic epitopes within a fixed tissue section. The specificity of the antibody is usually determined using liquid- and solid-phase adsorption with the homologous antigen, as well as with some closely related heterologous antigens (Sternberger, 1986). Using this method, various neurotransmitter, neuromodulator (i.e., peptides), and growth factor receptors, including nicotinic, muscarinic, dopaminergic, β-adrenergic, benzodiazepine/GABA, opioid, nerve growth factor, and epidermal growth factor receptors have been studied in the nervous system (Richards et al., 1987; Schroder et al., 1990; Birecree et al., 1991; Hill et al., 1993; Mansour et al., 1995a; Yung et al., 1995; Lee et al., 1998; Chabot et al., 1999).

Recently, using a polyclonal antibody against the IGF-II receptor, we reported intense IGF-II receptor immunoreactivity on neuronal cell bodies within the olfactory bulb, striatum, cortex, pyramidal, and granule cell layers of the hippocampus, selected thalamic nuclei, Purkinje cells of the cerebellum, pontine nucleus and motoneurons of the brainstem, as well as in the spinal cord. Moderate neuronal labeling was evident in the basal forebrain areas, hypothalamus, superior colliculus, midbrain areas, and in the intermediate regions of the spinal gray matter. We also observed dense neuropil labeling in many regions, suggesting that this receptor is localized on dendrites and/or axon terminals (Hawkes and Kar, 2003). The receptor immunoreactivity observed in the adult rat brain is very compatible with the autoradiographic distribution of the receptor studied using [^{125}I]IGF-II (Lesniak et al., 1988; Couce et al., 1992; Kar et al., 1993a; Hawkes and Kar, 2003) (➋ *Figures 11-4* and ➋ *11-5*). Our double immunolabeling studies further indicated that a subset of IGF-II receptors colocalized with cholinergic cell bodies and fibers in the septum, striatum, nucleus basalis, cortex, hippocampus, and motoneurons of the brainstem (Hawkes and Kar, 2003). In spite of the limitations associated with raising an antibody specific for a given receptor, immunohistochemical methods have the advantage of providing higher cellular resolution. Furthermore, with the use of electron microscopy, it is possible to determine whether the receptor-like immunoreactivity is localized in the plasma membrane or in subcellular compartments (Hill et al., 1993). However, unlike receptor autoradiography, the difficulties in quantifying the labeling is a major drawback of immunohistochemical approaches.

7 Receptor in Situ Hybridization

Receptor autoradiography and immunohistochemistry are used to demonstrate the final location of receptor proteins. In the nervous system, these receptor proteins are synthesized within soma, but are generally transported to dendrites or axons, distant from cell bodies where they are originated. Thus, demonstration of a receptor population by receptor autoradiography or immunohistochemistry is not sufficient to directly discriminate which perikarya synthesize the receptors (Kuhar et al., 1986; Quirion et al., 1993; Chabot et al., 1999). Knowledge of the sequence of the mRNA coding receptor proteins has made the development of nucleic acid probes possible, which can be used in combination with the

◘ Figure 11-4

Photomicrographs showing widespread IGF-II receptor expression in transverse sections of the adult rat brain. The profile of IGF-II receptor-immunoreactive neurons (a, d, g) is in general agreement with the distribution of [^{125}I]IGF-II binding sites (b, e, h) in all major brain areas, including the olfactory bulb and cortex (a–c), striatum, nucleus accumbens, and septal areas (d–f), as well as in the hippocampal, thalamic, and hypothalamic areas (g–i). The labeled schematic diagrams (c, f, i) demonstrate landmark nuclei/brain regions. AI, insular cerebral cortex; Arc, arcuate nucleus; Aov, anterior olfactory nucleus, ventral part; CC, corpus callosum; Cg, cingulate cerebral cortex; CPu, caudate putamen; DB, diagonal band complex; DG, dentate gyrus; Fr, frontal cerebral cortex; LS, lateral septal nucleus; Par, parietal cerebral cortex; Pir, piriform cerebral cortex; VPM, ventromedial thalamic nucleus. Scale bar = 0.1 cm. (Photomicrographs are from Hawkes and Kar. 2004. Brain Res Rev 44: 117–140.)

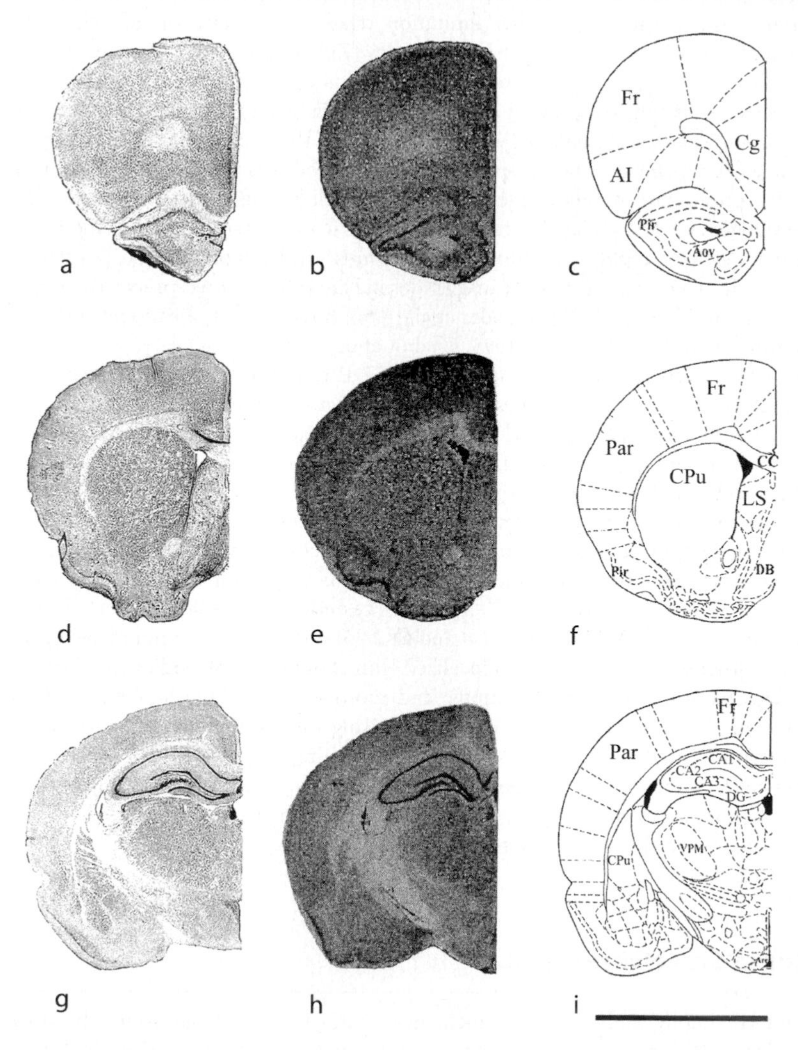

technique of in situ hybridization to visualize perikarya that express mRNA for a particular receptor. In this method, two single-stranded nucleic acid chains are allowed to bind together by hydrogen bonding of complementary base pairs, and the hybrids formed are subsequently detected using autoradiographic or histological procedures (Baldino et al., 1989; Mengod and Vilaro, 1993). Although both radioactive and

Figure 11-5
Photomicrographs showing widespread IGF-II receptor expression in transverse sections of the adult rat brain. The profile of IGF-II receptor-immunoreactive neurons (a, d) is in general agreement with the distribution of [^{125}I]IGF-II binding sites (b, e) in all major brain areas such as midbrain regions (a–c), cerebellum, and in the upper brainstem regions (d–f). Note the high density of IGF-II receptor expression in the pyramidal and granular cell layers of the hippocampal formation and in the granular cell layer of the cerebellum. g, represents a hippocampal section processed for IGF-II/M6P receptor immunoreactivity following preabsorption with 10 μM of purified IGF-II/M6P receptor antigen. h, demonstrates [^{125}I]IGF-II binding in the presence of 100 nM unlabeled human IGF-II. The labeled schematic diagrams (c, f) demonstrate landmark nuclei/brain regions. 7n, facial nucleus; LC, locus ceruleus; Oc, occipital cerebral cortex; pcuf, preculminate fissure; prf, primary fissure; py, pyramidal tract; R, red nucleus; SuG, superior colliculus; Tel, temporal cerebral cortex. Scale bar = 0.1 cm. (Photomicrographs are from Hawkes and Kar. 2004. Brain Res Rev 44: 117–140.)

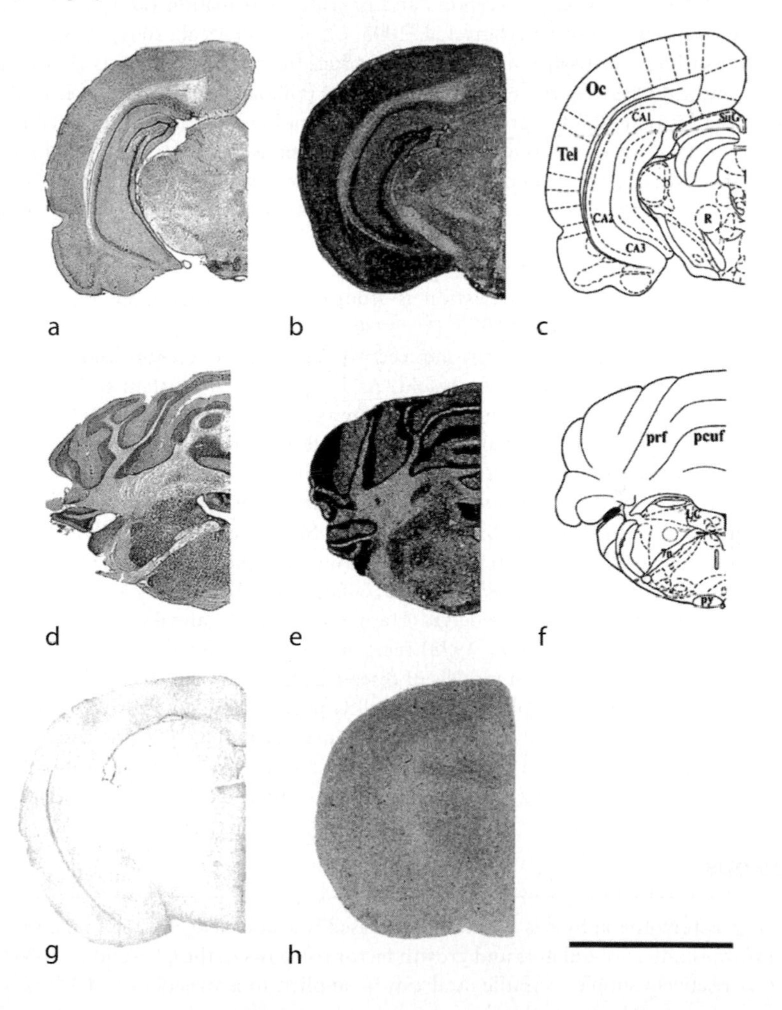

nonradioactive probes can be used to detect sites of active synthesis of receptor proteins, it is important that appropriate controls are carried out to test the specificity of the hybridization signal.

The discrete localization of a number of receptor mRNAs in brain has been determined using in situ hybridization; these include muscarinic (m1, m2, m3, m4, m5) (Vilaro et al., 1994), nicotinic (Wada et al., 1989), dopaminergic (Weiner et al., 1991; Mengod et al., 1992), GABAergic (Montpied et al., 1988; Sequier et al., 1988), glutamatergic (Pellegrini-Giampietro et al., 1991), neurokinin (Gerfen1991), opioid (μ, κ, δ) (Mansour et al., 1995b), neurotrophin (Miranda et al., 1993), and IGF-I and IGF-II (Werther et al., 1990; Couce et al., 1992; Nagano et al., 1995) receptors. One major advantage of the in situ hybridization is that it can be carried out in combination with other histological methods, either in the same or in consecutive tissue sections. The combination of in situ hybridization and receptor autoradiography/immunohistochemistry allows one to determine whether receptors found in a given brain nucleus are synthesized by cells intrinsic to the nucleus or synthesized by neurons whose perikarya are located in distant brain regions.

8 Functional Receptor Autoradiography Using [^{35}S]GTPγS

Unlike traditional receptor ligand autoradiography, [^{35}S]GTPγS binding autoradiography enables the visualization of only potentially active receptors and provides information regarding the first step of the intracellular signal transduction system (Sim et al., 1995, 1997; Sovago et al., 2001). A large number (at least 80%) of neurotransmitters, hormones, and peptides produce their biological effects via structurally related seven-transmembrane domain G-protein coupled receptors (Gilman 1987; Birnbaumer et al., 1990). Their signal is transduced by a variety of G-proteins, which act as molecular switches, activated by GTP binding and switched off by hydrolyzation of bound GTP. Each G protein is a heterotrimer composed of α, β, and γ subunits. In the inactive state, the α-subunit of the heterotrimer binds GDP. Binding of an agonist induces a conformational change of the receptor, which in turn interacts with and activates the α-subunit of the G-protein. When the G-protein is activated, the α–subunit releases its bound GDP and binds GTP instead, resulting in the formation of two active components: Gα-GTP and Gβγ, which subsequently activate downstream signaling elements. The activation is stopped by the intracellular GTPase activity of Gα (Gilman, 1987; Hamm, 1998; Gether, 2000).

In the [^{35}S]GTPγS binding assay, agonist-induced activation of the receptor leads to exchange of GDP for [^{35}S]GTPγS in the α-subunit of the G-protein. As [^{35}S]GTPγS is resistant to hydrolysis, it remains bound and the increase in radioactivity can be detected by autoradiography. A unique feature of this assay is its ability to allow simultaneous measurement of receptor function and localization with high precision, which is of great importance in receptor studies of the CNS. This technique is reasonably fast due to short film exposure time and it also enables one to localize a number of different receptors in the absence of selective radiolabeled receptor ligands (Sim et al., 1997; Sovago et al., 2001). Functional [^{35}S]GTPγS binding autoradiography has been used in the CNS for several receptors, including D1 and D2 dopaminergic (Coronas et al., 1999), μ, δ, κ opioid (Platzer et al., 2000), 5HT1A and 5HT2A/2C serotonergic (Dupuis et al., 1999; Adlersberg et al., 2000), α_2 adrenergic (Happe et al., 1998), cannabinoid (Breivogel et al., 1999), and cholinergic muscarinic (Capece et al., 1998) receptors. Additionally, this method can detect changes following experimental treatments or in models of disease states. However, one of the main limitations of the [^{35}S]GTPγS binding autoradiography is that not all G-protein coupled receptors can be detected with this method due to low receptor density and/or low efficiency of coupling. Only pertussis toxin-sensitive G proteins (Go or Gi) have been labeled so far. Another limitation is the inability to establish the types of G-proteins that are activated following receptor stimulation (Sim et al., 1997; Sovago et al., 2001).

9 Conclusions

In vitro receptor autoradiography has markedly increased our knowledge of the precise localization of a variety of neurotransmitter/modulators and growth factor receptors in the CNS and peripheral tissues. The methodology is relatively simple, versatile, and can be applied to a variety of systems including second messengers and enzymes. Moreover, the cloning of various receptors and receptor subtypes has permitted the use of immunohistochemical and in situ hybridization methods to add significantly to the information generated from receptor autoradiography. The development and characterization of selective radioligands/drugs will also provide a new means to investigate the precise distribution and possible roles of a given receptor subtype under normal homeostatic conditions, as well as in various neurological and psychiatric diseases, thereby hopefully leading to better therapeutic approaches in the coming years.

Acknowledgments

The research reported here was supported by grants from the Natural Sciences and Engineering Research Council of Canada and Canadian Institutes of Health Research to SK. CH is a recipient of the Alzheimer Society of Canada studentship award. SK is a recipient of a Tier-II Canada Research Chair in Medicine and Psychiatry and a Senior Scholar award from the Alberta Heritage Foundation for Medical Research.

References

Adamo M, Raizada MK, Le Roith D. 1989. Insulin and insulin-like growth factor receptors in the nervous system, Mol Neurobiol 3: 71-100.

Adlersberg M, Arango V, Hsiung S, Mann JJ, Underwood MD, et al. 2000. In vitro autoradiography of serotonin 5-HT (2A/2C) receptor-activated G protein: guanosine-5'-(gamma-[(35)S]thio)triphosphate binding in rat brain, J Neurosci Res 61: 674-685.

Anderson MF, Aberg MA, Nilsson M, Eriksson PS. 2002. Insulin-like growth factor-I and neurogenesis in the adult mammalian brain. Dev Brain Res 134: 115-122.

Baldino F, Chesselet MF, Lewis ME. 1989. High resolution *in situ* hybridization histochemistry. Methods Enzymol 168: 761-778.

Birecree E, King LE, Nanney LB. 1991. Epidermal growth factor and its receptor in the developing human nervous system. Dev Brain Res 60: 145-154.

Birnbaumer L, Abramowitz J, Brown AM. 1990. Receptor–effector coupling by G proteins. Biochim Biophys Acta 1031: 163-224.

Bouchard P, Roman F, Junien JL, Quirion R. 1996. Autoradiographic evidence for the modulation of *in vivo* sigma receptor labeling by neuropeptide Y and calcitonin gene-related peptide in the mouse brain. J Pharmacol Exp Ther 276: 223-230.

Breivogel CS, Sim LJ, Childers SR. 1997. Regional differences in cannabinoid receptor/G-protein coupling in rat brain. J Pharmacol Exp Ther 282: 1632-1642.

Capece ML, Baghdoyan HA, Lydic R. 1998. Carbachol stimulates [^{35}S]guanylyl 5'-(gamma-thio)-triphosphate binding in rapid eye movement sleep-related brainstem nuclei of rat, J Neurosci 18: 3779-3785.

Chabot J-G, Dumont Y, St- Pierre JA, Tong Y, Dore S, et al. 1999. Anatomical approaches to study G-protein coupled peptide receptors. Peptidergic G-protein-coupled receptors: from basic research to clinical application. Geppetti P, Müller-Esterl W, Regoli D, editors. Amsterdam: IOS Press; pp. 11-28.

Compton DR, Bagley RB, Katzen JS, Martin BR. 1987. (+)-and (-)-N-allylnormetazocine binding sites in mouse brain: *in vitro* and *in vivo* characterization and regional distribution. Life Sci 40: 2195-2206.

Coronas V, Krantic S, Jourdan F, Moyse E. 1999. Dopamine receptor coupling to adenylyl cyclase in rat olfactory pathway: a combined pharmacological-radioautographic approach. Neuroscience 90: 69-78.

Couce M, Weatherington A, McGinty JF. 1992. Expression of insulin-like growth factor-II (IGF-II) and IGF-II/Mannose-6-phosphate receptor in the rat hippocampus: an in situ hybridization and immunocytochemical study. Endocrinology 131: 1636-1642.

Dahms NM, Hancock MK. 2002. P-type lectins. Biochim Biophys Acta 1572: 317-340.

Dechant G, Rodriguez-Tebar A, Barde Y. 1994. Neurotrophin receptors. Prog Neurobiol 42: 347-352.

Dore S, Kar S, Quirion R. 1997. Rediscovering an old friend, IGF-I: potential use in the treatment of neurodegenerative diseases. Trends Neurosci 20: 326-331.

Dreyfus CF. 1998. Neurotransmitters and neurotrophins collaborate to influence brain development. Perspect Dev Neurobiol 5: 389-399.

Dupont J, Le Roith D. 2001. Insulin and insulin-like growth factor I receptors: similarities and differences in signal transduction, Horm Res 55(suppl 2): 22-26.

Dupuis DS, Pauwels PJ, Radu D, Hall H. 1999. Autoradiographic studies of 5-HT1A-receptor-stimulated [^{35}S] GTPγS-binding responses in the human and monkey brain. Eur J Neurosci 11: 1809-1817.

Gerfen CR. 1991. Substance P (neurokinin-1) receptor mRNA is selectively expressed in cholinergic neurons in the striatum and basal forebrain. Brain Res 556: 165-170.

Gether U. 2000. Uncovering molecular mechanisms involved in activation of G-protein coupled receptors. Endorine Rev 21: 90-113.

Gilman AG. 1987. G proteins: transducers of receptor-generated signals. Annu Rev Biochem 56: 615-649.

Guan J, Williams CE, Skinner SJ, Mallard EC, Gluckman PD. 1996. The effects of insulin-like growth factor (IGF)-1, IGF-2, and des-IGF-1 on neuronal loss after hypoxic–ischemic brain injury in adult rats: evidence for a role for IGF binding proteins. Endocrinology 137: 893-898.

Guthrie KM, Nguyen T, Gall CM. 1995. Insulin-like growth factor-1 mRNA is increased in deafferented hippocampus:

spatiotemporal correspondence of a trophic event with axon sprouting. J Comp Neurol 352: 147-160.

Hamm HE. 1998. The many faces of G-protein signaling. J Biol Chem 272: 669-672.

Happe HK, Bylund DB, Murrin LC. 2000. Alpha(2)-adrenoceptor-stimulated GTPγS binding in rat brain: an autoradiographic study. Eur J Pharmacol 399: 17-27.

Hawkes C, Kar S. 2003. Insulin-like growth factor-II/mannose-6-phosphate receptor: widespread distribution in neurons of the central nervous system including those expressing cholinergic phenotype. J Comp Neurol 458: 113-127.

Hawkes C, Kar S. 2004. The insulin-like growth factor-II/mannose-6-phosphate receptor: structure, distribution and function in the central nervous system. Brain Res Rev 44: 117-140.

Hawkes C, Jhamandas JH, Harris K, Fu J, Mac Donald RG, et al. 2006. Single transmembrane domain insulin-like growth factor-II/mannose-6-phosphate receptor regulates central cholinergic function by activating a G-protein-sensitive, protein kinaseC-dependent pathway. J Neurosci 26: 585-596.

Henderson CE. 1996. Role of neurotrophic factors in neuronal development. Curr Opin Neurobiol 6: 64-70.

Hill JM, Lesniak MA, Kiess W, Nissley SP. 1988. Radioimmunohistochemical localization of type II IGF receptors in rat brain. Peptides 9(suppl): 181-187.

Hill JA Jr, Zoli M, Bourgeois JP, Changeux JP. 1993. Immunocytochemical localization of a neuronal nicotinic receptor: the β_2-subunit. J Neurosci 13: 1551-1568.

Hille-Rehfeld A. 1995. Mannose 6-phosphate receptors in sorting and transport of lysosomal enzymes. Biochim Biophys Acta 1241: 177-194.

Hokfelt T, Millhorn D, Seroogy K, Tsuruo Y, Ceccatelli S, et al. 1987. Coexistence of peptides with classical neurotransmitters. Experientia 43: 768-780.

Hokfelt T, Broberger C, Xu ZQ, Sergeyev V, Ubink R, et al. 2000. Neuropeptides—an overview, Neuropharmacology 39: 1337-1356.

Ichise M, Meyer JH, Yonekura Y. 2001. An introduction to PET and SPECT neuroreceptor quantification models. J Nucl Med 42: 755-763.

Jafferali S, Dumont Y, Sotty F, Robitaille Y, Quirion R, et al. 2000. Insulin-like growth factor-I and its receptor in the frontal cortex, hippocampus and cerebellum of normal human and Alzheimer's disease brains. Synapse 38: 450-459.

Jones J, Clemmons D. 1995. Insulin-like growth factors and their binding proteins: biological actions. Endo Rev 16: 3-34.

Kar S, Chabot J-G, Quirion R. 1993a. Quantitative autoradiographic localization of [^{125}I]insulin-like growth factor I, [^{125}I]insulin-like growth factor II and [^{125}I]insulin receptor binding sites in developing and adult rat brain. J Comp Neurol 333: 375-397.

Kar S, Baccichet A, Quirion R, Poirier J. 1993b. Entorhinal cortex lesion induces differential responses in [^{125}I]insulin-like growth factor I, [^{125}I]insulin-like growth factor II and [^{125}I]insulin receptor binding sites in the rat hippocampal formation. Neuroscience 55: 69-80.

Kar S, Seto D, Dore S, Hanisch U-K, Quirion R. 1997a. Insulin-like growth factors-I and -II differentially regulate endogenous acetylcholine release from the rat hippocampal formation. Proc Natl Acad Sci USA 94: 14054-14059.

Kar S, Seto D, Doré D, Chabot J-G, Quirion R. 1997b. Systemic administration of kainic acid induces selective time-dependent decrease in [^{125}I]insulin-like growth factor I, [^{125}I]insulin-like growth factor II and [^{125}I]insulin receptor binding sites in adult rat hippocampal formation. Neuroscience 80: 1041-1055.

Keen M, Macdermot J. 1993. Analysis of receptors by radioligand binding. Receptor autoradiography: principles and practice. Wharton J, Polak JM, editors. Oxford, England: Oxford University Press; pp. 23-55.

Kornfeld S. 1992. Structure and function of the mannose 6-phosphate/insulin-like growth factor II receptors. Annu Rev Biochem 61: 307-330.

Kuhar MJ, De Souza EB, Unnerstall JR. 1986. Neurotransmitter receptor mapping by autoradiography and other methods. Annu Rev Neurosci 9: 27-59.

Kuhar MJ, Unnerstall JR. 1990. Receptor autoradiography. Methods in neurotransmitter receptor analysis. Yamamura HI, Enna SJ, Kuhar MJ, editors. New York: Raven Press; pp. 177-218.

Lee TH, Kato H, Pan LH, Ryu JH, Kogure K, Itoyama Y. 1998. Localization of nerve growth factor, trkA and P75 immunoreactivity in the hippocampal formation and basal forebrain of adult rats. Neuroscience 83: 335-349.

Le Roith D. 2003. The insulin-like growth factor system. Exp Diabesity Res 4: 205-212.

Lesniak M, Hill J, Kiess W, Rojeski, M, Pert C, et al. 1988. Receptors for insulin-like growth factors I and II: autoradiographic localization in rat brain and comparison to receptors for insulin. Endocrinology 123: 2089-2099.

Lindsay RM. 1994. Neurotrophins and receptors. Prog Brain Res 103: 3-14.

Mansour A, Fox CA, Burke S, Akil H, Watson SJ. 1995a. Immunohistochemical localization of the cloned mu opioid receptor in the rat CNS. J Chem Neuroanatomy 8: 283-305.

Mansour A, Fox CA, Akil H, Watson SJ. 1995b. Opioid-receptor mRNA expression in the rat CNS: anatomical and functional implications. Trends Neurosci 18: 22-29.

Mengod G, Vilaro MT. 1993. Analysis of receptor expression by *in situ* hybridization histochemistry. Receptor autoradiography: principles and practice. Wharton J, Polak JM, editors. Oxford, England: Oxford University Press; pp. 159-170.

Mengod G, Vilaro MT, Landwehrmeyer GB, Martinez-mir MI, Niznik HB, et al. 1992. Visualization of dopamine D1, D2 and D3 receptor mRNAs in human and rat brain. Neurochem Int 20(suppl): 33S-43S.

Miranda RC, Sohrabji F, Toran-Allerand CD. 1993. Neuronal colocalization of mRNAs for neurotrophins and their receptors in the developing central nervous system suggests a potential for autocrine interactions. Proc Natl Acad Sci USA 90: 6439-6443.

Montpied P, Martin BM, Cottingham SL, Stubblefield BK, Ginns EI, Paul SM. 1988. Regional distribution of the $GABA_A$/benzodiazepine receptor (α subunit) mRNA in rat brain. J Neurochem 51: 1651-1654.

Nagano T, Sato M, Mori Y, Du Y, Takagi H, et al. 1995. Regional distribution of messenger RNA encoding in the insulin-like growth factor type 2 receptor in the rat lower brainstem. Mol Brain Res 32: 14-24.

Pablo FD, de la Rosa EJ. 1995. The developing CNS: a scenario for the action of proinsulin, insulin and insulin-like growth factors. Trends Neurosci 18: 143-150.

Palacios JM, Dietl MM. 1987. Regulatory peptide receptors: visualization by autoradiography. Experientia 43: 750-761.

Pellegrini-Giampietro DE, Bennett MV, Zukin RS. 1991. Differential expression of three glutamate receptor genes in developing rat brain: an *in situ* hybridization study. Proc Natl Acad Sci USA 88: 4157-4161.

Platzer S, Winkler A, Schadrack J, Dworzak D, Tolle TR, et al. 2000. Autoradiographic distribution of mu-, delta-and kappa 1-opioid stimulated [^{35}S]guanylyl-5′-O-(gamma-thio)-triphosphate binding in human frontal cortex and cerebellum. Neurosci Lett 283: 213-216.

Quirion R, Dam TV, Sarrieau A, Rostene W. 1989. Receptor autoradiography in neuropeptide research. Brain imaging techniques and application. Sharif NA, Lewis ME, editors. Chichester, England: Ellis Horwood; pp. 77-94.

Quirion R, Kar S, Chabot JG, Dumont Y. 1993. Neuropeptide and growth factor receptor autoradiography. Receptor autoradiography: principles and practice. Wharton J, Polak JM, editors. Oxford, England: Oxford University Press; pp. 259-279.

Qume M. 1999. Overview of ligand–receptor binding techniques. Methods Mol Biol 106: 3-23.

Reynolds BA, Weiss S. 1993. Central nervous system growth and differentiation factors: clinical horizons—truth or dare? Curr Opin Biotech 4: 734-738.

Richards JG, Schoch P, Haring P, Takacs B, Mohler H. 1987. Resolving $GABA_A$/benzodiazepine receptors: cellular and subcellular localization in the CNS with monoclonal antibodies. J Neurosci 7: 1866-1886.

Schroder H, Zilles K, Luiten PG, Strosberg AD. 1990. Immunocytochemical visualization of muscarinic cholinoceptors in the human cerebral cortex. Brain Res 514: 249-258.

Sequier JM, Richards JG, Malherbe P, Price GW, Mathews S, et al. 1988. Mapping of brain areas containing RNA homologous to cDNAs encoding the α and β subunits of the rat $GABA_A$ gamma-aminobutyrate receptor. Proc Natl Acad Sci USA 85: 7815-7819.

Sharif NA. 1996. Quantitative autoradiographic methods. Brain mapping: the methods. Toga AW, Mazziota JC, editors. New York: Academic Press; pp. 115-144.

Sim LJ, Selley DE, Childers SR. 1995. In vitro autoradiography of receptor-activated G proteins in rat brain by agonist-stimulated guanylyl 5′-[γ-[35S]thio]-triphosphate binding. Proc Natl Acad Sci USA 92: 7242-7246.

Sim LJ, Selley DE, Childers SR. 1997. Autoradiographic visualization in brain of receptor-G protein coupling using [^{35}S]GTPγS binding. Methods Mol Biol 83: 117-132.

Smith M, Clemens J, Kerchner GA, Mendelsohn LG. 1988. The insulin-like growth factor-II (IGF-II) receptor of rat brain: regional distribution visualized by autoradiography. Brain Res 445: 241-246.

Sovago J, Dupuis DS, Gulyas B, Hall H. 2001. An overview on functional receptor autoradiography using [^{35}S]GTPγS. Brain Res Rev 38: 149-164.

Sternberger LA. 1986. Immunocytochemistry, 4th edn. New York: John Wiley & Sons.

Stumpf WE, Roth LJ. 1966. High resolution autoradiography with dry mounted, freeze-dried frozen sections. Comparative study of six methods using two diffusible compounds ^{3}H-estradiol and ^{3}H-mesobilirubinogen. J Histochem Cytochem 14: 274-287.

Teng KK, Hempstead BL. 2004. Neurotrophins and their receptors: signaling trios in complex biological systems. Cell Mol Life Sci 61: 35-48.

Thoenen H. 1991. The changing scene of neurotrophic factors. Trends Neurosci 14: 165-170.

Vilaro MT, Palacios JM, Mengod G. 1994. Multiplicity of muscarinic autoreceptor subtypes? Comparison of the distribution of cholinergic cells and cells containing mRNA for five subtypes of muscarinic receptors in the rat brain. Mol Brain Res 21: 30-46.

Wada E, Wada K, Boulter J, Deneris E, Heinemann S, et al. 1989. Distribution of α2, α3, α4, and β2 neuronal nicotinic receptor subunit mRNAs in the central nervous system: a hybridization histochemical study in the rat. J Comp Neurol 284: 314-335.

Weiner DM, Levey AI, Sunahara RK, Niznik HB, O'Dowd BF, et al. 1991. D1 and D2 dopamine receptor mRNA in rat brain. Proc Natl Acad Sci USA 88: 1859-1863.

Werther GA, Abate M, Hogg A, Cheesman H, Oldfield B, et al. 1990. Localization of insulin-like growth factor-I mRNA in rat brain by *in situ* hybridization—relationship to IGF-I receptors. Mol Endocrinol 4: 773-778.

Young WS, Kuhar MJ. 1979. A new method for receptor autoradiography: [^{3}H]opioid receptors in rat brain. Brain Res 179: 255-270.

Yung KK, Bolam JP, Smith AD, Hersch SM, Ciliax BJ, et al. 1995. Immunocytochemical localization of D1 and D2 dopamine receptors in the basal ganglia of the rat: light and electron microscopy. Neuroscience 65: 709-730.

12 In Silico Molecular Homology Modeling of Neurotransmitter Receptors

M. Wang · D. R. Hampson · Lakshmi P. Kotra

Abstract: The goal of this chapter is to introduce the basic strategies used in building and interpreting homology models of neurotransmitter receptors. A general step-by-step description of the model building process is outlined and potential problems and pitfalls are noted. In addition, procedures for the docking of ligands and drugs into the binding sites of *in silico* generated protein structures are also described. Although our emphasis is on the study of G-protein coupled receptors, the basic principles outlined here can be applied to the study of any protein.

List of Abbreviations: GPCRs, G-protein coupled receptors; PDB, protein database; SCRs, structurally conserved regions; SVRs, structurally variable regions

1 Introduction

Very few of the proteins encoded by the approximately 25,000 genes in the human genome have been subjected to rigorous structural analysis via X-ray crystallography or nuclear magnetic resonance. Although the number of solved structures is expected to increase rapidly over the next few years due to high throughput automated crystallization endeavors and advances in nuclear magnetic resonance technology, the number of structurally characterized proteins, particularly membrane proteins, will likely remain relatively small within the foreseeable future. In addition to the typical expensive and labor-intensive efforts required for purification and crystallization, this conundrum is further exacerbated in part by the refractoriness of some proteins to produce suitable crystals for high-resolution X-ray crystallographic analyses. For the proteins whose structures are not known, homology modeling using a related protein whose structure is known as a template for model building will play an increasingly important role in molecular neurobiology, pharmacology, and drug design. The goal of this chapter is to introduce the basic strategies used in building and interpreting homology models, and the docking of ligands into the binding sites of in silico generated protein structures. Although our emphasis is on the study of G-protein coupled receptors, the basic principles outlined here can be applied to the study of any protein.

G-protein coupled receptors (GPCRs) are the largest single family of signaling molecules in mammals and represent approximately 3–4% of all genes in the human genome (Fredriksson et al., 2003; Yin et al., 2004). They mediate signaling of an extremely wide variety of ligands such as amino acids, ions, biogenic amines, peptides, glycoproteins, light, pheromones, and other odorants. GPCRs have been classified into five primary families using the GRAFS classification system (GRAFS: Glutamate, Rhodopsin, Adhesion, Frizzled, and Secretin families; Fredriksson et al., 2003). Estimates of the total number of GPCR genes in the human genome range from about 750 to 1,000. Unfortunately, there are presently only a handful of GPCRs, whose structures have been elucidated. Of these, the mGluR1 subtype of metabotropic glutamate receptor (mGluR) and rhodopsin are the most widely used in modeling studies of GPCRs. In the case of mGluR1, only the structure of the extracellular ligand-binding domain of the molecule has been solved in the presence and absence of the ligand glutamic acid and the antagonist MCPG (Kunishima et al., 2000; Tsuchiya et al., 2002), whereas in the case of rhodopsin, the crystallographic data encompass the whole protein in the ground state with bound 11-*cis*-retinal (Palczewski et al., 2000).

As a general strategy consideration, it is important to have in place an array of experimental techniques that will be used to test the validity of the model. Although procedures for testing models of GPCRs are not discussed in detail here, they typically include a suitable protein expression system, site-directed mutagenesis, a radioligand binding assay, and a functional assay such as the measurement of a second messenger or calcium that is capable of detecting and quantitating responses in live transfected cells. For GPCRs and other types of neurotransmitter or hormone receptors, the results from radioligand binding and functional assays provide quantifiable pharmacological parameters such as the dissociation constant (K_D), inhibitory 50 constant (IC_{50}), and the effective 50 constant (EC_{50}). The parameters from the unmutated wild-type receptor are compared to those from a series of mutants produced to test the model (Kuang etal., 2003).

In addition to the techniques listed earlier, it is also desirable to conduct microscopy on the mutants expressed in transfected cells. The purpose is to compare the cell surface expression of the mutants with the wild-type receptor. This is particularly important for mutant receptors that lack functional responses or ligand binding. Some mutations can cause protein misfolding and in many, but not all cases, this inhibits proper trafficking to the cell surface; this will obviously negate responses in a functional assay carried out in live cells, and may also induce misfolding such that ligand binding is prevented. Thus, if a the goal of a study is to characterize a ligand or drug-binding site, mutants with mutation-induced trafficking defects must be identified and eliminated from further consideration. From the experimental standpoint, microscopy is most expeditiously carried out on fixed cells labeled with an antibody to an extracellularly exposed epitope and visualized with a fluorescent-tagged secondary antibody on a confocal or deconvolution microscope.

2 Strategies for Generating Models and Docking Ligands

The process of generating a three-dimensional molecular homology model and docking a natural ligand or synthetic drug into the receptor can be broken down into several major steps: (1) identifying the most suitable template structure, (2) producing an accurate amino acid sequence alignment between the target protein and the template protein, (3) producing an initial homology model using one of several software programs and subjecting the structure to energy minimization using a force field, (4) energy minimizing the small ligand or drug, and (5) docking the energy minimized ligand or drug into the putative binding pocket of the target protein. The docking step requires prior knowledge of the location of the binding site; this may be obtained from the template, from other related proteins, or from experimental analyses. Each of these steps is described in detail below.

2.1 Selecting the Most Appropriate Template

The first step is to retrieve the molecular coordinates from the RSCB Protein Data Bank (PDB, http://www.resb.org, see *Table 12-1* for a list of Web site addresses) of the protein that will be used as the template on

Table 12-1
Alignment, modeling, and related programs and algorithms

Program	Web site	Reference
CLUSTALW	www.ebi.ac.uk/clustalw/	Thompson et al. (1994)
Modeling Programs		
COMPOSER	www.tripos.com/sciTech/inSilicoDisc/bioInformatics/composer.html	Sutcliffe et al. (1987)
MODELLER	salilab.org/modeller/modeller.html	Sali and Blundell (1993)
SWISS-MODEL	www.expasy.ch/swissmod/SWISS-MODEL.html	Peitsch (1996)
WHATIF	www.sander.embl-heidelberg.de/whatif/	Vriend (1990)
Simulation Programs		
AMBER	amber.scripps.edu	Pearlman et al. (1995)
CHARMM	www.charmm.org	Brooks et al. (1983)
GROMACS	www.gromacs.org	Lindahl et al. (2001)
NAMD	www.ks.uiuc.edu/Research/namd/namd.html	Kalè et al. (1999)
Miscellaneous evaluation programs		
PROCHECK	www.biochem.ucl.ac.uk/~roman/procheck/procheck.html	Laskowski et al. (1993)
Prosall	www.came.sbg.ac.at	Sippl (1993)
VERIFY3D	shannon.mbi.ucla.edu/DOE/Services/Verify_3D/	Luthy et al. (1992)
WHAT_CHECK	www.sander.embl-heidelberg.de/whatcheck/	Hooft et al. (1996)

which the model will be based. In many instances, the identification of the best choice of template will be obvious and is based on the PDB entry, whose primary amino acid sequence shows the highest identity to the target protein under study. If this is not already known, it can be determined by conducting a BLAST search of the NCBI databases or any other sequence alignment search of the target sequence, identifying the top hits, and then searching the PDB for these structures. The most critical parameter is the degree of sequence identity between the target and the potential template sequences, which typically should be at least 30% for a reasonable homology model generation. Other points to be considered are the degree and the number of gaps introduced into the sequences during homology model generation and the resolution of the three-dimensional structure of the reference protein. Although it is preferable to have a high level of sequence identity between the target and the template, we note that successful homology models have been generated with lower degrees of sequence identity. For example, Hampson et al. (1999) produced a homology model of the mGluR4 subtype of metabotropic glutamate receptor, which provided good predictive capabilities in terms of explaining the mechanism of the ligand binding, using a bacterial periplasmic binding protein that had only 20% sequence identity to mGluR4.

In some cases only a part of the whole protein structure may be available in the PDB. For example, as noted earlier, structural data in the PDB for the mGluR1 receptor encompass only the extracellular ligand-binding domain; structural information on the transmembrane heptahelical domain and the intracellular carboxyl terminus of the mGluR receptors is not yet available. Thus, to formulate a homology model based on mGluR1, only the region of a target protein that aligns with the ligand-binding domain portion of mGluR1 can be used. To model the heptahelical transmembrane region of a GPCR, the structure of bovine rhodopsin can be used.

2.2 Producing an Amino Acid Sequence Alignment Between the Target and the Template

An alignment can be produced by generating a pairwise alignment using only the target sequence and the template sequence, or by producing a multiple sequence alignment, which also includes the target, the template, and other members of the family. Since the modeling programs require inputting a pairwise alignment, a multiple sequence alignment can be conducted to assist in deciding on the optimal pairwise alignment that will be used to construct the model. The alignment step is absolutely critical because an incorrect sequence alignment will ultimately produce an incorrect three-dimensional model. The extent of the deleterious downstream effects of the misaligned sequences will depend on the extent of the sequence within the total sequence that is mismatched and the importance of the misaligned sequence in the functional activity of the protein. Although it is possible that the mismatched sequence will not affect the function of the protein under investigation, every effort should be made to produce the optimal alignment at the initial stage of the project.

Most proteins exist as members of a family. These family members, along with other proteins with a high level of amino acid sequence identity, can be used to generate the multiple alignments. A close analysis of a multiple sequence alignment before deciding on the final pairwise alignment between the template and target sequence has the advantage that it can often produce a more accurate alignment than a simple pairwise alignment (Baxevanis et al. 2001). Conserved residues in the family members and the related sequences are often crucial for producing the correct alignment as is the consideration of protein secondary structure.

ClustalW (Thompson et al., 1994) is one of the most widely used algorithms for producing multiple sequence alignments. Although it is easy to produce a multiple alignment with ClustalW, care should be taken before accepting the initial alignment output. In many cases, additional adjustments need to be carried out, including changing the default parameters of the program or manually manipulating segments of selected sequences within the alignment. Additional precautions include avoiding the alignment of sequences that are substantially different in length, removing sequences that disrupt the alignment, and visually checking to ensure that key residues are aligned (Barton, 2001). Gaps or insertions and deletions (INDELS) should be given close scrutiny and alignment visualization tools such as ALSCRIPT/JalView can be used to identify positions in the alignment that show conserved physiochemical properties across the multiple alignment.

2.3 Building the Homology Model

After obtaining the optimum alignment between the target and the template sequences, a homology model can be built using one of the software programs listed in ➋ *Table 12-1* (➋ *Figure 12-1*). These programs adopt different approaches for building a three-dimensional structure based on conserved sequences between the target and the template. The various algorithms used in the different software programs

Figure 12-1
A simplified flow chart for homology model building

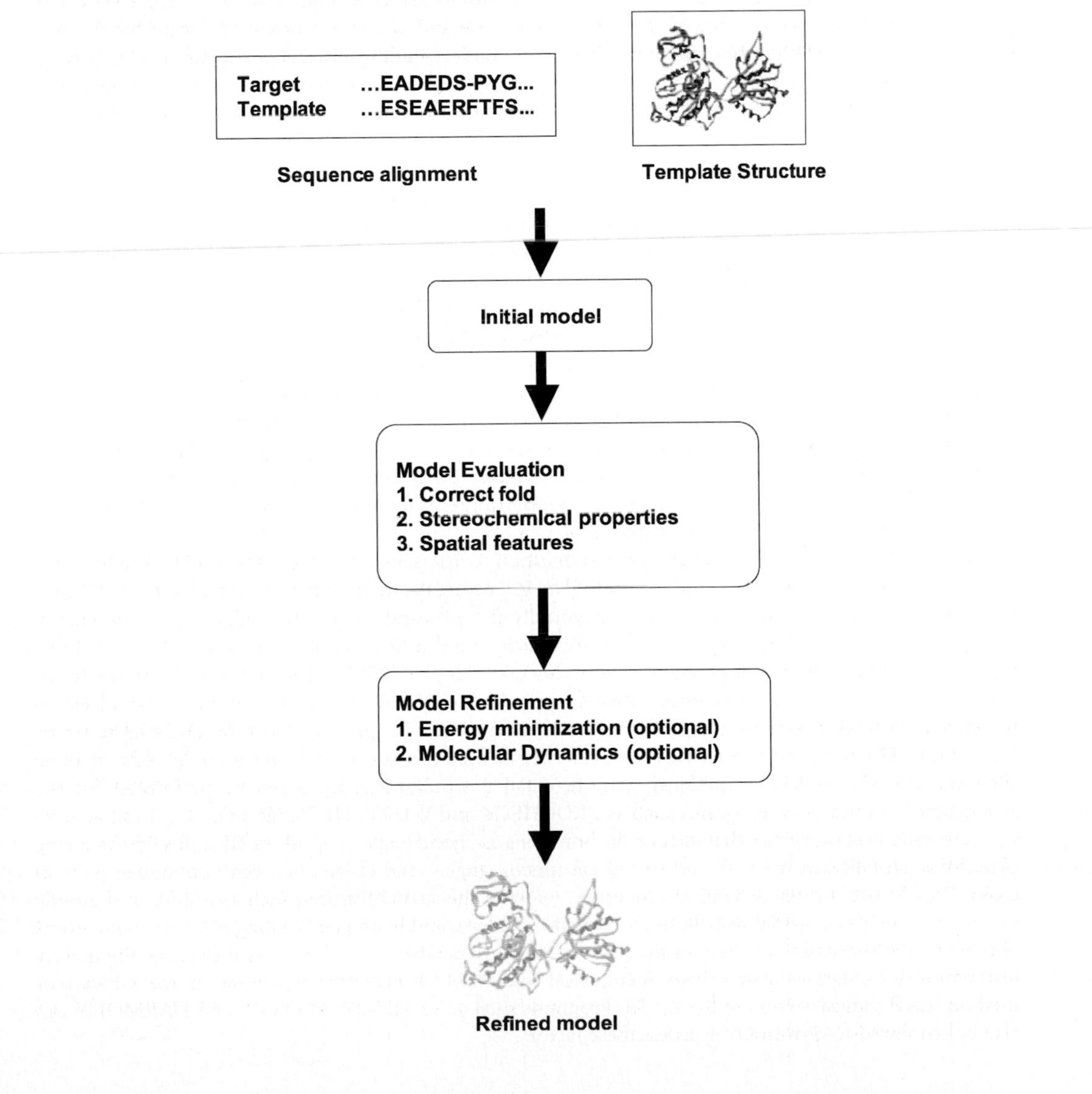

have been reviewed in detail by Sanchez and Sali (1997) and Marti-Renom et al. (2000). We focus on two of the most popular software tools, COMPOSER (Sutcliffe et al., 1987) and MODELLER (Sali and Blundell, 1993).

COMPOSER, which is marketed as part of a suite of tools from Tripos Software Inc., is a model-building program that uses a rigid assembly method, one of the earliest methods developed for homology modeling. The sequence is divided into two classes: structurally conserved regions (SCRs), which define the common tertiary folding pattern, and structurally variable regions (SVRs), which connect the SCRs and are generally loops that lie at the surface of the proteins. When multiple structural homologs are available, SCRs can be obtained by superposing the multiple template structures. In the case of a single template, SCRs are defined based only on the sequence alignment between the target and the template. The $C\alpha$ coordinates of the SCRs in the target model are built by first using the structure of the homolog; subsequently, the corresponding protein chain backbone and the side chain atoms of the target sequence are generated. Finally, the SCRs are connected by constructing SVRs using loop structure templates retrieved from the PDB.

MODELLER is another widely used program for homology modeling (Fiser and Sali, 2003). This program can be downloaded free of cost for academic use and is also a component within the Accelrys Software package (Accelrys Inc., ➋ *Table 12-1*). The model-building process is similar to that used in determining protein structures via NMR whereby the structures are generated by building models that satisfy the geometry restraints derived from experimental measurements. In MODELLER, restraints on the target sequence are determined from an alignment between the target and the template sequences, assuming that the aligned residues in the template and target structure adopt a similar orientation. The all-atom model is then constructed using this approach to satisfy the spatial restraints on distances and dihedral angles. All the above processes are invisible to the user. After inputting the sequence, the alignment procedure provides an output of the model coordinates. One advantage of MODELLER is that it has provisions for incorporating experimental information, if available, as constraints in the modeling process. For example, restraints obtained from NMR experiments, cross-linking experiments, image reconstruction in electron microscopy, and site-directed mutagenesis can be incorporated into the modeling process. By introducing such restraints, the model can be refined to make it consistent with the experimental data.

In homology modeling, loop regions are the most difficult portions to model because of the lack of a secondary structure. Loop regions are the most variable segments in a protein structure and usually exist on the surface where the mutation rates are high. In most cases, insertion of loops containing 3–5 amino acids can be modeled with relatively high confidence. If loop insertions longer than five amino acids are far from the regions of interest (e.g., the ligand-binding pocket in the study of ligand–receptor interactions), their influence on the ligand binding may be minimal.

After the initial model is obtained, the first question to ask is, how reliable is the model? The first step is to assess whether the model has the correct fold by inspecting the three-dimensional fold of the backbone. A high degree of amino acid sequence identity typically of $>30\%$ and conservation of known key functional or structural residues in the target sequence are usually good indicators of the correctness of the three-dimensional model. Statistical potential energy function (Sippl, 1995) has been used to discriminate the correct fold from the incorrect ones (Melo et al., 2002). This allows quantification of the observed preference of different residue types relative to others from known protein structures and can be transformed into a pseudoenergy score. This score can then be used to assess the fold of the model. After making sure that the correct fold is modeled, a more detailed structural analysis can be performed for the stereochemical properties. Programs such as PROCHECK and WHATCHECK (➋ *Table 12-1*) can analyze the stereochemical properties that include the bond lengths, bond angles, peptide bond, and side-chain ring planarities, chirality, main-chain and side-chain torsion angles, and clashes between nonbonded pairs of atoms. In addition to stereochemical properties, many of the spatial features, such as residue and atomic solvent accessibilities, spatial distribution of specific residues, and hydrogen bonding of main-chain atoms obtained from the statistical study of the protein structure database can also be used to judge the quality of the model. Deviations from statistical empirical values (which may vary depending on the software or method used) indicate errors in the model. Programs such as VERIFY3D, PROSAII, and HARMONY can also be employed to quantitatively assess these factors.

The results of these model evaluations permit one to choose the best model generated by the modeling program. This model can then be refined by molecular mechanics techniques. Energy minimization is the most popular means to relax geometric, angle, torsional, and other local structural strains in the homology model as a first step of refinement. A typical protein energy landscape is rugged and possesses many local energy minima. By performing an energy minimization, the protein structure can be optimized to a local minimum by eliminating bad contacts between side chains. However, this does not guarantee a global minimum for the structure in the energy landscape. To refine the structure further, the structure can be subjected to molecular dynamics simulations. After several hundred picoseconds or several nanoseconds of trajectory analysis during molecular dynamics simulations, a stable conformation can often be obtained. An average structure can also be obtained from the collection of snapshots taken after the molecular dynamics simulations and minimized again to obtain a relaxed structure of the target protein.

Software programs with force fields to perform energy minimization and molecular dynamics simulations include AMBER, CHARMM, GROMACS, and NAMD (➲ *Table 12-1*). These programs are usually developed using a specific force field that needs to be self-consistent (see Step 4). Employment of these techniques yields improved and refined structures by releasing the geometric strains and bad contacts; however, they will not eliminate errors inherited from an incorrect sequence alignment. It has been argued that molecular dynamics does not help in the refinement of model structure because the structure of the homology model may "drift" during the simulations, sometimes arriving at unreasonable structures. Therefore, restrained dynamics, in which the backbone of the model structure is constrained by applying forces to hold the atoms in position, is usually employed to prevent drifting from the starting structure. A recent study demonstrated that molecular dynamics can improve the accuracy in the presence of a solvent environment (e.g., water; Fan et al., 2004). Most simulation algorithms and software are capable of modeling environments that include water and/or lipids. Examples of analyses conducted on GPCRs in a lipid environment can be found in Trent et al. (2003) for the CXCR4 receptor and Varady et al. (2003) for the dopamine D3 subtype receptor.

2.4 Energy Minimization of Ligands and Drugs

Before proceeding to dock the ligand into the protein model, the structure of the ligand must be formulated. Ligands, drugs, and other small molecules can be constructed using sketch tools available within the modeling software. Alternatively, three-dimensional structures of small molecules may be obtained from the Cambridge Structural Database. The correct bond order, tautomeric states, and ionization states, which are necessary for building ligand molecules, are required before docking and molecular dynamics simulation. Structure Checker in the Accelrys software is an example of a program that can provide this information. The three-dimensional structures of the molecules can then be generated by programs such as CONCORD, CORINA, and COBRA (see ➲ *Table 12-2*; also see Sadowski and Gasteiger, 1994).

After the molecular structure is built, conformational analysis can be employed to explore the energy surface of the molecule. Most molecules exist in multiple conformations and transformations from one conformation to another are primarily associated with rotations about single bonds. The conformational space can be explored computationally by carrying out a conformational search. A conformational search will assist in identifying all possible conformations, and then this large set of conformations can be reduced to a small set, which may be of interest for other studies such as QSAR, pharmacophore modeling, and drug docking. Under most circumstances, molecular conformations of interest are subjected to energy minimization to release any constraints such as torsional strain or steric constraints on each of these conformers.

There are several ways to conduct conformational searches including systematic searches, random searches, genetic searches, and molecular dynamics simulations. A systematic search (also called grid search) emulates all the possible conformations in an exhaustive way by systematically changing the rotatable bonds in the molecules. The time scale of the systematic search depends on the rotatable bond in the molecule and the step length of the increment grid, which means that it may be very time consuming for large systems. Therefore, systematic searches are routinely restricted to molecules with less than

Table 12-2
Software for analysis of ligand molecules

Program	Web site	Company/ reference
Conformational analysis		
CONCORD	www.tripos.com/sciTech/inSilicoDisc/chemInfo/concord.html	Tripos Inc.
CORINA	www2.chemie.uni-erlangen.de/software/corina/corina.html	Gasteiger et al. (1990)
CATALYST	www.accelrys.com/catalyst/index.html	Accelrys Inc.
MACROMODEL	www.schrodinger.com/Products/macromodel.html	Schrodinger Inc.
CONFORT	www.tripos.com/sciTech/inSilicoDisc/moleculeModeling/confort.html	Tripos Inc.
Ab initio calculation		
GAUSSIAN	www.gaussian.com/	Gaussian Inc.
GAMESS	www.msg.ameslab.gov/GAMESS/GAMESS.html	Schmidt et al. (1993)

15 rotatable bonds. In addition to these methods, simulation methods such as molecular dynamics can also be employed to overcome the energy barriers between local minima and to explore different regions of the conformational space.

Programs that are capable of performing conformational analyses include Macromodel (Schrodinger Inc), Catalyst (Accelrys Inc), and Confort (Tripos Inc). These programs have been compared in terms of their ability to reproduce the X-ray structures (Sadowski and Gasteiger, 1994). A conformational search results in a large amount of conformers, which must be processed and analyzed. In comparison with a protein conformational space, small ligands have a relatively simple conformational space and therefore it is relatively easy to obtain a global minimum for a ligand. However, ligands do not necessarily bind in the lowest energy state. The bound conformation is usually within 3 kcal/mol of the global energy minimum state (Bostrom et al., 1998). Because a ligand molecule may not always bind to a receptor in its global energy minimum state (Perola and Charifson, 2004), it is desirable to select some representative conformations for subsequent docking studies. This is usually done using cluster analysis, which groups similar objects together. Two methods implemented in most of the commercial modeling packages are hierarchical clustering and principal components analysis (PCA).

Conformational analysis is especially useful for manual docking (see Step 5). If a pharmacophore model is available, it can be used as a constraint in a conformational search and thereby reduce the search time. Energy minimizations and other computational tasks are carried out using either *ab initio*, semiempirical, or molecular mechanics methods. In computer modeling of proteins and other large biomolecules, molecular mechanics methods are the most popular due to their efficiency. With these methods, molecular systems are treated based on a set of parameters that describe the molecular properties such as bond length, bond angle, etc. and have been specifically developed for application to a particular class of molecules. These sets of properties are collectively termed as the force field. For energy minimizations, molecular dynamics simulations, and other computational experiments, force field parameters for the small ligand may need to be developed because most of the simulation programs only include force field parameters for the building blocks such as amino acids. This parameter development process includes deriving the partial atomic charges and deriving the missing force field parameters; these tasks may require *ab initio* calculations and software tools such as Gaussian (Gaussian, Inc). A current trend is to automate these procedures as in the antechamber module of AMBER (Wang, 2004).

2.5 Docking Ligands into a Binding Pocket

Once a homology model of a receptor and the ligand structure are available, the ligand can be docked into the receptor. This procedure has been widely used in the structure-based drug design process to screen possible hits from a chemical database (Shoichet et al., 2002; Alvarez, 2004). Molecular docking can be defined as the process of identifying an optimal match within a binding pocket between two molecules, usually a ligand or drug and a protein in which electrostatic and steric properties between the two molecules are complementary. The process is accomplished in two stages; a conformational search process, which samples all possible conformations, and a ranking step which scores all the solutions to identify the most likely conformation of the complex. The binding pocket must be defined prior to the docking process. In most cases, the function of the receptor and its ligand(s) is known and the binding pocket has been previously elucidated from the experimental data. In cases where binding pocket is not known, software programs such as LIGSITE, PASS, and SITEID can be used to predict the binding site from the structure (Sotriffer and Klebe, 2002).

Several programs have been developed for docking ligands into receptors (see *Table 12-3*). These programs differ in the treatment of the ligands and the ranking methods. Efficiency and accuracy are

Table 12-3
Docking programs

Program	Web site	Reference
AUTODOCK	www.scripps.edu/pub/olson-web/doc/autodock/index.html	Morris et al. (1996)
DOCK	dock.compbio.ucsf.edu	Kuntz et al. (1982)
GLIDE	www.schrodinger.com/Products/glide.html	Friesner et al. (2004)
GOLD	http://www.ccdc.cam.ac.uk/products/life_sciences/gold/	Jones et al. (1997)
FLEXX	www.biosolveit.de/FlexX/	Kramer et al. (1999)
FLEXE	www.biosolveit.de/FlexE/	Claussen et al. (2001)
SLIDE	http://www.bch.msu.edu/~kuhn/projects/slide/home.html	Jacobs et al. (2001)

two factors considered in the development of docking software. When used in *in silico* screening, hundreds of thousands of compounds are docked; thus, speed and efficiency are key factors. Moreover, the determination of the accuracy of the mode of binding is important. In all docking programs, the scoring functions are completed within seconds. In such cases, empirical methods are usually adopted in the development of scoring functions for a quick and efficient ranking of the docked ligands (Gohlke and Klebe, 2002). It is very difficult to predict the ligand affinity accurately with the current technologies and methods. One way to overcome this limitation is to combine several scoring functions to judge the goodness of the docking results (Wang et al., 2003); this too works well at the qualitative level and experiments will have to be conducted to measure the actual affinities of the ligands.

In addition to the accuracy of scoring functions, the flexibility of the molecular system presents another challenge. The older rigid body docking methods in which both the ligand and protein are treated as rigid bodies are relatively easy to use and simplify the conformational space to search. However, most molecules are flexible and binding processes are associated with conformational changes in both the ligands and receptors. An analysis of ligand flexibility is an integral component of most docking programs. Receptor flexibility is more difficult to deal with because the complexity added to the system is astronomic. A workaround is to consider receptor flexibility as a cluster of structures that represent the flexibility (Carlson, 2002). These structures can be obtained either from experimental structure determination methods or computational methods such as molecular dynamics (Mangoni et al., 1999; Lin et al., 2002), simulated annealing (Verkhivker et al., 2001), or rotamer library (Frimurer et al., 2003). With continuing enhancements in computing power and advances in algorithms, programs that consider receptor flexibility are now under active development and will likely be more widely used in the near future. Several programs such as FlexE (Claussen et al., 2001), Glide (Friesner et al., 2004), and Slide (Jacobs et al., 2001) have emerged as useful programs capable of addressing the receptor flexibility issue.

In comparison with computer docking using a crystal structure, the docking of ligands into a homology model is more challenging because of the uncertainties in the homology model. When the homology model or the binding pocket is highly similar to the template, it is feasible to proceed with the same procedure as for crystal structure. Assessments of the performance of docking using homology models have been reported (Oshiro et al. 2004; Wang et al. 2006).

In many cases, manual docking is often preferred in studies of interactions between ligands and receptors. In the manual docking process, the minimized ligand structure is visually positioned into the binding pocket in three dimensions by exploring the shape complementary and the hydrogen bonding possibilities. If the template structure contains a similar ligand, this ligand structure can serve as a template to dock the new ligand. For example, in the case of L-glutamate activated GPCRs (metabotropic glutamate receptors), glutamate analogs have been docked manually using the crystal structure of mGluR1 with glutamate bound in the binding pocket (Yao et al., 2003; Rosemond et al., 2004). Manual docking can be performed using appropriate software such as Sybyl, InsightII, or other molecular modeling programs. The shape complementary can be visualized by constructing a Connolly surface of the binding pocket (Connolly, 1983). Although manual docking is somewhat subjective, it has the advantage of using experimental information and ligand-receptor interactions can be monitored during the docking process. If any atoms of the ligand clash with atoms of the receptor, the side chains of receptor residues can be adjusted so that the steric clashes are eliminated. Subsequently, energy minimization and molecular dynamics can be used to refine the structure of the complex; the refinement procedure is the same as that described earlier in Step 3.

In some cases, the information gleaned from a homology model of a ligand–receptor complex can be extended to include estimates of ligand affinity. If several ligand–receptor complexes are analyzed in parallel, they can be assessed using scoring functions for the binding affinity. More sophisticated and computing intensive molecular mechanics techniques such as thermodynamic integration and free energy perturbation can also be employed to predict the binding affinity by calculating the free–energy difference in the binding process. A more economic and yet efficient approach, MM-PBSA (molecular mechanics Poisson–Boltzmann surface area), adopts a continuum model to calculate solvation and has gained popularity in calculating free energies for complex molecular systems (Kollman, et al. 2000; Massova and Kollman, 2000). This method has also been applied to study the binding mode between HIV inhibitors and HIV reverse transcriptase (Wang et al., 2001). Several additional computational techniques and related applications are discussed in further detail by Jorgensen (2004).

In summary, we have described the complex process of building an in silico model for receptors using various computational tools in conjunction with experimental data. Ultimately, a successful model is one that holds up under rigorous experimental scrutiny and is capable of providing useful predictions for drug–receptor interactions.

Acknowledgments

This work was supported by operating grants to D.R.H. and L.P.K. from the Canadian Institutes for Health Research, a PDF to M.W. from the Canadian Institute of Health Research Strategic Training Grant on Membrane Proteins Linked to Disease, and the resources of the Molecular Design and Information Technology Centre, University of Toronto.

References

Alvarez JC. 2004. High-throughput docking as a source of novel drug leads. Curr Opin Chem Biol 8(1): 1-6.

Barton GJ. 2001. Creation and analysis of protein multiple sequence alignments. Bioinformatics: a practical guide to the analysis of genes and proteins. Baxevanis AD, Ouellette FF, editors. New York: Wiley-Liss Inc.

Baxevanis AD, Francis Ouellette BF. (editors) 2001. Bioinformatics: a practical guide to the analysis of genes and proteins 2nd edn. John Weley & Sons.

Bostrom J. 2001. Reproducing the conformations of protein-bound ligands: a critical evaluation of several popular conformational searching tools. J Comput Aided Mol Des 15(12): 1137-1152.

Bostrom J, Norrby PO, Liljefors T. 1998. Conformational energy penalties of protein-bound ligands. J Comput Aided Mol Des 12(4): 383-396.

Brooks BR, Bruccoleri RE, Olafson BD, States DJ, Swaminathan S, et al. 1983. CHARMM: a program for macromolecular energy, minimization, and dynamics calculations. J Comp Chem 4: 187-217.

Carlson HA. 2002. Protein flexibility and drug design: how to hit a moving target. Curr Opin Chem Biol 6(4): 447-452.

Claussen H, Buning C, Rarey M, Lengauer T. 2001. FlexE: efficient molecular docking considering protein structure variations. J Mol Biol 308(2): 377-395.

Connolly ML. 1983. Solvent-accessible surfaces of proteins and nucleic acids. Science 221(4612): 709-713.

Fan H, Mark AE. 2004. Refinement of homology-based protein structures by molecular dynamics simulation techniques. Protein Sci 13(1): 211-220.

Fiser A, Sali A. 2003. Modeller: generation and refinement of homology-based protein structure models. Methods Enzymol 374: 461-491.

Fredriksson R, Lagerström MC, Lundin LG, Schiöth HB. 2003. The G-protein-coupled receptors in the human genome form five main families. Phylogenetic analysis, paralogon groups, and fingerprints. Mol Pharmacol 63(6): 1256-1272.

Friesner RA, Banks JL, Murphy RB, Halgren TA, Klicic JJ, et al. 2004. Glide: a new approach for rapid, accurate docking and scoring. 1. Method and assessment of docking accuracy. J Med Chem 47(7): 1739-1749.

Frimurer TM, Peters GH, Iversen LF, Andersen HS, Moller NP, et al. 2003. Ligand-induced conformational changes: improved predictions of ligand binding conformations and affinities. Biophys J 84(4): 2273-2281.

Gasteiger J, Rudolph C, Sadowski J. 1990. Automatic generation of 3D-atomic coordinates for organic molecules. Tetrahedron Comp Method 3: 537

Gohlke H, Klebe G. 2002. Approaches to the description and prediction of the binding affinity of small-molecule ligands to macromolecular receptors. Agnew Chem Int Ed Engl 41(15): 2644-2676.

Hampson DR, Huang XP, Pekhletski R, Peltekova V, Hornby G, et al. 1999. Probing the ligand binding domain of the mGluR4 subtype of metabotropic glutamate receptor. J Biol Chem 274: 33488-33495.

Hooft RW, Vriend G, Sander C, Abola EE. 1996. Errors in protein structures. Nature 381(6580): 272.

Jacobs DJ, Rader AJ, Kunh LA, Thorpe MF. 2001. Protein flexibility predictions using graph theory. Proteins 44(2): 150-165.

Jones G, Willett P, Glen AR, Taylor R. 1997. Development and validation of a genetic algorithm for flexible docking. J Mol Biol 267(3): 727-748.

Jorgensen WL. 2004. The many roles of computation in drug discovery. Science 303(5665): 1813-1818.

Kalé L, Skeel R, Bhandarkar M, Brunner R, Gursoy A, et al. 1999. NAMD2: greater scalability for parallel molecular dynamics. Comput Phys 151: 283-312.

Kollman PA, Massova I, Reyes C, Kuhn B, Huo S, et al. 2000. Calculating structures and free energies of complex molecules: combining molecular mechanics and continuum models. Acc Chem Res 33(12): 889-897.

Kramer B, Rarey M, Lengauer T. 1999. Evaluation of the FLEXX incremental construction algorithm for protein-ligand docking. Proteins 37(2): 228-241.

Kuang D, Yao Y, Wang M, Pattabiramann N, Kotra L, et al. 2003. Molecular similarities in the ligand binding pockets of an odorant receptor and the metabotropic glutamate receptors. J Biol Chem 278(43): 42551-42559.

Kunishima N, Shimada Y, Tsuji Y, Sato T, Yamamoto M, et al. 2000. Structural basis of glutamate recognition by a dimeric metabotropic glutamate receptor. Nature 407: 971-977.

Kuntz ID, Blaney JM, Oatley SJ, Langridge R, Ferrin TE, 1982. A geometric approach to macromolecule–ligand interactions. J Mol Biol 161(2): 269-288.

Laskowski RA, Mac Arthur MW, Moss DS, Thornton JM. 1993. PROCHECK: a program to check the stereochemical quality of protein structures. J Appl Crystallogr 26(2): 283-291.

Lin JH, Perryman AL, Schames JR, McCammon JA. 2002. Computational drug design accommodating receptor flexibility: the relaxed complex scheme. J Am Chem Soc 124(20): 5632-5633.

Lindahl E, Hess B, Spoel Van der D. 2001. Gromacs 3.0: a package for molecular simulation and trajectory analysis. J Mol Med 7: 306-317.

Luthy R, Bowie JU, Eisenberg D. 1992. Assessment of protein models with three-dimensional profiles. Nature 356(6364): 83-85.

Mangoni M, Roccatano D, Di Nola A. 1999. Docking of flexible ligands to flexible receptors in solution by molecular dynamics simulation. Proteins 35(2): 153-162.

Marti-Renom MA, Stuart AC, Fiser A, Sanchez R, Melo F, et al. 2000. Comparative protein structure modeling of genes and genomes. Annu Rev Biophys Biomol Struct 29: 291-325.

Massova I, Kollman PA. 2000. Combined molecular mechanical and continuum solvent approach (MM-PBSA/GBSA) to predict ligand binding. Perspect Drug Discov Des 18: 113-135.

Melo F, Sanchez R, Sali A. 2002. Statistical potentials for fold assessment. Protein Sci 11(2): 430-448.

Morris GM, Goodsell DS, Huey R, Olson AJ. 1996. Distributed automated docking of flexible ligands to proteins: parallel applications of AutoDock 2.4. J Comput Aided Mol Des 10(4): 293-304.

Oshiro C, Bradley EK, Eksterowicz J, Evensen E, Lamb ML, et al. 2004. Performance of 3D-database molecular docking studies into homology models. J Med Chem 47(3): 764-767.

Palczewshi K, Kumasaka T, Hori T, Behnke CA, Motoshima H, et al. 2000. Crystal structure of Rhodopsin: a G Protein-coupled receptor. Science 289: 739-745.

Pearlman DA, Case DA, Caldwell JW, Ross WS, Cheatham TE, III et al. 1995. AMBER, a package of computer programs for applying molecular mechanics, normal mode analysis, molecular dynamics and free energy calculations to simulate the structural and energetic properties of molecules. Comput Phys Commun 91: 1-41.

Peitsch MC. 1996. ProMod and Swiss-Model: internet-based tools for automated comparative protein modelling. Biochem Soc Trans 24(1): 274-279.

Perola E, Charifson PS. 2004. Conformational analysis of drug-like molecules bound to proteins: an extensive study of ligand reorganization upon binding. J Med Chem 47(10): 2499-2510.

Rosemond E, Wang M, Yao Y, Storjohann L, Johnson EC, et al. 2004. Molecular basis for the differential agonist affinities of group III metabotropic glutamate receptors. Mol Pharmacol 66: 834-842.

Sadowski J, Gasteiger J. 1994. Comparison of automatic three-dimensional model builders using 639 X-ray structures. J Chem Inf Comput Sci 34: 1000-1008.

Sali A, Blundell TL. 1993. Comparative protein modelling by satisfaction of spatial restraints. J Mol Biol 234(3): 779-815.

Sanchez R, Sali A. 1997. Advances in comparative protein-structure modelling. Curr Opin Struct Biol 7(2): 206-214.

Schmidt MW, Baldridge KK, Boatz JA, Elbert ST, Gordon MS, et al. 1993. General atomic and molecular structure system. J Comput Chem 14: 1347-1363.

Shoichet BK, McGovern SL, Wei B, Irwin JJ. 2002. Lead discovery using molecular docking. Curr Opin Chem Biol 6(4): 439-446.

Sippl MJ. 1993. Recognition of errors in three-dimensional structures of proteins. Proteins 17(4): 355-362.

Sippl MJ. 1995. Knowledge-based potentials for proteins. Curr Opin Struct Biol 5(2): 229-235.

Sotriffer C, Klebe G. 2002. Identification and mapping of small-molecule binding sites in proteins: computational tools for structure-based drug design. Farmaco 57(3): 243-251.

Sutcliffe MJ, Haneef I, Carney D, Blundell TL. 1987. Knowledge based modelling of homologous proteins, Part I: Three-dimensional frameworks derived from the simultaneous superposition of multiple structures. Protein Eng 1(5): 377-384.

Thompson JD, Higgins DG, Gibson TJ. 1994. CLUSTAL W: improving the sensitivity of progressive multiple sequence alignment through sequence weighting, position-specific gap penalties and weight matrix choice. Nucleic Acids Res 22(22): 4673-4680.

Trent JO, Wang ZX, Murray JL, Shao W, Tamamura H, et al. 2003. Lipid bilayer simulations of CXCR4 with inverse agonists and weak partial agonists. J Biol Chem 278(47): 47136-47144.

Tsuchiya D, Kunishima N, Kamiya N, Jingami H, Morikawa K. 2002. Structural views of the ligand-binding cores of a metabotropic glutamate receptor complexed with an antagonist and both glutamate and Gd+. Proc Natl Acad Sci 99: 2260-2665.

Varady J, Wu X, Fang S, Min J, Hu Z, et al. 2003. Molecular modeling of the three-dimensional structure of dopamine 3 (D3) subtype receptor: discovery of novel and potent D3 ligands through a hybrid pharmacophore- and structure-based database searching approach. J Med Chem 46(21): 4377-4392.

Verkhivker GM, Bouzida D, Gehlhaar DK, Rejto PA, Schaffer L, et al. 2001. Hierarchy of simulation models in predicting molecular recognition mechanisms from the binding energy landscapes: structural analysis of the peptide complexes with SH2 domains. Proteins 45(4): 456-470.

Vriend G. 1990. WHAT IF: a molecular modeling and drug design program. J Mol Graph 8(1): 52-56, 29.

Wang J, Morin P, Wang W, Kollman PA. 2001. Use of MM-PBSA in reproducing the binding free energies to HIV-1 RT of TIBO derivatives and predicting the binding mode to HIV-1 RT of efavirenz by docking and MM-PBSA. J Am Chem Soc 123(22): 5221-5230.

Wang J, Wolf RM, Caldwell JW, Kollman PA, Case DA. 2004. Development and testing of a general amber force field. J Comput Chem 25(9): 1157-1174.

Wang R, Lu Y, Wang S. 2003. Comparative evaluation of 11 scoring functions for molecular docking. J Med Chem 46(12): 2287-2303.

Wang M, Hampson DR. 2006. An evaluation of automated in silico ligand docking of amino acid ligands to family C G-protein coupled receptors. Bioorg Med Chem 14: 2032-2039.

Yao Y, Pattabiraman N, Michne WF, Huang X-P, Hampson DR. 2003. Molecular modeling and mutagenesis of the ligand binding pocket of the mGluR3 subtype of metabotropic glutamate receptor. J Neurochem 86: 947-957.

Yin D, Gavi S, Wang H, Malbon CC. 2004. Probing receptor structure/function with chimeric g-protein-coupled receptors. Mol Pharmacol 65: 1323-1332.

13 Flow Cytometry

Wendy P. Gati

Abstract: Flow cytometry is a technique for rapidly examining multiple characteristics of individual cells, by recording fluorescence signals emitted from cell-associated reporter molecules, and measuring cellular light scattering properties. This chapter introduces the principles and practice of flow cytometry, and reviews examples from the literature that highlight applications of this experimental tool in the neurosciences. The chapter concludes with protocols for three basic procedures that illustrate some practical aspects of analytical flow cytometry.

1 Introduction

Flow cytometry is a technique that uses fluorescence and light scatter to examine the biochemical and biophysical characteristics of cells in a fluid suspension. By probing cellular components with fluorescent molecules (fluorochromes), this unique tool allows the investigator to record data for individual cells at rates of thousands of cells per second, and to identify subsets of cells in which specified properties are correlated. Although this technique was introduced in its simplest form more than 70 years ago, flow cytometry as we know it today began in the 1960s with the development of instruments capable of fluorescence activated cell sorting (FACS), a preparative mode in which defined cell subsets can be physically isolated from cell populations. Since that time, improvements in the electronic and optical components of flow cytometers and the availability of a wide choice of suitable fluorochromes have greatly extended the realm of applications. Flow cytometry has become the method of choice for detection and quantification of biochemical heterogeneity within cell populations, and for the resolution and isolation of single cells or rare subsets of cells according to specified characteristics.

Although the concept of flow cytometry originated in studies of blood cells and tumor cells (for review, see Darzynkiewicz et al., 2004), flow cytometric procedures are now used routinely in disciplines as diverse as immunology, the neurosciences, nutritional sciences, pharmacology, parasitology, and marine biology. This chapter offers an introduction to the theory and practice of analytical flow cytometry, with emphasis on applications in the neurosciences.

2 Principles

2.1 Instrumentation

Most flow cytometers are designed to be used as either analyzers or cell sorters, although some offer both capabilities. The newest instruments are compact and semi-automated, incorporating solid state lasers and digital electronics. However, the basic principles of operation have not changed significantly during the past few decades. ➋ *Figure 13-1* shows the elements of an analytical flow cytometer in which scattered light signals and two fluorescence parameters are resolved. Cells in suspension are delivered under pressure into a stream of sheath fluid within a flow cell, where they are hydrodynamically focused into a single file pattern. As the cells individually intercept a laser beam, cell-associated fluorescent molecules are excited, and scattered light is produced according to the physical properties of the cells. Both scattered light and fluorescence signals are focused and resolved by an optical system, which ensures that light of specified wavelengths reaches the appropriate detector. Two kinds of scattered light signals, namely, forward scatter (FSC) and side scatter (SSC, orthogonal light scatter), yield, respectively, information about relative cell size and cellular complexity or granularity. Fluorescence signals, which are emitted from intrinsic cellular components (autofluorescence) and from cell-associated fluorochromes, provide biochemical information.

The number of cellular parameters that can be evaluated simultaneously depends on the optical and electronic configuration of a given instrument. A flow cytometer may be a single-laser analyzer that can measure three fluorescence colors and two light scatter parameters (five parameters), or a complex instrument with multiple lasers and detectors capable of recording signals for as many as 13 fluorescence colors (15 parameters) in individual cells. For example, two widely used single-laser instruments, the BD

■ Figure 13-1
Principal components of a single-laser, analytical flow cytometer. Focused within a stream of sheath fluid, cells flow in single file past a laser beam, which excites cell-associated fluorescent molecules and is scattered according to structural features of the cells. Scattered light and fluorescence signals are focused and filtered before reaching detectors, which convert them to electronic pulses. *FSC*, forward light scatter; *SSC*, side (orthogonal) light scatter

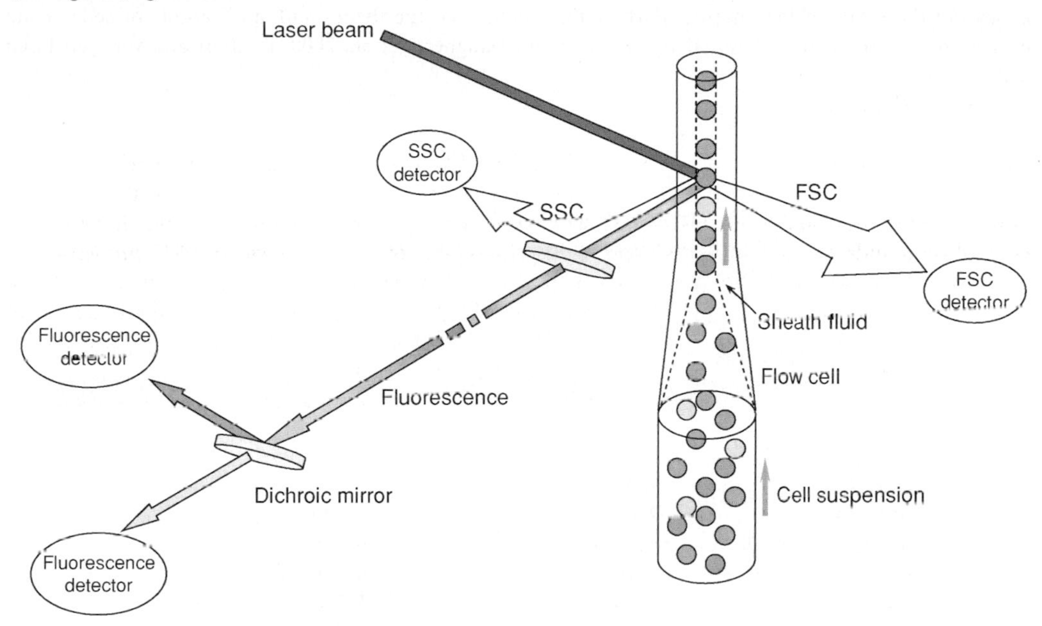

FACScan (BD Biosciences, San Jose, CA) and the Coulter Epics XL (Beckman Coulter Inc, Fullerton, CA) are equipped with 488-nm, argon-ion lasers, and are capable of recording three or four fluorescence colors (five or six parameters). Two-laser analyzers, such as the BD FACSCalibur, the Cytomics FC500 (Beckman Coulter Inc), and the BD FACSCanto are configured additionally with a red solid-state diode (635-nm) laser or helium-neon (HeNe, 633-nm) laser, and can detect four to six colors (five to seven parameters). Three and four-laser instruments, such as the BD LSR II analyzer and the Cytomation MoFlo (DakoCytomation Colorado Inc, Fort Collins, CO) and BD FACSAria high-speed cell sorters, are equipped with UV-emitting and/or violet diode (405-nm) lasers together with argon-ion and red diode lasers.

Other properties that define instrument capability include sensitivity, limits of particle size resolution, and rate of data acquisition. For example, detection limits for green fluorescence range from ~600 molecules of equivalent soluble fluorochrome (MESF)/cell to <50 MESF/cell, and particle sizes of 0.5 μm to 50 μm may be resolved in some instruments. Acquisition rates in analytical instruments vary from <3,500 cells (events)/s in older instruments, to ~10,000 events/s in the newest models. High acquisition rates and the availability of automatic sampling devices for microplate-based assays have enhanced the efficiency of analyzing large sample sets.

Fluorescence activated cell sorting (preparative flow cytometry) is a means of physically isolating individual cells or cell subsets on the basis of cell-associated fluorescence signals and/or light scattering properties. Cell sorters provide a unique interface between flow cytometry and other analytical methods, as the isolated cells can be further studied by using microscopic, chemical, electrophysiological, or molecular techniques, or may be expanded in cell culture. Live cell sorting under sterile conditions is the most efficient means for recovering transfectants that harbor fluorescent genetic markers, or rare subsets of cells, such as stem cells, from mixed populations. High-speed instruments such as the Coulter Epics ALTRA (Beckman Coulter Inc) and the Cytomation MoFlo are capable of sorting at rates of 25,000 to 50,000 cells/s. Four-way sorting options that are available on some high-speed instruments, such as the Cytomation MoFlo and the BD FACSAria, allow the investigator to recover several fractions simultaneously. Instruments that are equipped with single-cell deposition modules are capable of sorting individual cells onto microscope slides or into multi-well plates with high precision, offering a sophisticated and efficient method for cloning, and for the subsequent genetic analysis of single cells by using PCR techniques.

A detailed description of cell-sorting procedures, which are usually performed by a trained operator, is beyond the scope of this chapter. Further information on the theory and applications of cell sorting may be found in recent reviews (Battye et al., 2000; Daugherty et al., 2000; Ibrahim and van den Engh 2003, 2004).

2.2 Fluorochromes

Fluorochromes are stains that interact directly with cellular components, or are used to form conjugates with antibodies or ligands, yielding fluorescent reporter molecules. *Table 13-1* lists a selection of fluoro-

Table 13-1
Fluorochromes for flow cytometry

Fluorochrome	Laser line for excitation, nm	Nominal λ_{max} of emission, nm (colour)
Fluorescein (FITC)	488	519 (green)
Alexa Fluor 488	488	519 (green)
R-Phycoerythrin (PE)	488	578 (orange)
Propidium iodide (PI)	488	617 (red)
7-Aminoactinomycin D (7-AAD)	488	647 (red)
Allophycocyanine (APC)	633/635	660 (red)
Alexa Fluor 647	633/635	665 (red)
PE-Cy5 tandem	488	667 (red)
Cyanine dye 5 (Cy5)	633/635	670 (red)
Peridinin chlorophyll protein (PerCP)	488	670 (red)
APC-Cy7 tandem	633/635	760 (infrared)
PE-Cy7 tandem	488	785 (infrared)
4′,6-Diamidino-2-phenylindole (DAPI)	UV, 405	455 (blue)
Hoechst 33342	UV	478 (blue)
Indo-1	UV	475 (blue)/401 (violet)

chromes that are commonly used in flow cytometry. Allowable combinations of fluorescent agents for a particular multiparameter experiment will be limited by the availability of laser lines and optical filtering capability of the analyzer, and will depend on experimental conditions and goals. Many of the fluorochromes listed in *Table 13-1* are excitable with the widely available 488-nm, argon-ion laser. Thus, combinations of two or three such agents may be used in a single-laser instrument, provided that fluorescence emission maxima (λ_{max}) are sufficiently well separated and spectral overlap is minimal. In addition, fluorochrome brightness, which is a function of the extinction coefficient and quantum yield, may be a consideration if rare subsets of cells are to be discerned, or if the cellular abundance of target molecules is low. Fluorochromes that emit fluorescence in a pH-dependent manner may not be compatible with all experimental conditions.

Fluorescein (fluorescein isothiocyanate, FITC), which has had a long history of use in the preparation of fluorescent conjugates, is still adequate for many applications, although newer green-fluorescing agents such as Alexa Fluor 488 have superior quantum yields and fluorescence emission that does not vary with pH. Phycoerythrin (PE), which is valued for its brightness, has been one of the most useful orange-fluorescing agents. Red-fluorescing stains, such as Cyanine dye 5 (Cy5), and allophycocyanine (APC), are excited by the longer wavelength lines of HeNe or red solid-state diode lasers, which are less damaging to cells and induce minimal cellular autofluorescence. Tandem dyes that emit fluorescence in the infrared region, such as PE-Cy7 and APC-Cy7, are most easily combined with popular green- and orange-fluorescing agents.

A UV laser is needed for exciting the blue-fluorescing agents, 4′,6-diamidino-2-phenylindole (DAPI) and Hoechst 33342, which are DNA-intercalating stains, and for indo-1, a fluorescent calcium chelator dye. Violet diode lasers that are offered in some newer instruments accommodate fluorochromes such as Cascade Blue, Pacific Blue, and cyan fluorescent protein, and are also capable of exciting DAPI (Shapiro and Perlmutter 2001; Telford et al., 2003).

3 Practical Considerations

3.1 Sample Preparation and Staining

It is essential that samples for flow cytometry are prepared as single-cell suspensions. Whereas nonadherent cells need only be washed free of medium before staining, attached cells must first be released from their growth surface and prepared as suspensions that are stable and free of clumps. This is most often accomplished by treating cell cultures with trypsin and/or EDTA, or by mechanically disaggregating cells that have been teased from tissues or scraped from growth surfaces (see, e.g., Barrett et al., 1998; Lecoeur et al., 2004). Cell suspensions prepared in this way should be passed through a 50- to 100-μm nylon mesh filter before flow cytometric analysis, to avoid flow cell clogging.

Fixation procedures are necessary when biohazardous samples are analyzed, and are sometimes used to allow access of membrane impermeant fluorochromes, or to stabilize samples for short-term storage. Optimal fixatives are those that have low autofluorescence and do not significantly affect staining. Paraformaldehyde, at concentrations of 0.5–2%, and ethanol (70%, 4°C) are widely used fixatives for flow cytometry. Combinations of paraformaldehyde with Triton X-100 or saponin have been employed in procedures that fix and permeabilize cells.

Most manufacturers of fluorochromes and fluorescent conjugates that are intended for flow cytometry offer technical bulletins that describe optimal product use. These provide a good starting point from which adjustments for specific cell types and experimental conditions may be made. Sample sets should include unstained samples for measuring autofluorescence, single-stained standards for setting spectral compensation, and in some instances, controls that measure nonspecific staining. For example, matched isotype controls are often used to measure nonspecific fluorescence when cells are stained with fluorochrome-conjugated monoclonal antibodies.

3.2 Data Acquisition

Acquisition parameters, such as detector voltages and amplification modes, are selected in instrument setup procedures according to anticipated ranges of fluorescence intensities and intrinsic properties of the stained cell samples. Flow cytometer detectors convert fluorescence signals to electronic pulses, which are collected with either linear or logarithmic amplification. For example, light scatter signals (FSC, SSC) and signals that measure DNA content, which vary over a limited range, are usually collected with linear amplification and plotted on a scale of channel numbers. Other parameters, such as the abundance of a cell surface marker that may vary by orders of magnitude within a heterogeneous cell population, will be measured with logarithmic amplification to resolve dimly fluorescent cell subsets, and plotted on a logarithmic scale of relative fluorescence intensity units. Although it is appropriate in most instances to record pulse height values, which are proportional to signal intensity, some situations require the measurement of pulse width and/or area (see ➲ Section 13.5.3). Instrument setup procedures may be expedited by using acquisition templates, designed with instrument software, which also serve to ensure the consistency of data acquisition conditions in replicate experiments. Reference standard fluorescent beads, such as Quantum MESF kits

(Bangs Laboratories, Inc., Fishers, IN) or QuantiBRITE beads (BD Biosciences), are useful for the absolute quantification of fluorescence signals and for day-to-day monitoring of instrument performance.

When fluorochromes are combined in a multiparameter assay, it is almost always necessary to correct for signals from overlapping portions of emission spectra that have not been eliminated by optical filtration. This is accomplished through a spectral compensation procedure that is performed according to instrument-specific instructions, and involves the adjustment of detector voltages to electronically subtract extraneous fluorescence.

3.3 Display and Analysis of Data

Flow cytometry software is designed to store data in list-mode files, as sequential values for multiple parameters for individual cells, allowing the investigator to perform retrospective analyses. Fundamental to flow cytometric analysis is the establishment of gates to electronically isolate data for particular subsets of cells, and to probe within those subsets multiple correlations among specified parameters. Several formats are available for the display of data and extraction of statistical information. Fluorescence histograms (see, e.g., *Figures 13-3* and *13-7*), which are used in univariate analysis, can reveal multiple subpopulations of cells that are distinguishable according to a single characteristic. Bivariate plots, which allow the investigator to recognize cell subsets in which two characteristics are correlated, are usually displayed as dot distributions or dot plots (*Figure 13-2, left panel*). In this format, signals from single cells are displayed as data points, for which x- and y-coordinates correspond to values for two parameters. Other bivariate plot options that show frequency distributions within cell subsets are contour plots (*Figure 13-2, center panel*), density plots (*Figure 13-2, right panel*), and surface plots (*Figure 13-4, top*). Three-dimensional plots, such as the tomograms of *Figure 13-4*, are useful for multivariate analysis.

4 Applications

Flow cytometry is a rapid and sensitive means of identifying, characterizing, and isolating individual cells or cell subsets, according to specified characteristics. Important applications of this technology are phenotypic analysis, the determination of kinetic and equilibrium binding constants for ligands at receptors (Bednar et al., 1997; Waller et al., 2001) and membrane transporters (Buolamwini et al., 1994; Gati et al., 1997), and the measurement of cellular thiol content (Sen et al., 1999), intracellular free calcium (Burchiel et al., 2000), intracellular pH (Ringel et al., 2000), and mitochondrial membrane potential (Shapiro 2000). Fluorescence resonance energy transfer (FRET) analysis of membrane receptor interactions may be accomplished by flow cytometry (Chan et al., 2001). Flow cytometry techniques complement microscopic analysis in the characterization of cell death processes by providing quantitative information for cell populations (Darzynkiewicz et al., 1997; Lecoeur et al., 2004). Furthermore, flow cytometry has played a central role in studies of the cell cycle (Darzynkiewicz et al., 2004), and in the identification and characterization of rare subsets of cells such as stem cells (Kim and Morshead 2003) and transfectants (Dell'Arciprete et al., 1996), which can be isolated by high-speed cell sorting.

4.1 Phenotypic Analysis

Flow cytometric analysis is the method of choice for recognizing cell-surface or intracellular phenotypic markers that define cell subsets, by using fluorescently labelled antibodies to detect cell surface antigens or cytoplasmic epitopes. For example, Gylys et al., (2004) identified and characterized synaptosomes in a crude homogenate from human cortex by gating on events that stained positively for the presynaptic marker, SNAP-25. By using a similar approach, amyloid-β levels were measured in synaptosomes in a crude preparation from rat brain (Gylys et al., 2003). Nonyl acridine orange (NAO), which binds to cardiolipin in the inner mitochondrial membrane (Ratinaud et al., 1988), was used by Mattiasson et al. (2003) to identify mitochondria in a crude preparation from rat hippocampus. McLaren et al. (2001) used antibodies against nestin, glial

◘ Figure 13-2

Bivariate analysis of a mononuclear cell-enriched human leukemic bone marrow sample, after staining with antiCD45-PE. The data are shown as a dot plot (*upper left panel*), in which gates were set to define major cell populations, a contour plot (*upper right panel*), and a density plot (*lower panel*). The data were acquired with a BD FACScan and analyzed using Cytomics RXP software (Beckman Coulter Inc.)

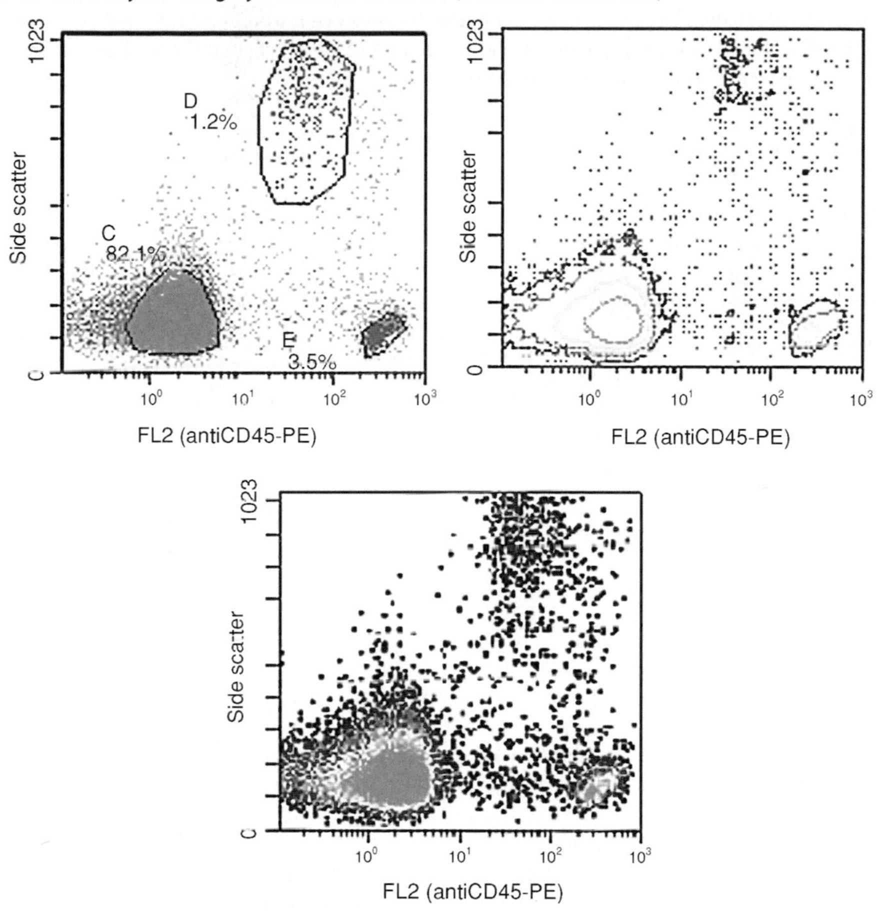

fibrillary acidic protein (GFAP), β-tubulin, and galactocerebroside C to monitor cellular differentiation in rat neurosphere cell cultures. Furthermore, neural stem cells from mouse forebrain have been purified to high levels of homogeneity by sorting cells according to the intensity of staining with fluorochrome-conjugated peanut agglutinin and heat stable antigen (CD24) (Rietze et al., 2001), or according to differential efflux of the DNA binding fluorochrome, Hoechst 33342 (Kim and Morshead 2003).

4.2 Receptor Quantitation and Characterization

Analytical flow cytometry offers a rapid and facile means of monitoring cellular receptor content. For example, multiparameter flow cytometry techniques were used to monitor expression of $GABA_A$ receptor subunits during neurogenesis in embryonic rat brain (Maric et al., 2001). The content of the cell surface p75 neurotrophin receptor was measured in a heterogeneous population of mouse dorsal root sensory neurons, from which high and low p75 subsets were subsequently isolated by cell sorting (Barrett et al., 1998).

Figure 13-3

Determination of equilibrium binding constants for SAENTA-fluorescein at *es* nucleoside transporter sites in CEM cells. Cells were stained with graded concentrations of SAENTA-fluorescein (0.5–20 nM), in the absence or presence of 5 μM nitrobenzylthioinosine, which was used to measure nonspecific binding. Fluorescence signals were acquired with a BD FACScan and the data were analyzed using Cytomics RXP software (*left panel*). Mean fluorescence intensity values were determined from the fluorescence histograms (*left panel*), which are shown as an overlay plot. Nonlinear regression analysis of the data (*right panel*) yielded the binding constants shown. *AF*, autofluorescence; *RFI*, relative fluorescence intensity

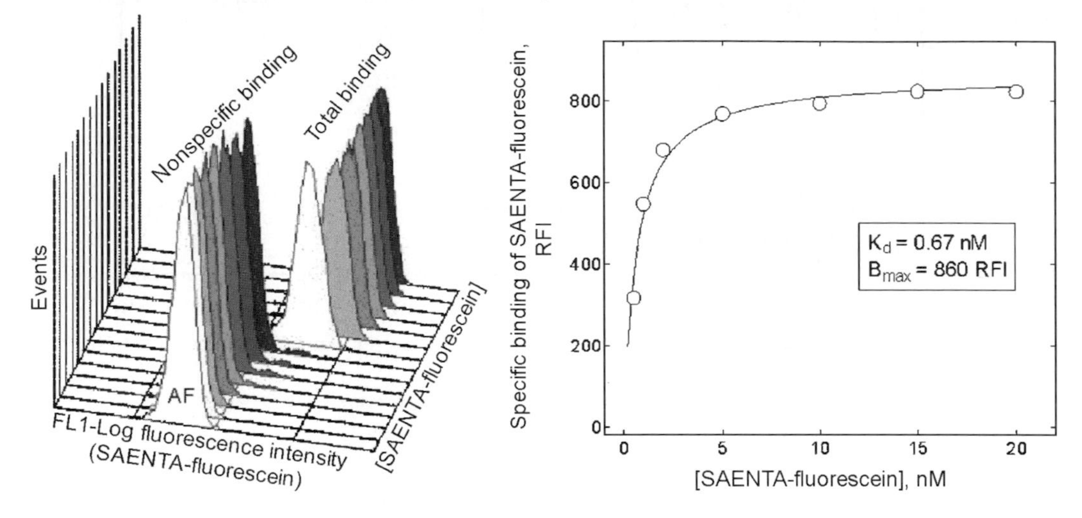

In some instances, flow cytometry assays are a superior alternative to conventional procedures for the determination of equilibrium binding constants (Stein et al., 2001). In contrast to assays that employ radiolabelled ligands, which measure population mean values for binding constants, flow cytometry methods can measure those values in individual cells, revealing heterogeneity in receptor expression within a population of cells or membrane vesicles. Furthermore, small samples can be characterized in a short period of time (hours). This approach to receptor-binding analysis may be limited only by the availability of a properly characterized fluorescent ligand.

The experiment in *Figure 13-3* illustrates the use of flow cytometry to determine K_d and B_{max} values for binding of the *es* nucleoside transporter ligand, SAENTA-fluorescein, at sites in human CEM leukemic lymphoblasts. The *es* nucleoside transporter is widely expressed in mammalian cells, and is concentrated in regions in the CNS that coordinately express adenosine A_1 receptors (Jennings et al., 2001).

Absolute values for transporter abundance (B_{max}) have been obtained with this assay by using fluorescent calibration beads to convert fluorescence signals to MESF (Wiley et al., 1994), or by calibrating the assay with a standard cell population in which B_{max} was determined by radioligand binding (Gati et al., 1997). *Figure 13-4* illustrates an extension of the SAENTA-fluorescein assay in which the correlation of *es* transporter content with the expression of immunophenotypic markers was tested in a clinical sample.

4.3 Cell-Cycle Analysis

The analysis of cell-cycle progression was one of the earliest applications of flow cytometry (for review, see Darzynkiewicz et al., 2004). In this assay, fluorescence signals from cells stained with DNA-binding fluorochromes are plotted as DNA content histograms that may be analyzed by using histogram deconvolution software to quantify cell-cycle phase distributions (Rabinovitch 1994). Fluorochromes that are useful for this purpose are the plasma membrane-impermeant DNA stains, propidium iodide (PI),

Figure 13-4

Correlative analysis of the expression of immunophenotypic markers (CD34, CD45) and *es* nucleoside transporter abundance in the human leukemic bone marrow sample of *Figure 13-2*. The three-dimensional plots allow recognition of the coordinate positivity of cells for various combinations of these characteristics, which were measured by staining with antiCD34-CyChrome, antiCD45-PE, and SAENTA-fluorescein. The *upper* surface plot shows CD45 (x-axis), *es* nucleoside transporter content (y-axis), and frequency (z-axis). In the lower tomograms, CD34 is shown on the z-axis. An unstained sample (*lower left*) shows autofluorescence, whereas the triple-stained sample (*lower right*) shows the pattern of cell positivity for the three markers. Data were acquired with a BD FACScan and analyzed with Cytomics RXP software

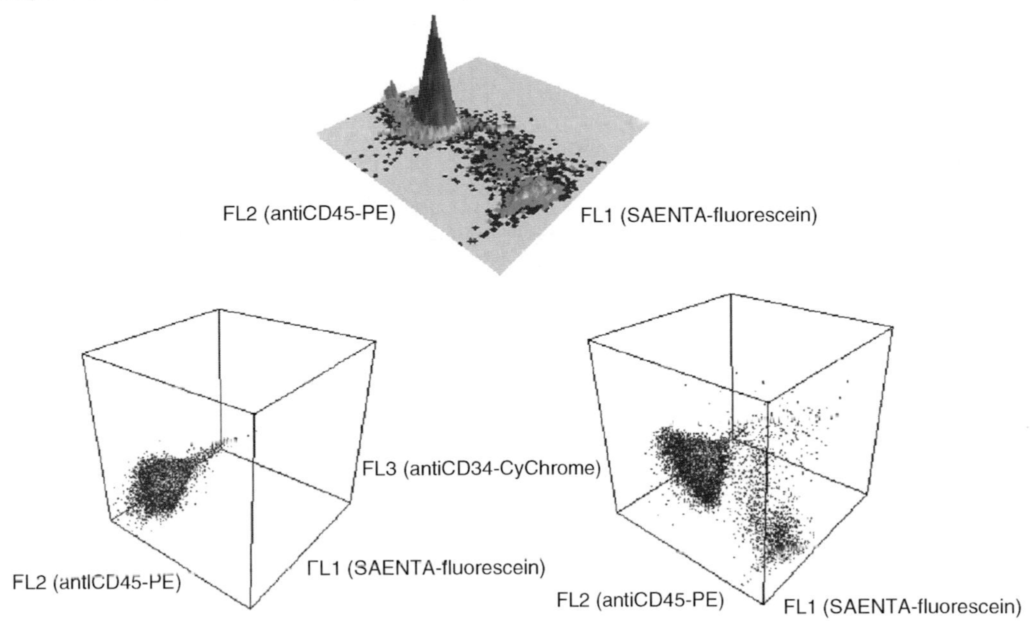

7-aminoactinomycin D (7-AAD), and DAPI for fixed cells, and the membrane-permeable Hoechst 33342 for live cells. For example, Inokoshi et al. (1999) examined the relationship between cell-cycle arrest and neuronal differentiation in Neuro 2a cells, by flow cytometric analysis of PI-stained cells. Other agents, such as pyronin Y (Toba et al., 1996) and antibodies against cyclins A and B_1 (Gong et al., 1993), have been useful in combination with DNA stains to discriminate G_0 and G_1 cells, and G_2 and M cells, respectively.

4.4 Cytotoxicity and Characterization of Cell Death

The membrane-impermeant DNA intercalating dyes, PI and 7-AAD, are useful for recognizing and gating out dead (necrotic) cells, which have leaky membranes and are selectively stained with those agents (Lyons et al., 1992; Schmid et al., 1992). Enumeration of cells that are negatively stained under these conditions can be used to monitor cell viability in simple, flow cytometric assays of cytotoxicity. Alternatively, viable cells can often be distinguished from apoptotic cells without the use of fluorochromes, according to decreases in forward scatter and increases in side scatter, which characterize cells that are undergoing early stages of apoptosis (Ormerod et al., 1995; Darzynkiewicz et al., 1997). The light scatter assay has been useful for measuring cytotoxicity in many cell types in which viable cells exhibit a homogeneous scatter profile (see, e.g., *Figure 13-5*), and may be used in conjunction with the measurement of phenotypic or functional characteristics, yielding information about the processes underlying cell death.

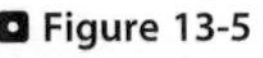

Figure 13-5

Determination of cytotoxicity by measuring light scatter signals with a flow cytometer. Human leukemic myeloblasts from a peripheral blood sample were exposed to graded concentrations of cytarabine (araC) for 96 h. The data were acquired with a BD FACScan and analyzed with CellQuest software (BD Biosciences), according to the protocol in Section 13.5.1. In the bivariate analysis of FSC versus SSC, a viable (healthy) cell gate was set to exclude dying cells and debris, and the number of events within that gate was plotted for each concentration of araC, yielding the concentration-effect curve shown

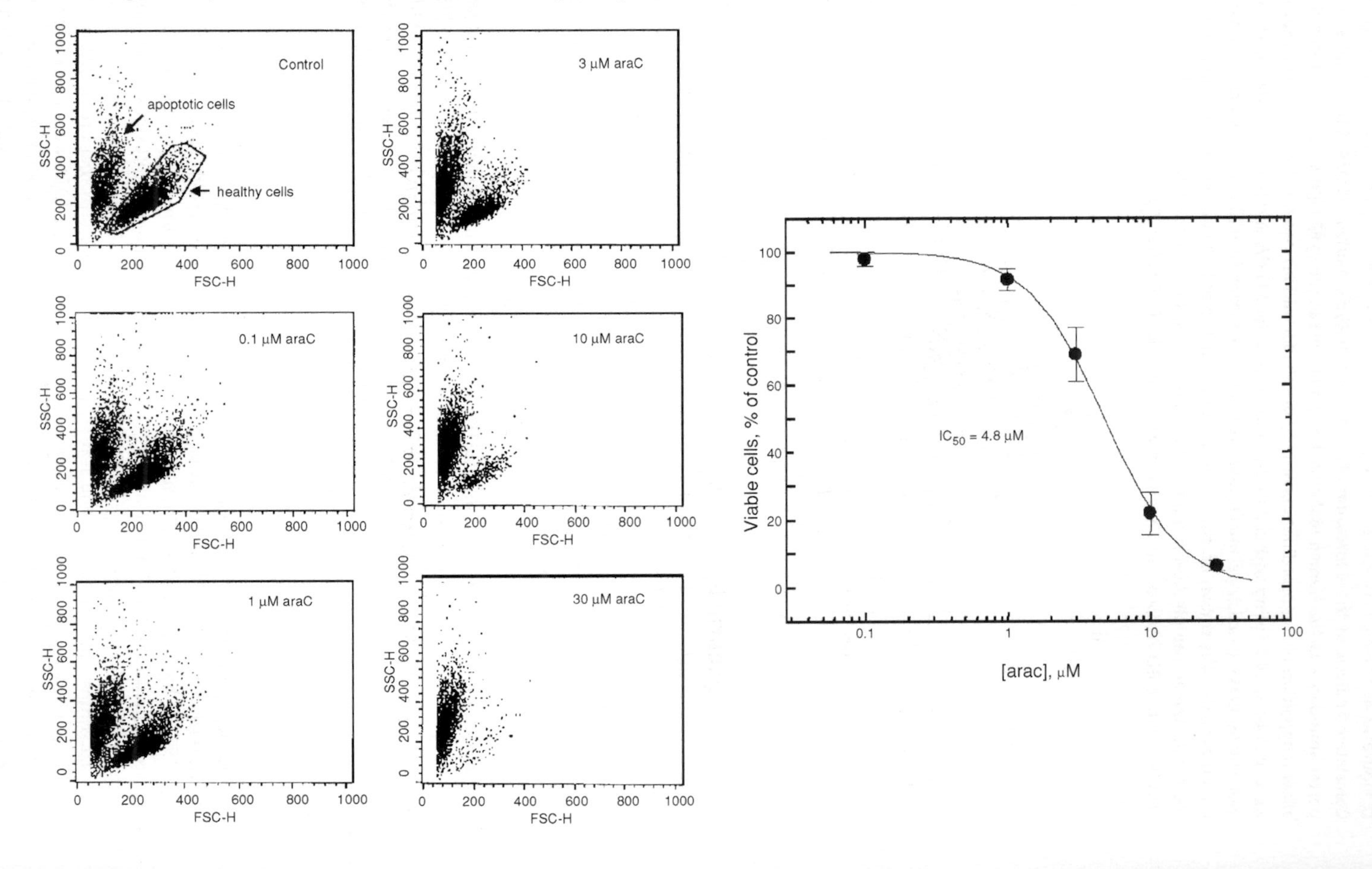

Several methods for characterizing cell death processes employ flow cytometric techniques. For example, fluorochrome-labeled annexin V has been used to recognize the translocation of phosphatidylserine to the external surface of the plasma membrane in early apoptotic cells (Koopman et al., 1994), and to distinguish those cells from late-stage apoptotic or necrotic cells, by counterstaining with PI or 7-AAD (Lecoeur et al., 1997). Other assays for apoptotic events monitor changes in mitochondrial membrane potential (Salvioli et al., 1997) or in caspase activity (Köhler et al., 2002). DNA fragmentation and loss from cell nuclei during apoptosis can be recognized as a quantifiable sub-G_1 peak on DNA content histograms of cells stained with PI (Nicoletti et al., 1991) or DAPI (Migheli et al., 1999).

Lecoeur et al. (2004) used flow cytometry to distinguish early- and late-stage apoptotic, primary cortical neurons by staining cells with fluorescein-labeled annexin V and 7-AAD. They measured mitochondrial membrane potential and caspase 3 activity in those cells by staining with 5,5′,6,6′-tetrachloro-1,1,3,3′-tetraethylbenzimidazolylcarbocyanine iodide (JC-1), and fluorochrome-labeled caspase inhibitors, respectively. Mattiasson et al. (2003) measured Ca^{2+}-induced changes in levels of reactive oxygen species (ROS) and mitochondrial membrane potential simultaneously in crude preparations from rat hippocampus by counterstaining NAO-labeled mitochondria with 2′,7′-dichlorodihydrofluorescein-diacetate and 1,1′,3,3,3′,3′-hexamethylindodicarbocyanine iodide ($DiIC_1(5)$), respectively. The role of Akt1 kinase in reducing ROS production in rat pheochromocytoma PC12 cells treated with an apoptosis-inducing Alzheimer amyloid fragment was examined in a multiparameter flow cytometry assay by transfecting cells with an enhanced green fluorescent protein (EGFP)-Akt1 fusion protein and counterstaining with hydroethidine to measure ROS levels (Martín et al., 2001). Those investigators also measured the sub-G_1 peak in DNA content histograms of PI-stained cells to determine the percentage of apoptotic cells in the population.

5 Experimental Protocols

The following basic protocols may be used alone or combined with other staining procedures in multiparameter flow cytometry experiments. Although they are illustrated with data from cells that proliferate in suspension, these protocols may be easily modified for the analysis of cells isolated from tissues or adherent cells in culture, by incorporating an initial step for the preparation of single cell suspensions. The assays are conducted at room temperature, unless otherwise noted.

5.1 Determination of Cytotoxicity by Measuring Light Scatter Signals

The flow cytometric measurement of light scatter signals from cells is a facile means of enumerating viable cells that are suspended at low concentrations in small volumes. This technique may be applied in assays of cytotoxicity (see Section 13.4.4), in which viable cells are enumerated after exposure of cell populations to agents that cause cell death. The present protocol has been designed for determining IC_{50} values for a toxicant, after exposing cells to graded concentrations of that agent in 24-well (1.5 ml/well) or 48-well (1.0 ml/well) culture plates for specified time intervals. Cells should be seeded at concentrations that yield $\sim 4 \times 10^5$ cells in control cultures (no toxicant) at the time of harvesting.

1. Mix and transfer the contents of each microwell to a 1.5-ml microcentrifuge tube.
2. Centrifuge the cell suspensions at 238 × *g* (2000 rpm in an Eppendorf microcentrifuge) for 5 min.
3. Aspirate supernatants, taking care not to disturb the cell pellets, and leaving about 30–50 μl in each tube. Add about 1.3 ml of a suitable buffer, such as phosphate-buffered saline or HEPES-buffered saline.
4. Close tube and vortex gently for 1 s to resuspend the cells. Repeat the centrifugation step (238 × *g* for 5 min).
5. Remove each supernatant as in step 3, leaving no more than 50 μl above the pellet.
6. Use a micropipette to add exactly 1.0 ml buffer to each tube, and mix the cell suspension thoroughly by drawing it in and out of the pipette tip.
7. Transfer each suspension to a flow cytometer tube, and cap the tube.

8. Set up the flow cytometer with fluorescence detectors turned off and the acquisition terminator set for TIME, so that a constant volume from each sample will be analyzed. A time interval (e.g., 1 min) that will be sufficient for the acquisition of approximately 10,000 events in the control samples should be used.
9. Acquire light scatter signals (FSC and SSC) for each sample, vortexing the cell suspensions briefly but vigorously before introducing each sample into the flow cytometer.
10. Analyze the data by setting a gate around the viable (healthy) cell population in a control sample, and recording the number of events within that gate in each of the test samples. Calculate the number of viable cells as a percentage of the viable event number in control samples.
11. Construct a concentration-effect plot as in *Figure 13-5*, and determine the IC_{50} value by fitting a logistic model to the data.

The light scatter assay may be used to determine absolute numbers of viable cells if flow cytometry data from cell suspensions of known concentration are used to construct a standard curve. For that purpose, cell concentrations should be determined in a series of graded, "standard" cell suspensions with the use of a Coulter counter. A plot of those standard concentrations versus the number of events (light scatter signals) acquired during a specified acquisition interval in the flow cytometer may then be used to interpolate cell concentrations for test samples that have been assayed by the light scatter procedure.

5.2 Characterization of Cell Death by Staining with Annexin V and 7-AAD

This procedure, which complements other methods for distinguishing apoptotic and necrotic cell death, employs annexin V-PE as a marker for early apoptotic cells, and 7-AAD for late apoptotic or necrotic cells. Although other versions of this assay have used annexin V-fluorescein together with PI, that combination precludes the use of a third fluorescence color to measure an additional parameter, such as a phenotypic marker, because PI, unlike 7-AAD, has a broad emission spectrum that includes both orange and red fluorescence.

1. Dilute the annexin V-PE solution (BD Biosciences), just before use, by adding an equal volume of HSC buffer (140 mM NaCl containing 2.5 mM $CaCl_2$ and 10 mM HEPES, pH 7.4).
2. Dissolve 7-AAD (Molecular Probes Inc.) in buffer at a concentration of 100 μg/ml. The solution should be stored as small aliquots at 4°C or -20°C, and brought to room temperature before use.
3. Prepare a series of cell samples that includes control samples as well as test samples in which cell death has been induced by exposure to a toxicant. Each sample should contain at least $\sim 10^5$ cells.
4. Samples in tubes 2 and 3 (see table below) are single-stained standards that will be used for setting compensation. These should contain a substantial number of apoptotic and/or necrotic cells, so that subsets that are positively stained with annexin V-PE and/or 7-AAD may be readily recognized, and are best prepared from designated replicates of test samples that have been exposed to a high concentration of toxicant.
5. Wash and resuspend each sample in 1 ml of HSC buffer, and transfer the washed cells to a flow cytometer tube.
6. Add annexin V-PE and 7-AAD to each tube according to the table, and gently vortex the samples.

Tube #	Sample	Annexin V-PE (μl)	7-AAD (μl)
1	Autofluorescence	---	---
2	Annexin V-PE Standard	5	---
3	7-AAD Standard	---	10
4	Control sample 1	5	10
5	Control sample 2	5	10
6	Test sample 1	5	10
7	Test sample 2	5	10

7. Set up the flow cytometer to acquire orange (annexin V-PE) and red (7-AAD) fluorescence signals, and adjust the compensation settings by using the samples in tubes 1–3.
8. Proceed with acquisition of data from tubes 4 to 7, until signals from 10,000 cells have been collected from each sample.
9. Analyze the data by constructing bivariate plots in which quadrants are set to enclose viable cells (annexin V-negative, 7-AAD-negative), early apoptotic cells (annexin V-positive, 7-AAD-negative), and late apoptotic or necrotic cells (annexin V-positive, 7-AAD-positive), as described in the legend of *Figure 13-6*.

Figure 13-6

Characterization of cell death in human leukemic myeloblasts after a 96-h exposure to 10 μM cytarabine. The cells were stained with annexin V-PE and 7-AAD, according to the protocol in Section 13.5.2. Data were acquired with a BD FACScan and analyzed using CellQuest software. The *upper left panel* is a light scatter plot of the data, in which gates B (excluding debris) and D (live myeloblast gate) are defined. The *upper right* and *lower panels* show bivariate fluorescence plots of the stained cells within gates B and D, respectively. In each plot, the relative numbers of viable cells (lower left quadrant), apoptotic cells (lower right quadrant), and late apoptotic or necrotic cells (upper right quadrant), are given by the quadrant statistics. The experiment shows that cells within gate B (*upper right panel*) included viable, apoptotic, and late apoptotic or necrotic cells, whereas virtually all of those within gate D (*lower panel*) were viable, in agreement with the light scatter data (*upper left panel*)

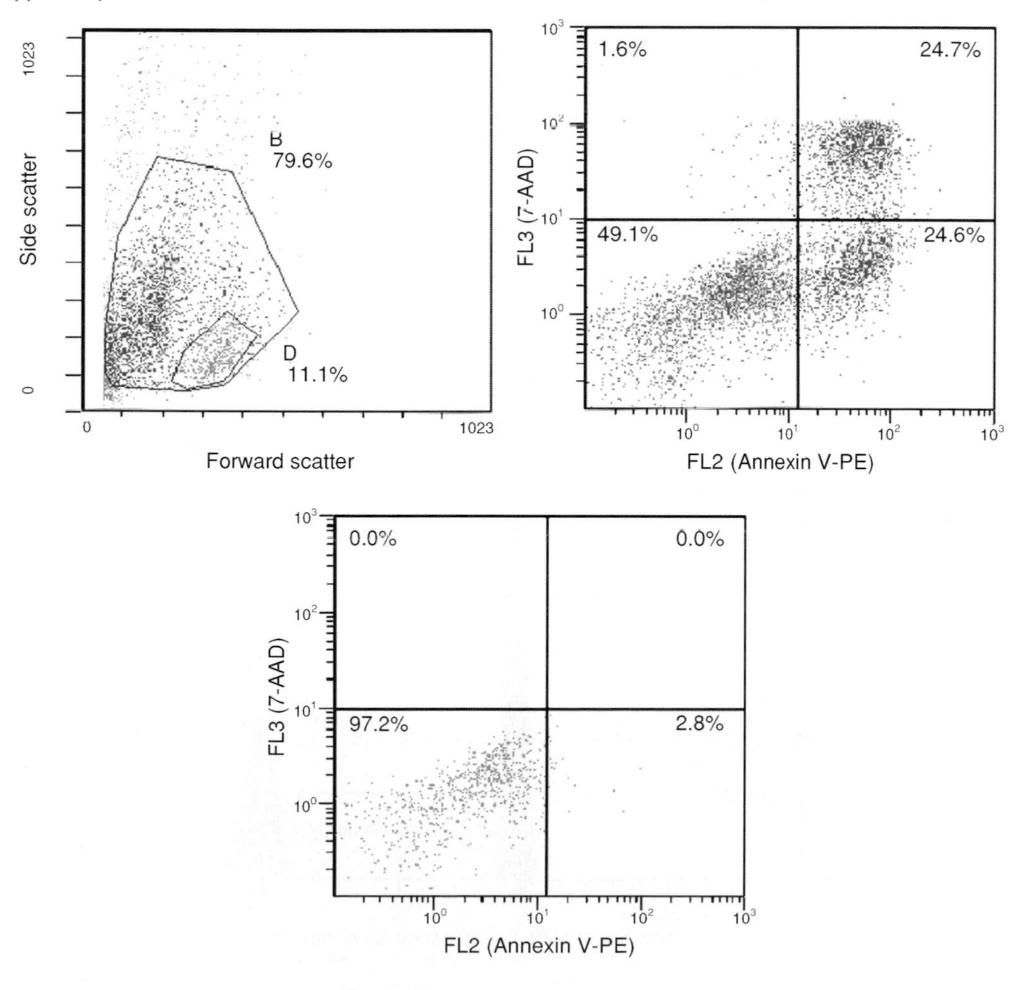

5.3 Determination of Cell-Cycle Phase Distributions by Staining with PI

This protocol describes one of several useful procedures for constructing DNA content frequency histograms that yield information about cell-cycle phase distributions. The method uses ethanol-fixation to prepare cells for staining with PI.

Fixation procedure

1. Each sample should contain at least 10^6 cells that have been washed and resuspended in 0.75 ml phosphate-buffered saline in a polypropylene tube.
2. Place the tube in an ice bath, and add 1.9 ml ice-cold 70% ethanol, dropwise, while gently agitating the sample.
3. Close the tube tightly, and store at 4°C overnight or for several days.

Propidium iodide staining procedure

1. A "Working Solution" of the stain should be prepared just before use, by combining 10 ml of 0.1% Triton X-100 in phosphate-buffered saline, 2 mg RNase (DNase-free, Sigma), and 200 μl PI solution (Sigma, 1 mg/ml in distilled water). Stock solutions of Triton X-100 and PI should be stored at 4°C.
2. Centrifuge the ethanol-fixed cells (200 × *g* for 5 min) and discard the supernatant, removing it completely.
3. Resuspend the pellet in 5 ml phosphate-buffered saline, by vortexing gently. After a 60-s pause to allow residual ethanol to diffuse from the cells, centrifuge the sample (200 × *g* for 5 min). Discard the supernatant.

Figure 13-7
Cell-cycle phase distribution in an asynchronous population of CEM cells that were fixed in ethanol and stained with PI. The data, acquired with a BD FACScan, are presented as a histogram of PI fluorescence (a measure of DNA content), recorded in units of channel numbers. Analysis of the DNA frequency content histogram was performed with Modfit LT software

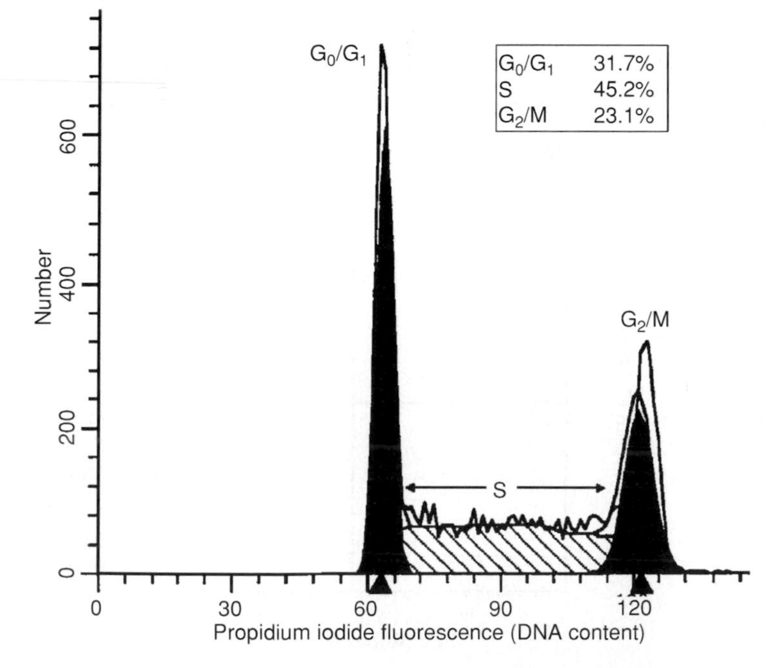

4. Resuspend the pellet in 1 ml Working Solution. Keep at room temperature, protected from light, for 30 min.
5. Filter the suspension through a nylon mesh (50 μm) and prepare to begin data acquisition within 30 min.
6. Set up the flow cytometer to acquire red fluorescence signals as pulse width and pulse area (integral) values, so that doublets can be discerned and gated out before analyzing the DNA content frequency histogram (see Wersto et al., 2001, for a discussion of the doublet discrimination procedure).
7. Analyze the data (as in ➋ *Figure 13-7*) by using DNA content frequency histogram deconvolution software, such as ModFit LT (Verity Software, Inc., Topsham, ME).

6 Conclusion

The availability of a wide array of fluorescent probes and versatile, high-speed instruments has established flow cytometry as a preferred scientific tool with applications that seem limited only by the imagination of the investigator. Excellent discussions of the theory and practice of flow cytometry may be found in the reviews cited herein, in the comprehensive textbook by Shapiro (2003), and at a dedicated web site for flow cytometry (Purdue University Cytometry Laboratories, 2005, http://www.cyto.purdue.edu/), which provides many useful links to additional technical resources.

Acknowledgments

Experimental data presented in this chapter were obtained with the expert technical assistance of Dorothy Rutkowski, Rena Pawlik, and Carmen Rieder, and with financial assistance from the Alberta Heritage Foundation for Medical Research, the Alberta Cancer Board, and the University of Alberta Hospital Foundation.

References

Barrett GL, Georgiou A, Reid K, Bartlett PF, Leung D. 1998. Rescue of dorsal root sensory neurons by nerve growth factor and neurotrophin-3, but not brain-derived neurotrophic factor or neurotrophin-4, is dependent on the level of the p75 neurotrophin receptor. Neuroscience 85: 1321-1328.

Battye FL, Light A, Tarlinton DM. 2000. Single cell sorting and cloning. J Immunol Methods 243: 25-32.

Bednar B, Cunningham ME, McQueney PA, Egbertson MS, Askew BC, et al. 1997. Flow cytometric measurement of kinetic and equilibrium binding parameters of arginine-glycine-aspartic acid ligands in binding to glycoprotein IIb/IIIa on platelets. Cytometry 28: 58-65.

Buolamwini JK, Craik JD, Wiley JS, Robins MJ, Gati WP, et al. 1994. Conjugates of fluorescein and SAENTA (5′-*S*-(2-aminoethyl)-N^6-(4-nitrobenzyl)-5′-thioadenosine): flow cytometry probes for the *es* nucleoside transporter elements of the plasma membrane. Nucleosides Nucleotides 13: 737-751.

Burchiel SW, Edwards BS, Kuckuck FW, Lauer FT, Prossnitz ER, et al. 2000. Analysis of free intracellular calcium by flow cytometry: Multiparameter and pharmacologic applications. Methods 21: 221-230.

Chan FK-M, Siegel RM, Zacharias D, Swofford R, Holmes KL, et al. 2001. Fluorescence resonance energy transfer analysis of cell surface receptor interactions and signaling using spectral variants of the green fluorescent protein. Cytometry 44: 361-368.

Darzynkiewicz Z, Juan G, Li X, Gorczyca W, Murakami T, et al. 1997. Cytometry in cell necrobiology: Analysis of apoptosis and accidental cell death (necrosis). Cytometry 27: 1-20.

Darzynkiewicz Z, Crissman H, Jacobberger JW. 2004. Cytometry of the cell cycle: Cycling through history. Cytometry 58A: 21-32.

Daugherty PS, Iverson BL, Georgiou G. 2000. Flow cytometric screening of cell-based libraries. J Immunol Methods 243: 211-227.

Dell'Arciprete R, Stella M, Fornaro M, Ciccocioppo R, Capri MG, et al. 1996. High-efficiency expression gene cloning by flow cytometry. J Histochem Cytochem 44: 629-640.

Gati WP, Paterson ARP, Larratt LM, Turner AR, Belch AR. 1997. Sensitivity of acute leukemia cells to cytarabine is a correlate of cellular *es* nucleoside transporter site content measured by flow cytometry with SAENTA-fluorescein. Blood 90: 346-53.

Gong J, Traganos F, Darzynkiewicz Z. 1993. Simultaneous analysis of cell cycle kinetics at two different DNA ploidy levels based on DNA content and cyclin B measurements. Cancer Res 53: 5096-5099.

Gylys KH, Fein JA, Tan AM, Cole GM. 2003. Apolipoprotein E enhances uptake of soluble but not aggregated amyloid-β protein into synaptic terminals. J Neurochem 84: 1442-1451.

Gylys KH, Fein JA, Yang F, Cole GM. 2004. Enrichment of presynaptic and postsynaptic markers by size-based gating analysis of synaptosome preparations from rat and human cortex. Cytometry 60: 90-96.

Ibrahim SF, Engh van den G. 2003. High-speed cell sorting: fundamentals and recent advances. Curr Opin Biotech 14: 5-12.

Ibrahim SF, Engh van den G. 2004. High-speed chromosome sorting. Chromosome Res 12: 5-14.

Inokoshi J, Katagiri M, Arima S, Tanaka H, Hayashi M, et al. 1999. Neuronal differentiation of Neuro 2a cells by inhibitors of cell cycle progression, trichostatin A and butyrolactone I. Biochem Biophys Res Commun 256: 372-376.

Jennings LL, Hao C, Cabrita MA, Vickers MF, Baldwin SA, et al. 2001. Distinct regional distribution of human equilibrative nucleoside transporter proteins 1 and 2 (hENT1 and hENT2) in the central nervous system. Neuropharmacology 40: 722-731.

Kim M, Morshead CM. 2003. Distinct populations of forebrain neural stem and progenitor cells can be isolated using side-population analysis. J Neurosci 23: 10703-10709.

Köhler C, Orrenius S, Zhivotovsky B. 2002. Evaluation of caspase activity in apoptotic cells. J Immunol Methods 265: 97-110.

Koopman G, Reutelingsperger CPM, Kuijten GAM, Keehnen RMJ, Pals ST, et al. 1994. Annexin V for flow cytometric detection of phosphatidylserine expression on B cells undergoing apoptosis. Blood 84: 1415-1420.

Lecoeur H, Ledru E, Prévost M-C, Gougeon M-L. 1997. Strategies for phenotyping apoptotic peripheral human lymphocytes comparing ISNT, annexin-V and 7-AAD cytofluorometric staining methods. J Immunol Methods 209: 111-123.

Lecoeur H, Chauvier D, Langonné A, Rebouillat D, Brugg B, et al. 2004. Dynamic analysis of apoptosis in primary cortical neurons by fixed- and real-time cytofluorometry. Apoptosis 9: 157-169.

Lyons AB, Samuel K, Sanderson A, Maddy AH. 1992. Simultaneous analysis of immunophenotype and apoptosis of murine thymocytes by single laser flow cytometry. Cytometry 13: 809-821.

Maric D, Liu Q-Y, Maric I, Chaudry S, Chang Y-H, et al. 2001. GABA expression dominates neuronal lineage progression in the embryonic rat neocortex and facilitates neurite outgrowth via $GABA_A$ autoreceptor/Cl^- channels. J Neurosci 21: 2343-2360.

Martín D, Salinas M, López-Valdaliso R, Serrano E, Recuero M, et al. 2001. Effect of the Alzheimer amyloid fragment Aβ(25-35) on Akt/PKB kinase and survival of PC12 cells. J Neurochem 78: 1000-1008.

Mattiasson G, Friberg H, Hansson M, Elmér E, Wieloch T. 2003. Flow cytometric analysis of mitochondria from CA1 and CA3 regions of rat hippocampus reveals differences in permeability transition pore activation. J Neurochem 87: 532-544.

McLaren FH, Svendsen CN, Van der Meide P, Joly E. 2001. Analysis of neural stem cells by flow cytometry: cellular differentiation modifies patterns of MHC expression. J Neuroimmunol 112: 35-46.

Migheli R, Godani C, Sciola L, Delogu MR, Serra PA, et al. 1999. Enhancing effect of manganese on L-DOPA-induced apoptosis in PC12 cells: Role of oxidative stress. J Neurochem 73: 1155-1163.

Nicoletti I, Migliorati G, Pagliacci MC, Grignani F, Riccardi C. 1991. A rapid and simple method for measuring thymocyte apoptosis by propidium iodide staining and flow cytometry. J Immunol Methods 139: 271-279.

Ormerod MG, Cheetham FPM, Sun X-M. 1995. Discrimination of apoptotic thymocytes by forward light scatter. Cytometry 21: 300-304.

Purdue University Cytometry Laboratories, West Lafayette, IN (February 14, 2005); http://www.cyto.purdue.edu/

Rabinovitch PS. 1994. DNA content histogram and cell-cycle analysis. Methods Cell Biol 41: 263-296.

Ratinaud MH, Leprat P, Julien R. 1988. In situ flow cytometric analysis of nonyl acridine orange-stained mitochondria from splenocytes. Cytometry 9: 206-212.

Rietze RL, Valcanis H, Brooker GF, Thomas T, Voss AK, et al. 2001. Purification of a pluripotent neural stem cell from the adult mouse brain. Nature 412: 736-739.

Ringel F, Chang RCC, Staub F, Baethmann A, Plesnila N. 2000. Contribution of anion transporters to the acidosis-induced swelling and intracellular acidification of glial cells. J Neurochem 75: 125-132.

Salvioli S, Ardizzoni A, Franceschi C, Cossarizza A. 1997. JC-1, but not $DiOC_6(3)$ or rhodamine 123, is a reliable fluorescent probe to assess ΔΨ changes in intact cells: implications for studies on mitochondrial functionality during apoptosis. FEBS Lett 411: 77-82.

Schmid I, Krall WJ, Uittenbogaart CH, Braun J Giorgi JV. 1992. Dead cell discrimination with 7-amino-actinomycin

D in combination with dual color immunofluorescence in single laser flow cytometry. Cytometry 13: 204-208.

Sen CK, Roy S, Packer L. 1999. Flow cytometric determination of cellular thiols. Methods Enzymol 299: 247-258.

Shapiro HM. 2000. Membrane potential estimation by flow cytometry. Methods 21: 271-279.

Shapiro HM, Perlmutter NG. 2001. Violet laser diodes as light sources for cytometry. Cytometry 44: 133-136.

Shapiro HM. 2003. Practical flow cytometry, Fourth edition, New York: Wiley-Liss.

Stein RA, Wilkinson JC, Guyer CA, Staros JV. 2001. An analytical approach to the measurement of equilibrium binding constants: application to EGF binding to EGF receptors in intact cells measured by flow cytometry. Biochemistry 40: 6142-6154.

Telford WG, Hawley TS, Hawley RG. 2003. Analysis of violet-excited fluorochromes by flow cytometry using a violet laser diode. Cytometry 54A: 48-55.

Toba K, Kishi J, Koike T, Winton EF, Takahashi H, et al. 1996. Profile of cell cycle in hematopoietic malignancy by DNA/RNA quantitation using 7AAD/PY. Exp Hematol 24: 894-901.

Waller A, Pipdorn D, Sutton KL, Linderman JJ, Omann GM. 2001. Validation of flow cytometric competitive binding protocols and characterization of fluorescently labelled ligands. Cytometry 45: 102-114.

Wersto RP, Chrest FJ, Leary JF, Morris C, Stetler-Stevenson MA, et al. 2001. Doublet discrimination in DNA cell-cycle analysis. Cytometry 46: 296-306.

Wiley JS, Cebon JS, Jamieson GP, Szer J, Gibson J, et al. 1994. Assessment of proliferative responses to granulocyte-macrophage colony-stimulating factor (GM-CSF) in acute myeloid leukaemia using a fluorescent ligand for the nucleoside transporter. Leukemia 8: 181-185.

14 Expression and Study of Ligand-Gated Ion Channels in *Xenopus laevis* Oocytes

A. Kapur · J. M. C. Derry · R. S. Hansen

Abstract: The study of neurotransmitter ion channels has been greatly facilitated by use of the *Xenopus laevis* oocyte expression system. The reliable expression of exogenous receptors in the *Xenopus* oocyte, including those belonging to the ligand-gated ion channel family, has provided a powerful means for investigating structure-function relationships in these receptors using a variety of methods such as two-electrode voltage clamp electrophysiology, patch-clamp studies, and radioligand binding assays. The following chapter outlines morphology of *Xenopus laevis* frogs and oocytes, and reviews the advantages and limitations of using *Xenopus* oocytes to study exogenously expressed proteins. Techniques for oocyte isolation and preparation, *in vitro* preparation and injection of RNA transcripts, and subsequent functional analysis of expressed ligand-gated ion channel receptors which include the nAChR, the $GABA_A$ receptor, the 5-HT_3 and glycine receptors, are discussed in detail. The use of the two-electrode voltage clamp setup to characterize whole-cell receptor currents, and the types of concentration-effect experiments that can be performed using this arrangement, are reviewed. Additional uses of *Xenopus* oocytes in characterizing the structure-activity relationship of these receptors are also explored.

List of Abbreviations: RNA, ribonucleic acid; nAChR, nicotinic acetylcholine receptor; $GABA_A$, γ-amino-butyric type A; 5-HT_3, 5-hydroxytryptamine type 3

1 Introduction

Oocytes from the South African clawed frog, *Xenopus laevis,* provide a reliable and powerful system for the transient heterologous expression of proteins. The use of this expression system has become very popular as oocytes have a high translational capacity and they are able to express multi-subunit proteins derived from exogenously introduced RNA or DNA. Furthermore, the expressed receptors frequently appear to be correctly assembled, post-translationally modified and oriented to the appropriate site. The relative scarcity of endogenous ion channels in the oocyte membrane makes it a versatile tool for the study of a range of heterologously expressed ion channel proteins.

Gurdon and colleagues (1971) first demonstrated the ability of *Xenopus* oocytes to synthesize the globin protein from foreign globin mRNA. The first of the ligand-gated ion channel (LGIC) superfamily to be successfully expressed in *Xenopus* oocytes was the nicotinic acetylcholine receptor (nAChR) (Sumikawa et al., 1981). Since then, the system has been used to study many different ion channels and receptors. The large size of the *Xenopus* oocyte (~1.2 mm) makes it amenable to electrophysiological analysis and, in some cases, recombinant receptors expressed in oocytes have also proved to be useful in radioligand binding assays. The oocyte expression system has been extensively used in the study of structure–function relationships of membrane proteins whereby the properties of a mutant receptor (injecting cRNA transcribed from a site-directed mutagenesis modified cDNA) are compared with the wild-type receptor protein. Thus oocytes have become a model for studying exogenously expressed ion channels that have no natural role in the development and working of the oocyte. In this chapter, we discuss the morphology of *Xenopus* oocytes, the expression and study of neurotransmitter ion channels, and how this system has helped to provide a better understanding of the LGIC family which includes the nACh, $GABA_A$, 5-HT_3, and glycine receptors.

2 *Xenopus* Frogs and Oocytes

The *Xenopus laevis* frog is native to South Africa but the frogs can be purchased from several commercial suppliers across the globe (e.g. Nasco®, Fort Atkinson, WI). Four subspecies of *Xenopus laevis* are known, namely, *X. laevis laevis, X. laevis borealis, X. laevis peteresi,* and *X. laevis victorianus* (Goldin, 1992). The oocyte is the germ cell or gamete in the ovary of the frog. It is in the meiotic arrest prophase stage (Dascal, 1987). During the process of oogenesis, significant amounts of proteins, enzymes and organelles are acquired and accumulated; these serve as the maternal reserve for use during early embryonic development. This maternal reserve (histones, nucleoplasmin, RNA polymerases, tRNAs, and ribosomes) makes the oocytes an attractive system for studying transcription, replication, assembly, and translation of injected foreign cRNA (Colman, 1984).

2.1 Morphology

The oocytes have six different stages of development, I to VI (Dumont, 1972), and the ovaries contain a mixture of these. The last two stages (V and VI) are large fully grown cells (1.0–1.3 mm in diameter). Oocytes from stages I to III are small (0.5–0.6 mm) and there is no clear differentiation between the nucleus-containing animal pole (which is dark brown due to melanin-containing pigment granules) and the light-colored vegetal pole (yellowish). Stage IV shows a clear demarcation between the two poles, although the vegetal pole still has some melanin pigment. The oocytes from stage V and VI, which are usually selected for expression studies, are well defined by an unpigmented equatorial band separating the two poles and are clearly distinguishable from the smaller, immature oocytes (Goldin, 1992; Yao et al., 2000).

Several layers of membranes surround the fully mature oocyte. From inside out, these are the vitelline membrane, the follicular layer, and the connective tissue layer (Dascal, 1987; Stühmer and Parekh, 1995). To study channels and receptors expressed in oocytes, it is imperative to remove these surrounding structures. This will not only provide better access of solutions to the oocytes, but will also assist in impaling the cells with low-resistance (large diameter) electrodes for use in electrophysiological experiments. Further, the follicular layer forms gap junctions with the oocyte membrane that allows passage of molecules such as Ca^{2+} and second messengers. In addition, follicular cells contain certain receptors (e.g., angiotensin II) and channels (e.g., delayed-rectifier K^+ type) that may interfere in the analysis and interpretation of results (Stühmer and Parekh, 1995). When viewed under a microscope, tiny blood vessels are sometime visible in the vegetal pole, an indication that the follicular layer is still covering the oocyte (Quick and Lester, 1994). All outer layers except the vitelline membrane can be removed by collagenase (described later) or similar proteolytic treatment.

During maturation of the oocyte, its surface area increases significantly resulting in a change of its resistance and capacitance. This change is attributed to the appearance of microvilli. The specific capacitance increases from $\sim$2 $\mu F/cm^2$ in stage I to $\sim$6–7 $\mu F/cm^2$ in stage IV and V and reduces to $\sim$4 $\mu F/cm^2$ in stage VI oocytes (Dascal 1987). Also, there is a high density of the endogenous Ca^{2+}-activated Cl^- channel in the animal pole of a mature oocyte, as shown to be activated by inositol 1,4,5-triphosphate (IP_3)-induced release of Ca^{2+} stores (Lupu-Meiri et al., 1988). These factors have been suggested to be of some concern for the use of stage VI oocytes (Stühmer and Parekh, 1995); however, in our studies of members of the LGIC family, oocytes from both stage V and VI show robust receptor expression and are readily amenable to two-electrode voltage clamp studies.

3 Heterologous Expression in *Xenopus* Oocytes

Xenopus oocytes have proved very useful for the study of exogenously expressed proteins, especially those that are expressed on the cell surface. These include members of the LGIC family (nAChR, $GABA_A$ and $5\text{-}HT_{3A}$ receptors), which we study in our laboratory. Receptors in their native membrane environment can also be reconstituted in the oocyte membrane by directly injecting receptor-enriched membrane fragments into the oocyte. This method, which was originally described for the membrane-bound *Torpedo* nAChR (Marshal et al., 1995), excludes *de novo* protein synthesis and has the advantage of reconstituting low abundance native receptors in a system that facilitates their electrophysiological analysis.

3.1 Pros and Cons of the *Xenopus* Oocyte Expression System

3.1.1 The Advantages of Using Oocytes as a Heterologous Expression System that Make it a Favorable Model are Numerous and Well Accepted

1. The translation of exogenous RNA occurs in a normal living cell, and is therefore devoid of the artifacts that might be associated with a cell-free (*in vitro*) system.

2. The ovaries of a fully mature *Xenopus* frog contain upward of 10,000 oocytes. Hundreds of viable oocytes can be isolated from a donor frog and the same frog can be reused.
3. *Xenopus* oocytes show little species specificity for the type of foreign RNA or DNA they can finally translate into membrane proteins.
4. Oocytes have a low level of background endogenous channels.
5. The robustness and large size (diameter ~ 1.2–1.3 mm) of the oocytes make them tolerant to repeated impalement of microelectrode and injection pipettes, permitting functional analysis by a number of electrophysiological techniques.
6. Although the expression in oocytes is transient, we have successfully used the cells for up to 14 days for electrophysiological studies.
7. Foreign proteins are correctly post-translationally modified (e.g., by glycosylation and phosphorylation), and oocytes correctly orient multi-subunit proteins that acquire native activity.

3.1.2 The *Xenopus* Oocyte System has Some Limitations as Well

1. Oocytes exhibit seasonal variations affecting the levels of protein expression. We have experienced such difficulties, especially in the summer months.
2. The physiological functions of oocytes are optimal at 16–22°C. The biochemical properties of mammalian proteins studied in the system may therefore be questioned.
3. The properties of the lipids in oocyte membranes are different from those seen in mammalian cells and this may affect the functional properties of the expressed proteins.
4. Expression is transient and there can be significant variation in receptor protein levels seen within (and across) different batches of oocytes.
5. Since oocytes are a transient, small-scale expression system, they are not the best suited for radioligand binding studies as (usually) hundreds of oocytes must be injected to harvest sufficient membranes for such measurements. This is both time-consuming and expensive. However, with high affinity and highly specific activity ligands, the expression of binding sites on the membranes of intact oocytes can be readily quantified (see later).

3.2 Endogenous Channels in Oocytes

One limitation of the *Xenopus* oocyte expression system is the presence of endogenous channels. Although these are few in number, they can interfere with the analysis and interpretation of small currents mediated by exogenously expressed ion channels. Such interference can be identified by carrying out control studies of un-injected or water-injected oocytes in parallel with experiments on oocytes expressing the receptor of interest.

The predominant endogenous channel present in oocytes is a Ca^{2+}-activated Cl^- channel. Depolarization of voltage-clamped *Xenopus* oocytes results in an outward Cl^- current that is dependent upon calcium entry (Miledi, 1982). An intracellular injection of Ca^{2+} elicits a large ~4.1 μA outward chloride current (Miledi and Parker, 1984). A voltage jump from −100 to +30 mV produces 1.5 μA of this type of current (White and Aylwin, 1990). If the presence of these channels poses a problem, one should consider injecting Ca^{2+} chelators such as EGTA (Miledi and Parker, 1984) or nonsteroidal anti-inflammatory agents such as niflumic and flumenanic acid, for instance, procedures that have been shown to abolish these currents (White and Aylwin, 1990). However, these compounds have been shown to also affect $GABA_A$ receptor-mediated currents (Woodward, et al., 1994). Thus the appropriate conditions must be verified for each receptor of interest.

The oocyte membrane has been shown to contain muscarinic acetylcholine receptors (Kusano et al., 1982; Barnard et al., 1982). In naive oocytes, acetylcholine causes a concentration-dependent depolarization of the membrane, which is blocked by atropine but not by curare or α-bungarotoxin. In our laboratory, we routinely add 1 μM atropine to any buffer used in studying *Torpedo* nAChR expressed in *Xenopus* oocytes in order to block any muscarinic AChRs that are present. Other examples of endogenous channels present in *Xenopus* oocytes include voltage-dependent Ca^{2+} channels, slow voltage dependent Na^{+} channels, and a delayed rectifier K^{+} current. The total current carried by other endogenous channels is small and does not generally interfere with the currents seen with exogenous channels. It is beyond the scope of this chapter to discuss this aspect in detail and the authors suggest considering other sources including the extensive review by Dascal (1987).

3.3 Methods for Studying *Xenopus* Oocyte Expression

The following techniques are commonly used to investigate heterologously expressed proteins in oocytes: (1) whole cell current analysis by two-electrode voltage clamp, (2) single channel properties by patch-clamp studies, and (3) radioligand binding assays on oocyte membranes or intact oocytes. Two-electrode voltage clamp is the most routinely employed technique (see later). Patch clamping involves more labor since devitellinized oocytes have to be used. The vitelline membrane surrounding the oocyte will obstruct the formation of a high resistance seal on to the oocyte membrane and therefore must be removed manually prior to clamping. Single-channel analysis with the patch-clamp technique is not the focus of this review and hence is not discussed.

3.4 Precautions for Electrophysiological Studies

The defolliculation of oocytes can result in a change of the resting membrane potential (Wallace and Steinhardt, 1977) as a consequence of damage to the membrane and this may take several hours to heal. Healthy defolliculated oocytes have been reported to have a resting membrane potential between −45 to −60 mV (Dascal et al., 1984). When injecting oocytes, one should avoid using freshly defolliculated cells. Another area of concern is the trauma caused to the oocyte membrane while performing electrophysiological studies. Subsequent to electrode penetration, oocytes undergo hyperpolarization (Kusano et al., 1982) and gradually recover. This can result in an underestimation of the resting membrane potential and membrane resistance. It is therefore advisable to wait several minutes after clamping an oocyte prior to the study of ion channels.

4 *Xenopus* Frog Handling and Oocyte Isolation

Numerous protocols describing the care and maintenance of *Xenopus laevis* frogs and the isolation of oocytes from their ovaries have been published (Colman, 1984; Goldin, 1992; Quick and Lester, 1994; Stühmer and Parekh, 1995; Theodoulou and Miller, 1995). The following methods are based primarily on the procedures described by Yao and colleagues (2000).

4.1 Maintenance of *Xenopus* Frogs

Housing environment for the frogs should ideally constitute a temperature-controlled setting at 16–18°C, and an alternating 12-h light-dark cycle (Yao et al., 2000). The containers in which the frogs are kept can be fairly simple structures. The animals are aquatic and will die of dehydration in the absence of water; however, they must be able to reach the surface in order to breathe (Colman, 1984). Covered tanks containing 3 L of water per frog with a depth of 15–25 cm is sufficient (Yao et al., 2000). As the frogs are sensitive to chlorine and chloramine, it is important to ensure that the water in the tanks is free of these reagents (Goldin, 1992; Stühmer and Parekh, 1995). A number of different feeding protocols exist with regard to the laboratory breeding of *X. laevis*. A feeding schedule of 0.2 g of pellet diet (Tetra, Blacksburg,

VA) per frog once a day is adequate (Yao *et al.*, 2000). An additional consideration may be the need to induce ovulation by injection of human chorionic gonadotropin hormone (hCG, see Goldin, 1992).

In the event that accommodation and care for the frogs is not possible, *X. laevis* surgically extracted ovaries containing oocytes can be ordered through Nasco® (Fort Atkinson, WI); however the quantity and quality of the oocytes cannot be guaranteed upon shipping.

4.2 Surgery

Small sections of ovary (containing the oocytes to be isolated) can be removed surgically from the frog without sacrificing its life. As such, the same frog can be used multiple times; however, with subsequent surgeries, oocyte quality may deteriorate (Goldin, 1992; Yao et al., 2000).

For ovary removal where the frog is to be kept alive, an anesthetic reagent is used. A combination of cold water treatment to slow down frog metabolism, followed by exposure to anesthetic is effective (Quick and Lester, 1994; Stühmer and Parekh, 1995). MS-222 (tricaine, methane sulfonate salt of 3-aminobenzoic acid ethyl ester; Sigma, Oakville, ON, Cat. # A 5040) at a 0.15–0.20% concentration is the most common choice of anesthetic; 15–30 min of exposure is usually sufficient for effect (Goldin, 1992; Quick and Lester, 1994; Stühmer and Parekh, 1995). The extent of anesthesia is measured by assessing the loss of righting reflex (Goldin, 1992; Stühmer and Parekh, 1995). For larger-scale experiments, the frog can be sacrificed to obtain a greater supply of oocytes. The animal is placed directly in ice water containing anesthetic for 15 min, stumped, and then pithed (Theodoulou and Miller, 1995; Yao et al., 2000).

A small, 1–2 cm abdominal incision is made through the skin and the underlying muscle layer. A section of, or the entire, ovarian lobe is removed using sterile forceps and is placed into a sterile dish containing Modified Barth's Medium (MBM; 88 mM NaCl, 1 mM KCl, 0.33 mM $Ca(NO_3)_2$, 0.41 mM $CaCl_2$, 0.82 mM $MgSO_4$, 2.4 mM $NaHCO_3$, 2.5 mM sodium pyruvate, 0.1 mg/mL penicillin, 0.05 mg/mL gentamicin sulfate, and 10 mM HEPES, pH 7.5, filter-sterilized). For frogs under anesthetic, the remaining ovary is replaced into the abdomen and the incision site closed using a few sutures, preferably at both the muscle and the skin layer to minimize the chance of infection (Colman, 1984; Stühmer and Parekh, 1995). As *X. laevis* are sensitive to antibiotics and secrete magainin (protective antimicrobial peptides) from the skin, surgery conducted in clean but not necessarily sterile conditions is adequate (Goldin, 1992).

4.3 Isolation and Preparation of Oocytes

The collected oocytes need to be individually isolated from the connective ovarian tissue and the layer of follicular cells that surround each oocyte (Goldin, 1992). For a detailed description, see Yao et al. (2000). Briefly, the excised ovarian lobe is washed in MBM and dissected using sterile forceps into smaller sections containing approximately 5–10 oocytes each. The sectioned clumps of oocytes are treated with MBM supplemented with 2 mg/mL type I collagenase (Worthington, Lakewood, NJ) for ~2 h at room temperature with constant shaking to separate the oocytes from the follicular cell layer. The oocytes are subsequently washed five times with a BSA solution (bovine serum albumin, 1 mg/mL in MBM), followed immediately by five wash cycles in MBM.

Collagenase over-treatment can affect oocyte viability (Goldin, 1992). For isolation of a small number of oocytes, manual defolliculation can be performed as an alternative. The individual oocytes are separated with care, and the follicular cell layer removed, using forceps or microscissors (Colman, 1984; Goldin, 1992; Stühmer and Parekh, 1995). For large-scale isolation using collagenase, as described earlier, the oocytes should be carefully monitored to prevent over-treatment. In addition, treatment carried out in Ca^{2+}-free medium can minimize the activation of proteases and prevent resultant damage to the oocytes (Goldin, 1992; Stühmer and Parekh, 1995).

For ligand-gated ion channel receptor expression studies, stage V and VI oocytes are preferred (see earlier). Under a dissecting microscope, oocytes of this stage that are in good condition are sorted from the rest and then incubated at 16–18°C in MBM overnight. Incubation in hypertonic phosphate solution (1 mg/mL BSA, 100 mM K_2PO_4, pH 6.5) for 1 h with constant shaking is subsequently performed to remove any remaining follicular cell layer. The oocytes are then washed five times each first in BSA solution and then in MBM. Oocytes in good condition are once again selected, and those chosen for microinjection are incubated in MBM at 16–18°C for several hours before proceeding.

The isolated oocytes are still surrounded by the vitelline membrane, an inner glycoprotein matrix layer (Yao et al., 2000). This layer does not interfere with whole-cell recording experiments. It can be removed manually however, for patch-clamp experiments requiring the formation of high-resistance seals.

5 Nucleic Acid Injection into Oocytes

The heterologous expression of proteins, including ligand-gated ion channels, is most commonly achieved by injection of cDNA or cRNA encoding the protein of interest into the oocyte.

5.1 Different Injection Techniques

1. Injection of recombinant vaccinia virus carrying cDNA of the protein of interest driven by an early promoter has proven successful (Yang et al., 1991). This method precludes the need to synthesize cRNA *in vitro.*
2. More recently, coinjection of T7- or SP6-driven cDNA and T7- or SP6-RNA polymerase, respectively, into the cytoplasm achieved high levels of expression for some proteins, and may serve as an alternative to other techniques (Geib et al., 2001).
3. Liposomes reconstituted with exogenous plasma membrane proteins can be microinjected into the cytoplasm, resulting in the expression of functional proteins in the oocyte membrane (Le Caherec et al., 1996).
4. Direct injection of cDNA into the oocyte nucleus can lead to high levels of expression (note that this technique may require centrifugation of the oocytes to orient the nuclei as injection targets; see Goldin, 1992).
5. Cytoplasmic injection of RNA is the most widely used approach. Examples of the types of RNA that are used include mRNA isolated from tissue and cRNA synthesized by *in vitro* transcription.

5.2 Benefits of Cytoplasmic cRNA Injection

High levels of protein expression are generally achieved with cRNA injection. This technique requires the *in vitro* synthesis of the appropriate cRNA from the template cDNA (see later). Although this approach can be time-consuming and costly, it remains the most common technique used to ensure the robust expression of receptors and ion channels in the oocyte membrane. In our laboratory, we routinely use the cRNA injection technique to promote high expression levels of LGIC receptors ($GABA_A$ receptor subtypes, nACh and $5\text{-}HT_3$ receptors).

6 *In Vitro* cRNA Preparation

A prerequisite for cRNA transcript synthesis is the presence of the cDNA of interest within a suitable expression vector containing a bacteriophage polymerase promoter (e.g. T3, T7 or SP6). The vector must be linearized downstream of the cDNA insert prior to its use as the template in the *in vitro* reaction.

6.1 DNA Template Linearization

A restriction digestion is set up using the target plasmid DNA template and an appropriate restriction enzyme. Enzymes are available through numerous companies including Invitrogen (Burlington, ON), New England

Biolabs (Pickering, ON), and Promega (Madison, WI). It is important to note that restriction enzymes producing 3′-overhangs should be avoided if possible (e.g., *Pst*I, *Sfi* I, *Kpn*I). The use of such enzymes has been reported to result in the production of additional, nonspecific transcripts (Schenborn and Mierendorf, 1985). If these enzymes must be used, an exonuclease such as DNA Polymerase 1 Large (Klenow) Fragment can be utilized to convert the overhang to a blunt end before the template is transcribed.

We set up a restriction digest using 10 μg of DNA in a final reaction volume of approximately 10 μL. The reaction is mixed and incubated in a 37°C water bath for 1 h (temperature and incubation time may vary with type of enzyme used; check supplier's protocol).

6.2 cRNA Synthesis

Subsequent to DNA template linearization, the procedures for *in vitro* RNA transcription reaction are set up. It is essential to avoid RNase contamination by using gloves, sterile glassware, and water devoid of RNase activity [treated with 0.1% diethyl pyrocarbonate (DEPC) and autoclaved].

6.2.1 Components Required for the Transcription Reaction

The reagents required for transcript preparation (*Table 14-1*) are available through numerous companies, including those suppliers listed earlier for DNA template linearization.

Table 14-1
***In vitro* transcription reagents**

5× transcription buffer (1:5 final reaction volume)
60 U RNase Inhibitor
10 mM DTT (dithiothreitol)
0.5 mM 7-methyldiguanosine triphosphate [mG(5′)PPP(5′)G] RNA capping analog
0.5 mM NTP mix (UTP, ATP, GTP, CTP)
5.0 μg linearized DNA template
80–100 U of T3-, T7- or SP6-RNA polymerase
0.1% DEPC-treated water to bring reaction to 50 μL final volume

6.2.2 Method of cRNA Transcript Preparation

1. The reaction is set up on ice by adding first the DEPC-treated water to a 1.5 mL microfuge tube and subsequently all other components listed earlier, except the polymerase.
2. Next, add 40–50 U of the RNA polymerase to the reaction mix. Cap the tube, invert to mix, centrifuge for a few seconds, and incubate in a 37°C water bath for 1 h.
3. Add an additional 40–50 U of RNA polymerase and incubate the reaction for another 1 h in a 37°C water bath.
4. Add 5.0 U of RNase-free DNase to the reaction mix and incubate at 37°C for 15 min. (RNase-free DNase is added to degrade the template DNA following the transcription reaction.)
5. Extract the RNA transcript from solution using 25:24:1 (v/v) phenol/chloroform/isoamyl alcohol (conduct in a fumehood). First, add DEPC-treated water to bring the final volume to 400 μL; subsequently, add 400 μL of phenol/chloroform/isoamyl alcohol to the reaction mix. Briefly vortex the tube and microfuge at 13,000 rpm (~15,000 *g*) for 2 min at room temperature to separate the

aqueous from the organic phase. Collect the upper (aqueous) layer and carefully transfer into another 1.5 mL microfuge tube.

6. Precipitate the RNA transcript by the addition of 1:10 volume of 3M sodium acetate (pH 5.2) and 2.5 volumes of cold 100% ethanol. Mix and place at –80°C for at least 1 h.
7. Spin the tube at 13,000 rpm for 15 min at 4°C (to collect RNA pellet). Carefully discard the supernatant so as not to dislodge the pellet.
8. Add 500 μL of 70% ethanol (in DEPC-treated water) to the tube and spin again for 5 min at 13,000 rpm at room temperature. Allow pellet to air dry for about 5 min. Resuspend in 10 μL of DEPC-treated water.
9. The final concentration of the cRNA transcript can be determined by its absorbance. The absorbance of a 1:1000–2000 dilution of the transcript is read at a wavelength of 260 nm (one unit A_{260} is equivalent to 40 μg/mL of RNA). Alternatively, densitometry on ethidium bromide-stained agarose gels or colorimetric stains (e.g., RiboGreen, Molecular Probes, Inc., Eugene, OR) can be used to estimate RNA concentration.
10. cRNA transcripts are diluted to a concentration of 1 μg/μl in DEPC-treated water and stored at –80°C until required.

6.3 cRNA Injection into *Xenopus laevis* Oocytes

With respect to a multi-subunit LGIC receptor, a common approach is to inject oocytes with an equivalent molar ratio of each transcript. However, it should be noted that the ratio of injected transcripts might have an effect on the final stoichiometry of subunits in the expressed receptors (e.g., Boileau et al., 2002). This may lead to an expression of a heterogeneous population of receptors that can complicate functional analysis.

RNA transcript combinations are prepared for injection by adding a small volume (0.5–1.0 μL) of the cRNA(s) encoding each required subunit to a 1.5 mL microfuge tube. Individual subunit cRNAs are combined and pulse spun for a few seconds and kept on ice until ready for injection. Preparation of a suitable pipette for microinjection of transcripts is imperative to ensure proper injection with minimum damage to the oocyte membrane.

6.3.1 Microinjection Procedure

1. Pull a glass pipette tip (10 μL Drummond Microdispenser 100 replacement tubes, Broomall, PA, Cat. # 3–000–210-G) using a one-stage magnetic puller with a heating element.
2. Score the end of the injecting tip using flame-heated forceps.
3. Fill the glass pipette with mineral oil (dyed with 4-amino-3-nitrotoluene; MP Biomedicals, Irvine, CA, Cat. # 154757) to check that the tip is patent.
4. Attach the glass pipette to a microdispenser (10 μL digital, Drummond, Broomall, PA, Cat. # 3–000–510-X), and mount the microdispenser onto a micromanipulator.
5. Under a microscope, carefully draw the transcript solution droplet (placed on Parafilm) into the microinjector tip. (It is essential to avoid drawing air bubbles into the glass pipette tip, as these will interfere with RNA injection.)
6. Transfer oocytes into a petri dish containing, for example, ND96 buffer (see below) supplemented with gentamicin (see ➲ Section 6.3.2). The dish should have a small piece of mesh attached to the bottom to hold oocytes in place.
7. Use the manipulator to gently puncture the membrane and inject the required volume of transcript solution into the oocyte vegetal pole. We normally inject 20–50 μL to optimize expression levels. Wait a few seconds before slowly withdrawing the tip to minimize leakage of oocyte contents and RNA transcript(s) through the puncture site.
8. Store injected oocytes in 200 μL of ND96 buffer in a 96-well plate and incubate at 14–16°C for 48 h before use.

6.3.2 Buffer Considerations

For transcript injection and oocyte incubation, a suitable buffer is required. ND96 (96 mM NaCl, 2 mM KCl, 2.5 mM sodium pyruvate, 1.8 mM $CaCl_2$, 1 mM $MgCl_2$, and 5 mM HEPES, pH 7.4) is supplemented with 0.1 mg/mL of gentamicin sulfate (Invitrogen, Burlington, ON, Cat. # 15710–064). Antibiotic addition to the buffer medium prevents microbial contamination (Elsner et al., 2000). However, for the study of ligand-gated ion channels, it should be noted that antibiotic-mediated effects have been reported, in particular at nicotinic acetylcholine receptors. Gentamicin has been shown to accelerate the decay of whole cell currents, an effect that cannot be reversed by subsequent switching to a gentamicin-free buffer (Okamoto and Sumikawa, 1991. Nishizaki et al., 1994). In our studies of recombinant nicotinic acetylcholine receptors, we decrease the gentamicin sulfate concentration to 0.05 mg/mL.

7 Expression of Functional LGIC Receptors in *Xenopus* Oocytes

Characterization of the properties of LGIC receptors expressed in *Xenopus* oocytes has been facilitated by the variety of studies that can be carried out on recombinant receptors in this system.

The landmark paper by Sumikawa et al. (1981) established that the multi-subunit nAChR could be expressed in *Xenopus* oocytes. Extracellular, but not intracellelular, application of acetylcholine (ACh) was shown to activate these receptors, suggesting that they were correctly oriented within the membrane. Furthermore, the expressed receptors retained the pharmacological attributes of native *Torpedo* receptors (Barnard et al., 1982). The functional characterization of the other homologous members of the receptor family soon followed. The recombinant expression of the nAChR (Mishina et al., 1985), the $GABA_A$ (Schofield et al., 1987), glycine (Grenningloh et al., 1987) and 5-HT_3 (Maricq et al., 1991) receptors laid the foundation for studying LGICs in the *Xenopus* oocyte system.

Initial characterization of LGIC receptors relied mainly on biochemical techniques including affinity and photoaffinity labeling and radioligand binding to study ligand interaction. In addition, characterization of structure function relationships in LGIC receptors has been furthered by the use of chimeric and mutant proteins. Therefore, receptor expression in *Xenopus* oocytes has proven to be a convenient and useful method for examining specific aspects of both wild-type and mutant receptor functions. The study of receptor currents in oocytes permits analysis of key LGIC features including subunit specificity of ligand interaction, ligand potency, receptor activation, ion channel properties and desensitization.

Oocyte expression of LGIC receptors incorporating mutations has proven useful for examining the functional significance of binding site residues (Sigel et al., 1992) as well as identifying subunit protein domains involved in forming the receptor ion channel (e.g., Akabas and Karlin, 1995). The characterization of novel receptor subunit isoforms (e.g., Davies et al., 1999) as well as an examination of the desensitization properties of whole-cell currents of wild-type and mutant LGIC receptors (e.g., Newell and Dunn, 2002) has also been carried out in oocytes. More recently, expression in *Xenopus* oocytes has been used to examine receptors that incorporate site-specific mutations in an attempt to detect which residues are accessible to the aqueous environment and those which may line receptor binding pockets (e.g., Wagner and Czajkowski, 2001). Additionally, mutant LGICs in oocytes have been used to make predictions about receptor subunit structure (e.g., Holden and Czajkowski, 2002) in comparison to a recently identified molluscan acetylcholine binding protein (Brejc et al., 2001) that bears sequence homology to the extracellular, ligand binding domain of LGIC receptor subunits. Many varied aspects of LGIC function can therefore be observed following their robust expression in *Xenopus* oocytes. Results from these types of studies have contributed to the body of knowledge that shapes our current understanding of LGIC structure and function.

A major benefit of using oocyte expression of these receptors is the ability to select for the specific receptor subtype by injecting the cRNA for a defined combination of subunits or subunit isoforms. As these receptors assemble into anionic- (e.g., $GABA_A$) and cationic- (e.g., nAChR) selective ion channels, a standard two-electrode voltage clamp arrangement is a convenient method of analyzing agonist-evoked currents passing through the expressed receptors in the oocytes.

8 Two-Electrode Voltage Clamp

In order to record and analyze ligand-gated ion channel currents in oocytes as a measure of receptor function, the two-electrode voltage clamp arrangement is used to observe electrical changes that occur upon receptor activation. With the oocyte membrane held steady under voltage clamp, whole-cell currents are measured in response to agonist activation of expressed receptors as current that must be passed to the oocyte to maintain its command holding potential (V_{hold}). This response is taken as the drug-evoked current.

8.1 Preparation of the Perfusion System

In our laboratory, *Xenopus* oocytes are placed into a small-volume bath (~0.5 mL) through which there is a constant flow of perfusion buffer supplied with a gravity-flow drip system (see *Figure 14-1*). A peristaltic pump (e.g., Minipuls 3, Gilson, Middleton, WI) drives the constant flow of buffer from a reservoir to a

Figure 14-1

Schematic diagram illustrating a typical two-electrode voltage clamp setup for the study of currents of ligand-gated ion channel receptors expressed in *Xenopus* oocytes. The oocyte is voltage clamped in the bath and drug-evoked currents are recorded (see text)

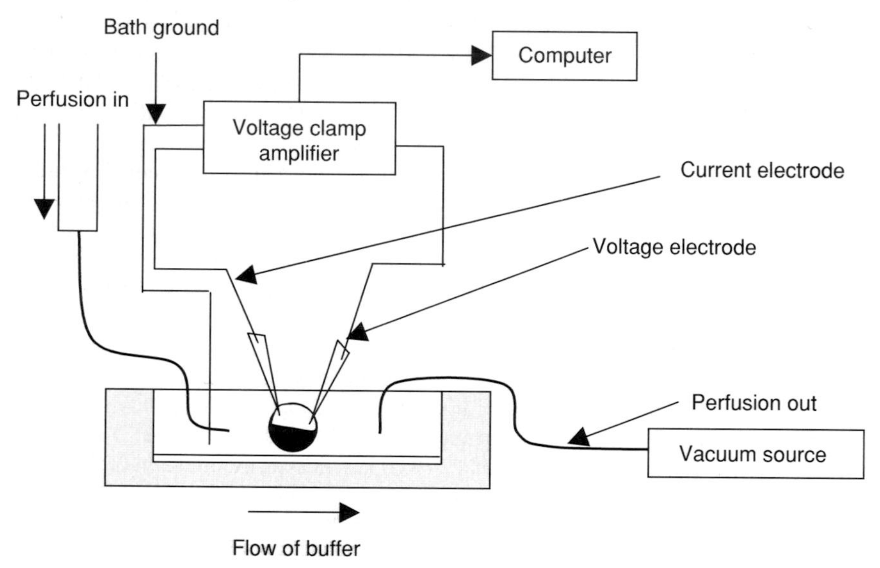

10-mL syringe. Flow of buffer from the syringe into the oocyte bath via a small-bore flexible tube perfuses the oocyte. Buffer is removed from the bath with a small tube connected to a vacuum source. An adequate rate for continuous flow of buffer is ~3–5 mL/min. For the application of drugs to be tested, the pump-driven flow is stopped, and a drug solution is introduced into the syringe to perfuse the bath. Care must be taken not to permit the introduction of air bubbles while switching solutions.

A stable setup for oocyte recording is essential for the proper acquisition of electrical currents. The oocyte bath is placed under a microscope on a vibration isolation table (e.g., Micro-g, Technical Manufacturing Corporation, Peabody, MA) to dampen mechanical vibrations that can lead to a less stable voltage clamp and therefore electrical noise. The bath chamber is a hollowed groove in a small block of Lucite, partially filled with a silicone elastomer (e.g., Sylgard, Dow Corning) to provide a smooth bed surface on which the oocyte can be deposited prior to impalement with the recording electrodes.

8.2 Composition of the Perfusion Buffer

The composition of the perfusion buffer is critical to the recording of ligand-gated ion channel currents. In our laboratory, we have achieved reliable currents from $GABA_A$ and 5-HT_3 receptors with ND96 buffer. This buffer is made as a 20X stock, from which the working strength is prepared fresh daily to a final concentration of 96 mM NaCl, 2 mM KCl, 1.8 mM $CaCl_2$, 1 mM $MgCl_2$ and 5 mM HEPES, pH 7.4. Alternatively, we have also used frog Ringer's solution: 120 mM NaCl, 2 mM KCl, 1.8 mM $CaCl_2$, 5 mM HEPES, pH 7.4. In order to examine *Torpedo* nAChR currents, we use a modified ND96 buffer to reduce the calcium ion concetration to 0.1–0.3 mM (extracellular calcium is known to enhance receptor desensitization; Miledi, 1980) and also supplement the perfusion buffer with 0.3–1.0 μM atropine to block endogenous muscarinic acetylcholine receptor responses (see ➋ Section 3.2).

8.3 Voltage Clamp Setup

With the oocyte in the bath, a voltage clamp amplifier (e.g., GeneClamp 500B, Axon Instruments, Foster City, CA) is used to hold the membrane potential at –60 mV. First, small borosilicate glass capillary tubing (outer diameter: 1.5 mm; inner diameter: 0.86 mm) is pulled into electrodes with a two-stage electrode puller (e.g., Model PP-830, Narishige, Japan). Pulled electrodes are filled with 3 M KCl such that they have a resistance of 0.5–3.0 MΩ in recording buffer. The electrodes are attached to the Ag/AgCl wire in the amplifier headstages (e.g., HS-2A, Axon Instruments) and are mounted on fine manual manipulators to permit the smooth movement of the electrodes to the oocyte membrane. Both the voltage-clamp and the current-passing electrodes are positioned onto the oocyte membrane and slight pressure is applied until the resting membrane potential of the oocyte is detected. The membrane potential of the oocyte is then artificially held at a potential of –60 mV via voltage clamp.

The analog electrical signal from the amplifier is transduced into a digital signal that is then fed into a computer by a 16-bit data acquisition system (e.g., Digidata 1322A, Axon Instruments). In this manner, the electrical response of the oocyte to drug application (a change in the amplifier-driven current that must be supplied to maintain V_{hold}) is digitally detected. The current response to drug over a time interval (i.e., a current trace) can be obtained with an acquisition program (e.g., Axoscope 9.0, Axon Instruments) for the analysis of the trace parameters.

8.4 Experiments on Ligand-Gated Ion Channel Function

A range of experiments can characterize LGICs that are expressed in *Xenopus* oocytes. Application of a drug solution into the perfusion system allows for exposure of the oocyte to that concentration of ligand. In these studies, it is usually assumed that the bath volume is relatively small and that there is minimal dead-space in the flow tubing.

As all biological tissues will have varied responses, it is paramount that the current response measured is stable. That is, repeated application of the same concentration of test drug to the oocyte yields quantitatively similar current amplitudes upon each successive application (e.g., ±5%). An oocyte with a stable response is suitable for use in concentration-dependent experiments to examine the properties of the expressed receptor.

An intrinsic property of LGICs is that they undergo desensitization (see later). These receptors, upon exposure to a sufficiently high concentration of activating ligand, will show an attenuation of the current passed by the receptor, even in the continued presence of the agonist. Recovery from desensitization is not simply achieved upon removal of the agonist since the receptor, once desensitized, must undergo conformational changes to return to a resting state. The time-course for this process should be established for each receptor combination used to determine the refractory period and therefore the time interval needed between subsequent drug applications to allow for an accurate characterization of the full response to the test ligand. In our laboratory, we have found that a washout period of 12–15 minutes between subsequent agonist challenges permits recovery of most subtypes of $GABA_A$ receptors, the *Torpedo* nACh and 5-HT_3 receptors.

8.5 Concentration–Effect Relationships for Agonists

A fundamental experiment that characterizes the function of an expressed receptor is to determine the concentration–effect relationship. By measuring the response of the receptor population to a range of drug concentrations, the potency of a drug can be determined (e.g., *Figure 14-2*). In these experiments, an

Figure 14-2

Concentration–effect curves of agonists at *Torpedo* nAChR expressed in oocytes. Potency of ligands: suberyldicholine (□) > acetylcholine (●) > PTMA (○) (unpublished observations)

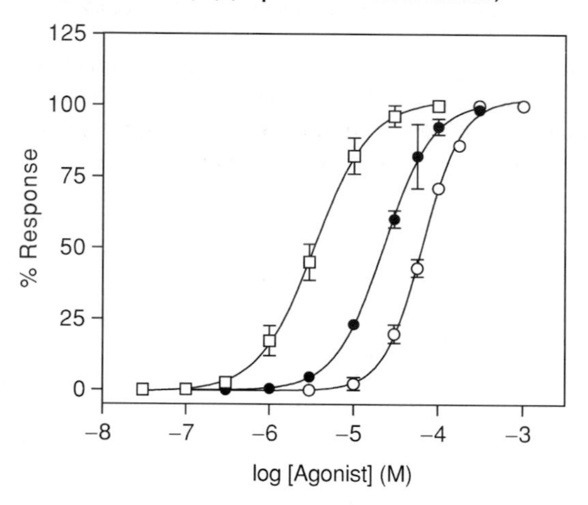

oocyte under voltage-clamp, is exposed to a range of drug concentrations (usually spanning at least four orders of magnitude). Successive drug applications should be applied at intervals that are sufficiently long to ensure that the receptor population has had sufficient time to recover from any desensitization process. The magnitude of currents can be normalized to the effect giving the maximum response. Data can be fit by a three-parameter logistic equation

$$I = \frac{I_{max}{}^{*}[L]^n}{EC_{50}{}^n + [L]^n}$$

where I is the measured agonist-evoked current, $[L]$ is the agonist concentration, EC_{50} is the agonist concentration that evokes half the maximal current (I_{max}) and n is the Hill slope. Data analysis can be performed using software such as GraphPad Prism 4.0 (GraphPad, San Diego, CA). In this manner, the EC_{50} value defines the potency of the drug in activating the expressed receptor. These studies are useful in defining properties of receptor subtypes. Additionally, functional characterization of receptors bearing specific mutations can shed light on the role of the mutated residues in channel activation; these studies can be useful for examination of channel gating phenomena (see Colquhoun, 1998).

8.6 Characterizing Receptor Antagonists and Allosteric Modulators

The effect of antagonists and allosteric modulators can also be characterized at LGIC receptors expressed in *Xenopus* oocytes. The potency of an antagonist (IC_{50}) can be determined by testing a range of concentrations for its ability to block the agonist-evoked current in the oocyte. Additionally, the potency and efficacy of allosteric modulators acting at distinct binding sites (e.g., benzodiazepine compound modulation of $GABA_A$ receptor function; see Sigel and Buhr, 1997) can be observed and characterized.

The effect of these compounds on agonist-evoked current is studied with a pre-perfusion protocol. Oocytes are pre-exposed to a concentration of antagonist or modulator (test drug) followed by coapplication of the same concentration of test drug with agonist. The evoked current in the presence of the test drug is compared to the amplitude of current from agonist alone. The percentage change in the current magnitude can be determined this way, normalized to the response to agonist alone, and the effect of a range of test drug concentrations defines the potency of that drug.

With respect to pre-exposing oocytes to the test compound, sufficient time is required to permit the drug to bind and occupy its binding sites and approach equilibrium prior to agonist exposure. In the case of each test drug and its receptor, it is essential that the time used for pre-perfusion is optimized to ensure that equilibrium has been reached. For studies of antagonist inhibition, we normally use an agonist concentration that is close to its predetermined EC_{50} for channel activation (see *Figure 14-3*). For the characterization of modulators, the efficacy of the test compound dictates the agonist concentration to be used. Submaximal agonist concentrations (e.g., EC_{10-20}) are used for studying positive modulatory compounds whereas higher agonist concentrations (e.g., EC_{50}) are used for negative modulators.

Figure 14-3
Concentration–effect curve of an antagonist (■, d-tubocurarine + agonist) and agonist alone (□, suberyldicholine) at *Torpedo* nAChR expressed in oocytes (unpublished observations)

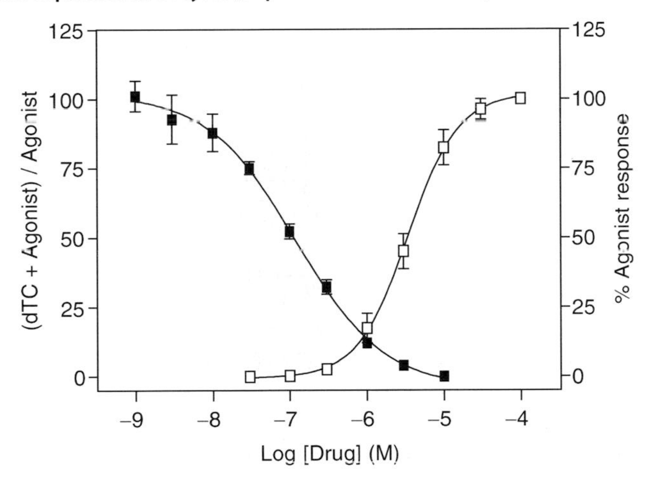

8.7 Studying Ligand-Gated Ion Channel Desensitization

Although receptor activation occurs on millisecond timescales that are too rapid to be readily resolved using two-electrode voltageclamp techniques, the macroscopic current elicited by an agonist can be analyzed to provide valuable information on receptor–ligand interactions. An intrinsic property of desensitization is that it is concentration-dependent and the threshold for the phenomenon can be determined in a manner analogous to characterizing agonist concentration–effect relationships. The kinetics of the conductance changes can be examined during the time of exposure to agonist and any attenuation of current (desensitization) can be readily monitored. Receptor desensitization kinetics can be analyzed using commercially available data analysis software (e.g., Clampfit 8.2, Axon Instruments). Such analysis can be very useful in characterizing the properties of different receptor subtypes and the effect of specific mutations on desensitization processes.

9 Additional Uses of *Xenopus* Oocytes

9.1 Radioligand Binding on Receptors Expressed in *Xenopus* Oocytes

In addition to functional experiments, LGIC receptors can be expressed in *Xenopus* oocytes for biochemical studies. For example, oocytes have been used to express the nAChR for the characterization of radioligands that bind to the receptor. In analogy to binding performed on membranes harvested from cell lines, membranes harvested from oocytes expressing the nAChR have been used for binding studies (e.g., Xie and Cohen, 2001). Furthermore, binding to intact oocytes has been used to study surface receptor expression levels (see ➋ *Figure 14-4*).

Figure 14-4
Correlation of surface receptor binding on intact oocytes (■, ^{125}I-α-bungarotoxin) and maximum ACh response (□) at *Torpedo* nAChR (unpublished observations)

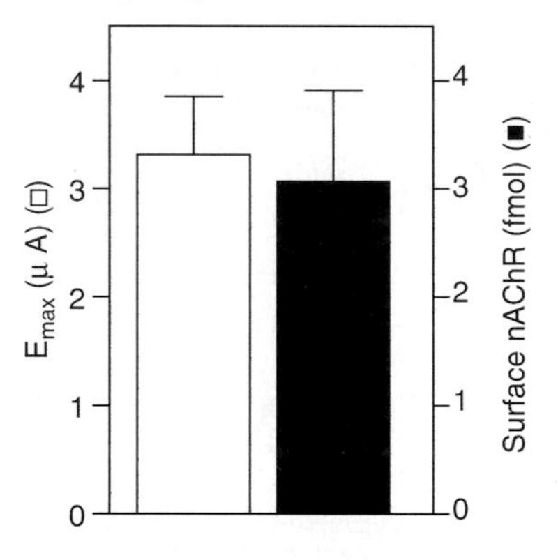

Xenopus oocyte expression has been used in parallel studies of ligand binding and receptor expression in a single population of receptors. The first example of such a study was the expression of the ρ1 subtype of the $GABA_C$ receptor in a single oocyte that permitted the parallel examination of both the binding parameters and functional properties of receptor ligands (Chang and Weiss, 1999). This was later extended to attempts to correlate ligand apparent affinities and efficacies at the same discrete population of $GABA_C$ receptors (Chang et al., 2000).

9.2 Novel Uses of *Xenopus* Oocytes

A few recent technical advances have capitalized on the reliable LGIC expression in the *Xenopus* system to study receptor function. In one such study, $GABA_C$ receptors that had been covalently tagged with a fluorescent group were expressed in oocytes allowing the parallel monitoring of agonist-induced conductance changes with measurements of fluorescence signals that monitored transitions between receptor conformational states (Chang and Weiss, 2002). Studies of this type will be essential for our improved understanding of the relationship between receptor structure and function.

The use of large-scale automated perfusion systems has allowed for several oocytes to be tested in parallel and presents an opportunity for high-throughput drug screening. For example, the Robooctye system (see

Schnizler et al., 2003) permits the automation of both cRNA injection and two-electrode voltage clamp recordings from multiple oocytes in standard 96-well plates. Although the cost and maintenance of such an apparatus is likely to be restricted to industrial research environments, other automated oocyte perfusion systems have been described that would be suitable for smaller scale laboratories (e.g., Joshi et al., 2004).

10 Conclusions

The *Xenopus* oocyte can reliably express LGIC receptors. In our laboratory, we have seen robust expression of the *Torpedo* nAChR, 5-HT_3 receptors and various $GABA_A$ receptor subtypes in oocytes. Injection of cRNA transcripts is a convenient and reproducible way to achieve the expression levels needed for functional analysis of receptor subtypes. We have found that functional characterization with this system complements biochemical experiments conducted on native receptors or those that have been expressed in mammalian cells. A combination of these approaches is essential for furthering our understanding of structure–function relationships in these receptors.

References

Akabas MH, Karlin A. 1995. Identification of acetylcholine receptor channel-lining residues in the M1 segment of the alpha-subunit. Biochemistry 34 (39): 12496.

Barnard EA, Miledi R, Sumikawa K. 1982. Translational of exogenous messenger RNA coding for nicotinic acetylcholine receptors produces functional receptors in *Xenopus* oocytes. Proc R Soc Lond B 215: 241.

Boileau AJ, Baur R, Sharkey LM, Sigel E, Czajkowski C. 2002. The relative amount of cRNA coding for $\gamma 2$ subunits affects stimulation by benzodiazepines in $GABA_A$ receptors expressed in *Xenopus* oocytes. Neuropharmacology 43: 695.

Brejc K, van Dijk WJ, Klaassen RV, Schuurmans M, van Der Oost J, et al. 2001. Crystal structure of an Ach-binding protein reveals the ligand-binding domain of nicotinic receptors. Nature 411: 269.

Chang Y, Weiss DS. 1999. Channel opening locks agonist onto the $GABA_C$ receptor. Nat Neurosci 2 (3): 219.

Chang Y, Weiss DS. 2002. Site-specific fluorescence reveals distinct structural changes with GABA receptor activation and antagonism. Nat Neurosci 5 (11): 1163.

Chang Y, Covey DF, Weiss DS. 2000. Correlation of the apparent affinities and efficacies of γ–aminobutyric $acid_C$ receptor agonists. Mol Pharmacol 58 (6): 1375.

Colman A. 1984. Translation of eukaryotic messenger RNA in *Xenopus* oocytes. Transcription and translation: a practical approach. Rickwood D, Hames BD, editors. Washington, DC: Oxford University Press; pp. 271–302.

Colquhoun D. 1998. Binding, gating, affinity and efficacy: the interpretation of structure–activity relationships for agonists and of the effects of mutating receptors. Br J Pharmacol 125 (5): 924.

Dascal N. 1987. The use of *Xenopus* oocytes for the study of ion channels. CRC Crit Rev Biochem 22 (4): 317.

Dascal N, Landau EM, Lass Y. 1984. *Xenopus* oocyte resting potential, muscarinic responses and the role of calcium and cyclic GMP. J Physiol 352: 551.

Davies PA, Pistis M, Hanna MC, Peters JA, Lambert JJ, et al. 1999. The 5-HT3B subunit is a major determinant of serotonin receptor function. Nature 397 (6717): 359.

Dumont JN. 1972. Oogenesis in *Xenopus laevis* (Daudin). I. Stages of oocyte development in laboratory maintained animals. J Morphol 136: 153.

Elsner HA, Honck HH, Willmann F, Kreienkamp HJ, Iglauer F. 2000. Poor quality of oocytes from *Xenopus laevis* used in laboratory experiments: prevention by use of antiseptic surgical technique and antibiotic supplementation. Comp Med 50 (2): 206.

Geib S, Sandoz G, Carlier E, Cornet V, Cheynet-Sauvion V, De Waard M. 2001. A novel *Xenopus* oocyte expression system based on cytoplasmic coinjection of T7-driven plasmids and purified T7-RNA polymerase. Receptors Channels 7 (5): 331.

Goldin AL. 1992. Maintenance of *Xenopus laevis* and oocyte injection. Methods in Enzmology, Vol. 207. Ion Channels. Rudy B, Iverson LE, editors. London: Academic Press; pp. 266–279.

Grenningloh G, Rienitz A, Schmitt B, Methfessel C, Zensen M, et al. 1987. The strychnine-binding subunit of the glycine receptor shows homology with nicotinic acetylcholine receptors. Nature 328: 215.

Gurdon JB, Lane CD, Woodland HR, Marbaix G. 1971. Use of frog eggs and oocytes for the study of messenger RNA and its translation in living cells. Nature. 233: 177.

Holden JH, Czajkowski C. 2002. Different residues in the GABA(A) receptor alpha 1T60–alpha K70 mediate GABA and SR-95531 actions. J Biol Chem 277 (21): 18785.

Joshi PR, Suryanarayanan A, Schulte MK. 2004. A vertical flow chamber *for Xenopus* oocyte electrophysiology and automated drug screening. J Neurosci Methods 132: 69.

Kusano K, Miledi R, Stinnakre J. 1982. Cholinergic and catecholaminergic receptors in the *Xenopus* oocyte membrane. J Physiol 328: 143.

Le Caherec F, Bron P, Verbavatz JM, Garret A, Morel G, et al. 1996. Incorporation of proteins into (*Xenopus*) oocytes by proteoliposome microinjection: functional characterization of a novel aquaporin. J Cell Sci 109 (Pt 6): 1285.

Lupu-Meiri M, Shapira H, Oron Y. 1988. Hemispheric asymmetry of rapid chloride responses to inositol triphosphate and calcium in *Xenopus* oocytes. FEBS Lett 240 (1,2): 83.

Maricq AV, Peterson AS, Brake AJ, Myers RM, Julius D. 1991. Primary structure and functional expression of the 5HT3 receptor, a serotonin-gated ion channel. Science 254: 432.

Marshal J, Tigyi G, Miledi R. 1995. Incorporation of acetylcholine receptors and chloride channels in *Xenopus* oocytes injected with *Torpedo* electroplaque membranes. Proc Natl Acad Sci USA 92: 5224.

Miledi R. 1980. Intracellular calcium and the desensitization of acetylcholine receptors. Proc R Soc Lond B Biol Sci 209: 447.

Miledi R. 1982. A calcium-dependent transient outward current in *Xenopus laevis* oocytes. Proc R Soc Lond B Biol Sci 215: 491.

Miledi R, Parker I. 1984. Chloride current induced by injection of calcium into *Xenopus* oocytes. J Physiol 357: 173.

Mishina M, Tobimatsu T, Imoto K, Tanaka KI, Fujita Y, et al. 1985. Location of functional regions of acetylcholine receptor α-subunit by site-directed mutagenesis. Nature 313: 364.

Newell JG, Dunn SMJ. 2002. Functional consequences of the loss of high affinity agonist binding to gamma-aminobutyric acid type A receptors. Implication for receptor desensitization. J Biol Chem 277 (24): 21423.

Nishizaki T, Morales A, Gehle VM, Sumikawa K. 1994. Differential interactions of gentamicin with mouse junctional and extrajunctional ACh receptors expressed in *Xenopus* oocytes. Brain Res Mol Brain Res 21 (1–2): 99.

Okamoto T, Sumikawa K. 1991. Antibiotics cause changes in the desensitization of ACh receptors expressed in *Xenopus* oocytes. Brain Res Mol Brain Res 9 (1–2): 165.

Quick MW, Lester HA. 1994. Methods for expression of excitability proteins in *Xenopus* oocytes. Methods in Neuroscience. London: Academic Press, Vol. 19; pp. 261-279.

Schenborn ET, Mierendorf RC. 1985. A novel transcription property of SP6 and T7 RNA polymerases: Dependence on template structure. Nucl Acids Res 13: 6223.

Schnizler K, Kuster M, Methfessel C, Fejtl M. 2003. The roboocyte: Automated cDNA/mRNA injection and subsequent TEVC recording on *Xenopus* oocytes in 96-well microtiter plates. Receptors Channels 9: 41.

Schofield PR, Darlison MG, Fujita N, Burt DR, Stephenson FA, et al. 1987. Sequence and functional expression of the GABAA receptor shows a ligand-gated receptor superfamily. Nature 328: 221.

Sigel E, Buhr E. 1997. The benzodiazepine binding site of GABAA receptors. Trends Pharmacol Sci 18 (11): 425.

Sigel E, Baur R, Kellenberger S, Malherbe P. 1992. Point mutations affecting antagonist affinity and agonist dependent gating of GABAA receptor channels. EMBO J 11 (6): 2017.

Stühmer W, Parekh AB. 1995. Electrophysiological recordings from *Xenopus* oocytes. Single-channel recording, Second edition. Sakmann B, Neher E, editors. New York: Plenum Press; pp 341-356.

Sumikawa K, Houghton M, Emtage JS, Richards BM, Barnard EA. 1981. Active multi-subunit ACh receptor assembled by translation of heterologous mRNA in *Xenopus* oocytes. Nature 292: 862.

Theodoulou FL, Miller AJ. 1995. *Xenopus* oocytes as a heterologous expression system. Methods Mol Biol 49: 317.

Wagner DA, Czajkowski C. 2001. Structure and dynamics of the GABA binding pocket: a narrowing cleft that constricts during activation. J Neurosci 21 (1): 67.

Wallace RA, Steinhardt J. 1977. Maturation of *Xenopus* oocytes. II. Observations on membrane potential. Dev Biol 57: 305.

White MM, Aylwin M. 1990. Niflumic and flufenamic acids are potent reversible blockers of Ca^{2+}-activated Cl^- channels in *Xenopus* oocytes. Mol Pharmcol 37: 720.

Woodward RM, Polenzani L, Miledi R. 1994. Effects of fenamates and other nonsteroidal anti-inflammatory drugs on rat brain GABAA receptors expressed in *Xenopus* oocytes. J Pharmacol Exp Ther 268 (2): 806.

Xie Y, Cohen JB. 2001. Contributions of Torpedo nicotinic acetylcholine receptor γTrp-55 and δTrp-57 to agonist and competitive antagonist function. J Biol Chem 276 (4): 2417.

Yang XC, Karschin A, Labarca C, Elroy-Stein O, Moss B, et al. 1991. Expression of ion channels and receptors in *Xenopus* oocytes using vaccinia virus. FASEB J 5 (8): 2209.

Yao SYM, Cass CE, Young JD. 2000. The *Xenopus* oocyte expression system for the cDNA cloning and characterization of plasma membrane transport proteins. In: *Membrane transport: A Practical Approach.* Baldwin SA, editor. Oxford University Press, Oxford, pp. 47-78.

15 Nucleic Acid Quantitation Using the Competitive Polymerase Chain Reaction

J. Auta · Y. Chen · W. B. Ruzicka · D. R. Grayson

Abstract: Since the advent of the polymerase chain reaction as a method for amplifying and detecting DNA and RNA, researchers have attempted to draw quantitative conclusions based on signals that not only vary exponentially but that amplify at different rates depending on the primer and template compositions. For many years, it was clear that because these signals saturate quite readily, attempts at quantification remained elusive. At best, the investigator had to rely on establishing a linear amplification range for the amplicon of interest and then make comparisons to known external standards that were independently shown not to vary with the experimental paradigm. This so-called semi-quantitative approach was not usually quantitative as the assumptions needed to stay within the confines of linearity were largely ignored when comparisons were made to the external standard (which often times was β-actin or one of a group of house-keeping genes). Here we describe a method that allows for absolute quantitation of DNA and/or RNA using internal standards that amplify at the same rate as the target template. Refinements in the use of this competitive PCR approach no longer require radio-isotope and labor intense efforts like cutting bands out of gels, etc. Moreover, the method is applicable to diverse types of experiments from quantitating micro amounts of RNA to determining gene copy number in a transgenic cell lines or animals. The method for making the internal standard template is easy, reproducible and time effective. Moreover, the experimental approach can be readily adapted for making site-directed mutants in any template of interest.

List of Abbreviations: PCR, polymerase chain reaction; RT-PCR, reverse transcription polymerase chain reaction; DNA, deoxyribonucleic acid; RNA, ribonucleic acid; RNase, ribonuclease; mRNA, messenger RNA; $GABA_A$, γ-aminobutyric acid type A; cRNA, copy RNA; dNTPs, deoxy nucleoside triphosphates; MMLV, Mouse Moloney murine leukemia virus; RT, reverse transcriptase; bp, base pair; Tm, melting temperature; DEPC, diethylpyrocarbonate; OD, optical density; mL, milliliter; SA–PMPs, streptavidin paramagnetic particles; dT, deoxy thymidine; DTT, dithiothreitol; DNase, deoxyribonuclease; RNasin, ribonuclease inhibitor; UV, ultraviolet; TBE, Tris-borate, 1 mM EDTA; EDTA, ethylenediaminetetraacetic acid; Buffer RLT, guanidium thiocyanate lysis buffer; PBS, phosphate buffered saline; NT2, Ntera 2 neural progenitor cells

1 Introduction

The polymerase chain reaction (PCR) amplification is an extremely powerful technique for obtaining readily detectable and manipulable amounts of DNA or RNA from starting material that may contain only a few molecules of the desired sequence in a complex mixture of other molecules. The importance and power of this method is apparent in applications such as the diagnosis of inherited diseases, studies of gene expression during development, and forensic medicine. The employment of a heat-stable DNA polymerase isolated from the thermostable *Thermus aquaticus* bacterium was seminal in the development of the technique as we know it today (Saiki et al., 1988). Shortly after its inception, it became clear that a popular application of PCR would be in quantitating mRNA levels by reverse-transcription PCR (RT-PCR). Conventional PCR analysis is plagued by false positives due to contamination and, often, a band with the size of the desired product is counted as the signal, regardless of the intensity. The nature of amplification reactions are such that they are limited by saturation effects that tend to limit the linearity of the response at the upper ends of the cycle numbers. To work around this problem, researchers have employed a variety of approaches to avoid this and other problems associated with using PCR to quantify signal intensity. The two most common approaches include competitive PCR (Grayson et al., 1993; Grayson and Ikonomovic, 1998; Freeman et al., 1999), and more recently, real-time RT-PCR quantitation (Higuchi et al., 1993 reviewed in Ginzinger, 2002; Bustin and Nolan, 2004).

Unlike other methods currently employed for quantitative transcript measurements, including cDNA microarrays and real-time RT-PCR, competitive RT-PCR is amenable to quality control, which is critical for clinical diagnostic and pharmaceutical industry applications. Furthermore, microarray approaches are limited to generating "snap-shot" like profiles, but they do not control for differences in hybridization efficiencies of different gene probes with their corresponding cDNAs. That is, cross comparisons are relative and not absolute. Real-time PCR has gained acceptance recently largely due to the reduced cost associated with the instrumentation needed for the analyses. However, successful application of real-time PCR for quantitation is not trivial and many variables can limit the usefulness of the technology (Ginzinger, 2002). In contrast, quantitative competitive RT-PCR has the capability to achieve hybridization-independent transcript quantification through the incorporation of internal competitors virtually identical to the native cDNA templates in the reaction mixture (Pagliarulo et al., 2004). The competitive RT-PCR approach uses an internal standard molecule that is coamplified with the target template using the same primer pair, and the origin of the amplification products are determined in a postamplification step (Wang et al., 1989; Gilliand et al., 1990a, b; Buck et al., 1991; Bovolin et al., 1992a, b; Grayson et al., 1993). Whereas the same degree of quantitation can be achieved by conventional Northern analyses or by RNase protection assays using appropriate internal standards, the sensitivity accessible by PCR allows the investigator to examine both abundant and rare transcripts by adjusting the concentrations of the internal standard and the number of cycles used.

2 Strategy

There is a need to develop analytical techniques that can be used in the quantification of small variations in transcript expression because many important cellular events are modified or regulated by small changes in the expression of certain genes. The competitive RT-PCR technology provides quantitative information that can reflect altered function with very small amounts of RNA source materials. Therefore, an acceptable sensitivity limit would be the ability to quantify changes in gene expression from a single sample with sufficient resolution to detect differences of less than 10% between two populations, using as little as 20–50 cells per sample. Thus, as an example, one would like to be able to quantitatively compare absolute amounts of mRNA expression for related members of a gene family to determine the dominant isoforms likely contributing to the formation of native receptor assemblies. As mentioned here, semiquantitative measurements of individual mRNAs can be attained using conventional Northern blots, RNase protection assays, and RT-PCR with external standards. Each of these approaches yields reproducible measurements or relative amounts of mRNA. However, accurate comparisons among different mRNAs require the use of known amounts of internal standards that can be coamplified in the same reaction tube using a single pair of primers, which provides for an end point PCR product that is used for absolute quantitation. The desire for determining absolute amounts of mRNAs is predicated by the need to obtain an accurate and reproducible quantitation of mRNA under conditions in which known mRNA is likely to vary or in RNA isolated from different tissue sources (e.g., different brain regions, different neuronal cell types, or cell populations following pharmacological manipulations).

Several years ago, we adopted the use of a competitive PCR based assay using internal standard cRNAs to quantify mRNAs encoding subunits of the $GABA_A$ receptor superfamily (Bovolin et al., 1992a, b; Grayson et al., 1993; Harris et al., 1994). Our interest at that time was to devise an assay that used the sensitivity of PCR and at the same time allowed for a comparison of molar amounts of individual subunit mRNAs. The overall experimental approach is illustrated in ❯ *Figure 15-1*. In brief, we designed and made internal standards corresponding to each of the 14 known rat $GABA_A$ receptor subunit mRNAs using known sequences and introduced a restriction enzyme cleavage site (*Bgl*II) into the center of the target sequence (Bovolin et al., 1992b), as described here. The cRNA is prepared from the linearized internal standard and the amounts are quantitated. Decreasing amounts of cRNA are added to constant amounts of RNA isolated from the tissue of choice and containing the mRNA of interest. The approach allows for the coreverse transcription of both templates (mRNA and cRNA) in the same tube. Both the native mRNA and internal standard cRNA are coamplified using the same primer pair. The template can also be genomic DNA for some applications, in which case the internal standard would also be a DNA molecule and the RT step would be eliminated. Subsequently, the origins of the amplification products are determined in a post-amplification step. This is done either by a size separation of the amplification products or by an added restriction enzyme cleavage step. That is, digestion of the amplification products with *Bgl*II, as illustrated in the internal standard example shown in ❯ *Figure 15-2*, cleaves only those molecules arising from the internal standard cRNA, whereas those arising from the target mRNA are uncut by the enzyme. These two

Figure 15-1
Overview of Competitive RT-PCR. This depicts the general strategies utilized in the competitive PCR reaction to quantitate mRNA levels. The cloned internal standard is used to make cRNA using T7 polymerase and increasing amounts are added to a fixed amount of RNA (or genomic DNA) isolated from the source of interest. The templates are reverse transcribed in the same tube using random hexamers and reverse transcriptase and then subject to the polymerase chain reaction. By performing all steps in the same tubes, internal variations are minimal and affect each template to the same extent. The amplification products are subsequently separated and their origins are determined in a postamplification step. The gels are scanned using any of a number of software programs and the intensities of the bands are used to extrapolate the amount of the mRNA present in the RNA sample isolated

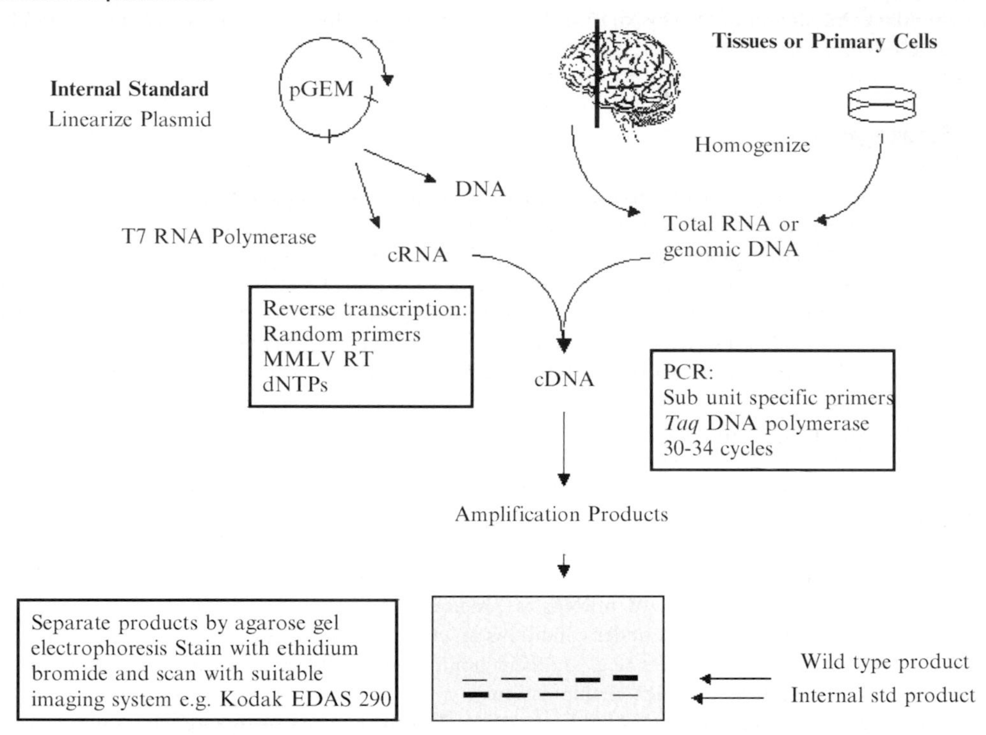

products are easily distinguished and quantitated by adding trace amounts of radioactivity in the PCR reaction and by excising the respective bands from agarose gels following electrophoretic separation.

Another assay for quantitating $GABA_A$ subunit mRNAs that has been described uses degenerate primer pairs to coamplify multiple receptor subunit mRNAs and to differentiate their relative amounts using a post-PCR hybridization step (Buck et al., 1991). Numeric signals are obtained by densitometric scanning of the hybridized blots. Whereas this multiplexing type of approach allows for a comparison of different subunits within the same RNA sample, it is more difficult to compare between samples because of inherent variations in reverse transcriptase efficiencies and sample-to-sample PCR variations. However, a variation of this approach has been used successfully to compare the levels of individual $GABA_A$ receptor subunit mRNAs present in pharmacologically defined individual cerebellar granule cells (Santi et al., 1994).

During recent years, many of our applications of competitive RT-PCR have involved the analysis of RNA from very small tissue samples obtained by laser capture microdissection. We have therefore adopted a modified reamplification protocol (competitive Nested-RT-PCR) for the quantitation of very small tissue samples (Costa et al., 2002). The procedure involves the use of very low concentrations of cRNA (internal standard) in the presence of total RNA isolated from a measured and microdissected tissue sample or a

Figure 15-2
General Design of an Internal Standard. These internal standards were originally designed to amplify individual subunits of the GABA-A receptor subunit family. Each subunit spans the membrane 4 times as indicated (M1-M4). The corresponding mRNA transcript is shown beneath the subunit primary structure. Most proteins are fairly well conserved between species in the coding region (shown in green) and vary considerably in the 5′ and 3′ untranslated regions. The location of the external primers (1 and 4) and the internal primers (2 and 3) are shown. In constructing the internal standard, primers 1 and 3, and also 2 and 4, are used to generate the intermediate species shown as the primary PCR product by overlap extension PCR. The region of overlap is shown in red. PCR with the two templates in equal proportions and the outside primers (1 and 4) result in a amplification product that contains the introduced mutation (either point mutations as shown, or deletion mutants as in *Figure 15-3*). The outside primers (O1 and O2) are then used to amplify the internal standard and the mRNA of interest in parallel. The presence of a restriction enzyme cleavage site or a deletion allows determination of the origin of the amplification products in a postamplification step. This protocol can be used in a nested PCR reaction with additional primers (N1 and N2) as shown. Digestion of the PCR product with the enzyme *Bgl* II will cut the product as indicated, which can be resolved using conventional agarose gel electrophoresis

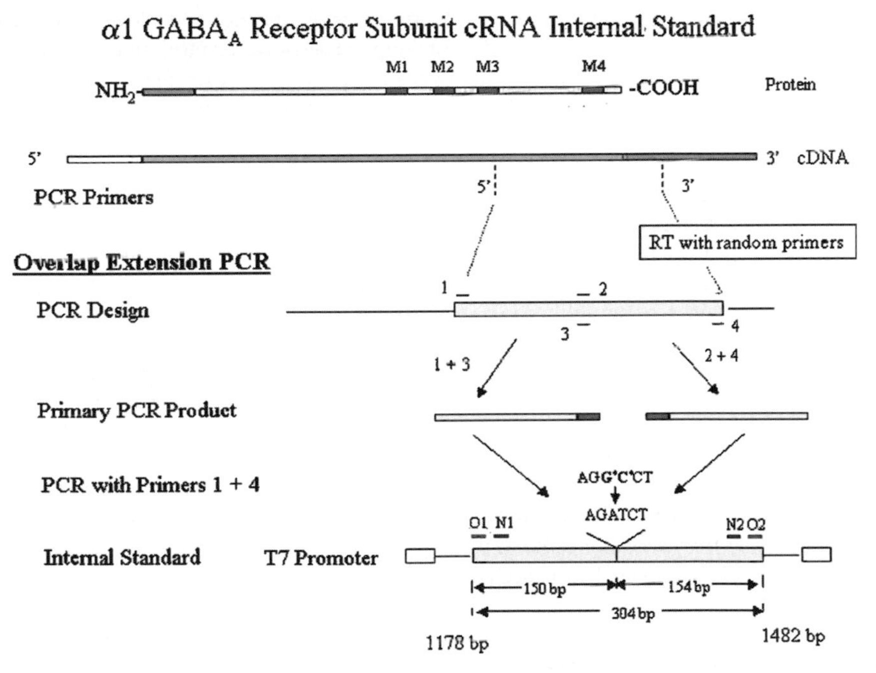

specific number of cell types. Because the samples used in this analysis are obtained by a laser capture microdissection technique, the amount of RNA extracted from them is extremely low. Therefore, we have adopted a modified NIH RNA extraction protocol (described later) by Arcturus (http/www.arctur.com) for RNA extraction. The competitive Nested-RT-PCR consists of two successive amplification protocols (nested PCR). The first amplification protocol is the standard protocol that amplifies the transcript of interest using specific oligonucleotide primers. In the second amplification protocol, the initial amplification products are amplified with a second set of nested primers designed such that the primer pair is interior to the first primer pair (see for example the outside (O1, O2) and inside primers (N1, N2) in *Figure 15-2*). This application is discussed in more detail in Section 5.

Over the past 2 years, we have made further advances in the design of the internal standard used in our competitive RT-PCR measurements. We have adopted a deletion method for the design and synthesis of internal standards corresponding to the transcript to be studied (see a detailed description later). For example, by deleting 150 bp from the Dnmt1 internal standard, we can differentiate the amplicons by size (wild type and the internal standard). By adopting this technique, we have eliminated the need for introducing an enzyme-restriction site midway between the targeted sequences as we have done in the past. This avoids the postamplification enzyme-digestion step that requires additional time and involves added costs. Moreover, the extra enzyme-digestion step introduced some uncertainty into the protocol, as the investigator had to assume complete digestion for accuracy. The deleted internal standard amplifies with the same kinetics as the wild-type cDNA as determined by a series of rate measurements.

3 Design and Construction of Competitive Internal Standards

Quantitative PCR has been widely used to determine the amount (number of molecules) of DNA molecules in a test sample. The best quantitative PCR method involves the addition of known amounts of a similar DNA or RNA fragment, such as one containing a short deletion or specific mutation, to the test sample before amplification. Such internal standards must be precisely calibrated to ensure that they are amplified and detected in a form and manner that are similar to the test sample. The ratio of the internal standard and the targeted template will depend on the amount of internal standard added and allows for the determination of the amount of the targeted molecule in the test sample. Therefore, the ideal standard for quantitative amplification based assays should have a structure that is comparable to the template of interest and which allows for the simultaneous amplification of both template and standard using a single primer pair.

For a competitive internal standard to be accurately used for the quantitative analyses, it is essential that its rate of amplification and that of the target template be identical. For this reason, it is desirable that the internal standard and the targeted template share common features that critically affect their rate of amplification. These factors include primer annealing efficiency, template length, and sequence identity. Among these factors, the primer annealing efficiency appears to be perhaps the most critical. For this reason, it is most often desirable to amplify both templates using a single primer pair and to differentiate the origin of the templates in a postamplification step. Several variations of this approach have been used successfully. If the internal standard template and the targeted template are sufficiently different in length, then electrophoretic separation of the amplification products will be suitable for determining their origin. The design of this type of internal standard template can be achieved using two different approaches: (1) by introducing a specific mutation into the targeted sequence in a manner that generates an internal standard template with a specific enzyme restriction site midway through the sequence (see ➋ *Figure 15-2, mutated internal standard*) or (2) by creating an internal deletion within the sequence of interest such that the internal standard template generated is several base pairs shorter than the targeted template (see ➋ *Figure 15-3, deletion internal standard*). In both cases, it is imperative that the amplification primers are identical in the targeted sequence and the internal standard template.

3.1 Mutated Internal Standard

As described above, a restriction cleavage site is introduced into the middle of the internal standard sequence by site-directed mutagenesis. This method of internal standard design allows the determination of the origin of the amplification products by postamplification digestion followed by electrophoretic size separation of the products. ➋ *Figure 15-2* is a schematic representation of the construction of the internal standard we used to quantitate the expression levels of the $\alpha 1$ $GABA_A$ receptor subunit mRNA. A linear representation of the $\alpha 1$subunit primary sequence is shown at the top. This structure is typical of many of the subunits of the ligand-gated ion channel family and is representative of all $GABA_A$ receptor subunits. The amino terminal signal peptide is shaded; each subunit contains a large extracellular amino-terminal domain. There are four hydrophobic transmembrane spanning regions with a large intracellular loop separating the third and the fourth transmembrane segments. Considering the $GABA_A$ receptor subunit gene family (i.e., α, β, δ, and γ) as one example of the gene families we have extensively studied, we observed a high degree of sequence identity within this family, thus making it difficult to design primers that

Figure 15-3
Design and construction of the reelin promoter internal standard. The reelin promoter and its characterization have been described in some detail previously (Chen et al., 2002). The top bar shows a restriction enzyme cleavage map of the 5′ flanking sequence relative to the location of the RNA start site and first exon. Just prior to the ATG initiation codon, there is a CGG repeat present in the first exon. Below this first line is the diagram depicting the internal standard construction. Bp's indicated by the green arrow were deleted as described. The full size of the internal standard amplicon is 524 bp and that corresponding to the intact promoter is 734 bp. In the photograph of the gel shown at the bottom of the Fig. are the bands that were obtained after the individual amplifications used in constructing the standard. Lane 1 shows the PCR products of the 5′ half (227 bp) and Lane 2 shows the products of the 3′ half (297 bp). Lanes 3 and 4 show the amplification obtained after mixing the two halves together and amplifying with only the outside primer pairs. Lane 5 shows the products obtained using the outside primers with human genomic DNA as the template. Lane 6 shows the amplification of the cloned internal standard using these same outside primer pairs. Two different DNA markers are shown in the flanking lanes for sizing purposes

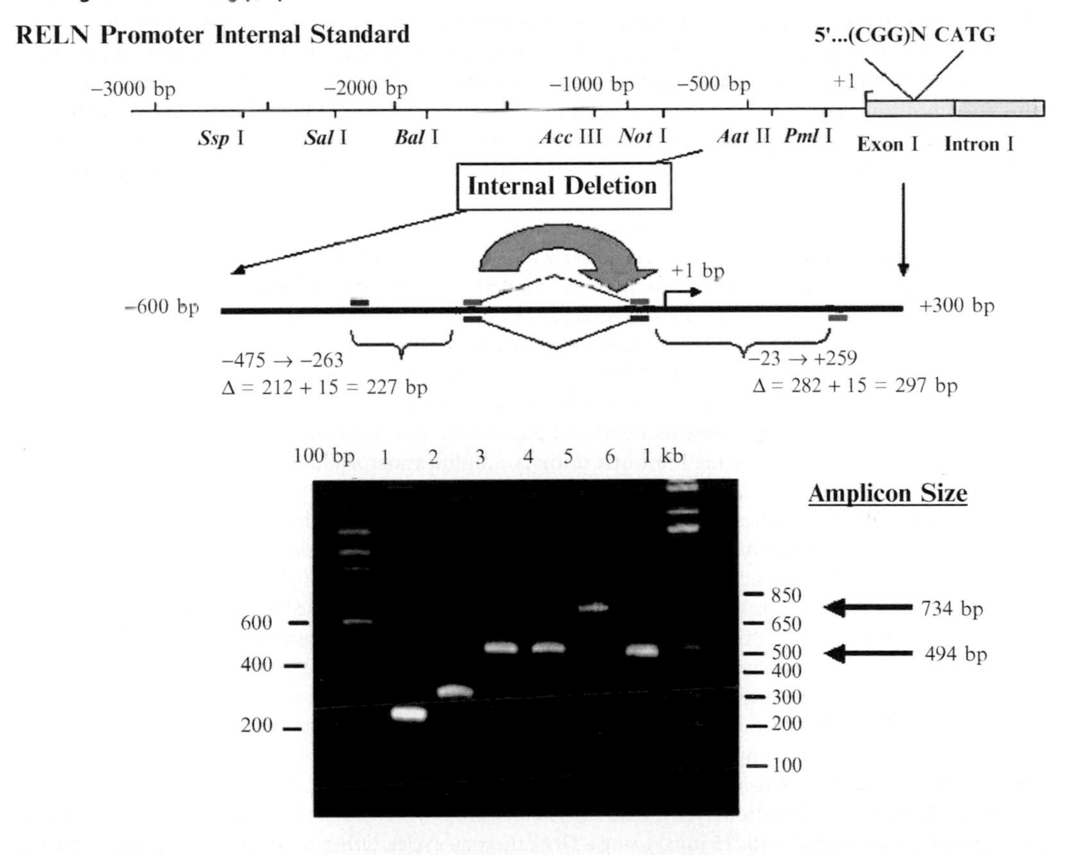

uniquely amplify individual subunit mRNAs. The second line in Figure 15-2 shows a representation of the mRNA encoding the α1 subunit mRNA with the region targeted for amplification bracketed by the internal primers. The primers are designed to selectively amplify only this subunit and have been chosen such that the 5′ primer lies within the cDNA sequence that corresponds to the intracellular loop domain, whereas the 3′ primer extends within the 3′ untranslated region. Because of the highly variable sequences present within the untranslated regions, designing unique amplification primers within this region is simplified for genes encoding homologous receptor subunits. One disadvantage of the above approach is that it is unlikely that these primers will also be useful in amplifying the homologous mRNA from different species, as these sequences are likely to diverge. In many situations, it is preferable to design a single internal standard that is useful in quantitating mRNAs from multiple species. To do this, it is easiest to perform several two-sequence Blast analyses (http://www.ncbi.nlm.nih.gov/blast/bl2seq/wblast2.cgi) using the sequences from each of the species of interest and then to design the internal standard based on regions in which there is sequence conservation among these species. It is generally quite easy to find regions in which there is sufficient sequence overlap to accomplish this, particularly for the regions that correspond to the primers that will be used. The degree of sequence identity in between the primers is less critical provided that there is a fairly normal distribution of bases and no large sequence gaps. Constructing internal standards in this way then avoids confusion when it comes to particular internal standards and the primers that were used in their design.

To construct the mutated internal standard, we adopted the overlap extension PCR method originally described by Ho et al. (1989). This method is a two-step protocol that allows the creation of selected point mutations and variable size insertions or deletions at targeted sites in the cDNA within the desired template. For a more detailed description of this protocol, readers are referred to Grayson et al. (1993). In brief, external primers are chosen to ensure selective amplification of a unique region within the targeted sequence. Once these primers are designed, the sequences midway between the external primer pairs are examined so as to design internal primers that contain a palindrome (e.g., AGATCT) flanked on both sides by sequences that match the region being modified. In the representative example depicted in ➲ *Figure 15-2*, this modification involved the alteration of wild-type AGGCCT to contain the AGATCT site that is recognized by the *Bgl*II restriction enzyme. Internal primers are complimentary and contain 9 bp both 5′ and 3′ to the restriction site that matched the wild-type sequence. In addition, the internal primers are designed such that there is a 24-bp overlap of the primary PCR products (i.e., the products of the initial amplification). It is also important to maintain the G/C content of both the internal mutating primers and the external amplification primers at 50%. By keeping both the internal and external primers 24 bp in length and maintaining the same proportion of G/C content, we consistently observed that an annealing temperature of ~60°C results in fairly specific amplification with minimal nonspecific contamination (Grayson et al., 1993). This is a very general guideline and when testing a new primer pair with mRNAs and internal standard, it is recommended to optimize the annealing temperature by performing a temperature gradient profile to find the appropriate conditions.

As depicted in ➲ *Figure 15-2*, two sequential amplifications are carried out. Primers 1 and 3, and 2 and 4 were used separately to amplify the indicated primary amplification products. Reversed transcribed RNA isolated from rat frontal cortex (a brain area in which the $\alpha 1$ subunit is highly expressed) was used as the starting template. An aliquot of the cloned template could also be used, although the amount of material and numbers of cycles would have to be reduced accordingly. From our experience both types (RT vs. clone) of starting material worked equally well, although the products of the initial amplification step tend to be "cleaner" (absence of nonspecific extraneous fragments) if the clone is used as the initial template. This is because the cloned cDNA represents a single molecular species, while reverse transcribed mRNA is a heterogenous population of molecules only one type of which represents the intended target. When cDNA (RNA following RT) is the starting template, we routinely use random hexamers to prime the reverse transcription reaction, although oligo dT primed cDNA will work as well. The reverse transcription is carried out as described for the competitive RT-PCR assay using 1–2 μg of purified RNA. The primary PCR amplification is performed using the heat denatured cDNA (either RT cDNA or cloned cDNA) template and from 32 to 35 amplification cycles, each cycle consisting of 94°C/45 s, 60°C/1 min, and 72°C/1 min (with a final 72°C extension for 15 min) using a DNA thermal cycler. Other protocols will work equally well and the reaction should be optimized. Prior to the next amplification, the primary amplification products are examined by electrophoresis through an agarose gel. The primary PCR products are purified using Quick spin PCR purification columns as described by the manufacturer (Qiagen, Chatsworth, CA) to separate the products of the internal and external primers.

As shown in ➲ *Figure 15-2*, the two separate amplification products obtained from the primary amplification contain the desired mutation and correspond to one-half of the targeted template. These purified fragments also contain a 24-bp region at each end that is complementary to the other half. During the secondary PCR, equal amounts of each primary PCR product are mixed and only the external primers (numbers 1 and 4) are used. Hybrid structures form between the denatured templates within the region of the 24-bp overlap, which primes the remainder of the two strands. This secondary amplification completes the construction of the two halves of the internal standard such that the mutation is now introduced into the template. The external primers then amplify this structure during subsequent cycling rounds. The specificity of the secondary amplification is checked by agarose gel electrophoresis before proceeding to subsequent steps. If the amount of the resulting product from this amplification is low or if there are nonspecific DNA bands contaminating the reaction, the process can be refined by adjusting the amplification cycles, annealing temperature, and the amounts of starting templates, or by adjusting the Mg^{2+} concentration. The product of this reaction is purified using PCR cleanup columns (Qiagen, Valencia, CA) or extracted from low-melting-point agarose after electrophoretic size separation using the carbohydrate-digesting enzyme Gelase (Epicenter Technologies, Madison, WI) as described by the manufacturer. This is a low-cost, efficient alternative to other purification techniques and typically gives excellent yields with high-molecular-size DNAs. The fragment is ligated into an appropriately digested and gel-purified vector for cloning purposes (Sambrook et al., 1989). During our initial experiments, we used the pGEM-1 (Promega/Fisher, Madison, WI) for cloning all our internal standards, although the commercial distribution of this vector has been discontinued. Newer versions of this vector include additional features that are useful for a variety of applications. More recently, we have used TA cloning (Invitrogen) to symmetrically clone the internal standard templates. We use the resulting internal standards for large-scale cRNA preparation for quantitative competitive RT-PCR. Protocols from AMBION and PROMEGA routinely yield high amounts of cRNA from 5 μg of starting template. We have also found these subclones convenient for generating ^{35}S-labeled sense and antisense probes for in situ hybridization (Zheng et al., 1993).

3.2 Deletion Internal Standard

The use of an internal standard containing a deletion mutation has several advantages over the use of an internal standard containing an engineered enzyme restriction site. As described above, the main advantage includes the elimination of the need for postamplification digestion to differentiate the internal standard amplicon, which, in addition, saves time, money, and resolves any concerns regarding incomplete digestion. It does, however, introduce concern that the internal standard template cRNA could amplify at a rate faster than that of the longer wild-type template. As stated earlier, the primer annealing efficiency is most likely the critical determinant of amplification rate. As the internal and wild-type templates are amplified by the same set of primers, their amplification rates should be similar. This is indeed what we have determined in the laboratory (Dennis Grayson, unpublished data). Amplifications performed in parallel with the two types of templates show comparable rates of product formation.

The generation of an internal standard containing a deletion mutation is similar to the procedure described for constructing the mutated internal standard. Similar to the mutated internal standard, we use a two-step overlap extension PCR, differing only in the design of the internal primers. When designing the primers for this protocol, one must carefully consider their placement along the template. The deletion mutation should be long enough so that the internal standard amplicon will be sufficiently different in size from the wild type to be easily resolved by gel electrophoresis size separation. The external primers must also be spaced far enough apart so that the internal standard will not be unreasonably short after a portion of it is deleted, especially if the internal standard is to be used in nested PCR. In the example described in ➋ *Figure 15-3*, the wild-type amplicon is 734 bp long and the internal standard amplicon is 494 bp long, containing a deletion of 341 bp. Aside from this additional consideration of size, the external primers are designed as described for mutated internal standard construction.

The design of the internal primers for generation of a deletion mutation is also similar to that for a point mutation. The two primers are complementary to one another and consist of a region on each end that is complementary to the template with the deletion mutation at the center. In other words, the 5′ half of the forward primer is identical to the sequence of the template found upstream of the 5′ end of the desired deletion. The 3′ half is identical to the sequence found downstream of the 3′ end of the desired deletion. The primers are designed to flank the deletion, generally 15 bp before the deletion point and then the 100–200 bp deletion followed by the 5′ terminal 15-bp primer.

The length of the internal primers is another factor to be considered. The internal primers should be long enough to ensure specific annealing during the first round of amplification, where only half of the primer anneals to the template. Our laboratory has achieved good success with internal primers of 30–34 bp or from 15 to 17 bp flanking the internal deletion mutation. The GC content, which defines the annealing temperature, should be kept constant at around 50%.

After designing the primers, the rest of the procedure is the same as that for the mutated internal standard construction. In brief, the first amplification consists of two separate PCR protocols to generate the two halves of the internal standard. In the first PCR protocol, primers 1 and 3 are used and the upstream half of primer 3 anneals to the site just 5′ to the intended deletion. In the second PCR, we employ primers 2 and 4 for amplification. Each of these amplification procedures results in one-half of the internal standard containing the deletion mutation in the middle of a 34-bp sequence at one end, which will allow the halves to anneal to each other. The products of these separate PCRs are separated on a 1.2% low melting temperature gel to check the size of the amplicons, which are then purified by gel extraction as described for the mutated internal standard construction.

In the second round of amplification, a PCR is performed using the products of the two individual PCRs from the first amplification as templates in the presence of only primers 1 and 4. The complementary ends of the two internal standard halves anneal to each other, priming the production of a complete double-stranded internal standard. This product is then further amplified by repeated (PCR) amplification with primers 1 and 4.

The deleted internal standard has now been created and should be checked for the expected size by gel electrophoresis and purified using a PCR cleanup column. Symmetrical cloning of the internal standard for amplification can be achieved following the protocol described earlier for the mutated internal standard after the TA tailing step. However, one disadvantage with using the pGEM-T Easy or other symmetrical cloning systems is that the construct's orientation is not known and therefore requires additional sequencing or restriction mapping site steps to identify the construct's orientation.

4 Routine Competitive RT-PCR

Competitive reverse transcriptase (RT)-PCR has evolved over the years as a versatile tool for the quantitation of gene transcripts of interest and has been adopted as a reliable strategy for comparing amounts of RNAs arising from different genes. This sensitive method has been standardized by mixing known amounts of competitors for multiple cDNAs (internal standard) with the native templates in the PCR reaction. It is noteworthy that this technique is more labor intensive than either the Northern blot analysis or RNAse protection assays. In brief, this technique involves titrating the transcript of interest in a mixed population of molecules by spiking the mix with known amounts of the corresponding internal standard. The internal standard competes with the endogenous RNA during the PCR as both templates are amplified using the same primer pair (➋ *Figure 15-1*). When the abundance of the native or wild-type RNA is high relative to the amount of the internal standard, the resulting amplification products are dominated by those arising from the wild-type RNA. In contrast, when the internal standard cRNA has exceeded the levels of the wild-type RNA, the product arising from the cRNA predominates. Thus, the initial reaction is generally performed over a broad internal standard concentration range. When appropriately designed, competitive RT-PCR is capable of estimating the actual number of molecules of each specific message present in the same sample, permitting comparison on the basis of the number of molecules not only across samples but also across genes encoding different proteins (Hayward et al., 1999). Furthermore, when competitive RT-PCR is carefully designed and applied, both precision and accuracy are high, even when microscopic tissue samples (e.g., laser microdissection) are analyzed (Hayward-Lester et al., 1997).

4.1 Oligonucleotide Primer Design and Synthesis

We design primers by aligning homologous sequences from as many taxa as possible. These sequences can be readily retrieved from large sequence databases such as Genbank (http://www.ncbi.nlm.nih.gov/web/

Genbank). In selecting for homologs within an archaea, primers specific for the archaeal domain are the most effective. However, it is not always possible to design primers that span the entire archaeal domain; therefore, primers may be restricted to a single group. Homologous regions that are conserved throughout the selected organisms serve as sites for primer design. In many cases, these primers will only amplify a portion of the gene.

For several reasons, we select our primers manually rather than relying on the use of commercially available computer software that will pick primers based on user-defined criteria. This is, in part, biased as we began making internal standards and PCR primers before the current accessibility of these programs and because we have had reasonable success following the criteria that were somewhat self-imposed by the limits of the experimental design. However, the Primer Express program, provided to all users of the 7700 SDS software (Applied Biosystems), can assist with primer design issues. If suitable exon/exon gene boundaries cannot be identified, we design primers to an exonic sequence within the gene-coding region. In this case, the potential effects of nonspecific, contaminating genomic signals can be assessed by comparing results obtained without RT and with RT in the reactions.

We typically examine sequence alignments for sequence discordance or identity (depending on the experimental end point desired) and design primers such that they are 24 bp in length with a 50% G/C content and a narrow melting temperature (T_m) difference (0.2–1.0) between the primers to enhance amplification efficiency. Preliminary semiquantitative RT-PCR experiments (Memo et al., 1991) suggested that this design would routinely allow us to set the annealing temperature for each internal standard at 60°C. To facilitate directional cloning of the amplification products, we routinely include a 6-bp recognition sequence of a selected restriction enzyme on both external primers (e.g., 5′ *Eco*RI, and 3′ *Hind*III). A 3-bp clamp of random nucleotide precedes the enzyme recognition site to ensure efficient digestion with the selected enzyme. For example, at the 5′ external primer the sequence may read 5′-**TGC**GAATTC-24 bp. In this case, nucleotides TGC represent the 3-bp clamp that is followed by the recognition site for digestion by *Eco*RI and the 24-bp region unique to the $GABA_A$ receptor subunit of interest. We have applied similar principles of oligonucleotide primer design described for $GABA_A$ receptor subunits here for all other gene transcripts of interest. Alternatively, the templates can be cloned symmetrically using commercially available cloning systems to avoid these steps.

Once an internal standard is made and subcloned, we generally prepare subsequent primers lacking the restriction enzyme cleavage site and the 3-bp clamp. In this way, we amplify both target templates with the same degree of overlap throughout the length of the primers. Primers can be made commercially or with available oligonucleotide synthesizers. We routinely prepare primer stocks by diluting the concentrated stock primer solutions to 20 μM with nuclease-free water.

4.2 RNA Isolation

Isolation of high-quality, intact RNA is less important for cloning experiments but is essential to analyzing mRNA expression from either tissues or cultured cell samples. The critical factor in all RNA experiments is the isolation of full-length RNA. The major source of failure in any attempt to isolate RNA is contamination by ribonucleases, enzymes that are very stable and generally require no cofactors. Thus, a small amount of RNase in an RNA preparation can create a serious problem. To minimize RNase contamination, it is desirable to ensure that all solutions, glassware, and plasticware used for RNA isolation are autoclaved (although autoclaving will not fully inactivate many RNases). Other routine precautions for handling RNA include the use of a sterile technique when handling reagents used for RNA isolation or analysis; wearing hand gloves at all times to avoid introducing “fingertip” RNases; treating all solutions used in RNA preparation with diethylpyrocarbonate, DEPC (0.1%); the use of DEPC-treated nondisposable glassware or sterile disposable plasticware and RNase-free water. In addition, dissected brain tissues should be frozen immediately on dry ice; cells grown in monolayer should be lysed directly in strong denaturants while in the culture dishes. Frozen cryosections are fixed for 2 min in 70% ethanol at −20°C, air dried for 2–3 min, stained with 0.1% toluidine blue (dissolved in DEPC-treated water) for about 5–10 s, and then air dried before microdissection.

Conventional (total RNA) and magnetic separation of poly A^+ mRNA are methods routinely used in our laboratory. The method of choice used for RNA separation will depend on the sample size and the abundance of the RNA transcript under study. Specifically, we use the conventional total RNA isolation method when the amount of starting material is not limited. This would include the isolation of total RNA from postnatal and adult dissected brain regions and primary cerebellar granule cell cultures. For isolation of RNA from tissue punches or other primary neuronal cultures, the use of magnetic bead isolation technology offers the advantage of increased yield from small sample size. Recently, we have collected microscopic tissue samples using the laser-capture microdissection technique. For isolating RNA from laser-microdissected samples, we use a modified NIH total RNA extraction protocol provided by Arcturus (http://www.arctur.com) or RNeasy total RNA isolation kit (QIAGEN Inc., Valencia, CA). We will describe the two main approaches we use for isolating total RNA and the use of magnetic beads for mRNA isolation.

The first step in all RNA isolation protocols involves lysing the cell in a chemical environment that denatures ribonucleases. The RNA is then fractionated from other cellular macromolecules by either homogenizing the tissue (dissected brain tissue) or simply vortexing the sample (very small tissues and laser-microdissected sample) without further homogenization. The cell type from which the RNA is to be isolated, the sample size, and the eventual use of the RNA will determine which procedure described here is appropriate.

Frozen dissected brain tissue is homogenized in 5 M guanidium isothiocyanate, 100 mM Tris–HCl, pH 7.4, and 1 mM EDTA, and total RNA is isolated following $CsCl_2$ ultracentrifugation as described previously (Chirgwin et al., 1979). The RNA pellet is resuspended in 400 μL nuclease-free water, extracted with equal volumes of phenol:chloroform:isoamyl alcohol (Sambrook, 1989), and precipitated using 0.3 M sodium acetate, pH 5.2, by adding ice-cold ethanol (2.5 vol). The total RNA yield is determined by measuring the optical density (OD) of an aliquot of the precipitated stock at 260/280 nm. RNA prepared by the ultracentrifugation is typically free of genomic DNA contamination, but it can be a lengthy procedure. We have also found that total RNA extraction with the TRI Reagent (Molecular Research Center, Cincinnati, OH) is reliable and efficient and does not require the overnight ultracentrifugation step. This approach utilizes phenol and guanidium thiocyanate for a rapid disruption and denaturation of cells or tissue. The frozen tissue is homogenized in TRI Reagent (1 mL for 50–100 mg tissue) or cells grown in monolayer are lysed directly in a culture dish (1 mL TRI Reagent/10 cm^2 area culture plate). Total RNA is then extracted using a modification of the originally published method (Chomczynski, 1993) as described by the manufacturer. Following vigorous mixing and centrifugation at 12,000 $\times$ *g* for 15 min at 4°C, RNA from the aqueous phase is concentrated by precipitation with isopropanol and centrifuged at 12,000 $\times$ *g* for 10 min at 4°C. The RNA pellet is washed with 1 mL of 75% ice-cold ethanol, dried under reduced pressure, and resuspended in nuclease-free water. The total RNA yield is determined by measuring the OD of an aliquot as described earlier. The entire procedure can be completed in 2 h depending on the sample size.

For low amounts of RNA (as low as 1.0 ng/mL RNA in solution) and for accurate determination of amounts of internal standard, we use the RiboGreen RNA quantitation reagent and Kit (Molecular Probes, Eugene, OR). The RiboGreen RNA quantitation reagent is an ultrasensitive fluorescent nucleic acid stain that alleviates many of the problems (the relatively large contribution of proteins and free nucleotides to the signals) encountered with the OD method (for details, see Jones et al., 1998). We routinely use this method to determine the concentrations of our diluted internal standard cRNAs. As the precision of our method depends on knowing the quantities of standard accurately, this is perhaps the most important aspect of the quantitation.

Whereas many applications for the analysis of RNA may be performed using total RNA, greater sensitivity is obtained using purified mRNA samples, especially polyA^+ mRNA. Magnetic based isolation of mRNA from a broad range of tissues and cells has been used successfully. These systems use a biotinylated oligonucleotide probe to hybridize the targeted nucleic acids in solution, and the hybrids are captured using covalently coupled streptavidin paramagnetic particles (SA-PMPs). The streptavidin paramagnetic particles have a high binding capacity for biotinylated oligonucleotides and low nonspecific binding for nucleic acids. The isolation of polyA^+ mRNA is performed according to the manufacturer's protocol (Promega/Fisher, PolyATract System), which allows for direct isolation of mRNA from tissues and cells, without intermediate purification of total RNA. With some modifications of the protocol, the procedure can be used to extract mRNA from total RNA as the starting template. The amount of the extraction buffer used depends on the tissue sample (20 μL per 2.5 mg of tissue or 1×10^5 for cultured cells). If the maximum sample concentration is exceeded, the increased lysate viscosity may lead to irreversible clumping of the particles during magnetic capture and result in lower RNA yields. The tissues of interest or cultured cells are homogenized in the extraction buffer (4 M guanidium thiocyanate, 25 mM sodium citrate, and 2% b-mercaptoethanol) at a high speed for 15–30 s. Forty μL of the preheated (at 70°C) hybridization buffer (containing 3 pmol of biotinylated oligo[dT] is added to 20 μL of homogenate. The mixture is incubated for 5 min at room temperature and 35 μL of blocking particles are added to each tube; the tubes are then centrifuged at 1700 × *g* for 10 min to clear the homogenate of cell debris and precipitated proteins. During this centrifugation, SA-PMPs are resuspended and transferred to a sterile tube (63 μL/sample) and captured using the magnetic stand for approximately 30 s. The particles are washed three times by holding the tube in the magnetic stand and then carefully pouring off the storage buffer so that the solution runs over the captured particles. Each time an equal volume of wash solution (7.5 mM sodium citrate, 75 mM NaCl) is used and the particles are resuspended in 78 μL of the washing buffer. The homogenate from the centrifugation step is added to the washed SA-PMPs and incubated at room temperature for 10 min. The SA-PMPs are then captured using the magnetic stand, and the supernatant is carefully removed and particles washed five times with 180 μL of the wash solution. After the final wash, mRNA is eluted by the addition of nuclease-free water to the particles. The particles are gently resuspended by flicking and are magnetically captured again and the eluted mRNA in the aqueous phase is transferred to the sterile RNase-free tube. The elution step is repeated once more and the mRNA remaining in the solution is now purified and ready for reverse transcription.

4.3 In Vitro cRNA Synthesis

In vitro transcription of RNA from a DNA insert cloned into a vector containing the appropriate RNA promoter is widely used to generate large amounts of transcripts. When a transcript of a defined size is desired, the plasmid containing the cloned DNA segment is linearized with an appropriate restriction endonuclease before the transcription reaction and only discrete "runoff" transcripts are obtained, virtually free of vector sequences. The presence of cohesive ends generated by specific restriction endonucleases, particularly those with 3′ overhangs or protruding ends (e.g., *Pvu*I, *Sac*I, *Sph*I), often leads to the production of a contaminated product with transcripts arising from the opposite strand. In addition, this can be avoided by blunting the ends following digestion using the Klenow fragment of DNA polymerase (Promega/Fisher). This is readily achieved by incubating the template with Klenow (5 U/μg template) for 15 min at room temperature in the absence of dNTPs.

cRNA is synthesized on a large scale from 5 μg of the linearized template and T7 RNA polymerase according to the manufacturer's technical manual (Promega/Fisher). The reaction mixture contains 5 μg DNA, 40 mM Tris–HCl, pH 7.9, 6 mM $MgCl_2$, 2 mM spermidine, 10 mM NaCl, 10 mM dithiothreitol (DTT), 100 U RNasin ribonuclease inhibitor, 0.5 mM each of rNTP, and from 30 to 40 U T7 RNA polymerase in a 100-μL reaction volume. The reagents are added to the template at room temperature and the transcription reaction is carried out at 37°C for 2 h. Mixing the reagents at room temperature is critical as the presence of spermidine will precipitate the DNA template if the reagents are cold. This will reduce the yield of the RNA considerably and is readily avoided by pipetting the reagents to the side of the tube before mixing. Once the enzyme is added, the mixture is spun down and incubated as described above. Removal of the DNA template following transcription is achieved by digestion of the residual plasmid with 5 U of RQ1 RNase-free DNase at 37°C for 1 h to ensure complete digestion of the DNA template.

Following DNase 1 digestion, the synthesized RNA is extracted with 1 volume of TE-saturated phenol/chloroform, vortexed for 1 min and centrifuged in a microcentrifuge (12,000 × *g*) for 5 min. The aqueous phase is transferred to a fresh tube, 1 volume of chloroform:isoamyl alcohol (24:1) is added, and the tube is vortexed for 1 min and centrifuged (12,000 × *g*) for 5 min. The aqueous phase is transferred into a fresh tube and cRNA precipitated by adding an equal volume of 5 M ammonium acetate and 2 volumes of ice-cold ethanol or an equal volume of isopropanol. The precipitation is carried on dry ice for a minimum of

30 min or overnight at −20°C. RNA is then pelleted in a refrigerated (4°C) microcentrifuge for 15 min at 12,000 × *g* and resuspended in 100 μL of 1 M ammonium acetate. The ethanol precipitation is repeated as before, and following centrifugation, the pellet is washed with 500 μL of ice-cold 70% ethanol, vacuum dried, and resuspended in 20 μL of nuclease-free water. We use about 10% of this material to determine the absorbance at 260/280 nm and subsequently the cRNA yield. For a more accurate determination of the cRNA yield, we use the RiboGreen RNA quantitation reagent described earlier. Typically, we obtain from 25 to 35 μg of cRNA as determined by these methods of measurement. Kits for preparing larger amounts of cRNA are also commercially (Ambion) available.

4.4 cDNA Synthesis

The first step in competitive RT-PCR is the reverse transcription of total RNA and the appropriate internal standard cRNA into complementary (c)DNA. We routinely use the thermal cycler (GeneAmp PCR system 9600, Perkin Elmer Cetus, and more recently, Mastercycler 5333, Eppendorf Scientific Inc., Westbury, NY) for all manipulations starting with cDNA synthesis. Decreasing amounts of cRNA prepared from the appropriate standard template are added to a constant amount of total RNA (1 μg or less) isolated from the source of interest (see later). This approach allows the coreverse transcription of both the internal standard and the wild-type RNA template in the same tube and thus control for variations in reverse transcription efficiency. In addition, the incorporation of internal standards in the reaction has permitted important advances in the accuracy and reliability of the competitive RT-PCR reaction (Becker-Andre and Hahbrock, 1989; Wang et al., 1989; Gilliand et al., 1990; Grayson et al., 1993). In general, we first try a broad range of internal standards to establish the appropriate concentration range (e.g., 800 pg-50 pg) prior to a more restrictive titration. This allows us to define the approximate abundance of a particular message first and then to determine more precisely the concentrations to be used for a more accurate quantitation. The RNA/cRNA mixtures are transcribed with cloned Moloney murine leukemia virus (MMLV) reverse transcriptase using random hexamer primers. The reverse transcriptase (200 U) is added to the RNA/cRNA in 50 mM Tris–HCl (pH 8.3), 75 mM KCl, 3 mM $MgCl_2$, 1 mM dNTPs, 2.5 μM random hexamers, and 40 U RNasin in a final volume of 20 μL. The reverse transcription reaction is carried out in the thermal cycler using this temperature/time settings: 25°C for 10 min and then at 37°C for 1 h, heat-denatured at 98°C (to inactivate the reverse transcriptase and to denature the starting template) for 5 min, and rapidly chilled to 4°C.

4.5 Competitive RT-PCR Amplification

To the 10 μL of cDNA reaction mixture (described earlier) is added 40 μL of a PCR master mix, resulting in the following final concentrations: 0.5 μM of subunit-specific primer pairs, 200 μM dNTPs, 2 mM $MgCl_2$, 10 mM Tris–HCl (pH 9.0), 0.1% Triton X-100, 50 mM KCl, and 2.5 U Hot Tub DNA polymerase (Amersham/Pharmacia). In previous years, we added trace amounts of [^{32}P]dCTP 90.3 μCi per tube) to the PCR reaction mixture for subsequent quantification. However, with the advent of the Kodak Electrophoresis Documentation and Analysis System described in the strategy section, we have discontinued the use of trace amounts of radioactive nucleotide for quantification purposes. Each amplification cycle consists of a denaturing step (94°C, 30 s), an annealing step (60°C, 45 s), and an elongation step (72°C, 45 s) with a final 5-min elongation step and rapid chilling to 4°C. The cycling parameters are adjusted according to the particular primer pair melting temperatures and the targeted template.

Amplification products arising from both the mRNA and cRNA templates are analyzed after enzymatic digestion. All of the amplification mixture is digested with the appropriate restriction enzyme (mostly *Bgl*II) introduced in the sequence of interest in a 60-μL reaction volume, and aliquots are electrophoresed in triplicate on 1.6% (w/v) agarose gels prepared in TBE buffer (0.04 M Tris–borate and 1 mM EDTA). The digested PCR products are visualized and photographed following ethidium bromide (0.5 μg/mL) staining under UV light and the corresponding bands are excised from the gel for radioactive [^{32}P]dCTP counts incorporated into each band. Alternatively, the gel is stained with SYBR gold, scanned using the Storm phosphoimager to quantify the respective band intensities for subsequent analysis, as described earlier. The ethidium bromide-stained gel may also be analyzed without excising the bands by using the Kodak Electrophoresis Documentation and Analysis system described earlier. The later two approaches greatly reduce the postamplification time needed to acquire data and also reduce the exposure of the researcher to excessive UV light and accidental exposure to excessive material.

5 Quantitation of RNA Transcripts from Laser-Capture Microdissection Material

As described earlier, we have adopted the use of a reamplification protocol (Nested-RT-PCR) in our competitive assay to improve the sensitivity of quantitative RNA measurements from a microdissected tissue sample while maintaining the specificity of the technique (Costa et al., 2002).

5.1 RNA Isolation from Laser-Capture Microdissected Samples

For the preparation of total RNA from microdissected tissues, samples are catapulted by laser microscopy into the cap of a 500-μL Eppendorf tube containing 40 μL of lysis buffer (4 M guanidium isothiocyanate and β-mercaptoethanol). The cap is then transferred into a sterile 500-μL Eppendorf tube and lysed in 200 μL denaturing buffer (4 M guanidium isothiocyanate, 50 mM Tris–HCl (pH 7.5), 25 mM EDTA, 1% Triton X-100, and 0.1 M β-mercaptoethanol) by gently mixing for 2 min. Total RNA is then extracted with the addition of 0.1$\times$ volume 3 M sodium acetate (pH 4.0) and 1$\times$ volume of phenol:chloroform:isoamyl alcohol (22:24:1, v/v) and vortexed vigorously for 1 min. The aqueous and organic phases are separated by incubation on ice for 15 min followed by centrifugation at 14,000 rpm for 30 min at 4°C. The upper aqueous layer is carefully pipetted into a sterile, fresh tube and RNA is precipitated at −80°C for 1 h in a 1$\times$ volume of isopropanol containing 2 μg of Pellet Paint (Novagen) as a coprecipitant. Total RNA is then pelleted by centrifuging in a microcentrifuge at 14,000 rpm for 30 min at 4°C. The RNA pellet is washed with 400 μL of cold 70% ethanol and air-dried on ice to remove any residual ethanol. The final pellet is reconstituted in a final volume of 13 μL. As the amount of total RNA extracted from laser-microdissected samples is not sufficient to be measured by UV detection, we routinely compare the relative abundance of mRNA among samples by referring to the volume of microdissected samples or the number of microdissected cells. We measure sample volumes using the UTH SCSA Image Tool provided by Uthscsa (http://ddsdx.uthscsa.edu). However, the updated version of the software supplied by the manufacturer of the Leica Laser Microdissection System (DMLMD) has provisions for measuring the area or volume of microdissected tissues. Microdissected RNA is often contaminated with DNA, and it is therefore important to DNAase treat the RNA for certain protocols. We have also used the RNeasy procedure (QIAGEN Inc.) for the isolation of total RNA from microdissected samples. The samples are disrupted and lysed with guanidium thiocyanate containing buffer (Buffer RLT) by vortexing for 30 s. This is followed by homogenization of the lysate by adding 1 volume of 70% ethanol and mixing by pipetting to shear genomic DNA and reduce the viscosity of the lysate. Total RNA and digested DNA are isolated by adsorption to the RNeasy MinElute Spin Column membrane. For the preparation of total RNA, the column is treated with DNase using RNase-Free Buffer RDD followed by a 15-min incubation at room temperature (RT). The column is then washed using 500 μL of buffer RPE followed by a 15-s centrifugation at 10,000 rpm in a microcentrifuge. The total RNA is then eluted from the column by adding 14 μL of RNase-free water directly onto the center of the spin column silica-gel membrane and centrifuging for 1 min at maximum speed.

5.2 Quantitative Nested-RT-PCR

The routine competitive RT-PCR protocol described earlier is modified for the quantitative nested RT-PCR analysis of RNA isolated from laser-microdissected samples. The low amount of starting template and the corresponding internal standard cRNA calls for more amplification cycles, thus the need for a two-step PCR amplification protocol to improve sample detection. The protocol for the reverse transcription of RNA to cDNA is the same as described earlier. In brief, the reversed transcribed RNA/cRNA mixture is subjected to 30 cycles of first-round amplification using a specific primer pair corresponding to the target of interest. For the second amplification (nested PCR), the first-round PCR amplification products are diluted 20-fold with RNase-free water. Aliquots (5 μL) from this dilution are then used for the second-round PCR amplification using nested primers designed in such a manner that they are inner to the first set of the primer pair. The second-round amplification is carried out in a 50-μL final volume and contains the same reagents in the same final concentrations described earlier and 20 cycles of amplification. The final PCR products are then digested with *Bgl*II or an appropriate restriction enzyme, separated by agarose gel electrophoresis, stained with SYBR gold, and scanned using the Storm phosphoimager to quantify the respective band intensities.

The data are analyzed by plotting the ratio of the counts incorporated into the product arising from the cRNA to that of the targeted RNA as a function of the amount of added external standard. If the SYBR gold/ Storm phosphoimager or the Kodak electrophoresis documentation and analysis system is used to quantify the respective band intensities, the ratio of the intensities (OD) of the amplified cRNA standard bands to that of the targeted RNA is also plotted as a function of the external standard used. Thus, irrespective of the method used for gel analysis, the determined ratios are plotted as a function of the known amount of internal standard cRNA that competed against the targeted RNA in the test sample. A linear regression analysis of these data yields the so-called point of equivalence (that is, the concentration at which the internal standard cRNA/sample ratio is 1. The point of equivalence represents the absolute amount of targeted RNA present in the sample. This is illustrated in the example shown in ➋ *Figure 15-4b* (see red and blue arrows) and ➋ *Figure 15-4c* (corresponding bar graphs).

6 Application of Genomic Internal Standard for Quantifying Gene Copy Number

There are also many applications for the use of competitive PCR in the context of quantitating amounts of genomic DNA. One example that we have recently adapted this technology for is the characterization of gene copy number. This can be particularly useful when engineering cell lines or mice with transgenes to determine the copy number relative to the normal diploid genome of the nonengineered cell line or mouse. For example, we have generated multiple cell lines of neural progenitor cells that contain additional copies of a reelin promoter (Chen et al., 2002) that is designed to drive expression of a β-galactosidase reporter gene. The goal of the exercise was to determine the accuracy of various amounts of 5′ flanking sequences to direct the expression of β-gal activity in these cells, which do not normally express the endogenous gene. NT2 cells are human progenitor cells, which will differentiate into neurons when treated for extended periods with retinoic acid. Transient transfection of these cells with reelin promoter constructs showed the promoter to be promiscuous relative to the normal gene. That is, promoter constructs containing 5′ flanking sequences drive high levels of expression of a luciferase reporter when introduced into the cells by transient transfection (Chen et al., 2002). However, the endogenous human gene is silent or expressed at very low levels in these cells as determined by quantitating the corresponding mRNA using competitive RT-PCR. We chose three different constructs that contained varying amounts of 5′ flanking sequence (from −300 bp up to −1700 bp) to determine whether integration of the DNA was a prerequisite for biologically accurate transcription. To test this, we cotransfected the NT2 cells with the promoter/reporter constructs and a gene conferring hygromycin resistance. After selecting for antibiotic resistance, we maintained several lines from each parental transfection for the various experiments. However, prior to making expression comparisons, we needed to define the gene/promoter copy number in each cell line so that we could normalize the reporter data obtained either before or after retinoic acid treatment based on the gene copy number present.

Genomic DNA was isolated from each cell line using standard techniques, which involved sequential washing in PBS and lysis in the presence of proteinase k/ 0.4% SDS (Zuccotti and Monk, 1995). Following multiple phenol/chloroform extractions and ethanol precipitations, the amount from each line was

Figure 15-4

Quantitative nested RT-PCR assay of α1 GABA$_A$ receptor subunit mRNA in layer V pyramidal neuronal somata (900 cells) microdissected from 10-μm slices of frontoparietal motor cortex. A pool of 900 cell somata from 14 day vehicle (V) or diazepam (D) treated rat brains (300 somata each from three separate V or D treated rats) were catapulted by laser microscopy into a 500-μL Eppendorf tube containing 40 μL of lysis buffer (guanidine isothiocyanate and β-mercaptoethanol). Total RNA was extracted as described in the text and reconstituted in a final volume of 13 μL. Aliquots (3 μL) were used for competitive nested RT-PCR in the presence of decreasing competing amounts (0.2, 0.1, 0.05, 0.025 pg, lanes 1, 2, 3 and 4 respectively) of the corresponding α1 GABA$_A$ receptor subunit cRNA internal standard (I.S.) containing midway in the structure a base substitution to create a *Bgl* II restriction site (Grayson and Ikonomovic, 1998). Panel A is the SYBR gold staining of the final nested RT-PCR products separated by agarose gel electrophoresis after *Bgl* II digestion. Panel B is a plot of the ratio of the I.S. band O.D. versus the sample O.D. against the respective I.S. amounts used for competitive analysis. The I.S. concentration at which the I.S./sample ratio is 1 represents the amount of α1 GABA$_A$ receptor subunit mRNA in the sample aliquot. Panel C bar graph represents the absolute amounts of α1 GABA$_A$ subunit mRNA measured in V and D treated brain samples. The gel was scanned using a Storm phosphoimager to quantify the respective band intensities

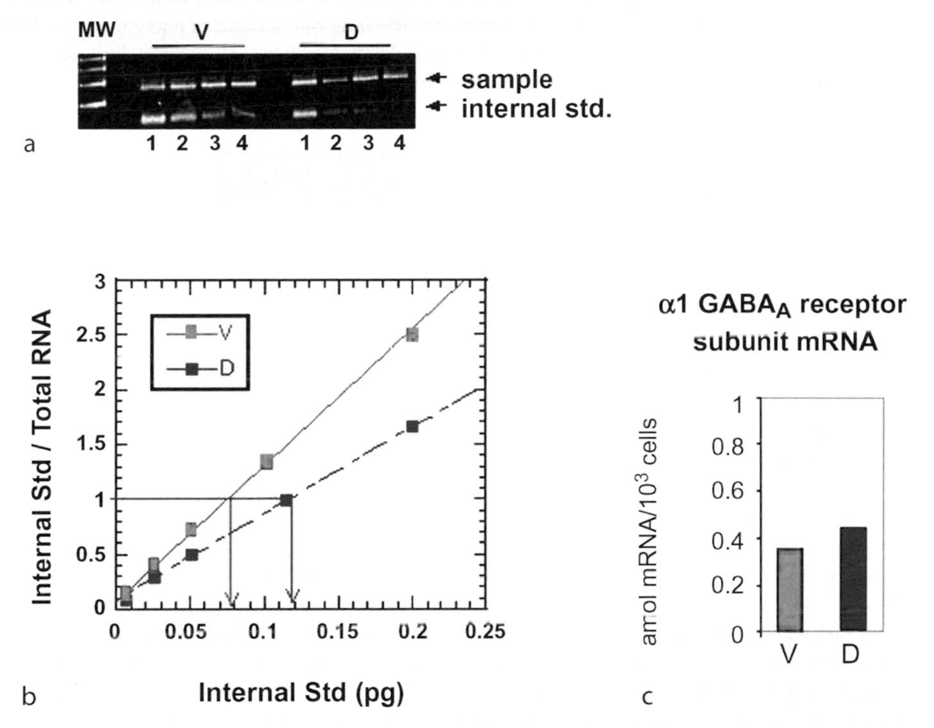

determined using optical density measurements. Although the absorbance at 260 nm (OD_{260}) is not particularly sensitive, we obtained large amounts of genomic DNA from each source for accurate concentration measurements. The internal standard used was that illustrated in *Figure 15-3* and the indicated amplicon is present in each of the three constructs used and in the human genome of the cells. As shown in *Figure 15-3*, we used this as a deletion internal standard so that we were able to separate the products of amplification without restriction enzyme digestion. To prepare the internal standard, we linearized the plasmid and gel-purified the ~500-bp insert DNA. The purification utilized Gelase (Epicenter Technologies) enzyme digestion of agarose, which is described earlier. The amounts of the internal standard insert were also calculated from the measured OD_{260} reading. To obtain the appropriate range, we first performed a broad-range PCR reaction with amounts of internal standard that varied from low fg (femtogram) to ng (nanogram) amounts. Once these results were available, we were able to narrow the range to determine that which was used in *Figure 15-5*. A calculation of the ratio of the amount present in each cell line was an integral proportion to that number obtained with the genomic DNA isolated from nonengineered NT2 cells. This is what was anticipated and allowed us to determine the gene copy for each line that we had derived.

Figure 15-5

Use of competitive PCR to determine copy number. The internal standard described in *Figure 15-3* was used to coamplify genomic DNA isolated from NT2 cells and NT2 cells engineered to contain the reelin promoter driving expression of a β-galactosidase reporter gene. We used the internal standard to determine the numbers of copies of the promoter/reporter construct were integrated in the genome of the cell line. The internal standard was linearized and increasing amounts were added to 1 μg genomic DNA isolated from either NT2 cells (right) or the reelin promoter containing NT2 cell line (left). The amplification products were separated using a 1.2% agarose gel and the stained gel was scanned and analyzed with the KODAK 1D image analysis software. The optical density of each band was determined. The graphs show representations of the ratio of the internal standard DNA to the target DNA and plotted as a function of the amount of internal standard DNA added. The point of equivalence is shown below each graph. The ratio of the point of equivalence in the case is the ratio of the copy number in the two cells lines, which is nearly 2. This means that in the engineered cell line there is twice the diploid number of copies of the reelin promoter as in the nonengineered cells

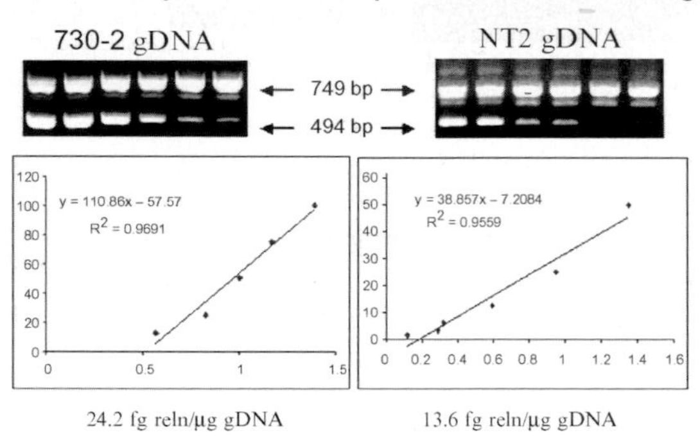

7 Optimization of Amplification Conditions

The efficiency of PCR reactions varies with fragment size, primer pair, and even the nature of the template (e.g., genomic DNA, cDNA library, plasmid, or products of first-strand cDNA synthesis); therefore, it is prudent to vary reaction conditions to improve reaction efficiency. In most instances, if the reaction conditions are not optimal, the PCR reaction often generates a smear of end products rather than a defined band on agarose gel. Amplification reactions are not 100% efficient because numerous experimental parameters affect the rate and specificity of the reaction. The usual parameters that require monitoring include magnesium (Mg^{2+}) concentration and annealing temperature. Most DNA polymerases are inactive in the absence of free Mg^{2+}, because Mg^{2+} permits the enzymes to bind to the primer template when they otherwise would not. The optimal Mg^{2+} concentration varies with each sequence and possibly the type of DNA polymerase used, but it is usually between 1 and 4 mM. For instance, we have observed in our hands that adjusting the Mg^{2+} concentration to 7 mM during our experiments with the γ_{2L} $GABA_A$-receptor subunit primers reduced the number of nonspecific bands. In addition, we found that 2 mM $MgCl_2$ works sufficiently well with Taq DNA polymerase from Promega/Fisher for most templates. Today's manufacturers supply their DNA polymerases with Mg^{2+}-free 10× reaction buffer and a tube of 25 mM $MgCl_2$ that allow the researcher to adjust the Mg^{2+} concentration to optimize each reaction. In light of these, we optimize our experiments using different Mg^{2+} concentrations with different enzymes from different manufacturers to increase our amplification yields. However, the variations in optimum conditions for a given enzyme with different fragments or primers are frequently less significant than the variations between enzymes. It is noteworthy that increasing the amount of enzyme also increases the possibility of generating artifacts that result in smearing on agarose gel. In general, low annealing temperature favors mispriming, whereas higher temperatures produce fewer mispriming errors. Thus, changes in the temperature of annealing result primarily in changes in mispriming.

The specificity of a PCR reaction can also be improved by nested PCR. As described earlier, the nested PCR involves two amplification protocols. During the second amplification, a new set of primers is used that anneal within the target amplified by the first primer pair, resulting in a shorter final PCR product. If the first-round amplification results in some nonspecific products (resulting in a smear or numerous bands on agarose gel), the nested primers ensure that only the desired product is amplified from the mixture and should be the only sequence present containing both sets of primer-binding sites. However, it is important to note that a large dilution of the first PCR product is necessary for a specific amplification product to be obtained at the end of the second-round amplification. It is therefore advisable that a serial dilution of the first PCR product be performed to determine the optimum dilution that results in a specific and good yield during the reamplification step.

8 Conclusions

In the current chapter, we have detailed the competitive RT-PCR reaction as originally derived for quantitating mRNA amounts isolated from either tissue, cell lines or primary neurons, and/or glia in culture. Since the original descriptions of the method, several advances have streamlined the approach, which has allowed more user friendly or less-labor-intense assays. It should be recognized that any quantitative information is not readily attained and precautions and care must be taken at each step of the assay to obtain both reliable and reproducible numbers. Of the more significant advances, switching to nonradioactive quantitation has provided the biggest improvement in terms of reducing labor and cost. In addition, a safety benefit comes from not having to expose the researcher to radioactivity. Most labs these days have access to gel photographic or scanning equipment, which then obviates the need for using isotopes and cutting out bands or exposing these to film. A second improvement in the process is the more general use of deletion standards as opposed to mutated standards. This latter advance avoids the time-consuming restriction enzyme cleavage step. This also prevents any uncertainties associated with incomplete enzyme digestion, which could complicate data interpretation. Although the assay was originally developed for quantitating RNAs from the GABA-A receptor subunit family of mRNAs, the method is obviously generally applicable to any mRNA of interest. It is more common these days to design the internal standards based on homology rather than on regions of sequence discordance. That is, it is desirable to develop standards that are applicable to each species that the lab is currently investigating as opposed to devising a standard for each independently.

The need for obtaining quantitative information from very small amounts of starting material has spawned the development of new techniques for performing single-cell PCR types of analyses. For the most part, these techniques have enjoyed only limited success. The use of T7 polymerase amplification based protocols coupled with microarray type screens has been useful in the development of molecular fingerprints in terms of expression profiling (Glanzer et al., 2004). Although these studies are quite informative, this type of information is not quantitative in the sense that we have discussed in this article. We have devised a modification to the competitive RT-PCR assay that works with small amounts of material that includes the use of nested primer pairs. As presented earlier, we have been able to show that this approach can be used to obtain quantitative information from laser-captured material, which amounts to defining the expression profiles of mRNAs present in small populations of cells or neurons. Although we have not performed this assay at the single-cell level, the ability to approach this is in development and only awaits the advent of newer and better technology. It may be that a nested approach coupled with the T7 amplification system may provide the added sensitivity required for the desired end point.

Finally, as we show in the last section of the chapter, the method is generally applicable to quantitating nucleic acids rather than just mRNAs. For example, establishment of gene copy number in engineered cell lines and in animals expressing a transgene is readily achieved as outlined earlier. The assay is also readily extended to Chromatin immunoprecipitation assays in which antibodies are used to pull down genomic DNA to which a specific factor is bound (Das et al., 2004). The amounts of DNA obtained from genomic DNA using this assay are limited, but the use of a nested PCR approach should provide sufficient material for this to be successful. For example, when performing ChIP assays with antibodies prepared against modified histone proteins, we obtain enough DNA such that with a nested PCR approach we can perform as many as 40–50 titrations in the second round of amplification (see Mitchell et al., 2005). This approach provides one of only a few ways to quantitate the amounts of DNA obtained with this research tool and this is currently largely ignored by those working in the field. Another genomic application of competitive PCR would be to determine copy numbers of specific viruses present in a genome following infection. Although the reduced cost associated with real-time PCR machines may make this technology more approachable for many investigators, this technique also requires rigor and care when making quantitative measurements. However, at the same time, though real-time PCR can provide quantitative information, it will not provide absolute quantitation in the way that the competitive PCR reaction will.

References

Becker-Andre M, Hahlbrock K. 1989. Absolute quantification using the polymerase chain reaction (PCR). Nucleic Acid Res 17: 9347-9446.

Bovolin P, Sanit M-R, Memo M, Costa E, Grayson DR. 1992a. Distinct developmental patterns of expression of the rat a1, α5, γ2S and γ2L $GABA_A$ receptor mRNAs in vivo and in vitro. J Neurochem 59: 62-72.

Bovolin P, Santi M-R, Costa E, Grayson DR. 1992b. Differential patterns of expression of γ-aminobutyric acid type A receptor subunit mRNAs in primary cultures of neurons and astrocytes from cerebella of the neonatal rat. Proc Natl Acad Sci USA 89: 9344-9348.

Buck KJ, Harris RA, Sikela JM. 1991. A general method for quantitative PCR analysis of mRNA levels for members of gene families: application to $GABA_A$ receptor subunits. Biotechniques 11: 636-641.

Bustin SA, Nolan T. 2004. Pitfalls of quantitative real-time reverse-transcription polymerase chain reaction. J Biomol Tech 15: 155-166.

Chen Y, Sharma RP, Costa RH, Costa E, Grayson DR. 2002. On the epigenetic regulation of human reelin promoter expression. Nuc Acids Res 30: 2930-2939.

Chirgwin JM, Pryzbyla AE, MacDonald RJ, Rutter WJ. 1979. Isolation of biologically active ribonucleic acid from sources enriched in ribonuclease. Biochemistry 18: 5294-5299.

Chomczynski P. 1993. A reagent for the single-step simultaneous isolation of RNA, DNA and proteins from cell and tissue samples. Biotechniques 15: 532-535.

Costa E, Auta J, Grayson DR, Matsumoto K, Pappas GD, et al. 2002. $GABA_A$ receptors and benzodiazepines: a role for dendritic resident subunit mRNAs. Neuropharmacology 43: 925-937.

Das PM, Ramachandran K, vanWert J, Singal R. 2004. Chromatin immunoprecipitation assay. Biotechniques 37: 961-969.

Freeman WM, Walker SJ, Vrana KE. 1999. Quantitative RT-PCR: pitfalls and potential. Biotechniques 26: 112-124.

Gilliand G, Perrin S, Bunn HF. 1990a. Competitive PCR quantitation of mRNA in PCR Protocols. Innis MA, Gelfand DH, Sninsky JJ, White TJ, editors. San Diego: Academic Press; pp. 60-69.

Gilliand G, Perrin S, Blanchard K, Bunn HF. 1990b. Analysis of cytokine mRNA and DNA: detection and quantitation by competitive polymerase chain reaction. Proc Natl Acad Sci USA 87: 2725-2729.

Ginzinger DR. 2002. Gene quantification using real-time quantitative PCR: an emerging technology hits mainstream. Exp Hematol 30: 503-512.

Glanzer JG, Haydon PG, Eberwine JH. 2004. Expression profile analysis of neurodegenerative disease: advances in specificity and resolution. Neurochem Res 29: 1161-1168.

Grayson DR, Ikonomovic S. 1998. Competitive RT-PCR quantitate steady state mRNA levels. Neuromethods 34 (In vitro neurochemical techniques). Boulton AA, Baker GB, Bateson A, editors. Totowa, NJ: Humana Press; pp. 127-151.

Grayson DR, Bovolin P, Santi M-R. 1993. Absolute quantitation of γ-aminobutyric acid A receptor subunit mRNAs by competitive polymerase chain reaction. Methods in neuroscience, vol 12. Academic Press; San Diego: pp. 191-208.

Harris BT, Charlton M, Costa E, Grayson DR. 1994. Quantitative changes in the α1 and α5 $GABA_A$ receptor subunit mRNAs and proteins following a single treatment of cerebellar granule neurons with either NMDA or glutamate. Mol Pharmacol 45: 637-648.

Hayward AL, Hinojos CA, Nurowska B, Hewetson A, Sabatini S, et al. 1999. Altered sodium pump alpha and gamma subunit gene expression in nephron segments from hypertensive rats. J Hypertens 17(8): 1081-1087.

Hayward-Lester A, Chilton BS, Underhil PA, Oefner PJ, Doris PA. 1997. Quantitation of specific nucleic acids, regulated RNA processing and genomic polymorphism using reversed-phase HPLC. Gene quantification. Ferre F, editor. Boston: Birkhauser.

Ho SN, Hunt HD, Horton RM, Pullen JK, Pease LR. 1989. Site-directed mutagenesis by overlap extension using polymerase chain reaction. Gene 77: 51-59.

Jones LJ, Yue ST, Cheung C-Y, Singer VL. 1998. RNA quantitation by fluorescence-based solution assay: ribogreen reagent characterization. Anal Biochem 265: 368-374.

Memo M, Bovolin P, Costa E, Grayson DR. 1991. Antagonists of the NMDA sensitive glutamate receptor down regulate mRNAs encoding various $GABA_A$ receptors in cerebellar granule neurons. Mol Pharmacol 39: 599-603.

Mitchell CP, Chen Y, Kundakovic M, Costa E, Grayson DR. 2005. Histone deacetylase inhibitors decrase reelin promoter methylation in vitro. J Neuochem 93: 483-92.

Pagliarulo V, George B, Beil SJ, Groshen S, Laird PW, et al. 2004. Sensitivity and reproducibility of standardized competitive RT-PCR for transcript quantitation and its comparison with real time RT-PCR. Mol Cancer 3: 5

Saiki RK, Gelfand DH, Stoffel S, Scharf SJ, Higuchi R, et al. 1988. Primer-directed enzymatic amplification of DNA with thermostable DNA polymerase. Science 239: 487-491.

Santi MR, Ikonomovic S, Wroblewski JT, Grayson DR. 1994. Temporal and depolarization-induced changes in the absolute amounts of mRNAs encoding metabotropic glutamate receptors in cerebellar granule neurons in vitro. J Neurochem 63: 1207-1217.

Sambrook J, Fritsch EF, Maniatis T. 1989. Molecular cloning: a laboratory manual. Cold Sping Harbor, NY: Cold Spring Harbor Laboratory.

Wang AM, Doyle MV, Mark DF. 1989. Quantitation of mRNA by the polymerase chain reaction. Proc Natl Acad Sci USA 86: 9717-9721.

Zheng T, Santi MR, Marlier LNJ-L, Grayson DR. 1993. Expression of the mRNA encoding the (6 subunit of the $GABA_A$ receptor occurs in cerebellar granule cells only after completion of migration to the internal granule cell layers. Dev Brain Res 75: 91-103.

Zuccotti M, Monk M. 1995. Methylation of the mouse Xist gene in sperm and eggs correlates with imprinted Xist expression and paternal X-inactivation. Nat Genet 9: 316-320.

16 In Situ Hybridization

Alan N. Bateson

Abstract: In situ hybridisation is a powerful technique for determining the distribution of specific mRNA species in tissues such as brain that comprise different cell types. Its utility in molecular neurobiology has driven the development of a variety of variations of this technique will be reviewed.

List of Abbreviations: BCIP, 5-bromo-4-chloro-3-indolyl phosphate, toluidine salt; CCD, charged-coupled device; CNS, central nervous system; cDNA, copy or complimentary RNA; cRNA, copy or complimentary RNA; DNase, deoxyribonuclease; FITC, fluorescein isothiocyanate; mRNA, messenger RNA; NBT, nitro blue tetrazolium choride; PCR, polymerase chain reaction; RNase, ribonclease

1 Introduction

The unique identity of a cell type is determined by the profile of genes it expresses through its development and during its fully differentiated state. Thus, the vast cellular heterogeneity of the central nervous system is specified by differential gene expression. Indeed, of the approximately 30,000 genes present in the human genome, 50% are thought to be expressed in brain (Sandberg et al., 2000). As a consequence, central to our understanding the properties of neurons is our ability to determine which genes they each express.

One of the most widely used techniques in neurobiology for the examination of neuronal gene expression is in situ hybridization histochemistry. This relatively straightforward technique allows the determination of the spatial and temporal distribution of the initial product of expression of a specific gene, namely its messenger ribonucleic acid (mRNA), at different levels of resolution, from regional distribution patterns through cellular to subcellular mRNA localization. In situ hybridization histochemistry utilizes the single-stranded nature of mRNA and the ability of nucleic acid probes of complementary sequence to specifically hybridize. Appropriate labeling of the nucleic acid probe and detection of that label allows the distribution of the mRNA species corresponding to the probe to be determined. This review is not intended to provide detailed protocols for conducting in situ hybridization histochemistry experiments; these can be found elsewhere. Instead, we give an overview of common variations of the basic methodology, examples of standard applications of the technique, and some of its more novel uses, illustrated by using examples from the primary literature.

2 In Situ Hybridization Probes

Two types of molecules are most commonly used as in situ hybridization probes, namely RNA probes and DNA oligonucleotides probes. RNA probes are synthesized in vitro from cDNA sequences that correspond to the mRNA of interest. cDNA sequences are cloned into plasmid vectors adjacent to bacterial RNA polymerase promoters such as T7 or Sp6 such that RNA molecules can be synthesized in vitro using the appropriate RNA polymerases driving transcription from these promoters. These constructs are initially linearized by restriction enzyme digestion prior to the in vitro generation of RNA with bacterial RNA polymerases using the cDNA insert as the template. Complementary RNA, sometimes called cRNA, is synthesized, which is the complement (or anti sense) to the sense strand and therefore has the ability to hybridize with the corresponding mRNA sequence. These probes may be full length, covering the whole transcript including the 5′-untranslated, coding, and 3′-untranslated regions, or they may only correspond to part of the target mRNA transcript. The latter approach is most often used when examining the distribution patterns of mRNA species that are encoded by members of a gene family in which there are regions of significant sequence homology among different family members. In these situations, there is the possibility that a full-length probe may detect not only the transcript of interest but also transcripts of closely related members of its gene family. In the central nervous system (CNS), there are many examples of closely related gene families that are expressed with discrete or overlapping patterns (e.g. $GABA_A$ receptor subunit genes), however, with due consideration to the possibility of cRNA probe cross-hybridization and the use of appropriate controls, false positive signals can be avoided.

Most protocols that use RNA probes recommend stringent procedures to ensure the absence, as far as possible, of contaminating RNase during the probe preparation and hybridization procedures. RNase is resistant to heat denaturation and is therefore not destroyed by standard autoclaving. Removal of RNase entails the treatment of water with diethylpyrocarbonate prior to autoclaving for the preparation of all solutions, the baking of glassware at 180°C, the use of RNase-free sterile plastic ware and the inclusion of RNase inhibitors during probe or tissue preparation (see, e.g., Suzuki et al., 1999). Too stringent a control of RNase levels may not be necessary, as it has been reported that provided contaminating RNase levels are kept below 0.1 μg/ml under less stringent conditions, which are cheaper and easier to perform, it is sufficient (Tongiorgi et al., 1998).

Oligonucleotide probes are synthesized chemically rather than by in vitro biological systems. Short DNA oligonucleotides are relatively cheap, readily available commercially, and can be used "off the shelf," requiring no further manipulation upon receipt other than dilution to appropriate working concentrations. They can be stored almost indefinitely and being DNA, are not susceptible to RNase degradation whereas contaminating DNase is destroyed by standard autoclaving. Oligonucleotide probes are normally between 30 and 60 nucleotides in length, with 45 being most common. In designing oligonucleotide probes (Bateson and Darlison, 1992; Erdtmann-Vourliotis et al., 1999; Ambesi-Impiombato et al., 2003) it is important to avoid sequences in which guanine and cytosine are either under- or overrepresented (AT or GC rich) or which have runs of a single base. Further, sequences that display intramolecular complementarity (hairpin formation) or intermolecular complementarity (probe–dimer formation) should also be avoided. A number of software packages are available to aid this process, including freeware such as Primer3 (http://frodo.wi.mit.edu/cgi-bin/primer3/primer3_www.cgi).

As they are short, it is relatively easy to identify in silico regions of a transcript that are unique to a target mRNA sequence, thereby avoiding the possibility of cross hybridization to related gene family members (see, e.g., Wisden et al., 1992). Caution should be taken, however, to ensure that part of the probe does not have a contiguous region of homology with an unrelated sequence. This can also be achieved in silico by performing BLAST searches of the relevant genome database (http://www.ncbi.nlm.nih.gov/BLAST/) under conditions that look for stretches of sequence identity with no gaps rather than a high degree of overall sequence identity. There is, however, no foolproof method of choosing the most appropriate sequence for an oligonucleotide probe, and a degree of empirical testing must be anticipated. The low cost of commercial oligonucleotides makes this feasible, and the most rational approach to take when investigating the distribution of a new mRNA sequence is to design at least three separate oligonucleotide probes, taking into account the parameters mentioned earlier, and use these in separate experiments. If all three recognize the same transcript specifically, this will be apparent in the distribution pattern visualized. Any probes that either show a high signal-to-noise ratio or reveal a different distribution pattern should be discarded and a new probe designed and tested. Even when examining mRNA species of supposedly well-known gene families, artifacts can become apparent (Mladinic et al., 2000), although it is likely that these experiences are most often not reported in the literature.

An examination of the in situ hybridization literature suggests that the choice of control used depends to a certain extent upon the particular type of probe employed, although this appears to be largely for historical reasons. The use of multiple probes that correspond to nonoverlapping sections of the target mRNA are widely, although not always, incorporated into oligonucleotide probe protocols as indicated earlier. This approach can also be used taken with cRNA probes, although this is less common. The widespread application of the polymerase chain reaction (PCR) to the isolation and cloning of specific cDNA sequences means that there is little to prevent the same strategy being applied when using cRNA probes. In contrast to this positive control, negative controls can be performed by employing sense probes. Again, these can be incorporated into either cRNA or oligonucleotide protocols, and are often considered useful in assessing the extent of background signal obtained. A problem arises with this type of control, however, because it may be the probe's sequence itself that is the cause of a high background signal. Thus, although these controls can be useful in identifying optimal conditions, such as probe density or washing conditions that will minimize background, this will only be useful if the sense probe is not specifically hybridizing to an unrelated transcript. A more useful negative control is that of RNase pretreatment of the sample. This allows assessment of the degree of nonspecific "sticking" of the probe to the considerable amount of non-mRNA components present in the tissue section, that is, the background signal.

3 Probe Labeling and Detection

Three types of detection methods can be used for in situ hybridization: radioactive, fluorescent, and chromogenic. Radioactive detection techniques are the most commonly used primarily because they are more sensitive, but also because the other techniques are relatively more recent in terms of the development of the appropriate chemistry and the techniques that enable in vitro labeling of probes by these nonradioactive means.

Radioactive probes are synthesized in vitro by either incorporating radiolabeled ribonucleotides into the nascent cRNA during its synthesis or by adding radiolabeled deoxyribonucleotides to the 3′ end of oligonucleotide probes with terminal transferase. In performing the latter technique, often referred to as "tailing," due cognizance must be taken of the fact that a longer tail means greater specific activity but decreased specificity. ^{32}P-, ^{33}P-, or ^{35}S-labeled ribonucleotides can be used for cRNA probes, but oligonucleotide probes are normally labeled with ^{35}S. cRNA probes can be labeled to a higher specific activity than oligonucleotide probes and hence are considered more sensitive.

For many the choice of isotope employed for radioactive labeling of in situ hybridization probes is determined by the competing needs of sensitivity and resolution. ^{32}P-labeled probes give the best sensitivity due to the high energy β particles emitted by this isotope, whereas ^{35}S-labeled probes emit β particles of lower energy, giving rise to better resolution. Detection of β particles can be augmented by the use of intensifying screens that incorporate a scintillant, thereby allowing an amplification of the signal when detected by film autoradiography. This process is carried out at −80°C, and many workers believe that this is because the scintillation screens only work at this temperature. In fact, it is because the film is only sensitive to the wavelength of light emitted by the screens when bombarded with β particles at −80°C. Because β particles emitting from a point source of a ^{32}P atom can exit the tissue diagonally (due to their high energy they are not stopped by the thickness of tissue used), the signal detected is more diffuse than a weaker β particle-emitter such as ^{35}S. Consequently, ^{35}S-labeled probes tend to be more frequently used for cellular and subcellular localizations of mRNA transcripts compared to ^{32}P-labeled probes. The isotope ^{33}P emits β particles with an energy that is intermediate between that of ^{32}P and ^{35}S and has found favor with some workers; however, ^{35}S-labeled probes (whether cRNA or oligonucleotide) are probably the most widely used radiolabeled probes used for in situ hybridization.

The nature of the radioactive label needs to be taken into consideration when choosing the detection procedure. Traditionally, X-ray film or direct emulsion autoradiography were the choices; however, phosphor imaging systems (see, e.g.,Vizi and Gulya, 2000; Key et al., 2001) are also now available. The expense of these latter systems makes X-ray film or direct emulsion autoradiography still the most common methods used in academia. X-ray film autoradiography provides gross anatomical localization and works with both ^{32}P- and ^{35}S-labeled probes, the former being more sensitive. The greater resolution afforded by ^{35}S-labeled probes does result in considerable advantage when combined with X-ray film autoradiography but is much more evident when direct emulsion visualization is employed.

Despite the continued popularity of radiolabeled probes for in situ hybridization, fluorescent and chromogenic methods are increasingly being used, particularly for medium to high abundance transcripts. These alternatives to radiolabeling also allow dual- and triple-labeling protocols to be used for the simultaneous analysis of specific transcripts. Essentially, these methods borrow heavily from methods used to detect proteins by immunohistochemistry. Thus, nucleic acid probes are synthesized with modified nucleotides that can be detected by labeled antibodies. Consequently, unlike radioactively labeled probes, fluorescent and chromogenic methods are indirect. Many variations of these methods exist, but only a few general examples will be mentioned. cRNA molecules are synthesized by in vitro transcription under conditions that allow the incorporation of digoxigenin- (Kerner et al., 1998; Meltzer et al., 1998; Tongiorgi et al., 1998; Luo and Jackson, 1999) or biotin-conjugated (Grino and Zamora, 1998; Tongiorgi et al., 1998) ribonucleotides. Oligonucleotide probes can be similarly tailed with modified deoxyribonucleotides (Tata, 2001). Following hybridization, antibodies to digoxigenin or biotin are bound to the hybridized probes, with the detection method determined by the reporter molecule conjugated to the antibody. Alkaline phosphatase-conjugated antibodies are incubated with chromogenic substrates (Kerner et al., 1998;

Tongiorgi et al., 1998) nitro blue tetrazolium chloride (NBT) and 5-bromo-4-chloro-3-indolyl phosphate, toluidine salt (BCIP), which gives rise to a blue precipitate, although other chromogens, which can be detected by light microscopy, have been reported (Stern, 1998). Biotin-labeled probes can be detected by streptavidin conjugated to horseradish peroxidase and the signal subsequently amplified by a tyramide reaction (Grino and Zamora, 1998; Tata, 2001). This tyramide amplification step initially found success for the boosting of immunohistochemical signals. Tyramide, conjugated to a fluorophore (Grino and Zamora, 1998) or biotin (Tata, 2001), is activated by peroxidase to form a short-lived radical that couples to tyrosine residues. The short half-life of the activated tyramide means that deposition is localized. Tyramide is then revealed by a streptavidin-fluorophore (e.g., streptavidin-FITC), if tyramide-biotin is used, followed by fluorescence microscopy detection, or directly by fluorescence microscopy if a tyramide-fluorophore conjugate is used. Although it is possible to detect fluorescence signals using film autoradiography, this is rarely used, and fluorescence microscopy is the method of choice.

The complexities of protocols for fluorescent and chromogenic in situ hybridization necessarily entail careful attention to controls. In particular, the possibility of native enzyme activity or the presence of endogenous biotin in the experimental tissue should be considered, though this can be addressed by exposing control tissue to the detection system in the absence of probe. The relative merits of digoxigenin versus biotin, and some of the technical problems associated with each, have been previously discussed (Chevalier et al., 1997; Luo and Jackson, 1999).

Compared to detection by radioactivity, fluorescent and chromogenic detection of oligonucleotide probes is less sensitive, again presumably due to the lower specific activity that can be obtained with this sort of probe relative to that which can be obtained with cRNA probes (Kerner et al., 1998). Although the sensitivity of radioactively-labeled probes is greater than that achievable with nonradioactive probes, the difference is not as large as it once was. In situ hybridization with nonradioactively labeled probes is increasingly being used, particularly for medium- to high-abundance transcripts and they have clear safety advantages over the traditional radiolabeled probes. The use of nonradioactive probes is only likely to increase because as the techniques for incorporation of modified nucleotides becomes more efficient, newer non-radioactive chemistries with greater amplification power are developed and detection systems improved.

4 Localization

Regional localization of specific mRNA species is achieved with radioactively-labeled probes and X-ray film autoradiography (Erdtmann-Vourliotis et al., 1999; Ambesi-Impiombato et al., 2003) or phosphor image acquisition (Vizi and Gulya, 2000; Key et al., 2001). Subsequent counterstaining of sections with standard histochemical techniques allows the assignment of autoradiographic signals to previously defined anatomical regions, such as those of Paxinos and Watson (1997) for the rat brain.

The localization of transcripts to individual cell bodies can be determined using all types of probe labeling procedures, although the detection methods differ. Liquid film emulsion is used to directly coat sections hybridized with radioactively labeled probes. Following exposure and development, light- (Kerner et al., 1998) or dark-field (Bateson et al., 1991) illumination microscopy can be used to analyze the resultant silver grains. Chromogenic (Kerner et al., 1998; Tata, 2001) and fluorescence (Grino and Zamora, 1998) detection are achieved by light and fluorescence microscopy, respectively. Further, by labeling probes that detect different mRNA species either with different fluorophores, or with a fluorophore and a chromogenic or radioactive label, multiple mRNA species can be detected at the single cell level (Grino and Zamora, 1998; Kerner et al., 1998). Indeed, confocal microscopy has been used to detect such multiply labeled mRNAs with much success (Grino and Zamora, 1998).

Perhaps one of the most exciting areas in neuroscience to which in situ hybridization has made significant contributions in recent years is the discovery that specific mRNA transcripts can be localized to certain compartments of neuronal cells (Steward and Schuman, 2003), such as axons (Tohda, 2003), dendrites (Ma and Morris, 2002; Böckers et al., 2004), cell bodies (Lu et al., 1998), and growth cones (Zhang et al., 1999). Neurons exhibit a polarized morphology, with different parts of a neuron displaying specific functional roles. The discovery of localized protein synthetic machinery suggested that part of a neuron's ability to rapidly respond to a changing extracellular signaling or environment lay in the ability to synthesize particular proteins close to the sites of the cell they are required, rather than in the cell body, followed by transport along neurites (Steward and Schuman, 2003). In situ hybridization at the subcellular level, either by light (Böckers et al., 2004) or electron (Ma and Morris, 2002) microscopy, has demonstrated that specific mRNA species are targeted to these sites where the mRNA molecules are thought to be held "in reserve" for translation upon the appropriate signal (Steward and Schuman, 2003).

5 Sample Preparation

In situ hybridization can be applied to cultured neuronal or glial cells, often grown on treated slides (e.g., poly-D-lysine) that are fixed directly on the slides on which they are grown (Luo and Jackson, 1999). Wholemounts of dissected tissue, such as cochlear (Judice et al., 2002), spinal cord (Shifman and Selzer, 2000) or dorsal root ganlia (Tata, 2001), and whole embryos (Luque et al., 1998) have been used. Two methods of fixation of brain tissue are widely used. Under deep, terminal anesthesia, perfusion of brain tissue with fixative is performed via the aorta (Tongiorgi et al., 1998). Alternatively, animals are sacrificed, brains quickly removed, immediately frozen on dry ice (Grino and Zamora, 1998; Erdtmann-Vourliotis et al., 1999) or subzero isopentane (Kerner et al., 1998), and stored frozen at -80°C. Fixing of this tissue takes place after sections are cut and thaw-mounted onto glass slides (Grino and Zamora, 1998; Erdtmann-Vourliotis et al., 1999). Brains fixed by perfusion are also cut to produce sections that can either be mounted onto glass slides (Cloez-Tayarani and Fillion, 1997) or can be subjected to probe hybridization as free-floating sections (Tongiorgi et al., 1998).

Perfusion of brain tissue is not as easy to perform as immediate freezing following removal of fresh tissue from the animal. The former technique has long been preferred for immunohistochemistry, but it has been argued that fresh-frozen sections are better for analysis of mRNA distribution and can be adapted for immunohistochemical analysis (Newton et al., 2002).

Postmortem human brain tissue is obviously not subject to as rigorous control as experimental tissue from laboratory animals. As such much care needs to be taken when attempting in situ hybridization with human brain. In addition to the normal factors such as age and gender that need to be controlled with human samples, other factors, such as the immediate premortem state of the individual, the time taken for tissue recovery, and the storage conditions, must also be considered (for review, see Hynd et al., 2003).

6 Quantification

Although in situ hybridization is predominantly used in the qualitative assessment of mRNA localization patterns many workers attempt to quantify their results. Often this merely takes the form of what is sometimes called "semi-quantitative" analysis in which descriptors, such as weak, strong, very strong, and so on, are assigned to the signal intensities observed (see, e.g., Wisden et al., 1992). True quantification of signal strength is easily obtained using automated imaging detection methods but can also be achieved with film detection, using appropriate standards, combined with digital capture (CCD camera or high-quality scanner) of the resulting autoradiographic image and subsequent analysis with appropriate software such as NIH Image (http://rsb.info.nih.gov/nih-image/) or ImageJ (http://rsb.info.nih.gov/ij/). An issue that is sometimes overlooked, however, is what the nature of the mathematical relationship is between the signal obtained and the true level of target transcript. Other factors that also affect the ability to accurately quantify the target transcript include the ability of the probe to uniformly penetrate a tissue and the dynamic range of the detection system, which can be relatively small for autoradiographic film-based detection.

Prior to data analysis, some preprocessing may be desirable or necessary. Normalization of acquired images to regions that should contain no mRNA (e.g., corpus callosum), which can reduce signal variance (see Ambesi-Impiombato et al., 2003). Algorithms have also been developed that allow the comparison of signals generated by multiple probes (Zhao et al., 1999) or the reconstruction of 3-dimensional in situ hybridization patterns and the averaging of image sets from multiple brain samples (Ginsberg et al., 1996).

7 Developing Areas

A number of advancements have been made recently that are promising to improve in situ hybridization techniques or widen its applicability. The analysis of the rate of gene expression has been reported by the use of intron-specific probes that allow the examination of nascent mRNA levels, that is, before processing of the immature transcript has occurred (Chang et al., 2000). Combining the polymerase chain reaction (PCR) into in situ hybridization protocols has always been an attractive goal, but the very sensitivity of PCR can lead to significant background problems (Nuovo, 2001). Three-dimensional reconstruction of confocal images of fluorescence in situ hybridization allows automated analysis of hybridization signals in specific brain nuclei and should allow the analysis of large data sets (Chawla et al., 2004). High-throughput applications are spreading throughout biological research and in situ hybridization is not being excluded. Following on from a microarray analysis of gene expression in the mouse hippocampus, Fred Gage's group conducted in situ hybridization in this brain region with over 100 probes (Lein et al., 2004).

8 Conclusion

The CNS is the most complex organ in the body and contains a wider range of different cells than any other organ. The identity of each cell is determined by its gene expression pattern, which itself is subject to alteration as the organism develops and responds to its environment. In situ hybridization techniques have proved to be powerful tools in the hands of neurobiologists for the study of neuronal and glial gene expression. Further improvements in sensitivity, resolution, reliability, and throughput of this technology will no doubt be forthcoming and help to provide greater insight into the structure and function of the CNS.

References

Ambesi-Impiombato A, D'Urso G, Muscettola G, de Bartolomeis A. 2003. Method for quantitative in situ hybridization histochemistry and image analysis applied for Homer1a gene expression in rat brain. Brain Res Prot 11: 189-196.

Bateson AN, Darlison MG. 1992. The design and use of oligonucleotides. Longstaff A, Revest P, *A Laboratory Manual in Molecular Biology,* Protocols in Molecular Neurobiology. Vol. 13: , Humana Press, Totowa, New Jersey: pp. 55-66. editors. Chapter 4;

Bateson AN, Harvey RJ, Wisden W, Glencorse TA, Hicks AA, et al. 1991. The chicken $GABA_A$ receptor α1-subunit: cDNA sequence and localization of the corresponding mRNA. Mol Brain Res 9: 333-339.

Böckers TM, Segger-Junius M, Iglauer P, Bockmann J, Gundelfinger ED, et al. 2004. Differential expression and dendritic transcript localization of Shank family members: identification of a dendritic targeting element in the 3V untranslated region of Shank1 mRNA. Mol Cell Neurosci 26: 182-190.

Chang M-S, Hahn MK, Sved AF, Zigmond MJ, Austin MC, et al. 2000. Analysis of tyrosine hydroxylase gene transcription using an intron specific probe. J Neurosci Meth 94: 177-185.

Chawla MK, Lin G, Olson K, Vazdarjanova A, Burke SN, et al. 2004. 3D-catFISH: a system for automated quantitative three-dimensional compartmental analysis of temporal gene transcription activity imaged by fluorescence in situ hybridization. J Neurosci Meth 139: 13-24.

Chevalier J, Yi J, Michel O, Tang X-M. 1997. Biotin and digoxigenin as labels for light and electron microscopy in situ hybridization process: where do we stand? J Histochem Cytochem 45: 481-491.

Cloez-Tayarani I, Fillion G. 1997. The in situ hybridization and immunocytochemistry techniques for characterization of cells expressing specific mRNAs in paraffin-embedded brains. Brain Res Prot 1: 195-202.

Erdtmann-Vourliotis M, Mayer P, Riechert U, Handel M, Kriebitzsch J, et al. 1999. Rational design of oligonucleotide probes to avoid optimization steps in in situ hybridization. Brain Res Prot 4: 82-91.

Ginsberg MD, Zhao W, Singer JT, Alonso OF, Loor-Estades Y, et al. 1996. Computer-assisted image-averaging strategies for the topographic analysis of in situ hybridization autoradiographs. J Neurosci Meth 68: 225-233.

Grino M, Zamora AJ. 1998. An in situ hybridization histochemistry technique allowing simultaneous visualization by the use of confocal microscopy of three cellular mRNA species in individual neurons. J Histochem Cytochem 46: 753-759.

Hynd MR, Lewohl JM, Scott HL, Dodd PR. 2003. Biochemical and molecular studies using human autopsy brain tissue. J Neurochem 85: 543-562.

Judice TN, Nelson NC, Beisel CL, Delimont DC, Fritzsch B, et al. 2002. Cochlear whole mount in situ hybridization: identification of longitudinal and radial gradients. Brain Res Prot 9: 65-76.

Kerner JA, Standaert DA, Penney JB Jr, Young AB, Landwehrmeyer GB. 1998. Simultaneous isotopic and nonisotopic in situ hybridization histochemistry with cRNA probes. Brain Res Prot 3: 22-32.

Key M, Wirick B, Cool D, Morris M. 2001. Quantitative in situ hybridization for peptide mRNAs in mouse brain. Brain Res Prot 8: 8-15.

Lein ES, Zhao X, Gage FH. 2004. Defining a molecular atlas of the hippocampus using DNA microarrays and high-throughput in situ hybridization. J Neurosci 24: 3879-3889.

Lu Z, McLaren RS, Winters CA, Ralston E. 1998. Ribosome association contributes to restricting mRNAs to the cell body of hippocampal neurons. Mol Cell Neurosci 12: 363-375.

Luo L-G, Jackson IMD. 1999. Advantage of double labeled in situ hybridization for detecting the effects of glucocorticoids on the mRNAs of protooncogenes and neural peptides TRH in cultured hypothalamic neurons. Brain Res Prot 4: 201-208.

Luque JM, Adams WB, Nicholls JG. 1998. Procedures for whole-mount immunohistochemistry and in situ hybridization of immature mammalian CNS. Brain Res Prot 2: 165-173.

Ma D, Morris JF. 2002. Protein synthetic machinery in the dendrites of the magnocellular neurosecretory neurons of wild-type Long-Evans and homozygous Brattleboro rats. J Histochem Cytochem 23: 171-186.

Meltzer JC, Sanders V, Grimm PC, Stern E, Rivier C, et al. 1998. Production of digoxigenin-labelled RNA probes and the detection of cytokine mRNA in rat spleen and brain by in situ hybridization. Brain Res Prot 2: 339-351.

Mladinic M, Frederic Didelon F, Cherubini E, Bradbury A. 2000. 'Specific' oligonucleotides often recognize more than one gene: the limits of in situ hybridization applied to GABA receptors. J Neurosci Meth 98: 33-42.

Newton SS, Dow A, Terwilliger R, Duman R. 2002. A simplified method for combined immunohistochemistry and in-situ hybridization in fresh-frozen, cryocut mouse brain sections. Brain Res Prot 9: 214-219.

Nuovo GJ. 2001. Co-labeling using in situ PCR: a review. J Histochem Cytochem 49: 1329-1339.

Paxinos G, Watson C. 1997. *The rat brain stereotaxic coordinates.* New York: Academic.

Sandberg R, Yasuda R, Pankratz DG, Carter TA, Del Rio JA, et al. 2000. Regional and strain-specific gene expression mapping in the adult mouse brain. Proc Natl Acad Sci USA 97: 11038-11043.

Shifman MI, Selzer ME. 2000. In situ hybridization in whole-mounted lamprey spinal cord: localization of netrin mRNA expression. J Neurosci Meth 104: 19-25.

Stern CA. 1995. Detection of multiple gene products simultaneously by in situ hybridization histochemistry and immunohistochemistry in while mounts of avian embryos. Curr Top Dev Biol 36: 223-243.

Steward O, Schuman EM. 2003. Compartmentalized synthesis and degradation of proteins in neurons. Neuron 40: 347-359.

Suzuki T, Ogata A, Tashiro K, Nagashima K, Tamura M, et al. 1999. A method for detection of a cytokine and its mRNA in the central nervous system of the developing rat. Brain Res Prot 4: 271-279.

Tata, AM. 2001. An in situ hybridization protocol to detect rare mRNA expressed in neural tissue using biotin-labelled oligonucleotide probes, Brain Res Prot 6: 178-184.

Tohda C. 2003. Comprehensive identifying method for localized mRNAs in single neuronal axons. J Biochem Biophys Meth 57: 57-63.

Tongiorgi E, Righi M, Cattaneo A. 1998. A non-radioactive in situ hybridization method that does not require RNAse-free conditions. J Neurosci Meth 85: 129-139.

Wisden W, Laurie DJ, Monyer H, Seeburg PH. 1992. The distribution of 13 $GABA_A$ receptor subunit mRNAs in the rat brain. I. Telencephalon, diencephalon, mesencephalon. J Neurosci 12: 1040-10462.

Vizi S, Gulya K. 2000. Calculation of maximal hybridization capacity (Hmax) for quantitative in situ hybridization: a case study for multiple calmodulin mRNAs. J Histochem Cytochem 48: 893-904.

Zhang HL, Byrd AL, Singer RH, Bassell GJ. 1999. Neurotrophin regulation of β-actin mRNA and protein localization within growth cones. J. Cell Biol. 147: 59-70.

Zhao W, Truettner J, Schmidt-Kastner R, Belayev L, Ginsberg MD. 1999. Quantitation of multiple gene expression by in situ hybridization autoradiography: accurate normalization using Bayes classifier. J Neurosci Meth 88: 63-70.

17 Analysis of Gene Expression in the Brain Using Differential Display

K. L. Gilby · E. M. Denovan-Wright

Abstract: The utility of differential display as a method of screening for differences in gene expression between a number of control and experimental conditions using small amounts of RNA and standard equipment and reagents found in most molecular biology laboratories has been demonstrated over the last 15 years. This chapter will discuss the advantages and limitations of differential display compared to some other methods used to study gene expression patterns and discuss some parameters of experimental design that should be considered prior to initiating a differential display screen. Methods for differential display reactions and verification of differential gene expression are presented.

List of Abbreviations: cDNA, complementary DNA; ddH$_2$O, double-distilled H$_2$O; dNTP, deoxyribonucleotide triphosphate; EDTA, ethylenediaminetetraacetic acid; MgCl$_2$, magnesium chloride; mRNA, messenger ribonucleic acid; NaOH, sodium hydroxide; PCR, polymerase chain reaction; qRT PCR, quantitative reverse transcriptase polymerase chain reaction; RNase, ribonuclease; RT PCR, reverse transcriptase polymerase chain reaction; UTR, untranslated region

1 Introduction

Even slight variations in gene expression impact cellular function. Thus, examination of differential gene expression and its cellular repercussion—dubbed expression genetics or functional genomics—has become an essential tool in the study of brain function. In the past decade, an arsenal of different technologies has become available to study and compare gene expression on a cellular, structural, or systems level. Differential display employs two of the most simple and commonly used molecular biological tools, namely PCR and DNA sequencing, to systematically amplify, visualize, and identify differences in mRNA levels among cell types or tissues (Liang, 2002). Since its introduction in 1992, differential display has been applied to nearly all areas of biological and biomedical research (Matz and Lukyanov, 1998; To, 2000). More than 4,000 articles have been published that have made use of differential display to identify changes in gene expression, which exceeds the number of articles that have used alternative gene expression analysis technologies including subtractive hybridization, representational difference analysis (RDA), serial analysis of gene expression (SAGE) and DNA microarrays (Liang, 2002; Liang and Pardee, 2003).

2 Advantages and Limitations of Differential Display

The majority of "gene screening" technologies are based on the same, conceptually simple, principle whereby total or polyA RNA is isolated from cell populations of interest and converted to cDNA by reverse transcription and the levels of specific cDNAs are compared between samples. As such, many of the same technical considerations for the preservation and isolation of tissues of interest, which impact the quality of isolated RNA, apply to all methods used for analysis of differential gene expression (Sandberg et al., 2000). However, once the RNA is extracted and purified, the technical demand, sensitivity, and ultimate capability of these techniques diverge. Advances in molecular and computational biology have provided high-throughput screening methods such as microarray analysis of gene expression. Although these methods are increasing in popularity and contributing significantly to our understanding of gene expression, differential display still has considerable utility. Comprehensive comparisons of various screening methods and the distinctive, and potentially advantageous, qualities of differential display have been reviewed (Ahmed, 2002; Stein and Liang, 2002; Bartlett, 2003; Liang and Pardee, 2003; Ding and Cantor, 2004).

The major differences between various screening methods include differences in the amount of starting RNA required, sensitivities to high and low abundance RNAs, and the detection/visualization methodology used. Due to the recent predilection toward microarray technology, advantages of differential display will primarily be referenced to microarray procedures. Differential display can be used to efficiently compare samples derived from small amounts of RNA (10 ηg of total RNA per reaction) whereas microarray technology requires significantly larger amounts of polyA RNA per comparison ($\geq$10 μg) (Eberwine et al., 2001; Stein and Liang, 2002). Moreover, differential display is capable of detecting expression differences that are as subtle as 1.1–1.5-fold (Liao and Freedman, 1998; Kuno et al., 2000) whereas the criterion for defining differences between samples in microarray analysis is generally set at greater than twofold. In response to the concern that differential display is heavily biased to high copy number mRNAs, Wan and colleagues (1996) have demonstrated that differential display can detect mRNA species with a prevalence of 1 in 20,000 transcripts. Thus, differential display is a more sensitive procedure that can be applied to systems for which biological samples are limited without the need to combine (pool) individual samples. In fact, many studies that have used microarrays as their screening technique have had to introduce an additional, potentially confounding, RNA amplification step to obtain the required amount of starting RNA for analysis. Moreover, unlike microarray analysis, differential display can be used to detect mRNAs that differ by size or sequence. For example, small deletions were detected in the regulatory 3′UTR of the human alpha-tropomyosin gene using differential display which manifested as a size shift instead of an absence or presence of cDNA band when RNA samples from different individuals were compared (Planitzer et al., 1998).

DNA microarray analysis utilizes hybridization of complex cDNA probes to a mostly incomplete set of cDNA templates (Liang, 2002). In contrast, the number of primer combinations and reactions that an investigator is willing to perform is the limiting factor in differential display screening. Results generated via differential display are also not subject to variability introduced by probe labeling biases (Cy3 or Cy5 incorporation levels), which require additional reciprocal labeling comparisons, or the variability introduced by hybridization and washing conditions (Brown and Botstein, 1999; Geschwind, 2000; Lockhart and Winzeler, 2000). Further, the complexity of the cDNA probes used in microarray analysis has recently been called into question as the rate of cross-hybridization between complex cDNA probes and cDNA sequences that share regions of sequence similarity has yet to be determined (Liang, 2002, Liang and Pardee, 2003). In light of the fact that many genes share numerous conserved sequences among family members, cross-hybridization is a valid concern. It is also important to note that cDNA sequence error rates on arrays have been shown to be as high as 30% (Halgren et al., 2001; Liang, 2002).

Perhaps, the most compelling reason for using differential display as a primary screening strategy is the fact that it allows for simultaneous direct comparison of gene expression among individual samples. In light of the fact that gene expression is regulated developmentally, spatially, temporally, environmentally, and pathologically, the ability to compare gene expression patterns simultaneously in a large number of individual samples is a powerful advantage when screening to generate testable hypotheses regarding genes that are important in particular biological situations. In contrast, the majority of screening technologies, including DNA microarrays and subtractive hybridization, are limited to the direct comparison of only two RNA populations. If the research goal of a given experiment is to identify genes critical to a specific biological process, comparisons between control conditions and a number of samples per treatment group are required. Such equivalent broad comparison screening is neither easy nor cost-effective using current DNA microarray technology. Ultimately, however, it is important to note that all gene-expression profiling strategies should be considered prone to error and require further confirmation using more traditional RNA expression comparison methodologies such as Northern blot, in situ hybridization, RNase protection assays and quantitative PCR (Liang and Pardee, 2003).

3 Experimental Design

Execution of the differential display procedure is relatively simple and financially feasible for most laboratories. It is believed by some, however, that differential display is an inefficient screening technique because the reactions generate a large number of false positive bands. A number of technical factors can influence the rate of false positive results generated by differential display including reagent and enzyme quality, reaction setup, criterion for picking candidate bands, types of tubes and thermal cyclers, pipetting errors and most importantly primer design (Cho et al., 2002). To address the methodological concerns, several modifications to original differential display protocols have been made (Liang et al., 1993; Liang and Pardee, 1995; Simon and Oppenheimer, 1996; Matz et al., 1997; Jurecic et al., 1998; Matz and Lukyanov, 1998; Cho et al., 2002). If the quality of cDNA is consistent among samples and positive control reactions are included in the first sets of differential display reactions, the techniques can be readily optimized and the number of false positive bands will be minimized.

Appropriate experimental design will also limit the number of false positive products observed and increase the probability of finding differential display products that are relevant to the biological paradigm (Stein and Liang, 2002). As such, several guidelines should be adhered to when conducting differential display experiments. For instance, drastically different conditions should not be compared, as many genes will be differentially expressed making evaluation of the contribution of any one gene difference difficult (Stein and Liang, 2002). In addition, differential display reactions for each sample should always be replicated in order to quickly distinguish artifacts, generated through reverse transcription and PCR, from differences in gene expression. Most importantly, multiple within-group samples should be used for each differential display experiment to distinguish within-population variation from expression differences induced by the condition or treatment (Stein and Liang, 2002). An example that illustrates the need for multiple within-group comparisons is provided in *Figure 17-1*. In this example, differential display

Figure 17-1

Differential display analysis of hippocampal RNA isolated from individual out-bred Long Evans Hooded (LEH), seizure-prone (SP), and seizure-resistant (SR) rats

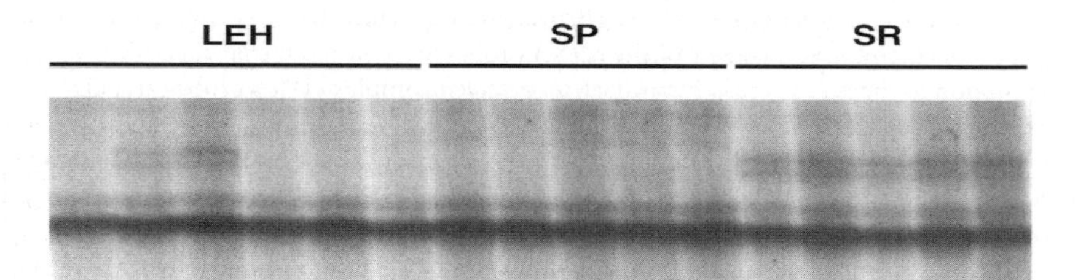

was used to examine differences in constitutive gene expression related to seizure-susceptibility in rats. Seizure-prone and seizure-resistant rat strains examined in this study were selectively bred over 40 years in order to establish stable seizure-susceptibility phenotypes from initial matings of Long Evans Hooded and Wistar rats (Racine et al., 1999). Results of this study showed considerable variability in constitutive gene expression within limbic structures of individual out-bred Long Evans Hooded rats (*Figure 17-1*). Although individual Long Evans Hooded rats exhibit varying degrees of seizure-susceptibility, it was not possible to identify changes in gene expression that directly related to such susceptibility within the heterogeneous population of animals. In contrast, seizure-prone and seizure-resistant rat strains showed a very high degree of homogeneity in constitutive gene expression within limbic structures and major differences in gene expression between strains could, therefore, be identified. This finding illustrates that, when using out-bred animals, within-population variability is an issue that must be factored into gene expression analysis. The inclusion of multiple individual (not pooled) samples per condition, brain structure, or time-point will increase the rigor when selecting positive differential display bands for analysis. As a general rule, design of a differential display experiment should be questioned when more than 5% of the transcripts are differentially expressed or if no differences in expression are identified after ten or more sets of differential display reactions using various primer combinations are performed (Stein and Liang, 2002). If the complexity of the experimental design is sufficient, then any cDNA band that is consistently observed in a group of individual samples should be selected for further analysis. Moreover, the first differential display reactions should include, if possible, an internal positive control. Primers that anneal to the cDNA of a gene that is known to be differentially expressed in one of the tissues or time-points included in the experimental set can be selected. Moreover, to demonstrate that clear differences between samples can be identified, a sample of RNA from another tissue, such as liver, may be included as a positive control.

Generally differential display is used to compare levels of gene expression among experimental groups. However, this technique has also been used successfully to reveal mRNA sequence variation between samples. For example, differential display demonstrated that a cDNA fragment could be amplified from samples derived from hippocampi of seizure-resistant but not seizure-prone rats (*Figure 17-2*). A smaller

Figure 17-2

Differential display analysis of hippocampal RNA isolated from individual seizure-resistant (SR) rats and seizure-prone (SP) rats

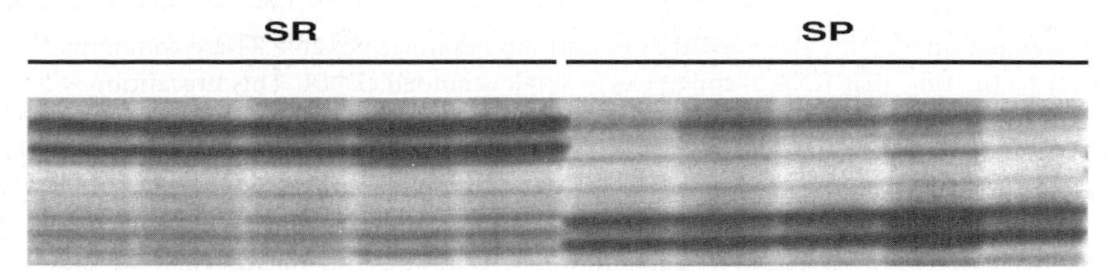

cDNA fragment was produced from both structures of seizure-prone but not seizure-resistant rats. Ultimately, each of these fragments were sequenced and identified as sequence variants of N-Chimaerin. Individual out-bred Long Evans Hooded rats expressed one of the two isoforms of N-Chimaerin (data not shown). The difference in size and sequence of differential display bands isolated from seizure-prone and seizure-resistant rats suggests that splice variants or genetic alleles of N-Chimaerin exist between the strains and may contribute to the difference in seizure susceptibility. The precise role of N-Chimaerin in seizure-susceptibility is currently under investigation. Findings such as these demonstrate the utility of differential display for comparisons of complex mRNA populations such as those found in the brain. It has been suggested that alternative splicing of mRNA, which permits a single coding sequence to create multiple proteins with different functions, lies at the root of higher order complexity (Brett et al., 2002). In fact, recent estimates based on alignment of expressed sequence tags and genomic DNA indicate that >50% of human genes may be alternatively spliced (Hanke et al., 1999; Mironov et al., 1999; The international genome sequencing consortium, 2001; Venter et al., 2001). As such, splice variants could conceivably explain the diversity of function achieved by the brain. Thus, strategies used to identify differential gene expression within the brain must incorporate this possibility into their search strategies. Most other gene screening techniques, including microarrays, are far less sensitive to such differences. However, one advantage of using DNA microarrays over differential display is that microarrays contain target sequences for functionally linked genes and thus can be used to ascertain the involvement of entire molecular pathways in a biological paradigm, whereas differential display produces random cDNA fragments. In response to this apparent disadvantage, it is a good idea to use small "pathway arrays" (Superarray, Inc.) that contain the gene found to be differentially expressed via differential display along with genes that are functionally linked to the candidate gene as a primary follow-up strategy.

In some cases, positive differential display results are later dismissed as "false positive" bands. Following reamplification, cloning, and screening of individual clones within a complex mixture, occasionally none of the clones tested correspond to an mRNA that is present at a different level in the experimental tissue. The reason for this may be that low abundance cDNA fragments present in the differential display band become enriched during reamplification. As such, the clone of interest may be overlooked during subsequent screening. For this reason, the reamplified differential display product can be used directly as a hybridization probe in Northern blot to determine the number of products present and whether any of them correspond to differentially expressed genes prior to cloning and extensive screening of individual clones. Differential display bands that correspond to small insertion/deletion differences or sequence variation in primer binding sites may also be dismissed as "false positive" bands because the sequences cross-hybridize to each isoform such that no significant difference in mRNA levels are observed following Northern hybridization analysis. Alternative methods for gene expression confirmation following differential display such as RNase protection assays or quantitative RT PCR are, in this case, more likely to confirm the differential expression of such mRNAs. As shown in ➋ *Figure 17-2*, the insertion/deletion difference in N-Chimaerin observed in seizure-resistant and seizure-prone rats would likely not have been detected if either probe were used in Northern blot analysis. The probe would have hybridized to mRNAs that do not differ significantly in size and appear equally abundant in both rat strains.

4 Differential Display Protocol

4.1 Preparation of Solutions and Equipment

To prevent degradation of RNA, all standard procedures to minimize possible RNase contamination should be used prior to the time that RNA is converted to single-stranded cDNA. This precaution will minimize differences in RNA populations caused by nucleases that have been introduced during tissue extraction and RNA manipulation. It is convenient to use disposable sterile plasticware whenever possible. Glassware can be soaked in 0.1 M NaOH for 1 h at room temperature and then extensively rinsed in water, which is free from any RNase. Following sodium hydroxide treatment, glassware should be autoclaved and baked in a 160°C oven overnight. RNase-free water or other solutions can be prepared by adding diethyl pyrocarbonate to a final concentration of 0.05% (v/v), stirring the solution at room temperature for 12 to 16 h and then autoclaving the solution to eliminate diethyl pyrocarbonate. It is important to eliminate all traces of diethyl pyrocarbonate as it will inhibit downstream enzymatic reactions. Similarly, it is important to eliminate all traces of sodium hydroxide from glassware as residual sodium hydroxide can lead to RNA hydrolysis.

4.2 Isolation of Tissue

The tissue of interest should be removed from the animals using equipment that has been treated to ensure that it is RNase-free. The surface hair of the animals can be soaked in 70% ethanol to minimize the inclusion of hair or dander in the isolated tissues. A sterile disposable Petri plate, placed on top of a bed of crushed ice, is a suitable RNase-free surface for microdissection. Immediately after isolation from the animal, individual samples of tissue can be flash frozen by immersion in liquid nitrogen and stored in cryovials in liquid nitrogen or at –70°C until all samples from an experimental set have been obtained. RNA extraction and preparation of single-stranded cDNA can then be performed simultaneously on all samples. This will minimize differences between RNA populations that result from differences in sample preparation.

4.3 Isolation of Total Cellular RNA and Removal of Genomic DNA

Although there are many different commercially available methods to isolate RNA, we have found that TRIzol reagent (Invitrogen), based on the standard acid phenol–guanidinium thiocyanate method, is a rapid, economical and reliable method to isolate high-quality RNA from many different tissues. Up to 100 mg of frozen tissue can be added directly to 1 ml of TRIzol reagent and dispersed using a dounce homogenizer with a tight fitting pestle or an electric homogenizer. If tissue samples weigh less than 10 mg, the initial volume of TRIzol can be reduced to 100 μl and the volumes of all reagents reduced proportionately. The tissue is homogenized in TRIzol reagent until it is dissociated into homogeneous solution. The homogenate can be stored at –70°C until all samples are processed to minimize the difference in the amount of time each tissue is exposed to TRIzol reagent. Tissue homogenates in TRIzol reagent are stable for months at –70°C. When processing multiple samples, the homogenizer can be cleaned between samples by sequentially rinsing the probe with RNase-free H_20, 1% (w/v) SDS, H_20, 70% (v/v) ethanol, followed by a final rinse in H_20. Small handheld homogenizers with interchangeable tips that can be pretreated to be RNase-free are convenient for tissues such as brain that are easily homogenized. In this case, the tissue can be disrupted in 100 μl of TRIzol in a microfuge tube and then diluted with 900 μl of TRIzol reagent. Bring all TRIzol-dissociated samples back to room temperature and vortex well. Chloroform (200 μl) is added and the solution is shaken vigorously by hand for 15 s and allowed to incubate at room temperature for 3 min. The homogenates are subjected to centrifugation at 12,000 × *g* for 15 min at 4°C. Genomic DNA and lipid will partition to the acidic organic phase and a dense layer of protein will be present between the organic and aqueous phases following centrifugation. RNA will be present in the upper aqueous phase. Carefully remove the colorless upper aqueous phase, avoiding the protein material collected at the interface and place in an RNase-free microfuge tube. Generally, 600 μl of aqueous phase can be removed following homogenization of up to 100 mg of tissue in 1 ml of TRIzol reagent. Isopropanol (500 μl) is added to the aqueous supernatant to precipitate total RNA. The supernatant and isopropanol are mixed thoroughly and allowed to incubate at room temperature for 15 min. Following centrifugation at 12,000 × *g* for 15 min at 4°C, the RNA pellet should be visible. The supernatant is removed and the RNA pellet can be rinsed by adding 1 ml of 75% ethanol, vortexing, and recentrifuging the solution containing the RNA pellet at 7,500 × *g* for three min at 4°C. After removing the supernatant, pulse-spin the tubes, remove any remaining traces of supernatant, and then allow the RNA pellet to dry at room temperature for approximately 10 min. It is not advisable to dry the RNA in a lyophilizer or extend the period of drying, as it makes the resuspension of RNA difficult. Although the amount of RNA that can be isolated from different tissues varies, resuspension of the RNA pellet in 1 μl of ddH_20 per 1 mg of tissue should yield sufficiently concentrated RNA for further manipulations and accurate spectrophotometric quantification. To resuspend the RNA, heat the solution at 60°C for 10 min, vortex, and place on ice or store at –70°C. The concentration of RNA can be determined spectrophotometrically. Generally, the ratio of absorbance at 260 and 280 nm is between 1.5 and 1.7 after TRIzol extraction.

4.4 Evaluation of Quality and Quantity of RNA

Although TRIzol reagent is extremely efficient for isolating total RNA, there can be trace amounts of contaminating genomic DNA in some of the RNA samples. Amplification of contaminating genomic DNA will produce differences in the distribution or intensity of differential display products between samples that are not due to differences in gene expression. To reduce the probability that genomic DNA will produce false positive differential display bands, each RNA sample is treated with DNase to remove any potentially contaminating genomic DNA prior to the generation of single-stranded cDNA. There are a number of commercially available RNase-free DNases that are suitable. An RNase inhibitor such as Rnasin (Promega) is included to prevent degradation of RNA during DNase treatment. Of each RNA sample, 10 μg is treated with RNase-free DNase for 60 min at 37°C using the buffers recommended by the manufacturer in a volume of 50 μl. Following DNase treatment, the RNA samples are re-extracted using 100 μl of TRIzol reagent to remove the DNase. The pellet is resuspended in 9 μl of RNase-free H_20. The concentration of RNA should be ~1 μg/μl before dilution to a final concentration of 500 ηg/μl.

Although DNase treatment should eliminate any traces of genomic DNA, each sample should be tested prior to the generation of single-stranded cDNA to eliminate the possibility that differential display bands are the result of amplification of genomic DNA. Any primer pair that supports amplification of genomic DNA can be used. We have found that amplification of cyclophilin is a highly sensitive method to determine if contaminating genomic DNA is present in total RNA isolated from mouse, rat, human, and zebrafish. Based on the estimate that 2 μg of RNA will be reverse transcribed, the reverse transcriptase reaction will be diluted to a final volume of 100 μl, and that 0.5 μl of cDNA will be used in each subsequent differential display reaction, the amount of contaminating genomic DNA from any one sample in a differential display reaction set would be equivalent to that found in 10 ηg of input RNA. The inclusion of only 10 ηg of RNA as the template in PCR reactions followed by agarose gel electrophoresis and ethidium bromide staining, however, may lead to an underestimation of the levels of contaminating genomic DNA because the differential display reaction products are radio-labeled during synthesis. For this reason, at least 500 ηg of input RNA should be included in the initial PCR analysis. Although only a few of the original RNA samples likely contained traces of genomic DNA, all samples in an experimental set are treated with DNase to ensure that differences in cDNA populations are not introduced during manipulation of RNA. The primers for cyclophilin (Sense, 5′ TGG TCA ACC CCA CCG TGT TCT T 3′; Antisense, 5′ GCC ATC CAG CCA CTC AGT CTT G 3′) amplify a 371 bp fragment from genomic DNA from tissue obtained from human, mouse, rat, and zebrafish. To amplify cyclophilin, any standard PCR reaction reagents can be employed using an annealing temperature of 50°C and a minimum of 35 PCR cycles. Samples that have genomic DNA should be removed and replaced prior to synthesis of cDNA.

4.5 cDNA Synthesis

Following isolation of high-quality RNA and testing to determine that genomic DNA is absent from the RNA, single-stranded cDNA is prepared. An oligonucleotide with the sequence 5′-T_{18} MN-3′ (M = C, G, or A; N=C, G, A, or T) is used as the primer in reverse transcription reactions. Theoretically, this primer should anchor the start of the reverse transcription reaction to the junction of the encoded 3′ ends of mRNA and polyA tails. Total RNA is reverse transcribed by combining 2 μg of DNA-free RNA at a concentration of 500 ηg/μl with 1 μl of 1 μM 5′-T_{18} MN-3′. The oligo dT/RNA mixture is denatured at 70°C for 3 min, vortexed briefly, and then placed directly on ice for a minimum of 2 min. A cDNA synthesis master mix is prepared that includes 2 μl of 5X first-strand cDNA synthesis buffer, 2 μl of 5 mM dNTP, and 1 μl of M-MLV reverse transcriptase (200 units/μl). Alternatively, we have used Superscript reverse transcriptase and have not found that it significantly alters the distribution or length of molecules that are amplified during subsequent differential display reactions. After the oligo dT/RNA mixture has cooled on ice, 5 μl of the first-strand cDNA master mix is added to each sample and the solution is mixed by pipetting. The reverse transcriptase reaction is allowed to incubate at 42°C for 60 min after which the reverse transcriptase is heat-inactivated at 80°C for 5 min. Each single-stranded cDNA mixture is diluted to a final volume of 100 μl and stored at −20°C in 20 μl aliquots to minimize the number of freeze–thaw cycles for the cDNA.

Prior to undertaking a large number of differential display reactions, we determine whether the concentration of cDNA is approximately equivalent in each sample. If an individual sample has a markedly different concentration of cDNA, the concentration can be adjusted by dilution or the sample can be replaced with a biologically equivalent sample. If three or more replicates of each experimental point are included in differential display analysis, differences among a group of samples will be immediately obvious. Any set of primers that amplify cDNA from a gene that is constitutively expressed in the tissues of interest can be used. It is known that β-actin levels are developmentally and tissue-specifically regulated in the brain. For this reason, we use primers that anneal with cDNA generated from the mRNA encoded by the hypoxanthine ribosyl transferase gene. The primers for hypoxanthine ribosyl transferase (Sense, 5′ GCT GGT GAA AAG GAC CTC T 3′; Antisense, 5′ CAC AGG ACT AGA ACA CCT GC 3′) are specific to mouse. As a simple test to determine the relative abundance of the housekeeping gene transcript, standard PCR reactions are assembled and aliquots of the reaction are removed after 24, 26, and 28 complete PCR cycles. After agarose gel electrophoresis, the optical density of ethidium bromide stained products can be determined by densitometry and the relative amount of product can be estimated. Generally, one of the cycles will correspond to the exponential amplification phase and comparisons among samples will be informative. If a cDNA sample is twofold or greater in abundance than other samples in the set, the concentration should be adjusted by dilution. It is not necessary to adjust the concentration of cDNA if there are only slight differences in the amount of product observed following PCR amplification and ethidium bromide staining, as these differences may reflect minor differences caused during the setup of individual reactions.

4.6 Differential Display Reactions

There are a number of differential display protocols that are modifications of the original protocol described by Liang and Pardee (Liang et al., 1993; Liang and Pardee, 1995; Simon and Oppenheimer, 1996; Matz et al., 1997; Jurecic et al., 1998; Matz and Lukyanov, 1998; Cho et al., 2002). The protocol presented here is based on the Clontech Delta Fingerprinting protocol. Equivalent amounts of cDNA are used as templates in PCR reactions that include different combinations of primers. Primers P1 to P10 (➋ *Table 17-1*) contain 16 nucleotides of the T3 RNA polymerase recognition sequence at their 5′ end. P1 to P10 have unique sequences at their 3′ end. The unique terminal 9 nucleotides anneal, at low temperature, to complementary sequences found in the single-stranded cDNA during the first rounds of amplification generating double-stranded cDNA. The 5′ end of primers T1 to T9 (➋ *Table 17-1*) are identical in sequence and contain the recognition sequence for T7 polymerase followed by nine T residues. Primers T1 to T9

Table 17-1
Primers used in differential display reactions

Anchored primers (5′ to 3′)	Upstream primers (5′ to 3′)
T1	P1
CATTATGCTGAGTGATATCTTTTTTTTTAA	ATTAACCCTCACTAAATGCTGGGGA
T2	P2
CATTATGCTGAGTGATATCTTTTTTTTTAC	ATTAACCCTCACTAAATGCTGGAGG
T3	P3
CATTATGCTGACTCATATCTTTTTTTTTAG	ATTAACCCTCACTAAATGCTGGTGG
T4	P4
CATTATGCTGACTCATATCTTTTTTTTTCA	ATTAACCCTCACTAAATGCTGGTAG
T5	P5
CATTATGCTGACTCATATCTTTTTTTTTCC	ATTAACCCTCACTAAAGATCTGACTG
T6	P6
CATTATGCTGACTCATATCTTTTTTTTTCG	ATTAACCCTCACTAAATGCTGGGTG
T7	P7
CATTATGCTGACTCATATCTTTTT TTTTGA	ATTAACCCTCACTAAATGCTGTATG
T8	P8
CATTATGCTGACTCATATCTTTTTTTTTGC	ATTAACCCTCACTAAATGGAGCTGG
T9	P9
CATTATGCTGACTCATATCTTTTTTTTTGG	ATTAACCCTCACTAAATGTGGCAGG
	P10
	ATTAACCCTCACTAAAGCACCGTCC

differ with respect to the last two nucleotides on the 3′ end of each primer. The T primers, anneal with the polyA sequence of the cDNA generated after the first linear amplification of single-stranded cDNA. Following the low temperature annealing of the terminal 11 nucleotides of each primer, the subsequent rounds of annealing occur at much higher temperatures, taking advantage of the specificity and stability of primer/target interactions due to the T3 and T7 sequences. In theory, the use of different T primers that contain oligo dT followed by the nucleotides MN at the 3′ end should reduce the complexity of the differential display products by anchoring the primer next to the polyA sequence. In practice, the PCR primers amplify regions throughout mRNA due to internal polyA repeats. In addition, many products contain the same primer at each end. As such, not all differential display products represent the 3′ end of mRNAs. To increase the longevity of the T and P primers, resuspend the synthetic oligonucleotides in 10 mM Tris-HCl, pH 8.0 containing 1 mM Na_2EDTA to a concentration of 100 μM (stock concentration) and store at −70°C. Dilute the oligonucleotide stock solutions in ddH_2O to a final concentration of 20 μM and store at −20°C.

The differential display reactions themselves are very simple PCR reactions, which use combinations of T and P primers (*Table 17-1*). The differential display products are radio-labeled during synthesis by the incorporation of [α-^{33} P]dATP. Although this isotope is relatively expensive, it has a half-life of 28 days and we have found that even after four half-lives, the isotope can be used to generate good results in differential display reactions. A master mix of the differential display PCR reagents is prepared, ensuring that there is enough material for the entire set of experimental reactions. For each differential display reaction, combine 3.3 μl of ddH_20, 0.5 μl of 10X cDNA PCR reaction buffer, 0.05 μl of 5 mM dNTP, 0.05 μl of [α-^{33} P]dATP (3000 Ci/mmol), 0.25 μl of 20 μM T primer, 0.25 μl of 20 μM P primer, and 0.1 μl of 50X cDNA polymerase mix. The Clontech 50X cDNA polymerase mix includes two thermostable DNA polymerases, which have limited 5′ exonuclease activity and 3′ to 5′ proofreading activity. The polymerases are bound to an antibody, which is denatured and released from the polymerase prior to amplification. The polymerase activity, therefore, is blocked during the assembly of the PCR reactions. We have found that "hot start" PCR is absolutely necessary for the production of highly reproducible differential display banding patterns. Aliquots (0.5 μl) of each single-stranded cDNA are placed in the bottom of separate PCR tubes and 4.5 μl of the differential display master mix is added directly to the cDNA. Very little evaporation will occur if high-quality thin-wall PCR tubes with tight fitting lids and a thermocycler with a heated lid are used. The relative volume of each sample can be checked after PCR by visual inspection. Use a larger volume PCR reaction if evaporation of the sample is observed. During the first rounds of amplification, the low annealing temperature allows the very short unique sequences in the primers to anneal and generate a subpopulation of double-stranded cDNA molecules. Following two rounds of amplification using low temperature annealing, the temperature of annealing is significantly increased to take advantage of the specificity conferred by the T3 and T7 primer sequences. PCR conditions are as follows: 94°C for 5 min, 40°C for 5 min, 68°C for 5 min, 94°C for 2 min, 40°C for 5 min, 68°C for 5 min. This sequence of steps is repeated one time and followed by 26 cycles of 94°C for 1 min, 60°C for 1 min, and 68°C for 2 min plus 4 s per cycle. The final step is at 68°C for 7 min before holding the reaction at 0°C.

4.7 Gel Electrophoresis

The radiolabeled PCR products are fractionated in a denaturing 6% polyacrylamide gel using standard conditions (Sambrook and Russell, 2001). One of the plates should be siliconized and rinsed thoroughly with 70% ethanol before pouring the gel. This will allow for the removal of one glass plate while leaving the gel firmly attached to the non-siliconized plate after electrophoresis. The number of products that can be resolved by gel electrophoresis will be increased if a buffer gradient or acrylamide concentration gradient is used. However, gradients based on gel thickness (wedge-shaped spacers) are not appropriate because it is difficult to elute and reamplify differential display bands that are located in the thicker part of the gel. Following PCR, an equivalent volume of denaturing loading dye is added to each differential display reaction and the samples are heated at 92° for 3 min. After mixing the samples and collecting all material at the bottom of the tube by subjecting the PCR tubes to a brief pulse-spin, the samples are placed on ice until they are loaded on the denaturing acrylamide gel. It is important to remove all traces of acrylamide and thoroughly flush the top of the gel with running buffer to remove excess urea before positioning the sharks tooth comb. Four μl of denatured PCR product in loading dye is loaded into each well. The gel is run until the xylene cyanol exits from the bottom of the sequencing gel. The resolved differential display products that remain on the gel will be greater than 100 base pairs in length. The conditions for individual sequencing apparatus may need to be optimized. We routinely subject the gel to 3,000 V, 100 W for 5 h at a temperature of 50°C. The remainder of the differential display reaction can be stored at –20°C for up to 2 days and subjected to a longer period of electrophoresis to resolve higher molecular weight fragments. The samples do not need to be heat-denatured a second time prior to loading on a second gel.

Excess urea must be removed from the gel prior to autoradiography so that it does not interfere in subsequent reamplification of isolated differential display bands. Also, if the urea is not removed, the gel will be highly hygroscopic and is likely to adhere to the autoradiography film such that the gel is damaged when the film is removed. A simple method to remove excess urea is to separate the plates of the sequencing apparatus and use a handheld hairdryer to dry the gel directly on one plate. The urea will crystallize throughout the gel and can be removed by rinsing the gel thoroughly with H_20. After rinsing, excess H_20 is drained from the gel and a dry piece of Whatman 3 MM paper is placed directly on the gel. One end of the Whatman paper is placed flush with the edge of the plate and the filter paper is slowly lowered on the gel to avoid trapping pockets of air. The gel will be firmly attached to the filter paper and can be readily removed from the plate and placed in a gel dryer. The gel/filter paper must be completely dry before exposing the gel directly to maximum resolution autoradiography film (Kodak BioMax MR) at room temperature. The signal from the incorporated [α^{33}-P]dATP is extensively attenuated by plastic wrap, therefore, plastic wrap used during the drying process must be removed. An overnight exposure of the dried gel is generally sufficient to clearly visualize the differential display bands distributed throughout the gel.

4.8 Isolation of Differentially Displayed Fragments

We found that it is necessary to run several sets of differential display primers prior to an analysis of the distribution of differential display bands. This allows for a comparison between different independent reactions using different PCR primers to assess the quality of individual cDNA samples and discriminate between sample-to-sample variability and potential positive bands that are consistently found in different replicates. The presence or absence of a specific band in lanes corresponding to independent experimental samples indicates a reproducible difference in the relative amount of cDNA in a given sample, which should reflect differences in mRNA levels. However, the interpretation of the differential display results is not always straightforward. For example, a thick band can reflect quantitative differences in the initial concentration of a specific cDNA between samples or can represent comigration of two bands. Replication of the PCR reactions for samples that have differences in banding pattern will eliminate a significant number of false positive differential display differences. Also, in some cases, it may be informative to alter the electrophoresis conditions to maximize resolution of a band of interest prior to isolation, reamplification, and further analysis of potential positive bands.

After identification of potential differential display bands on the film, the samples must be located on the gel and excised. The easiest way to locate the band of interest on the gel is to outline the band on the film by placing opaque removable tape around the band. The gel and autoradiography film can be carefully aligned using landmarks that are visible on the film provided by trace amounts of radioactivity at the bottom and edges of the dried gel. The edges of the dried gel will correspond to this radioactive signal. Alternatively, the film and gel can be pierced at the edges in an asymmetric pattern using an 18 gauge needle before the film is removed for developing. The film and gel can then be realigned using the holes in the gel and film as landmarks. Lastly, a dilution of radio-label in ink (1:10,000 to 1:100,000) can be prepared and spotted on the edges of the filter paper containing the gel prior to exposure of the film. Invert the aligned film and gel on a light box and outline the region of the gel containing the band on the back of the dried gel/filter paper in the area of the gel corresponding to the window created by the opaque tape. Remove the film and excise the band from the gel/filter paper using a scalpel blade. The gel slice and adherent filter paper can be placed in a microfuge tube and stored at room temperature. The gel can be reexposed to ensure the correct band was removed. The dried gels can be stored, if kept dry and flat, for extended periods of time and differential display bands can be removed for further analysis when required. It is preferable, however, to isolate any potential bands of interest and store them separately in microfuge tubes prior to reamplification and further analysis as the dried gels will tend to curl or pick up moisture over time. In addition, the radioactive signal will decay preventing reexposure of the gel to ensure that the correct band was removed. Due to the number of bands that may be isolated and processed during a complete differential screen, it is important to establish a simple coding scheme for isolated bands at the beginning of the experiment.

4.9 Reamplification of cDNA Fragments

To isolate the cDNA from the excised gel slice, add 40 μl of ddH_20 to the tube and incubate the submerged gel slice at room temperature for 10 to 30 min. The solution is then heated at 95°C for 5 min and cooled on ice. For PCR re-amplification of the differential display band, combine 10.75 μl of ddH_20, 8.5 μl of the solution from the rehydrated band, 1.25 μl each of the P and T primers (20 μM) used in the original differential display reaction, 2.5 μl of 10X PCR buffer containing 1.5 mM $MgCl_2$, 0.5 μl of 5 mM dNTP, and 0.25 μl of *Taq* polymerase (5 U/μl). It is not necessary to use the high-fidelity enzyme mix used for the original differential display reaction. Any inexpensive thermostable polymerase and compatible buffer can be used for reamplification. The PCR reamplification conditions include 94°C for 1 min followed by 20 cycles of 94°C for 30 sec, 58°C for 30 sec, 68°C for 2 min, and a final extension at 68°C for 10 min. If a "hot-start" polymerase is used, alter the time of the original denaturation according to the directions provided by the manufacturer. After reamplification, the products are subjected to agarose gel electrophoresis and ethidium bromide staining and the product is gel purified. Generally, 20 cycles of PCR are sufficient to generate a visible band, although some reamplification reactions must be subjected to more cycles to generate sufficient product for subsequent analysis.

4.10 Identifying Differential Display Fragments within Complex Mixtures

Even if the isolated differential display band contains a majority of one type of molecule, it is likely that it will also contain trace amounts of other PCR products of similar size. After reamplification, there is a distinct possibility that cDNA bands, which do not represent changes in gene expression, become relatively abundant within the reamplified products. In this case, analysis of independent clones derived from the reamplified differential display band will demonstrate that a significant number of clones contain sequences that have no relationship to the "positive" differential display band. Unless very large numbers of individual clones are analyzed, a differential display band may be discarded as a "false positive" band simply because the appropriate clone was not chosen for analysis. For this reason, we isolate the reamplified product, radio-label the complex mixture, and use the population of cDNA molecules as a probe in Northern hybridization analysis prior to cloning and DNA sequence analysis of individual products. In some cases, the reamplified differential display band will contain only one or a few unique sequences, which anneal with a limited number of mRNA molecules in Northern blots. If this is the case, and it appears that there is a difference between experimental samples in the relative levels of one or more of the fractionated RNAs, the reamplified differential display products can be cloned and individual clones can be used as hybridization probes. Clones that hybridize to specific mRNAs can be isolated and sequenced and the clone corresponding to the differential display band can be easily identified. However, if the reamplified band contains a number of different products, the number of hybridizing RNAs will be large, making the identification of the differential display band time-consuming. We have developed a method for eluting the hybridizing "positive" differential display band directly from Northern blot and directly cloning such fragments (Denovan-Wright et al., 1999). This method was used to identify a gene that is differentially expressed during the development of heart failure in a model of cardiomyopathy (Denovan-Wright et al., 2000). This initial analysis precedes cloning and sequencing of the differential display bands and is followed by one or more complementary methods to determine the spatial and temporal distribution of mRNAs produced by differentially expressed genes. As discussed previously, differential display bands may result from amplification of splice variants or cDNAs that have differences in sequence at the primer binding sites. These differential display products may be overlooked following Northern blot hybridization. In such cases, it may be necessary to design in situ hybridization, quantitative PCR, or RNase protection experiments to investigate the differences in mRNA that generated the differential display product.

5 Verification of Differential Display Products

Ultimately, individual positive clones are generated and the sequence of the cDNA can be determined using standard methods (Sambrook and Russell, 2001). The identity of the differentially expressed gene can then be determined by comparing the partial cDNA sequence to databases, such as the annotated genomic Sanger Ensembl database available through the Wellcome Trust Sanger Institute (Cambridge, UK). Below is a brief description of methods that are commonly used to verify that a differential display cDNA corresponds to a gene that is truly differentially expressed.

5.1 Northern Blot

Traditionally, Northern blot analysis of isolated total RNA or mRNA has been used to determine differences in levels of mRNA between and among different animal groups, tissues or time-points. Standard protocols for performing Northern blot hybridization are available (Sambrook and Russell, 2001). Radioisotopic and colorimetric methods are available for probe labeling. The advantage of Northern blot analysis, especially for determining the relative levels of brain mRNAs, is that the information derived from such analyses provides both quantitative evaluations of the levels of mRNA, as well as an accurate description of the size of the mRNAs. As the complexity of mRNA populations in the brain is frequently due to splice variants, the ability to discriminate between size of an mRNA is valuable and cannot be observed following other types of analysis such as in situ hybridization, quantitative PCR, or RNase protection assays unless the conditions are specifically set to identify particular isoforms of mRNA. However, splice variants that are not significantly different in size or expression of genes that have undergone small insertion/deletion differences may not appear to be different in Northern blot analysis. Slot or dot blot hybridization can be used to determine whether there is a difference in the concentration of hybridizing mRNAs among different samples, however, this type of hybridization analysis will not provide information regarding the size of the message. Slot or dot blots also do not provide information as to whether or not different mRNA isoforms have been detected by the hybridizing probes.

5.2 In Situ Hybridization

In situ hybridization is a highly sensitive method to determine the temporal and anatomical resolution of differentially expressed genes. This technique is especially useful to analyze mRNA levels and distribution in the brain, which have distinct differences in expression patterns among closely spaced but functionally and phenotypically distinct tissues. In situ hybridization analysis is particularly suited to the study of gene expression in mice and rats as several complete coronal or sagital tissue sections can be placed on one slide and subjected to identical in situ hybridization conditions for comparative analysis. There are two basic ways to label probes for in situ hybridization. RNA probes (riboprobes) can be synthesized in vitro, which are complementary to the mRNA of interest, or synthetic oligonucleotide probes can be produced. Either type of probe can be radiolabeled or labeled for colorimetric detection. Detailed protocols for in situ hybridization analysis of brain mRNAs are available (Wilkinson, 1999). One of the major advantages of using in situ hybridization to verify that a band is differentially expressed is that the expression of specific genes can be simultaneously analyzed in a large number of different tissues and animals if the probes are radio-labeled and all sections are exposed to the same autoradiography film. Slight variations in the distribution of hybridizing probes are immediately obvious and the interpretation of the data is straightforward. Densitometry can then be employed to determine the relative levels of mRNA in a given tissue and statistical analysis of relative mRNA levels can indicate subtle changes in mRNA levels within different experimental conditions or at different time-points. If the hybridizing in situ probes are detected colorimetrically, this technique can provide cellular resolution of the distribution of mRNA but the levels of mRNA cannot be quantitated. One disadvantage of in situ hybridization is that low abundance messages may not be detected because the mRNA-specific hybridization signal is not significantly higher than background nonspecific binding of the probe in the tissue.

5.3 RNase Protection Assays

RNase protection assay is a relatively sensitive and accurate way to determine mRNA levels. The accuracy of determining the concentration of a specific mRNA by quantifying the amount of probe that hybridizes with the mRNA target and is protected from RNase digestion is sensitive over several orders magnitude of mRNA copy number. Riboprobes can be synthesized that contain a radiolabeled or biotin-linked NTP. The limitation of RNase protection assays is that, like Northern blot and quantitative PCR, this method utilizes isolated RNA populations and, therefore, differences in dissection or quality of RNA can impact the results. Moreover, to detect rare mRNAs, it may be necessary to include up to 50 μg of total RNA in the liquid hybridization reaction. However, RNase protection assays can provide valuable information about 5′ transcription start sites, variability in the position of 3′ ends of messages and, if the probe spans exon/intron boundaries, information regarding splice variants within an RNA population. For this reason, RNase protection assays can be used to detect different mRNA isoforms that generate differential display bands.

5.4 Quantitative RT PCR

Quantitative RT PCR (qRT PCR) can be used to accurately determine the levels of messages within given preparations of RNA. qRT PCR thermocyclers provide rapid online detection and quantification of mRNA, however, the initial purchase cost and the cost of reagents may be prohibitive for some laboratories. Methods of semiquantitative RT PCR have been used and good descriptions of these techniques are available (Sambrook and Russell, 2001). However, the same cDNA populations should not be used for differential display reactions and verification that a potential differential display band represents a differentially expressed gene. For this reason, independent cDNA samples should be prepared if both the screening and verification methods rely on PCR. qRT PCR, therefore, should be used in conjunction with other methods to verify that a differential display band represents a differentially expressed gene.

6 Conclusions

The utility of differential display as a primary gene expression screening tool has been demonstrated by the number of manuscripts that have appeared since the method was first described in the early 1990s. This method of gene expression analysis is relatively simple and financially feasible for most laboratories and requires standard molecular biology equipment. Because differential display enables simultaneous evaluation of a number of control and experimental samples using small amounts of input RNA and because this technique can detect different mRNA isoforms, differential display will continue to be of use in understanding changes in gene expression, especially in complex tissues such as brain.

References

Ahmed FE. 2002. Molecular techniques for studying gene expression in carcinogenesis. J Environ Sci Health C20(2): 77.

Bartlett JMS. 2003. Differential display: a technical overview. Methods Mol Biol 226: 217.

Brett D, Pospisil H, Valcarcel J, Reich J, Bork P. 2002. Alternative splicing and genome complexity. Nat Genet 30: 29.

Brown PO, Botstein. 1999. Exploring the new world of the genome with DNA microarrays. Nat Genet 21: 33.

Cho Y, Prezioso VR, Liang P. 2002. Systematic analysis of intrinsic factors affecting differential display. Biotechniques 32: 762.

Denovan-Wright EM, Ferrier GR, Robertson HA, Howlett SE. 2000. Increased expression of the gene for α-interferon inducible protein in cardiomyopathic hamster heart. Biochem Biophys Res Comm 267: 103.

Denovan-Wright EM, Howlett SE, Robertson HA. 1999. Direct cloning of differential display products eluted from Northern blots. Biotechniques 26: 1046.

Ding C, Cantor CR. 2004. Quantitative analysis of nucleic acids—the last few years of progress. J Biochem Mol Biol 37(1): 1.

Eberwine J, Kacharmina JE, Andrews C, Miyashiro K, McIntosh T, et al. 2001. mRNA Expression analysis of tissue sections and single cells. J Neurosci 21: 8310.

Geschwind DH. 2000. Mice, microarrays and the genetic diversity of the brain. Proc Natl Acad Sci USA 97: 10676.

Halgren RG, Fielden MR, Fong CJ, Zacharewski TR. 2001. Assessment of clone identity and sequence fidelity for 1189 IMAGE cDNA clones. Nucleic Acids Res 29(2): 582.

Hanke J, Brett D, Zastrow I, Aydin A, Delbruck S, et al. 1999. Alternative splicing of human genes: more the rule than the exception. Trends Genet 15: 389.

Jurecic R, Nachtman RG, Colicos SM, Belmont JW. 1998. Identification and cloning of differentially expressed genes by long-distance differential display. Anal Biochem 259: 235.

Kuno N, Muramatsu T, Hamazato F, Furuya M. 2000. Identification by large-scale screening of phytochrome-regulated genes in etiolated seedlings of *Arabidopsis* using a fluorescent differential display technique. Plant Physiol 122: 15.

Liang P. 2002. A decade of differential display. Biotechniques 33: 338.

Liang P, Averboukh L, Pardee AB. 1993. Distribution and cloning of eukaryotic mRNAs by means of differential display: refinements and optimization. Nucleic Acids Res 21:3269.

Liang P, Pardee AB. 1995. Recent advances in differential display. Curr Opin Immunol 7: 274.

Liang P, Pardee AB. 2003. Analyzing differential gene expression in cancer. Nat Rev 3: 869.

Liao VH, Freedman JH. 1998. Cadmium-regulated genes from the nematode *Caenorhabditis elegans*. J Biol Chem 273: 31962.

Lockhart DJ, Winzeler EA. 2000. Genomics, gene expression and DNA arrays. Nature 405: 827.

Matz MV, Lukyanov SA. 1998. Different strategies of differential display: areas of application. Nucleic Acids Res 26: 5537.

Matz M, Usman N, Shagin D, Bogdanova E, Lukyanov S. 1997. Ordered differential display: a simple method for systematic comparison of gene expression profiles. Nucleic Acids Res 25: 2541.

Mironov AA, Fickett JW, Gelfand MS. 1999. Frequent alternative splicing of human genes. Genome Res 9: 1288.

Planitzer SA, Machl AW, Schindler D, Kubbies M. 1998. Small deletions in the regulatory 3′UTR of the human alpha-tropomyosin gene identified by differential display. Mol Cell Probes 12(1): 35.

Racine RJ, Steingart M, McIntyre DC. 1999. Development of kindling prone and kindling resistant rats: selective breeding and electrophysiological studies. Epilepsy Res 35(3): 183.

Sambrook J, Russell DW. 2001. Molecular cloning: a laboratory manual. Cold Spring Harbor: Cold Spring Harbor Laboratory Press.

Sandberg R, Yasuda R, Pankratz D, Carter T, Del Rio J, et al. 2000. Regional and strain-specific gene expression mapping in the adult mouse brain. Proc Natl Acad Sci USA 97: 11038.

Simon H-G, Oppenheimer S. 1996. Advanced mRNA differential display: isolation of a new differentially regulated myosin heavy chain-encoding gene in amphibian limb regeneration. Gene 172: 175.

Stein J, Liang P. 2002. Differential display technology: a general guide. Cell Mol Life Sci 59: 1235.

The International Genome Sequencing Consortium. 2001. Initial sequencing and analysis of the human genome. Nature 409: 860.

To K. 2000. Identification of differential gene expression by high throughput analysis. Comb Chem High Throughput Screen 3: 235.

Venter C, et al. 2001. The sequence of the human genome. Science 16: 1304.

Wan JS, Sharp SJ, Poirer GMC, Wagaman PC, Chambers J, et al. 1996. Cloning differentially expressed mRNAs. Nat Biotechnol 14: 1685.

Wilkinson DG. 1999. In situ Hybridization: a Practical Approach 2nd. Edition. New York: Oxford University Press; pp 1-221.

18 Gene Arrays: A Practical Approach to Studying Stroke with Microarray

R. W. Gilbert · W. J. Costain · H. A. Robertson

Abstract: Cell death from cerebral ischemia is a dynamic process. In the minutes to days following an ischemic insult, progressive changes in cellular morphology are observed in ischemic tissues. Many of these changes are believed to be associated with the regulation of competing programs of gene expression; some of which are protective against ischemic insult and facilitate cell survival, and others that contribute to delayed cell death. In the past, paradigms for stroke genomic research have focused on the identification and investigation of individual gene candidates thought to underlie these events. While this approach has identified many genes we continue to lack truly effective strategies for the treatment of stroke. It is generally believe that effective strategies for managing the consequences of stroke will come only after details of the coordinated patterns of gene expressions underlying its pathogenesis have been described. Such investigations have until recently been hampered by a lack of reliable and efficient methods for large-scale gene expression screening.

Today the stage has been set for describing, on a global level, the integration of gene expressions pathways following cerebral ischemia. This approach has largely been made possible by the development of DNA microarray technology, an approach that provides a large-scale strategy for monitoring differential gene expression. This chapter discusses the potential applications of cDNA microarray-based technologies for furthering our understanding of diseases like stroke. A discussion of issues pertaining to sample preparation, labeling and purification, hybridization, data analysis and the validation of findings is provided. Important considerations for achieving a successful experiment are discussed.

Abbreviation: Bp, nucleotide base pairs; cDNA, complementary DNA; ChIP, chromatin Immunoprecipitation; Cy5, cyanine 5-dCTP; Cy3, cyanine 3-dCTP; ESTs, expressed sequence tags; FDR, false discovery rate; MIAME, minimum information about a microarray experiment; mRNA, RNA, messenger; NIA, National Institutes of Aging; RFUs, relative fluorescence units; RT-PCR, reverse transcriptase polymerase chain reaction; SAGE, serial analysis of gene expression; SAM, significance analysis of microarrays

1 Introduction

Current developments in DNA microarray technology provide a powerful tool for investigating large-scale changes in gene expression. DNA microarray technology has made it possible to simultaneously analyze gene expression for tens of thousands of genes from a single sample. Expression profiles are proving instrumental in the characterization of temporal and spatial changes in gene expression and are providing insight into cellular functions and underlying mechanisms of disease pathogenesis. To date this approach has been used to investigate disease-related changes in gene expression for cancer (Luo et al., 2003), psychiatric disorders (Bunney et al., 2003), brain injury (Matzilevich et al., 2002), and neurodegenerative diseases including Parkinson's disease (Grunblatt et al., 2001), Huntington's disease (Sipione et al., 2002), and Alzheimer's disease (Marcotte et al., 2003). The adoption of DNA microarray technologies to stroke research is likewise advancing. Current studies are identifying novel and complex patterns in gene expression in the brain after experimental cerebral ischemia, patterns that might provide novel avenues for treatment. In this chapter we explain the technological principles that underlie DNA microarrays and highlight the limitations and utility of the technology for investigating the molecular basis of neurobiological diseases like stroke. The potential of microarray technology in accelerating our understanding of this complex genetic disease will be described.

2 Gene Expression in the Cell Biology of Stroke

Stroke is a debilitating disease with an enormous societal cost ($53.6 billion/year in the US), and is therefore a very active area of research. Stroke research has many inherent obstacles, not the least of which is establishing good experimental models. Beyond this, stroke researchers are presented with the tremendously difficult task of pinpointing the key molecular events in a disease that is characterized by disruptions in complex and divergent processes. This difficulty of the task is emphasized by the number of failed attempts at creating a stroke therapy based on our current understanding of the disease (Wahlgren and Ahmed, 2004). Because of this, it is apparent that our current understanding of the molecular events surrounding stroke is insufficient. To better treat stroke we need a better picture of the events, including gene expression, that occur after stroke injury.

2.1 Gene Expression in Stroke Pathology

Cell death from cerebral ischemia is a dynamic process. In the minutes to days following an ischemic insult, progressive changes in cellular morphology are observed in ischemic tissues. Many of these changes are now known to be associated with the regulation of competing programs of gene expression, some of which are protective against ischemic insult and facilitate cell survival, and others that contribute to delayed cell death. In the past, paradigms for stroke genomic research promoted the identification of individual gene candidates. Many of these gene candidates were further investigated to determine their potential role in the pathogenesis of cerebral ischemia. Although this paradigm has identified a wealth of genes, individually these studies have provided only limited insight into the molecular pathways involved in these processes (Read et al., 2001). As a consequence, we continue to lack truly effective strategies for the treatment of stroke. Today it is generally believed that an appreciation of the sequence of events underlying stroke pathogenesis will come only after the coordinated patterns of gene expressions have been described. Such investigations have until recently been hampered by a lack of reliable and efficient methods for large-scale gene expression screening.

Today the stage has been set for describing, on a global level, the integration of gene expression pathways following cerebral ischemia. This approach has largely been made possible by two recent developments; the completion of mammalian genome projects and the development of DNA microarray technology. The completion of sequencing for the human, mouse (Waterson et al., 2002) and rat genomes (Rat genome sequencing project consortium, 2004) have provided an astounding resource of gene sequence data for both known genes and for thousands of expressed sequence tags (partial cDNA sequences) representing as yet unknown genes. This information has been deposited in gene sequence databases that today contain vast amounts of readable and accessible information. Collectively, these databases form the cornerstone of functional genomics (the analysis of the expression of mRNA under defined conditions) setting the stage for the analysis of gene expression on the genomic scale. However, the development of systematic approaches for identifying biological functions of these genes remains a major challenge. Microarray technology represents a major advance in this regard, providing a large-scale strategy for monitoring differential gene expression. The obvious application of these genome sequences will be in accelerating the identification of candidate genes whose expression patterns are associated with human disease and the modification of disease processes in both animal models of disease and human disorders. An alteration in the expression of a given mRNA in pathological tissue is at least circumstantial evidence that the gene may play a role in the disease.

2.2 Gene Expression Profiling in Stroke Models Using Microarrays

Cerebral ischemia is suitable for DNA microarray analysis because it is a complex process, dependent in part on differential RNA expression. Soriano et al. (2000) pioneered the use of this technology in the investigation of cerebral ischemia. Using an oligonucleotide array they provided the first large-scale profile of the differential mRNA expression associated with permanent focal ischemia. Their study profiled 750 genes and showed impressive changes in gene expression. Subsequent investigations have characterized differential gene expression in rodent models of global brain ischemia (Jin et al., 2001), focal ischemia reperfusion (Schmidt-Kastner et al., 2002), hypoxia-ischemia (Gilbert et al., 2003) and a model of hypoxic preconditioning (Stenzel-Poore et al., 2003). MacManus et al. (2004) recently published a novel translation-state analysis of gene expression in mouse brain following transient ischemia. An important feature of these studies is that they have each confirmed the expression of many known ischemia responsive genes and connected many new genes to cerebral ischemia. Significantly, despite using a variety of gene array formats (cDNA, oligonucleotide arrays, and Affymetrix.) these studies have produced data with a high degree of consistency between studies. Consistent observations in these studies included genes involved in molecular chaperoning, metabolism, development, and cell structure. Such initial findings illustrate the validity and robustness of DNA microarray technology as an approach for developing hypotheses regarding molecular pathways in cerebral ischemia. Although many of these studies were performed using microarrays that contained oligonucleotides for only a small number of genes, a number of differentially expressed genes novel to cerebral ischemia have been confirmed using techniques such as in situ hybridization.

By revealing the identity of a spectrum of molecules involved in different cellular functions, such results provide a means for mechanistic speculation and the subsequent development of neuroprotective strategies. Results such as these are now commonly used as a reference for predicting the functions of cells under given conditions and in assigning putative biological process annotation to unknown genes.

3 The Microarray Revolution

In a span of a very few years the use of microarray technologies for the high-throughput analysis of gene expression has become a standard laboratory technique. In biological science today, the term microarray refers to a group of technologies that uses micro-fabricated arrays of DNA to quantify the characteristics or levels of nucleic acids in a highly parallel manner. The specific application of microarrays of DNA has evolved from measuring mRNA levels (gene expression) to applications including but not limited to:

1. gene expression analysis, for the purpose of identifying differential gene expression across tissues types or for cells exposed to different experimental conditions
2. genotyping (analyzing the genomic content of an organism)
3. promoter/transcription factor analysis (ChIP on chip)
4. drug discovery and development
5. sequencing

In this review we focus on the use of DNA microarrays for the purpose of differential gene expression analysis. Discussions are intended to provide an overview of DNA microarray technology focusing on themes currently believed important to ensuring a successful DNA microarray experiment. Specifically, we address practical issues surrounding the use of DNA microarray technology with emphasis placed on its utility in the investigation of experimental stroke.

4 DNA Microarray Technologies

DNA microarray technology has empowered individual laboratories with the ability to examine biological processes with a speed and comprehensiveness unimagined a scant 10 years ago. As discussed later, this technology has an enormous capacity for data generation and lends both time and resource benefits. In comparison with other techniques, they offer ease of use and rapidly growing commercial availability. Furthermore, as our comprehension of the genomes of model organisms matures, the time when we will be able to examine the whole genome of a species in a single experiment is growing near.

The quest for a technology capable of detecting global changes in gene transcription within a single experiment is not new. For example, early studies saw the re-probing of Northern blots with multiple probes for the purpose of detecting multiple gene expressions in one experiment. More recently, we have seen the development of powerful techniques such as differential display of mRNA (Liang and Pardee, 1992; Gilby and Denovan-Wright, 2004), serial analysis of gene expression (SAGE; Velculescu et al., 1995) and subtractive hybridization (Sagerstrom et al., 1997), each capable of detecting multiple transcripts from a single RNA sample. Although differential display and subtractive hybridization have made significant contributions to the field of stroke genomics (Read et al., 2000), they are laborious and do not provide the incentive for shifting biological research toward a nonreductionist approach. Although the generation of SAGE libraries is arguably more comprehensive than microarrays and has the added advantage of revealing previously unidentified transcripts, its expense and resource demands make it prohibitive for all but a few laboratories.

The current response to the demand for high-throughput analysis of gene expression is the DNA microarray. By combining standard molecular biology techniques, precision robotic printing technology, and high-throughput screening methodologies, DNA microarrays now provide an efficient and expedient platform for large-scale analysis of changes in gene expression. Unlike previous technologies that are limited to the analysis of small numbers of mRNAs, DNA microarrays enable the simultaneous analysis of tens of thousands of genes in a single experiment.

Two basic platforms of DNA microarray technologies currently lead the charge to large-scale gene analysis; spotted microarrays (cDNA or oliognucleotide, Schena et al., 1995) and the in situ synthesized oligonucleotides microarrays (GeneChip) marketed by Affymetrix (Lockhart et al., 1996). These two technologies, which have developed in parallel, are methodologically distinct yet share several essential features.

In terms of similarities, all DNA microarray platforms are based upon the principles of nucleic acid hybridization. The amount of each transcript is determined by measuring the amount of fluorescently labeled cDNA hybridizing to a spot (containing DNA immobilized on a solid surface) on a microarray. Both platforms use DNA fragments spotted at high density to a structural base, such as a glass slide or nylon membrane. Each spot on a microarray contains a unique DNA species that is representative of a single gene. The spots on a microarray are analogous to probes that are used in Northern and Southern blotting. DNA fragments may be for characterized genes or ESTs (expressed sequence tags, the mRNA of genes known to be expressed in a tissue but not yet characterized). Finally, both formats require detection and quantification of the hybridized molecules to determine mRNA expression levels across different experimental conditions. Although these two platforms for gene expression analysis have their respective advantages and disadvantages, collectively their developments have led to the current standards for large-scale gene expression measurement. The following section summarizes the basic characteristics of these two platforms and provides a brief commentary on their applications.

4.1 GeneChip Microarrays

The best-known commercial microarray platform was developed by Fodor et al. (1991) and uses photolithographic techniques for the in situ synthesis of DNA fragments (oligonucleotides) on microarray slides (Lockhart et al., 1996; Lipshutz et al., 1999; Lockhart and Winzeler, 2000). This method has traditionally been referred to as DNA chip technology and is proprietary to Affymetrix, Inc. (Santa Clara, CA). Affymetrix sells its photo lithographically-fabricated microarrays under the trademark name GeneChip although a number of other companies now manufacture oligonucleotide-based microarrays using alternative in situ synthesis technologies.

The typical GeneChip is a high density microarray containing hundreds of thousands of oligonucleotides (25-mer probes). The expression of each gene represented on a GeneChip is determined by a number (11 or more) of distinct oligonucleotide probes (probe sets). Each probe set is made up of perfect match (PM—exactly complementary to target) and mismatch (MM—single nucleotide substitutions in probe) probes. Comparison of the PM and MM signals is used to determine the amount of specific hybridization to a given probe set. In a typical microarray experiment, mRNA is isolated from the experimental sample, labeled with and then hybridized to oligonucleotides spotted on a GeneChip. The abundance of an mRNA present in the original sample is determined by recording the fluorescent intensity levels for each probe set. In contrast to the cDNA microarrays (described later) only a single sample can be hybridized to each microarray and comparisons are made among multiple arrays (one color hybridization).

GeneChip technology possesses a number of strengths. It provides a strategy to monitor differential gene expression on a genomic scale but what is of particular value is its ability to detect single nucleotide polymorphisms and mutations. These features are well suited to epidemiological studies. The specificity in design of oligonucleotide probe pairs minimizes the potential for confounding effects due to cross hybridizations between highly homologous members of a gene family. This technology also has a number of drawbacks. Cost is the primary limitation as the technology is very expensive to establish, especially in an academic setting. Furthermore GeneChip microarrays are highly standardized with limited flexibility. For these practical reasons the remainder of this chapter focuses on spotted cDNA microarray technologies.

4.2 Spotted Microarrays

Spotted microarray technology is widely considered to have originated at Stanford University (Stanford, CA) under the direction of groups led by Ronald Davis and Patrick Brown (Schena et al., 1995, 1996; Brown and Botstein, 1999). Using an in-house automatic printer these researchers spotted (arrayed), on glass slides, cDNAs encoding for 1,046 distinct human genes. They next isolated mRNA transcripts from their experimental samples and reverse transcribed them to labeled cDNA, using nucleotides conjugated with a fluorescent label. The labeled cDNAs were then competitively hybridized to the cDNAs spotted on the glass slide (the microarray). Finally the intensity of labeled cDNA hybridization at each cDNA spot was measured using a computerized array reader, providing a measure of the abundance of target mRNA in the experimental sample.

Today spotted microarray manufacturing follows a similar format. Small glass slides or nylon membranes are arrayed at high density with DNA fragments (PCR amplified cDNA or synthesized oligonucleotides). In this chapter we use the term "probe" when referring to the DNA that is spotted on the fixed substrate (glass slides or nylon membranes). The probes may originate from a variety of sources. Oligonucleotide probes (60–70 mer) are commonly used to create small custom microarrays that can be used to examine a subset of genes that may be species, tissue, or pathway specific. Similarly, companies such as Agilent manufacture oligonucleotide-based high-density spotted microarrays that offer very good coverage of the genome. Similarly, Qiagen sells oligonucleotide sets for a variety of organisms that can be printed onto glass slides. Oligonucleotide base microarrays offer the theoretical advantage that all of the spots on the microarray have relatively consistent hybridization kinetics, thereby increasing the quality of the data derived from each microarray experiment.

The probes used in cDNA microarrays are PCR products amplified from cloned mRNA transcripts that range from 80–500+ base pairs in length. The cloned cDNAs are typically either full-length transcripts or ESTs. EST-based cDNA microarrays have been popular because EST libraries can be created for any organism in an unbiased manner without any prior knowledge of the genetic makeup of the organism. EST libraries are easily sequenced and amplified. When compared to genomic sequences ESTs frequently cluster (UniGene) in regions that contain characterized genes, thereby confirming the identity and origin of the original transcripts. ESTs offer an additional advantage that they often cluster in regions of the genome that have not been identified as encoding a gene.

cDNA microarray technology exploits the basic principles described in the Watson–Crick model of nucleic acid base pairing (hybridization). In a typical two-condition experiment, the mRNA transcripts isolated from two samples (e.g., experimental and control) are reverse transcribed to labeled cDNA (herein referred to as targets) using fluorescently labeled nucleotides. A different fluorescent dye is used for each sample (usually a red fluorescent dye (Cy5) and a green fluorescent dye (Cy3). The two targets are then mixed in equal proportions and competitively hybridized to probe cDNA sequences on a single microarray (this type of experiment is commonly referred to as a two-color hybridization). Competitive hybridization occurs between the labeled targets with the known printed cDNA probes. The hybridization step is then followed by a series of washes to eliminate unbound nucleotides and nonspecific binding, leaving only the appropriate target--probe complexes intact. The microarray is then imaged using a scanner and measurements made for each fluorescent dye at each spot on the microarray. The fluorescent intensity measured at each probe reflects the abundance of target mRNA in the original sample. With two-color hybridizations, a measurement of both samples is made on the same probe (spot). This eliminates any variability in the ratios that may result from variability in the spotting procedure. Differential expression is then defined as any ratio that significantly deviates from unity. Experiments are then replicated with fluorophores interchanged to control for the effect of biases introduced during dye incorporation (Liang et al., 2003).

cDNA microarray technologies have a number of strengths and weaknesses. When contrasted with GeneChip technology the cDNA platform allows the researcher to perform parallel analyses of differential gene expression using the same target set (two samples compared on the same chip). Furthermore, a growing availability of cDNA clone sets for use in the large-scale production of microarrays combined with a relatively low cost and high level of flexibility makes them well suited for academic institutions.

However, there are also a number of limitations to the use of cDNA-based microarrays. As with GeneChip, oligo and cDNA microarrays, the user is limited to detection of entities (genes or ESTs) that are present on the array. Importantly, detection of novel "genes" remains possible with microarrays that probe for uncharacterized EST clusters. Cross-hybridization is also a problem inherent to microarray experiments. This can be due to sequence homology, particularly among targets and probes from members of highly homologous gene families, poor probe design, or poor stringency hybridization. Cross-hybridization is more of a concern for EST-based microarrays, because there is no rational approach to the design of the probes as there is with oligonucleotide-based microarrays. Introns, promoters, and intergenic sequences involved in gene regulation have not been represented on cDNA and EST microarrays, limiting their utility in epigenetic studies. However, many labs working on the characterization of bacterial genomes have begun producing genome-chips that facilitate examination of these issues (Carrillo et al., 2004). And finally, the sensitivity for detection of low-level gene expression changes requires further improvement.

This section provides only a basic outline of concepts surrounding cDNA microarrays. For further reading a number of excellent detailed descriptions of this technology are recommended (Geschwind, 2000; Luo and Geschwind, 2001; Marcotte et al., 2001; Li et al., 2002).

5 Experimental Design of cDNA Microarray Experiments

Goals:

1. To provide a clear definition of the experimental questions and/or hypothesis to be addressed
2. To ensure that all steps in the experimental process are optimized to reduce variation and ensure reproducibility
3. To ensure that the experiment incorporates an appropriate level of replication for statistical validation of the data
4. To ensure that the experimental plan is in compliance with standards for microarray information collection

For every microarray experiment the first and most important step is experimental design. A badly designed experiment can render microarray data unsuitable for addressing the experimental questions or worse, lead the investigator to draw false conclusions. Furthermore, failed microarray experiments can be very costly both in terms of resources and time. There are many issues that must be addressed when planning a cDNA microarray experiment, some intuitive, others requiring considerable thought.

5.1 Hypothesis Testing

The basis of experimental science is hypothesis testing. Microarray experiments are no exception to this rule. Although high-throughput experimentation is often criticized as being nothing more than a "fishing trip," the reality is that a hypothesis is tested when a microarray experiment is conducted. Like all experimental techniques, microarray experiments are as good or bad as the hypothesis being tested. The only distinction that could be made for array experiments is that the experimenter is testing a complex hypothesis that examines the biological state on a systematic level. It is necessary for the experimenter to understand what the limitations of the microarray platform being used are. Specifically, any hypothesis is limited to what is being tested on the array. Therefore, a necessary first step is to ensure that the genes hypothesized to be involved in the biological phenomenon being tested are present on the microarray. Furthermore, if a given pathway is known to be involved in the experimental manipulation, then you would want the array platform chosen to have a good representation of the genes in that pathway. An important benefit of gaining a thorough understanding of the array platform prior to use is that you will be better able to interpret the reliability of the data.

A number of excellent reviews provide detailed discussions pertaining to cDNA microarray experimental design (Kerr and Churchill, 2001; Churchill, 2002; Foster and Huber, 2002; Yang and Speed, 2002; Li et al., 2002; Pan et al., 2002; Simon et al., 2002; Simon and Dobbin, 2003). Here we provide only a brief discussion of selected issues in cDNA microarray experiment design. We strongly encourage readers to develop an expertise in design issues prior to beginning a microarray experiment.

5.2 Experimental Controls

Good experimental practice includes the use of positive and negative controls. This can be done very easily with microarrays. If known alterations in gene expression (knock-outs, over-expression) are part of the experimental paradigm, then the manipulated genes should be used as positive controls. Alternatively, positive controls can be identified from the literature. By checking for the presence/regulation of sentinel genes, the experimenter can gain an appreciation for the validity of the microarray data that has been collected. The use of sentinel genes may also be of assistance in determining the necessity for additional replication or methodological alterations. Negative controls commonly take the form of alien spots (gene not expressed in the model organism) or genes that are not expected to be expressed in a given tissue or time frame.

5.3 Controlling Variability

Experiments that are designed well provide good primary data (spot intensity measurements), with which robust statistical analyses can be performed. Of utmost importance to collecting good data is the ability to control variability. It is now established that replication in microarray experiments is essential both for increasing the robustness of findings and for minimizing the influence of inherent variability in gene expression data. Numerous investigations have shown that even under tightly controlled conditions a single microarray experiment is subject to substantial variability and that the pooling of data from replicates is necessary to ensure statistical validity (Lee et al., 2000). The sources of variability for cDNA microarray experiments can be grouped into three broad categories; *biological variation, technical variations* and *measurement error* (Churchill 2002; Dudoit et al., 2002). Biological variations are common among individuals within a population and may be the result of environmental influence, genetics, or both. Biological variance can be managed by incorporating biological replications in the experimental design (Simon and Dobbin, 2003). Biological replication within a microarray experiment refers to using mRNA from different samples (obtained from different individuals, tissues, or versions of a cell line). Inclusion of biological replicates enables the determination of biological variability rather than technical variability. Current recommendations suggest that a minimum of three biological replicates be performed for cDNA microarray experiments (Lee et al., 2000). Technical variation can be considered the sum of all possible variations arising from the microarrays technology (Simon et al., 2002). Substantial variation among individual microarrays, at the individual probe level, is not uncommon (even when the arrays are obtained from the same producer and lot). Irregularities in manufacturing (variation in the size, shape, and amount of a spotted probe) can influence image analysis and confound comparisons. It would be considered good practice to perform experiments designed to determine the amount of technical variability (control versus. control) in the array platform that is chosen. Technical replicates of mRNA from the same pool help to ensure that procedures, reagents, and equipment are operating correctly. However, since most replicates will be performed by the same person using the same protocols and equipment, they generally reveal a smaller degree of variation in measurement than would be seen in the biological replicates described earlier.

Biases introduced during the isolation or labeling of the mRNA transcripts or in the hybridization process are also commonplace. These biases can be minimized by ensuring consistent RNA quality and quantity prior to labeling, as well as ensuring that the efficiency of labeling is consistent among samples.

This can be accomplished using standard electophoretic and spectrophotometric methods. Researchers frequently note substantial variation in the quality of hybridizations obtained. Heavy background levels, scratches, and fingerprints are common contributors to hybridization variation. Once again, replicates for each RNA specimen permits discarding of bad arrays and serves to increase the precision of the data collected.

A final source of variation in microarray experiments is derived from measurement errors. Measurement errors may occur during the processes of image acquisition and normalization or during the multifactorial data analysis required to extract biological relevance from the collected data. The effect of measurement error can be minimized by ensuring consistency in all aspects of microarray experimentation. If possible, experiments should be performed by the same technician, and subsequent data analyses be applied to all datasets consistently.

5.4 Statistical Analysis

Statistical analysis of microarrays has been an area of active research in recent years. The fundamental problem associated with microarray data is that they are highly dimensional. Typical statistical analyses are designed for use in situations where there are a few variables with many replicates. However, microarray experiments produce data composed of many variables with few replicates. This makes the use of standard statistical tests prone to type I error (false negative findings). Although this is not a new problem, the methods designed to deal with this, such as the Bonferroni correction, do not scale well for use in cases where there are >10,000 variables. Fortunately, a number of methods have been developed to provide researchers with a method for determining the statistical significance of microarray data. Each of these is a modification of an existing method and they all have limitations. Therefore, it is incumbent upon the researchers to determine the most appropriate method for their experimental design.

An additional experimental design consideration is the type of comparisons needed to answer the experimental question. Currently cDNA microarrays are used almost exclusively to make relative comparisons between two experimental conditions generally via direct design comparisons (e.g., a comparison between a tissue sample exposed to different experimental conditions). Simple pair-wise comparisons like these are common and issues pertaining to statistical analysis and data mining of these comparisons have been extensively described in the literature. As DNA microarrays grow in their ability to measure gene expression on a genomic scale, researchers will be tempted to design increasingly complex experiments. For a discussion of the potential gains from multiple comparisons of microarray studies and issues surrounding their experimental design, see Townsend (2003).

5.5 Standards

Early on in the process it was realized that to facilitate sharing of microarray data a community infrastructure needed to be developed. Global initiatives have established a series of standards that enable the faithful replication, verification, and comparison of data between similar experiments. The Microarray Gene Expression Data Society (MGED, http://www.mged.org) was created to facilitate the sharing of microarray data. In 2001, MGED published the first comprehensive description of international standards for microarray experiments. Referred to as Minimum Information About Microarray Experiments (MIAME) (Brazma et al., 2001) this six part protocol sets guidelines for (1) experimental design, (2) array design, (3) samples, (4) hybridizations, (5) measurements, and (6) normalization controls. Most research journals currently require compliance with MIAME for publication. Detailed descriptions of the six parts are available on the MIAME web site (http://www.mged.org/Workgroups/MIAME/miame.html).

Although microarray experiments generate vast amounts of data, typically, the experimental question can be answered with only a small fraction of this information. By sharing complete datasets with the research community (published results or results that will not be published) the full utility of microarray results can be realized. By conforming to the MIAME standards, microarray data become more interpretable and extensible.

Today a number of gene expression data repositories have been created to facilitate the data sharing process. Services such as ArrayExpress at the European Bioinformatics Institute (http://www.ebi.ac.uk/arrayexpress), the Gene Expression Omnibus at the NCBI (http://www.ncbi.nlm.nih.gov/geo/), and the Stanford Microarray Database (SMD, http://genome-www5.stanford.edu/) have been created for the storage of microarray gene expression data for all platforms. These services act both as a repository and as an online resource for the retrieval of gene expression data.

6 The cDNA Microarray Process

The growing commercial availability and relative affordability of cDNA microarrays combined with well-defined protocols for hybridization has made functional genomics a reality for many laboratories. However microarray experiments produce massive quantities of gene expression and functional genomics data, the analysis of which is complicated and involves many steps, each requiring careful consideration.

The typical cDNA microarray study can be described in nine steps: (1) establishing an appropriate experimental design; (2) isolation and conversion of mRNA to labeled cDNA; (3) hybridization of labeled cDNA to the microarray slide; (4) image acquisition, (5) data storage, (6) normalization; (7) statistical analysis; (8) data mining; and (9) validation of the results. Each of these steps is multifaceted and the introduction of error at any point in the process can lead to costly loss of data. The following section describes the steps followed in experimental design.

6.1 Isolation and Labeling of RNA (making labeled cDNA probes)

Goals:

1. To ensure the proper isolation and collection of the biological samples
2. Isolation of RNA from biological samples
3. Characterization of isolated RNA in terms of integrity, purity, and concentration
4. Preparation of labeled cDNA probes from characterized mRNA

A variety of standard molecular biology techniques are used in performing a DNA microarray experiment. These include techniques for handling biological samples (the source of the mRNA), isolating RNA from samples, characterizing the purity, integrity, and concentration of isolated RNA, and conversion of mRNA to fluorescently labeled cDNA. This process is based upon comparison of labeled cDNA samples so it is essential that equal quantities of RNA are used when making the probes. The integrity and purity of the starting RNA can affect the accurate quantification of RNA samples and therefore the precision and reproducibility of microarray data obtained. Poor quality RNA typically results from improper sample collection and handling. Protocols for the proper collection and handling of RNA and for assessing its purity, concentration, and integrity are well established and can be found in any text of molecular biology techniques. The recent interest in micro RNA has prompted the development of microarrays that survey theexpression of these small RNAs (Krichevsky et al., 2003). It should be noted that some commercial kits for isolating total RNA exclude RNA below a certain size (100 bp). Therefore, if one intends to study these important smaller RNA species, it is necessary to ensure that the RNA isolation protocol is appropriate.

Fluorescent labeling of cDNA can be a potential source of technical variability. In a typical two-color experiment, fluorescently labeled cDNA probes are transcribed from separate mRNA populations (e.g., cerebral ischemia versus sham). One set of cDNA probes is labeled with one fluorescent dye (typically Cy5) and the second set with a different fluorescent dye (Cy3). A number of methods for making labeled cDNA from the RNA samples have been tested and reviewed (Stears et al., 2000; Vernon et al., 2000; Li et al., 2002) and a number of potential sources for variation must be appreciated. First, the molecular structure of the fluorescent dyes used in making labeled cDNA can affect efficiency of dye incorporation. Second the mode of dye incorporation (direct verses indirect labeling) can affect subsequent hybridization kinetics (Stears et al., 2000; Li et al., 2002). Fortunately, commercial DNA microarray manufacturers provide optimized protocols for making labeled cDNA probes that are suited to their products and provide information on ways to control for dye incorporation problems. A good practical approach to ensuring a balance of the amount of dye/cDNA used during hybridization is to ensure the use of equivalent amounts of starting material and equivalent amounts of fluorescence following labeling (see Section 18.6.3).

6.2 DNA Microarray Hybridization

Goals:

1. To facilitate high-quality hybridization that demonstrates both specificity of nucleotide base pairing and minimum background noise
2. To ensure the hybridization protocol chosen affords minimum variability and a high level of reproducibility

In the typical two-color hybridization, labeled cDNA is hybridized to DNA probes fixed to the surface of the microarray. Although this sounds straightforward there are in fact no universal hybridization protocols that can be applied to all experiments. This is due to the fact that the basic hybridization conditions, including ionic strength, probe concentrations, and temperature, are largely influenced by the length of the DNA fragments spotted on the microarray. Therefore, there are considerable differences in optimal hybridization conditions for oligonucleotide arrays compared with cDNA arrays. The optimization strategy used should be appropriate for the type of microarray that is being used. Optimizing the hybridization conditions provides the greatest opportunity to minimize variability and to maximize the experimental reproducibility. If one is using commercial microarrays, it is advisable to use the methods recommended by the manufacturer. However, if the results are unsatisfactory, there is a wealth of alternative protocols available in the literature and the public domain. The benefit of using standardized protocols, where appropriate, is that the data are likely to be more reproducible and interpretable by other laboratories. Additionally, optimizing the washing conditions can increase the signal-to-noise ratio and decrease the number of artifacts.

Recently, the issue of hybridization kinetics has been examined (Wetmur, 1991; Dai et al., 2002; Sartor et al., 2004). Standard microarray hybridization protocols suggest hybridization times of 24 h or less. However, there was no empirical evidence to suggest that microarray hybridizations can reach equilibrium in 24 h. An important observation made by Dai et al. (2002) was that specific hybridization requires more time than nonspecific to reach equilibrium. Furthermore, these authors indicate that there are differences in the time to equilibrium between nonfragmented and fragmented (as used in GeneChip experiments) samples due to differences in the mobility of the samples. Experiments conducted by Dai et al. (2002) and Sartor et al. (2004) indicated that specific hybridization and spot intensity continued to increase beyond 48 and 72 h. Sartor et al. (2004) found that spot intensity and number were also dependent on the amount of RNA in the sample. An interesting observation presented in this paper is that the number of spots detected with 72 h hybridization was roughly equivalent for all concentrations of RNA used. This suggests that the reliability of low-intensity measurements can be improved by increasing the duration of hybridization.

6.3 Image Acquisition and Data Processing

Goals:

1. To extract high-quality raw intensity data from the images on the microarray
2. To remove unacceptable spots
3. To adjust (normalize) the data for biases that may have arisen during the technical process thereby ensuring that meaningful biological comparisons and conclusions can be made

Obtaining an image, by scanning the slide to produce an image, is the first step in the process of data acquisition. The image is then analyzed in order to identify the spots on the slide. An intensity measurement and a local background measurement are then made for each spot. The raw intensity values (foreground and background) are then preprocessed, to exclude poor quality spots, and corrected (normalized) to remove any bias in the data. The processed, normalized data are then ready for the lengthy, computationally intensive procedures that follow.

6.3.1 Image Acquisition

The steps involved in the image acquisition process for cDNA microarrays have been reviewed and a number of image analysis programs are available to mediate this process. The process can be described in four basic steps (1) scanning, (2) spot recognition, (3) segmentation (4) intensity measurement, and (5) ratio calculation (Leung and Cavalieri, 2003).

Although image acquisition is an important step in the DNA microarray experiment, it is limited by the quality of the hybridization. The quality of the hybridization affects all of the subsequent steps and image acquisition provides an opportunity to critically assess the quality of the hybridization. As with each step, a poorly acquired image will require subsequent manipulation to improve its quality, a process that decreases the validity of analysis. Choosing the correct scanning parameters is essential to good image acquisition. Although some scanners offer the option of scanning slides in batches, it is our experience that the scanning parameters require optimization for each slide. This is true even for technical replicates. There are several factors that should be optimized during scanning (1) dynamic range, (2) saturation, (3) balance, and (4) merging. The dynamic range of an image refers to the maximum and minimum intensity values. To maximize the dynamic range, without saturating spots, the PMT should be adjusted so that the brightest spots are just below threshold. The dynamic range is adjusted for each channel (Cy3 and Cy5). A balanced image is one in which the average spot intensity is equal in both channels. Obtaining balanced images with the greatest dynamic range will minimize the effect/need for normalization and provide data with greater reliability.

In some instances, particularly in vertebrate tissue samples, there may be a population of genes that is expressed at levels that are far in excess of the majority of the genes. When scanning parameters are set so that the brightest spots are not saturated, it can become very difficult to obtain data for genes expressed at low or even medium levels. Under this circumstance, the choice is either to use the scan with the majority of the genes being undetected, or to saturate the bright spots and capture more data at the low end. Either option results in some data loss. Fortunately, methods that allow for the merging of low- and high-intensity scans have been developed. This is a very useful feature of the BASE (BioArray Software Environment) microarray software package (http://base.thep.lu.se/).

Another issue associated with microarray scanning is the mathematical limits of $\log_2$ ratios (Sharov et al., 2004). In this paper, it was pointed out that optimal results are obtained when the boundaries of the dynamic range are avoided. Furthermore, it was observed that the maximal $\log_2$ ratios are possible when the background intensity is only 256 RFUs. Targeting the spot intensities to such a value would be impractical, and the authors suggested a target range of 1,024–4,096 for average spot intensity. Incidentally, in our hands, microarray studies of mouse brain hippocampus produced scans, using the maximized dynamic range method, with average spot intensity in the range that would provide the maximal $\log_2$ ratios (Gilbert et al., 2003).

There are several microarray scanners on the market, all of which utilize a laser (and which may or may not be confocal) to excite the fluorescent dyes. Each of these machines generates a 16-bit grayscale tiff image for each channel. Certain scanners have as many as 5 lasers and can be used with a wider variety of fluorescent dyes. Microarray scanners come with software for spot recognition, segmentation, and determination of spot intensity. However, because each software package is capable of importing tiff images created by any scanner, the researcher is free to use any software to perform spot recognition, segmentation, and spot intensity quantification. In addition, free software packages are available that perform these tasks (TM4 package from TIGR, http://www.tigr.org/software/tm4/).

6.3.2 Normalization of Ratios

Normalization of cDNA microarray data is a very important step in the process of data analysis. With current technology, systematic bias is unavoidable and must be dealt with in a sensible manner. Furthermore, normalization methods need to be consistently applied to all raw data. Using different normalization methods on different datasets may introduce bias and thereby decrease the validity of the data. Normalized data should be free of systematic bias and should thereby provide a truer representation of the biological variance. Furthermore, normalized data increases the validity of slide to slide comparisons.

There are many approaches for normalizing ratio data (Li et al., 2002; Quackenbush 2002; Park et al., 2003). The various methods can be viewed as having increasing resolution on the data. Normalization can be performed on all of the data by simply applying an algebraic factor to one channel so that the total average intensities in the two channels are equalized. This is the most primitive method and does not account for known sources of systematic bias (e.g., print tips, intensity) or regional differences in image quality due to printing abnormalities or hybridization artifacts. The best approach is to group spots according to a common biasing characteristic and normalize those spots separately. This is the approach that is used when normalization is done according to print tip or sub-grid. The use of nonlinear normalization methods provides the ability to correct for systematic biases that cause ratio plots to deviate from linearity. The lowess method, as an example, is very useful in correcting intensity dependent biases. Although normalization is capable of correcting suboptimal data, the researcher needs to be objective about the application of normalization. The temptation to over-normalize low-quality data should be avoided. In our experience, time and effort is better spent repeating experiments, rather than attempting to interpret and confirm low-quality experiments.

7 Analysis and Interpretation of DNA Microarray Data

Goals:
1. To use analysis tools to organize the gene expression data (gene expression profiling) in a way that will facilitate the exploration of changes in gene expressions
2. To examine the microarray data from the perspective of functional pathways

By this stage, the raw data have been processed and normalized, enabling subsequent statistical and biological analysis of gene expression. At this point the amount of data begins to accumulate and the task of making inter-hybridization comparisons is at hand. The challenge therefore is to organize the data in a way that will facilitate statistical analysis, data mining, and biological interpretation. Attempting to organize vast amounts of data (often in excess of 1 million data points and all associated meta data) outside of an SQL framework would be inefficient and ill advised. Fortunately, there are commercial and free solutions to this problem. Once the data have been consolidated, it is now possible to perform statistical analysis aimed at identifying differentially expressed genes. The statistical analysis employed will depend on the type of comparison that has been determined by the experimental design. The opportunity to reduce the dataset that is used in subsequent data mining procedures is presented with the identification of the significantly altered genes. Data mining is a general term that refers to a large number of methods that act as exploratory tools for extracting meaning from complex datasets. Concurrently, analysis and interpretation of the biological annotation of the significant altered genes occurs at this stage. Recently, efforts have been made to create tools that enable researchers to explore their data in the context of biological/biochemical pathways. Further advances in the area of literature mining have enabled researchers to examine their data in the context of biological associations that are reported in PubMed.

7.1 Data Management

As mentioned, microarray experiments generate tremendous amounts of data. Further complicating matters, each data point can have numerous meta data associated with it. Meta data (data about data) refers to the biological annotations that are typically associated with a given gene. Meta data originates from a number of databases that describe features of the genes, mRNA, and protein. Further information regarding protein–protein interaction, gene and protein structure, and gene ontology can be utilized (e.g., Entrez, SwissProt, GO, Kegg, Bind). The necessity of having the ability to easily access all of the data associated with a microarray project was realized early on. Today there are a number of commercial and open source/free software packages that suit the needs of both large and small laboratories. A notable option in this field is the BASE package that is freely available and open-source. BASE has the same capabilities of commercial packages (including an integrated LIMS system), and also allows for the development of customizable plug-ins.

7.2 Statistical Analysis

The consolidation of normalized microarray data sets the stage for the subsequent statistical analyses. As discussed in Section 18.5.4, performing statistical analyses on microarray data presented a challenge to early researchers. Due to practical considerations, experiments are often designed so that direct comparisons of conditions are not made in a single hybridization. A frequently cited example of this is the common reference design (Yang and Speed, 2002). Experiments using the common reference design are performed with the RNA from each condition being hybridized with RNA from a common source (typically from large batch of RNA). Use of this design requires a data transformation step that enables indirect comparisons between conditions. This design is popular because it is much simpler than other indirect comparison designs, and is often cheaper than directly comparing all conditions. The mathematics used to transform the data are straightforward, and are explained in detail in Smyth et al. (2003).

One approach to identifying differentially expressed genes involves the use of an arbitrary threshold (cutoff value), below which differences in gene expression were considered insignificant (Draghici, 2002). Although practical in nature, this strategy lacked a statistical basis and typically used a threshold value of a twofold difference in gene expression to define a significant change. From a biological perspective this practice poses the question of what degree of expression change constitutes a relevant biological event. In other words, how does one choose a cutoff value that will allow the separation of true experimentally induced changes in gene expression from natural variability in the data? The importance of such questions is exemplified by the fact that many biologically relevant changes in genes expression involve change in small amounts of mRNA. In such cases conservative cutoff values increase considerably the chance of missing important genes (Yang et al., 2002). To date, issues surrounding the use of arbitrary thresholds for determining significance change remain unresolved. Few researchers have taken up the task of proving the validity of such strategies and work that has been done suggests that true cutoff values are heavily influenced by the personnel, facilities and protocols employed (see Li et al., 2002). Furthermore, because systematic variance (the sum of random, biological, and technical/measurement variance) is not accounted for when using this method, it is impossible to discriminate between true and spurious findings.

Recently, statistical procedures have been described that allow researchers to identify alterations in gene expression more objectively. The development of good statistical tools is essential from the perspective of managing the false discovery rate (FDR) associated with DNA microarray experiments. Early experiments were plagued by high FDR. Although technological advances have reduced this value considerably, if we consider that the typical cDNA microarray may be spotted with 40,000 probes then even an FDR of 0.5% could generate upward of 200 false positives. Since most microarray experiments have revealed only a few hundred gene expression changes, false values of this magnitude can be an important confounding problem. Therefore, it has been important to develop strategies that allow the determination of real signals from false ones.

Current practice in microarray experimentation suggests that a balance design with adequate replication be used. Good experimental design and execution will produce data that minimize technical variance, allowing the statistical analyses to evaluate biological variance more effectively. Still, the nature of the data requires that an estimate of the FDR be included in the statistical analysis. This enables the researcher to assess the reliability/validity of the results of the statistical analysis. As discussed earlier, cDNA microarray experiments are subject to multiple sources of experimental variation. Therefore, replications within an experiment are a necessity to increase the robustness of findings and minimize systemic variation in the data. It has been recommended that a minimum of three replicates be performed for every cDNA microarray experiment (Lee et al., 2000). Although replicates lend statistical strength to microarray data, they are expensive and create the need for statistical algorithms capable of analyzing large data sets. There are currently many tools for the statistical analysis of the results of DNA microarray experiments. Although there is yet no single universally accepted method for primary statistical analysis of microarray data, approaches such as CyberT (Baldi and Long, 2001), and *s*ignificance *a*nalysis of *m*icroarrays (SAM) (Tusher et al., 2001; Storey and Tibshirani, 2003) have proved useful in analyzing microarray data. For example, the SAM method uses data permutations to provide estimates of FDR for multiple testing. For a comprehensive review of the issues pertaining to the statistical validation, causes of error, and variations in microarray data (see Kerr and Churchill, 2001; Pan, 2002; Krajewski and Bocinowski, 2002; Nadon and Shoemaker, 2002; Brody et al., 2002; Dudoit et al., 2002; Smyth et al., 2003). If the user is so inclined, a wealth of statistical methods used to analyze microarray data can be accessed, and contributed to, through the BioConductor project (http://www.bioconductor.org/).

7.3 Data Mining

In the present context, data mining can be considered the automated extraction of hidden predictive information from microarray data and meta data. Data mining is typically performed on a dataset that has been reduced as the result of statistical analysis. By focusing on expression profiles, the analyses used in data mining will have their greatest predictive power. There are many powerful techniques for data mining including algorithms designed to facilitate the organization and interpretation of large-scale gene expression data. These also include tools for dimensionality reduction (principal component analysis and singular value decomposition) and clustering (hierarchical clustering, K-means clustering, and self-organizing map clustering). Currently there are no clear guidelines for choosing one particular method of data mining. This is due to the difficulty in determining which method provides the best results. What is clear, however, is that no one technique can suit all situations and that the various types of analysis or even different parameters within the same analysis can reveal unique aspects of the gene expression data (Leung and Cavalieri, 2003). Today cluster analysis is one of the most commonly used approaches for exploring relationships (similarities and dissimilarities) in gene expression data. Cluster analysis is an unsupervised method that assigns (segments) genes to a given cluster according to their similarity to other genes. This is achieved by first calculating the distances between genes based upon their expression patterns, under various conditions, followed by the merging genes whose expression levels are close to each other under these conditions to a cluster. The assumption is that there will be correlation between gene expression and its function. Therefore, genes assigned to the same cluster might be coregulated, might perform a similar function, or may be involved in the same biological pathway. Comparing clustered genes of known function with ones of unknown function will provide information on those novel genes, whereas analyzing the promoter's elements of clustered genes may reveal common regulatory motifs and provide a better understanding of the pathways within a biological system (Pilpel, 2001). There are several excellent reviews on data mining techniques (Burke, 2000; Sherlock, 2000; Hughes et al., 2000; Quackenbush, 2001; Valafar, 2002).

A different application of clustering methods for interpreting microarray data is the problem of classification. Classification methods have been used to assess the potential of gene expression data in making clinical diagnoses. Classic examples of this type of problem include the classification of human tumors or tumor subclasses (Sorlie et al., 2001; Lossos et al., 2004). Sorlie and colleagues used clustering of samples (rather than genes) to discriminate between subtypes of breast cancer. This is an example of how a given pattern of gene expression can be used to identify or define a subtype of cancer. Lossos and coworkers (2004) studied a subset of 36 genes that had been reported to predict cancer survival, and through the use of a multivariate model, found that measuring the expression of six of these was sufficient to predict survival. Through the use of this type of analysis, it becomes clear how gene expression data could be used clinically to expedite diagnosis and treatment.

7.4 Pathway Analysis and Literature Mining

DNA microarray technology has the potential to change the way we approach experimental design. In the past, we explored possible biological pathways prior to designing an experiment. In the microarray era, we explore biological pathways revealed during the process of microarray data analysis. It is widely perceived that genes do not work alone in a biological system but instead function within a cascade of pathways. The wealth of data provided by a cDNA microarray experiment provides us with the opportunity to gain novel understanding of biological systems through the exploration of functional pathways. Earlier we discussed the utility of cluster analysis in the identification of clusters of genes that share common expression characteristics. However, cluster analysis does not provide information on the biological relatedness of these genes. Recently, a number of methods for analyzing gene expression data from a pathway perspective have emerged. Each of these methods has the capability of loading microarray results with biological meaning and includes probabilistic relational models (Segal et al., 2001). Certain tools, such as GenMAPP (http://www.genmapp.org), allow the researcher to overlay their gene expression data in the context of known biological pathways or ontologies (Dahlquist et al., 2002). This enables the recognition of the importance or biological context of the observed changes in gene expression. Another approach to this problem is to utilize a combination of literature mining (currently limited to PubMed abstract scanning) and protein interaction databases to construct biological associations between observed alterations in gene expression. PubGene (http://www.pubgene.org), PathwayAssist (http://www.iobion.com), and Osprey (http://biodata.mshri.on.ca/osprey/servlet/Index) are examples of extremely useful tools that use this approach. A key development has been the development of natural language processors that enable the extraction of meaningful functional data from PubMed (i.e., literature mining). An alternative approach has been taken by Ingenuity Systems who have created a curated pathway knowledge base that is utilized by a pathway analysis tool (http://www.ingenuity.com). Collectively, these methods facilitate the unbiased integration of biological information with microarray results.

8 Validation of Microarray Experiments

Goals:

1. To provide confirmation of the results of the experiment by follow-up experiments.

Technological advances have significantly reduced the incidence of false discoveries in microarray experiments. Today the majority of published confirmatory studies have shown that DNA microarrays and reverse transcription-polymerase chain reaction (RT-PCR)/Northern blot analysis support each other qualitatively. Yet, in spite of these findings most publishers of microarray studies continue to require the use of high-resolution methods for the verification of gene expression levels and the debate on whether corroboration of statistically significant microarray data is necessary remains unsettled. A strong argument for the continued validation of gene expression data can be made from the recent woes of Affymetrix. When numerous sequence identity errors were detected on their mouse chips it became clear that the public sequence databases used in the construction of oligonucleotides contained inaccuracies. The fact that these sequence databases are used in both the making of DNA microarrays and the subsequent analysis of information obtained from them exemplified the uncertainties often associated with new technology and the importance of caution in the interpretation of their results.

At present there are no standards to describe the methods to be used for corroboration. Similarly the microarray community lacks consensus regarding the extent to which corroboration must be undertaken. From the perspective of publishing, many journals require extensive confirmation of microarray data whereas others limit corroboration to the small number of genes that are of interest or those whose expression levels are in question. Obviously, confirmation of many genes is time-consuming and defeats the purpose of a high-throughput system.

Currently two approaches are used in the verification of DNA microarray results; in silico based analysis and laboratory-based analysis. From the laboratory-based perspective a number of proven and accurate techniques have provided confirmation of gene expression data including RT-PCR, the most popular method, Northern analysis, in situ hybridization, and ribonuclease protection assays (for review, see Marcotte et al., 2001; Chuaqui et al., 2002). Cellular localization of differentially expressed genes has been facilitated by in situ hybridization and changes in protein expression can be determined with immunoblot and immunohistochemical technology. With the in silico based methods, validation of microarray results is achieved by corroborating gene expression results with expression information published in the scientific literature or expression databases. Agreement between array results from other groups and/or with expression information from the literature can be used to validate the data.

Another issue pertaining to validation is deciding what should be done about poorly characterized ESTs. Many array platforms have a considerable number of spots that probe the expression of unannotated ESTs. In many studies, dramatic alterations in the expression of these ESTs are observed. This presents an interesting research opportunity, the possibility of discovering and characterizing a novel gene that is potentially of fundamental importance to the biological phenomenon being studied. However, the factor that prevents people from pursuing uncharacterized ESTs is that gene expression studies often produce long lists of uncharacterized ESTs that display similar patterns of gene expression. This makes it very difficult to identify the best candidate for further study.

Ultimately the final answer as to whether the product of a gene is of biological significance must be determined at the biological level, typically verified through a change in a physiological measure in vivo. Measurement of protein changes can also be an important confirmation of gene change based as it is on transformed genetic information.

9 Problems Inherent to Microarray Studies in Stroke

Although the data obtained from expression profiling are by no means an endpoint, they provide, in the critical hand, a way of defining genes or pathways acting in the pathogenesis of stroke. Like all methodologies however, the limitations of this technology must be appreciated in order to prevent unrealistic expectations or, worse, erroneous conclusions.

1. One limitation to cDNA microarray experiments is that relatively large quantities of mRNA are required. This can be difficult if, for example, the structure under investigation is small (e.g., the CA1 region of the hippocampus or the suprachiasmatic nucleus in the hypothalamus) or when RNA is to be obtained from postmortem brain tissue. A number of strategies have been used to address these concerns. Pooling tissue from biological replicates can provide the necessary amount of RNA for an experiment. In situations where replicates are cost prohibited PCR amplification of the mRNA population may be performed. Whereas this strategy has been used successfully by several investigators (Luo et al., 1999; Luzzi et al., 2001), PCR amplification has been associated with introducing bias in transcript levels (Emmert-Buck et al., 2000; Brown et al., 2003). For a review of these issues see Ginsberg et al., (2004).
2. Microarray data obtained from in vivo experiments are by virtue of design representations of transcripts expressed by all cell types of the tissue under investigation. For example, gene profiles obtained from investigation of hippocampal tissue would represent the sum of expression change across several cell types (e.g., neurons, interneurons, glial cells, vascular epithelial cells, or cells of the immune system that may have migrated to the ischemic region). Although the nature of the regulated genes may aid in defining the parent cells, additional experiments to localize and verify the differential expression of these transcripts (e.g., in situ hybridization, RT-PCR, Northern analysis) are imperative. Attempts to profile the differences in gene expressions within specific cell populations can be difficult. One approach is to isolate distinct populations of cell types within a given brain structure (e.g., separating astrocytes from neurons) or even distinct populations of neurons from within the same brain region. Recent publications have demonstrated the feasibility of integrating laser capture microdissection technology with DNA microarrays (Luo et al., 1999). Laser capture can be used to isolate mRNA from homogeneous cell populations, or even single cells. One drawback of this technology is that it may be difficult to isolate sufficient RNA from small tissue sections. In these situations the pooling of tissue

from multiple biological replicates may help to overcome this concern. As this field develops we will reach the point where it will be possible to map at the cellular level the functional pathways mediating stroke pathogenesis.

3. Another limitation that is common to all platforms of gene expression analysis is that these methods investigate the intermediate molecules in the process to protein synthesis. Although the identification of mRNA is instructive, ultimately it is the protein expression in cells that would mediate apoptotic related pathogenesis following cerebral ischemia. As increased transcription can, but does not always, result in increased protein synthesis, measured changes in RNA expression may lack biological significance.

   In a related vein, it is important to keep in mind that a lack of differential expression of a gene does not signify a lack of functional importance. The role of some genes may be regulated at the level of translation rather than transcription. Post-transcriptional and post-translational modifications (e.g., alternative splicing, mRNA $t_{1/2}$, phosphorylation, glycosylation, protein $t_{1/2}$) can all influence biological function (Luo and Geschwind, 2001). These alterations would remain undetected by DNA microarray analysis. Finally, DNA microarrays do not identify genes (and their protein products) that although not differentially expressed, provide supportive functions for genes undergoing induction/repression.
4. Even under strict experimental control, animal models of stroke are notoriously variable. As such, microarray experiments require special consideration during the experimental design. A number of protocols have been proposed for controlling the influence of variability to stroke-related experiments. The logical approach would be to include a level of biological replication known to suffice for variability (e.g., as established in morphological studies). However, this practice, when applied to cDNA microarray experiments, can be quite costly. A second more practical approach has been to pool tissue of interest from a number of animals (Gilbert et al., 2003).
5. The current dogma suggests that following a cerebral-ischemic event gene expression occurs in a defined sequence (Roth et al., 2003). For example, immediate early genes followed by heat-shock proteins. In other words genes involved in the pathogenesis of stroke may be differentially expressed at different times and the time points chosen for gene expression analysis may miss the window for detection of certain genes of interest. Similarly, factors that can affect the expression of a gene or pathway might include the developmental stage of the organism, sex, age, the cell type under investigation, as well as a large number of environmental factors. All of these confounding factors further illustrate the need for good experimental design and control.
6. Does the microarray chosen contain probes for genes required to answer your experimental question/hypothesis? Before a gene can be represented on a cDNA microarray the manufacturer must possess a cDNA clone for that gene or have knowledge of the gene sequence. Although the day is approaching when an entire species genome may be represented on a single microarray slide, we are not there yet. Estimates suggest the 15K mouse NIA clone set, which forms the basis of many commercially available microarrays in fact represents $\approx$7500 genes (Kargul et al., 2001). Many functionally important genes are expressed at low levels and consequently are not represented in cDNA libraries. Furthermore, many of the cDNA clone sets used in microarray manufacturing contain only constitutively expressed genes. Since many of the genes that have been implicated in stroke are induced they may not be represented on the slide.

10 Developing an Unbiased Approach to the use of DNA Microarray Results

Pathogenesis following stroke most likely reflects the integrative actions of multiple functional pathway acting in concert. Today, DNA microarray technology provides the opportunity for even the small-scale research laboratory to study complex interactions, like those occurring following cerebral ischemia. By enabling the observation of simultaneously occurring events, DNA microarrays now provide a platform for a nonreductionist approach to hypothesis generation. Unfortunately, for many researchers the wealth of results provided in these experiments becomes intimidating, and rather than trying to organize the results into a meaningful biological framework they tend to focus mainly on what is already known. Consequently, the majority of DNA microarray-related stroke research that has been published has focused on the same subset of genes leaving the results obtained far short of reaching their full potential as a hypothesis-generating tool. Only a handful of publications have attempted to describe microarray results from a pathway perspective, and yet an understanding of the interplay between pathways may be vital for a complete understanding of the pathogenesis of stroke. As described earlier we currently have at our disposal a number of methods developed for the automatic loading of microarray results with biological meaning. There are already many examples in the literature that have highlighted specific pathways that might have otherwise been overlooked. With time and increased familiarity with these methods, researchers will hopefully begin to place stroke gene expression results within biological context, increasing the development of new hypotheses.

11 Exploiting Public Databases and Repositories of Gene Expression Data for Stroke Research

Annotation of mammalian genomes is an evolutionary process. With each build of public databases, the amount of information available to researchers increases. An example of this is the UniGene database. Having used EST-based microarrays in the past (Gilbert et al., 2003), it has been our observation that a large number of the ESTs that were altered following experimental stroke were either poorly characterized, or completely uncharacterized. This situation also holds true for oligonucleotide-based microarrays. Commercial manufacturers (including Affymetrix) of oligonucleotide microarrays design a large proportion of their probes to recognize ESTs. Although the continued annotation of previously uncharacterized ESTs is a slow process, it warrants revisiting data (published or otherwise) occasionally to see if any new information has been made available that would be of interest.

The principles of open access are expanding the scope of experimental science (Brown et al., 2003). Today, it is possible to access, analyze, and publish findings derived from publicly available datasets (Cheadle et al., 2003). A number of sources of data are freely available to researchers (see Section 18.5.5) who wish to augment/extend their own data (reducing costly replication of previous research). Similarly, publicly available datasets are rich sources of information and, coupled with current data mining tools, could be a goldmine for hypothesis generation. Indeed, there are probably enough data available in public repositories to drive an in silico biology program.

References

Baldi P, Long AD. 2001. A Bayesian framework for the analysis of microarray expression data: regularized t-test and statistical inference of gene changes. Bioinformatics 17: 509.

Brazma A, Hingamp P, Quackenbush J, Sherlock G, Spellman P, et al. 2001. Minimum information about a microarray experiment (MIAME)-toward standards for microarray data. Nature Genetics 29: 365.

Brody JP, Williams BA, Wold BJ, Quake SR. 2002. Significance and statistical errors in the analysis of DNA microarray data. Proc Natl Acad Sci USA 99: 12975.

Brown PO, Botstein D. 1999. Exploring the new world of the genome with DNA microarrays. Nature Genetics 21: 33.

Brown PO, Eisen MB, Varmus HE. 2003. Why PLoS became a publisher. PLoS Biol 1: E36.

Bunney WE, Bunney BG, Vawter MP, Tomita H, et al. 2003. Microarray Technology: A review of new strategies to discover candidate vulnerability genes in psychiatric disorders. Am J Psychiatry 160: 657.

Burke HB. 2000. Discovering patterns in microarray data. Mol Diagn 5: 349.

Carrillo CD, Taboada E, Nash JH, Lanthier P, Kelly J, et al. 2004. Genome-wide expression analyses of *Campylobacter jejuni* NCTC11168 reveals coordinate regulation of motility and virulence by flhA. J Biol Chem 279: 20327.

Cheadle C, Cho-Chung YS, Becker KG, Vawter MP. 2003. Application of z-score transformation to Affymetrix data. Appl Bioinformatics 2: 209.

Chuaqui RF, et al. 2002. Post-analysis follow-up and validation of microarray experiments. Nature Genetics 32: 509.

Churchill GA. 2002. Fundamentals of cDNA microarray design. Nature Genetics 32: 490.

Dahlquist KD, Salomonis N, Vranizan K, Lawlor SC, Conklin BR. 2002. GenMAPP, a new tool for viewing and analyzing microarray data on biological pathways. Nature Genetics 31: 19.

Dai H, Meyer M, Stepaniants S, Ziman M, Stoughton R. 2002. Use of hybridization kinetics for differentiating specific from non-specific binding to oligonucleotide microarrays. Nucleic Acids Res 30: e86.

Draghici S. 2002. Statistical intelligence: effective analysis of high-density microarray experiments. Drug Discovery Today 7: 555.

Dudoit S, Yang YH, Callow MJ, Speed TP. 2002. Statistical methods for identifying differentially expressed genes in replicate cDNA microarray experiments. Statistica Sinica 12: 111.

Emmert-Buck MR, et al. 2000. Molecular profiling of clinical tissue specimens: feasibility and applications. Am J Pathol 156: 1109.

Foster FG, Huber R. 2002. Current themes in microarray experimental design and analysis. Drug Discovery Today 7: 290.

Geschwind DH. 2000. Mice, microarrays, and the genetic diversity of the brain. Proc Natl Acad Sci USA 97: 10676.

Gilbert RW, Costain WJ, Blanchard ME, Mullen K, Currie RW, et al. 2003. DNA microarray analysis of hippocampal gene expression measured 12 hours after hypoxia-ischemia in the mouse. J Cereb Blood Flow Metabol 23: 1195.

Gilby KL, Denovan-Wright EM. 2004. Analysis of gene expression in the brain using differential display. Handbook of neurochemistry and molecular neurobiology (this volume).

Ginsberg SD, Elarova I, Ruben M, Tan F, Counts SE, et al. 2004. Single-cell gene expression analysis: implications for neurodegenerative and neuropsychiatric disorders. Neurochem Res 29: 1053.

Grunblatt E, Mandel S, Maor G, Youdim MB. 2001. Gene expression analysis in N-methyl-4-phenyl-1,2,3,6-tetrahydropyridine mice model of Parkinson's disease using cDNA microarray: effect of R-apomorphine. J Neurochem 78: 1.

Hughes TR, et al. 2000. Functional discovery via a compendium of expression profiles. Cell 102: 109.

Jin K, Mao XO, Eshoo MW, Nagayama T, Minami M, et al. 2001. Microarray analysis of hippocampal gene expression in global cerebral ischemia. Ann Neurol 50: 93.

Kargul GJ, Dudekula DB, Qian Y, Lim MK, Jaradat SA, et al. 2001. Verification and initial annotation of the NIA mouse 15K cDNA clone set. Nature Genetics 28: 17.

Kerr MK, Churchill GA. 2001. Statistical design and the analysis of gene expression microarray data. Genet Res 77: 123.

Krajewski P, Bocinowski J. 2002. Statistical methods for microarray analysis. J Appl Genet 43: 269.

Krichevsky AM, King KS, Donahue CP, Khrapko K, Kosik KS. 2003. A microRNA array reveals extensive regulation of microRNAs during brain development. RNA 9: 1274.

Lee MI, Kuo FC, Whitmore GA, Sklar J. 2000. Importance of replication in microarray gene expression studies: statistical methods and evidence from repetitive cDNA hybridizations. Proc Natl Acad Sci USA 97: 9834.

Leung YF, Cavalieri D. 2003. Fundamentals of cDNA microarray data analysis. Trends in Genetics 19: 649.

Li X, Gu W, Mohan S, Baylink DJ. 2002. Microarrays – their use and misuse. Microcirculation 19: 13.

Liang M, Briggs AG, Rute E, Greene AS, Cowley AW. 2003. Quantitative assessment of the importance of dye switching and biological replication in cDNA microarray studies. Physiol Genomics 14: 199.

Liang P, Pardee AB. 1992. Differential display of eukaryotic messenger RNA by means of the polymerase chain reaction. Science 257: 967.

Lipshutz RJ, Fodor SP, Gingeras TR, Lockhart DJ. 1999. High density synthetic oligonucleotides arrays. Nature Genetics 21: 20.

Lockhart DJ, Winzeler EA. 2000. Genomics, gene expression and DNA arrays. Nature 405: 827.

Lockhart DJ, Dong H, Byrne MC, Follettie MT, Gallo MV, et al. 1996. Expression monitoring by hybridization to high density oligonucleotides arrays. Nature Biotechnol 14: 1675.

Lossos IS, Czerwinski DK, Alizadeh AA, Wechser MA, Tibshirani R, et al. 2004. Prediction of survival in diffuse large-B-cell lymphoma based on the expression of six genes. N Engl J Med 350: 1828.

Luo L, Salunga RC, Guo H, Bittner A, Joy KC, et al. 1999. Gene expression profiles of laser-capture adjacent neuronal subtypes. Nature Med 5: 117.

Luo J, Isaacs WB, Trent JM, Duggan DJ. 2003. Looking beyond morphology: cancer gene expression profiling using DNA microarrays. Cancer Invest 21: 937.

Luo Z, Geschwind DH. 2001. Microarray applications in neuroscience. Neurobiol Dis 8: 183.

Luzzi V, Holtschlag V, Watson MA. 2001. Expression profiling of ductal carcinoma in situ by laser capture microdissection and high density oligonucleotides arrays. Am J Pathol 158: 2005.

Marcotte ER, Srivastava LK, Quirion R. 2001. DNA microarrays in neuropsychopharmacology. Trends Pharmacol Sci 22: 426.

Marcotte ER, Srivastava LK, Quirion R. 2003. cDNA microarray and proteomic approaches in the study of brain diseases: focus on schizophrenia and Alzheimer's disease. Pharmacol Ther 100: 63.

Matzilevich DA, Rall JM, Moore AN, Grill RJ, Dash PK. 2002. High-density microarray analysis of hippocampal gene expression following experimental brain injury. J Neurosci Res 67: 646.

Mac Manus JP, Graber T, Luebbert C, Preston E, Rasquinha I, et al. 2004. Translation-state analysis of gene expression in mouse brain after focal ischemia. J Cereb Blood Flow Metab 24: 657.

Nadon R, Shoemaker J. 2002. Statistical issues with microarrays: processing and analysis. Trends Genet 18: 265.

Pan W, Lin J, Le C. 2002. How many replicates of arrays are required to detect gene expression changes in microarray experiments? A mixture model approach. Genome Biology 3(5): research0022.1.

Pan W. 2002. A comparative review of statistical methods for discovering differentially expressed genes in replicated microarray experiments. Bioinformatics 18: 546.

Park T, Yi SG, Kang SH, Lee S, Lee YS, et al. 2003. Evaluation of normalization methods for microarray data. BMC Bioinformatics 4: 33.

Pilpel Y. 2001. Identifying regulatory networks by combinatorial analysis of promoter elements. Nature Genetics 29: 153.

Quackenbush J. 2001. Computational analysis of microarray data. Nature Review Genetics 2: 418.

Quackenbush J. 2002. Microarray data normalization and transformation. Nature Genetics 32: 496.

Rat Genome Sequencing Project Consortium. 2004. Genome sequence of the Brown Norway rat yields insights into mammalian evolution. Nature 428: 493.

Read SJ, Parsons AA, Harrison DC, Philpott K, Kabnick KS, et al. 2001. Stroke genomics: approaches to identify, validate and understand ischemic stroke gene expression. J Cereb Blood Flow Metab 21: 755.

Roth A, Gill R, Certa U. 2003. Temporal and spatial gene expression patterns after experimental stroke in a rat model and characterization of PC4, a potential regulator of transcription. Mol Cell Neurosci 22: 353.

Sagerstrom CG, Sun BI, Sive HL. 1997. Subtractive cloning: past present and future. Annu Rev Biochem 60: 751.

Sartor M, Schwanekamp J, Halbleib D, Mohamed I, Karyala S, et al. 2004. Microarray results improve significantly as hybridization approaches equilibrium. Biotechniques 36: 790.

Schena M, Shalon D, Davis RW, Brown PO. 1995. Quantitative monitoring of gene expression patterns with a complementary DNA microarray. Science 270: 467.

Schena M, Shalon D, Heller R, Chai A, Brown PO, et al. 1996. Parallel human genome analysis: microarray-based expression monitoring of 1000 genes. Proc Natl Acad Sci USA 93: 10614.

Schmidt-Kastner R, Zhang B, Belayev L, Khoutorova L, Amin R, et al. 2002. DNA microarray analysis of cortical gene expression during early recirculation after focal brain ischemia in rat. Brain Res Mol Brain Res 108: 81.

Segal E, Taskar B, Gasch A, Friedman N, Koller D. 2001. Rich probabilistic models for gene expression. Bioinformatics 17 Suppl 1: S243.

Sharov V, Kwong KY, Frank B, Chen E, Hasseman J, et al. 2004. The limits of log-ratios. BMC Biotechnol 4: 3.

Sherlock G. 2000. Analysis of large-scale gene expression data. Curr Opin Immunol 12: 201.

Simon R. Dobbin K. 2003. Experimental design of DNA microarray experiments. Bio Techniques 34: S16.

Simon R, Radmacher MD, Dobbin K. 2002. Design of studies using microarrays. Genetic Epidemiology 23: 21.

Smyth GK, Yang YH, Speed T. 2003. Statistical issues in cDNA microarray data analysis. Methods Mol Biol 224: 111.

Sipione S, Rigamonti D, Valenza M, Zuccato C, Conti L, et al. 2002. Early transcriptional profiles in huntingtin-inducible striatal cells by microarray analyses. Hum Mol Genet 11: 1953.

Soriano MA, Tessier M, Certa U, Gill R. 2000. Parallel gene expression monitoring using oligonucleotide probe arrays of multiple transcripts with an animal model of focal ischemia. J Cereb Blood Flow Metab 20: 1045.

Sorlie T, Perou CM, Tibshirani R, Aas T, Geisler S, et al. 2001. Gene expression patterns of breast carcinomas distinguish tumor subclasses with clinical implications. Proc Natl Acad Sci USA 98: 10869.

Stears RL, Getts RC, Gullans SR. 2000. A novel, sensitive detection system for high-density microarrays using dendrimer technology. Physiol Genomics 3: 93.

Stenzel-Poore MP, Stevens SL, Xiong Z, Lessov NS, Harrington CA, et al. 2003. Effect of ischaemic preconditioning on genomic response to cerebral ischaemia: similarity to neuroprotective strategies in hibernation and hypoxia-tolerant states. Lancet 362: 1028.

Storey JD, Tibshirani R. 2003. SMA thresholding and false discovery rates for detecting differential gene expression in DNA microarrays. The Analysis of Gene Expression Data: Methods and Software. Parmigiani G, et al. editors. Heidelberg: Springer-Verlag.

Townsend JP. 2003. Multifactorial experimental design and the transactivity of ratios with spotted DNA microarrays. BMC Genomics 4: 41.

Tusher VG, Tibshirani R, Chu G. 2001. Significance analysis of microarrays applied to ionizing radiation response. Proc Natl Acad Sci USA 98: 5116.

Valafar F. 2002. Pattern recognition techniques in microarray data analysis: a survey. Ann NY Acad Sci 980: 41.

Velculescu VE, Zhang L, Vogelstein B, Kinzler KW. 1995. Serial analysis of gene expression. Science 270: 484.

Vernon SD, Unger ER, Rajeevan M, Dimulescu IM, Nisenbaum R, et al. 2000. Reproducibility of alternative probe synthesis approaches for gene expression profiling with arrays. J Mol Diag 2: 124.

Wahlgren NG, Ahmed N. 2004. Neuroprotection in cerebral ischaemia: facts and fancies – the need for new approaches. Cerebrovasc Dis 1: 153.

Waterson RH, et al. 2002. The Mouse Genome Sequencing Consortium, Initial sequencing and comparative analysis of the mouse genome. Nature 420: 520.

Wetmur JG. 1991. DNA probes: applications of the principles of nucleic acid hybridization. Crit Rev Biochem Mol Biol 26: 227.

Yang IV, Speed T. 2002. Design Issues for cDNA Microarrays. Nat Rev Genet 3: 579.

Yang IV, Chen E, Hasseman JP, et al. 2002. Within thee fold: assessing differential expression measures and reproducibility in microarray assays. Genome Biol 3: research0062.1.

19 Yeast Two-Hybrid Studies

M. J. Smith · K. Pozo · F. A. Stephenson

Abstract: The yeast two-hybrid system is a high throughput method for the study of protein-protein interactions. The use and application of two GAL4 based and the CytoTrap Sos recruitment yeast two-hybrid systems are described. This includes a detailed consideration of appropriate baits together with their generation and characterization for use in the screening of cDNA libraries for the detection of interacting proteins, the characterization of positive clones resulting from cDNA library screens and strategies for the validation of protein-protein interactions using alternative experimental paradigms.

Abbreviations: Activation domain, AD; Bioluminescence resonance energy transfer, BRET; DNA binding domain, DNA-BD; Fluorescence resonance energy transfer, FRET; γ-aminobutyric acid type A, $GABA_A$; $GABA_A$ receptor interacting factor, GRIF; Golgi-specific DHHC zinc finger protein, GODZ; N-methyl-D-aspartate, NMDA; O-GlcNAc transferase interacting protein 106, OIP106; SDS-polyacrylamide gel electrophoresis, SDS-PAGE

1 Introduction

The yeast two-hybrid system is an elegant, high-throughput method for the study of protein–protein interactions. Its main use is in identifying protein partners of defined proteins of either known or unknown function. The identification of these proteins thus yields insight into the function and/or dynamics of proteins within the cell. It is a genetic-based methodology and has the distinct advantage that protein–protein interactions are detected and analyzed in situ in yeast cells negating the need for the isolation of large amounts of proteins as may be necessary for other methods. It can also be used to study the association between two known proteins to identify the specific respective binding site domains. However, it should be noted that the association between two proteins as determined by yeast two-hybrid technology is not sufficient evidence to prove that two proteins are associated in vivo. Protein–protein interactions that have been identified by yeast two-hybrid screens need to be validated by alternative, supplementary experimental approaches. The first yeast two-hybrid system to be developed was the GAL4, transcription-based assay (Fields and Song, 1989). The GAL4 system has been available commercially for over 10 years. It has been used successfully by many laboratories in all areas of biochemistry for the study of protein–protein interactions. More recently, it has been adapted to improve the efficiency of use. Also, alternative two-hybrid systems have been introduced to overcome limitations of the GAL4-based system. In this chapter, we describe our experience in the use of the GAL4, a LexA, and the CytoTrap yeast two-hybrid systems focusing on the studies of the γ-aminobutyric acid type A ($GABA_A$) receptors, ($GABA_A$) receptor associated protein, GRIF-1, a protein isolated via a yeast two-hybrid cDNA library screen and thought to function in the anterograde trafficking of β2 subunit containing $GABA_A$ receptors to synapses (Beck et al., 2002), and the association of GRIF-1 and the GRIF-1 homolog, O-GlcNAc transferase interacting protein 106 (OIP106) with kinesin heavy chain, KIF5C (Brickley et al., 2005). General detailed methods for the culturing, transformation, and recombinant analysis of yeast can be found in Bartel and Fields (1997) and Zhu and Hannon (2000).

2 Yeast Two-Hybrid Systems

2.1 The GAL4 Yeast Two-Hybrid System

The theoretical basis of the GAL4 yeast two-hybrid system is shown in ➋ *Figure 19-1*. Transcription factors are modular proteins composed of separable, functional domains. These are the DNA binding domains (DNA-BD) that bind the upstream activation sequence (UAS) of a gene promoter and the acidic activation domain (AD) that initiates assembly of gene transcription apparatus. When separated, each of these domains retains its independent function but alone, is not capable of activating gene transcription. Transcription factor function can be restored if the two domains are brought within close proximity to each other. In the GAL4 yeast two-hybrid system, the AD and the DNA-BD of the GAL4 transcription factor are encoded separately by

Figure 19-1
Schematic diagrams outlining the principles of the GAL4 yeast two-hybrid system. A depicts the activation (AD) and the DNA binding domain (DNA-BD) of the transcription factor fused to the bait and fish proteins, respectively. When these associate with each other, the AD and DNA-BD are brought into close proximity reconstituting transcription factor activity with the subsequent activation of reporter genes. UAS is the upstream activation sequence. B depicts the cotransformation of the bait plasmid; here, it is the intracellular loop domain of the β2 subunit (β2-IL) of the inhibitory $GABA_A$ neurotransmitter receptor, and the fish plasmid, rat brain cDNA, into competent yeast cells, strain AH109. The DNA-BD vector carries the *TRP1* gene, the AD vector, the *LEU2* gene thus yeast cells successfully transformed with both vectors will show colony growth but β-galactosidase reporter gene activity is seen only when bait and fish proteins interact

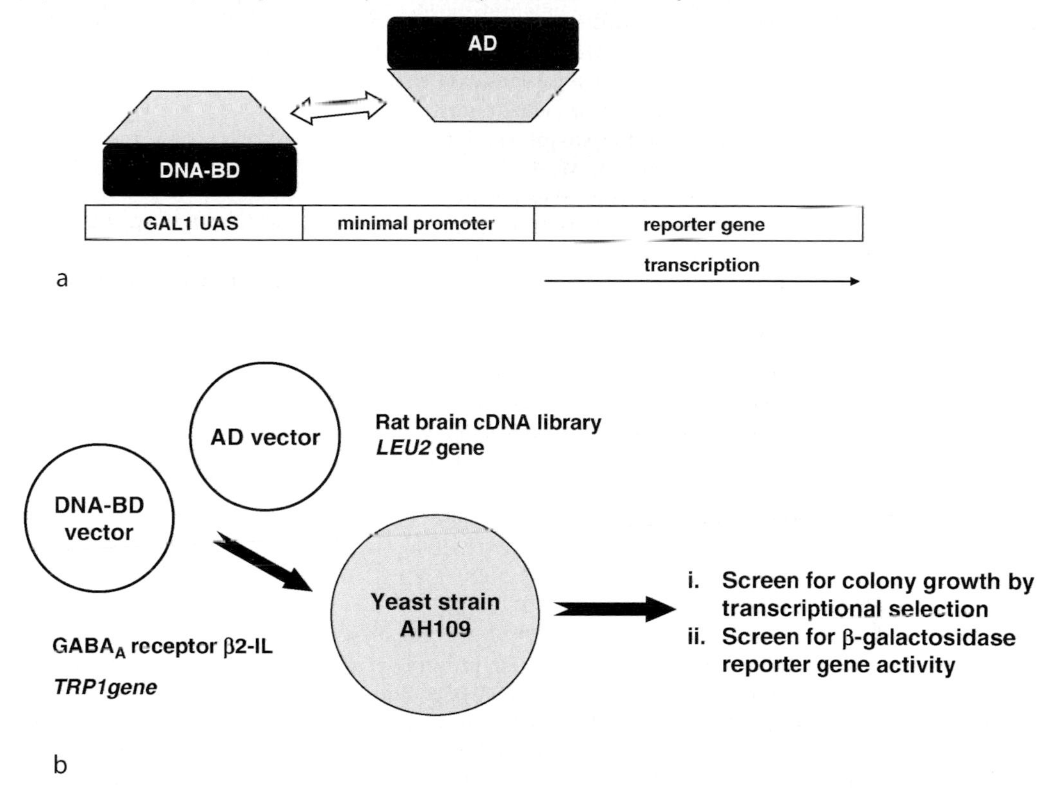

two yeast expression vectors. Each vector contains a multiple cloning site at the C-terminal end of the encoded AD or the DNA-BD domain. DNAs encoding the test proteins are subcloned in frame into the respective vectors to generate AD or DNA-BD fusion proteins. The two vector constructs are cotransformed into competent yeast cells that contain an inactivated GAL4 gene. Association between the two test proteins brings the AD and DNA-BD domains together thus reconstituting the transcription factor activity. The reconstituted GAL4 transcription factor induces transcription of reporter genes under the control of the GAL4 promoter. Detection of the reporter gene activities demonstrates an association between the two test proteins.

There are two types of reporter genes. The first of these are the nutritional selection marker genes that encode enzymes that are involved in the biosynthesis of leucine and tryptophan, i.e., the *LEU2* and *TRP1* genes. These nutritional marker genes are plasmid-based. They are important because unlike bacteria, successful transformation of yeast cells cannot be identified by antibiotic resistance. The yeast strain AH109 used in the GAL4 yeast two-hybrid system contains inactive leucine (L) and tryptophan (W) selection genes. If AH109 cells are grown in the absence of L or W in the growth media, i.e., -W or -L synthetic defined (SD) media, AH109 yeast cells do not grow. The DNA-BD and AD plasmids contain the W or L gene, respectively, both under the control of the GAL4 promoter enabling the yeast cells transformed with both plasmids to grow in -W, -L SD media. The second type of reporter gene is a gene within the yeast genome that is engineered to be under the control of the GAL4 transcription factor. They include *MEL1* and *LacZ* that encode α- and β-galactosidase enzymes, respectively and the nutritional selection marker genes *HIS3* and *ADE2* that encode enzymes involved in the biosynthesis of histidine (H) and adenine (Ade). The *MEL1* and *LacZ* genes have been placed under the control of the GAL4 responsive promoters *GAL1*, *GAL2*, and *MEL1*, respectively. Functional transcription of the GAL4 responsive genes is restored only by the interactions of exogenously introduced proteins fused to the separate GAL4 AD and DNA-BD transcription domains. The introduction of the two nutritional selection markers, i.e., H and Ade, rather than H alone (as found in other yeast strains), provides added stringency to the assay and reduces the occurrence of false positives due to the leaky *HIS* activity. The *MEL1* promoter controls the *MEL1* and *LacZ* reporter genes. These genes encode the α-galactosidase and β-galactosidase enzymes and their activity can be readily determined by filter lift assays followed by colorimetric assays.

Thus, to summarize and to show that the two proteins interact, AH109 yeast cells transformed with the appropriate DNA-BD and AD vectors grow in -W, -L SD media and also in -W, -L, -H, -Ade SD media and yield α-galactosidase and β-galactosidase reporter gene activities.

The GAL4 yeast two-hybrid system is available commercially from Clontech (Palo Alto, California, USA). It is marketed as the Matchmaker System with the current version being Matchmaker 3. The major difference between the earlier Matchmaker Systems 1 and 2 and System 3 is in the design of the vectors. In System 3, both DNA-BD and AD vectors have an incorporated epitope tag thus negating the need for antibodies against the test proteins in further studies. Also the AD vector has been engineered to contain a nuclear localization signal to match the DNA-BD vector. This ensures that both AD and DNA-BD fusion proteins are targeted into the yeast nucleus thus overriding most potential subcellular sorting sequences within the test proteins. Additionally, the two new generation vectors have different antibiotic selectivities thus facilitating the isolation of either DNA-BD or AD plasmid DNA from bacteria. The Matchmaker 3 vectors are the DNA-BD vector, also known as the "bait" vector pGBKT7, and the AD vector, also known as the "fish" vector pGADT7.

2.2 The LexA Yeast Two-Hybrid System

The LexA yeast two-hybrid system works on the same principle as the GAL4 system, i.e., via reporter gene activation using transcription factor AD and DNA-BDs. But, the LexA system uses a prokaryotic rather than a eukaryotic transcription factor. This method is advantageous in the study of eukaryotic proteins since the use of a prokaryotic transcription factor reduces possible interaction between the test protein and the respective DNA-BD or AD of the transcription factor encoded by the vectors. The DNA-BD fusion protein is the full-length LexA transcription factor, and the AD fusion protein is the blob 42 (B42) protein. These protein domains are used to activate transcription of the *Escherichia coli LacZ* reporter gene. The DNA-BD "bait" vector is pEG202-NLS, the AD "fish" vector is pJG4-5, and an alternative yeast strain is used. A LexA yeast two-hybrid system is available from Clontech (Palo Alto, California, USA).

2.3 The CytoTrap, Sos Recruitment, Yeast Two-Hybrid System

The Sos recruitment yeast two-hybrid system was developed by Aronheim and colleagues (1997). It is now known as the CytoTrap yeast two-hybrid system and marketed by Stratagene (La Jolla, California, USA). The CytoTrap system differs from both the GAL4 and the LexA systems in that it is not dependent on transcription factor activation in the nucleus for the detection of protein–protein association. Instead, protein interactions are detected in the cytoplasm and involve the reconstitution of the Sos/Ras signaling pathway in conjunction with the temperature-sensitive yeast strain, cdc25H.

This strain has a point mutation in its *cdc25* gene. The *cdc25* gene encodes the protein, cdc25, which is a guanyl nucleotide exchange factor (GNF) and the yeast homolog of the human Sos (hSos) protein. Sos is responsible for the activation of the Ras signaling pathway and the regulation of cell growth. The mutation in the *cdc25* gene leads to a temperature-sensitive cdc25 protein that prevents cell growth at 37°C but not at the permissive temperature of 25°C. It is possible to rescue the Ras signaling pathway with the hSos protein such that cells can grow at both 37°C and 25°C. A myristylation signal is required to target the hSos protein to the cell membrane in which the interaction takes place. Thus in the CytoTrap system, the bait vector encodes hSos. Test protein A is cloned in frame such that the hSos protein is fused to its N-terminus. The fish vector contains the myristylation sequence. Test protein B is cloned in frame in this vector such that the myristylation signal is at its N-terminus. The interaction of bait and fish proteins results in the transport of the hSos protein to the membrane as a result of association with the myristylated protein. hSos can now associate with endogenous Ras resulting in the activation of the cell division signaling pathway enabling the cdc25H yeast cells to grow at 37°C. The selection of transformants is again based on the nutritional SD media. Further details can be found in the legend to *Figure 19-2* and in the Stratagene CytoTrap Vector Kit User Manual. The figure also shows the basis for the CytoTrap system.

Figure 19-2

Schematic diagram outlining the principle of the CytoTrap, Sos recruitment, yeast two-hybrid system. The figure shows the bait and fish fusion constructs. The bait vector encodes a Sos fusion protein. The fish vector encodes a myristylation sequence that targets the fusion protein to the cell membrane. Interaction between the fish and bait proteins targets the Sos fusion protein to the intracellular face of the cell membrane where it can interact with Ras and rescue cell growth at 37°C

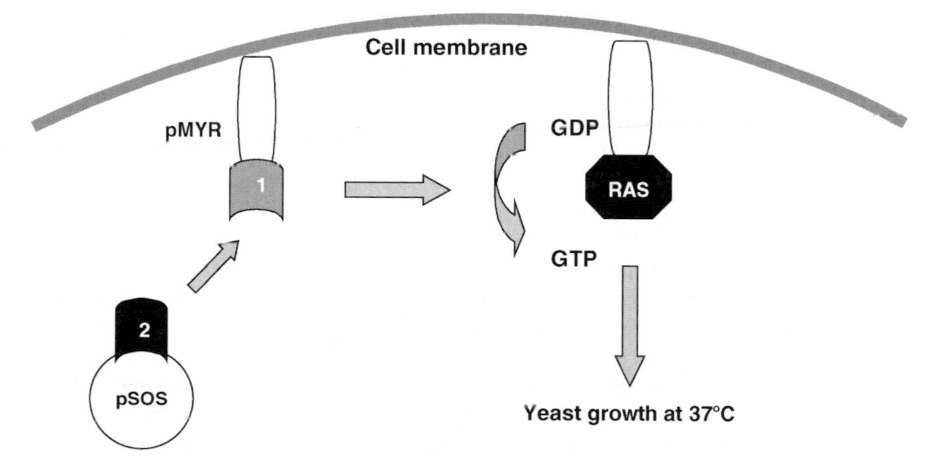

3 Identification of Interacting Proteins Using Transcription Factor-Based Yeast Two-Hybrid Systems

By far, the most widely used application of the yeast two-hybrid system as intimated in Introduction is the identification of protein partners for a test protein of either known or unknown function. Here, the DNA encoding the test protein or the domain of a test protein is cloned in frame into the bait vector. The fish vector contains cDNA and it is constructed so that there is one cDNA molecule per vector. Fish and bait vectors are cotransformed into the appropriate strain of competent yeast, and the resultant transformed yeast cells are screened for growth on SD media and for reporter gene activities. Putative positive clones are then isolated and characterized further. In the next section, each of these stages is discussed in detail.

3.1 Selection and Characterization of Bait Constructs

The assumption in yeast two-hybrid protein–protein interaction assays is that the test protein or a defined domain of the test protein, when fused to the BD of the transcription factor will fold independently of the DNA-BD. For complete protein sequences, it is assumed that the bait protein will adopt the conformation found in vivo, and where the bait is a protein domain, it is assumed that the domain will adopt the same or a similar conformation to the native protein. A second point to consider is the posttranslational modifications of proteins. Posttranslational modifications, such as N-glycosylation and phosphorylation differ between yeast and mammalian systems, which may affect the protein–protein association. Further, consideration should be given to disulfide bond formation within proteins. In eukaryotic cells, protein disulfide isomerase catalyses the formation of disulfide bonds and this occurs in the lumen of the endoplasmic reticulum. In yeast, the bait and fish fusion proteins are synthesized in the cytoplasm before being transported into the cell nucleus thus bypassing the endoplasmic reticulum. The environment in the cell cytoplasm is generally reducing thus one would predict that disulfide bond formation does not occur. However, some proteins that contain disulfide bonds, which are essential for the formation of their native structure, e.g., prolactin, growth hormone, and vascular endothelial growth factor, have been successfully shown to interact with their respective ligands in yeast (Ozenberger and Young, 1995), suggesting that in some cases, disulfide bond formation in the cytoplasm is possible. The thiol-disulfide redox buffer, glutathione, which is found in the cytoplasm, is competent to mediate correct disulfide linkage formation and may explain the above observations. Our own experience attempting to study the assembly motifs within the N-termini of $GABA_A$ receptor subunits, where a critical disulfide bridge is important for structural integrity, would suggest that disulfide bridge formation does not occur efficiently (Smith et al., 2003). Finally, because in the transcription-based systems, the fusion proteins need to be soluble for transport into the nucleus, it is inappropriate to use membrane proteins as the bait. In these cases, hydrophilic extramembranous regions of integral membrane proteins are selected as the bait. Examples of baits for the identification of neurotransmitter receptor associated proteins include the intracellular C-terminus of the excitatory ionotropic L-glutamate neurotransmitter receptors (e.g., for *N*-methyl-D-aspartate (NMDA) receptors, Kornau et al., 1995 and for non-NMDA receptors, Nishimune et al., 1998) and the intracellular loop located between transmembrane domains 3 and 4 of $GABA_A$ receptor subunits (e.g., Wang et al., 1999; Beck et al., 2002). Membrane proteins can be used as baits and conversely can be fished in the CytoTrap system. Indeed, recently Keller and coworkers (2004) used a full-length $GABA_A$ receptor $\gamma2$ subunit as a bait to identify Golgi-specific DHHC zinc finger protein (GODZ), an integral membrane protein, which is a palmitoyltransferase enzyme, as a $GABA_A$ receptor associated protein. An additional advantage of the CytoTrap system is that it permits the study of proteins that have intrinsic transcription factor activity (see the discussion on autoactivation later).

For the GAL4 and CytoTrap systems, the vectors are constructed such that the AD and DNA-BD of the GAL4 transcription factor or the hSos and the myristylation sequences are fused to the N-terminus of the test proteins. For the LexA system, there is an option to use a commercial vector that will permit construction of bait proteins where their C-termini are fused to the DNA-BD. The DNA encoding the bait protein is subcloned in frame into the bait vector. This should be verified by nucleotide sequencing of the resultant construct. The successful expression in the yeast of the fusion protein can be verified by transforming competent yeast cells with the bait construct, preparing yeast lysates, and analyzing these by immunoblotting using appropriate specificity antibodies. We have used the protocol within the CytoTrap Vector Kit User Manual for the preparation of yeast lysates. An example for the characterization of several GRIF-1 constructs is shown in ➋ *Figure 19-3.*

Another important requirement for the chosen bait protein is that it should not yield autoactivation of transcription factor activity. This may occur in the GAL4 and LexA systems if the bait actually is a transcription factor or if the bait mimics the transcription factor activity. Autoactivation is evident if the DNA-BD bait construct is cotransformed into competent yeast with an empty AD vector and the resultant transformants show reporter gene enzyme activities and growth on SD media. Obviously, this needs to be avoided because cDNA library screening with such a bait will result in the detection of many false positives.

Figure 19-3

An example for the characterization of fusion proteins to be used in yeast two-hybrid studies. A illustrates the maintenance of the reading frame in the pGADT7GRIF-1 (1–913) construct following the subcloning of the DNA encoding GRIF-1 (1–913) into the *EcoRI* restriction enzyme site in the multiple cloning site of pGADT7. B is a schematic drawing to scale of the fusion proteins. NLS is the nuclear localization signal, GAL4-AD is the GAL4 activation domain and HA is an epitope tag. C is an immunoblot verifying the expression of the fusion constructs depicted in B. GRIF-1 and OIP106 are both members of a coiled-coil gene family. The DNAs encoding the fusion proteins in B were subcloned in frame into the fish, AD vector, and pGADT7; competent yeast cells were transformed; liquid cultures were grown; proteins were extracted using a denaturating SDS/urea buffer and analyzed by immunoblotting using antiGRIF-1 antibodies that recognize both GRIF-1 and OIP106. The gel lanes are: 1, nontransformed yeast; lane 2, yeast transformed with empty pGADT7; lane 3, yeast transformed with pGADT7GRIF-1 (1–913); lane 4, yeast transformed with pGADT7GRIF-1 (8–633); lane 5, yeast transformed with pGADT7GRIF-1 (124–283); lane 6, yeast transformed with pGADT7OIP106 (1–953); lane 7, yeast transformed with pGADT7OIP106 (124–283). The positions of molecular weight standards kDa are shown on the left. It can be seen that the antibodies recognize the fusion protein with the correct molecular weight in the appropriately transformed yeast cells. Thus all the bait proteins are in frame and expressed. It is notable that the expression level of the lower molecular weight GRIF-1 and OIP106 fragments are apparently expressed at higher levels than the full-length proteins

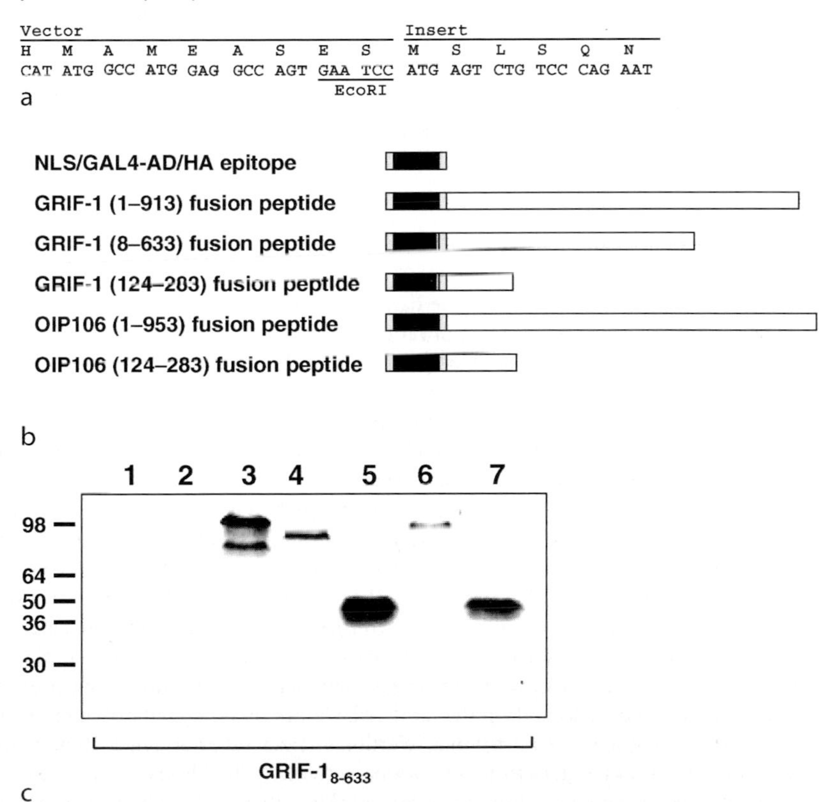

Autoactivation may be circumvented by modification of the bait protein usually by the construction of truncation or deletion bait proteins that do not yield autoactivation. The disadvantage is that this may result in the loss of protein–protein interaction domains. Again, the use of two different yeast two-hybrid systems in parallel would negate these potential problems. *Figure 19-4* shows an example of manipulation

Figure 19-4

Schematic diagram illustrating the engineering of GRIF-1 fish constructs to ensure that they do not result in autoactivation of transcription factor activity. The full-length cDNA encoding the protein, GRIF-1, was subcloned into the bait vector, pGBKT7, competent yeast strain AH109 cotransformed with pGBKT7GRIF-1 and pGADT7 and grown on SD -W, -L media. Resultant colonies were assayed for reporter gene activities by filter lift assays. It can be seen that despite the fact that the pGADT7 fish vector is empty, cotransformation with pGBKT7GRIF-1 results in robust reporter gene activity. Subsequent control experiments were carried out using bait constructs encoding GRIF-1 fragments as shown. It can be seen that the longest GRIF-1 fragment that did not yield reporter gene activity was GRIF-1 (545–913). This bait probe is now appropriate for use in the screening of a cDNA library to search for fish clones encoding GRIF-1 interacting proteins. Note the GRIF-1 associated proteins that bind to GRIF-1 within the N-terminal sequence GRIF-1 (1–544) will not be obviously detected by a cDNA library screen using GRIF-1 (545–913)

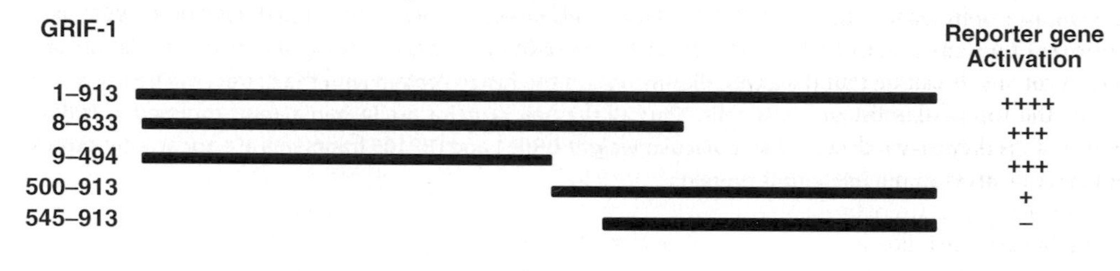

of a bait protein, GRIF-1, such that it does not yield autoactivation in the GAL4 system. In the CytoTrap system, the full-length GRIF-1 can be used as a bait because it shows no autoactivation thus permitting the identification of interacting protein partners that bind to the GRIF-1 (1–544) N-terminal domain.

3.2 cDNA Library Screening and Initial Characterization of Positive Clones

Following the selection and characterization of the bait as discussed earlier, the next consideration is the cDNA library. There are many commercial cDNA libraries currently available for yeast two-hybrid screening from various species and tissue types. This obviously enables specific screening for interacting proteins that are expressed in the native tissue of a test protein, thus reducing the likelihood of identifying interactions that would not normally occur in vivo, i.e., interactions between two proteins that are not expressed in the same tissues. First, the titer of the cDNA library should be determined. One should then ensure that enough DNA is used in the transformation assays to ensure complete cDNA library coverage, i.e., that all clones contained within the cDNA library are represented. Yeast can be either cotransformed with the bait plasmids and cDNA library plasmids or alternatively, yeast can be sequentially transformed with the bait plasmid followed by transformation with the cDNA library plasmid. In the latter case, a single transformant is selected, grown on SD-W media, and then transformed with the library fish plasmids. This sequential method has the advantage that the bait construct does not limit the efficiency of transformation, i.e., all cells transformed with the library cDNA plasmids already contain the bait plasmid. This is not necessarily the case for cotransformations, in which there is a risk that interacting proteins may go undetected due to the lack of both plasmids in the same cell. Clones encoding putative interacting proteins are identified for the GAL4 system by nutritional selection marker reporter gene activities followed by colorimetric assays for α- and/or β-galactosidase enzyme activities. The library plasmid DNAs are then isolated and characterized. Characterization includes checking to see if a single plasmid is present; sometimes multiple library plasmids can be contained within the same colony. In these cases, these colonies can be retained under appropriate nutritional selection, which promotes loss of the noninteracting clones. The size of the inserts should be determined by restriction enzyme digestion for multiple putative interacting clones, perhaps sorted according to the insert size, then DNA sequenced, and the sequence checked against appropriate databases. For a true interacting protein, nucleotide sequencing should show that any identified positive should be in frame with the AD. In some rare cases however, yeast will allow a translational frameshift; therefore, a clone with a large open reading frame that appears to be in the wrong reading frame may be expressed correctly.

As mentioned in the Introduction, yeast two-hybrid cDNA library screening is an elegant method used to identify protein–protein interactions. At this point, it should be stressed that a "positive" in a cDNA library screen is not necessarily a true reflection of an in vivo association. To have confidence that the "positive" is real, the optimum result of a cDNA library screen is that several fish plasmids that encode overlapping parts of the same protein are identified. A second best result would be that several "positives" are obtained and they are all the same clone. However, single positively interacting clones should not be dismissed since they may be the result of their representation in the cDNA library, they may reflect a low affinity or transient interaction between the bait and the fish proteins or they may be underrepresented due to the competition of the bait between different fish proteins or indeed from the bait protein itself if it exists in vivo as an oligomeric protein complex. Importantly, some proteins are well known to result in "false positives" thus any putative interacting protein should be screened against a public access database listing these proteins. It can be found at: http://www.fccc.edu/research/labs/golemis/main_false.html.

At the top of this list of "false positives" are the heat shock proteins. These could be termed "real positives" as they may reflect the fact that the bait protein does not fold appropriately and thus binds as in a normal cell stress response to the heat shock proteins.

Further, yeast two-hybrid specificity tests with the "positive" clones should also be carried out. These would include the following: (1) retransformation of yeast with the isolated identified fish plasmids and bait plasmid, (2) cotransformation of the identified fish plasmids with an empty bait vector, and (3) cotransformation of the identified fish plasmids with a bait vector containing an insert encoding a protein unrelated to the original bait protein. Further verification of putative interacting clones can also be obtained by yeast mating assays (but note that in general, these are not routine). These assays entail mating yeast strains on mating type a (i.e., CG1945 or Y190 yeast) containing the bait plasmid with yeast strains of mating type α (e.g., Y187) containing the interacting clones/fish plasmid. The diploid colonies formed can then be analyzed for reporter gene activation. If these additional controls yield the appropriate results, further experiments should be initiated to validate the protein–protein interaction using different experimental paradigms (see later). ➋ *Figure 19-5* summarizes the different stages of a yeast two-hybrid cDNA library screen.

3.3 Validation of Protein–Protein Interactions Using Alternative Experimental Paradigms

Protein–protein interactions that have been identified by yeast two-hybrid screens need to be confirmed by alternative, supplementary experimental strategies. These include in vitro protein interaction or fusion protein "pull-down" assays, far Western blotting, coimmunoprecipitations from mammalian cells transfected with the bait and fish proteins, and importantly from native tissues, colocalization either by fluorescence resonance energy transfer (FRET) or by bioluminescence resonance energy transfer (BRET) methods, and surface plasmon resonance studies. There are caveats, advantages, and disadvantages for each of these. Detailed protocols can be found in Golemis (2002).

Fusion protein "pull-down" assays involve the overexpression of bait and/or fusion proteins in bacteria. Often, the expressed fusion proteins are localized in occlusion bodies and not readily soluble under nondenaturing conditions. The expressed proteins can be extracted using urea, sonication, sodium dodecyl sulfate (SDS), or a combination of all the three. The net result is the denaturation of the recombinant protein and it may need to be refolded if the interaction domain is conformationally dependent. A major advantage of the "pull-down" assay is that high concentrations of proteins can be easily generated thus favoring protein association for a reversible equilibrium between two proteins.

For far Western blotting, native or recombinant proteins are separated by SDS-polyacrylamide gel electrophoresis (SDS-PAGE), transferred to the nitrocellulose membranes and the blotted proteins are

Figure 19-5
Schematic diagram illustrating the different stages in the identification of interacting proteins starting with a yeast two-hybrid cDNA library screen

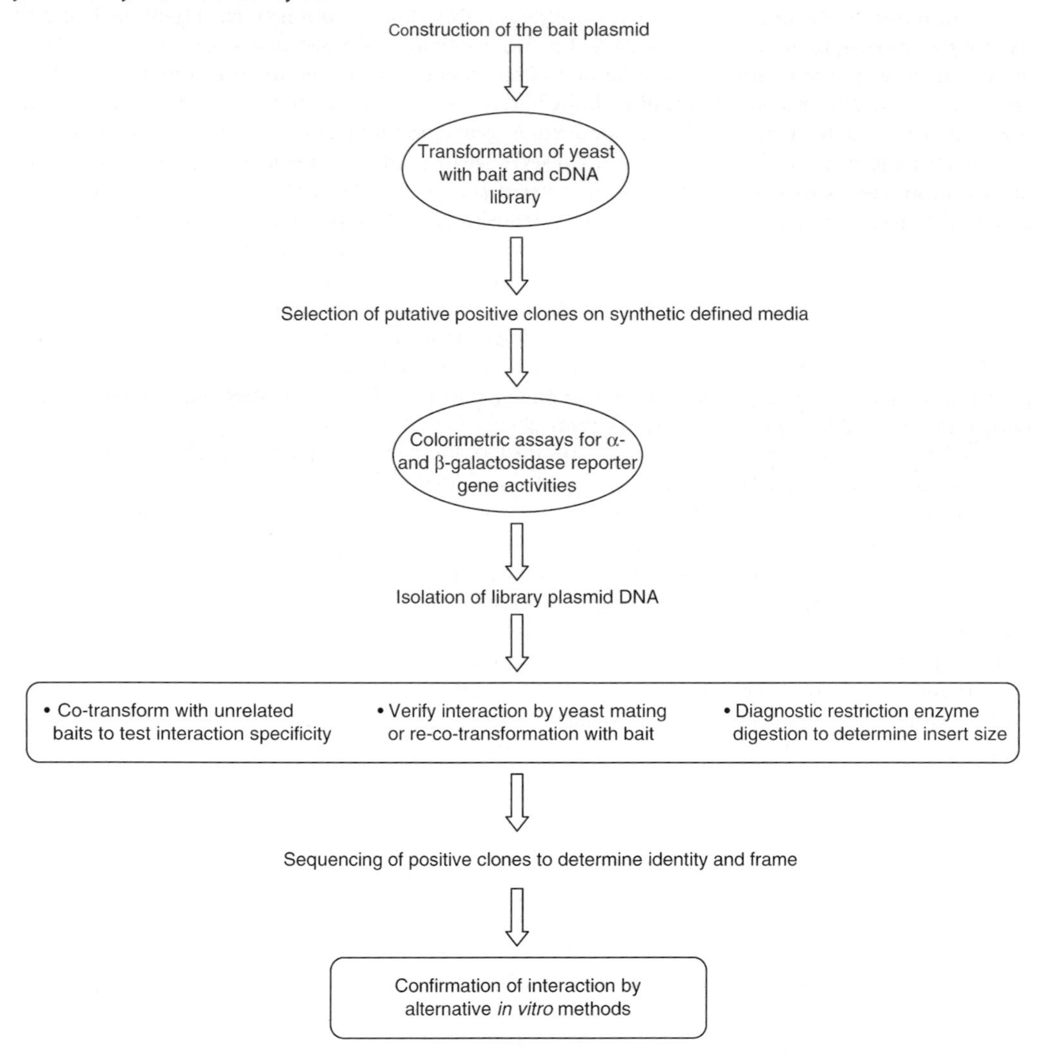

probed with ^{35}S-labeled fish protein. Again, following SDS-PAGE, proteins are denatured but they can be partially refolded in situ on the nitrocellulose membrane filters. Importantly, a positive result for a far western is the best method to demonstrate unequivocally that two proteins interact directly.

For immunoprecipitations from native tissues, one requires antibodies directed against both the fish and the bait proteins. Further, these antibodies should not bind to epitopes within the putative protein–protein BDs. It is technically difficult to determine low affinity or transient association among proteins by immunoprecipitation because low-affinity interactions may be lost by washing immune pellets to remove nonspecifically bound proteins. Also, one cannot manipulate protein concentrations to favor protein association as one can in a "pull-down" assay. Under these circumstances, probably the best method to use would be FRET.

4 Mapping of Protein BDs Using Yeast Two-Hybrid Interaction Assays

In addition to identifying protein partners, yeast two-hybrid technology can be used to identify and study in detail the interaction domains between two proteins. Here, bait and/or fish truncation or deletion constructs of the parent proteins are engineered and characterized as described earlier (see 3.1 Selection and characterization of bait constructs). These are then investigated for association in a yeast two-hybrid interaction assay. Once the BD has been identified, it can be further refined by mutagenesis. The same caveat applies to these studies as for the identification of associating proteins, i.e., it is assumed that the respective fusion proteins fold and adopt the same or a similar three-dimensional conformation to the native protein. This is not always the case and results should be interpreted with caution and if possible, always validated by an alternative experimental approach. ➋ *Table 19-1* shows an example of mapping the kinesin KIF5C heavy chain BD of GRIF-1 using a modified LexA yeast two-hybrid interaction assay. The table also serves to demonstrate how results of yeast two-hybrid assays can be presented. The results in ➋ *Table 19-1* are qualitative, i.e., the strength of the interaction between the two proteins is assessed by the number of colony forming units (cfu) and the qualitative assessment of the blue color generated by the β-galactosidase enzyme activity. The assay can be made semiquantitative. In this case, β-galactosidase reporter gene activity is determined in liquid culture assays using o-nitrophenyl β-D-galactopyranoside as substrate, measuring the resultant optical density (OD) at $\lambda = 420$ nm and expressing this as a percentage of wild-type Gal4p.

Table 19-1
Mapping the kinesin KIF5C interaction domain of GRIF-1 using a modified LEXA yeast two-hybrid interaction assay

AD Vector "Fish"	DNA-BD Vector "Bait"	Growth on -L, -W SD media	Growth on -L, -W, -H SD media	β-Galactosidase reporter gene activity
pGADT7GRIF-1_{1-913}	pMBL33KIF5C_{1-957}	++	(+)	(+)
pGAD10GRIF-1_{8-633}	pMBL33KIF5C_{1-957}	++	++	+
pGADT7GRIF-$1_{124-283}$	pMBL33KIF5C_{1-957}	++	+++	++
pGADT7OIP106_{1-953}	pMBL33KIF5C_{1-957}	++	–	–
pGADT7OIP$106_{124-283}$	pMBL33KIF5C_{1-957}	++	–	–
pGADT7GRIF-1_{1-913}	pMBL33	++	–	–
pGADT7GRIF-1_{8-633}	pMBL33	++	–	–
pGAD10GRIF-$1_{124-283}$	pMBL33	++	–	–
pGADT7OIP106_{1-953}	pMBL33	++	–	–
pGADT7OIP$106_{124-283}$	pMBL33	++	–	–
pGADT7	pMBL33KIF5C_{1-957}	++	–	–
pGADT7	pMBL33β2-$IL_{303-427}$	++	–	–
pGADT7	pMBL33	++	–	–

Yeast were cotransformed with bait and fish constructs as shown and transformants were grown on -2 SD, i.e., -L, -W and -3 SD, i.e., -L, -W, -H. Colonies from -2 plates were restreaked onto new plates and β-galactosidase activity was measured by filter lift assays. For colony growth, (+) = < 10 cfu; + = 10–100 cfu; ++ = 100–200 cfu; +++ > 200 cfu. OIP106 shares ~44% amino acid identity with GRIF-1 and is thought to be a member of the same coiled-coil family of proteins. The numbers in subscript refer to the amino acids of GRIF-1 and OIP106. Note the control in which each of the bait plasmids was cotransformed with an empty fish plasmid and *vice versa.* In all these controls, colony growth was observed in the SD -W, -L media because of the presence of the two plasmids but no reporter gene activation was observed, i.e., no growth on the -L, -W, -H SD plates and no detectable β-galactosidase activity was found. The results taken from

Brickley and coworkers (2005) show that KIF5C interacts with GRIF-1 and that the region, GRIF-$1_{124-283}$, contains the KIF5C BD. Further, OIP106 does not associate with KIF5C. It is concluded that the interaction between GRIF-1 and KIF5C is direct; however, one must be aware that in some rare cases proteins can associate because an intermediary adaptor protein may be supplied by the yeast cells themselves.

5 Variations of Two-Hybrid Systems

Since the introduction of the original GAL4 yeast two-hybrid system, a number of modified versions have been created. These include the yeast one-hybrid system, the yeast tribrid system, the reverse hybrid system, the mammalian two-hybrid system, and the bacterial two-hybrid system (for more detailed information see Zhu and Hannon, 2000). The yeast one-hybrid system is used to identify proteins that interact with DNA. In the tribrid system, a third protein is transformed into a yeast reporter strain, together with the two test proteins. This third protein may be a bridging or an adaptor protein that provides a link between two proteins, it may be required to stabilize a weak protein–protein interaction or it may be a modifying enzyme such as a kinase or phosphatase that is required to activate one or both of the interacting proteins that in the case of a kinase, may require phosphorylation for association. The reverse yeast two-hybrid assay enables the identification of proteins that dissociate protein–protein interactions through the use of a counter-selectable marker. Here, a positive interaction between two proteins initiates the expression of a marker that is toxic to the cell. Inactivation mutations in either of the two interacting proteins or the addition of a third protein that blocks a known interaction prevents expression of the toxic marker thus allowing cell growth on selective media. The mammalian two-hybrid system was developed to counter differences in posttranslational modification between yeast and mammalian cells that may be required for protein–protein association.

6 Conclusions

In this chapter, we have described three yeast two-hybrid interaction assay systems and our experience of their use in the identification of protein partners and in the localization of protein–protein interaction domains. Each of the three methods is distinct with respect to ease of use, merits, and disadvantages. The different systems are relatively easy to establish in the laboratory even without prior experience of working with yeast cultures. Commercial systems come with positive controls that aid in the validation of the method. The appropriate user manuals are very good with accompanying protocols and trouble shooting guides. The yeast two-hybrid system does indeed facilitate the identification of interacting proteins and it is a good starting point for the elucidation of the significance of particular protein–protein association. The system provides a means to an end. The physiological significance(s) of identified protein interactors must always be verified and consolidated via alternative experimental paradigms.

Acknowledgments

The authors would like to acknowledge and thank Dr. Mike Beck, Dr. Seema Sharma, and Dr. Helen Wilkinson for their contribution to the development of the yeast two-hybrid work in the laboratory. Work described in this chapter is funded by the Biotechnology and Biological Research Council (BBSRC) UK.

References

Aronheim A, Zandi E, Heinemann H, Elledge SJ, Karin M. 1997. Isolation of an AP-1 repressor by a novel method for detecting protein–protein interactions. Mol Cell Biol 17(6): 3094-3102.

Bartel PL, Fields S, editors. 1997. The yeast two-hybrid system. Oxford, UK: Oxford University Press.

Beck M, Brickley K, Wilkinson H, Sharma S, Smith M, et al. 2002. Identification, molecular cloning and characterization of a novel $GABA_A$ receptor associated protein, GRIF-1. J Biol Chem 277: 30079-30090.

Brickley K, Smith MJ, Beck M, Stephenson FA. 2005. GRIF-1 and OIP106, members of a novel gene family of coiled-coil domain proteins: association in vivo and in vitro with kinesin. J Biol Chem 280: 14723-14732.

Clontech MATCHMAKER GAL4 Two-hybrid system 3 and libraries user manual. Clontech yeast protocols handbook.

Fields S, Song O-K. 1989. A novel genetic system to detect protein–protein interactions. Nature 340: 245-246.

Golemis E, editor. 2002. Protein–protein interactions. New York, USA: Cold Spring Harbour Laboratory Press.

Keller CA, Yuan X, Panzanelli P, Martin ML, Alldred M, et al. 2004. The $\gamma 2$ subunit of $GABA_A$ receptors is a substrate for palmitoylation by GODZ. J Neurosci 24: 5881-5891.

Kornau HC, Schenker LT, Kennedy MB, Seeburg PH. 1995. Domain interaction between NMDA receptor subunits and the postsynaptic density protein PSD-95. Science 269: 1737-1740.

Nishimune A, Isaac JT, Molnar E, Noel J, Nash SR, et al. 1998. NSF binding to GluR2 regulates synaptic transmission. Neuron 21: 87-97.

Ozenberger BA, Young KH. 1995. Functional interaction of ligands and receptors of the hematopoietic superfamily in yeast. Mol Endocrinol 9: 1321-1329.

Smith MJ, Beck M, Stephenson FA. 2003. Can the yeast two-hybrid system be used to study assembly domains of GABA-A receptor subunits? British Neurosci Assoc Abstr 17: 26.12.

Wang H, Bedford FK, Brandon NJ, Moss SJ, Olsen RW. 1999. $GABA_A$-receptor-associated protein links $GABA_A$ receptors and the cytoskeleton. Nature 397: 69-72.

Zhu L, Hannon GJ, 2000. editors. Yeast hybrid technologies. Natick, MA, USA: Eaton Publishing.

20 Protein Engineering: Chimeragenesis and Site-Directed Mutagenesis

M. Davies

Abstract: The manipulation and alteration of proteins at the amino acid level currently presents many technical problems that are difficult to overcome. Because of this, protein engineering is best performed at the level of the gene that codes for the protein of interest. The two techniques described in this chapter, chimeragenesis and site-directed mutagenesis, are especially powerful tools when used in combination. By manipulating the cDNA of a protein of interest, chimeragenesis allows for the substitution of large sections of a protein while site-directed mutagenesis allows for the substituion of specific amino acids within a protein. This chapter describes how these techniques can be performed in a rapid, reliable manner.

List of Abbreviations: $GABA_A$, gamma-aminobutyric acid type A; $5HT_3$, 5-hydroxytryptamine type 3; SDM, site-directed mutagenesis; PCR, polymerase chain reaction; TRCP, targeted random chimera production; SDS, sodium dodecyl sulphate

1 Introduction

Protein engineering plays a crucial role in structure–function studies. Currently, there are no methods available by which the sequence of a protein can be reliably and specifically altered at the amino acid level. Instead, if a protein is to be altered, these changes must be introduced into the cDNA that encodes the protein. Two of the major techniques used to manipulate cDNA sequences are chimeragenesis (forming chimeras between proteins that are encoded by homologous cDNAs) and site-directed mutagenesis (SDM). In both cases, the general idea is to rearrange or alter the cDNA for a protein so that the mature protein has a primary sequence that is different from the wild-type version. The overall goal is to identify specific domains or amino acids that confer unique properties to the protein in question. These techniques have been used with great success to explore the architecture of the functional domains of many proteins in many different families. Much of our expertise lies in the manipulation of the cDNAs' coding for the subunits of gamma-aminobutyric acid type A ($GABA_A$), nicotinic acetylcholine (nACh), and 5-hydroxytryptamine type 3 ($5HT_3$) receptors. It is thus from the perspective of this family of receptors that we write this chapter. We have used SDM and chimera production extensively to investigate the structure and function of ligand binding domains in these proteins and have also used SDM to insert epitope tags and polyhistidine residues into receptor subunits. The latter approaches have helped us understand receptor quaternary structure and have aided in the purification of receptor protein. When coupled with information obtained using complementary biochemical techniques, it is possible to form a detailed map of receptor ligand binding and structural domains.

Typically, SDM and chimeragenesis are used in concert. As an example, consider two homologous receptor proteins, one of which has a high affinity for a specific drug, the other with a low affinity for the same drug. This is a perfect situation for using the techniques described in this chapter to identify the region(s) of the proteins that confer differential drug recognition. Because they are homologous, they lend themselves to the production of chimeric molecules consisting of portions of each of the cDNAs encoding the proteins. This will allow for the identification of a specific domain(s) in the proteins that confers drug sensitivity. Once that domain is identified, SDM can be used to identify specific amino acids within this region that are important for drug recognition. This chapter is divided into two sections. In the first, the technique of random chimera production and the method for generating a panel of chimeras from two homologous cDNAs are discussed. In the second section, our current methods for PCR-mediated SDM are outlined.

2 Random Chimeras

The production of chimeric cDNA molecules is typically used to localize an amino acid domain in a protein that gives rise to a specific property that is absent in a homologous protein. For example, this property could be the recognition of a specific drug, interaction with a specific protein, or localization to a specific cellular location. Chimeras are often a necessary first step in the path of identifying individual amino acids that confer these properties. The chimeras can be constructed so as to gain function in the protein that normally does not manifest the property of interest or to remove it from the protein that does (see *Figure 20-1*). Although both approaches are valid, gain-of-function studies generally carry more weight than do loss-of-function studies (see Section 4 given later).

Figure 20-1
Depiction of the use of chimeras to identify a region of cDNA that codes for a specific protein property, in this case recognition of drug Z. Top, two cDNAs, cDNA1 and cDNA2 code for homologous proteins that either recognize drug Z or do not recognize drug Z, respectively. To find the region of the cDNA that confers this property of drug recognition, chimeric molecules of the two cDNAs are constructed (chimeras 1–5). It is found that chimeras 1 and 2 do not recognize drug Z whereas chimeras 3–5 do. Therefore, the region delimited by the dashed lines between chimeras 2 and 3 codes for the property of drug recognition

cDNA1
cDNA1 codes for a homologous protein that recognizes drug Z
cDNA2
cDNA2 codes for protein that does not recognize drug Z
1
Chimera 1 does not recognize drug Z
2
Chimera 2 does not recognize drug Z
3
Chimera 3 recognizes drug Z
4
Chimera 4 recognizes drug Z
Chimera 5 recognizes drug Z

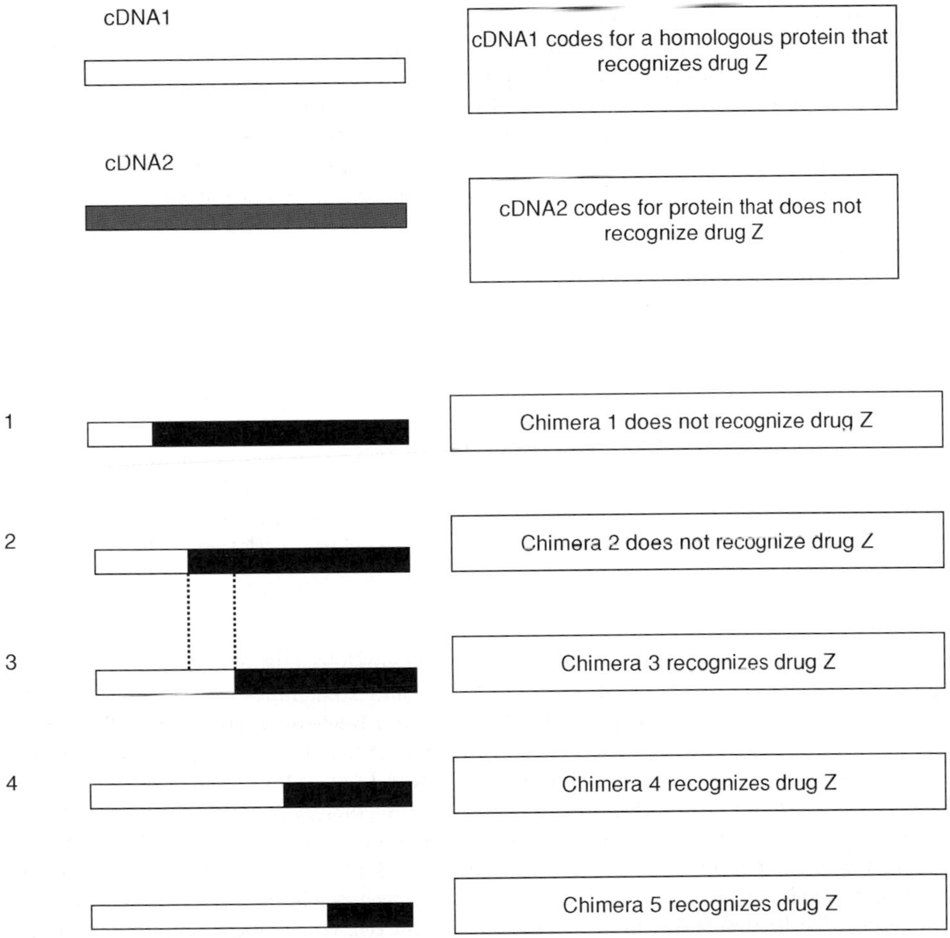

2.1 Methods of Chimeragenesis

Chimeras can be produced in one of two ways. In the first, restriction sites are chosen within the cDNA at positions where the experimenter wants the exchanges to take place. Typically, these restriction sites are not conserved between cDNAs and so they have to be introduced by SDM. This can be difficult because the sites must be "silent," i.e., they must not alter the amino acid sequence of the proteins. Once these sites are introduced, the cDNAs are cut and ligated to form chimeric cDNAs (see, i.e., Wieland et al., 1992). This approach has been used successfully in many instances, but does suffer from some drawbacks. First, as already mentioned, the introduction of restriction sites can be quite time-consuming because multiple sites in at least two separate cDNAs have to be introduced. Second, in our experience, it is often the case that many of the chimeras will not express in a heterologous expression system. This is obviously a major problem when performing structure–function studies and it is not easily resolved except by choosing different exchange sites in the cDNAs. Third, the exchanges are biased by the experimenter who chooses the location of the restriction sites.

An alternative technique introduced relatively recently by Moore and Blakely (1994) allows for the random production of chimeras without the need for restriction digests and ligations. We have used this approach with great success in our laboratory (Derry et al., 2004), as have several others. This is the technique upon which this section of the chapter focuses. The approach was first used to generate multiple chimeras from homologous serotonin transporter cDNAs (Moore and Blakely, 1994) and has since been used to generate many different chimeric receptors and enzymes (Satoh et al., 1999; Brock and Waterman, 2000; Derry et al., 2004; Tirona et al., 2004). The general idea is as follows: The cDNAs for the two proteins of interest are subcloned into a single vector. The construct is linearized and then introduced into competent bacteria. The chimeras are generated within the bacteria, but the process by which this occurs is unclear. Moore and Blakely (1994) speculated that it is not likely to be a recombination event given that they used a $recA^-$ strain (a strain lacking enzymes required for recombination) of bacteria in their experiments. Rather, they thought that it might be an annealing event resulting simply from the two cDNAs aligning with each other in areas of high homology. Indeed, when chimeras generated by this technique are sequenced, it appears that the exchange points are in areas of high homology. The advantages of this approach are that no preexisting restriction sites are needed, no ligation reactions are required, and there is no experimenter bias that dictates where the crossover points occur. Additionally, at least with the cDNAs that have been used in this laboratory to this point, the chimeras express well in mammalian cells. This technique can rapidly produce a panel of chimeric molecules in a short time and the major bottleneck is the screening of the putative chimeras. However, by using "colony cracking" (described later) even this step can be done in a single afternoon. A potential drawback is that it might not be possible to obtain a crossover in exactly the position that is required (e.g., in a region of low homology between the two cDNAs). Nevertheless, the ability to delimit the crossover area may increase the chances of obtaining the desired chimera [targeted random chimera production (TRCP), described later in ➋ Section 2.6].

2.2 Random Chimera Protocol

The process by which random chimeras are generated is quite straightforward. The cDNAs of the two homologous proteins are subcloned into a single vector in a head-to-tail fashion (see ➋ *Figure 20-2*). If there are plans to express the chimera in cells or to obtain cRNA, it is advantageous to subclone them directly into an expression vector. We typically use the pcDNA3.1 (+/–) (Invitrogen, USA) family of vectors as this allows for the direct expression of the chimeric cDNA in HEK293 cells or transcription of cRNA for expression in *Xenopus* oocytes for use in electrophysiological experiments. When subcloning the cDNAs into the vector, it is important to ensure that the appropriate leader sequences in the first cDNA and the appropriate stop codon in the second cDNA are kept intact as these will be needed for the correct expression and appropriate localization of the chimeric protein within the cell. When subcloning, also plan for the two cDNAs to be separated by a number of base pairs in the polylinker, which should be long enough to ensure that at least two restriction sites are left between them. This is necessary because this region must be removed by using restriction endonucleases to linearize the construct in order to enable the cDNAs to come into contact with one another so that exchange events can occur. Although a single restriction site can be used, the molecule tends to reform more easily through the resulting sticky ends and the efficiency of the chimera generation process decreases dramatically.

Figure 20-2
General procedure to produce random chimeras from two homologous cDNAs. The procedure is started by subcloning the desired cDNAs in a head to tail fashion in a single vector. The construct is linearized and introduced into competent bacterial cells. Colonies from the growth of construct-containing cells are then subjected to colony cracking and their DNA is visualized on an agarose gel. Depicted is a cartoon of an agarose gel on which several lanes have been used. The white bands represent supercoiled DNA of different sizes. In lanes one and two, controls have been run. The first is a sample obtained from a colony containing bacteria harboring a vector with the double insert, the starting point in the chimeragenesis procedure. The second lane is obtained from bacteria harboring a vector containing a single insert. In lane three, a sample has been run and the pattern of the bands indicates that this colony contains bacteria that have the vector with both inserts. No chimeragenesis took place in these bacteria. In lane four, the position of the bands indicates that an exchange has taken place between the two cDNAs and a single insert of the correct size (as compared to the control) remains in the vector. This putative chimera can now be further characterized by restriction digests and sequencing

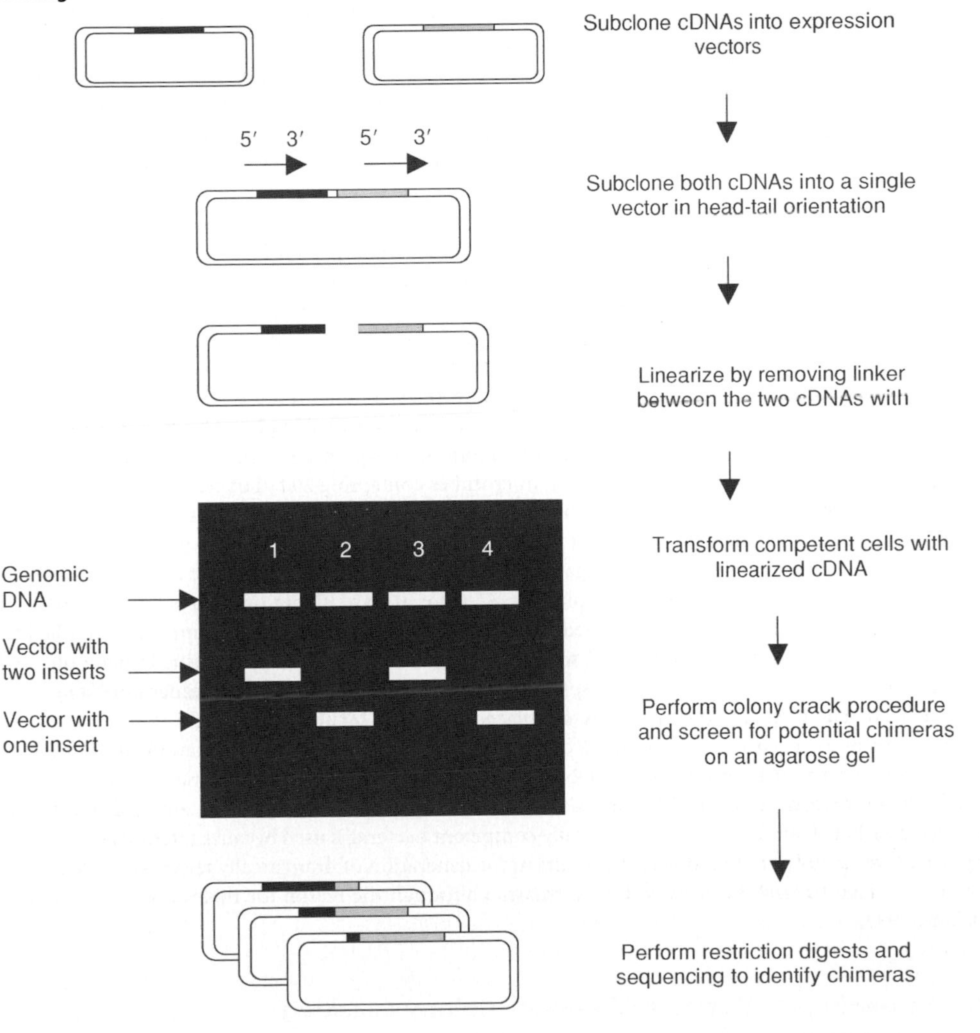

The construct is then linearized with a double restriction enzyme digestion procedure. As different enzymes require different incubation conditions, it is best to follow the manufacturer's instructions for the specific enzymes that are chosen. In the interest of saving time, the best-case scenario is to have two restriction enzymes that are equally active under the same conditions so that they can be used simultaneously in a single reaction. At this point, it is wise to digest a reasonable amount of the construct such that it can be used for a number of experiments if needed. We generally digest microgram amounts of DNA at this step and store whatever is not immediately required at –80°C. Following the restriction digestion, it is prudent to purify the linearized fragment from the enzymes, enzyme buffer, and the small fragment that was released from the construct by applying the products of the reaction to a low melting point gel and excising the linearized product. The linearized construct can then be purified from the excised gel portion using the Qiagen Gel Extraction Kit (Qiagen, USA). There are several commercially available kits that can be used to purify DNA from gel slices. In our experience, we have found the Qiagen kits to be the most consistent and they typically give a reasonable yield at the end of the procedure. The purified linearized construct can then be used to transform competent bacteria.

2.3 Bacterial Transformation

The transformation procedure that we use in the laboratory is the same for both chimeragenesis and PCR-based mutagenesis (described in the latter half of this chapter). Typically, JM109 cells are used for the former and DH5α are used for the latter. Commercial sources typically can provide highly efficient cells (as measured by transformants per μg of DNA) often in the region of 1×10^9. However, we have found that using competent cells prepared in-house with an efficiency ranging from 1×10^6–1×10^8 are quite suitable for carrying out the techniques described here. If using a commercial source, it is best to follow the instructions given by the supplier with respect to the amount of DNA to add, heat shock temperature, etc. Here, I only describe the procedure we have in place using our own prepared cells. 200 μl of cells in 1.5 ml microcentrifuge tube are thawed on ice. For the chimera technique, we typically add 25–50 ng of DNA. More or less might have to be added depending on how many colonies result from the procedure. The tube is capped and flicked with a finger to mix the contents. It is immediately placed back on ice and the DNA and cells mixture is allowed to incubate on ice for 20 min. At this point, the cells are subjected to a brief heat shock. We have found that with regular 1.5-ml microtubes containing 200 μl of cells, a heat shock of 90 s at 42°C gives optimal results. The tubes are then immediately returned to the ice for 2 min. 800 μl of Luria broth is added and the tubes are incubated in a shaking incubator at 37°C for 45 min. Following this, an aliquot of the mixture is plated out on agarose plates, containing an appropriate antibiotic to select for transformed bacteria. We preincubate the plates at 37°C for at least 30 min prior to plating, in order to dry any excess liquid that may have accumulated on the plate during storage. The volume that is applied to the plate can vary, but a smaller volume will soak into the plate more quickly. We usually plate out a 50 μl aliquot and a 250 μl aliquot. The liquid is spread around the plate with a sterile spreader until it is absorbed into the agar. The plates are then inverted and placed in a 37°C incubator overnight.

As controls are needed to assess the success of this protocol, parallel transformations using each of the vectors containing single inserts of the cDNAs of interest and one with the insertion of both the cDNAs should be performed (see later). The bacteria are then plated out on antibiotic-containing agarose plates as described earlier. Transformation of chemically competent bacteria is used because interestingly, it has been reported that electroporation of bacteria results in the generation of dramatically fewer chimeric constructs when compared to heat shock-based transformation although the reason for this is not clear (Moore and Blakely, 1994).

2.4 Screening for Putative Chimeras: Colony Cracking

If the transformation was successful, on the following day the plates will contain 50–100 colonies. The next step is to determine which of these colonies contain chimeric cDNAs. Since a large number of colonies must be screened, it is essential to use a technique that allows for simultaneous rapid screening of a large number of potential positives without requiring DNA amplification or using expensive restriction enzymes. A technique that fulfills these criteria is "colony cracking." With this technique, bacterial colonies are removed from a plate, the cells lysed, and the supercoiled contents assayed for their size relative to controls. This technique is a relatively crude but fast way to determine if the plasmid size corresponds to a vector with two inserts (where exchange has not taken place) or a vector with a single insert (where exchange events have taken place). The technique described below is based on a Promega protocol (Promega Corp., 1991). To "crack" the colonies, they are picked from the plate with a sterile toothpick. The colony is then smeared at the bottom of a 1.5-ml microtube. The same toothpick is then placed in a culture tube containing 3 ml of Luria broth. If a positive result is obtained with the colony cracking procedure, this tube will later be used to grow the bacteria harboring this particular chimera. In the meantime, it can be stored in a refrigerator. At this time, the three controls (see earlier) can be prepared in parallel. Two will be derived from colonies containing the single cDNAs in the vector and the third will be from a colony containing the construct with both cDNAs inserted. These will be used as markers to interpret the band patterns on the agarose gel (see later). Regular molecular weight markers are not useful for sizing the bands on the gels because the samples that are applied will run in a supercoiled form and will not therefore correspond to a linear DNA ladder. 50 μl of 10-mM EDTA (pH 8.0) is now added to the microtube and the bacteria are resuspended by vortexing rapidly for 1–2 min. To this mixture, 50 μl of freshly prepared cracking buffer is added (make the buffer as follows: 40 μl of 5N NaOH, 50 μl of 10% sodium dodecyl sulphate (SDS), 0.2 g of sucrose and 800 μl of H_2O; this can be scaled up if necessary). Vortex briefly to ensure that the contents are well mixed and then incubate the tubes in a 70°C waterbath for 5 min. This should lyse the bacteria and release both the genomic and plasmid DNA. Following this incubation, remove the tubes from the waterbath and let cool to room temperature. Add 1.5 ml of 4 M KCl and 0.5 ml of 0.4% bromophenol blue and vortex to mix the contents. Place the tubes on ice for 5 min to precipitate out proteins and then centrifuge for 3 min at 12,000 g. The supernatant should now contain the supercoiled genomic and plasmid DNA. The next step is to size the DNA components on an agarose gel. Because relatively large pieces of DNA must be resolved, it is wise to use a low percentage agarose gel (such as 0.5% or lower). It should be noted that the amount of DNA from a single colony is small. It is preferable, therefore, to use wide lanes with relatively large sample volume (35–50 μl) to allow for easier visualization of the DNA on the gel. Load the appropriate volume of supernatant in the microtubes directly into the sample lanes of the gel. Be sure to also load the control samples that have undergone the same manipulations as the putative chimeras. Run the gel until the dye front is a few centimeters from the bottom of the gel. The size of the plasmid in each lane is compared with that of the plasmid in the controls (see ➋ *Figure 20-2*). It should be relatively easy to identify those colonies that contained plasmids, which are consistent with a single insert, and those that correspond to the double insert. Any colonies with a size that is consistent with a single insert are, at this point, assumed to be chimeric. Because the toothpicks used to place the colonies in the eppendorf tubes were saved and placed in Luria broth, we can now grow the positive clones, perform minipreps on them, and subject the resulting DNA to further analysis.

2.5 Confirmation of the Presence of Chimeric Molecules

Once a miniprep or larger DNA preparation has been made from the putative positives, the basic first step in the identification of putative chimeras is to perform a restriction analysis on the DNA. Each cDNA will have unique restriction sites that act as landmarks. The presence or absence of these restriction sites should give some indication as to the general location of the exchange event and as to which sections of each cDNA are present in the chimeric molecule. Once appropriate crossovers have been identified with this technique, the vector containing the chimera should be sent for sequencing to confirm its primary sequence. Sequencing is important to determine that the crossover event has not disrupted the reading frame of the cDNA and to pinpoint the position of the crossover event. We have not yet found a chimera generated by this technique that has produced a shift in the reading frame. However, it is important to verify this before proceeding with additional experiments. Since the exchanges typically take place in areas of high homology, even after sequencing, it may be difficult to determine the exact site of crossover as these regions often have identical sequences in both cDNAs.

2.6 Targeted Random Chimera Production

A modification of the random chimera production technique was first described by Boileau et al. (1998). Targeted random chimera production (TRCP) allows the user to delimit the area of recombination by inserting a specific fragment of one of the cDNAs into the dual cDNA construct (➋ *Figure 20-3*). For this

Figure 20-3

For TRCP, the construct containing the two intact cDNAs (a) is modified so that only part of one of the constructs (in this case, the N-terminal half of the first cDNA) is present (b). This restricts the regions in which exchange events can take place, and the resulting chimeras will reflect this (c)

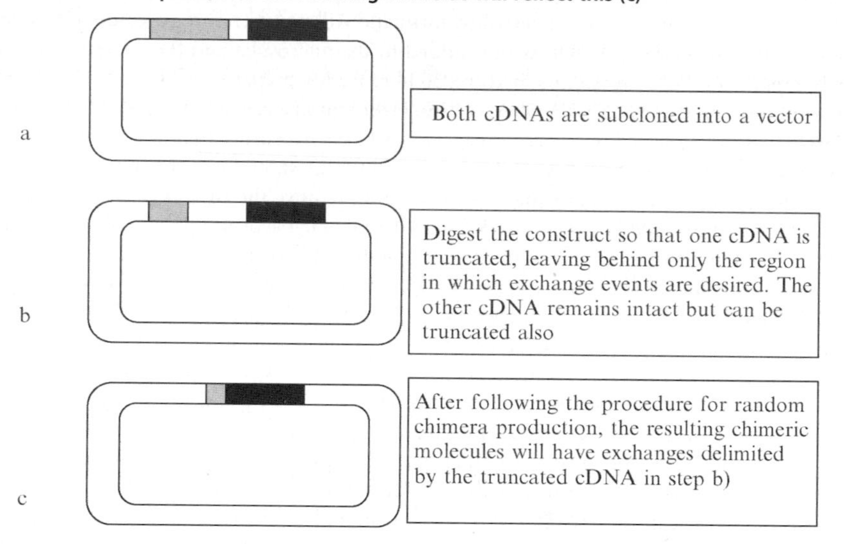

variation of the random chimera technique, two cDNAs are subcloned head-to-tail into a single vector. One is left essentially intact, but the second one is digested and only the section in which crossover events are desired is left in the construct. One drawback of this technique is that if a suitable restriction site for subcloning the truncated cDNA is not present, it may have to be introduced using SDM. However, by limiting the crossover area, time is not wasted in the characterization of chimeras that have been formed in areas of the cDNA that are not of interest.

3 PCR-Based Site-Directed Mutagenesis

This technique allows for the substitution, insertion, or deletion of specific amino acids within a peptide chain by the manipulation of cDNA. As with many molecular biological techniques, SDM was initially a cumbersome and time-consuming process that required multiple subcloning steps and the use of specialized vectors. However, it has since evolved into a streamlined procedure that can be completed in a day or two. There are many variations of PCR-based mutagenesis available and several of them have been used in this laboratory. We have found the Stratagene QuickChange protocol to be the easiest, fastest, and most reliable method to use. The general idea behind this technique is to use mutagenic oligonucleotide primers targeted to a cDNA in a closed, circular, double-stranded vector. In the PCR reaction, this allows the mutation to be incorporated into all subsequent copies of the DNA construct. The major advantage of this technique over many others is that no subcloning or specialized vectors are required. Previous approaches required that the cDNA of interest be subcloned into a vector in which the mutagenesis was to be performed. Following this, the cDNA was subcloned back into the appropriate expression vector. In addition, it was necessary to make single-stranded forms of the vector in order for the mutagenesis reaction to take place. This latter step is not needed in Thermocycler-based techniques because the double-stranded DNA is denatured as part of the cycle protocol. By eliminating many of the tedious steps once associated with SDM, the reaction can be completed in a day and preliminary verification of the mutations can be ascertained by the end of the following day.

Another problem inherent in most mutagenesis techniques is the isolation of mutated constructs from wild-type constructs that are used as a template for the mutagenesis reaction. The Stratagene QuickChange protocol involves the use of a restriction enzyme (*Dpn* I) that only digests methylated and hemimethylated DNA. Thus, the enzyme will degrade any DNA harvested from a bacterial source, but will not recognize DNA synthesized in vitro during the PCR. Thus, the template DNA that was grown in bacteria will be preferentially degraded. This is a fast and efficient way to remove contaminating wild-type DNA and this speeds up the whole process considerably.

In the following sections, some of the basic considerations and strategies used for SDM are discussed, as are the basic techniques that are currently in use in our laboratory and some of the caveats of which one must be aware.

3.1 Background

To perform SDM, the gene of interest must be in a suitable vector. We generally subclone all of our cDNAs into an expression vector so that, after mutagenesis, the cDNA can be used directly in cell transfection and/or cRNA production. The three basic types of mutagenesis are substitution, insertion, and deletion. Briefly, substitution is the targeting and replacing of one, or a number of amino acids with different amino acids so that the total number of amino acids in the protein is not changed. This is the most common type of mutagenesis. Insertion mutagenesis entails designing primers that will insert a number of amino acids at the desired position, adding to the total number of amino acids in the protein. This type of approach can be used for inserting epitope tags or histidine tags. With deletion mutagenesis, specific amino acids are targeted for removal, decreasing the total number of amino acids in the protein. In all three cases, the mutagenesis procedure is the same; the only difference is the way in which the mutagenic primers are designed. Because of space constraints, this chapter deals only with substitution mutagenesis.

3.2 Strategy

There are two basic considerations when attempting SDM. One is to determine the amino acids that should be mutated and the other is to decide what to replace them with. The first question is, of course, dependant upon information gathered from previous experimentation in order to target residues that are appropriate. Such information may be derived from biochemical techniques. For instance, in our binding site studies, we have specifically mutated amino acids that had previously shown to be covalently labeled by photoactive ligands. Additionally, we have used comparisons between the sequences of different receptor subunits that correlate with receptor function to identify domains of interest. Chimeragenesis, the technique described in the first half of this chapter, can provide important information in this regard. Obviously, those proteins for which a detailed structural model is available will lend themselves to more rational substitutions.

The second question, of which amino acid to use as a substitute, is perhaps more difficult to answer. There are a number of different strategies one can use (see Ward et al., 1990 for review) but, unfortunately, there are no set rules to predict the success of the substitution. It is critical to avoid any substitution that results in either a drastic topological change in the protein or a protein that can no longer be expressed but it is very difficult to predict the changes that might bring this about. One approach is to substitute individual amino acids with alanine. This can be quite a drastic change, but alanine is typically well tolerated by proteins. Indeed such "alanine scanning" mutagenesis has been used successfully to scan entire ligand binding domains Mingarro et al. 1996. Another tactic is to substitute a closely related amino acid; for example, a tyrosine may be substituted with a phenylalanine or vice versa. The reasoning here is that since these amino acids have similar molecular volumes, such substitutions may not have a drastic impact on protein structure. In addition, a conservative substitution such as this allows the experimenter to assess the contribution of a specific functional group to the parameter of interest. This is not possible with many amino acids, as changing functional groups usually necessitates altering the size and volume of the residue in that space. For example, changing from a basic to an acidic group introduces not only a change in charge but also a change in the size and volume of the residue and this will inevitably complicate the interpretation of results. The only way around this dilemma is to use unnatural amino acids, a topic that is beyond the scope of this chapter.

3.3 PCR Procedure

Outlined here is the basic procedure that we are currently using (➋ *Figure 20-5*). It is based on the Stratagene Quick Change protocol. This approach can be adapted to a wide range of applications depending on the particular needs of the experimenter and only the basic procedure is described here.

3.4 Template

The template is obviously an essential component of the PCR. Although only small amounts are needed for the reactions, it is essential that the DNA be clean and free of any protein or chemical contaminants. We typically use Qiagen column purification kits to harvest the template DNA. As mentioned earlier, any vector/cDNA construct can be used. However, it should contain some features that are needed to successfully accomplish the mutagenic procedures. One is that some form of antibiotic resistance gene should be present because, after transformation, this allows selection of the bacteria containing the mutated construct. Additionally, smaller constructs require less time during PCR cycling and are generally easier to manipulate. We typically use expression vectors of approximately 5–6 kb in size containing a 1.4–2.0 kb insert. The vector that we use in most cases is pcDNA3 (Invitrogen). This particular vector was chosen for a number of reasons. First, because it is an expression vector, it can be used directly after mutagenesis to transfect mammalian cells for testing the effects of the mutation on receptor structure. Second, cRNA encoding the protein of interest can be transcribed for expression in *Xenopus* oocytes for electrophysiological analysis of the impact that the mutation would have on receptor function. Generally, we harvest milligram amounts of construct DNA because the wild-type form of the construct is routinely required for expression experiments. In this case, we use Qiagen Mega or Giga kits to provide a stock of DNA. However, for harvesting smaller amounts of DNA to check the consequences of a mutagenic change, we typically use a smaller-scale Qiagen miniprep kit. It is important to note that the parental DNA must be methylated to permit recognition by *Dpn* I when selecting for mutagenized strands (see later). Some bacterial strains, therefore, should be avoided when used for DNA amplification. These include any strains that are dam^-. Most of the commonly used strains (JM109, XL-1 blue) are dam^+.

3.5 Primers

As there are many excellent reference materials regarding primer design, only the basic concepts are discussed in this section. The design of the primer that is to be used to introduce the substitution is the most important step in mutagenesis studies. The areas of the primer that mismatch with the template must be flanked by regions that are a perfect match with the surrounding template (➋ *Figure 20-4*). The primers should be designed such that the mismatched nucleotides are in the middle portion of the primer. Additionally, the primers must overlap such that they anneal to the same region of DNA. For a single base pair mismatch, the matching flanking regions should be of 12 base pairs in length. This typically

Figure 20-4

The design of primers to substitute a glutamate for a valine and to add a silent restriction site. Boxed are six base pairs found in the cDNA that is to be mutated. The bases that need to be mutated in the cDNA are bolded whereas the mismatches at these positions are bolded in the primers. The first three code for valine (GTX). To substitute valine for glutamate (GAA/G), the thymine residue in the second position of the codon in the upper strand of the cDNA must be replaced with an adenine. The corresponding residue in the lower cDNA strand must be changed to a thymine residue. Therefore, mutagenic primer 1 contains the adenine at this position whereas mutagenic primer 2 contains a thymine at this position. Additionally, a silent mutation will be introduced that will result in a novel restriction endonucleases site (GAGCTC, *Sst* II) in this group of six base pairs. To construct the *Sst* II recognition site, the thymine in the upper strand of the cDNA must be altered to a cytosine residue and the adenine in the lower strand must be changed to a guanine residue. Therefore, the cytosine residue is introduced into Primer 1 whereas the guanine residue is introduced in Primer 2. Note that there are 15 perfectly matched residues on either side of the mutations to ensure that the primers will anneal properly and that the primers overlap each other perfectly. Also note the orientation of the primers with respect to the 5′ to 3′ direction. It is crucial to take this into account when synthesizing the primers

Primer[1]
5′ C A G T T G C C A A C G T G **A** G **C** T C C A G T T A G A C G A C C 3′

5′ C A G T T G C C A A C G T | G **T** G **T** T C | C A G T T A G A C G A C C 3′
3′ G T C A A C G G T T G C A | C **A** C **A** A G | G T C A A T C T G C T G G 5′

3′ C A G T T G G G T T G C A C **T** C **G** A G G T C A A T C T G C T G G 5′
Primer[2]

provides sufficient hydrogen bonding to allow the primer to bind despite the presence of the mismatch. One additional consideration is to terminate the primer at the 3′ end with purines rather than pyrimidines since the former form three hydrogen bonds rather than two, and thus provide a more stable anchor for the polymerase. This may require that the primer is extended by a nucleotide or two in order to accommodate these anchor residues. The example in ➲ *Figure 20-4* shows the primer design for the introduction of a glutamate residue to replace a valine residue. Unfortunately, because the composition of the primers changes depending on the substitutions that are to be made, the reaction will need to be optimized for each different mutagenesis reaction (see Optimization of PCR ➲ Section 3.8). An additional consideration is that the primers should be designed so that their predicted annealing temperatures are reasonably close to one another. If they are drastically different and they anneal at markedly different temperatures, this may lead to PCR problems such as the preferential amplification of one strand when compared to the other.

To determine the appropriate annealing temperatures, a number of different programs that are available on the Internet can be used. Strategene has a program that allows for the design of optimal primers specifically for mutagenesis. This site can be found at http://labtools.stratagene.com/QC. Occasionally, for unknown reasons, certain sets of primers simply will not work in the reaction. If repeated trials with varying conditions fail to produce results, it is recommended that, if possible, the primers are redesigned. For the Stratagene protocol, the primers do not need to be phosphorylated. However, to maximize the efficiency of the reaction, they should be purified. When purchasing primers commercially, a purification option is often available for a minimal cost. A gel-purification procedure can be performed in the laboratory but this can be time-consuming and typically affords poor recovery. That being said, we have, on occasion forgone the purification step and used the primers directly. The desired mutations were still obtained but at a lower efficiency.

3.6 Verification of Mutagenesis: Additional Primer Considerations

After completing the reaction, only those double-stranded vectors containing a mutation should be present. However, this is not always the case as the process is never 100(efficient. One way to rapidly screen the final product of mutants is to build a silent unique restriction site into the primer along with the required

Figure 20-5
The basic steps of PCR-based mutagenesis. Primers are indicated as black bars, parental strands as solid lines, and newly-synthesized strands as dashed lines

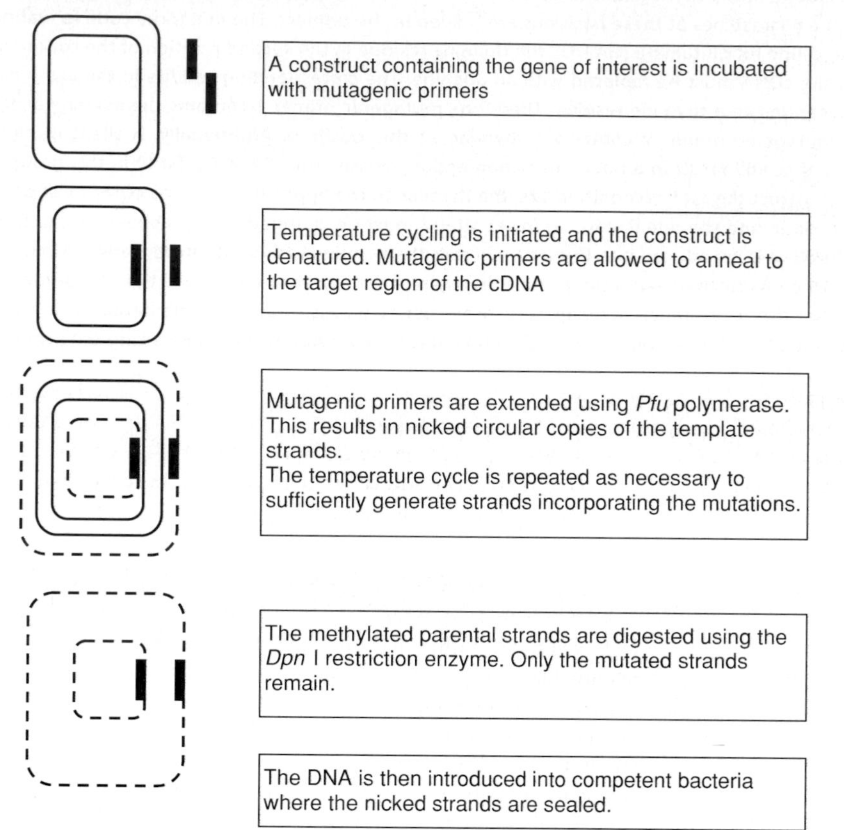

mismatch. Several tables and computer algorithms exist that greatly assist the user in determining which silent mutations will work in a given stretch of DNA (Shankarappa et al., 1992; see also the New England Biolabs Web site at www.neb.com/neb/tech/). It is usually best to introduce a site that is not present in the wild-type construct so that interpretation of the restriction digest is as simple as possible. In the example shown in *Figure 20-4*, an *Sst* II site has been introduced into the primers so that the final product will have this site incorporated into its strands. By digesting the DNA obtained in the final steps of the procedure, one can easily check for the presence of this cut site. This in turn is indicative of the mutagenic primers being incorporated into the DNA.

3.7 The PCR Protocol

The components listed in *Table 20-1* are typical for PCR. As can be seen, it is a fairly simple reaction to set up. The template and primer concentrations must be determined beforehand; the buffer, enzyme, and deoxynucleotide triphosphate (dNTP) mixture are commercially available.

After adding the first five components listed in *Table 20-1*, water is added to bring the reaction volume to a total of 50 μl. Ensure that the contents of the tube are mixed by pipetting the contents up and

Table 20-1
Components of a typical PCR

Component	Volume μl	Amount
10 X buffer	5	–
Primer 1	Depends on concentration	125 ng[a]
Primer 2	Depends on concentration	125 ng[a]
10 mM dNTP mix	1	200 μM final
Template	Depends on concentration	5–50 ng[a]
Double distilled water	To volume of 50	–
Pfu turbo polymerase	1	2.5 U

[a]These amounts may need to be altered if the reaction is not successful using these guidelines. The optimum reaction conditions may require different primer/template ratios and these can only be determined empirically

down several times. The polymerase should always be added last. *Pyrococcus furiosus* (*Pfu*) polymerase is used rather than *Taq* polymerase because of its proofreading capacity and because it does not displace the mutagenic primers. This enzyme greatly improves the yield of longer targets (Lasken et al., 1996; Hogrefe et al., 2002). Specifically, we use *Pfu* Turbo from Stratagene. This formulation contains a dUTPase that removes any dUTP that forms from cCTP during temperature cycling since cUTP has been shown to dramatically inhibit *Pfu* polymerase.

The samples are then overlaid with mineral oil (20 μl) to prevent evaporation of the contents and the PCR can proceed. If, however, the PCR machine being used has a hot-start lid, addition of mineral oil is not necessary.

The cycles used are as follows:

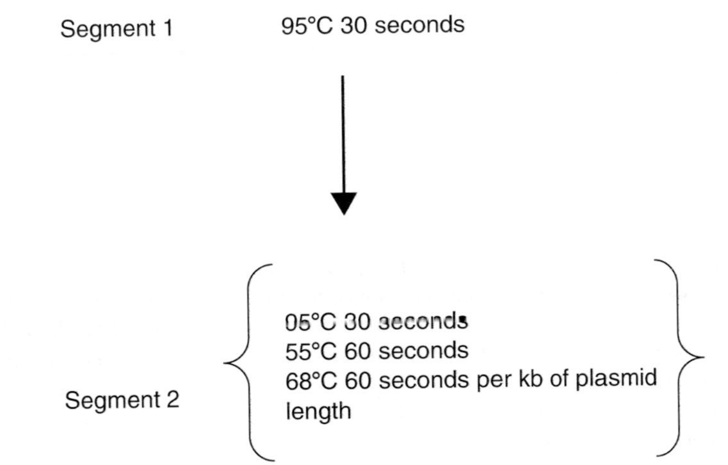

Segment 1 always cycles only once. The number of cycles for segment 2 is variable, depending on the type of reaction that is to be performed. Single nucleotide point mutations need 12 cycles; single amino acid changes (replacing two or three bases) should have 16 repeats of cycle 2 and mutations in which multiple amino acids are added or deleted should have at least 18 repeats of cycle 2. These numbers are not immutable; however, they are recommended by Stratagene and we have found that they work quite well. However, in some cases, especially when performing more complicated reactions, the numbers may need to be adjusted on an individual basis.

Following the reaction, it should be possible to visualize the product on an agarose gel. This is a fairly quick procedure that will allow the user to determine if the PCR reaction was successful and whether it is worth proceeding with the following steps. From the reaction tube used in the PCR, 10 μl of the reaction mixture is removed, mixed with loading buffer, and applied to the gel. We use a 1% agarose gel containing ethidium bromide. The sample is run along with a 1-kb DNA ladder as a marker in order to determine whether the size of the product matches with that of the expected size. Typically, excess primer will be seen close to the dye front. Once it appears that the reaction has proceeded as anticipated, selection of newly synthesized strands can begin.

3.8 Optimization of PCR Reaction

If, after an aliquot of the reaction is run on the gel, there is no product present or there is a product that is of the wrong size, it is necessary to optimize the reaction. This should be done for each new set of primers and template because each set of primers will have different properties. As the base content of the primers will be different, the annealing temperature must also be altered in each case. In most cases, we have found that optimizing this parameter is generally sufficient to obtain a successful reaction. Additional alterations can include varying the ratio of the template and primers and/or changing the concentration of magnesium ions in the buffer. Ratios are best altered by changing the amount of template in the reaction while keeping the primer amount constant. As it is sometimes difficult to obtain accurate measurements of small quantities of oligonucleotides, the ratio of these two components to the template in the reaction mixture can differ from that which had been anticipated.

3.9 Digestion of Parental Strands

The digestion of the parental strands is performed in the same tubes in which the PCR was carried out. 10 U of the restriction endonucleases *Dpn* I is added beneath the mineral oil (if present) and the contents are mixed by pipetting the solution up and down several times. At this point, it is wise to briefly centrifuge the tubes to ensure that all of the contents are at the bottom of the tube. The tubes can then be placed in a 37°C water bath for 1 h.

3.10 Competent Cells

Following *Dpn* I digestion, the remaining DNA must be grown in sufficient amounts for further manipulation and so it is introduced into competent bacteria. Once introduced into competent cells, the nicks in the duplex DNA are repaired and the intact construct can be replicated by bacterial machinery. As described earlier, we typically make our own competent cells, but supercompetent cells are available from a number of suppliers (e.g., Strategene and Promega).

3.11 Transformation of Competent Cells

To introduce the mutated construct into competent cells, we follow the following protocol:

1. Cells are removed from the –85°C freezer and thawed on ice.
2. An aliquot of the reaction mixture is removed from the PCR tube. If mineral oil was used, be sure that it is not transferred to the tube containing competent cells as it will drastically reduce transformation efficiency.
3. The transformation protocol is the same as described earlier in the chimera production section. In this case, 1–10 μl of the PCR mixture is used to transform the cells and, after transformation, the mixture is plated out on antibiotic plates.

3.12 Testing for the Presence of Mutant DNA

If the mutagenic primers have been incorporated, the newly synthesized double-stranded DNA should have both the mutation that will code for a new amino acid in the protein and the mutation that will give rise to a new restriction enzyme site. To test for the presence of the mutant, restriction digests are performed on the plasmid DNA that has been harvested from the colonies that grew on the agarose plates overnight. This is done using a basic restriction digest following the directions of the restriction enzyme supplier. If positive colonies are identified, these can be used to streak additional plates and grow the DNA further. Because the restriction digest pattern suggests the presence of the silent mutation but does not confirm the presence of the mutation that will change the amino acid composition, it is imperative that all putative positives be fully sequenced across the entire length of the cDNA. This will confirm not only that the expected mutation is present but also that no unexpected alterations have been introduced.

3.13 Other Applications

The technique described earlier can be modified for many different purposes. One way in which we regularly use a slight modification of the technique is to clean up "dirty" DNA. Often, when cDNAs are cloned from a library, additional extraneous DNA is associated with the cDNA of interest. After subcloning into a vector, this extra DNA can interfere with the expression of the clone and the extra size that it adds to the insert can make the DNA difficult to manipulate. Additionally, the cDNA may be in a vector that is not suitable for expression and may not have the correct restriction endonucleases cut sites to allow it to be subcloned into the vector of choice. PCR mutagenesis can be useful to remove the extraneous DNA and to insert restriction sites at convenient positions.

We have also used mutagenic primers to insert histidine tags and epitope tags into cDNAs of interest (Neish et al., 2003). Inserting several codons into a cDNA is a little more difficult than simply substituting one amino acid for another, but is quite feasible using this technique. Optimization of the reaction is absolutely critical when performing more difficult manipulations such as this.

Several laboratories have replaced consecutive amino acids with cysteine residues as part of a technique called the substituted cysteine accessibility method (SCAM) (Akabas et al., 1992). In this technique, which is described elsewhere in this volume, the accessibility of the inserted cysteine residues to an aqueous environment is measured by their accessibility to covalent chemical modifying agents. The effects of the addition of these bulky agents can then be used to explore protein functional domains such as those that line ion channels or ligand binding sites.

4 Interpreting the Impact of Chimeras and Mutations on Function

The results of protein engineering must always be interpreted with caution. Loss-of-function mutations or chimeras are especially problematic, mainly because it is often difficult to resolve the difference between a specific, local change and a global alteration of protein structure. For example, if the binding domain of a receptor has been targeted and the resulting mutant protein displays no specific binding for a radioligand, it cannot be assumed that the binding site has been selectively disrupted. A number of ligands having different efficacies at the receptor should be tested for their effects on the mutant protein. It would, for example, be significant if agonist binding was disrupted but the binding of inverse agonists or antagonists was unaltered. In addition, a decrease in binding may be due to an effect of the mutation on a step in the pathway of receptor maturation. This might include alteration of a process involved in the membrane insertion of the receptor or one that is required for the assembly of multiple subunits into a mature protein. The use of antibody-labeling techniques to visualize receptor expression on the cell surface or its retention within the cell can provide useful information on such effects. SDM has been extensively used to investigate structure–function relationships of ligand-gated ion channels and has led to many insights into the structure of ligand recognition domains and ion channels. Again, the interpretation of any mutations must be made with great caution. Especially problematic is discriminating between binding and gating effects. A comprehensive review on this subject covers many of the potential pitfalls in interpreting data from analysis of mutant proteins (Colquhoun, 1998).

5 Conclusions

Chimeragenesis and SDM are powerful techniques that can be used to investigate the complex relationships between protein structure and function. The methods detailed here are relatively simple to perform and can be carried out in a short period of time. They are applicable to any protein type for which the cDNA is available and can be modified for many different purposes in protein engineering.

References

Akabas MH, Stauffer DA, Xu M, Karlin A. 1992. Acetylcholine receptor channel structure probed in cysteine-substitution mutants. Science 258: 307.

Boileau AJ, Kucken AM, Evers AR, Czajkowski C. 1998. Molecular dissection of benzodiazepine binding and allosteric coupling using chimeric γ-aminobutyric $acid_A$ receptor subunits. Mol Pharmacol 53: 295.

Brock BJ, Waterman MR. 2000. The use of random chimeragenesis to study structure/function properties of rat and human P450c17. Arch Biochem Biophys 373(2): 401

Colquhoun D. 1998. Binding, gating, affinity and efficacy: the interpretation of structure–activity relationships for agonists and of the effects of mutating receptors. Br J Pharmacol 125(5): 924.

Derry JMC, Dunn SMJ, Davies M. 2004. Identification of a residue in the γ-aminobutyric acid type A receptor α subunit that differentially affects diazepam-sensitive and –insensitive benzodiazepine site binding. J Neurochem 88(6): 1431-1438.

Hogrefe HH, Hansen CJ, Scott BR, Nielson KB. 2002. Archael dUTPase enhances PCR amplifications with archael DNA polymerases by preventing dUTP incorporation. Proc Nat Acad Sci 99(2): 596.

Lasken RS, Schuster DM, Rashtchian A. 1996. Archaebacterial DNA polymerases tightly bind uracil-containing DNA. J Biol Chem 271(30): 17692-17696.

Moore KR, Blakely RD. 1994. Restriction site-independent formation of chimeras from homologous neurotransmitter–transporter cDNAs. BioTechniques 17: 130.

Mingarro I, Whitley P, Lemmon MA, Von Heijne G. 1996. Ala-insertion scanning mutagenesis of the glycophorin A transmembrane helix: a rapid way to map helix–helix interactions. Protein Science 5: 1339.

Neish CS, Martin IL, Davies M, Henderson RM, Edwardson JM. 2003. Atomic force microscopy of ionotropic receptors bearing subunit-specific tags provides a method for determining receptor architecture. Nanotechnology 14: 864.

Promega Corporation. 1991. Plasmid cloning and transcription in vitro. Promega protocols and applications guide 2nd edn, Titus DE, editor. Promega Corporation USA.

Satoh T, Takahashi Y, Oshida N, Shimizu A, Shinoda H, et al. 1999. A chimeric inorganic pyrophosphatase derived from Escherichia coli and Thermus thermophilus has an increased thermostability. Biochemistry 38(5): 1531.

Shankarappa B, Vijayananda K, Ehrlich G. 1992. Silmut: a computer program for the identification of regions suitable for silent mutagenesis to introduce restriction enzyme recognition sequences. Biotechniques 12: 882.

Tirona RG, Leake BF, Podust LM, Kim RB. 2004. Identification of amino acids in rat pregnane X receptor that determine species-specific activation. Mol Pharmacol 65(1): 36.

Ward WHJ, Timms D, Fersht AR. 1990. Protein engineering and the study of structure–function relationships. Trends Pharmacol Sci 11: 280

Wieland HA, Luddens H, Seeburg PH. 1992. A single histidine in $GABA_A$ receptors is essential for benzodiazepine binding. J Biol Chem 267(3): 1426

21 Cysteine Scanning Mutagenesis: Mapping Binding Sites of Ligand-Gated Ion Channels

J. G. Newell · C. Czajkowski

Abstract: The substituted cysteine accessibility method (SCAM) is a powerful technique that combines site-directed mutagenesis and chemical modification to address fundamental questions about the structure and function of proteins. In this chapter, we discuss SCAM theory and the application of the technique to the study of binding pockets of ligand-gated ion channels. SCAM entails the introduction of individual cysteines residues into protein regions (e.g. binding sites or channel lumen) and subsequent application of thiol (-SH)-specific reagents (e.g. derivatives of methanethiosulfonate) to probe the physico-chemical nature of the environment. Methanethiosulfonate (MTS) reagents react with cysteine residues in open, aqueous environments one billion times faster than cysteine residues in interior of a protein or at a protein-lipid interface. Cysteine residues can be introduced into a wild-type protein by using a standard method, after which cDNA sequencing is required to verify the correct insertion of the foreign DNA sequence. Thereafter, fundamental properties of the mutant receptors in an artificial expression system (e.g., Xenopus oocytes) must be characterized (e.g. EC_{50}) in order to carry out SCAM experiments using similar experimental conditions for each mutant receptor. Once the preliminary characterization has been completed, experiments with high MTS concentrations can be performed to elucidate the secondary structure of the region, which can be deduced on the basis of the pattern of cysteine accessibility. More sophisticated experiments with lower MTS concentrations in the presence or absence of antagonists and other drugs can be used to determine the rates of MTS modification of individual cysteine residues. This, in turn, can identify amino acid residues that play a role in ligand binding, allosteric modulation or channel opening.

List of Abbreviations: DMSO, Dimethylsulfoxide; EC_{50}, Concentration of agonist that produces a half-maximal response; $GABA_AR$, γ-Aminobutyric acid type A receptor; HEK, Human embryonic kidney; I, Current; IAF, 5-iodoacetamidofluorescein; K_I, Antagonist binding constant; k_1, Pseudo-first order rate constant; k_2, Second order rate constant; LGIC, Ligand-gated ion channel; M, Molar; MTS, Methanethiosulfonate; MTSEA, 2-aminoethylmethanethiosulfonate; MTSEA-biotin, *N*-biotinylaminoethyl-methanethiosulfonate; MTSEA-biotincap, *N*-biotinylcaproylaminoethylmethanethiosulfonate; MTSES, 2-trimethylammonioethyl-methanethiosulfonate; MTSET, 2-hydroxyethylthiosulfonate; nAChR, Nicotinic acetylcholine receptor; SCAM, Substituted cysteine accessibility method

1 Introduction

Arthur Karlin and Myles Akabas at Columbia University developed the substituted cysteine accessibility method (SCAM) to identify channel-lining residues of the nicotinic acetylcholine receptor (nAChR). SCAM entails the introduction of cysteines, one at a time, into a protein region and the subsequent application of thiol-specific reagents to the engineered residues to determine whether they are modified (Karlin and Akabas, 1998). Modification of the introduced cysteine is monitored using electrophysiological or biochemical assays. Since its development, SCAM has gained widespread use not only in examining channel domains but also in mapping ligand-gated ion channel (LGIC) binding sites (Boileau et al., 1999; Sullivan and Cohen, 2000; Teissére and Czajkowski, 2001; Reeves et al., 2001; Holden and Czajkowski, 2002; Torres and Weiss, 2002; Wagner and Czajkowski, 2001; Boileau et al., 2002; Sullivan et al., 2002; Newell and Czajkowski, 2003; Bali and Akabas, 2004; Newell et al., 2004). Unlike photoaffinity labeling and classical mutagenesis, SCAM not only identifies binding site residues but also provide important insight into binding site structure and dynamics.

Photoaffinity labeling is one of the most direct means of investigating ligand-binding site residues (Bayley and Knowles, 1977; Chowdry and Westheimer, 1979; Dormán and Prestwich, 2000). However, photoaffinity labeling is a labor-intensive method that requires protein sequencing to identify labeled residues and the synthesis of high affinity photoactive ligands. In addition, depending upon the photoactive moiety attached to the ligand, this method only identifies a subset of residues in the recognition site (Middleton and Cohen, 1991; Cohen et al., 1991; Smith and Olsen, 1994; Duncalfe et al., 1996; Chiara and Cohen, 1997; Chiara et al., 1998; Grutter et al., 2000; Sawyer et al., 2002). Finally, another important consideration in photolabeling is the possible alteration of the structure of the binding site by ultraviolet irradiation during experimentation (Guillory and Jeng, 1983).

Site-directed mutagenesis is widely used in the study of LGIC neurotransmitter-binding sites. However, the interpretation of data gathered using this method is difficult, especially in the absence of high-resolution crystal structures. Residues mutated outside of a binding site can alter binding through nonlocal effects on receptor structure, and even mutations of binding site residues may not necessarily alter binding properties. In addition, as the binding of agonists induces conformational changes in LGIC protein structures that promote gating transitions, it is difficult to determine whether a mutation affects binding and/or gating processes using macroscopic approaches such as concentration response studies and changes in EC_{50} values (Colquhoun, 1998). An important distinction between SCAM and site-directed mutagenesis is that SCAM does not rely on studying the functional effects of the mutations alone, but combines systematic mutagenesis of a series of residues and chemical modification to glean information about binding site structure and dynamics. It is important to recognize that, regardless of the experimental approach, the agonist-binding site itself undergoes a series of conformational changes during allosteric transitions among resting, active, and desensitized states (for review, see Changeux and Edelstein, 1998), suggesting that the complement of residues that forms the binding site is not static (Colquhoun, 1998).

2 Cysteine Substitution

Cysteine mutagenesis can be carried out using any number of commercially available systems. Cysteine substitution in LGIC proteins is remarkably well tolerated, e.g., 103/105 cysteine substitutions in the nAChR (Akabas et al., 1994; Akabas and Karlin, 1995; Zhang and Karlin, 1997; Zhang and Karlin, 1998) and 121/131 substitutions in the $GABA_AR$ have been tolerated. Expression of mutant receptors in heterologous systems such as *Xenopus laevis* oocytes or human embryonic kidney (HEK) 293 cells allows one to monitor functional modification of a given cysteine residue using electrophysiological techniques (e.g., two electrode voltage clamp recording) or radioligand-binding assays (Javitch et al., 1996; Boileau et al., 1999). Ideally, only cysteine mutations that have near wild-type function should be used for analysis because this is the best indication that the overall protein structure has not been disrupted (Wagner and Czajkowski, 2001; Newell and Czajkowski, 2003). In an electrophysiological assay, we assume that if the EC_{50} value (concentration of agonist that produces half of the maximum response) of the mutant and wild type receptors are similar, then the structure of the mutant receptor is similar to the wild-type protein (Cheung and Akabas, 1996; Boileau et al., 2002).

If sulfhydryl-specific reagents do not affect wild-type channel function, we assume that any effect observed after exposure of the mutant cysteine containing receptor to sulfhydryl-specific reagents is due to modification of the introduced cysteine. However, if endogenous cysteine residues are accessible and wild-type function is altered by sulfhydryl modification, then a pseudo (cysteine-less) wild-type protein must be engineered, the function of which must be tested prior to SCAM analysis. The most suitable substitutions for cysteine are alanine or serine residues. For their study of the $GABA_AR$ M2-M3 loop, Bera et al. (2002) replaced the endogenous cysteine residues in the M1 and M3 transmembrane regions of the $GABA_A$ receptor subunits with serine to facilitate data interpretation. These "cys-less" α_1, β_1, and γ_2 subunits, when coexpressed in *Xenopus* oocytes, retained normal function. It must be noted that endogenous cysteines can be problematic, even if they are not located on a water-accessible surface. This is due to the fact that mutagenesis may produce subtle changes in receptor structure that unmask inaccessible cysteine residues.

3 Premises Underlying SCAM Analysis of Binding Site Structure

SCAM is based on the following assumptions:

1. The introduced cysteine residues are found in one of three possible environments: (1) on the water-accessible surface, (2) within the protein interior, or (3) at the protein–lipid interface (Karlin and Akabas, 1998).

2. After cysteine substitution, if the mutant receptor is expressed and has close to wild-type functional properties, then its three-dimensional structure will be close to wild type, and the substituted cysteine acts as a reporter for the environment surrounding the wild-type side chain (Karlin and Akabas, 1998; Javitch et al., 2002).
3. Residues lining water-accessible binding sites will react with charged, hydrophilic, and sulfhydryl-specific reagents and with residues that face the interior of the protein or the lipid bilayer will not be accessible.
4. Modification of a binding site residue will alter agonist and antagonist binding irreversibly.
5. The effects of covalently attaching a sulfhydryl-specific reagent to an accessible cysteine are due to local steric or electrostatic effects in the binding site and not due to nonlocal effects.
6. If reaction with a sulfhydryl-specific reagent is prevented by agonists and antagonists, the residue lines the binding site. However, not every one of the identified binding site residues need contact ligand; ligand could protect residues located deeper in the binding site pocket by binding above them and blocking the passage of the sulfhydryl-specific reagents from the extracellular medium to this area of the receptor. In addition, we cannot rule out indirect effects through propagated structural changes caused by the agonists used in the protection assays, although these same changes are unlikely to be induced by antagonists. It seems unlikely that the alteration of binding by attachment of a sulfhydryl reagent to an engineered cysteine residue would be due to a nonlocal effect if the reaction were inhibited by the presence of agonists and antagonists.

4 Methanethiosulfonate Reagents

Sulfhydryl-specific reagents commonly used in SCAM analysis are derivatives of methanethiosulfonate (MTS). These include 2-aminoethylmethanethiosulfonate (MTSEA), 2-hydroxyethylthiosulfonate (MTSET), and 2-trimethylammonioethyl-methanethiosulfonate (MTSES). These compounds are roughly cylindrical in shape with dimensions of about 0.6 nm in diameter and 1 nm in length (Karlin and Akabas, 1998). These MTS reagents form mixed disulfides with free sulfhydryls, covalently linking the -SCH_2CH_2X group to the cysteine sulfhydryl (➋ *Figure 21-1*). The rank order of reactivity is MTSET > MTSEA > MTSES,

Figure 21-1
Chemistry of sulfhydryl-specific modification. Reaction of MTS ($CH_3SO_2SCH_2CH_2X$) with a cysteine will lead to specific modification of the SH group by MTS. Note that X = NH_3^+ (MTSEA), SO_3^- (MTSES), $N(CH_3)_3^+$ (MTSET), NH-biotin (MTSEA-biotin) or $NHCO(CH_2)_5$ (MTSEA-biotincap)

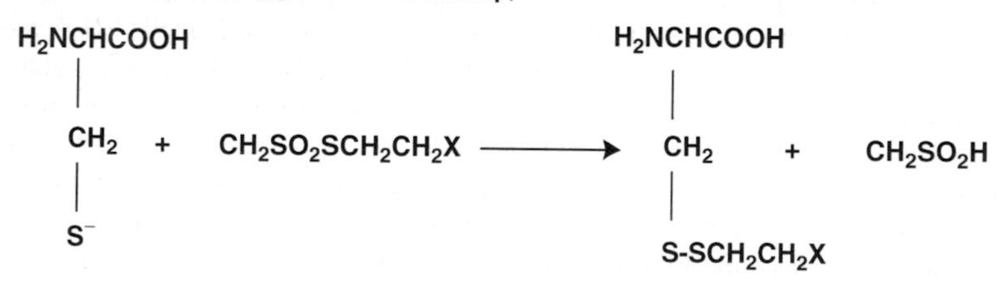

with MTSET being ~10–13 times and MTSEA 2.5 times more reactive than MTSES (Staufer and Karlin, 1994). When using MTSET, MTSEA and MTSES, the optimal concentrations (to correct for differences in intrinsic reactivity) have been determined to be 1 mM, 2.5 mM, and 10 mM, respectively (Zhang and Karlin, 1997). The MTS reagents react with ionized thiolates (S-) a billion times faster than with unionized thiols (SH) (Roberts et al., 1986). Since ionization of a sulfhydryl is much more likely in an aqueous environment than in a protein/lipid environment, these reagents should react with sulfhydryl groups accessible from the aqueous medium and not with sulfhydryls buried in the protein or facing the lipid bilayer. An additional advantage of these reagents is that they are available in a variety of sizes and charges and thus can be used to probe the electrostatic and steric environment of any residue mutated to cysteine. Charged MTS reagents are water-soluble, whereas uncharged reagents should be prepared in a vehicle such as DMSO. MTS reagents are available from Toronto Research Chemicals (Toronto, Ontario, Canada, www.trc-canada.com) or Biotium Inc. (Hayward, California, www.biotium.com). MTS reagents are hygroscopic and should be stored dessicated at −20°C.

4.1 MTSEA

At neutral pH, MTSEA is predominantly positively charged, though this depends on the degree of ionization of the amine (Zhang and Karlin, 1997). The half-life of MTSEA in buffer (pH 7.5, 22°C) is 15 min (Karlin and Akabas, 1998). MTSEA is membrane permeant (Holmgren et al., 1996); therefore, membrane topology of proteins cannot be determined using this reagent, unless MTSEA quenchers (e.g., cysteine or glutathione) are applied intracellularly (Pascual and Karlin, 1998). The length of MTSEA that covalently links to the cysteine sulfhydryl is ∼3.7 Å (Teissére and Czajkowski, 2001; Holden and Czajkowski, 2002).

4.2 MTSET

MTSET has a permanently charged quaternary ammonium ion derivative. The half-life of MTSET in buffer (pH 7.5, 22°C) is 10 min (Karlin and Akabas, 1998). MTSET is not membrane permeant (Holmgren et al., 1996). The length of MTSET that covalently links to the cysteine sulfhydryl is ∼4.7 Å (Holden and Czajkowski, 2002). MTSET is a low-affinity nAChR agonist, the effects of which can be blocked by 10 μM d-tubocurarine (Zhang and Karlin, 1997; Sullivan and Cohen, 2000).

4.3 MTSES

MTSES carries a net negative charge (Karlin and Akabas, 1998) and is membrane impermeant (Holmgren et al., 1996; Seal et al., 1998). The half-life of MTSES in buffer (pH 7.5, 22°C) is 20 min (Karlin and Akabas, 1998). The length of MTSES that covalently links to the cysteine sulfhydryl is ∼4.8 Å (Holden and Czajkowski, 2002).

4.4 MTSEA-biotin

N-biotinylaminoethylmethanethiosulfonate (MTSEA-biotin) is a relatively membrane-impermeant MTS derivative to which a biotin moiety is conjugated (Seal et al., 1998; Boileau et al., 2002). It is composed of two structural domains: a flexible tail ∼12–14 Å in length, 2.5 Å in diameter, and 4 × 5 Å planar head group (Teissére and Czajkowski, 2001; Wagner and Czajkowski, 2001). The reactive disulfide is near the end of the tail ∼12 Å from the head group.

4.5 MTSEA-biotincap

N-biotinylcaproylaminoethylmethanethiosulfonate (MTSEA-biotincap) is structurally related to MTSEA-biotincap, except that it contains a longer linker group (Seal et al., 1998). The length of MTSEA-biotincap that covalently links to the cysteine sulfhydryl is ∼20.0 Å (Teissére and Czajkowski, 2001).

5 Secondary Structure

The pattern of accessibility of the engineered cysteines that react with the sulfhydryl-specific reagents along the primary amino acid sequence reflects the secondary structure of the region. If one side of an α helix contributes to the binding site, every third or fourth residue should react with the sulfhydryl-specific reagents; by contrast, if a β strand contributes to the binding site, every other residue should be reactive (➋ *Figure 21-2*; Boileau et al., 1999). Turns, loops, and irregular secondary structures will not give a clear

◘ Figure 21-2

Summary of the inhibitory effects of application of MTSEA-biotin (3 mM) for a prolonged application period (3min) for cysteine mutants in the V93C-D101C region of the $GABA_AR$ $β_2$ submit. The peak amplitude of the current was measured before and after application of MTSEA-biotin. This pattern of accessibility is indicative of a β-strand. Reprinted from Boileau et al. (2002) with permission from the American Society of Biochemistry and Molecular Biology

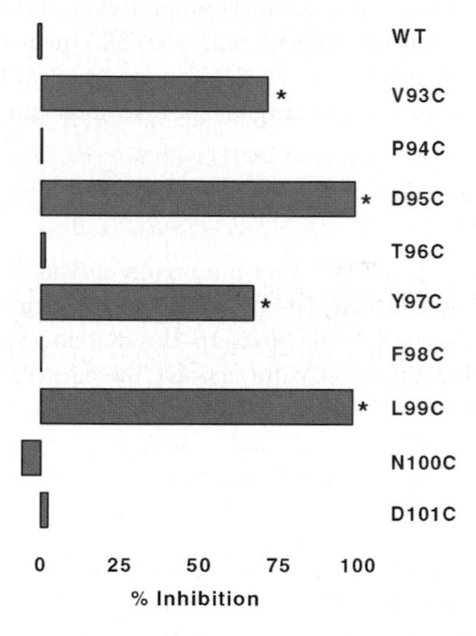

pattern of reactivity (Newell and Czajkowski, 2003). We generally use a high concentration (e.g., 2 mM) of a neutral reagent (MTSEA-biotin) applied over a prolonged period (2–5 min). The effect of MTS on the current response (I) to an EC_{50} concentration of agonist is calculated using the following equation: $(I_{POST\text{-}MTS}/I_{PRE\text{-}MTS})$-1 (see Section 10.1). A residue is accessible if the effect of MTS is statistically different from the effect of MTS on wild-type receptors. Thus, in cases where the effects of MTS are small (<30%), the effect may not be significant, depending on the number of mutations and the *post-hoc* test used (Bera et al., 2002). An alternative method for statistical analysis of large data sets has been reported (Newell and Czajkowski, 2003).

Alternative approaches to mapping secondary structure have also been reported. These include alanine- and lysine-scanning mutagenesis, both of which do not rely on chemical modification (Yan et al., 1999; Sine et al., 2002).

6 The Rate of Reaction

Modification of an introduced cysteine is both time- and concentration-dependent. It is therefore possible to measure the rate of sulfhydryl modification by applying low concentrations of MTS over short periods of time. The rate of reaction depends on the following factors: (1) the permeability of the access pathway to the substituted cysteine, (2) electrostatic potentials, (3) the degree of ionization of the thiol, and (4) local steric constraints (Karlin and Akabas, 1998). The rate constant provides important information about the physicochemical environment of the introduced cysteine residue, relative to other accessible residues within the same region. Thus, SCAM can be used to determine whether a residue is found in an open, aqueous environment or a restricted environment. The rates of reaction of cysteines in an aqueous environment are generally faster than cysteine residues buried within the protein interior (Reeves et al., 2001). In addition, the rates of reaction of a residue in an aqueous environment tend to be slower than the rates of reaction with cysteine in free solution (Karlin and Akabas, 1998). This, however, is not always the case. Since MTS reaction is influenced by the electrostatic potentials, it is possible to have faster rates than those measured in free solution if the intrinsic properties of the protein region permit. In fact, it has been demonstrated that the rate constants for MTSEA and MTSES at α_1F64C of the $GABA_AR$ are approximately 30-fold faster (Holden and Czajkowski, 2002) than those obtained in free solution (*Table 21-1*).

The rates of modification of an introduced cysteine are obtained by measuring the effect of successive low applications of MTS on agonist-gated current (*Figure 21-3*). The optimal MTS concentration to use

Table 21-1

Second-order rate constants for MTSEA-, MTSET-, and MTSES-modification of residues within the loop D region of the GABA-binding site. Second-order rate constants (k_2) represent the mean ′ standard deviation. NR, no reaction. The free solution rates were reported by Karlin and Akabas (1998) and reflect the rates of MTS reaction with 2-mercaptoethanol. Adapted from Holden and Czajkowski (2002) with permission from the American Society of Biochemistry and Molecular Biology

	MTSES		MTSET		MTSEA	
Receptor	$k_2(M^{-1}s^{-1})$	n	$k_2(M^{-1}s^{-1})$	n	$k_2(M^{-1}s^{-1})$	n
$\alpha_1(D62C)\beta_2$	NR	1	NR	1	16 ± 1	4
$\alpha_1(F64C)\beta_2$	$23{,}400 \pm 6000$	4	$5{,}500{,}000 \pm 2{,}800{,}000$	4	$2{,}475{,}000 \pm 235{,}000$	4
$\alpha_1(R66C)\beta_2$	50 ± 11	5	$116{,}000 \pm 13{,}000$	4	$13{,}000 \pm 1700$	4
$\alpha_1(S68C)\beta_2$	270 ± 80	7	2800 ± 1400	6	2000 ± 300	3
free sol.	17,000		212,000		76,000	

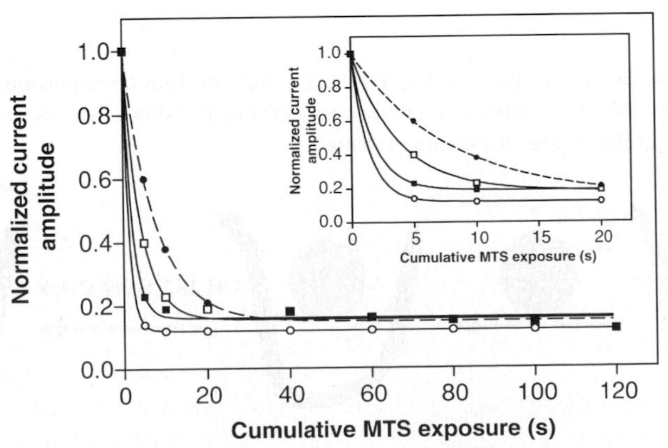

Figure 21-3

Theoretical rate of modification of an introduced cysteine using four different concentrations of MTS (10 μM, ●; 20 μM, □; 30 μM, ■ and 60 μM, □). The pseudo first-order rate constants (k_1) are 0.1285, 0.2445, 0.4849, and 0.7564, respectively. When corrected for the concentration-dependence of MTS modification, the second-order rate constants (k_2) are 12,900, 12,275', 12,122', and 12,606 $M^{-1}s^{-1}$, respectively. Although all k_2 values are similar, the early phase of modification (<20 s) is less well described when higher concentrations (>20 μM) of the reagents are applied (see inset). Based on these data, we would use 10 μM for control experiments and protection assays

for rate experiments can only be determined empirically by examining several MTS concentrations and by nonlinear regression fit of the data. In a well-designed experiment (see Figure 21-3), the curve will be defined with data points on the slope of the curve (early phase of modification <20s) and the plateau (i.e., maximum effect) will be the same as that obtained following application of high MTS concentrations (i.e., as in experiments to determine secondary structure). We recommend starting experiments with a neutral MTS reagent such as MTSEA-biotin. In our functional assays, we use a test concentration of agonist that is equivalent to the EC_{50} for the mutant receptor. To calculate the pseudo first-order rate of reaction (k_1), the decrease in current amplitude is plotted as a function of cumulative MTS exposure and fit using an exponential equation, $y=span^{*}e^{-kt}+plateau$, where span=1-plateau. From this, the second-order rate (k_2) constant is determined as follows: $k_2=k_1/M$, where M is the molar MTS concentration (see Section 10.2). Second-order rate constants should be determined using at least two different concentrations of MTS. The calculated rates should be similar if the concentration of MTS reagent does not change considerably over the time course of an experiment (Holden and Czajkowski, 2002).

Comparison of $MTSET^{+}$ and $MTSES^{-}$ rates of reaction makes it possible to determine if there is charge selectivity within a protein region. This is due to the fact that the two reagents (1) have opposite charges, (2) are similar in size, and (3) share a common reaction mechanism (Holden and Czajkowski, 2002). Within one region of the GABA-binding site of the α_1 subunit, Holden and Czajkowski (2002) determined that the second-order rate constants for $MTSET^{+}$ modification of α_1F64C and α_1R66C were 235- and 2,320-fold faster than that for $MTSES^{-}$. To factor out the intrinsic differences in the reactivities of the two MTS reagents, the ratio of the rates of the two reagents at each substituted cysteine is divided by the ratio of the rates for the reagents modification of 2-mercaptoethanol in free solution. A ratio significantly different from this one indicates that there is an electrostatic potential experienced near the thiol. For α_1F64C and α_1R66C, the ratios were 19 and 186, from which it was concluded that a negative subsite exists within the GABA binding pocket near α_1F64 and α_1R66 (Holden and Czajkowski, 2002).

7 The Protection Assay

7.1 Identification of Binding Site Residues Measuring Rates

One of the fundamental assumptions of SCAM is that sulfhydryl modification of binding site residues will alter the binding of agonists and antagonists irreversibly. Agonists and antagonists that recognize the binding should therefore alter the rate of MTS modification (Figure 21-4). In order to determine whether the presence of a reversible ligand alters the rate of reaction, we coapply a reversible binding-site specific ligand and MTS.

There are three possible outcomes in this type of experiment: (1) a slowing of the rate, (2) an acceleration of the rate, or (3) no change in the rate in the presence of ligand. If both antagonists and

Figure 21-4

Schematic diagram of the modification of protection from modification by application of a reversible ligand (agonist or antagonist, ●). The reversible binding site specific ligand slows the rate at which the MTS reagent modifies the thiolate group of the introduced cysteine

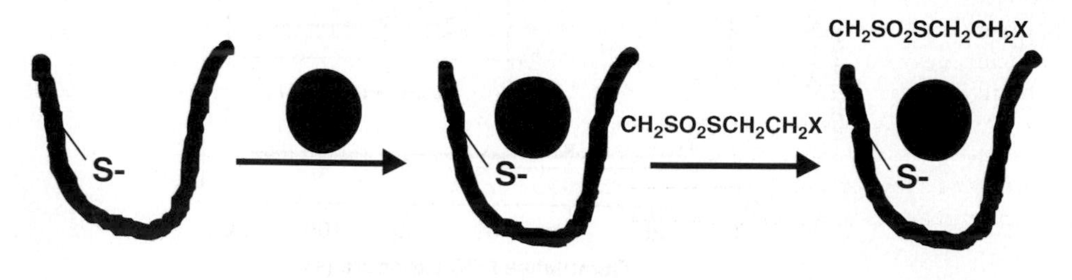

Figure 21-5

Theoretical rate of MTS modification (10 μM) in the absence or presence of agonist (5 × EC_{50}) or antagonist (4 × K_1). Sequential application of MTS reduced the amplitude of subsequent agonist-evoked (EC_{50}) currents. Data were normalized to the current measured at t = 0 for each experiment and plotted as a function of cumulative MTS exposure. Data were fitted to a single exponential function to get a pseudo first-order rate constant (k_1). Second-order rate constants (k_2) were calculated by dividing the pseudo first-order rate constant by the concentration of MTS used. Data points represent the mean ± S.E. for control (●), antagonist (□), or agonist (■), for at least three independent experiments. The calculated second-order rate constants are 12,900 $M^{-1}s^{-1}$ (control, ●), 7,294 $M^{-1}s^{-1}$ (in the presence of antagonist, □), and 3,804 $M^{-1}s^{-1}$ (in the presence of agonist, ■). According to our criteria, this residue would be classified as a binding site residue

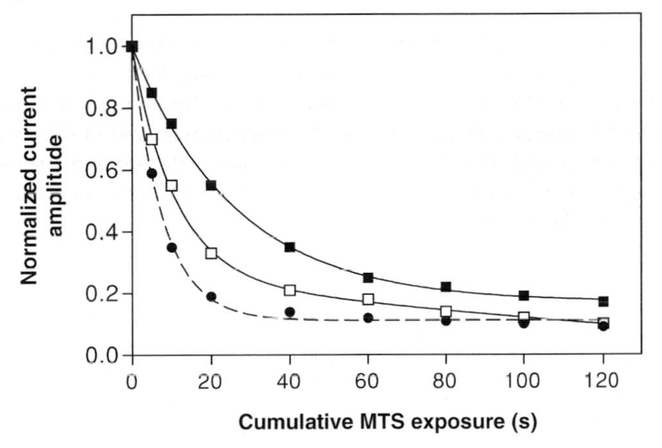

agonists slow the rate of reaction, we infer that "protection" has occurred (*Figure 21-5*). Protection refers to the ability of reversible ligands to impede modification of the introduced cysteine by MTS. The mechanism is likely to be steric hindrance because agonists and antagonists promote different conformational changes within the binding site. When both classes of ligand confer protection, we conclude that the residue at this position lines the core of the binding site. This does not indicate that the residue is a contact point for the ligand. It is equally plausible that the agonist or antagonist may block the access pathway of the MTS reagent to the engineered cysteine. Other mechanisms of protection include changes in electrostatic potentials and allosteric structural changes. If increases in the rate of reaction occur only in the presence of ligand, we conclude that the ligand causes allosteric movements that increase the accessibility of the cysteine.

The concentration of reversible ligand is arbitrary and must be optimized. We generally start with a concentration of agonist equivalent to $5\times EC_{50}$ (Newell and Czajkowski, 2003) and a concentration of antagonist that is $4\times K_1$. We have also used EC_{60}–EC_{95} for GABA and IC_{90}–IC_{95} for the competitive antagonist, SR95531 (Wagner and Czajkowski, 2001; Holden and Czajkowski, 2002). If these concentrations slow covalent modification to the extent that it is impossible to determine an accurate rate over the course of the experiment, they must be reduced. The concentrations of "protectant" should be kept constant for each cysteine mutant in the region (see Section 10.3). Alterations in the maximum effect of modification (plateau) in rate experiments occur occasionally. We have observed these effects when studying agonists (GABA) or modulators (e.g., Ro-15-1788). If the maximum MTS effect returns to control levels on washout of the "protectant" (as a function of time), it may suggest that coapplication of efficacious ligands drives receptors into a (partially) desensitized or blocked/nonconducting state (unpublished observations) that changes accessibility of the cysteine.

7.2 Identification of Binding Site Residues Without Measuring Rates

In lieu of measuring rates of modification, one can examine the ability of reversible ligands to confer protection by coapplication of MTS and a reversible binding site-specific ligand (➋ *Figure 21-6*). In this case, one chooses the lowest concentration of MTS that produces the maximal reduction (Boileau et al., 1999; Torres and Weiss, 2002) or half-maximal reduction (Boileau et al., 2002). High concentrations of "protectant" (e.g., 500 × EC_{50}) are used to measure changes in the extent of modification (Boileau et al., 1999; Boileau et al., 2002; Torres and Weiss, 2002).

Figure 21-6

Identification of binding site residues in the V93C-D101C region of the $GABA_AR$ β_2 subunit. Peak responses to GABA (EC_{50}) of accessible cysteine mutants (V93C, D95C, Y97C, and L99C) were stabilized and cells were exposed to a concentration of MTS that reduced the response by ~50%. Half-maximal inhibition was normalized to the maximum inhibition at each position. The same experiments were carried out in the presence of GABA (~500 × EC_{50}, grey bars) and SR95531 (~10 × K_1, white bars). Data represent the mean ± S.E. for at least three independent experiments. Reprinted from Boileau et al. (2002) with permission from the American Society of Biochemistry and Molecular Biology

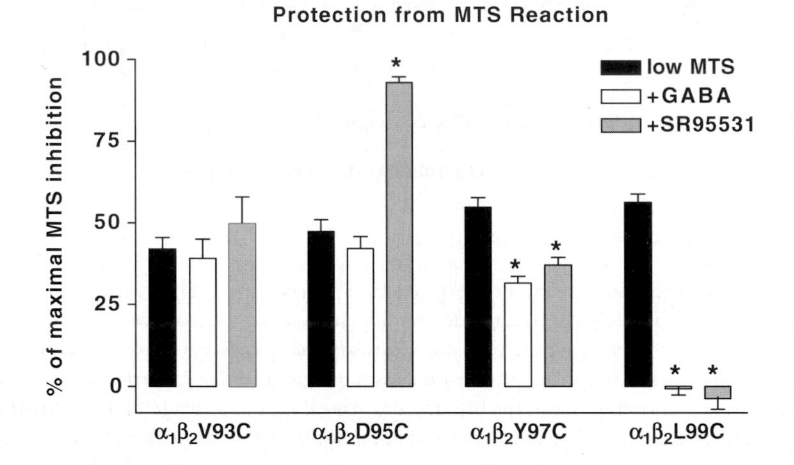

7.3 Detection of Conformational Changes

SCAM is useful not only for binding site studies, but also for detecting structural movements within LGIC. The use of MTS reagents as reporter molecules for distant conformational changes has been described for benzodiazepines, general anesthetics, and barbiturates, all of which modulate $GABA_AR$ function (Wagner and Czajkowski, 2001; Holden and Czajkowski, 2002; Newell and Czajkowski, 2003; Williams and Akabas, 2000). These types of studies provide evidence that ligands that bind at distant sites induce movements that stabilize alternate conformations of the protein. Although reciprocal cross talk between the benzodiazepine and GABA-binding sites of the $GABA_A$ receptors had been described for many years, SCAM analysis of the loop D region of the $GABA_A$ receptor benzodiazepine-binding site (of the γ_2 subunit) provided evidence that GABA was able to accelerate the MTSEA-biotin-mediated modification of γ_2A79C (Teissére and Czajkowski, 2001). The benzodiazepine, diazepam, has been shown to alter the pattern of $pCMBS^-$ accessibility in the M3 region of the $GABA_A$ receptor (Williams and Akabas, 2000). Further, it has also been demonstrated that concentrations of the general anaesthetic, propofol, produce two different patterns of $pCMBS^-$ accessibility within the M3 region of the $GABA_A$ receptor during allosteric modulation and direct gating (Williams and Akabas, 2002). Similarly, barbiturates have been shown to induce structural rearrangements within the GABA-binding site during modulation (Wagner and Czajkowski, 2001; Holden and Czajkowski, 2002) or gating (Wagner and Czajkowski, 2001; Holden and Czajkowski, 2002; Newell and Czajkowski, 2003). Recently, this approach has revealed that the competitive $GABA_AR$ antagonist, SR95531, induces movements within the GABA binding site (Boileau et al., 2002). This marked one of the first observations that competitive antagonists do indeed cause conformational movements within the binding site, contrary to classical pharmacological dogma. SCAM has been used to delineate the importance of the extracellular M2–M3 loop of $GABA_A$ and GlyR in the gating machinery of the receptors as the patterns of modification differ in the absence and presence of high concentrations of agonist (Lynch et al., 2001; Bera et al., 2002). Studies of the channel lining region of nAChR and $GABA_AR$ have used SCAM to characterize movements within the regions in the presence and absence of agonist (Xu and Akabas, 1996; Zhang and Karlin, 1997).

The observed changes in rate of modification/patterns of accessibility can be the result of movements of the introduced cysteine or movements of nearby residues that alter accessibility or ionization of the sulfhydryl. A variant of this approach is to modify the introduced cysteine with sulfhydryl-specific fluorescent reagents that can report changes in environment during gating (Chang and Weiss, 2002). This method requires expensive equipment to monitor changes in function and fluorescence simultaneously (Chang and Weiss, 2002). Although site-specific changes in fluorescence measure conformational changes in real-time, they, like SCAM, cannot determine whether the introduced cysteine or a nearby residue is moving.

8 Interpretations and Limitations of SCAM

8.1 Lack of Functional Effect of Sulfhydryl Modification

As described earlier, SCAM relies on measuring an alteration in ligand binding or function to determine the accessibility of a cysteine to reaction with a sulfhydryl-specific reagent. However, there are occasions when application of an MTS reagent will produce no change in function, suggesting (1) that modification has not occurred or (2) that modification has taken place, but is functionally silent (Boileau et al., 2002). It is plausible that microscopic parameters such as single channel conductance and open probability are reciprocally altered so as to produce no net change in macroscopic function (Cheung and Akabas, 1996). Additional information could be obtained if one could detect the incorporation of the sulfhydryl-reagent directly onto the engineered cysteine. By using MTSEA-biotin, which covalently links $-SCH_2CH_2$ biotin to a cysteine, one can directly determine whether a cysteine substituted residue reacts when using a combination of avidin-bead precipitation and Western blotting (Seal et al., 1998). Given the relatively low protein yields from heterologous expression systems, availability of appropriate antibodies and the labor-intensive nature of the experiments, it is often impractical to detect MTSEA-biotin modification of an entire series of cysteine residues directly by avidin in Western blotting experiments (Javitch et al., 2002). Thus, if a cysteine residue appears to be unreactive, other MTS reagents can be used. For example, modification of the $GABA_AR$ α_1D62C was only achieved using MTSEA and MTSEA-biotin, but not MTSET or MTSES (Holden and Czajkowski, 2002). In a functional assay, it is therefore possible to use the functionally unreactive MTS reagent as a tool to determine whether it prevents the modification of the cysteine by functionally reactive MTS reagents (Holden and Czajkowski, 2002).

8.2 Agonist/Antagonist Subsites

It must also be noted that ligands of divergent chemical structure (agonists and antagonists) likely have different contact points within a common binding site. It is therefore possible to observe protection by one class of ligand, but not with the other. This could be indicative of subsites for agonists and antagonists (Holden and Czajkowski, 2002). A study of the dopamine D_2 receptor illustrated that the affinities of agonists or antagonists were differentially altered following MTSEA modification of Cys^{118}, depending on the chemistry of the ligand. Specifically, the affinity of substituted benzamide antagonists such as YM-09151-2 was decreased 50–2,800 fold, whereas the affinity of antagonists such as N-methyl spiperone was decreased less than sixfold (Javitch et al., 1996).

8.3 Concentration–Response Studies

Several studies have examined the effects of modification on concentration–response curves. This provides information about the mechanisms underlying the functional effects of modification in terms of agonist sensitivity and efficacy (Sullivan and Cohen, 2000; Sullivan et al., 2002; Teissére and Czajkowski, 2001). A reduced maximum response to agonist without changes in sensitivity is indicative of irreversible antagonism. This was observed for MTSEA and MTSET modification of γE57C and αY93C in the nAChR acetylcholine binding site (Sullivan and Cohen, 2000), whereas modification of αP97C produced a tenfold increase in the sensitivity of agonist after MTSET modification (Sullivan et al., 2002). Teissére and Czajkowski (2001) noted that MTSEA-biotin or MTSEA-biotincap modification of γ2A79C of the benzodiazepine binding site shifted the concentration response curve for GABA to the left by 1.5- and 2.6-fold, similar to the degree of leftward shift on application of flurazepam (1 μM). These findings illustrate that MTS reagents can alter the function of a given channel in several ways. The microscopic mechanisms of MTS modification are unknown unless one studies its effects at the single-channel level to determine its effects on conductance, mean open time, and open probability (Cheung and Akabas, 1996).

9 Crystal Structure and SCAM

The SCAM methodology has been validated as a powerful means of determining secondary structure and for identifying crucial functional elements by a study of the 1.85Å crystal structure of the periplasmic domain of the aspartate receptor (Danielson et al., 1997). As a positive control for mapping the unknown structure of helix α2, Danielson et al. (1997) introduced cysteine residues into the Thr-95 to His-103 region, for which the crystal structure had been solved. The pattern of chemical reactivity of the introduced cysteines with the alkylating agent, 5-iodoacetamidofluorescein (IAF), was strongly correlated with the pattern of solvent exposure determined from the crystal structure. This was later confirmed using another sulfhydryl-specific probe, IANBD (Bass and Falke, 1998). In addition, SCAM analysis of the $GABA_AR$ correctly predicted the extended α-helical structure of the M2 domain (Goren et al., 2004) recently seen in Nigel Unwin's 4Å structure of the related nAChR (Miyazawa et al., 2003). Also, our SCAM analysis of the $GABA_AR$ ligand-binding domain determined that the binding site is composed of β-strands and loop structures (Boileau et al., 1999; Wagner and Czajkowski, 2001; Holden and Czajkowski, 2002; Boileau et al., 2002; Newell and Czajkowski, 2003). Our secondary structure prediction and our identification of binding site residues agree well with the published molluscan AChBP crystal structure (Brejc et al., 2001) and homology models of the $GABA_AR$ extracellular ligand-binding domain (Cromer et al., 2002).

10 Protocols

10.1 Patterns of Accessibility: Secondary Structure

In order to determine the pattern of accessibility, we determine the effect of a high concentration of MTS on a functional parameter. We most often use the evoked agonist current ($I_{AGONIST}$) as our "readout."

1. We measure the amplitude of the agonist current (I) as our functional "readout." It is important to begin by determining the mean effective concentration of the neurotransmitter (EC_{50}) for each mutant in the series.

2. We apply agonist (EC_{50}) at regular time intervals (10 min) until $I_{AGONIST}$ varies <10% on 2–3 successive agonist applications.
3. We then allow the oocyte to recover from desentization (3 min) before applying a high concentration of MTSEA-biotin (2 mM) for 2 min, followed by a brief washout period (5 min).
4. We now reapply the same concentration of agonist and assess changes in $I_{AGONIST}$.
5. Finally, we plot the data as a function of position to determine the pattern of accessibility.

10.2 Rate of Methanethiosulfonate Modification

Once the accessible residues have been identified, the rate of covalent modification can be determined (1) to assess the water accessible environment of the cysteine residues, (2) to determine which residues (if any) may line the neurotransmitter binding site; and (3) to monitor structural rearrangements that occur within putative binding sites during agonist/antagonist or allosteric modulator binding.

1. As the application of high MTS concentrations may cause the reaction to go to completion too quickly, one needs to empirically determine the optimal MTS concentration (as described earlier).
2. $I_{AGONIST}$ must be stabilized prior to experimentation. Normally, we use a more rigid criterion for stability for this line of experimentation (i.e., <5% on three successive applications 3 min apart).
3. The sequence of reactions is as follows: at time 0, agonist (EC_{50}) is applied (5 s), followed by a brief period of recovery (25 s); MTS is then applied for 10 s, followed by a recovery period of 2 min and 20 s. This cycle is repeated until the reaction comes to apparent completion (i.e., further application of MTS produces no further alterations in $I_{AGONIST}$).
4. The data are then plotted as a function of cumulative MTS exposure and fit with an exponential equation to obtain the pseudo first-order rate (k_1) and plateau. The plateau achieved with lower MTS concentrations should approach that obtained with the application (2 min) of a high MTS concentration.
5. Given the concentration and time-dependence of MTS modification, several factors may need to be adjusted following initial attempts to determine a rate. This could include adjusting the concentration of the reagent or the duration of application (up to 20 s). We find that the shortest possible duration of agonist application when using an oocyte system is 5 s. The wash times can be adjusted accordingly.
6. The second-order rate constant can then be determined by dividing the pseudo first-order rate (k_1) by the molar concentration of the MTS reagent applied. This value corrects for the concentration-dependence of the effect and permits comparisons about the nature of the environment within the series of mutations.
7. Multiple MTS concentrations should be used to ensure that the concentration of MTS does not change significantly during the reaction, and that we can determine an accurate pseudo first-order rate constant.

10.3 The Protection Assay

The protection assay is carried out in much the same way as the control rate, the primary difference being that an agonist/antagonist/modulator is coapplied with the MTS reagent. The purpose of this kind of experiment is to determine if any such ligand will alter the rate of reaction.

1. $I_{AGONIST}$ must be stabilized prior to experimentation. As this experiment requires use of a "protectant," stabilization must be carried out in the presence of the drug to ensure that any changes in $I_{AGONIST}$ are attributed to MTS. To stabilize, agonist (EC_{50}) is applied (5 s), followed by a brief period of recovery (25 s); the reversible binding site specific ligand (protectant) is then applied for 10 s, followed by a recovery period of 2 min and 20 s. This cycle is repeated until stability is achieved. (i.e., <5% on three successive applications of test concentration).

2. The sequence of reactions is as follows: at time 0, agonist (EC_{50}) is applied (5 s), followed by a brief period of recovery (25 s); MTS and the other ligand are then co-applied for 10 s, followed by a recovery period of 2 min and 20 s. This cycle is repeated until the reaction comes to apparent completion (i.e., further application of MTS produces no further alterations in $I_{AGONIST}$). Depending on the concentration of protectant, the wash times may need to be adjusted accordingly.
3. The data are then analyzed as described earlier in Section 10.2.

11 Permissions

All figures and tables were adapted and reprinted with permission from the American Society for Biochemistry and Molecular Biology in format "Other Book" through Copyright Clearance Center.

Acknowledgements

We wish to thank Drs. Jessica Holden, Jeremy Teissére, David Wagner, and Andrew Boileau for their invaluable discussion on SCAM and its applications.

References

Akabas MH, Kaufmann C, Archdeacon P, Karlin A. 1994. Identification of acetylcholine receptor channel-lining residues in the entire M2 segment of the α subunit. Neuron 13: 919-927.

Akabas MH, Karlin A. 1995. Identification of acetylcholine receptor channel lining residues in the M1 segment of the α subunit. Biochemistry 34: 12496-12500.

Bali M, Akabas MH. 2004. Defining the propofol binding site location on the $GABA_A$ receptor. Mol Pharmacol 65: 68-76.

Bass RB, Falke JJ. 1998. Detection of a converved α-helix in the kinase-docking region of the aspartate receptor by cysteine and disulfide scanning. J Biol Chem 273: 25006-25014.

Bayley H, Knowles JR. 1977. Photoaffinity labelling. Methods Enzymol 46: 69-115.

Bera AK, Chatav M, Akabas MH. 2002. $GABA_A$ receptor M2–M3 loop secondary structure and changes in accessibility during channel gating. J Biol Chem 277: 43002-43010.

Boileau AJ, Evers AM, Davis AF, Czajkowski C. 1999. Mapping the agonist binding site of the $GABA_A$ receptor: evidence for a β-strand. J Neurosci 19: 4847-4854.

Boileau AJ, Newell JG, Czajkowski C. 2002. $GABA_A$ receptor β_2 Tyr^{97} and Leu^{99} line the GABA-binding site: insights into the mechanisms of agonist and antagonist actions. J Biol Chem 277: 2931-2937.

Brejc K, van Dijk WJ, Klaassen RV, Schuurmans M, van Der Oost J, Smit AB, Sixma TK. 2001. Crystal structure of an ACh-binding protein reveals the ligand-binding domain of nicotinic receptors. Nature 411: 269-276.

Chang Y, Weiss DS. 2002. Site-specific fluorescence reveals distinct structural changes with GABA receptor activation and antagonism. Nat Neurosci 5: 1162-1168.

Changeux JP, Edelstein SJ. 1998. Allosteric receptors after 30 years. Neuron 21: 959-980.

Cheung M, Akabas MH. 1996. Identification of the cystic fibrosis transmembrane conductance regulator channel-lining residues in and flanking the M6 membrane-spanning segment. Biophys J 70: 2688-2695.

Chiara DC, Cohen JB. 1997. Identification of amino acids contributing to high and low affinity d-tubocurarine sites in the *Torpedo* nicotinic acetylcholine receptor. J Bio Chem 273: 32940-32950.

Chiara DC, Middleton RE, Cohen JB. 1998. Identification of tryptophan 55 as the primary site of [^{3}H] nicotine photoincorporation in the γ-subunit of the Torpedo nicotinic acetylcholine receptor. FEBS Lett 423: 223-226.

Chowdry V, Westheimer FH. 1979. Photoaffinity labelling of biological systems. Annu Rev Biochem 48: 293-3

Cohen JB, Sharp SD, Liu WS. 1991. Structure of the agonist-binding site of the nicotinic acetylcholine receptor. [^{3}H] acetylcholine mustard identified residues in the cation binding subsite. J Biol Chem 266: 23354-23364.

Colquhoun D. 1998. Binding, gating and efficacy: the interpretation of structure–activity relationships for agonists and of the effects of mutating receptors. Br J Pharmacol 125: 925-947.

Cromer BA, Morton CJ, Parker MW. 2002. Anxiety over $GABA_A$ structure relieved by AChBP. Trends Biochem Sci 27: 280-287.

Danielson MA, Bass RB, Falke JJ. 1997. Cysteine and disulfide scanning reveals a regulatory α-helix in the cytoplasmic domain of the aspartate receptor. J Biol Chem 272: 32878-32888.

Dormán G, Prestwich GD. 2000. Using photolabile ligands in drug discovery and development. Trends Biotechnol 18: 64-77.

Duncalfe LL, Carpenter MR, Smillie LB, Martin IL, Dunn SMJ. 1996. The major site of photoaffinity labelling of the γ-aminobutyric acid type A receptor by $[^3H]$ flunitrazepam is histidine 102 of the α subunit. J Biol Chem 271: 9209-9214.

Goren EN, Reeves DC, Akabas MH. 2004. Loose protein packing around the extracellular half of the $GABA_A$ receptor β1 subunit M2 channel-lining segment. J Biol Chem 279: 11198-111205.

Grutter T, Ehret-Sabatier L, Kotzyba-Hibert F, Goeldner M. 2000. Photoaffinity labelling of Torpedo nicotinic receptor with the agonist $[^3H]$ DCTA: identification of amino acid residues which contribute to the binding of the ester moiety of acetylcholine. Biochemistry 39: 3034-3043.

Guillory RJ, Jeng SJ. 1983. Photolabeling: theory and practice. Fed Proc 42: 2826-2830.

Holden JH, Czajkowski C. 2002. Different residues in the $GABA_A$ receptor α_1 T60-α_1K70 region medicate GABA and SR-95531 actions. J Biol Chem 277: 18785-18792.

Holmgren M, Liu Y, Yellen G. 1996. On the use of thiol-modifying agents to determine channel topology. Neuropharmacology 35: 797-804.

Javitch JA, Shi L, Liapkis G. 2002. Use of the substituted cysteine accessibility method to study the structure and function of G protein-coupled receptors. Methods Enzymol 343: 137-156.

Javitch JA, Fu D, Chen J. 1996. Differentiating dopamine D_2 ligands by their sensitivities to modification of the cysteine exposed in the binding-site crevice. Mol Pharmacol 49: 692-698.

Karlin A, Akabas M. 1998. Substituted-cysteine accessibility method. Methods Enzymol 293: 123-144.

Lynch JW, Reena Han N-L, Haddrill J, Pierce KD, Schofield PR. 2001. The surface accessibility of the glycine receptor M2–M3 loop is increased in the channel open state. J Neurosci 21: 2589-2599.

Middleton RE, Cohen JB. 1991. Mapping of the acetylcholine binding site of the nicotinic acetylcholine receptor: $[^3H]$ nicotine as an agonist photoaffinity label. Biochemistry 30: 6987-6997.

Miyazawa A, Fujiyoshi F, Unwin N. 2003. Structure and gating mechanisms of the acetylcholine receptor pore. Nature 423: 949-955.

Newell JG, Czajkowski C. 2003. The $GABA_A$ receptor α_1 subunit Pro^{174}-Asp^{191} segment is involved in GABA binding and channel gating. J Biol Chem 278: 13166-13172.

Newell JG, McDevitt RA, Czajkowski C. 2004. Mutation of Glutamate 155 of the $GABA_A$ receptor β_2 subunit produces a spontaneously open channel: a trigger for channel activation. J Neurosci. 24: 11226-35.

Pascual JM, Karlin A. 1998. State-dependent accessibility and electrostatic potential in the channel of the acetylcholine receptor: Inferences from the rates of reaction of thiosulfonates with substituted cysteines in the M2 segment of the α subunit. J Gen Physiol 111: 717-739.

Reeves DC, Goren EN, Akabas MH, Lummis SCR. 2001. Structural and electrostatic properties of the $5\text{-}HT_3$ receptor pore revealed by substituted cysteine accessibility mutagenesis. J Biol Chem 276: 42035-42042.

Roberts DD, Lewis DD, Ballou DP, Oldon ST, Shafer JA. 1986. Reactivity of small thiolate anions and cuysteine-25 in papain toward methyl Methanethiosulfonate. Biochemistry 25: 5595-5601.

Sawyer GW, Chiara DC, Olsen RW, Cohen JB. 2002. Identification of the bovine γ-aminobutyric acid type A receptor a subunit residues photolabeled by the imidazobenzodiazepine $[^3H]$ Ro15-4513. J Biol Chem 51: 50036-50045.

Seal RP, Leighton BH, Amara SG. 1998. Transmembrane topology mapping using biotin-containing sulfhydryl reagents. Methods Enzymol 296: 318-331.

Sine SM, Wang HL, Bren NH. 2002. Lysine scanning mutagenesis delinates structural model of the nicotinic receptor ligand binding domain. J Biol Chem 277: 29210-292

Smith GB, Olsen RW. 1994. Identification of a $[^3H]$muscimol photoaffinity substrate in the bovine γ-aminobutyric $acid_A$ receptor α subunit. J Biol Chem 269: 20380-20387.

Staufer DA, Karlin A. 1994. Electrostatic potential of the acetylcholine binding sites in the nicotinic receptors probed by reactions of binding-site cysteines with charged methanethiosulfonates. Biochemistry 33: 6840-6849.

Sullivan D, Cohen JB. 2000. Mapping the agonist binding site of the nicotinic acetylcholine receptor: orientation requirements for activation by covalent agonist. J Biol Chem 275: 12651-12660.

Sullivan D, Chiara DC, Cohen JB. 2002. Mapping the agonist binding site of the nicotinic acetylcholine receptor by cysteine scanning mutagenesis: antagonist footprint and secondary structure predictions. Mol Pharmacol 61: 463-472.

Teissére JA, Czajkowski C. 2001. A β-strand in the γ_2 subunit lines the benzodiazepine binding site of the $GABA_A$

receptor: structural requirements detected during channel gating. J Neurosci 21: 4977-4986.

Torres VI, Weiss DS. 2002. Identification of a tyrosine in the agonist binding site of the homomeric ρl γ-aminobutyric acid (GABA) receptor that, when mutated, produces spontaneous opening. J Biol Chem 277: 43471-43748.

Wagner DA, Czajkowski C. 2001. Structure and dynamics of the GABA binding pocket: a narrowing cleft that constricts during activation. J Neurosci 21: 67-74.

Williams DB, Akabas MH. 1999. γ-Aminobutyric acid increases the water accessibility of M3 membrane-spanning segment residues in γ-aminobutyric acid type A receptors. Biophys J 77: 2563-2574.

Williams DB, Akabas MH. 2000. Benzodiazepines induce a conformational change in the region of the γ-aminobutyric acid type A receptor α_1-subunit M3 membrane-spanning segment. Mol Pharmacol 58: 1129-1138.

Williams DB, Akabas MH. 2002. Structural evidence that propofol stabilizes different $GABA_A$ receptor states at potentiating and activating concentrations. J Neurosci 22: 7417-7424.

Xu M, Akabas MH. 1996. Amino acids lining the channel of the γ-aminobutyric acid type A receptor identified by cysteine substitution. J Biol Chem 268: 21505-21508.

Yan D, Schulte MK, Bloom KE, While MM. 1999. Structural features of the ligand-binding domain of the serotonin $5HT_3$ receptor. J Biol Chem 274: 5537-5541.

Zhang H, Karlin A. 1997. Identification of acetylcholine receptor channel-lining residues in the MI segment of the beta-subunit β-subunit. Biochemistry 36: 15856-15864.

Zhang H, Karlin A. 1998. Contribution of the β-subunit M2 segment to the ion-conducting pathway of the acetylcholine receptors. Biochemistry 37: 7952-7964.

22 Protein X-Ray Crystallography

D. A. R. Sanders

Abstract: Neuroscientists may wish to examine the structure of a protein to aid in characterizing function or for mechanistic/design purposes. In order to do so, the researcher must have an understanding of what is required to determine a protein structure and how to interpret structures. Furthermore, such an understanding is vital if structural information is to be used in a research program. This chapter outlines the general process by which structural information can be obtained using protein X-ray crystallography and the type of information that it is possible to obtain.

1 Introduction

Protein X-ray crystallography is becoming a fundamental technique in all disciplines of the biological sciences. The increase in our knowledge of genomes, and our increasing ability to clone and purify any protein of choice have opened up protein structure determination to all biological disciplines. As the field of neuroscience moves towards the molecular level for understanding the function of receptors, their agonists/inhibitors, and their cellular effects, the ability to visualize receptors and other important proteins at the atomic level is becoming increasingly important.

This chapter is not intended for experienced crystallographers, rather it is for those who are wondering what the crystallographers down the hall are doing, and whether they can be of any use to their research. It also offers advice for those who are considering a crystallographic project. Although ultimately collaboration with an expert in the field of crystallography will be the likely route for most scientists, an understanding of the technique will increase the productivity of such collaborations. While crystallography does require a certain level of expertise, there are a number of aspects that can be initiated in any laboratory setting. In addition, with the increase in commercial companies catering to crystallography and the increasing power of the average laboratory PC, more and more aspects of crystallography are becoming accessible to all laboratories.

Trying to cover the theory behind crystallography, or even merely diffraction, would require a much larger chapter, or even a book, of which there are a number of excellent choices. For an in-depth understanding of the theory behind crystallography, readers are encouraged to investigate the texts listed at the end of this chapter, although it is hoped that this chapter will give scientists an understanding of what is required to carry out crystallographic experiments, and a basic understanding of how to interpret results and to view structures.

1.1 Is Crystallography for You?

Is protein X-ray crystallography a technique that can be used in your laboratory? The easiest way to answer this question is to discuss what crystallography can do and the information that can (and cannot) be extracted from crystal structures.

Protein X-ray crystallography gives a snapshot of the structure of a protein as it exists in a crystal. This technique provides a complete and unambiguous three-dimensional (3-D) representation of a protein molecule. It is important to note that the model generated from a crystallographic study is a static or time-averaged view of the molecular structure. Information about molecular motions can be obtained from precise diffraction data; however, the motions of molecules within a crystal are usually severely restricted in comparison to the motions of molecules in solution.

1.1.1 Identification of the Function of Novel Proteins

Proteins can be classed into groups based on their overall 3-D shapes, known as protein folds (➔ *Figure 22-1a*). In general, proteins that have similar functions have similar folds. This means that if you are the proud parent of an unknown protein whose structure is solved, it may be possible to make educated guesses as to the function of the protein based on its overall fold. There are a number of well-known exceptions to this [notably, the serine protease family, subtilisin and trypsin/chymotrypsin (Hartley, 1979)], but the

■ Figure 22-1

(a) Structures of two proteins unrelated by sequences (12% identity), but related by fold. The protein on the left is ActVA-Orf6, an enzyme involved in actinorhodin biosynthesis [(PDB code: 1LQ9) (Sciara et al., 2003)]. The protein on the left is from PA3566, a protein of unknown function from *Pseudomonas aureginosa* (unpublished results; PDB code: 1X7V). (b) Binding of 1-2-dimannose (arrowed) to Concanavalin A, showing residues involved in binding [(PDB code: 1I3H) (Sanders et al., 2001a, b)]

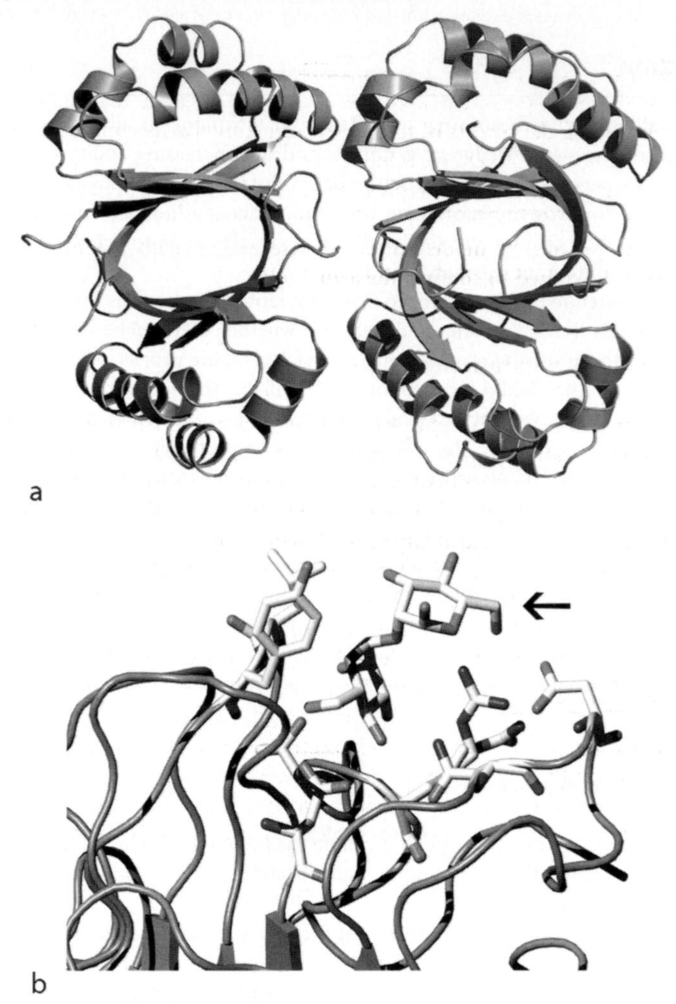

so-called protein structural super-families will likely offer clues as to function. For example, the so-called TIM-barrel fold is found in a large family of proteins that are all enzymes (Wise and Rayment, 2004), so a novel protein with a TIM-barrel fold is likely to be an enzyme. However, enzymes containing the TIM-barrel fold catalyze a diverse variety of reactions. Additional information about substrates/reactions can be gathered from a closer examination of potential active sites, but separating out closely related activities by examining protein structures without the actual ligands bound is not likely to be successful. It will allow you to narrow down the functional experiments that should be investigated.

There are numerous folds for which overall functions are not well characterized, and many folds are reused for different functions; so relying on a protein fold to determine function will not work. This approach can only be useful as a guide to further experiments.

1.1.2 Mechanisms

Understanding how enzymes catalyze their reactions, how receptors bind to their ligands and pass their signals onwards, and how proteins interact with each other is the most commonly used and greatest strength of protein crystallography. Crystallography allows identification of the amino acids that are critical to binding or activity. Once residues have been identified, they can be altered to confirm their roles in protein function. Proceeding in this way will dramatically improve the ability to choose amino acids for mutagenesis.

There are many examples of enzyme mechanisms and protein/ligand-binding studies that have been carried out using protein crystallography. Two of the earliest and best-known examples are: the mechanism of serine proteases, including the aspects of the protein structure that determines specificity (Steitz et al., 1969; Ding et al., 1994); and the cooperative binding of oxygen to hemoglobin, with the structural changes that occur during that process (Case and Karplus, 1979; Gelin et al., 1983; Ackers and Hazzard, 1993).

The use of crystallography for identification of key residues of proteins is a well-established technique. Crystallography merely provides a model for how these amino acids are involved in function, and biochemical tests are still required to confirm the actual roles.

1.1.3 Inhibitor Design/Modification

Protein crystallography allows the researcher to obtain a snapshot of the way a protein is binding to an inhibitor or agonist (or natural substrate). Not only does this allow identification of important amino acid residues for binding, it also allows identification of important features of the substrate for binding (➋ *Figure 22-1b*). Once an active site or binding site has been identified, it is possible to ether screen libraries of small molecules for potential ability to bind to the protein. In theory, it would be possible to use a small molecule to screen protein libraries for potential binding partners; however, this would require enormous computing resources and is probably still out of reach as a usable technique.

Often, crystallography is used as a tool for improving inhibitors that already exist. It can be quite difficult to rationally design an inhibitor from scratch, but improvements to known inhibitors can be easily and rapidly modeled. Potential improved inhibitors can then be designed and tested. This can be an iterative process with subsequent rounds of inhibitors designed from each subsequent improvement.

Currently, this is a major application of protein crystallography in most of the major drug companies. One of the best examples of this approach is the design of inhibitors for HIV protease (Dash et al., 2003). In brief, once the 3-D structure of HIV protease was determined, the active site was identified and used to screen small molecule libraries for potential compounds that could bind to HIV protease. These compounds were then tested for their ability to inhibit the protease. Lead compounds were then used to iteratively improve the inhibitors, using crystallographic studies, computational modeling, and biochemical tests.

2 Theory

The theory behind protein X-ray crystallography is probably of limited interest to the majority of readers of this chapter, and this text will not go into a great deal of depth on the theory behind this technique. Only theory sufficient to discuss the practical side of crystallography will be introduced. For the few who are interested in a more in-depth understanding of the theory, there are many excellent texts that deal with this subject in great detail. Some of these texts are listed under ➋ Section 5.

2.1 X-Ray Diffraction

Crystallography is in many ways analogous to light microscopy. Light of a particular wavelength is shone onto the object to be examined. The object causes the light to diffract and the diffracted light is then focused with a lens to show the image. In the case of X-ray crystallography, the light used is X-ray radiation (a wavelength of about 0.1 nm or 1 Å). The diffraction is caused by the electron clouds around each atom, and the "lens" that is used to refocus the image is a computer that, with some luck, can reconstitute the diffraction pattern into a usable image. Light microscopy is useful for resolving particles with diameters in the order of 0.2 mm; whereas X-rays can be used for resolving atoms that are 1–2 Å (0.1–0.2 nm) apart, equivalent to a 10^6 increase in magnification. This is based on Bragg's Law (➋ Eq. 22.1):

$$2d\sin\theta = \lambda \tag{1}$$

where d is the minimum spacing between the diffracting planes that can be resolved at wavelength, λ, and theta (θ) is the angle of diffraction. While this is the theoretical limit for resolution, most protein crystals are much more limited in resolution.

2.1.1 Generation of X-Rays

X-rays are commonly generated in three ways; using a sealed tube, a rotating anode, or a synchrotron source. The use of sealed tubes is very uncommon, as both rotating anodes and synchrotron sources have greatly superior performance; sealed tubes are thus not discussed here.

Rotating anode X-ray generators are used in "in-house" X-ray generators, commonly using copper as the metal target for the production of X-rays. A large amount of heat is generated with the production of X-rays from a metal target, and this metal "anode" is continually rotated to minimize the heat buildup on any single part of the anode, and cooled with water. The wavelength of X-rays generated from a rotating anode is fixed at 1.54 Å. This is acceptable for most general crystallography requirements, but for more modern techniques for solving structures, a "tunable" X-ray source is required.

Synchrotron radiation sources are currently the instruments of choice for collecting X-ray data from protein crystals. Synchrotrons are actually electron storage rings. The electron beam in this ring travels around the storage ring at very high speeds, directed by a series of "bending magnets." The X-ray radiation is generated when the paths of these electrons are turned by the magnets, or oscillated between a series of magnets called insertion devices, often referred to as undulators or "wigglers." There are a number of advantages for using synchrotron radiation instead of an in-house source. The two major ones are intensity of radiation and the "tunability" of the wavelength.

The X-rays that are emitted by an electron storage ring come out as a "white" beam, spanning a large range of wavelengths. This beam can be filtered down to a single wavelength through the use of different prisms and mirrors. Because of the strength of the X-ray beam, a very narrow wavelength can be chosen while still maintaining a very intense signal. The intensity of the beam means that there is a great increase in the strength of the diffraction signal when compared to a rotating anode X-ray generator. This means that more and better data can be collected in a much shorter time period.

Since the X-rays are emitted at a large number of wavelengths, the actual wavelength that is used can be adjusted through the use of monochromators. This allows the wavelength to be adjusted according to the desires of the user. This becomes important for determining de novo protein structures using techniques such as multiwavelength anomalous dispersion (MAD) discussed in ➋ Section 2.3.

2.1.2 Diffraction

Light waves, including X-rays, can be diffracted by objects, in the same way that diffraction of water waves is observed when a wooden block is placed in a ripple tank. The resulting diffracted waves interact with other waves to give a summed diffracted wave. In brief, the resulting diffracted wave is a summation of the constructive and destructive interference caused by a series of waves that are shifted slightly with respect to each other. The degree of the shift is known as the *phase angle* and is a critical value that needs to be determined.

In crystallography, the diffraction of the individual atoms within the crystal interacts with the diffracted waves from the crystal, or *reciprocal lattice.* This lattice represents all the points in the crystal (x,y,z) as points in the reciprocal lattice (h,k,l). The result is that a crystal gives a diffraction pattern only at the lattice points of the crystal (actually the reciprocal lattice points) (*Figure 22-2*). The positions of the spots or reflections on the image are determined by the dimensions of the crystal lattice. The intensity of each spot is determined by the nature and arrangement of the atoms with the smallest unit, the unit cell. Every diffracted beam that results in a reflection is made up of beams diffracted from all the atoms within the unit cell, and the intensity of each spot can be calculated from the sum of all the waves diffracted from all the atoms. Therefore, the intensity of each reflection contains information about the entire atomic structure within the unit cell.

Figure 22-2
Diffraction pattern of a typical protein

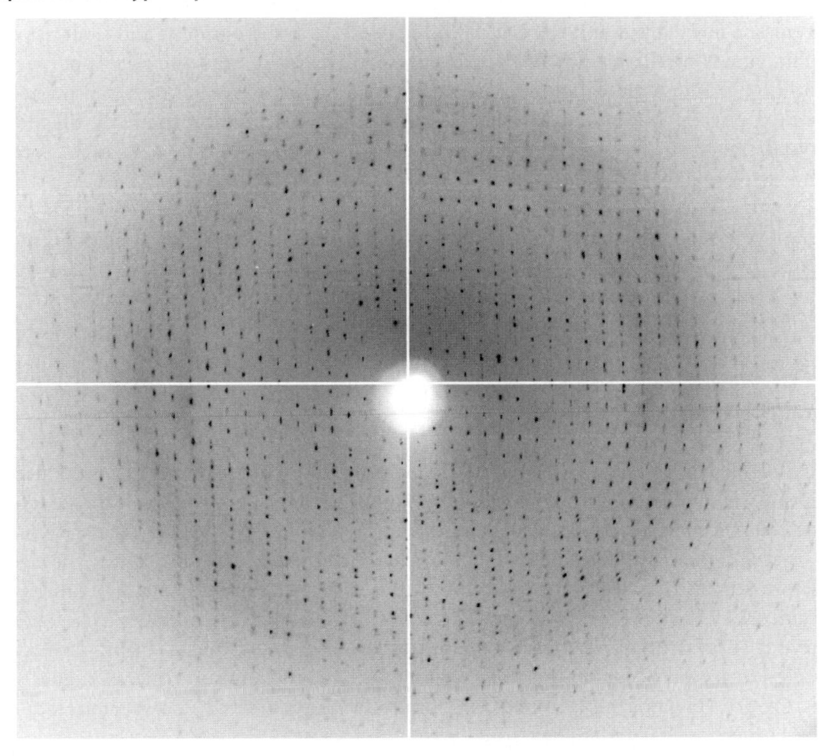

We wish to obtain an image of the scattering elements in three dimensions (the electron density). To do this, we perform a 3-D Fourier synthesis (summation). Fourier series are used because they can be applied to a regular periodic function; crystals are regular periodic distributions of atoms. The Fourier synthesis is given in Eq. 22.2:

$$\rho(\mathrm{x\,y\,z}) = 1/V\,\Sigma\Sigma\Sigma\,|F(\mathrm{h\,k\,l})|\cos[2\pi(\mathrm{hx}+\mathrm{ky}+\mathrm{lz}) - \alpha(\mathrm{h\,k\,l})] \quad (2)$$

ρ(x y z) refers to the electron density at any given (x y z) point, V is the volume of the cell, F(h k l) is the structure factor [the amplitude of the diffracted wave—proportional to the square root of the intensity (I^2)] at a reciprocal lattice point h k l, and α is the phase angle.

Although the intensities of the spots can be easily measured experimentally, the phases cannot. This is the so-called "phase problem" of crystallography. Determining an estimate of the phases is critical to solving structures, but must be done by one of the several second-hand techniques, outlined later (Section 2.3).

2.2 Crystallization

The electron cloud of protein atoms does not diffract X-rays very well. Only a small fraction of the X-ray beam will be diffracted by any atom. To increase the ability of proteins to diffract X-rays, we make use of protein crystals to amplify the number of X-rays diffracted, and hence the diffraction signal. For example, a crystal with dimensions of $0.3 \times 0.3 \times 0.3$ mm^3 contains approximately 10^{15} copies of an average-sized protein, which gives a corresponding amplification to the X-ray diffraction signal.

Crystals arise from molecules (in our case proteins) aggregating into a regular arrangement of molecules, precipitating in the crystalline form. Crystals are characterized by a high degree of internal order. This enables the diffraction of X-rays to become an additive phenomenon. The high degree of internal order means that a crystal can be broken down into small repeating units. The smallest unit repeated throughout the crystal is known as the *unit cell.* The unit cell is the region that will be examined by crystallographic studies, as the reflection generated will contain information from the entire unit cell. Unit cells do not necessarily have to contain only a single protein. Many unit cells contain multiple copies of protein molecules. Knowing the unit cell is a critical portion of crystallography, and is always described in crystallographic papers by the lengths of the cell edges or axis (a,b,c) and the angles between the edges (α,β,γ). In addition, although these unit cells are packed together in repeating units, they do not necessarily have to be packed directly on top of each other. There are many orientations that will still allow a regular, repeating crystal structure. The way the unit cells are arranged with respect to each is called crystal symmetry and is referred to as the *space group* of the crystal.

Crystal growth is the most difficult portion of crystallography. Because how protein crystals grow, and predicting the conditions that will promote crystal growth are not very well understood. Crystal growth contains three separate events that follow one another and either allow or disallow crystal growth. Understanding how the three events relate to one another and potential ways of manipulating these events are critical for designing crystallization experiments. The three events are nucleation, growth, and cessation.

2.2.1 Crystal Nucleation

To form crystals, a seed or *nucleus* must be formed to act as a scaffold for growth. How protein crystal nuclei are formed is not well understood. In small molecules, nuclei are formed when the surface tension energy barrier is crossed. The energy barrier is much easier to cross at high levels of supersaturation. We assume that protein crystals behave much in the same fashion as small molecules and therefore the classic strategy for crystallization studies is to bring the protein solution to a supersaturated state. A solubility diagram, such as ➋ *Figure 22-3*, is useful for understanding crystallization. A protein that is in a region above the solubility curve is supersaturated and may spontaneously form crystal nuclei. In practice, the optimum region for nuclei formation occurs in the *labile* region of the supersaturation zone. At higher supersaturation than this, the protein tends to precipitate as amorphous aggregates, and at lower supersaturation levels, the surface tension energy barrier is not crossed (Feher and Kam, 1985).

2.2.2 Crystal Growth

Crystal growth will spontaneously occur in the supersaturated region of the solubility diagram. In this region, the solution is still in a supersaturated state and the protein will still be driven towards coming out of solution. Crystal growth occurs at "steps" in the crystal surface, since binding energies are higher at these interfaces than on flat surfaces. These steps occur either at locations of defects in the crystal packing arrangement, or by the formation of a new nucleus on the crystal surface. The best region for crystal growth is the *metastable* region, as this will limit the formation of new nuclei and will therefore promote fewer crystals to grow to larger sizes rather than numerous, small crystals. In addition, the lower degree of supersaturation in this region will promote slow crystal growth, which will aid in larger and better crystal formation (Feher and Kam, 1985).

Figure 22-3
Theoretical protein solubility diagram

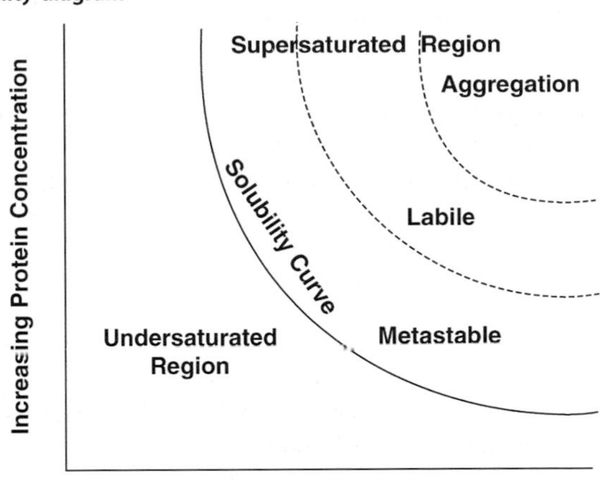

2.2.3 Cessation of Growth

The cessation of crystal growth is not well understood. One cause is likely to be that as the crystal grows, the solution moves towards the undersaturated region of the protein solubility curve. When the solution reaches the solubility curve, it reaches equilibrium and no further growth will be seen. This is not the only reason for cessation of growth, as crystals moved into a new, supersaturated solution may not continue to grow. It is likely that errors incorporated into the crystal during growth eventually reach a critical point and these defects or surface impurities eventually limit crystal growth (Feher and Kam, 1985).

2.3 Solving Structures

Once a diffraction dataset of a crystal has been collected, these data have to be retranslated back into an electron density map using the "lens" of computers. Remember in ➲ Eq. 22.1 (➲ Section 2.1.2) we discussed the phase angle (α), or phase. Determining the phase of each structure factor is critical to solving a structure, which is a nontrivial exercise, as they cannot be measured directly. This is the so-called "phase problem" of X-ray crystallography. Rather than determine what the exact phase angles for each structure factor are, crystallographers use a variety of techniques to estimate the phases. Once a good estimate is made, the phases are then refined by building a model and comparing the model with the experimental data. This section briefly outlines the three major ways that are currently in use for estimating the phases.

The method we will not spend much time discussing, as it is currently not of much use in protein crystallography, is the direct method. This is the method of choice for determining the structures of small molecules, but as yet is only of limited value in protein crystallography. Much work is being carried out on this particular problem and advances have been made, but are not enough to make it practical for most protein crystallography applications.

The three methods discussed are:

1. Molecular replacement (MR): A similar, known structure is used as a model for the unknown protein.
2. Isomorphous replacement: Heavy atoms (atoms with high atomic numbers, often metals) are attached to the protein molecules.
3. Multiwavelength anomalous diffraction: Makes use of atoms within a protein that diffract X-rays anomalously. Commonly uses seleno-methionine residues in place of methionines.

Once the phase problem is solved, a model of the protein can be built and "refined," so that the model best fits the experimental data. A refined model will accurately represent the positions of the atoms in the unit cell.

2.3.1 Molecular Replacement

The model protein is used to search the crystal space until an approximate location is found. This is, in a simplistic way, analogous to the child's game of blocks of differing shapes and matching holes. Classical molecular replacement does this in two steps. The first step is a rotation search. Simplistically, the orientation of a molecule can be described by the vectors between the points in the molecule; this is known as a Patterson function or map. The vector lengths and directions will be unique to a given orientation, and will be independent of physical location. The rotation search tries to match the vectors of the search model to the vectors of the unknown protein. Once the proper orientation is determined, the second step, the translational search, can be carried out. The translation search moves the properly oriented model through all the 3-D space until it finds the proper "hole" to fit in.

2.3.2 Isomorphous Replacement (Heavy Atom/Metal Derivatives)

A diffraction dataset from a crystal containing heavy metals is compared with the diffraction data from a crystal containing no heavy atoms. Since the heavy metals, commonly platinum, mercury, or gold, contain many more electrons than the other atoms in the protein, they diffract the X-rays much better and so have a large effect on the intensities of the reflections. The two datasets are subtracted and the only information left should be the diffraction from the heavy metals. If enough heavy metals are present, this information can be used to get a reasonable estimate of the phases. These phases can then be used to estimate the phases for the rest of the protein. Because the critical information is the difference between the intensities of the native crystal and the derivative crystals, the crystals must have the same unit-cell dimensions and space group (isomorphous). This can be done using either in-house X-ray generators or synchrotrons.

2.3.3 Multiwavelength Anomalous Diffraction

In most diffraction experiments, any reflection has a symmetry mate, known as Friedel (or Bijvoet) pairs. The pairs are a reflection at (h k l) and its mirror image at (–h –k –l). At certain wavelengths, atoms diffract the X-rays "anomalously." In other words, the Friedel pairs are no longer equal. This is due to atoms absorbing the radiation at particular wavelengths. All atoms exhibit this phenomenon, but not all atoms do this either at wavelengths that are useful, or have a large enough absorption component to be detectable. This absorption component results in a slight shift in the phase of the anomalous scatterer, compared with the other atoms. This slight phase shift can be used to estimate the phases of the atoms. Commonly, the experiments are carried out at three wavelengths, thus making a tunable X-ray source (a synchrotron) essential for these experiments.

2.3.4 Density Modification

Often the initial phases from one of the above techniques (particularly MIR and MAD) are not sufficiently accurate to build in all the atoms and further improvement of the phases must be carried out. The standard approach for this is called density modification, or solvent-flattening. The idea is that if a unit cell contains a sufficient amount of solvent in it (usually around 50%, which is reasonable for most protein crystals), you will be able to use this information to improve the phases for the protein. The region of the unit cell that contains the solvent is set to an average value and new structure factors and phases are then calculated for this "solvent-flattened" map (known as map inversion or solvent flipping." This is a very powerful tool for improving phases, and the benefits increase as the degree of solvent increases (Wang, 1985).

2.3.5 Refinement

Once the initial phases have been estimated, a model must be built and corrected to give the best fit of the model to the experimental data. Refinement is an iterative process, cycling between mathematical adjustments to the protein model and examination of the model versus the experimental data and manual rebuilding.

The mathematical portion of refinement consists of adjusting the positional parameters of the model and the "temperature factor (B value)" for each of the atoms in the model. The agreement between the model and the experimental data is a value known as the *R-factor.* The lower the *R*-factor, the better the agreement between the model and data. A completely random structure will have an *R*-factor of 59% (Wilson, 1950), while a fully refined structure will have an *R*-factor (usually) less than 25% and often less than 20%. The number of observations (reflections) versus the number of parameters to be adjusted is often in the order of 3:1, which is a low value of overdetermination. Crystallographers get around this problem by incorporating restraints based on knowledge of chemical bonds and angles as well as any other information available. Accurate details of bond distances and angles have been determined by small molecule crystallography for a large number of organic molecules and even small peptides. Proteins generally conform to the same bond characteristics as other molecules.

Mathematically, all this information is used to calculate the "best fit" of the model to the experimental data. Two techniques are currently used, least squares and maximum likelihood. Least-squares refinement is the same mathematical approach that is used to fit the best line through a number of points, so that the sum of the squares of the deviations from the line is at a minimum. Maximum likelihood is a more general approach that is the more common approach currently used. This method is based on the probability function that a certain model is correct for a given set of observations. This is done for each reflection, and the probabilities are then combined into a joint probability for the entire set of reflections. Both these approaches are performed over a number of cycles until the changes in the parameters become small. The refinement has then converged to a final set of parameters.

The second approach is to use Fourier methods to calculate the electron density based on the model (using calculated *F*s and phases, the vector $\boldsymbol{F}_c$) and compare this with the electron density based on the observations (with calculated phases, the vector $\boldsymbol{F}_o$). An electron-density map is calculated based on $|\boldsymbol{F}_o| - |\boldsymbol{F}_c|$. This so-called "difference" map will give an accurate representation of where the errors are in the model compared with the experimental data. If an atom is located in the model where there is no experimental observation for it, then the difference map will show a negative density peak. Conversely, when there is no atom in the model where there should be, then a positive peak will be present. This map can be used to manually move, remove, or add atoms into the model.

3 In Practice

3.1 Crystallization

Crystallization is still the bottleneck point in any protein crystallography study. With the vast improvements in genetic techniques, the ability to express and purify large amounts of recombinant proteins is now fairly routine. There are, of course, exceptions to this, and membrane-bound proteins still remain among the most difficult of proteins to purify. The advances in computers and the crystallographic programs that are available make the solving of most structures more or less routine and quite rapid. Crystallization, on the other hand, remains a technique that requires a large number of repetitive trials to determine optimum conditions.

Crystallization is the technique that most "noncrystallography" laboratories can easily carry out in the initial stages of a structural study. Crystallization experiments, for the most part, can be carried out without much expense or technical knowledge. Many proteins give preliminary crystals following the techniques outlined below, and anyone who approached a crystallographer with crystals in hand would quickly find a welcome reception.

3.1.1 Protein Requirements

Every protein that is used in crystallographic studies is unique. The generalities that are offered in this section are just that, generalities. The amount of protein that is probably required when starting a crystallization experiment is about 10–20 mg/mL, and probably close to 1 mL for a reasonably complete survey of crystallization conditions. The actual concentration of protein needed for crystallization trials varies quite widely. Some reported crystallization conditions use less than 5 mg/mL of protein, while others use more than 20 mg/mL. The only way to determine the protein concentration required for a specific individual protein is through trial and error.

Having sufficient protein to carry out crystal trials is certainly important, but more important is the quality of that protein. The key to crystallization is to start from a homogeneous protein solution. Since the formation of crystals is dependent both on the ability of a protein to form crystal nuceii and on the even addition of protein molecules to the surface of the crystal, the fewer obstructions that exist in a solution, the better the chance there is of forming large crystals. We can think of homogeneity in two general terms: protein purity and solution homogeneity. Purity refers to physical contaminants within the solution. The major impurities that are usually considered are other proteins left over from the protein purification. Homogeneity refers to the state of the molecules in the solution, whether they are monomer, dimer, or some other oligomeric state.

Protein purity: The general answer to "how pure does a protein have to be?" is "the purer the better." The usual consensus is 95% or better. Monitoring by SDS-PAGE is sufficient for this evaluation. There are examples of proteins that are less pure than those giving diffraction quality crystals, and there are examples of proteins that have to be much purer to form good crystals. Protein purity is often the first thing that is adjusted to improve crystallization results. There are many examples of proteins whose crystals were improved through the simple expedient of improving the purity of the sample.

Solution homogeneity: Having a pure protein is not necessarily enough. A protein that is in a variety of oligomeric states in a solution (a polydispersed solution) is much less likely to form consistent crystals than a protein that is all in a single state (monodispersed). The usual technique for monitoring protein homogeneity is called dynamic light scattering (DLS). This technique is useful for determining if a solution contains particles of one size (monodispersed), or more than one [polydispersed (Zulauf and D'Arcy, 1992; Ferre-D'Amare and Burley, 1994, 1997)]. In practice, solutions are considered polydispersed if the polydispersivity is greater than 30%. Polydispersed solutions do not generally yield crystals (Zulauf and D'Arcy, 1992; Ferre-D'Amare and Burley, 1994, 1997). Polydispersed solutions can be altered by changing the pH, the salt concentration, buffer type, or adding compounds such as glycerol.

It is useful to make as large a batch of protein as possible to carry on crystallization trials. This makes it easier to maintain reproducible results as a single batch should have the same purity and degree of homogeneity. Any other alterations that occur during the purification process should be consistent within a single batch.

3.1.2 Crystallization Experiments

As described earlier, there are a number of different ways of crystallizing proteins. By far, the most common approach is the vapor-diffusion method, as mentioned earlier. Approximately, 70% of the crystal structures reported have been crystallized through variations of the vapor-diffusion method. The technique can be carried out in a number of ways; the simplest two being the "hanging drop" method, and the "sitting drop" method. The two approaches are similar, but due to the different shapes of the crystallization drops, and therefore the different rates of equilibration within the drop, different results can be obtained from them.

Microbatch crystallization has recently become a popular choice, for a number of reasons. First, microbatch generally uses much less protein and reagents than vapor diffusion. In addition, microbatch is much more amenable to using robotics, and there are a number of proteomics groups that have turned to microbatch for their high-throughput programs. Lastly, microbatch will often yield different crystallization conditions than vapor diffusion (Baldock et al., 1996), so the two methods can be complementary to each other.

Both methods require the use of a broad range of crystallization solutions for the initial screening of crystals. These screens usually come in two types, grid screens and sparse matrix screens:

1. Grid screens: These screens usually change two parameters in a grid approach. Commonly, salt or precipitant concentration and pH are altered in a 2-D grid approach. Common grid screens include ammonium sulfate/pH and PEG/pH grid screens.
2. Sparse matrix screens: These screens are based on the factorial or incomplete factorial statistical approach for designing the screening experiment (Jancarik and Kim, 1991). An extended version of this approach, called sparse matrix sampling has been developed to cover a very large number of conditions for initial screening.

Hanging drop vapor diffusion: These experiments are typically carried out using 24-well tissue-culture-style plates and glass cover slips. The top of the wells must be coated with oil or grease to make a seal between the well and the cover slips (➋ *Figure 22-4a*). The cover slips are usually siliconized to aid proper drop formation. The crystallization trials are prepared by pipetting the well solution into the well and preparing the drop on the cover slip by mixing the well solution and protein solution together in a 1:1 ratio drop. The drops are usually 2–10 μL in size. The crystal plates are viewed under a microscope on a daily basis. This

Figure 22-4

(a) Typical set-up for hanging drop vapor-diffusion experiments. (b) Typical set-up for sitting drop vapor-diffusion experiments. (c) Typical set-up for microbatch experiments

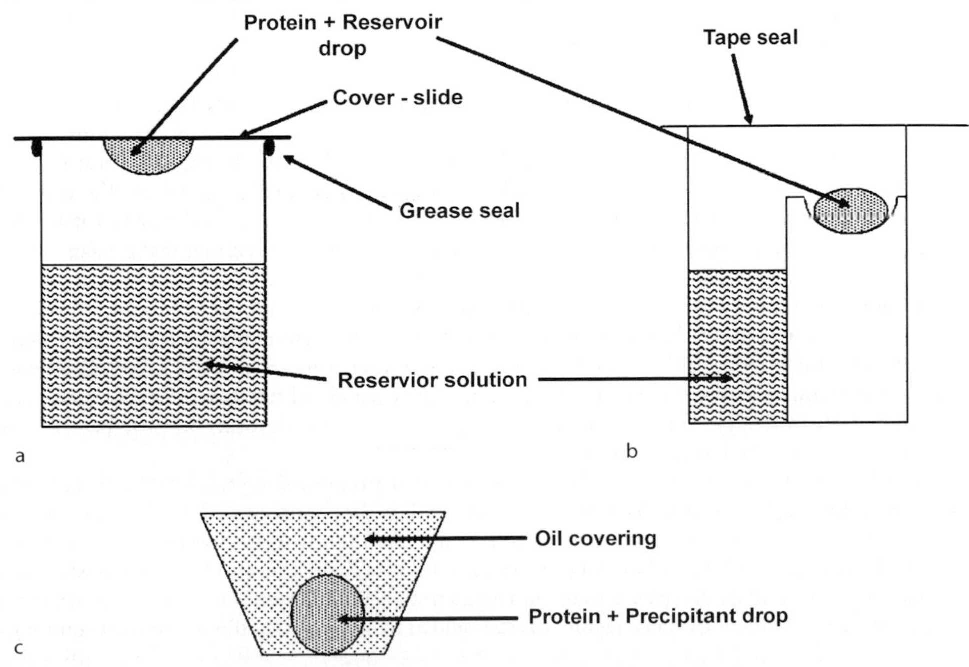

approach tends to use a large amount of reagents per trial (typically 1 mL of well solution), but has the advantage of being easy to see and easy to manipulate crystals.

Sitting drop vapor diffusion: These experiments are very similar in design to the hanging drop experiments. Instead of the mixed drop being hung over the mother liquor from a glass cover slip, the drop is placed on a platform (➋ *Figure 22-4b*). The experiments can use the same style plates as the hanging drop, with the addition of the platform for the drop; however, these experiments can also be carried out on smaller plates, with 96 wells per plate. This method has several advantages over hanging drop, including more trials per plate, smaller amounts of reagents required (typically 100 μL of well solution), and drops with low surface tension do not spread out over the surface.

Microbatch: These experiments are set up without a defined well solution and drop. The protein solution and precipitant solution are mixed together in a small well (such as that found on microtiter plates). The drops are then covered with oil to prevent evaporation (➋ *Figure 22-4c*). The drops are typically 1–2 μL in size. Current robotic techniques allow much smaller drops (100 nL of protein per drop). This method differs from vapor diffusion as vapor diffusion assumes that the starting conditions are well away from being supersaturated and the diffusion out of the drop increases the concentrations of protein and precipitants until supersaturation is reached. Microbatch assumes that the drop enters a supersaturated state immediately on mixing protein with precipitant solution. This method has the advantage of using much less protein and reagent than the vapor-diffusion methods. The Hauptmann–Woodward Medical Research Institute (SUNY, Buffalo, NY) operates a crystallization facility that accepts samples from outside users. This facility uses the microbatch method of crystallization.

3.1.3 Optimizing Crystals

Once preliminary crystallizing conditions have been found, they usually have to be optimized to grow large, diffraction quality crystals. There are several methods that are commonly used for improving the conditions. Diffraction quality crystals can only be judged by diffraction experiments, and the requirements will vary depending on what is required.

The first method is fine-tuning of the current conditions. Most crystals can be improved by adjusting any one of the variables from the original condition. This includes altering the pH (usually only slight variations are appropriate), changing buffers, salt concentrations, protein concentrations, or the concentrations of any other components of the drops. Another variable that can be adjusted is the size of the drop. On the theory that a bigger drop contains more protein and thus makes bigger crystals. An adjustment, like finding the initial crystallization conditions, is a trial-and-error procedure. It is best done with a systematic approach, varying one or two conditions at a time and then reoptimizing based on the improved conditions. This could be a never-ending process, and it is best to stop once good quality crystals have been found, rather than continuously improving the crystals in the hope of getting that 0.1–0.2 Å better resolution.

The second method is a continuation of the above fine-tuning, except this time using additives. In the simplest sense, anything can be added to a crystallization drop and it might improve the crystals. Common additives are salts, metals, various kinds of detergents, and reducing agents. These will have variable effects on crystals. For example, in one protein in our laboratory that crystallized into two different crystal forms, the addition of 10 mM *L*-cysteine improved the crystals, whereas in the other form it had no effect (McMahon et al., 1999; Sanders et al., 2001a, b).

The third method is called seeding. The theory is that if preformed crystals can be placed into a crystallization drop in the "metastable" region (see above), then these preformed crystals will be the only crystals to grow in that drop. This is a very powerful technique for improving the size of crystals. The major difficulty with seeding is finding the "metastable" region. This is done by trial and error. If you have a drop that is close to the conditions that gave crystals, but it has not produced crystals itself, then this drop has a good likelihood to be at the metastable region. Crystals added to this drop should not dissolve—if they do dissolve, then the drop is not in the metastable region, but is unsaturated instead. There are two major techniques used for seeding, macroseeding and microseeding. In macroseeding, visible crystals, usually around 0.5–0.1 mm are moved from the drop they grew in to the new metastable drop, usually using a loop. The crystals are usually washed several times in a stabilizing buffer (usually the well buffer) prior to transfer to remove any small, microscopic crystals. The second technique is called microseeding. In this technique, microcrystals are used in the seeding procedure. As they are too small to be seen, there are two options for moving them. The first option is to take a macrocrystal, crush it using a needle tip, and then transfer the shattered crystal fragments (microseeds) into a larger volume of well buffer. Small aliquots are then pipetted into the new drops. This is known as the dilution method of microseeding. The other option is to use a hair or fiber to touch a crystal. This hair should pick up some microseeds on it from the contact. The hair is then streaked through the new drops; and in theory, some of these microseeds will be deposited into the drops. This is known as "streak seeding."

3.1.4 Tools of the Trade

Setting up a laboratory to run crystallization experiments will not require a huge investment of resources for the initial trials. This will allow to carry out the routine crystallization trials on a protein. If you are planning this to become a major portion of your research program, then a more significant investment will become necessary. The basic components required for crystallization are:

1. Plates: Crystallization plates come in numerous shapes and sizes. The first decision to be made is what sort of crystallization experiment you will carry out. For hanging drop experiments, you require plates, cover slips, and a grease to seal the two together. For sitting drop experiments, you require plates and sealing tape. For microbatch, you require the plates and an oil (usually paraffin) to cover the drops. There are a number of companies that sell crystallization plates. Hampton Research (CA) probably has the largest supply and would be a good place to start, though there are other companies that sell specialized plates at cheaper prices.
2. Screens: Crystal screens can be obtained in a number of ways. The simplest screens are grid screens, with varying salt/precipitant concentration and pH. These can be easily produced in a laboratory and the most popular screens are ammonium sulfate/pH or PEG 6000/pH screens. There are numerous screening kits that are commercially available. These vary from grid screens to sparse matrix screens. Hampton Research, Emerald Biostructures, Molecular Dimension, Jena Biosciences, and Sigma/Aldrich all offer premade crystallization screens. Specialized screening kits also exist for membrane proteins and nucleic acids. Additive screens can also be purchased commercially from Hampton Research. The advantage of buying commercial screens is reproducibility and convenience.
3. Microscope: A microscope is essential for examining the crystals trials. The brand and type of microscope is purely down to personal preference. High magnification is not usually necessary, as crystals will be 10 mm or larger (smaller than that would not be useful). Depth of field is a more critical requirement, as scanning drops is a fairly laborious and time-consuming endeavor and a larger depth of field will improve efficiency. Most crystallographers tend to favor stereo zoom microscopes, with transmission light stages. These are available from all the major microscope manufacturers, including Olympus, Nikon, Zeiss, and Leica.

The following components are for a more advanced laboratory that is pursuing crystallography as a major component of a research program and is more independent of any established crystallography laboratory.

4. Incubator: Crystallization trials are heavily dependent on the temperature and maintaining a constant temperature is critical to reproducible trials. In addition, crystallization trials are often carried out at different temperatures, the three most common temperatures being around room temperature (20–22°C), 12–15°C, and 4°C. Using a vibration-free incubator to maintain a constant temperature

can be quite important. Some crystallographers have rooms that are temperature controlled, while others rely on smaller incubators.

5. Camera: It is important to have the pictures of crystals. Aside from the obvious reason of publication and grant writing, pictures give an on-going record of crystallization conditions and make it easy to see improvements in crystals over time. Cameras can be either film or digital and are usually easily mounted onto most microscopes. Olympus and Nikon offer very good digital camera systems.
6. Other supplies: There are a number of other supplies that are often necessary for a crystallography laboratory. These include a selection of loops and capillary tubes for mounting crystals (frozen or room temperature), microprobes for manipulating crystals, and goniometer heads (always best to have one on hand). The most complete source for these supplies is Hampton Research. Additionally, if the laboratory is planning on storing frozen crystals and transporting them to a synchrotron, then at least one cryostorage/shipping Dewar will be required.
7. Robotic systems: Robotic crystallization systems are probably beyond the scope of most laboratories, but they are worth at least mentioning. These systems tend to be quite expensive, but for a laboratory with suitable resources, they can be extremely useful. Current robotic systems are available for vapor-diffusion (hanging and sitting drop) and microbatch methods. It is with the microbatch method that robotic systems have their potential fully realized. Systems are capable of accurately pipetting protein drops of 50 nL or less which means that much less protein required, and can give very reproducible results when operated correctly. High-throughput crystallization laboratories depend on robotic systems for processing their samples. Gilson/Cyberlab Supplies and Douglas Instruments produce two commonly used systems.

3.1.5 The Crystal Blues

Many proteins will not yield crystals in initial crystal trials. Unfortunately, crystallization is still a trial-and-error procedure, with no real way of predicting success or failure. If you fail to get crystals in your initial trials, it need not be the end of your structural studies (although it could be!). Many of the targets for neuroscientists are going to be membrane-bound or membrane-spanning proteins, and these are notoriously difficult to crystallize. Techniques are continuously being developed and refined to improve our ability to crystallize difficult protein examples.

Several approaches have been successfully used. This chapter is far too short and general to cover all the possible methods of altering proteins for crystallization. This section is intended as an introduction to the most commonly used, or most striking examples of techniques (Derewenda, 2004).

1. Repurify or reclone the protein: Proteins often do not crystallize due to impurities and contaminants in the solutions. These may not be visible in any of the tests that can be done. One way to get around this is to use different techniques to purify the protein. This can be as simple as using different buffers or pHs for the columns, or as complicated as recloning the protein into a different expression vector. Many proteins are expressed with tags to aid purification. Cleaving or not cleaving these tags off the target protein can have a great effect in crystallization studies. Additionally, the cleavage sequence (i.e., the residues left over after cleavage) may have negative effects on crystallization. Altering the cleavage site may improve crystallization. The best options for proteolysis (and thus cleavage site) are to use specific viral proteases, such as tobacco etch virus protease (TEV) or human rhinovirus 3C protease [PreScission protease (Amersham) (Walker et al., 1994)]. Thrombin and Factor X have been commonly used, but can undergo secondary, nonspecific cleavage (Doskeland et al., 1996) which results in heterologous protein solutions.
2. Mutagenesis: Proteins are flexible molecules and many proteins contain highly mobile loops that can affect the ability of proteins to pack into crystals. Two common locations for flexible domains are at the N and C terminus. It is common for either or both termini to be disordered in a protein

crystal. If this disorder is extreme, it may prevent crystal formation. Removal of a number of residues from either end of the protein can be attempted to aid in purification. Internal flexible loops may also prevent proper crystal formation. These loops are difficult to remove, as they are often important for function, but sometimes altering the loops can be effective (Dale et al., 1999; Mazza et al., 2002).

Since crystals form through contacts between protein molecules, it can be expected that altering the surface residues on a protein will alter the ability to form crystals. Altering surface residues to allow better crystal surface contacts can work extremely well for improving crystals (D'Arcy et al., 1999; Lorber et al., 2002). Two residues in particular, lysine and glutamic acid, seem to be particularly good residues for mutation. Both are found predominantly on the surface of proteins, and both seem to be disfavored in protein–protein interaction surfaces. Since crystals depend on forming contacts between proteins, this approach may improve the crystal contact surfaces of the proteins in the crystal (Mateja et al., 2002).
This mutagenesis is very tricky to carry out in a predictable way, unless you have additional structural information that may aid you in this (for instance a homologous structure).

3. Complex formation: Complex formation for enhancing crystallization can take two forms. The first is to use substrates, inhibitors, ligands, or cofactors to cocrystallize with the protein. It is well known that proteins often undergo conformational changes on the binding of ligands or cofactors. The protein in a different conformation will often crystallize under completely different conditions (or give crystals where none was found previously). Since a cocrystal structure is often more important than protein by itself, this is often a good step to take early in crystallization trials.

The second way of using complexes is to use a "crystallization scaffold" (Derewenda, 2004). Some proteins will not crystallize on their own, but will form crystals as part of a complex. This complex does not have to be of any biological relevance, except to promote crystallization. The first way of doing this is to make use of the affinity tag used to purify a protein (if one was used). His-tagged proteins are very commonly used for purifying proteins, and could be left on for crystallization studies. This is not usually a useful scaffold, as the poly-His plus linker sequence is usually very flexible.
Proteins expressed as fusion proteins have also been crystallized. Maltose binding protein (MBP) (Kobe et al., 1999; Liu et al., 2001), thioredoxin (Stoll et al., 1998) and GST (Kuge et al., 1997) have all been used as fusion proteins in structure studies.

An alternative approach is to use antibody fragments (the Fab domian) raised against a particular protein. This technique has been used with some success for crystallizing a number of membrane proteins, including the ClC chloride channel (Dutzler et al., 2002), a K+ voltage-dependent channel (Jiang et al., 2003), and a cytochrome c oxidase (Iwata et al., 1995).

4. Limited proteolysis. Flexible regions of proteins can sometimes be removed by digestion of the protein with different proteases. This technique is based on the techniques that were used to determine the core folded regions of proteins, most notably antibodies (Porter, 1973). Limited proteolysis can be used to remove flexible loops of proteins, or separate multidomain proteins into separate domains and has been used successfully in a number of instances (Noel et al., 1993; Sondek et al., 1996; Mazza et al., 2002).

Limited proteolysis tends to introduce heterogeneity into the protein sample, and if the protein can be overexpressed using recombinant techniques, it is better to engineer these deletions into the protein at the gene level.

5. Homologous proteins. Homologous proteins usually crystallize under very different conditions. For example, in our laboratory two homologous proteins crystallize in either 10 mM sodium acetate, pH 3.8, 5 mM DTT, and 50% (v/v) MPD (Weichsel et al., 1996), or 100 mM ammonium sulfate, 30% PEG 6000, and 10 mM DTT (Filson et al., 2003). This technique was first used by Kendrew for solving the structure of sperm whale myoglobin (Kendrew et al., 1954) and has been used in many other structural studies. Currently, this technique is heavily exploited in membrane protein crystallography (Wiener,

2004). This technique is not useful if a specific structure is needed, as will often be the case in neurochemistry.

3.2 X-Ray Diffraction

3.2.1 Choices of Sources

The choice of the X-ray generator that you use will almost certainly be dictated to you by location. The major choice you will have is using an in-house generator or a synchrotron. In-house generators have the convenience of easy access, expertise available and (hopefully) willing to aid you in diffracting crystals. These sources are good and improving rapidly all the time. The intensity of the beam is less than at synchrotrons and the wavelength is fixed, but for initial datasets and possibly final publication datasets, in-house systems can be used very well.

Synchrotrons have the advantage of brighter beams, multiple wavelength availability, and a more focused beam, all of which can greatly improve the quality of data and may be essential for the experiments that are to be carried out. The synchrotrons I have worked at are staffed by excellent people; however, they do assume a certain level of user competence and are probably not the place to go to learn how to diffract crystals. Synchrotrons will grant time to users in a peer-reviewed process and generally have quite rapid access times (usually booked per 24-h slot).

A number of synchrotrons (including the National Synchrotron Light Source, New York, the Advanced Light Source, Berkley, and soon the Canadian Light Source, Saskatoon) operate "mail-in" crystallography services where a scientist can mail in crystals (prefrozen and mounted on loops) and data will be collected, processed (sometimes), and returned. This is becoming a method of choice, as it eliminates the need to travel to a synchrotron and speeds up the data collection procedure at the synchrotron also.

3.2.2 To Freeze or Not to Freeze

Traditionally, all protein crystal data were collected at room temperature. The increased intensity of X-ray beams (both from in-house generators and synchrotrons) has made cryocrystallography (or "frozen" crystals) the method of choice for data collection. Freezing of crystals requires using a cryoprotectant, usually glycerol, that is added to the crystal solution. This causes the crystal and the liquid surrounding it to freeze as a transparent glass rather than ice, which would show up as rings on the diffraction pattern.

Frozen crystals have several advantages over unfrozen crystals. Radiation damage is greatly reduced with frozen crystals, which means that one or more datasets can be collected from a single crystal. This is of particular importance when conducting MAD experiments, where the best results are gained by collecting all three datasets from a single crystal. The quality of general data is also often improved with frozen crystals because of reduction in motion within the crystal.

Disadvantages of freezing crystals center around the difficulty in determining the proper freezing buffer. Incorrect buffers can cause crystals to dissolve or crack, which are obvious problems to detect. More subtle problems also exist, such as a reduction or loss in diffraction through freezing. It is always a good idea to try room temperature diffraction before freezing crystals.

3.3 Solving Structures

Solving protein structures requires a significant investment in time and expertise. If you have made it to this point without the assistance of an established crystallographer, then you deserve praise. Attempting to solve and refine structures without the initial assistance and help of an expert crystallographer is not advisable. You may desire to eventually carry out all this work in your own laboratory, but initially the assistance of an established crystallographer will be invaluable.

3.3.1 Requirements for MR/MIR/MAD

Solving the phase problem in protein crystallography is a requirement for any structural study. The three common methods for determining phases are MR, multiple isomorphous replacement (MIR), and MAD. There are also a number of lesser used methods that are generally derivatives of the other three.

1. MR: Molecular replacement requires a reasonable starting model for the search algorithm. Defining a starting model is less clear. The general consensus is that proteins with greater than 30% identity can be used as a search model. This obviously varies on a case-by-case basis; variations on this can also be tried. Converting the model structure to a polyalanine model can be successful, as can using homologous domains rather than the entire protein. The computer programs and algorithms are being improved continuously and becoming increasingly powerful. Our laboratory has successfully performed MR using less than 30% homology (Filson et al., 2003) with some of these newer programs. The current best program for this is PHASER (Storoni et al., 2004), which is available for the CCP4 program suite (see later).
2. MIR: Multiple isomorphous replacement requires datasets to be collected on a native crystal and two or more crystals soaked with a heavy metal derivative. The most common heavy metals are platinum, mercury, and gold. The usual method is to grow crystals the same way, and then add varying amounts of metal (usually as salts) and soak for different lengths of times. There is no hard and fast rule regarding the length of time required for the metal soaks, with examples varying from minutes to days. This technique requires that the space group and unit cell of the derivative crystals are the same as that of the native crystal. Often many conditions/heavy metals must be screened to find suitable derivative crystals. This technique is very powerful and has a long, successful history for solving structures. It is becoming less used as MAD increasingly becomes the method of choice for solving novel structures.
3. MAD: Multiwavelength anomalous diffraction/dispersion requires datasets at different wavelengths, usually from the same crystal. The three wavelengths are chosen to maximize the differences in the anomalous signal. This is usually determined by prescanning the crystal to determine the optimum wavelengths. MAD experiments can be carried out using metals attached to proteins (especially useful for proteins that naturally contain metals), or by incorporating selenium [usually in the form of seleno-methionine (SeMet)] into the protein during overexpression. This is usually done either by expressing the protein in a Met-auxotroph [e.g., B834 cells (Novagen)], or by inhibiting the Met-biosynthetic pathway (Doublie, 1997). In the case of SeMet incorporation, it is important to keep the seleniums in the same state (either fully reduced or fully oxidized) during crystallization and data collection. Many of the high-throughput crystallization laboratories grow all their proteins as SeMet derivatives and solve the structures exclusively using MAD techniques.

3.3.2 Computers and Programs

If this had been written 10 years ago, then the choices and prices of computers for this work would be the major stumbling block for any laboratory to join the crystallographic world. Fortunately, with the speed of processors, and the wonderful work being carried out by computer programmers around the world, most crystallographic applications can be used on the average laboratory PC.

Computer requirements are currently not excessive. A mid- to top-end PC or Macintosh is capable of most crystallography work. For best functionality, a computer running Linux is the best option for running all crystallography programs, though a number of them will run under Windows or Mac operating systems. Other operating systems such as Sun are also able to perform most tasks, although like all such systems, they are occasionally limited by the programmer's desires to work on any particular system.

The biggest concern is disk space. A dataset of diffraction images can take more than 3 GB of disk space. This does not seem like an excessive amount, but with multiple datasets and refinement files, a single structure can very quickly require over 10 GB of disk space. Fortunately, 100-GB hard drives are not very expensive any more.

Graphics is also a very important component for building protein models and there are several options available. Silicon Graphics Inc. (SGIs) is the biggest name in selling computer systems that are capable of good quality graphics for protein structures. Unfortunately, these systems are quite expensive. Many PC graphics cards are capable of displaying the graphics at sufficiently high resolution and speed, and these are quickly supplanting SGIs as the systems of choice amongst crystallographers.

Crystallographic software comes in two categories, free software, and commercial software. In most cases, these software packages exist quite comfortably side by side and are often compatible with each other. There are advantages to each type of software, but as a member of the academic community and operating on a limited budget, my personal experience with commercial software is quite limited.

Software of particular value to laboratories interested in protein structures without wanting to become a crystallography laboratory will center around programs for viewing protein structures, of which there are a number of both free and commercial programs. These will be discussed in ➋ Section 4.2.

For a laboratory that is starting a crystallographic project and is interested in solving and refining structures independently, establishing a computer system will seem to be a huge undertaking, and in many ways, it is. It is best to make use of the advice and guidance of an experienced crystallographer, hopefully the one that you have been turning to for advice already. As many of the software packages have their own quirks, it would be best to use the software that your experienced crystallographer is familiar with.

If you are in the very odd situation of having no crystallographer to turn to, then an organization called CCP4 (Collaborative Computing Project Number 4) (CCP4, 1994) is recommended. This is a group of crystallographers who have been developing an integrated system for solving and refining crystal structures for more than 15 years. Their software package is free to academic users, runs easily on most computer systems, and the people associated with CCP4 are helpful in answering questions and solving problems.

4 Crystal Structures

4.1 Structure Papers

Structure papers generally contain a lot of information that will be unfamiliar to the majority of non-crystallographers. This section is offered as a very brief primer on what some of the more important terms mean and their importance.

Crystallographic papers do not show most of the data for the experiments. This is found in the raw structure factor files that are often deposited along with the coordinates for the model (➋ Section 4.3). The papers report numbers designed to inform the reader as to the accuracy of the data and the validity of the model. This is usually contained in one or two tables (➋ *Figure 22-5*).

The two common tables seen in crystallographic papers are combined into one table in this example. Crystallographic data and refinement statistics are measures of how good the diffraction data is and the "goodness" of the model, respectively.

The crystallographic data contain all the information about the data that were used for determining the model. The most important information is the resolution. This refers to the minimum d spacing (➋ Eq. 22.1) and indicates the smallest distance between two atoms that can be "resolved," i.e., completely separated based on electron density. The table also contains space group ($P2_12_12_1$) and unit-cell information along with the statistical measurements for the reflection data.

The refinement statistics show how well the model matches up with the experimental data. The most commonly discussed numbers are the *R-factor* and R_{free}. These numbers (*R*-factor was discussed in ➋ Section 2.3.5) are a measure of how well the model Fs match with the experimental Fs. The R_{free} value

Figure 22-5
Example of data table from a crystallographic structure paper (Sanders et al., 2001a,b)

Summary of structural data

Crystallographic data	Native data set ($P2_12_12_1$)
Wavelength (A)	0.934
Resolution[1] (A)	40–2.4 (2.52–2.4)
Cell constants	
a (A)	56.6
b (A)	98.1
c (A)	132.5
$\alpha = \beta = \gamma$ (°)	90
Unique Reflections	28.288
I/σ[1]	10.9 (3.3)
Average redundancy[1]	3.9 (3.7)
% Completeness[1]	98 (93)
R_{merge}[1,2]	5.9 (22.4)
Mosaic spread	0.5
Refinement statistics	
Number of protein atoms	6,084
Number of FAD molecules	2
Number of water molecules	321
R-factor (%)[3]	18.5
R_{free} (%)[3]	24.5
R.m.s. deviations from ideal	
Bond lengths (A)	0.012
Bond angles (°)	1.51
Mean temperature factor	
Main chain	26.6
Side chain	28.7
FAD molecules	18.9
Water molecules	31.8

[1]Values in parentheses refer to the highest resolution shell.
[2]$R_{merge}\ \Sigma = \Sigma\Sigma|I(h)j - \langle I(h)\rangle| / \Sigma\Sigma I(h)j$, where I(h) is the measured diffraction intensity and the summation includes all observations.
[3]*R*-factor $= \Sigma|F_D - F_D| / \Sigma|F_n|$. R_{free} is the R-factor calculated using 5% of the data that were excluded from the refinement.

is based on a subset of the data that is kept out of the refinement process to minimize bias. These values should be close to each other, with the R_{free} value slightly higher than the *R*-factor. A good rule of thumb is that the R_{free} value should be approximate to the resolution (resolution 2.4 Å should have an R_{free} of around 24%).

4.2 Viewing Crystal Structures

For many, examining crystal structures will be the extent of their crystallography experience. Although looking at figures in journal articles or occasionally online is useful, there is a large range of options that would allow a researcher to easily view a protein structure of interest in a much more interactive manner. This list is not intended to be an exhaustive list of options for viewing structures, merely a suggested sample of programs that may be useful.

Many of these programs are available for PC and Mac computers and require no additional software or hardware. To view protein models in stereo mode (enabling 3-D viewing), the choice is more limited and additional hardware will be required.

4.2.1 Basic Viewing Programs

Basic viewing programs are available for basic viewing of structures, including rotating the molecule, zooming in on interesting regions, highlighting of interesting residues, and coloring of residues or regions. All these programs are available as free downloads from various websites. In general, they are easy to use and have reasonable manuals available. The comments associated with these programs are my personal views about the programs.

Rasmol: It was one of the first Protein Databank (PDB) compatible viewers available as freeware. Upgrades have now been stopped, but the program is still widely available. Versions exist for both PC and Mac.

Protein explorer: This program is the expanded version of Rasmol and works very well. It is available at http://proteinexplorer.org and has better mouse control than Rasmol.

VMD: This program is available at http://www.ks.uiuc.edu/Research/vmd/. This is a good program for viewing structures. It can be used in stereo mode and it produces publication quality graphics.

PyMOL: This program is avalibale at http://pymol.sourceforge.net/ and is another effective program for viewing structures. It is more user friendly than VMD And is also, a good program for making movies of protein structures. It can be used in stereo mode. It displays electron-density maps reasonably well. It is another program that is constantly evolving.

Swiss-PDB viewer: This program is available at: http://www.expasy.org/spdbv/.

4.2.2 Advanced Programs

O: This program is one of the standards used for modeling of protein structures to experimental data. It is freely available to academic users. It is very good for manually editing structures, including altering residues and moving them. This program is constantly being updated and is available at http://xray.bmc.uu.se/~alwyn/.

XtalView: This is another excellent program for model building. It is available at: http://www.sdsc.edu/CCMS/Packages/XTALVIEW/xtalview.html.

Insight II: Available at: http://www.accelrys.com/insight/.

Sybyl: Available at: http://www.tripos.com.

These programs are commercially available ones. They are excellent for a number of tasks, depending on the modules. These programs are suitable for structure-based drug design applications as well as protein–ligand docking and protein homology modeling. Both these programs require significantly greater computer resources than the other programs on this list.

4.3 Crystallography Databases

For protein crystallography, the repository of most protein crystal structures is the PDB hosted at: http://www.rcsb.org/pdb/ (Berman et al., 2000). This database contains the 3-D coordinates (and sometimes the structure factor files) for almost all protein crystal structures. Most journals currently require deposition of the coordinates when publishing structure papers. Each structure is given a unique identification code that will be listed in the paper (see ➋ *Figure 22-1* for examples of PDB codes). Structures can be accessed using this code, or using various other search criteria. The PDB also contains structural information for NMR structures.

5 Further Reading Material

This list of books is by no means considered an exhaustive list on protein X-ray crystallography. These are merely books that I have found useful, or have been recommended to me by colleagues. Other suggestions for additions to this list will be gratefully received.

5.1 Introductory Crystallography Texts

Rhodes G. 1993. Crystallography made crystal clear. New York: Academic Press.
Drenth J. 1993. Principles of protein X-ray crystallography. New York: Springer.
Primer A, Glusker JP, Trueblood KN. 1985. Crystal structure analysis. New York: Oxford University Press.

5.2 Crystallization

McPherson A. 1982. Preparation and analysis of protein crystals. New York: Wiley Interscience.
Ducruix A, Giege R. 1999. Crystallization of nucleic acids and proteins. New York: Oxford University Press.
McPherson A. 1999. Crystallization of biological macromolecules. CSH Laboratory Press.
Michel H. 1990. Crystallization of membrane proteins. London: CRC Press.
Methods. 2004. Vol. 34, Issue 3, McPherson A, editor. pp. 251–423.

5.3 Protein Crystallography—Advanced Texts

Blundell TL, Johnson LN. 1976. Protein crystallography. London: Academic Press.
Stout GH, Jensen LH. 1989. X-ray structure determination: a practical guide. New York: Wiley.
McRee DE. 1999. Practical protein crystallography. London: Academic Press.
Methods in enzymology. Vol. 276, 277, 368, and 374. New York: Academic Press.

References

Ackers GK, Hazzard JH. 1993. Transduction of binding energy into hemoglobin cooperativity. Trends Biochem Sci 18: 385-390.

Baldock P, Mills V, Stewart PS. 1996. A comparison of microbatch and vapor diffusion for initial screening of crystallization conditions. J Cryst Growth 168: 170-174.

Berman HM, Westbrook J, Feng Z, Gilliland G, Bhat TN, et al. 2000. The protein data bank. Nucl Acids Res 28: 235-242.

Case DA, Karplus M. 1979. Dynamics of ligand binding to heme proteins. J Mol Biol 132: 343-368.

CCP4. 1994. The CCP4 suite: programs for protein crystallography. Acta Cryst D50: 760-763.

D'Arcy A, Stihle M, Kostrewa D, Dale G. 1999. Crystal engineering: a case study using the 24 kDa fragment of the DNA Gyrase B subunit from *Escherichia Coli.* Acta Cryst D55: 1623-1625.

Dale GE, Kostrewa D, Gsell B, Stieger M, D'Arcy A. 1999. Crystal engineering: deletion mutagenesis of the 24 kDa fragment of the DNA Gyrase B subunit from *Staphylococcus Aureus.* Acta Cryst D55: 1626-1629.

Dash C, Kulkarni A, Dunn B, Rao M. 2003. Aspartic peptidase inhibitors: implications in drug development. Crit Rev Biochem Mol Biol 38: 89-119.

Derewenda ZS. 2004. The use of recombinant methods and molecular engineering in protein crystallization. Methods 34: 354-363.

Ding X, Rasmussen BF, Petsko GA, Ringe D. 1994. Direct structural observation of an acyl-enzyme intermediate in the hydrolysis of an ester substrate by elastase. Biochemistry 33: 9285-9293.

Doskeland AP, Martinez A, Knappskog PM, Flatmark T. 1996. Phosphorylayion of recombnant human phenylalanine hydroxylase: effect on catalytic activity, substrate activation and protection against non-specific cleavage of the fusion protein by restriction protease. Biochem J 313 (Pt 2): 409-414.

Doublie S. 1997. Preparation of selenomethionyl proteins for phase determination. Methods Enzymol 276: 523-530.

Dutzler R, Campbell EB, Cadene M, Chait BT, MacKinnon R. 2002. X-ray structure of a ClC chloride channel at 3.0 A reveals the molecular basis of anion selectivity. Nature 415: 287-294.

Feher G, Kam Z. 1985. Nucleation and growth of protein crystals: general principles and assays. Methods Enzymol 114: 77-112.

Ferre-D'Amare AR, Burley SK. 1994. Use of dynamic light scattering to assess crystallizability of macromolecules and macromolecular assemblies. Structure 2: 357-359.

Ferre-D'Amare AR, Burley SK. 1997. Dynamic light scattering in evaluating crystallizability of macromolecules. Methods Enzymol 276: 157-166.

Filson H, Fox A, Kelleher D, Windle HJ, Sanders DA. 2003. Purification, crystallization and preliminary X-ray analysis of an unusual thioredoxin from the gastric pathogen *Helicobacter pylori.* Acta Cryst D59: 1280-1282.

Gelin BR, Lee AW-M, Karpls M. 1983. Hemoglobin tertiary structural change on ligand binding. J Mol Biol 171: 489-559.

Hartley BS. 1979. Evolution of enzyme structure. Proc R Soc Lond B Biol Sci 205: 443-452.

Iwata S, Ostermeier C, Ludwig B, Michel H. 1995. Structure at 2.8 A resolution of cytochrome c oxidase from *Paracoccus Denitrificans.* Nature 376: 660-669.

Jancarik J, Kim SH. 1991. Sparse matrix sampling: a screening method for crystallization of proteins. J Appl Cryst 24: 409-411.

Jiang Y, Lee A, Chen J, Ruta V, Cadene M, Chait BT, MacKinnon R. 2003. X-ray structure of a voltage-dependent K+ channel. Nature 423: 33-41.

Kendrew JC, Parrish RG, Marrack JR, Orlans ES. 1954. The species specificity of myoglobin. Nature 174: 946-949.

Kobe B, Center RJ, Kemp BE, Poumbouris P. 1999. Crystal structure of human T cell leukemia virus type 1 gp21 ectodomain crystallized as a maltose binding protein chimera reveals structural evolution of retroviral transmembrane proteins. Proc Natl Acad Sci USA 96: 4319-4324.

Kuge M, Fujii Y, Shimizu T, Hirose F, Matsukage A, et al. 1997. Use of a fusion protein to obtain crystals suitable for X-ray analysis: crystallization of a GST-fused protein containing the DNA-binding domain of DNA replication-related element-binding factor, DREF. Protein Sci 6: 1783-1786.

Liu Y, Manna A, Li R, Martin WE, Murphy RC, Cheung AL, Zhang G. 2001. Crystal structure of the SarR protein from *Staphylococcus Aureus.* Proc Natl Acad Sci USA 98: 6877-6882.

Lorber B, Theobald-Dietrich A, Charron C, Sauter C, Ng JD, et al. 2002. From Conventional Crystallization to Better Crystals from Space: A Review on Pilot Crystallographic Studies with Aspartyl-tRNA Synthetases. Acta Cryst D58: 1674-1680.

Mateja A, Devedjiev Y, Krowarsch D, Longenecker K, Dauter Z, et al. 2002. The Impact of Glu-Ala and Glu-Asp mutations on the crystallization properties of RhoGDI: the structure of RhoGDI at 1.3 A resolution. Acta Cryst D58: 1983-1991.

Mazza C, Segref A, Mattaj IW, Cusack S. 2002. Co-crystallization of the Human Nuclear Cap-binding Complex with a m7GpppG Cap Analogue Using Protein Engineering. Acta Cryst D58: 2194-2197.

McMahon SA, Leonard GA, Buchanan LV, Giraud MF, Naismith JH. 1999. Initiating a crystallographic study of UDP-galactopyranose mutase from *Escherichia coli.* Acta Cryst D55: 399-402.

Noel JP, Hamm HE, Sigler PB. 1993. The 2.2 A crystal structure of transducin-α complexed with GTP-γ-S. Nature 366: 654-663.

Porter RR. 1973. Structural studies of immunoglobulins. Science 180: 713-716.

Sanders DAR, Moothoo DN, Raftery J, Howard AJ, Helliwell JR, Naismith JH. 2001a. The 1.2 A resolution structure of the ConA–dimmanose complex. J Mol Biol 310: 875-884.

Sanders DAR, Staines AG, McMahon SA, McNeil MR, Whitfield C, Naismith JH. 2001b. UDP-galactopyranose mutase has a novel structure and mechanism. Nat Struct Biol 8: 858-863.

Sciara G, Kendrew SG, Miele AE, Marsh NG, Federici L, et al. 2003. The structure of ActVA-Orf6, a novel type of monooxygenase involved in actinorhodin biosynthesis. EMBO J 22: 205-215.

Sondek J, Bohm A, Lambright DG, Hamm HE, Sigler PB. 1996. Crystal structure of a G-protein βγ dimer at 2.1 A resolution. Nature 379: 369-374.

Steitz TA, Henderson R, Blow DM. 1969. Structure of crystalline α-chymotrypsin. 3. Crystallographic studies of substartes and inhibitors bound to the active site of α-chymotrypsin. J Mol Biol 46: 337-348.

Stoll VS, Manohar AV, Gillon W, Mac Farlane EL, Hynes RC, et al. 1998. A thioredoxin fusion protein of VanH, a D-lactate dehydrogenase from *Enterococcus Faecium*: cloning, expression, purification, kinetic analysis, and crystallization. Protein Sci 7: 1147-1155.

Storoni LC, McCoy AJ, Read RJ. 2004. Likelihood-enhanced fast rotation functions. Acta Cryst D60: 432-438.

Walker PA, Leong LE, Ng PW, Tan SH, Waller S, Murphy D, Porter AG. 1994. Efficient and rapid affinity purification of proteins using recombinant fusion proteases. Biotechnology 12: 601-605.

Wang B. 1985. Resolution of phase ambiguity in macromolecular crystallography. Methods Enzymol 115: 90-112.

Weichsel A, Gasdaska JR, Powis G, Montfort WR. 1996. Crystal structures of reduced, oxidized, and mutated human thioredoxins: evidence for a regulatory homodimer. Structure 4: 735-751.

Wiener MC. 2004. A pedestrian guide to membrane protein crystallization. Methods 34: 364-372.

Wilson AJC. 1950. Largest likely values for the reliability index. Acta Cryst 3: 397-398.

Wise EL, Rayment I. 2004. Understanding the importance of protein structure to nature's routes for divergent evolution in TIM barrel enzymes. Acc Chem Res 37: 149-158.

Zulauf M, D'Arcy A. 1992. Light scattering of proteins as criterion for crystallization. J Cryst Growth 122: 102-106.

Index

Printing: Krips bv, Meppel
Binding: Stürtz, Würzburg